Production, Properties, and Applications of High Temperature Coatings

Amir Hossein Pakseresht
University of Tehran, Iran & Materials and Energy Research Center, Iran

A volume in the Advances in Chemical and Materials Engineering (ACME) Book Series

IGI Global
DISSEMINATOR OF KNOWLEDGE

Published in the United States of America by
IGI Global
Engineering Science Reference (an imprint of IGI Global)
701 E. Chocolate Avenue
Hershey PA, USA 17033
Tel: 717-533-8845
Fax: 717-533-8661
E-mail: cust@igi-global.com
Web site: http://www.igi-global.com

Library of Congress Cataloging-in-Publication Data

Names: Pakseresht, Amir Hossein, 1981- editor.
Title: Production, properties, and applications of high temperature coatings / Amir Hossein Pakseresht, editor.
Description: Hershey, PA : Engineering Science Reference, [2018] | Includes bibliographical references.
Identifiers: LCCN 2017030321| ISBN 9781522541943 (hardcover) | ISBN 9781522541950 (eISBN)
Subjects: LCSH: Materials at high temperatures--Protection. | Ceramic coating. | Heat resistant paint.
Classification: LCC TA417.7.H55 P7575 2018 | DDC 620.1/1217--dc23 LC record available at https://lccn.loc.gov/2017030321

This book is published in the IGI Global book series Advances in Chemical and Materials Engineering (ACME) (ISSN: 2327-5448; eISSN: 2327-5456)

British Cataloguing in Publication Data
A Cataloguing in Publication record for this book is available from the British Library.

For electronic access to this publication, please contact: eresources@igi-global.com.

Advances in Chemical and Materials Engineering (ACME) Book Series

J. Paulo Davim
University of Aveiro, Portugal

ISSN:2327-5448
EISSN:2327-5456

Mission

The cross disciplinary approach of chemical and materials engineering is rapidly growing as it applies to the study of educational, scientific and industrial research activities by solving complex chemical problems using computational techniques and statistical methods.

The **Advances in Chemical and Materials Engineering (ACME) Book Series** provides research on the recent advances throughout computational and statistical methods of analysis and modeling. This series brings together collaboration between chemists, engineers, statisticians, and computer scientists and offers a wealth of knowledge and useful tools to academics, practitioners, and professionals through high quality publications.

Coverage

- Thermo-Chemical Treatments
- Ductility and Crack-Resistance
- Numerical Techniques
- Computational methods
- Wear of Materials
- Toxicity of Materials
- Nanomaterials
- Statistical methods
- Biomaterials
- Coatings and surface treatments

IGI Global is currently accepting manuscripts for publication within this series. To submit a proposal for a volume in this series, please contact our Acquisition Editors at Acquisitions@igi-global.com or visit: http://www.igi-global.com/publish/.

The Advances in Chemical and Materials Engineering (ACME) Book Series (ISSN 2327-5448) is published by IGI Global, 701 E. Chocolate Avenue, Hershey, PA 17033-1240, USA, www.igi-global.com. This series is composed of titles available for purchase individually; each title is edited to be contextually exclusive from any other title within the series. For pricing and ordering information please visit http://www.igi-global.com/book-series/advances-chemical-materials-engineering/73687. Postmaster: Send all address changes to above address.

Titles in this Series

For a list of additional titles in this series, please visit: www.igi-global.com/book-series

Emergent Research on Polymeric and Composite Materials
R. Somashekar (University of Mysore, India) and Thejas Urs G. (University of Mysore, India)
Engineering Science Reference • copyright 2018 • 341pp • H/C (ISBN: 9781522530237) • US $235.00 (our price)

3D Printing and Its Impact on the Production of Fully Functional Components Emerging Research and Opportunities
Petar Kocovic (Union – Nikola Tesla University, Serbia)
Engineering Science Reference • copyright 2017 • 115pp • H/C (ISBN: 9781522522898) • US $150.00 (our price)

Advanced Applications of Supercritical Fluids in Energy Systems
Lin Chen (Tohoku University, Japan & Japan Society for the Promotion of Science (JSPS), Japan) and Yuhiro Iwamoto (Nagoya Institute of Technology, Japan)
Engineering Science Reference • copyright 2017 • 682pp • H/C (ISBN: 9781522520474) • US $235.00 (our price)

Innovative Applications of Mo(W)-Based Catalysts in the Petroleum and Chemical Industry Emerging Research and Opportunities
Hui Ge (Chinese Academy of Sciences, China) Xingchen Liu (Chinese Academy of Sciences, China) Shanmin Wang (Oak Ridge National Laboratory, USA) Tao Yang (China University of Petroleum, China) and Xiaodong Wen (Synfuels China, China)
Engineering Science Reference • copyright 2017 • 162pp • H/C (ISBN: 9781522522744) • US $135.00 (our price)

Sustainable Nanosystems Development, Properties, and Applications
Mihai V. Putz (West University of Timişoara, Romania & Research and Development National Institute for Electrochemistry and Condensed Matter (INCEMC) Timişoara, Romania) and Marius Constantin Mirica (Research and Development National Institute for Electrochemistry and Condensed Matter (INCEMC) Timişoara, Romania)
Engineering Science Reference • copyright 2017 • 794pp • H/C (ISBN: 9781522504924) • US $245.00 (our price)

Computational Approaches to Materials Design Theoretical and Practical Aspects
Shubhabrata Datta (Calcutta Institute of Engineering and Management, India) and J. Paulo Davim (University of Aveiro, Portugal)
Engineering Science Reference • copyright 2016 • 475pp • H/C (ISBN: 9781522502906) • US $215.00 (our price)

Green Approaches to Biocomposite Materials Science and Engineering
Deepak Verma (Graphic Era Hill University, Dehradun, India) Siddharth Jain (College of Engineering Roorkee, India and University of Alberta, Canada) Xiaolei Zhang (Queens University, Belfast, UK) and Prakash Chandra Gope (College of Technology, G.B.Pant University of Agriculure and Technology, Pantnagar, India)
Engineering Science Reference • copyright 2016 • 322pp • H/C (ISBN: 9781522504245) • US $165.00 (our price)

IGI Global
DISSEMINATOR OF KNOWLEDGE

701 East Chocolate Avenue, Hershey, PA 17033, USA
Tel: 717-533-8845 x100 • Fax: 717-533-8661
E-Mail: cust@igi-global.com • www.igi-global.com

Editorial Advisory Board

Table of Contents

Section 3

Detailed Table of Contents

Concurrent with the development of new generation of gas turbines, many attempts have been made to introduce advanced thermal barrier coatings with lower thermal conductivity and higher temperature stability. Most of the research to improve TBCs performance are based on two general approaches: structural modifications and chemical modifications. In most cases, the improvements in some properties are at the expense of loss of some other properties. Changing in the TBCs architecture and the application of multilayer coatings, consisting of layers with engineered properties based on the requirements, is a solution to achieve a combination of desired properties. In all of these development methods it is to be understood that the principle is reducing the possibility of formation of cracks, but, once formed, all such cracks can grow under and thermal cycles and eventually lead to coating delamination and spallation. Self-healing is the most precious phenomenon to overcome this problem.

The zirconates of rare earth elements, such as Sm2Zr2O7, can be an alternative material for zirconia modified by yttria (8YSZ) usually used as an insulation layer in thermal barrier coatings (TBC) systems. This chapter describes the morphology of feedstock zirconate powder, internal morphology and selected properties of different samarium zirconate TBC systems. These included: composite TBC coatings of Sm2Zr2O7 + 8YSZ type with different ratio of both used to coatings deposition powders (25/75, 50/50 and 75/25) as well as the TBC of double ceramic layer (DCL) type with an 8YSZ internal layer and an outer layer of Sm2Zr2O7 type, and a monolayer TBC based on Sm2Zr2O7. Another presented subject is related to the oxidation resistance of TBC systems during static oxidation test at temperature 1100°C. In this case, the special emphasis was taken on the characterization of thermally grown oxides (TGO) zone thickness where the most important phenomena related with overall live-time of TBC systems usually take place.

This chapter describes how for power generators like gas turbines and aero engines, the economic and environmental challenges are increasing day by day for producing electricity more efficiently. The efficiency of power generators can be increased by changing its operating conditions like inlet temperature and procedure. Currently, the inlet temperature to the industrial gas turbine is reaching up to 1400°C. Also, in aero engines, the ring temperature reaches around 1550°C. Therefore, the coatings used in aero

This chapter describes how the aim of duplex surface engineering includes chronological application of two surface modification technologies for the production of a surface, with collective properties. Duplex treatment of nitriding and carbonitriding of austenitic stainless steel is of high technical importance owing to its capability to increase hardness, corrosion and wear resistance of treated surface. Duplex treatment has been utilized to enhance the surface mechanical properties of austenitic stainless steel (AISI 304). The microstructure of nitrided surface indicates the development of nitride phases, Fe4N, Fe2N, CrN, Cr2N and γN whereas, duplex treated films shows the formation of FeC, Fe3C, Fe7C3, Cr3C2, Cr7C3, along with nitride phases like Fe3N. Both nitrided and duplex treated samples show the formation of cauliflower like grains. Surface micro hardness of treated substrates has been dependent on the variation of crystallite size and increased by 1.26 times the hardness of nitrided sample and 4.60 times the hardness of the untreated substrates.

Foreword

The most important area of the high temperature coatings is application of thermal-barrier coating (TBC) that consist of usually zirconia-based refractory oxide and Ni-based intermetallic bonding layer, applied by thermal spraying (TS) or electron beam vapor deposition (EBPVD). This multilayer system is widely used as a protection of metallic components operating at high-temperatures, which include for example combustion chambers and other static components of engines which are manufactured in various international companies. Research and development of TBC is driven by need to increase the service temperature and or life of jet engines and auxiliary power units. This book offers an insight into current trends and approaches towards the research and development of TBC and their performance testing.

During last years the TBC production expanded and majority of the global production of engine parts cannot be realized without proper design of TBC systems. The increasing requirements on material properties which are caused by more aggressive service conditions (increased temperature and mechanical loading) consequently lead to development of new advanced materials and protective coatings.

The production is most economically performed with air plasma spray (APS) and therefore newly developed TBC have to be tailored precisely for APS technology where particles of powder either ceramic (top coat) or intermetallic (bond coat) are injected into plasma flame, melted, and sprayed on substrate surface – usually combustor liner made of nickel based superalloy. As was mentioned above, the current development within thermal spraying field is mainly focused on the advanced TBC with improved performance properties. Development is divided to three domains which are concentrated on: i) reduction of oxides and porosity level in bond coat (BC); ii) development of top coat (TC) with low-thermal conductivity and high strain tolerance and iii) CMAS resistant TBC. The key of the development is testing of advanced TBC which are sprayed by APS and tested on durability with modern testing equipment following acceptable degree of worldwide unification.

With the increased components lifetime the necessary service interruptions (which consist mainly from diagnosis, disassembly, repair of defective parts and the subsequent re-assembly and final inspection) are less frequent. The increased TBC life is closely connected with cost saving thanks to reduction of damaged parts. As was indicated, the TBC mostly consist of two separate layers: BC and TC. The BC is a metallic/intermetallic layer (MCrAlY, M = Ni, Co or both) and its main task is to ensure strong bond between substrate and TC and protect the substrate material against oxidation and hot-corrosion at high temperature. BC also eliminates thermal expansion coefficient mismatch between ceramic and metal substrate. Ceramic TC (as the most reliable is considered up to now ZrO2 + Y2O3, labeled YSZ) is designed to provide thermal insulation and therefore it has low thermal conductivity. In service, component as the combustor is exposed to high temperatures of flue gas (800–1300°C) combined with thermal cycling (takeoff/landing of the airplane). High service temperature causes chemical and phase changes occur in all materials, including ceramic TC, which consequently leads to degradation and failure of the coating (called spallation).

There are several possible ways of lowering thermal conductivity of a TBC. One way is to increase porosity of a TBC layer, another way is to prepare new ceramic materials either by doping YSZ by rare-earth elements or using new family of materials for example with the perovskite structure (ABO_3). It might also possible to lower thermal conductivity by introducing nanocrystalline structure or by designing advanced functionally graded materials. The aim of the Low Thermal Conductivity TBC development is to attempt top coat TBC with lower thermal conductivity than those of the conventional YSZ. Materials identification is the necessary first step. These certainly include rare-earth elements as dopants to YSZ (such as La, Nd, Sm, Gd, Dy, Er) or other elements (Sc). Some of these materials are commercially available. Others should be synthesized by using ball milling and other powder chemistry modification routes. The acquired powders might arrive in various conditions thus care must be taken to prepare the powders for the involved plasma spraying techniques. This step includes milling, sieving or possible agglomeration to obtain the right powder size distributions for plasma spraying techniques. Moreover, the novel designed approach termed Liquid Feedstock using nanometric powder colloidal suspensions of supplementary material fed into the plasma jet would also be used. There must be an essential step of organic chemistry with reproducible dispersant agent development before this technique have to be transferred into an industrial scale.

Standard MCrAlY bond coated samples are be used as substrates. The plasma spraying of the top coat involves optimization of the spraying parameters based on in-flight particle diagnostics and feed-back from basic evaluation of coating properties (porosity, materialography, etc.). In addition to newly selected materials, combination of old and new materials is be plasma sprayed to form a functionally graded material structure that will utilize higher temperature stability at its surface. Development of highly porous ceramic top coats will require special tuning of the spraying parameters.

Chemically resistant TBC coatings are desirable. With the increasing turbine inlet temperatures, the CMAS (calcium-magnesium-aluminium-silicate) attack of TBC from the environment is becoming a critical issue in the development of next generation of aerospace engines. Three basic types of protective coatings are classified as: impermeable, sacrificial and non-wetting to CMAS. The impermeable coatings protect the TBC by inhibiting the liquid CMAS from infiltrating into or reacting with the TBC at the operating temperature of turbine. This protective coating is defined as a dense, non-cracked, nonporous and non-oxide layer. The principle of sacrificial oxide layer is based on increasing the melting point or the viscosity of the CMAS deposit. This phenomenon is caused by chemical or physical changes of sacrificial oxide layer in contact with CMAS at operating temperatures, when the CMAS is dissolved or react with the sacrificial layer. Consequently, the non-liquid by product material or at least more viscous compound is formed. The result is, that the CMAS does not flow or form reactive liquid.

The non-wetting coatings are outer layers that minimize contact between underlying layers and the molten CMAS. The integrity of TBC is enhanced due to reduction of CMAS ability to penetrate the TBC via capillary action. The research of CMAS resistant part is focused on different kinds of new un-conventional and conventional CMAS resistant coatings. Different conditions of deposition techniques and their process parameters are under development due to tailoring anti-CMAS layers.

The test samples sometimes include both coatings and free-standing parts. The free-standing parts enable more precise measurement of thermal conductivity of the coatings be the known laser flash method. The complex characterization of TBC must include precise porosity determination, thermal conductivity measurements, isothermal oxidation, cyclic oxidation and JET tests. This specific JET test is able to partially simulate real service conditions of parts like combustion chambers and others. Beside TBCs, this book is about anti wear, erosion-corrosion, and self-lubricant coatings, which withstand in high temperature environments.

Pavel Ctibor
Academy of Sciences of the Czech Republic, Czech Republic

Pavel Ctibor *was born in 1971. In 1994, Dr. Ctibor graduated from Faculty of Mechanical Engineering, Czech Technical University (CTU), Prague. During 1995-2000, Dr. Ctibor underwent post-graduate studies at the Faculty of Nuclear Sciences and Physical Engineering, CTU, Prague. Since 1994, Ctibor has been an employee of the Institute of Plasma Physics (IPP), Academy of Sciences of the Czech Republic, Prague. Since 2013, Dr. Ctibor has been regularly teaching subjects related to materials at Faculty of Electrical Engineering (FEE) CTU, Prague. Since 2016, Dr. Ctibor has been an Associate professor at the Department of Electrotechnology, FEE CTU, Prague. Between April 2003 and February 2004, Dr. Ctibor was a post-doctoral researcher at Higher Engineering School, Limoges, France. In 2009, Dr. Ctibor spent a one month stay at University of Limoges, France, upon invitation. Dr. Ctibor has focused on research on plasma sprayed materials, namely by a Water-stabilized torch and sintered materials, namely by Spark Plasma Sintering technique. Dr. Ctibor is the author of about 100 scientific publications in reviewed journals and conference proceedings and of several patents. Dr. Ctibor is a member of the Scientific committee of the European Materials Research Society Spring Meeting 2009. Dr. Ctibor is also a two times member of the Best paper award committee in the Journal of Thermal Spray Technology.*

Preface

INTRODUCTION

Materials are frequently exposed to environments which change their chemical and physical characteristics on account of a reaction going on between environment and the material surface.The science and technology of surface preparation or improve material properties have gained importance during the last decade in many industrial sections. A coating is defined as a material that is applied to cover the surface of an object to add a desired functionality to a material which otherwise would not be present. The coating has been considered a very efficient procedure for protecting or preserving substrate and therefore has been developed and optimized continuously.

During the past centuries, our society has witnessed a great development of coatings for a variety of purposes. With technological advancement and the need for new coatings, a wide range of engineering materials on different substrates are used for electrical, magnetic, optical, thermal and medical applications. Different ceramic, polymeric and metallic materials are in this category. This kind of new coatings plays an important role in the development of a variety of coating strategies. One of the main drivers of the intense development has clearly been the high-temperature coatings such as thermal barrier, hard coatings, anti-wear and corrosion coatings.

A large number of industrial processes operate in very aggressive environments characterized by increased temperature gradients, large stresses on individual components, high pressure and the presence of an oxidizing and corroding atmosphere. The metallic alloys cant operate in mentioned harsh environments.Therefore, the only choice for protection against corrosion and oxidation is applying the coatings. In this regards, the coatings can provide barriers between the outer environments and the metallic alloys. High-temperature-resistant coatings are designed to withstand temperatures ranging from 150°C to over 2000°C while providing protection against harsh environments. High-temperature coatings belong to a technologically important, economically lucrative, and fast evolving field. This kind of coatings are often found in the power plant, aerospace, petrochemical, and manufacturing industries.

There is a so many valuable information available on various aspects of high-temperature coatings. However, the information disperses in conference proceedings, journals, book chapters, handbooks, and published industrials reports. Therefore, the author tries to collect valuable information together about high temperature coatings in one collection.This book addresses the different aspects of high temperature coatings. It includes basic understanding of materials, environmental interactions, principles of physical ceramics and metallurgy, manufacturing and different characterization techniques. One of the most important parts of this book is thermal barrier coatings (TBCs).

This book elaborates the different aspects of TBCs. In gas turbines, the TBC surface may be exposed to hot gases with temperatures in excess of 3000°F. In order to survive such environments without melting, the TBCs system has been used.The function of TBC is to reduce the substrate temperatures to increase the lifespan.

TBCs are generally a combination of multiple layers of coatings with each layer having a specific requirement and function. The topmost layer typically Y_2O_3 stabilized ZrO_2 (YSZ) topcoat, with low thermal conductivity, provides thermal insulation and a bond coat layer, which is typically MCrAlY (M = Ni, Co, Fe or a combination of these elements) as an oxidation resistant layer. Bond coat is often first deposited onto a metallic substrate to protect the substrate from high temperature corrosion and oxidation.TBCs are usually deposited either by electron beam-physical vapor deposition (EB-PVD) or by air plasma spraying (APS). However, because of the high deposition efficiency and comparatively cost-effective deposition conditions, APS is a popular one. There are some important factors affecting TBCs lifetime, including high temperature oxidation, infiltration of detrimental factors such as molten salts, which lead to spalling upon thermal cycling and formation of thermally grown oxide (TGO) that are discussed in this book. Oxidation is another aspect of this book, metallic substrates used at high temperatures are mainly optimized with regard to their mechanical properties. In a corrosive atmosphere or oxygen containing, however, they suffer from poor performance. So, anti-oxidation protection films are developed for metaillc alloys. Therefore, an effective protection of super alloys is essential for safe operation under different conditions. Anti wear, erosion-corrosion and self lubricant coatings are another aspects of the book. There are different techniques such as sputtering, thermal evaporation, pulsed laser deposition, gas phase chemical deposition methods (chemical vapor deposition, atomic layer deposition), wet-chemical strategies (sol-gel techniques, electroplating) and thermal spray methods enable to deposit materials from gas, solution or solid phase and transfer their unique properties to the substrate to produce high temperature coatings. Each of these techniques has advantages and disadvantages over the other, which some of them discussed in the book.

As a mentioned, among different aspects of coating science, this book is focused on high temperature coatings. The reader will find examples of exciting applications that are enabled by various kinds of coatings and materials. Those span various emerging application fields including thermal barrier, anti-wear, corrosion protection, thermal shock and oxidation and so on.The book is not focused on a specific deposition technology which it is a value of the book. In addition, this book is a truly multidisciplinary reference for the reader interested in high temperature coating.

ORGANIZATION OF THE BOOK

This book covers key aspects of the production methods, modification, characterization, application of the high temperature coatings. The book can be divided into three sections. The first one (Chapter 1–8) is related to general aspects of modern thermal barrier coatings, synthesis of coatings and characterization of them and investigate their properties. The second part (Chapter 9-12) is related to specific applications of hard, anti wear and erosion-corrosion of coatings which are produced by thermal spray methods. The last section (Chapter 13-17) is related to the hard and anti wear coatings which are produced by sol -gel, plasma assisted chemical vapor deposition (PACVD) and magnetron sputtering methods. The last part of section 3 is related to ultra high temperature coating that can protect C-C composites against oxidation and diffusion coatings.

The objective of Chapter 1 is to determine the effect of substrate surface conditions and temperature on splat formation and morphology; the effect of plasma spray parameters on splat formation and morphology; and thermal spray process. In this chapter physical behavior, such as thermal conductivity and wetting ability at impact between the thermal spray particle and the substrate is reviewed. Also, different kinds of thermal spray such as flame spray, plasma spray, HVOF Processes and cold spray are discussed. Moreover, the recent developments and the future perspectives in the field of TBCs, with the attention on advanced materials, new methods (Suspension and Solution Precursor Plasma Spraying) and new architectures as well as explaining the mechanisms affecting the failure of TBC systems are described in Chapter 2.

It should be noted TBC coatings with new materials, and structures known as new TBCs are described in the Chapters 3, 4 and 5. For better flow and ease of understanding, chapter 3 is physically separated into three sections starting with an introduction and nano TBC, multilayer TBC in the middle, and self healing TBC at the end. In all of these development methods it is to be understood that the principle is reducing the possibility of formation of cracks, but, once formed, all such cracks can grow under the thermal cycles and eventually lead to coating delamination and spallation. Self-healing is the most precious phenomenon to overcome this problem. Chapter 4 introduced samarium zirconate TBC systems: composite TBC coatings of $Sm_2Zr_2O_7$ + 8YSZ as well as the TBC of double ceramic layer (DCL) type, and monolayer TBC based on $Sm_2Zr_2O_7$. Another presented subject is related to the oxidation resistance of TBC systems during static oxidation test at 1100°C. lanthanum-titanium-aluminium oxide based thermal barrier coating are described in Chapter 5. Recently, lanthanum-titanium-aluminium oxide that is $LaTi_2Al_9O_{19}$ (LTA) is proposed as TBC material, which is having excellent phase stability from ambient to 1600°C. It is the combination of rare earth element and ceramic materials and shows better physical and mechanical properties.

The main physical and chemical concepts related to the performance of engineering coatings at high temperature. Especially, thermal shock and hot corrosion are described in Chapter 6. Several innovations for improving the performance of high temperature coatings have been introduced in the Chapter 7. Nanocrystallization has been demonstrated to be a practical way to prolong the lifetime of high temperature coatings. Enamel coatings with enhanced strength, toughness and thermophysical properties were developed for improving the hot corrosion resistance of superalloys are described in this chapter. The last chapter of the first section (8) is dedicated to Slurry spray technique (SST). This chapter reports the findings of the corrosion studies in simulated industrial corrosive environments and characterization results of the six sets of slurry sprayed mullite-nickel based coatings. The developed coatings are recommended for use in applications to endure the elevated temperature and inflict corrosion.

The demand for high temperature solid lubrication in the range 300-1000 °C has led to the development of self-lubricating wear resistant coatings. Chapter 9 describes the preparation, functionalization, analysis and usages of the high temperature self lubricant coatings. This chapter introduces typical nanostructured ceramic matrix coatings produced by plasma spraying for self-lubricating wear resistance purposes in a wide temperature range. Also, the lubrication mechanisms at moderate and high temperatures are discussed. Chapter 10 provides an overview of the role of various thermal spray coatings for erosion-corrosion protection of industrial components, feedstock materials, and their characteristics. Influence of process variables, test parameters, coating thickness and working temperature on the performance of thermal spray coatings for erosion-corrosion have been discussed, whereas Chapter 11 focuses on the effect of critical plasma spray parameter (CPSP), on microstructural and wear characteristics of conventional and nanostructured alumina-titania coatings. Chapter 12 presents reactive plasma spraying (RPS) technology as a promising solution for the in situ fabrication of several nitride ceramic coatings. The main attractive prospects of RPS for fabricating nitride coatings are specifically highlighted. Successful development of various high-temperature nitride coatings is presented in this chapter. Moreover, process optimization, the relationship between reaction and process parameters and the influence on coatings formation are comprehensively discussed.

The first part of the last section of the book (Chapter 13) is related to introduce the fundamentals of the sol-gel method to produce organic-inorganic hybrid (O-IH) protective thin coatings and discuss the methodologies used to apply these materials onto different metallic substrates for erosion and corrosion protection. Chapter 14 describes titanium nitride (TiN), titanium carbide (TiC), diamond like carbon (DLC) single and multilayer coatings that utilize in harsh environments. These hard coatings were usually produced by PACVD method as a modern technique. PACVD is used to deposit thin coatings for different usages such as automobile and aerospace parts. Mechanical features of these coatings were perused, whereas Chapter 15 focuses on the Thin-film deposition by magnetron sputtering. Magnetron sputtering is a state-of-the-art thin film deposition technique versatile for several unique applications. High temperature coatings for protection against oxidation of C-C composites are covered in Chapter 16 in reasonable detail including processing, characterization, and properties. Various methods used to protect C-C composite against oxidation (such as surface treatment, adding inhibitors in the matrix, and outer coating) are discussed. High-temperature outer coatings are the most effective method to the protection of C-C composites oxidation. Finally, Chapter 17 discusses duplex surface engineering. The aim of duplex surface engineering includes the chronological application of two surface modification technologies, for the production of a surface, with collective properties. Duplex treatment of nitriding and carbonitriding of austenitic stainless steel is of high technical importance owing to its capability to increase hardness, corrosion and wear resistance of the treated surface.

CONCLUSION

Academic societies involved with material science, microtechnology and nanoscience, consisting of researchers, scientists, engineers, and students around the globe with the focus on materials, their properties and applications, are considered the target audience of this book. The book aims to light the way to understand the basic and professional concepts of the high temperature coatings and hopefully help the target audience to find new advances in this field of study. It also guides them through the path to open new possibilities that are not imaginable for now. Also, the book covers new information around the topic with the aim to explain the research history in this area, to pass the knowledge on and to encourage other researchers and student's efforts on this ongoing subject.

Amirhossein Pakseresht
University of Tehran, Iran & Materials and Energy Research Centre, Iran

Acknowledgment

The editor would like to acknowledge the help of all the people involved in this project and, more specifically, to the authors and reviewers that took part in the review process. Without their support, this book would not have become a reality.

First, the editor would like to thank each one of the authors for their contributions. Our sincere gratitude goes to the chapters' authors who contributed their time and expertise to this book.

Second, the editor wishes to acknowledge the valuable contributions of the reviewers regarding the improvement of quality, coherence, and content presentation of chapters. Most of the authors also served as referees; we highly appreciate their double task.

Special thanks are also due to Marianne Caesar of IGI Global for her kind help and cooperation from the very beginning of this project. The author is deeply grateful to Iran National Science Foundation: INSF.

Amirhossein Pakseresht
University of Tehran, Iran & Materials and Energy Research Centre, Iran

Section 1

Chapter 1
Thermal Spray Methods and Splat Formation

Wei Xie
Seaway Advanced Materials Co. Ltd., China

ABSTRACT

This chapter describes how surface coatings play a significant role in manufacturing industries due to improvements in the properties of the engineering components that lead to longer service life and reliability. Thermal spray is well suited for large structures and can be easily repaired. Previous publications concerning the principle of thermal spray processing; powder and their fabrication methods; the effect of substrate surface conditions on splat formation and morphologies; the effect of process parameters on splat formation and morphology; and thermal spray coating morphology are reviewed. Also described in detail is physical behaviour, such as thermal conductivity and wetting ability at impact between the thermal spray particle and the substrate, influences the individual splat geometry and coating build-up and, consequently, the coating properties.

INTRODUCTION

Thermal spray processes hold a unique position in the family of available surface engineering technologies since coatings in excess of 20 µm can be deposited over large areas at high deposition rates (Petrovicova & Schadler, 2002). The maximum thickness deposited by plasma spray was reported (Sweet, 1993) to be 50.8 mm. However, thick coatings have technical concerns since substantial residual stresses may evolve and these need to be controlled by substrate cooling. Such processing needs are critical considerations for the implementation of thermal spray; especially when they exhibit advantages over alternative methods such as electroplating, physical vapor deposition (PVD) and chemical vapor deposition (CVD) (Kuroda, Kawakita, Watanabe, & Katanoda, 2008).

Coating materials available for thermal spray include metals, alloys, ceramics, plastics and composites. They are fed into a spray torch in the form of powder or wire and heated to a molten or semi molten state and accelerated towards substrates in the form of particles and droplets. The particles are deposited onto the substrates in the form of splats with a size range of several micrometers to ~100 µm (Kuroda et al., 2008).

DOI: 10.4018/978-1-5225-4194-3.ch001

The coating formation process is generally determined by the prime process parameters, which include substrate roughness, substrate temperature, impact velocity, stand-off distance (SOD), and nature of the local atmospheric environment (Bussmann, Chandra, & Mostaghimi, 2000; Herman, Sampath, & McCune, 2000; Sweet, 1993; Vardelle, Moreau, & Fauchais, 2000). Many of these factors are inter-related. For example, SOD will influence impact velocity and substrate temperature due to heat transfer from the thermal spray torch. The above processing variables influence the physical nature of the intrinsic building blocks of thermal spray coatings; i.e., the splat morphology (Fauchais, Fukumoto, Vardelle, & Vardelle, 2004; Sampath, Jiang, Matejicek, Leger, & Vardelle, 1999).

PRINCIPLE OF THERMAL SPRAYING

Thermal spray belongs to the family of granular methods of deposition. Fine particles of polymer or ceramic are propelled using a flow of carrier gas through a flame. The particles are heated, softened and ejected towards an often heated substrate where they collide, flatten and assemble to form a coating that consists of hundreds of thousands of well-adhered flattened particles.

Thermal Spray Techniques

Plasma, wire arc, and combustion spraying, including conventional flame and high velocity oxygen fuel (HVOF) processes, constitute the major thermal spray processes (Figure 1).

Flame Spray Processes

FS was first introduced by Dr. M. V. Schoop in the early 1900s (Fauchais, Vardelle, & Dussoubs, 2001), and this process was principally used for low melting point materials such as tin and lead. Later the wire-flame spray process was developed to spray higher melting point materials such as steel. However, the melting temperature of materials that could be so-sprayed was limited to 2,800 °C.

Schematic of flame spray torch can be found on Figure 2. and Flame Spray Technology website (Flame Spray Technologies, 2007). The powder feeder is located on the top of the torch so that the spray materials can be feed under gravity. The torch can spray a wide range of materials such as nickel, cobalt or ceramics. The chemical reaction between oxygen and acetylene produces the heat source. The powders are conveyed into the flame and compressed air is used to fragment and accelerate the molten particles towards the substrate. The schematic diagram of a flame wire spray torch also can be found on Flame Spray Technology website (Flame Spray Technologies, 2007). The feedstock is fed into the flame in the form of a wire and compressed air is used to break up the molten metal droplets and accelerate these more finely divided particles onto the substrate.

Plasma Processes

The torch for plasma spraying is based on the Gerdien-type plasma generator (Gerdien and Lotz, 1922). The first plasma spray coating process was introduced by Reinecke in 1939 (Davis, 2004). The process of spraying with the use of a plasma was patented by Gage *et al.* (1962), as well as by Giannini and Ducati (1960).

Figure 1. The thermal spray process family
Xie, 2013.

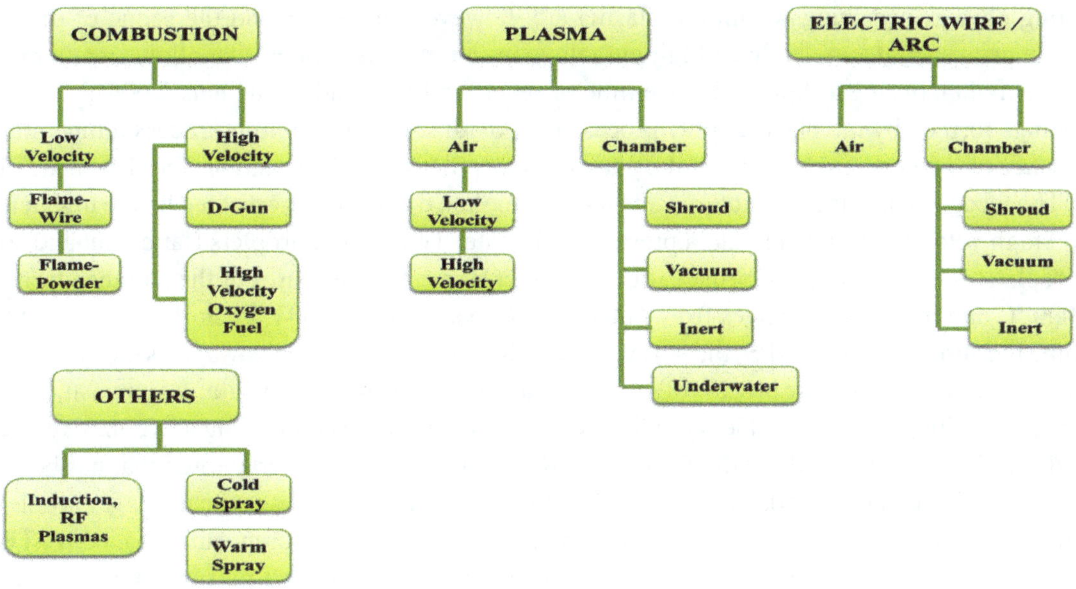

Figure 2. Scheme of flame spray
Xie, 2013.

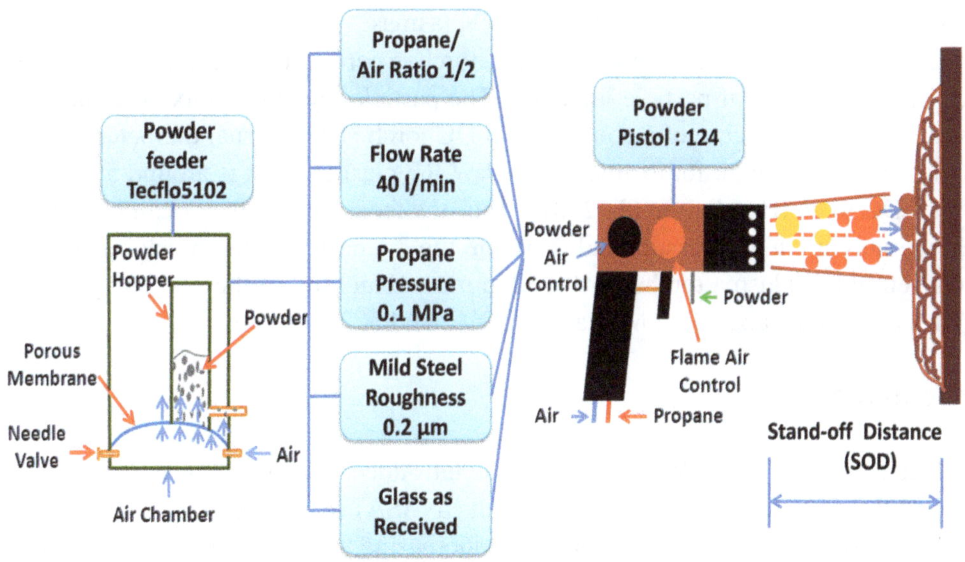

The use of the plasma jet has extended the technology possibilities to any material that can melt, including many refractory materials. The schematic of a plasma torch can be seen form University of Toronto website (University of Toronto). The process is commonly described as "atmospheric plasma spray", and is abbreviated to "APS". The direct current (DC) plasma torch consists of an anode made of copper and cathode made of thoriated-tungsten. The electric arc that is produced between the two electrodes heats the gas flowing between them. The plasma jet is generated by the gas that is partially ionized by the electric arc. The feedstock is injected into the plasma torch during spraying. There is quite interesting physics occurring within the plasma arc; however, this is beyond the scope of the current work. The primary outcome is that a high-enthalpy environment is created that allows many materials to become molten during a short residence time of several milliseconds (Fauchais, 2004).

Coatings formed by the plasma spray process may oxidize due to the high process temperature and the presence of oxygen. A schematic of the plasma spray process can be found on TSET website (TSET, 2008). The feedstock is injected into a plasma jet. The molten droplets are projected by the high temperature and high velocity plasma jet onto a prepared substrate. The molten droplets flatten out and solidify to form splats. A thick coating is formed after repeated passes. Temperatures of the jet range from 6,000 to 15,000 °C and the velocity ranges from 600 to 2,300 $m.s^{-1}$ (Davis, 2004; Lech Pawlowski, 2008). To overcome the shortcoming of APS, the soft vacuum (Kohzaki, Tsujita, Takizawa, & Kinoshita, 1987) or controlled (neutral) atmosphere process (Tsujita, Shibayama, Takizawa, Kinoshita, & Uematsu, 1987) was introduced; resulting in dense and less oxidized coatings with improved bonding and cohesive strength.

Boulos (1992) developed the radio frequency (RF) induction plasma spray torch that has been commercialized by Teckna in Canada. The RF torch exhibits low gas velocities and can spray particles with diameters over 100 μm. The schematic of a RF induction plasma spray torch can be found on TEKNA website (TEKNA, 2009). The RF induction plasma, or induction coupled plasma (ICP), is generated through the electromagnetic coupling of electrical energy into a discharge cavity. A plasma jet generated in an induction plasma torch exhibits a high specific enthalpy of 1–10 MJ/kg; depending on the plasma gas composition. It is claimed that the central axial feeding system provides a flexible and efficient approach compared to DC plasma torches. The feedstock particles are heated thoroughly because the residence time of the particles is greater than for DC plasmas.

The prime distinctions between the RF and DC torches pertain to (i) the internal diameters of the torches and (ii) the direction of particle injection. The particle injection is axial in the case of the RF torch whilst it is radial in the case of a conventional DC torch. The internal diameter of the RF torch is larger than that of the DC torch; for example, a 50 kW RF plasma torch is around 50 mm compared to 6 to 8 mm for a DC torch (Fauchais et al., 2001). Thus the feedstock diameter can be greater. However, an RF torch operates at a high pressure and therefore the operating cost is much higher than that for a DC torch. Currently DC plasma torches are used in more than 90% of thermal spray processes within the industrial sector (Fauchais, 2004; Mostaghimi & Chandra, 2007).

HVOF Processes

The high-velocity oxygen fuel (HVOF) process was introduced in the 1980s under the trade name as Jet Kote®, by Doloro-Stellite Inc. located in Goshen (IN-ISA) (Fauchais et al., 2001). There are many designs of HVOF torches with combustion pressures increasing from 0.4 to 1.35 MPa (Thorpe & Richter, 1992). High-velocity air fuel (HVAF) torches were also developed (Browning, 1992) where a pressure of 0.8 MPa was achieved.

A schematic of a typical commercial HVOF spray apparatus can be found in Kuroda' paper (Kuroda et al., 2008). A mixture of fuel and oxygen is propelled into a combustion chamber and ignited. The combustion gas flows through a converging–diverging nozzle at a pressure of ~1 MPa and travels through a straight barrel. The fuels can be either gases, such as hydrogen, methane, propane, propylene, acetylene or natural gas; or liquids such as kerosene. The gas velocity at the end of the barrel is normally above

1,000 m/s. Therefore, after being fed into the jet at the feed ports and heated, the powder material will be ejected and deposited onto the substrate at sonic or supersonic velocity (Kuroda et al., 2008).

The figure which indicates the characteristics of the coatings sprayed using HVOF and plasma spray processes can be seen in Sidhu'paper (Sidhu, Agrawal, & Prakash, 2005). HVOF coatings exhibit higher density, hardness and bond strength; an increased thickness capability; smoother surface finish; lower oxide levels and low porosities, less effect of the environment during the spray process and demonstrate enhanced corrosion and wear resistance properties compared with plasma spray coatings (Sidhu et al., 2005).

The HVOF process has wide applications in industry for preparing WC/Co-based wear-resistant coatings for drilling tools, and nickel-based corrosion-resistant coatings for chemical reactors (Li & Christofides, 2009). The formation of nanostructured coatings *via* HVOF has been investigated since 1989 (Ardanuy, Velasco, Maspoch, Haurie, & Fernández, 2009). The coatings sprayed using nano agglomerated powders are claimed to exhibit excellent properties in terms of micro hardness, elastic modulus, wear resistance and thermal conductivity (Meure, Wu, & Furman, 2010; Tsunashima, Nishioji, Hirasawa, & Yano, 1992).

There are more and more advanced process and equipement developed with the development of morden technology (Ciubotariu, Frunzăverde, Mărginean, Şerban, & Bîrdeanu, 2016; Graziani, Bianchi, Sassoni, Russo, & Marcacci, 2017; Kumar & Kandasubramanian, 2016; Yue Xu, Hu, Xu, & Li, 2017). The detailed thermal spray facilities can be found in the document Thermal spray equipment guide (Metco, 2014). Metco has advanced the HVOF equipments with gas fuel and liquid fuel separately (Metco, 2014). Recently laser remelting process for HVOF-sprayed Stellite 6 wear resistant coatings proves a homogenous Stellite 6 layer with good adhesion to the substrate (Ciubotariu et al., 2016).

Cold Spray

Cold spray was originally developed in the early 1980s by Papyrin *et al.* (Papyrin, Kosarev, Klinkov, Alkhimov, & Fomin, 2007) at the Institute of Theoretical and Applied Mechanics of the Siberian Branch of the Russian Academy of Sciences in Novosibirsk. The schematic of a cold spray facility can be seen from De Freitas Cunha Lins' paper (De Freitas Cunha Lins, Cury, & Moreira, 2008) which consists mainly of a supersonic nozzle, the powder feeder, the gas heater and the source of compressed gas. Particles of size 1-50 μm in diameter are propelled to speeds of 300–1,200 m/s by the carrier gas and exit through a converging-diverging de Laval type nozzle (Papyrin et al., 2007). The ejected particles have sufficient energy to impact onto a substrate, deform plastically, bond mechanically and form a coherent and integrated coating. Powder particles are always sprayed at a temperature lower than the melting point of the material.

Cold spray is an all-solid-state process, thus making it suitable for the deposition of oxygen-sensitive materials, such as Aluminum, copper, and titanium; and for temperature sensitive materials, such as nanostructured and amorphous powders. Compared to other spray technologies, cold spray has unique advantages (Irissou, Legoux, Ryabinin, Jodoin, & Moreau, 2008), including (i) the coatings exhibit wrought-like microstructures with near theoretical density; (ii) the spray footprint is typically 1-25 mm^2 and well defined, allowing for precise control on the area of deposition; (iii) the coatings can be produced with compressive stresses, thus ultra-thick coatings (5-50 mm) can be built-up without adhesion failure (Karthikeyan, Kay, Lindeman, Lima, & Berndt, 2000); and (iv) three dimensional free standing structures can be formed at high deposition rates: attributes which all make cold spray promising for industrial applications.

Table1 summaries important thermal spray process attributes. The maximum temperature of conventional flame spray (FS), HVOF, and cold spray is 2,800 K, 3,600 K, and 1,173 K, respectively. The maximum temperature of wire arc, air plasma and vacuum plasma is around 4,000 K. The velocity of conventional FS, HVOF, cold spray, wire arc, air plasma, vacuum plasma and detonation torch is 40–100 ms^{-1}, 200–1,000 ms^{-1}, 300–1,200 ms^{-1}, 50–150 ms^{-1}, 50–2000 ms^{-1}, 1500–3500 ms^{-1} and 910–3000 ms^{-1}, respectively (Pawlowski, 2008).

The size of the powder should be around 10 μm—150 μm which can be checked via manufacturer's specification according to different thermal spray methods. The better the fluidity of the powder, the smoother the thermal spray process for all thermal spray methods. The fluidity of the powder can be measured with Freeman Technology FT4 Powder Rheometer.

POWDERS AND THEIR FABRICATION METHODS

Nanostructured Ceramic Agglomerated Powders

Nanostructured, also referred to as nanophased, materials exhibit a crystal grain size between 1 nm and 100 nm. Such crystals may contain tens of thousands of ordered atoms (Siegel, 1996). Nanostructured ceramics and metals have unique properties compared with their microstructure counterparts. Ceramics are slightly ductile due to the capacity of small grains that may permit sliding and deformation. Nanostructured metals are stronger, which can be deduced from the Hall-Petch law (De Sansal, Devincre, & Kubin, 2010). Dislocations that are mobile can reduce the strength of materials. The bigger the grain size, then the more dislocations that exist in the grain. Thus, the materials that consist of larger grains

Table 1. Typical attributes of thermal spray processes

Attribute	Conventional Flame Spray	HVOF	Wire Arc	Air Plasma	Vacuum Plasma
		Jet			
Temperature (K)	Up to 3,500	>3,000	Up to 6,300	13,000	22,000
Velocity (ms^{-1})	50–100	500–1,300	50–100	300–2,000	200–600
Gas flow rate, (m^3s^{-1})	0.5–3.2 × 10^{-3}	6.5–18 × 10^{-3}	8.3–50 × 10^{-3}	1.5–3.2 × 10^{-3}	2.5–4.2 × 10^{-3}
Gas types	O_2 C_2H_2 (acetylene) C_3H_6 C_3H_8	air O_2 H_2 CH_4 C_3H_6 C_3H_8 kerosene	air Ar N_2	Ar N_2 He H_2	Ar He H_2
Power input (kW)	20	150–300	2–5	40–200	40–120
		Sprayed Particles			
Maximum temperature (K)	2,800	3,600	~4,000	~4,000	~4,000
Velocity, (ms^{-1})	40–100	200–1,000	50–150	80–800	80–350
Feed rate, (gs^{-1})	0.5–0.85	Up to 2.5	2.5–34	0.85–2.7	Up to 5.6

Source: (Petrovicova & Schadler, 2002).

exhibit lower strength. Grain boundaries obstruct the movement of dislocations, with the result that small–grained materials impede dislocation movement. The industrial potential of nanotechnology was predicted in 1959 by Richard Feynman in his talk "There's Plenty of Room at the Bottom" (Feynman, 2011) .

Nanostructured coatings can be synthesized by methods based on vapor condensation, such as physical vapor deposition (PVD) and chemical vapor deposition (CVD), by controlling the process parameters of deposition (Pawlowski & Dépôts Physiques, 2003). Nanostructured coatings can also be deposited by the thermal spray method *via* cryogenic cooling of the substrate and using nano agglomerated powders (Lech Pawlowski, 2008).

Sampath (High, Painter, & Coleman, 1992) reported that plasma sprayed coatings of the Metco 700 alloy consisted of a large proportion of nanocrystalline amorphous material within the deposits. These nanocrystalline or amorphous phases increased the hardness (i.e., from 550 to 700 HV) and abrasive wear resistance of the coatings (C.C. Berndt, 1998).

Thermal spray ceramic coatings are commonly manufactured from a powder feedstock. These powder particles typically exhibit a particle size distribution varying from 5 to 100 μm. Fine particles, including nanosized particles smaller than 100 nm, cannot be thermal sprayed due to their low mass (Roukes, 2001). Thus, they do not have the inertia required to cross the streamlines in the spray jet and would be segregated to its periphery without depositing on the substrate. That is, the nanoparticles would be blown away from the nozzle and not reach the substrate. The fine nanoparticles may, as well, block the powder delivery hose that transports the powder particles from the powder feeder to the powder injection port of the thermal spray torch. In addition, (i) injecting individual nanoparticles into a spray jet would require high levels of carrier gas flow, which would tend to destabilize the thermal spray jet, and (ii) individual nanoparticles would tend to exhibit low inertial levels to penetrate the stagnation layer that exists close to the substrate surface, which would make the deposition process inefficient (Lima & Marple, 2007).

In order to spray nanoparticles using regular powder feeders, micro particles with a nanostructure are adopted. The nanosized particles are agglomerated *via* spray-drying and then sintered into micrometer sized particles (Ahn, Hwang, Song, Lee, & Kim, 2006; Liang, Liao, Ding, & Coddet, 2005; Lin, Zeng, Lee, & Ding, 2004; Lin, Zeng, Zhou, & Ding, 2003; Song, Ahn, Lee, & Kim, 2006). They are fabricated by creating a dispersion of nanosized particles in a colloidal suspension, followed by the addition of a binder and subsequent spray drying into granules and then sintering (Strutt, Kear, & Roland, 2000). The goal of this process is the agglomeration of nanoparticles into larger, micrometer sized, particles that are spherically shaped.

Liquid Feedstock

Nanostructured coatings formed by a liquid precursor can also be considered as a viable spray process (Fan & Bai, 2016; Karthikeyan, Berndt, Tikkanen, Reddy, & Herman, 1997; Karthikeyan, Berndt, et al., 1997a, 1997b). The schematics of the liquid delivery and injection systems can be found from Pawlowski's paper (Pawlowski, 2008). The liquid can be delivered from a pressurized vessel or from a liquid container by a peristaltic pump to a nozzle or atomizer. A pressurized vessel allows adjustment of the injection velocity by controlling the gas pressure in the system. Multilayer coatings can be achieved by controlling rotation of the two peristaltic pumps in the system. The liquid can be injected with two kinds of injector system: (i) a nozzle having a size ranging from 0.05–1.2 mm; or (ii) an atomizer.

The coatings sprayed using liquid feedstock are thinner than those sprayed using powders. The coating thickness is usually in the range of 10 to 50 µm (Pawlowski, 2008). Liquid feedstock methods of thermal spray demonstrate greater deposition efficiency than the thermal spray of agglomerated nanostructured powders fabricated by spray drying and sintering methods.

Fabrication of Nanozirconia Powders

Suspension plasma spraying (SPS) and solution precursor plasma spraying (SPPS) techniques are relatively recent trends, since the 1990's, as a means to obtain nanostructured coatings (Fan & Bai, 2016; Fauchais, 2008). Nano zirconia powders can be produced by a variety of techniques, including SPS and SPPS processes (Karthikeyan, Berndt, Tikkanen, Reddy, et al., 1997; Karthikeyan, Berndt, et al., 1997a, 1997b). In the SPPS process, a liquid precursor solution is atomized, and this aerosol is injected into a high-temperature plasma flame. The plasma synthesized nanoparticles are deposited on a substrate or electrostatic precipitator (Berndt, 1998). During this process, various phenomena may occur (Berndt, 1998; Fauchais, Rat, Coudert, Etchart-Salas, & Montavon, 2008): (i) evaporation of the solvent; (ii) condensation of the precursor materials and initiation of chemical reactions; (iii) nucleation and growth of grains, followed by formation of dense particles; and (iv) formation of large particles, aggregates and agglomerates. Recently the radio frequency inductively coupled thermal plasma synthesis process has been developed (Samal, 2017). This process provides a high processing rate and yields high purity products.

EFFECT OF SUBSTRATE SURFACE CONDITIONS ON SPLAT FORMATION AND MORPHOLOGIES

Effect of Substrate Surface Wettability

The surface condition (Fukumoto, Nagai, & Yasui, 2006b; Jiang, Wan, Herman, & Sampath, 2001; Tran, Hyland, Qiu, Withy, & James, 2008; Tran, Hyland, Shinoda, & Sampath, 2011) and the wetting characteristics of the substrate play an important role in the morphology of splats (Zhao, Yadian, Li, Liu, & Zhang, 2009). There are three factors affecting the wettability of the substrate. The first factor includes the substrate roughness, porosity and physical properties such as thermal conductivity. Higher surface roughness induces higher wettability of the substrate (Fukumoto, Nishioka, & Matsubara, 2002). High thermal conductivity of the substrate leads to poor wettability of the substrate (Tanaka & Fukumoto, 1999). The second factor includes the composition, thickness and morphology of the oxide layer on the substrate surface. These oxide features are affected by the preheating rate, temperature and time. The thickness of the oxide layer affects the substrate roughness (Christoulis, Pantelis, Borit, Guipont, & Jeandin, 2004), and thus modifies the wettability of the substrate (Fauchais et al., 2004). The third factor relates to the condensates or adsorbents on the surface of the substrates. A higher content of condensates or adsorbates on the surface of the substrates lower the wettability of the substrate; thus, more splash splats occur (Tran, 2010; Tran & Hyland, 2009, 2010; Tran et al., 2008; Tran et al., 2011).

Substrate wettability is one factor that influences the maximum diameter of a single splat. Other factors include the droplet initial diameter, impact velocity, surface tension and viscosity. The wettability of the substrate plays a significant role when the impact velocity becomes low. The higher the wettability of the substrate, then the larger the diameter of a single splat (Li, Ashgriz, & Chandra, 2010), and the less likelihood of a splash splat (Fukumoto & Huang, 1999) feature.

Transition Temperature

The morphology of a single splat is related to the critical preheating temperature, T_c, of the substrate (Fauchais et al., 2001). This temperature is also defined as the transition temperature (T_t) of the substrate by Fukumoto (Fukumoto, Yamaguchi, Yamada, & Yasui, 2007b). The splats are more disc-like when the substrate temperature is over T_c. The splats are extensively fingered when the temperature is below T_c. The value of T_c is quite sensitive to the splat material rather than to the substrate (Fauchais et al., 2001). For alumina and zirconia, T_c is around 200 to 250 °C (Bianchi, Grimaud, Blein, Lucchese, & Fauchais, 1995; Fauchais, Vardelle, Vardelle, Bianchi, & Leger, 1996) for disc splats deposited on low-carbon steel, stainless steel, Aluminum alloys, zirconia, and alumina substrates. For ethylene methacrylic acid (EMAA), T_c is around 90-100 °C (Jeffrey Alan Brogan, 1996) for deposition onto mild steel.

It was also reported that the adhesion strength of coatings on the substrate preheated above the transition temperature was two to five times higher than those on substrates preheated below the transition temperature. The dependency of coating adhesion strength and splat morphology on substrate temperature is presented in Fukumoto's paper (Fukumoto, Yamaguchi, Yamada, & Yasui, 2007a). There is good correlation between splat morphology and adhesion strength. Similar results are provided by other researchers (Pershin, Lufitha, Chandra, & Mostaghimi, 2003).

Table 2 presents the transition temperature for several powder materials on AISI 304 stainless steel substrate. The standard deviation of T_t is approximately 5-10 K. The table indicates that T_t varies widely with regard to the particle material and that oxide ceramics exhibit a lower T_t.

Effect of Substrate Surface Roughness

Surface roughness was evaluated using two parameters: the average surface roughness (R_a) and skewness (S_k). R_a represents the average roughness, defined as the arithmetic average of the absolute values of the amplitudes of the surface with regard to a mean line along the measured distance (Lech Pawlowski,

Table 2. Transition temperature for several powder materials on AISI 304 stainless steel substrates

Powder Material	Transition Temperature, K
Ni	610
Mo	474
Cu	394
Cr	387
Cu-3Zn0	505
Al_2O_3	318
TiO_2	350
YSZ	345
Substrate: AISI 304 stainless steel. Spraying device: APS	

Source: (Fukumoto et al., 2007a).

2008), but R_a does not discriminate very well between peaks and valleys. Skewness is used to measure the degree of asymmetry of the distribution.

It was found that the adhesion between the coating and the substrate was improved by roughening the substrates up to 17 μm by grit blasting, grooving and heat treatment (Chandra & Fauchais, 2009; Fauchais et al., 2004; Varacalle, Guillen, Deason, Rhodaberger, & Sampson, 2006). A droplet adheres well to a rough substrate due to mechanical bonding between the surface of the substrate and bottom of the splat (Mehdizadeh, Chandra, & Mostaghimi, 2002).

Effect of Low Roughness (Ra < 0.2 μm)

Thermal treatment can increase the roughness of polished metal substrates (Ra < 0.2 μm) by oxidation. The roughness varies according to the polishing technique and preheating procedure, as shown in Table 3. Fukumoto et al. (Fukumoto et al., 2006b), studied copper splats formed on a 304 stainless steel substrate and found that the thickness of the oxide layer increased with heating time; however the oxide layer composition did not change with increasing the preheating time at a constant preheating temperature. The increasing oxide layer thickness modified the roughness. The contact angle of the droplet on the rough surface decreases compared with that on a smooth surface. Therefore, the wettability of the impacting droplet increased on thermal roughened substrates. More disc splats deposited on thermal roughened substrates due to enhanced wettability of the droplet. Hence the adhesion between the coating and substrate was promoted when the surface was roughened.

Cedelle et al. (Cedelle, Vardelle, & Fauchais, 2006) reported that the formation of copper, nickel and YSZ splats on stainless steel 304 substrates improved with thermal treatment of the substrates. This result was reflected in the skewness value of surface roughness since it transitioned from negative values to positive values after preheating.

An oxide layer that is a few tens of nanometers thick can be melted by an impacting droplet and if a chemical reaction takes place between the molten oxide layer and droplet, then adhesion can be promoted on smooth substrates (Fauchais et al., 2004).

Later studies (Fukumoto et al., 2006b; Tanaka & Fukumoto, 1999) proposed that the nanoscale roughness was not a dominant factor in splat formation. It was observed that the disk splat morphology was not enhanced on gold coated preheated substrates even though the surface roughness had increased on the nanoscale. McDonald et al. (McDonald, Moreau, & Chandra, 2007) also claimed that the average surface roughness on a nanometer scale was not a critical factor in splat morphology. They concluded that the release of adsorbates on the substrate surface with heating played a more significant role on the splat morphology (McDonald et al., 2007).

Tran (2010) confirmed that splat formation and morphology were not directly correlated to the substrate surface roughness, but were sensitive to surface chemistry; particularly the thickness and concentration of the hydroxide layer at the outermost surface. Tran (2010) also demonstrated that lower amounts of surface hydroxide significantly improved splat formation and morphology. Tran (2010) observed that a thick oxide layer of up to 350 nm on the substrate surface had a negligible effect on splat formation and morphology. Cross sections of the splat-substrate interfaces established that the oxide layer was not disrupted or removed under the impact of droplets.

Table 3. Roughness of substrates after different mechanical and/or heat treatments

Substrate	Surfacing	Heating	*Ra*, nm	Reference
Aluminum	Polished	None	5	(Fukumoto, Nishiyama, & Nishioka, 2002)
Aluminum	Polished	In air at 673 K	13	(Fukumoto, Nishiyama, et al., 2002)
Aluminum	Polished	In vacuum at 673 K	13	(Fukumoto, Nishiyama, et al., 2002)
SS-304L	Polished	None	0.9	(Fukumoto, Nishiyama, et al., 2002)
SS-304L	Polished	In air at 673 K	3.2	(Fukumoto, Nishiyama, et al., 2002)
SS-304L	Polished	In vacuum at 673 K	5.5	(Fukumoto, Nishiyama, et al., 2002)
SS-304L	Ground	In air at 673 K	50	(Fauchais et al., 1996)
SS-304L	Electropolished	In air at 673 K	400	(Fauchais et al., 1996)

Effect of High Roughness (Ra > 0.2 μm)

The effect of high substrate roughness (Ra > 0.2 μm) on splat formation is different compared to the case of low substrate roughness (Ra < 0.2 μm). Droplet spreading on a rough substrate is restricted by surface irregularities that lead to small diameter and thick splats, which can be found in Clyne's paper (Clyne, 1984). Rough substrates, as well, lead to more splash splats. Although the splat contact area might be greater on rough substrates; the quality of this contact in comparison to smooth substrates can be questioned. Thus, smooth substrates are more likely to exhibit clean and unambiguous clean interfacial contact in comparison to roughened substrates.

The splat cooling rate is lower with an increase in surface roughness (Leger, Vardelle, Vardelle, Dussoubs, & Fauchais, 1995; Moreau, Gougeon, & Lamontagne, 1995) since the cooling rate is faster with a decrease in the splat thickness. Interfacial cracks appear at smooth regions of the splat surface/interface, but do not appear in the rough areas of the splat surface/interface where mechanical bonding strengthens the interface between the splats and the roughened substrates.

It is difficult to analyze the oxide layer thickness and composition formed at the rough substrate surface. The substrate roughness should be adapted to the mean particle size of the feedstock. Hence R_t, the distance between the highest peak and deepest undercut, should not be larger than the splat mean size. A common guide is that R_t is 7 to 8 times less than Ra (Fauchais et al., 2004).

Effect of Substrate Surface Chemistry

Many metals exhibit oxide layers of diversified composition and thickness as a result of their fabrication process and storage. Heat treatment of metals releases gases from the metal surface and modifies the relative amounts of oxide layers and composition (Alexander et al., 2003; Alexander, Thompson, & Beamson, 2000; Grosvenor, Kobe, & McIntyre, 2004a, 2004b). Thermal energy can also generate surface segregation of elements because the oxide formation free energies are different (Birks, Meier, & Pettit, 2006). Thus, an alternative explanation for the influence of the transition temperature and surface chemistry may revolve around gas release from the substrate after heat treatment.

Splat morphology is affected by adsorbates or condensates on the substrate surface (Jiang et al., 2001; Tran & Hyland, 2010; Tran et al., 2011). To study the effect of adsorbates on splat morphology, Li brushed a thin layer of organic compounds (i.e., xylene, glycol and glycerol mixed together) onto a polished stainless-steel substrate (Li & Li, 2004). The substrate was subsequently preheated and plasma sprayed with either Aluminum, nickel, alumina or copper particles. Li et al. (Li & Li, 2004) reported that contiguous splats were achieved when the substrate was preheated to 50 °C, which is above the boiling point of the organic compounds. On the other hand, fragmented splats developed when the substrate was preheated to a temperature below the boiling point of the organic compounds.

Li et al. (Li & Li, 2004) believed that the evaporation of the organic layer with the impact of molten droplets on the substrate surface generated splashing of splats. Heat was rapidly conducted to the substrate upon impact. The heat caused desorption of adsorbates and condensates on the substrate surface that prevented contact between the droplet and the substrate. Therefore, the droplet spreads out to its maximum extent and starts to disintegrate. At the same time, the vaporized adsorbates burst through the liquid layer and gave rise to a fragmented splat (Li, Li, & Liao, 2006; Morks, Tsunekawa, Okumiya, & Shoeib, 2003; Tanaka & Fukumoto, 1999). Substrate preheat treatment can remove the adsorbates or condensates, hence the contact between the splat and substrate was improved and the formation of a disk-shaped splat promoted (Jiang et al., 2001). However, most substrates are unlikely to be contaminated with xylene, glycol or glycerol. Thus, an open question concerns the likely adsorbates or condensates on mild steel and glass substrates and the temperature at which these species may desorb.

Fukumoto et al. (Fukumoto et al., 2006b) also proved that splat morphology was affected by the adsorbates or condensates on the substrate surface. The fraction of disk splats of plasma-sprayed copper particles on a stainless steel AISI 304 substrate was measured. The substrates were preheated to 450 °C, cooled and exposed to ambient conditions for various durations. The results indicated that the disk splat fraction was constant for the first 24 hours of exposure, but after that the disc splat fraction reduced with increased exposure time. However, splat deposition on gold substrates was not significantly different between the two extreme conditions of no preheating and preheating. In this circumstance, the effect of surface chemistry on splat morphology was negligible since gold is a noble metal that does not oxidize under these conditions (Fukumoto, Nagai, & Yasui, 2006a).

Tran (2010) investigated the effects of surface chemistry on splat formation during the plasma spray process. Aluminum substrates were heated thermally and hydrothermally to grow specific types of oxide and hydroxide layers on the surface. Nickel-chromium and nickel single splats were then plasma-sprayed onto the pre-treated Aluminum substrates. It was observed that splash splats formed on samples where a thin layer of hydroxide was present on the substrate surface. The splash splats were caused by water vapor released from the dehydration of surface hydroxide upon impact of the NiCr molten droplets. Conversely, a large volume of released water from the thick layer of surface hydroxide contributed to the complete absence of splat formation on the boiled samples. It was concluded that splat formation and morphology were not directly correlated to the substrate surface roughness, but were sensitive to surface chemistry; particularly the thickness and concentration of the hydroxide layer at the outermost surface. The lower the amount of surface hydroxide, the better the splat formation and morphology.

In addition to releasing adsorbates or condensates, thermal treatment of substrates also modified the surface oxide composition and thickness. McDonald et al. (2007) found that a duplex layer of oxide with iron oxide at the outmost surface and an inner layer of chromium oxide was generated under thermal treatment of stainless steel. The thicknesses of iron oxide and chromium oxide on the substrate preheated at 350 °C were 56 nm and 124 nm respectively, while their thicknesses on a substrate preheated at 650

°C was 84 nm and 204 nm, respectively. However, there were no changes in surface roughness between these samples (Pershin et al., 2003). McDonald et al. (2007) found that the splashed splats appeared on the stainless-steel substrate preheated at 650 °C whereas disk splats were formed on the stainless steel substrate preheated at 350 °C. Thus, McDonald et al. (2007) suggested that the difference in splat morphology on the substrates preheated at 350 °C and 650 °C arose due to differences in the oxide composition and thickness.

EFFECT OF THE PROCESS PARAMETERS ON SPLAT FORMATION AND MORPHOLOGIES

Many parameters influence the structure of a single splat, and therefore, the thermal spray coating and its properties. The parameters include particle velocity, temperature, molten state, oxidation state and the substrate tilting angle. The physical properties of thermal conductivity and the wettability at impact between the thermal spray particle and the substrate influence the individual splat geometry (Fauchais et al., 2004; Verdian, Salehi, & Raeissi, 2010) and coating build-up and, consequently, the coating properties. The coating formation process is generally determined by the prime process variables, which include substrate roughness, substrate temperature, impact velocity, stand-off distance (SOD), and the nature of the local atmospheric environment (Bussmann et al., 2000; Herman et al., 2000; Sweet, 1993; Vardelle et al., 2000). The spray angle also influences the splat morphology. Splats display disc shapes at 90°, elliptical shapes between 90 and 60°, and become fingered for low angles (Fauchais et al., 2001). Many of these factors are inter-related. For example, SOD will influence the impact velocity as well as the substrate temperature due to heat transfer from the thermal spray torch. The target efficiency of EMAA decreases dramatically with an increase of SOD. Increasing SOD results in more spherical splats (EMAA) (Xie, Wang, & Berndt, 2014).

Many spray process parameters correlate to each other. The splat spread factor decreases with an increase in substrate roughness (Leger et al., 1995). When the spray angle increases, splats have an elliptical shape, and the aspect ratio, i.e. the ratio of the length of the major axis to that of the minor axis of the splats, increases accordingly. The splats become fingered for spray angles less than 30° (Fauchais et al., 2001).

The lower wettability, higher roughness and higher thermal conductivity of mild steel substrates compared with glass substrates led to more fragmentation of EMAA splats deposited on mild steel substrate than that on glass substrate. However, the surface chemistry of substrates played a more important role than wettability, roughness and thermal conductivity in single splat morphology. Substrates heat treatment above transitional temperature significantly reduced the splashed splats (Xie, Wang, & Berndt, 2013).

The spray process variables influence the physical nature of the intrinsic building blocks of thermal spray coatings; i.e., the splat morphology (Fauchais et al., 2004; Sampath et al., 1999).

THERMAL SPRAY COATING MORPHOLOGY

The performance of coatings deposited by thermal spray depends on the quality of contact between the coating and the substrate and between deposited layers. Plasma sprayed and HVOF coatings are characterized by a highly anisotropic lamellar structure.

A typical structure of a thermal spray coating can be seen on Flame Spray Technologies website (Flame Spray Technologies, 2007). It includes porosity, oxide inclusions, debris, unmelted particles and lamellar splats.

Coating particles impinging on the substrate may be in a molten, semi-molten or non-molten state depending on the thermal spray processes. The particles are normally in a molten state in the plasma spray process due to the high temperature. In the HVOF process, the high kinetic energy of particles impinging on the substrate does not require the particles to be fully molten to form the coatings. The high velocity air fuel (HVAF) and cold spray processes are also high kinetic energy coating processes where the particles remain in the non-molten state.

Polymer spraying is a one-coat process that acts both as the primer and the sealant (Loustanau & Horton, 1994). Polymers can be sprayed by flame spray (FS) (Brogan, 1996; Brogan & Berndt, 1996; Brogan & Berndt, 1997), atmospheric plasma spray (APS) (Bao, Gawne, Vesely, & Bevis, 1994a, 1994b; Bao, Gawne, & Zhang, 1995), high velocity oxygen fuel (HVOF) (Petrovicova, Knight, Smith, & Schadler, 1997; Petrovicova, Smith, Knight, & Schadler, 1998; Twardowski, Reilly, & Knight, 2001), high velocity air fuel (HVAF) (Petrovicova & Schadler, 2002) and cold spray (Xu & Hutchings, 2006) processes: processes that are selected on the basis of the physical properties of the polymer. The FS process exhibits some advantages over APS and HVOF; even for high melting polymers such as polyetheretherketone (PEEK) that undergoes heat sensitive phase transitions at 343 °C. Polymer fusion occurs when the preheating temperature and the heat input are appropriately selected due to the lower flame enthalpy and lower stagnation pressure compared to the plasma or HVOF (Liao, Beche, Berger, & Coddet, 1998) processes. FS can be used to fabricate dense and smooth PEEK coatings with less porosity and unmelted particles compared with those prepared by plasma and HVOF processes (Petrovicova & Schadler, 2002). The EMAA copolymer containing sodium ions is a common polymer that can be flame sprayed (Brogan, Lamp, & Berndt, 1995). Such polymers adhere well to steel and Aluminum (Loustaunau & Horton, 1994), and withstand high humidity (Glass & Depay, 1991) and low temperatures (Glass & Depay, 1991). They are, therefore, widely applied for coating pipes, ladder racking, structural steels, bridges, handrails, screw propellers, wastewater clarifiers, light poles and many other diverse applications where high chemical resistance and high impact resistance are required (Brogan & Berndt, 1997; Glass & Depay, 1991; Loustaunau & Horton, 1994).

Polymer-ceramic composites, or polycers, have the advantage of maintaining the toughness and coat-forming properties of the plastic, and allow mechanical properties such as modulus and abrasion resistance to be modified (Yan, Gross, Simon, & Berndt, 2004). Polycers can be manufactured by the thermal spray process (Berndt, Brogan, Montavon, Claudon, & Coddet, 1998), die compaction followed by sintering (Wang, Li, & Stevens, 1992) and tape casting (Böhnlein-Mauß et al., 1992). In conventional thermal spray processes, a high temperature flame is produced by either electrical or chemical energy and this flame is used to process solid feedstock material. Ceramic-polymer composites formed via FS consolidation have been reported by Berndt et al. in 1998 (Berndt et al., 1998). It has been demonstrated that thermal spray consolidation is an effective process to fabricate polycer composites. The advantage of this technology is that it permits on-site manufacture and it is a repairable composite system with a minimal cure time and close to zero release of volatile organic compounds (Berndt et al., 1998). The FTIR of EMAA coatings indicated that there was no detectable deterioration of the EMAA feedstock during the FS process (Xie, Wang, & Berndt, 2012; Xie et al., 2014).

Concerning the characterization of mechanical properties of thermal spray coating and process, some standards have to be adopted: Thermal spraying - Determination of the deposition efficiency for thermal spraying (BSI, 2004), Standard Test Method for Melt Flow Rates of Thermoplastics by Extrusion Plastometer (ASTM, 2010), Standard Test Method for Adhesion or Cohesion Strength of Thermal Spray Coatings (ASTM, 2001), etc.

FUTURE RESEARCH DIRECTIONS

The area of nanostructured thermal spray coatings is relatively immature and no large markets have been developed. Higher deposition efficiencies are often obtained with nanostructured materials so the overall cost could be reasonable. Improved performance and/or longer product life span may lead to a substantial reduction in the overall cost, and consequently reaching the cost-effective goal.

The deposits sprayed using solutions and suspensions are still at the stage of research and development. Whether polymers can be applied in some context to such applications is a topic worthy of further investigation.

CONCLUSION

The principles of thermal spray processes have been introduced. In addition, the effect of process parameters; e.g., stand-off distance; the powder size and its distribution; and filler contents on the properties of single splats and coatings have been presented and discussed.

The properties of coatings and splats are affected significantly by the substrate surface conditions, such as surface wettability, surface roughness, transition temperature and surface chemistry. There are numerous studies concerning surface wettability, surface roughness and transition temperature but little research about the correlation between the surface chemistry and splat morphology.

REFERENCES

Ahn, J., Hwang, B., Song, E. P., Lee, S., & Kim, N. J. (2006). Correlation of microstructure and wear resistance of Al_2O_3-TiO_2 coatings plasma sprayed with nanopowders. *Metallurgical and Materials Transactions. A, Physical Metallurgy and Materials Science*, *37*(6), 1851–1861. doi:10.1007/s11661-006-0128-5

Alexander, M. R., Beamson, G., Bailey, P., Noakes, T. C. Q., Skeldon, P., & Thompson, G. E. (2003). The distribution of hydroxyl ions at the surface of anodic alumina. *Surface and Interface Analysis*, *35*(8), 649–657. doi:10.1002/sia.1585

Alexander, M. R., Thompson, G. E., & Beamson, G. (2000). Characterization of the oxide/hydroxide surface of aluminum using X-ray photoelectron spectroscopy: A procedure for curve fitting the O 1s core level. *Surface and Interface Analysis*, *29*(7), 468–477. doi:10.1002/1096-9918(200007)29:7<468::AID-SIA890>3.0.CO;2-V

Ardanuy, M., Velasco, J. I., Maspoch, M. L., Haurie, L., & Fernández, A. I. (2009). Influence of EMAA compatibilizer on the structure and properties of HDPE/hydrotalcite nanocomposites prepared by melt mixing. *Journal of Applied Polymer Science*, *113*(2), 950–958. doi:10.1002/app.30133

ASTM. (2001). *Standard Test Method for Adhesion or Cohesion Strength of Thermal Spray Coatings* (pp. 846–852). ASTM.

ASTM. (2010). *Standard Test Method for Melt Flow Rates of Thermoplastics by Extrusion Plastometer* (pp. 282–296). West Conshohocken, PA: ASTM.

Bao, Y., Gawne, D. T., Vesely, D., & Bevis, M. J. (1994a). Formation and microstructure of plasma sprayed polyamide coatings. *Surface Engineering*, *10*(4), 307–313. doi:10.1179/sur.1994.10.4.307

Bao, Y., Gawne, D. T., Vesely, D., & Bevis, M. J. (1994b). Production of polymer matrix composite coatings by thermal spraying. *Transactions of the Institute of Metal Finishing*, *72*(Pt 3), 110–113. doi:10.1080/00202967.1994.11871033

Bao, Y., Gawne, D. T., & Zhang, T. (1995). Effect of feedstock particle size on the heat transfer rates and properties of thermally sprayed polymer coatings. *Transactions of the Institute of Metal Finishing*, *73*(Pt 4), 119–124. doi:10.1080/00202967.1995.11871072

Berndt, C. C., Brogan, J. A., Montavon, G., Claudon, A., & Coddet, C. (1998). Mechanical properties of metal- and ceramic-polymer composites formed via thermal spray consolidation - Extended abstract. *Journal of Thermal Spray Technology*, *7*(3), 337–339. doi:10.1361/105996398770350819

Berndt, C. C., & Lavernia, E. J. (1998). Thermal spray processing of nanoscale materials - A conference report with extended abstracts. *Journal of Thermal Spray Technology*, *7*(3), 411–440. doi:10.1007/s11666-998-0058-4

Bianchi, L., Grimaud, A., Blein, F., Lucchese, P., & Fauchais, P. (1995). Comparison of plasma-sprayed alumina coatings by RF and DC plasma spraying. *Journal of Thermal Spray Technology*, *4*(1), 59–66. doi:10.1007/BF02648529

Birks, N., Meier, G. H., & Pettit, F. S. (2006). *Introduction to the high temperature oxidation of metals* Retrieved from http://swin.eblib.com.au/patron/FullRecord.aspx?p=258481

Böhnlein-Mauß, J., Sigmund, W., Wegner, G., Meyer, W. H., Heßel, F., Seitz, K., & Roosen, A. (1992). The function of polymers in the tape casting of alumina. *Advanced Materials*, *4*(2), 73–81. doi:10.1002/adma.19920040203

Boulos, M. I. (1992). RF induction plasma spraying: State-of-the-art review. *Journal of Thermal Spray Technology*, *1*(1), 33–40. doi:10.1007/BF02657015

Brogan, J. A. (1996). *Processing and property relationships of thermally sprayed polymer systems*. Stony Brook, NY: State University of New York.

Brogan, J. A., & Berndt, C. C. (1996). The mechanical properties of combustion-sprayed polymers and blends. *Paper presented at the Thermal Spray: Practical Solutions for Engineering Problems*, Cincinnati, Ohio.

Brogan, J. A., & Berndt, C. C. (1997). The coalescence of combustion-sprayed ethylene-methacrylic acid copolymer. *Journal of Materials Science*, *32*(8), 2099–2106. doi:10.1023/A:1018526906452

Brogan, J. A., Lamp, R., & Berndt, C. C. (1995). Thermal spraying of polymers. *Paper presented at the Fourth World Congress on Coating Systems for Bridges and Steel Structures: Bridging the Environment*, St. Louis, MO.

Browning, J. A. (1992). Hypervelocity impact fusion-a technical note. *Journal of Thermal Spray Technology*, *1*(4), 289–292. doi:10.1007/BF02647155

BSI. (2004). *Thermal Spraying - Determination of the Deposition Efficiency for Thermal Spraying* (pp. 1–10). European Committee for Standardization.

Bussmann, M., Chandra, S., & Mostaghimi, J. (2000). Modeling the splash of a droplet impacting a solid surface. *Physics of Fluids*, *12*(12), 3121–3132. doi:10.1063/1.1321258

Cedelle, J., Vardelle, M., & Fauchais, P. (2006). Influence of stainless steel substrate preheating on surface topography and on millimeter- and micrometer-sized splat formation. *Surface and Coatings Technology*, *201*(3-4), 1373–1382. doi:10.1016/j.surfcoat.2006.02.003

Chandra, S., & Fauchais, P. (2009). Formation of solid splats during thermal spray deposition. *Journal of Thermal Spray Technology*, *18*(2), 148–180. doi:10.1007/s11666-009-9294-5

Christoulis, D. K., Pantelis, D. I., Borit, F., Guipont, V., & Jeandin, M. (2004). Effect of substrate roughness and temperature on splat formation in plasma sprayed aluminium bronze. *Paper presented at the 18th International Conference on Surface Modification Technologies*, Dijon, France. Retrieved from http://www.scopus.com/inward/record.url?eid=2-s2.0-33646056723&partnerID=40

Ciubotariu, C.-R., Frunzăverde, D., Mărginean, G., Şerban, V.-A., & Bîrdeanu, A.-V. (2016). Optimization of the laser remelting process for HVOF-sprayed Stellite 6 wear resistant coatings. *Optics & Laser Technology*, *77*, 98–103. doi:10.1016/j.optlastec.2015.09.005

Clyne, T. W. (1984). Numerical treatment of rapid solidification. *Metallurgical Transactions. B, Process Metallurgy*, *15*(2), 369–381. doi:10.1007/BF02667341

Davis, J. R. (Ed.). (2004). *Handbook of thermal spray technology*. Materials Park, OH, USA: ASM International.

De Freitas Cunha Lins, V., Cury, F. M., & Moreira, R. (2008). Infrared and ultraviolet-visible spectroscopy study of the degradation of polyester and polyester/ethylene methyl acrylate copolymer blend coatings on steel. *Journal of Applied Polymer Science*, *109*(4), 2103–2112. doi:10.1002/app.28231

De Sansal, C., Devincre, B., & Kubin, L. (2010). Grain size strengthening in microcrystalline copper: A three-dimensional dislocation dynamics simulation. *Key Engineering Materials*, *423*, 25–32. doi:10.4028/www.scientific.net/KEM.423.25

Fan, W., & Bai, Y. (2016). Review of suspension and solution precursor plasma sprayed thermal barrier coatings. *Ceramics International*, *42*(13), 14299–14312. doi:10.1016/j.ceramint.2016.06.063

Fauchais, P. (2004). Understanding plasma spraying. *Journal of Physics. D, Applied Physics*, *37*(9), R86–R108. doi:10.1088/0022-3727/37/9/R02

Fauchais, P. (2008). Suspension and solution plasma or HVOF spraying. *Journal of Thermal Spray Technology*, *17*(1), 1–3. doi:10.1007/s11666-008-9160-x

Fauchais, P., Fukumoto, M., Vardelle, A., & Vardelle, M. (2004). Knowledge concerning splat formation: An invited review. *Journal of Thermal Spray Technology*, *13*(3), 337–360. doi:10.1361/10599630419670

Fauchais, P., Rat, V., Coudert, J. F., Etchart-Salas, R., & Montavon, G. (2008). Operating parameters for suspension and solution plasma-spray coatings. *Surface and Coatings Technology*, *202*(18), 4309–4317. doi:10.1016/j.surfcoat.2008.04.003

Fauchais, P., Vardelle, A., & Dussoubs, B. (2001). Quo vadis thermal spraying? *Journal of Thermal Spray Technology*, *10*(1), 44–66. doi:10.1361/105996301770349510

Fauchais, P., Vardelle, M., Vardelle, A., Bianchi, L., & Leger, A. C. (1996). Parameters controlling the generation and properties of plasma sprayed zirconia coatings. *Plasma Chemistry and Plasma Processing*, *16*(1), S99–S125.

Feynman, R. P. (2011). There's plenty of room at the bottom. *Resonance*, *16*(9), 890–905. doi:10.1007/s12045-011-0109-x

Flame Spray Technologies. (2007). Thermal spray process. Retrieved from http://www.fst.nl/en/page00031.asp

Fukumoto, M., & Huang, Y. (1999). Flattening mechanism in thermal sprayed nickel particle impinging on flat substrate surface. *Journal of Thermal Spray Technology*, *8*(3), 427–432. doi:10.1361/105996399770350386

Fukumoto, M., Nagai, H., & Yasui, T. (2006a). Influence of surface character change of substrate due to heating on flattening behavior of thermal sprayed particles. *Paper presented at the 2006 International Thermal Spray Conference: Building on 100 Years of Success*, Seattle, WA. Retrieved from http://www.scopus.com/inward/record.url?eid=2-s2.0-34548752700&partnerID=40&md5=38ae31c0418098efc96220bff9c2022f

Fukumoto, M., Nagai, H., & Yasui, T. (2006b). Influence of surface character change of substrate due to heating on flattening behavior of thermal sprayed particles. *Journal of Thermal Spray Technology*, *15*(4), 759–764. doi:10.1361/105996306X146776

Fukumoto, M., Nishioka, E., & Matsubara, T. (2002). Effect of interface wetting on flattening of freely fallen metal droplet onto flat substrate surface. *Journal of Thermal Spray Technology*, *11*(1), 69–74. doi:10.1361/105996302770348998

Fukumoto, M., Nishiyama, H., & Nishioka, A. (2002). Effect of surface morphology of substrate on flattening behavior of freely fallen metal droplet. *Paper presented at the International Thermal Spray Conference*, Essen, Germany.

Fukumoto, M., Yamaguchi, T., Yamada, M., & Yasui, T. (2007a). Splash splat to disk splat transition behavior in plasma-sprayed metallic materials. *Journal of Thermal Spray Technology*, *16*(5-6), 905–912. doi:10.1007/s11666-007-9083-y

Fukumoto, M., Yamaguchi, T., Yamada, M., & Yasui, T. (2007b). Splash splat to disk splat transition behavior in plasma-sprayed metallic materials. *Paper presented at the International Thermal Spray Conference on Thermal Spray 2007: Global Coating Solutions*, Beijing, China. Retrieved from http://www.scopus.com/inward/record.url?eid=2-s2.0-59349099839&partnerID=40&md5=8c8c7aaaafea2cc271b8687671fea83a

Glass, T. W., & Depay, J. A. (1991). Thermal spray coatings: properties, processes and applications. *Paper presented at the Fourth National Thermal Spray Conference*, Pittsburgh, PA.

Graziani, G., Bianchi, M., Sassoni, E., Russo, A., & Marcacci, M. (2017). Ion-substituted calcium phosphate coatings deposited by plasma-assisted techniques: A review. *Materials Science and Engineering C, 74*, 219–229. doi:10.1016/j.msec.2016.12.018 PMID:28254288

Grosvenor, A. P., Kobe, B. A., & McIntyre, N. S. (2004a). Examination of the oxidation of iron by oxygen using X-ray photoelectron spectroscopy and QUASES™. *Surface Science, 565*(2-3), 151–162. doi:10.1016/j.susc.2004.06.210

Grosvenor, A. P., Kobe, B. A., & McIntyre, N. S. (2004b). Studies of the oxidation of iron by water vapour using X-ray photoelectron spectroscopy and QUASES™. *Surface Science, 572*(2-3), 217–227. doi:10.1016/j.susc.2004.08.035

Herman, H., Sampath, S., & McCune, R. (2000). Thermal spray: Current status and future trends. *MRS Bulletin, 25*(7), 17–25. doi:10.1557/mrs2000.119

High, M. S., Painter, P. C., & Coleman, M. M. (1992). Polymer-polymer mutual diffusion using transmission FTIR spectroscopy. *Macromolecules, 25*(2), 797–801. doi:10.1021/ma00028a045

Irissou, E., Legoux, J. G., Ryabinin, A. N., Jodoin, B., & Moreau, C. (2008). Review on cold spray process and technology: Part I - Intellectual property. *Journal of Thermal Spray Technology, 17*(4), 495–516. doi:10.1007/s11666-008-9203-3

Jiang, X., Wan, Y., Herman, H., & Sampath, S. (2001). Role of condensates and adsorbates on substrate surface on fragmentation of impinging molten droplets during thermal spray. *Thin Solid Films, 385*(1-2), 132–141. doi:10.1016/S0040-6090(01)00769-6

Karthikeyan, J., Berndt, C. C., Tikkanen, J., Reddy, S., & Herman, H. (1997). Plasma spray synthesis of nanomaterial powders and deposits. *Materials Science and Engineering a-Structural Materials Properties Microstructure and Processing, 238*(2), 275-286.

Karthikeyan, J., Berndt, C. C., Tikkanen, J., Wang, J. Y., King, A. H., & Herman, H. (1997a). Nanomaterial powders and deposits prepared by flame spray processing of liquid precursors. *Nanostructured Materials, 8*(1), 61–74. doi:10.1016/S0965-9773(97)00066-4

Karthikeyan, J., Berndt, C. C., Tikkanen, J., Wang, J. Y., King, A. H., & Herman, H. (1997b). Preparation of nanophase materials by thermal spray processing of liquid precursors. *Nanostructured Materials, 9*(1-8), 137-140.

Karthikeyan, J., Kay, C. M., Lindeman, J., Lima, P. S., & Berndt, C. C. (2000). *Cold spray processing of titanium powder*. Paper presented at the International Thermal Spray Conference, Montreal, QUE, Canada. Retrieved from http://www.scopus.com/inward/record.url?eid=2-s2.0-1442353812&partnerID=40

Kohzaki, M., Tsujita, Y., Takizawa, A., & Kinoshita, T. (1987). Crystallization and formation of cluster of ethylene ionomer during physical aging. *Journal of Applied Polymer Science*, *33*(7), 2393–2402. doi:10.1002/app.1987.070330710

Kumar, V., & Kandasubramanian, B. (2016). Processing and design methodologies for advanced and novel thermal barrier coatings for engineering applications. *Particuology*, *27*, 1–28. doi:10.1016/j.partic.2016.01.007

Kuroda, S., Kawakita, J., Watanabe, M., & Katanoda, H. (2008). Warm spraying-a novel coating process based on high-velocity impact of solid particles. *Science and Technology of Advanced Materials, 9*(3). doi:10.1088/1468-6996/9/3/033002

Leger, A. C., Vardelle, M., Vardelle, A., Dussoubs, B., & Fauchais, P. (1995). Splat formation: ceramic carticles on ceramic substrates. *Paper presented at the Advances in Thermal Spray Science & Technology*, Houston, TX.

Li, C. J., & Li, J. L. (2004). Evaporated-gas-induced splashing model for splat formation during plasma spraying. *Surface and Coatings Technology*, *184*(1), 13–23. doi:10.1016/j.surfcoat.2003.10.048

Li, C. J., Li, W. Y., & Liao, H. (2006). Examination of the critical velocity for deposition of particles in cold spraying. *Journal of Thermal Spray Technology*, *15*(2), 212–222. doi:10.1361/105996306X108093

Li, M., & Christofides, P. D. (2009). Modeling and control of high-velocity oxygen-fuel (HVOF) thermal spray: A tutorial review. *Journal of Thermal Spray Technology.*

Li, R., Ashgriz, N., & Chandra, S. (2010). Maximum spread of droplet on solid surface: Low reynolds and weber numbers. *Journal of Fluids Engineering-Transactions of the Asme, 132*(6).

Liang, B., Liao, H., Ding, C., & Coddet, C. (2005). Nanostructured zirconia-30 vol.% alumina composite coatings deposited by atmospheric plasma spraying. *Thin Solid Films*, *484*(1-2), 225–231. doi:10.1016/j.tsf.2005.02.040

Liao, H., Beche, E., Berger, F., & Coddet, C. (1998). On the microstructures of thermally sprayed "PEEK" polymer. *Paper presented at the Thermal Spray: Meeting the Challenges of the 21st Century*, Nice, France. Retrieved from http://www.scopus.com/inward/record.url?eid=2-s2.0-1442354654&partnerID=40&md5=b271e3c393176e0a41ae34b584c15782

Lima, R. S., & Marple, B. R. (2007). Thermal spray coatings engineered from nanostructured ceramic agglomerated powders for structural, thermal barrier and biomedical applications: A review. *Journal of Thermal Spray Technology*, *16*(1), 40–63. doi:10.1007/s11666-006-9010-7

Lin, X., Zeng, Y., Lee, S. W., & Ding, C. (2004). Characterization of alumina-3 wt.% titania coating prepared by plasma spraying of nanostructured powders. *Journal of the European Ceramic Society*, *24*(4), 627–634. doi:10.1016/S0955-2219(03)00254-1

Lin, X., Zeng, Y., Zhou, X., & Ding, C. (2003). Microstructure of alumina-3wt.% titania coatings by plasma spraying with nanostructured powders. *Materials Science and Engineering A*, *357*(1-2), 228–234. doi:10.1016/S0921-5093(03)00164-3

Loustanau, P. J., & Horton, D. (1994). EMAA thermoplastic powder coatings in shop and field applications. *Materials Performance*, *33*(7), 32–36.

Loustaunau, P. J., & Horton, D. (1994). EMAA thermoplastic powder coatings in shop and field applications. *Materials Performance*, *33*(7), 32–36.

McDonald, A., Moreau, C., & Chandra, S. (2007). Effect of substrate oxidation on spreading of plasma-sprayed nickel on stainless steel. *Surface and Coatings Technology*, *202*(1), 23–33. doi:10.1016/j.surfcoat.2007.04.041

Mehdizadeh, N. Z., Chandra, S., & Mostaghimi, J. (2002). Effect of substrate temperature and roughness on coating formation. *Paper presented at the International Thermal Spray Conference*, Essen, Germany.

Metco, O. (2014). Thermal spray equipment guide. Retrieved from https://www.oerlikon.com/metco/en/products-services/coating-equipment/thermal-spray/?rd=1

Meure, S., Wu, D. Y., & Furman, S. A. (2010). FTIR study of bonding between a thermoplastic healing agent and a mendable epoxy resin. *Vibrational Spectroscopy*, *52*(1), 10–15. doi:10.1016/j.vibspec.2009.09.005

Moreau, C., Gougeon, P., & Lamontagne, M. (1995). Influence of substrate preparation on the flattening and cooling of plasma-sprayed particles. *Journal of Thermal Spray Technology*, *4*(1), 25–33. doi:10.1007/BF02648525

Morks, M. F., Tsunekawa, Y., Okumiya, M., & Shoeib, M. A. (2003). Splat microstructure of plasma sprayed cast iron with different chamber pressures. *Journal of Thermal Spray Technology*, *12*(2), 282–289. doi:10.1361/105996303770348393

Mostaghimi, J., & Chandra, S. (2007). Heat transfer in plasma spray coating processes. *Advances in Heat Transfer*, *40*(C), 143–204. doi:10.1016/S0065-2717(07)40002-8

Papyrin, A., Kosarev, V., Klinkov, S., Alkhimov, A., & Fomin, V. (2007). *Cold spray technology.*

Pawlowski, L., & Dépôts Physiques. (2003). *Techniques, microstructures et propriétés*. Lausanne, Switzerland Presse Polytechnique et Universitaires Romandes.

Pawlowski, L. (2008). Finely grained nanometric and submicrometric coatings by thermal spraying: A review. *Surface and Coatings Technology*, *203*(3-4), 397–397. doi:10.1016/j.surfcoat.2008.09.004

Pawlowski, L. (2008). The science and engineering of thermal spray coatings (2nd ed.). Chichester, UK: John Wiley & Sons. doi:10.1002/9780470754085

Pershin, V., Lufitha, M., Chandra, S., & Mostaghimi, J. (2003). Effect of substrate temperature on adhesion strength of plasma-sprayed nickel coatings. *Journal of Thermal Spray Technology*, *12*(3), 370–376. doi:10.1361/105996303770348249

Petrovicova, E., Knight, R., Smith, R. W., & Schadler, L. S. (1997). Structure and properties of HVOF sprayed ceramic/polymer nano-composite coatings. *Paper presented at the Thermal Spray: A United Forum for Scientific and Technological Advances*, Indianapolis, IN.

Petrovicova, E., & Schadler, L. S. (2002). Thermal spraying of polymers. *International Materials Reviews*, *47*(4), 169–190. doi:10.1179/095066002225006566

Petrovicova, E., Smith, R. W., Knight, R., & Schadler, L. S. (1998). Particulate nanoreinforced polymer composites processing and properties. *J. Therm. Spray Techn.*, *7*(3), 426–427.

Roukes, M. (2001). Plenty of room, indeed. *Scientific American*, *285*(3), 42. doi:10.1038/scientificamerican0901-48 PMID:11524969

Samal, S. (2017). Thermal plasma technology: The prospective future in material processing. *Journal of Cleaner Production*, *142*, 3131–3150. doi:10.1016/j.jclepro.2016.10.154

Sampath, S., Jiang, X. Y., Matejicek, J., Leger, A. C., & Vardelle, A. (1999). Substrate temperature effects on splat formation, microstructure development and properties of plasma sprayed coatings Part I: Case study for partially stabilized zirconia. *Materials Science and Engineering A*, *272*(1), 181–188. doi:10.1016/S0921-5093(99)00459-1

Sidhu, T. S., Agrawal, R. D., & Prakash, S. (2005). Hot corrosion of some superalloys and role of high-velocity oxy-fuel spray coatings - a review. *Surface and Coatings Technology*, *198*(1-3), 441–446. doi:10.1016/j.surfcoat.2004.10.056

Siegel, R. W. (1996). Creating nanophase materials. *Scientific American*, *275*(6), 74–79. doi:10.1038/scientificamerican1296-74

Song, E. P., Ahn, J., Lee, S., & Kim, N. J. (2006). Microstructure and wear resistance of nanostructured Al_2O_3-8wt.%TiO_2 coatings plasma-sprayed with nanopowders. *Surface and Coatings Technology*, *201*(3-4), 1309–1315. doi:10.1016/j.surfcoat.2006.01.052

StruttP. R.KearB. H.RolandR. F. (2000).

Sweet, G. K. (1993). Applying thermoplastic/thermoset powder with a modified plasma system. *Paper presented at the Thermal Spray Coatings: Research, Design and Applications*, Anaheim, CA.

Tanaka, Y., & Fukumoto, M. (1999). Investigation of dominating factors on flattening behavior of plasma sprayed ceramic particles. *Surface and Coatings Technology*, *120-121*, 124–130. doi:10.1016/S0257-8972(99)00348-5

TEKNA. (2009). Plasma state technology. Retrieved from http://www.tekna.com/technology/plasma-state/

Thorpe, M. L., & Richter, H. J. (1992). A pragmatic analysis and comparison of HVOF processes. *Journal of Thermal Spray Technology*, *1*(2), 161–170. doi:10.1007/BF02659017

Tran, A. T. T. (2010). *Effects of surface chemistry on splat formation during plasma spraying. (Ph.D. Ph.D.)*. Auckland: The University of Auckland.

Tran, A. T. T., & Hyland, M. M. (2009). The role of substrate surface chemistry on splat formation during plasma spray deposition by experiments and simulations. *Paper presented at the International Thermal Spray Conference*, Las Vegas, NV. Retrieved from http://www.scopus.com/inward/record.url?eid=2-s2.0-76749126654&partnerID=40&md5=830a0ccf95426d84effca4bd27fbb1f0

Tran, A. T. T., & Hyland, M. M. (2010). The role of substrate surface chemistry on splat formation during plasma spray deposition by experiments and simulations. *Journal of Thermal Spray Technology*, *19*(1-2), 11–23. doi:10.1007/s11666-009-9414-2

Tran, A. T. T., Hyland, M. M., Qiu, T., Withy, B., & James, B. J. (2008). Effects of surface chemistry on splat formation during plasma spraying. *Journal of Thermal Spray Technology*, *17*(5-6), 637–645. doi:10.1007/s11666-008-9237-6

Tran, A. T. T., Hyland, M. M., Shinoda, K., & Sampath, S. (2011). Influence of substrate surface conditions on the deposition and spreading of molten droplets. *Thin Solid Films*, *519*(8), 2445–2456. doi:10.1016/j.tsf.2010.11.047

TSET. (2008). Plasma spray process. Retrieved from http://www.tsetinc.com/ceramicoat.htm

Tsujita, Y., Shibayama, K., Takizawa, A., Kinoshita, T., & Uematsu, I. (1987). Thermal properties of ethylene ionomers. *Journal of Applied Polymer Science*, *33*(4), 1307–1314. doi:10.1002/app.1987.070330421

Tsunashima, K., Nishioji, H., Hirasawa, E., & Yano, S. (1992). EXAFS studies on Zn(II) complexes with 1,3-bis(aminomethyl)cyclohexane of ethylene-methacrylic acid copolymer. *Polymer*, *33*(9), 1809–1812. doi:10.1016/0032-3861(92)90475-C

Twardowski, T., Reilly, M., & Knight, R. (2001). Properties of HVOF sprayed multi-scale polymer/silica nanocomposite coatings. *Paper presented at the Thermal Spray 2001: New Surfaces for a New Millennium*, Singapore.

University of Toronto. (2006). DC plasma modeling. Retrieved from http://www.mie.utoronto.ca/labs/cact/Research/DCPlasma/DCPlasma.html

Varacalle, D. J. Jr, Guillen, D. P., Deason, D. M., Rhodaberger, W., & Sampson, E. (2006). Effect of grit-blasting on substrate roughness and coating adhesion. *Journal of Thermal Spray Technology*, *15*(3), 348–355. doi:10.1361/105996306X124347

Vardelle, A., Moreau, C., & Fauchais, P. (2000). Dynamics of deposit formation in thermal-spray processes. *MRS Bulletin*, *25*(7), 32–37. doi:10.1557/mrs2000.121

Verdian, M. M., Salehi, M., & Raeissi, K. (2010). Effect of feedstock particle size on microstructure of APS coatings prepared from mechanically alloyed nickel-titanium powders. *Surface Engineering*, *26*(6), 447–452. doi:10.1179/026708409X12490360425927

Wang, J., Li, S. H. P., & Stevens, R. (1992). Effects of organic binders on the sintering of isostatically compacted zirconia powders. *Journal of Materials Science*, *27*(1), 63–67. doi:10.1007/BF02403644

Xie, W. (2013). *Thermal Spray Processing of Ethylene Methacrylic Acid and Polymer-Ceramic Composites. (PhD Doctor of Phyloshophy)*. Australia: Swinburne University of Technology.

Xie, W., Wang, J., & Berndt, C. C. (2012). Spreading Behavior and Morphology of Ethylene Methacrylic Acid (EMAA) Deposits via the Flame Spray Process. *Coatings*, *2*(2), 76–93. doi:10.3390/coatings2020076

Xie, W., Wang, J., & Berndt, C. C. (2013). Ethylene Methacrylic Acid (EMAA) Single Splat Morphology. *Coatings*, *3*(2), 82–97. doi:10.3390/coatings3020082

Xie, W., Wang, J., & Berndt, C. C. (2014). Analysis of EMAA Splats on Glass and Mild Steel Substrates. *Journal of Thermal Spray Technology*, *23*(3), 317–324. doi:10.1007/s11666-013-9989-5

Xu, Y., Hu, X., Xu, F., & Li, K. (2017). Rare earth silicate environmental barrier coatings: Present status and prospective. *Ceramics International*, *43*(8), 5847–5855. doi:10.1016/j.ceramint.2017.01.153

Xu, Y., & Hutchings, I. M. (2006). Cold spray deposition of thermoplastic powder. *Surface and Coatings Technology*, *201*(6), 3044–3050. doi:10.1016/j.surfcoat.2006.06.016

Yan, F. Y., Gross, K. A., Simon, G. P., & Berndt, C. C. (2004). Mechanical and erosion properties of $CaCO_3$-EMAA thermal sprayed coatings. *Polymer Engineering and Science*, *44*(8), 1448–1459. doi:10.1002/pen.20141

Zhao, B., Yadian, B. L., Li, Z. J., Liu, P., & Zhang, Y. F. (2009). Improvement on wettability between carbon nanotubes and Sn. *Surface Engineering*, *25*(1), 31–35. doi:10.1179/026708408X334104

KEY TERMS AND DEFINITIONS

Polycers: Polymer-ceramic composites.

SOD: Stand-off distance.

Surface Chemistry: The relative amounts of oxide layers and composition.

Surface Roughness: Surface roughness was evaluated using two parameters: the average surface roughness (Ra) and skewness (Sk). Ra represents the average roughness, defined as the arithmetic average of the absolute values of the amplitudes of the surface with regard to a mean line along the measured distance, but Ra does not discriminate very well between peaks and valleys. Skewness is used to measure the degree of asymmetry of the distribution.

Surface Wettability: The ability of a liquid to maintain contact with a solid surface, it determined by the contact angel.

Thermal Spray: Thermal spray belongs to the family of granular methods of deposition. Fine particles of polymer or ceramic are propelled using a flow of carrier gas through a flame. The particles are heated, softened and ejected towards an often-heated substrate where they collide, flatten and assemble to form a coating that consists of hundreds of thousands of well-adhered flattened particles.

Transition Temperature: The critical preheating temperature, Tc, of the substrate. This temperature is also defined as the transition temperature (Tt) of the substrate by Fukumoto.

ENDNOTE

1 All the figures mentioned above can be found in the thesis Thermal Spray Processing of Ethylene Methacrylic Acid and Polymer-Ceramic Composites (Xie, 2013).

Chapter 2
Current Challenges and Future Perspectives in the Field of Thermal Barrier Coatings

Giovanni Di Girolamo
Italian National Agency for New Technologies, Energy, and Sustainable Economic Development (ENEA), Italy

ABSTRACT

This chapter describes how ceramic thermal barrier coatings (TBCs) are usually applied on metal components of aircraft engines and land-based gas turbines, with the purpose to extend their lifetime as well as to increase performance and durability, by increasing the operating temperature. The TBCs have to satisfy basic requirements in terms of low thermal conductivity, high stress compliance, high sintering resistance as well as high resistance to the environmental attack promoted by oxygen, molten salts and CMAS (calcium-magnesium-alumino-silicate) deposits. This chapter is aimed at analyzing the state-of-the-art, the recent developments and the future perspectives in the field of TBCs, focusing the attention on advanced materials and new architectures as well as explaining the mechanisms affecting the failure of TBC systems.

BACKGROUND

The components of aircraft engines and land-based gas turbines typically operate in harsh service environments because of the presence of extreme temperatures, as well as of oxidative, corrosive and erosive agents, that reduce their durability and lifetime, thus involving high costs for maintenance, dismantling and substitution (Boyce, 2002; Davis, 2004; Pawlowski, 2008).

This also implies any restrictions on the operating temperature, imposing any technological barriers toward the development of new systems with enhanced performance, reduced fuel consumption and lower emissions of NO_x and CO.

The application of thermal barrier coatings (TBCs) is able to protect the surface of metal components from the surrounding environment (stationary guide vanes, rotating blades, blade outer air-seals and

DOI: 10.4018/978-1-5225-4194-3.ch002

shrouds in the high-pressure section behind the combustor, afterburners in the tail section of jet engines). Their first aim was to extend the lifetime of stationary engine parts, then, at the end of 1980s, the TBCs were applied to the rotating blades.

These insulating ceramic coatings are able to reduce the temperature on the surface of the component, as well as to protect it from the attack promoted by environmental contaminants, thus increasing both the temperature capability and durability, without compromising the mechanical strength of the structural component (Zhu & Miller, 2000; Padture et al., 2002; Xu & Guo, 2011).

TBCs typically allow to reduce the substrate temperature between 100 and 300 °C for thicknesses of 120-400 μm.

The enhanced temperature capability allows to increase the inlet temperature and the efficiency of the turbines. TBCs are generally applied on internally air-cooled components, such as combustor parts, transition ducts of stationary and rotation airfoils.

Ceramic TBCs have thickness from several hundreds of micrometers to >2mm. A TBC system generally consists of a metallic substrate, a MCrAlY-type bond coat and a ceramic top coat (Miller, 1997). The application of the intermediate layer increases the adhesion of the upper ceramic layer, by reducing the thermal stresses arising from different thermal expansion of ceramic and substrate, and also provides high resistance against infiltration of oxygen and molten salts (Saremi et al., 2008; Richer et al., 2010). Ceramic materials with low thermal conductivity are potential candidates for developing next-generation TBCs, because of their high temperature capability and thermal gradient tolerance (Curry et al., 2011; Cao et al., 2004; Hetmanczyk et al., 2007; Bast & Schumann, 2002).

The industrial gas turbines have a market share of 25%, the total contribution of the aero held sums up to 35%. The global TBC market was estimated to be US$12.4 bn in 2015 and is expected to reach US$22.4 bn by the end of 2024. The rise of the global market between 2016 and 2024 has been measured at CAGR (Compound Annual Growth rate) of 6.6% (TMR, 2017).

The growth of the automotive industry is feeding the demand for TBCs. The aerospace industry, as main end-user of TBCs, is contributing significantly to the expansion of the market. The rising number of applications for gas turbines in energy, aerospace and defense is also expected to dramatically increase the demand. The rise in the price of raw materials, coupled with the high cost of using superior manufacturing technologies, is able to restrain the growth of the global market.

The competitive position of the main players in TBC market is well protected due to license restrictions by original equipment manufacturing companies for sharing TBC solutions for hot-section gas turbines for aircrafts.

The actual context encourages new researches in the field, with the purpose to increase the knowledge and the TRL (Technology Readiness Level) of the products developed, in order to promote large-scale technological transfer to the industry. Increasing innovation in gas turbine technology and in high-temperature engine materials is a key demand for the coming years.

This paper deals with the recent findings on TBC materials and new architectures, focusing the attention on the major damaging mechanisms experienced by TBC systems during their service.

Current Challenges

The state-of-the-art on the development of advanced TBCs has to be discussed and analysed in terms of the main challenges that can be identified in the field, such as:

- Novel TBC compositions and architectures, to withstand higher temperatures and flexible fuels as well as high thermomechanical load flexibility (better strain tolerance, higher toughness, enhanced resistance to sintering) (Cao et al., 2004; Clarke & Levi, 2003; Evans et al., 2008; Paul et al., 2009; Vassen et al., 2010).
- Development of new thermal spray processing technologies, such as suspension and solution precursor plasma spraying (Guignard et al., 2012; Pawlowski, 2008).
- Coating systems with improved resistance to: CMAS deposits, vanadium pentoxide (V_2O_5) and sodium sulfate (Na_2SO_4) contained in duty fuels; erosion and foreign object damage (Borom et al., 1996; Drexler et al., 2010; Mohan et al., 2007; Saremi & Habibi, 2011).

Failure of TBCs under thermal cycling is a complex process, influenced by high-temperature evolution of coating microstructure and by residual stress development due to TGO (thermally grown oxide) growth and thermal mismatch during cooling. In principle, two main types of fracture phenomena are known, segmentation and delamination. Segmentation cracks are oriented normal to the interface and increase the strain tolerance, even if they are preferential paths for infiltration of oxygen and contaminants.

Delamination, occurring normal to the spraying direction, is promoted by propagation of cracks formed because of thermal stresses associated to different processes, such as sintering, oxidation and TGO growth at the interface, degradation and phase changes promoted by high temperature and infiltration of contaminants. Once long enough, it can promote spallation of TBC segments, so that delamination cracking is a criterion for evaluating the failure of TBCs.

The current limitations of traditional YSZ coatings at temperature higher than 1200°C, due to phase changes and sintering of the microstructure, as well as their fast degradation at lower temperature promoted by infiltration of molten salts and CMAS, impose the development of alternative materials with enhanced thermo-physical and functional properties.

The manufacturers are also focusing their efforts on the implementation of solutions for improving the reproducibility and efficiency of thermal spraying processes, in order to optimize coating performance, as well as for the standardization of powder manufacturing processes, for either improving material deposition and coating architecture.

PROCESSES AVAILABLE

Various processes are available for manufacturing of TBCs and, among them, thermal spraying and electron-beam physical vapour deposition (EB-PVD) are the best available technologies.

This paper briefly analyses the characteristics of the processing technologies, highlighting the differences in the as-produced microstructures and related functional properties.

Thermal Spraying

Thermal spraying is a cost-effective technology for manufacturing of relatively thick coatings on large components, through melting and deposition of micron-sized powder particles. It is characterized by high deposition rate and low manufacturing time. Thermal spraying includes many deposition techniques, depending on the heat source employed and the spraying environment. It is suitable to produce coatings with reduced thermal conductivity and high porosity (Wang et al., 2011), whereas EB-PVD is particularly used for producing coatings with higher density and compliance, by means of the formation of highly strain-tolerant columnar structures.

In thermal spraying a heat source is employed to melt the powder particles. The particles are fed by a carrier gas to the torch, injected in the flame/plasma jet and heated above their melting point. The molten droplets are then accelerated by a gas stream toward the substrate where they impact, and are flattened and quenched to room temperature, producing the deposition of overlapped splats and, finally, the build-up of a coating. Both the morphology and the size of the powder particles affect their melting degree and then the microstructural, physical and mechanical properties of the as-deposited coatings. These are generally characterized by a lamellar microstructure, as shown in Figure 1.

The selection of the most suitable thermal spraying method depends on the intrinsic characteristics of the raw materials employed, as well as on the functional requirements and on the costs for the equipment. New emerging thermal spraying processes, such as solution and suspension plasma spraying, are presented as further solutions for obtaining tailored microstructures, partly composed of nanostructured features.

Plasma Spraying

1. **Atmospheric Plasma Spraying (APS) of Conventional and Nanostructured Powders:** A high-temperature plasma is generated by transferring energy into an inert gas until the energy level is sufficient to ionize the gas, allowing the electrons and ions to act independently of one another. The plasma-forming gas mixture is introduced through the gun and enters the arc chamber through a gas injector that imparts a spin or vortex flow. Then it is accelerated as it expands and comes out of the nozzle.

Figure 2 shows plasma jet and spray particle stream during operation. Plasma spraying is suitable for depositing high melting-point materials, such as ceramic oxides for porous TBCs.

Figure 1. Lamellar microstructure of a TBC system composed of an inner NiCrAlY bond coat and an upper CYSZ (ceria-yttria co-stabilized zirconia) layer

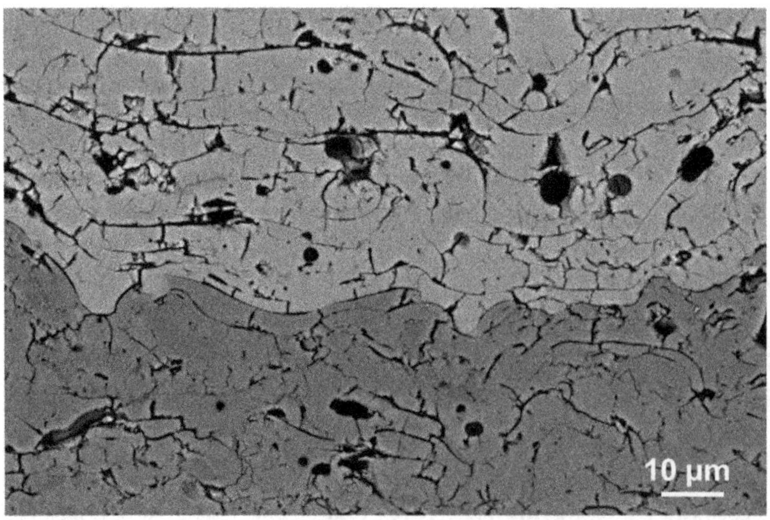

Figure 2. Plasma jet and spray particle stream during processing

The use of nanostructured powder feedstock allows to develop coatings with enhanced thermal and mechanical properties. By partially melting micronsized agglomerates of nanostructured particles it is possible to produce coatings with "bimodal" microstructure, characterized by melted areas with lamellar structure and semimolten particles preserving the original nanostructure, as shown in Figure 3.

The plasma spraying parameters have to be adjusted to minimize the grain growth and the detrimental changes associated to the high temperature experienced by the sprayed particles.

Figure 3. Cross-sectional microstructure of plasma sprayed nanostructured YSZ (yttria stabilized zirconia) coatings, showing melted areas surrounding partially melted nanozones

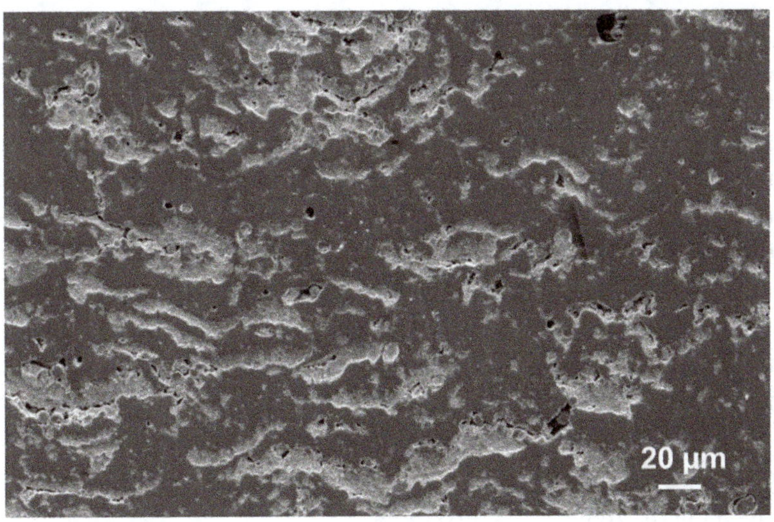

2. **SPS and SPPS:** More recently, atmospheric suspension and solution precursor plasma spraying (SPS and SPPS) processes have been introduced for fabrication of ceramic TBCs with very fine microstructure and enhanced service lifetime compared to that of the same coatings produced by conventional APS. The thermal spray community has a great interest and confidence in the development of these processes.

The suspensions are composed of fine solid particulates dispersed in liquid (water or organic solvent), whereas the solutions are inorganic salts and metalorganic compounds dissolved in water or organic solvent.

The suspensions are injected by a pressurized feed system in the plasma jet and atomized in a stream of fine droplets, that impinge on the substrate after a sequence of physical interactions (aerodynamic breaking up, evaporation of liquid, sintering and melting of fine solids and agglomerates). The suspension liquid evaporates after injection in the plasma jet and the solid may agglomerate, melt and impact the substrate. The suspension can be introduced to the plasma torch with a radial feedstock injector through an atomizer or a nozzle (Ahrens et al., 2012).

In turn, in SPS process the droplets go through several chemical and physical changes (breaking up, solvent evaporation, solute decomposition and melting), so that they impact on the substrate as fully melted or unpyrolized particles. The fine microstructure of these coatings is associated with the small size of the molten droplets as well as with their in-flight fragmentation in the plasma jet (Bhatia et al., 2002; Marchand et al., 2011).

EB-PVD

In EB-PVD process an electron beam is employed to vaporize an ingot of the coating material, to produce a vapor cloud that is accelerated toward a heated substrate, with the aid of high vapor pressure in vacuum atmosphere.

Coatings with high adhesion and strain tolerance are fabricated: they exhibit a columnar microstructure, composed of columns loosely connected, grown along the cooling direction, normal to the surface of the substrate. The columns are aligned along the heat flux, directing heat transfer by conduction during thermal exposure (Hass et al., 2001).

These coatings are able to follow the expansion of the metallic substrate accommodating thermal expansion mismatch, without developing further internal stresses. Increasing the intergrowth column separation distance reduces the strain energy driving delamination and increases the delamination resistance of TBCs.

EB-PVD coatings are more durable than APS ones, but the process is more expensive and time-consuming. They are generally used for severe applications in turbine blades and vanes (rotor components of aircraft engines), whereas low-cost APS process is suitable to coat large components of combustion liners, first-stage blades, stator vanes and after-burner flame holders of aircraft engines. APS is generally used to coat both the stationary and rotating hot-section parts of electricity-generation engines. APS is more efficient than EB-PVD because of lower equipment costs, relatively high performance-to-cost ratio and shorter manufacturing time. The same materials can be approximately processed using these deposition techniques.

Thermal conductivity of EB-PVD is generally in the range between 1.5 and 2 $Wm^{-1}K^{-1}$, whereas for APS coatings its value is in the range 0.8-1.7 $Wm^{-1}K^{-1}$.

MAIN FOCUS OF THE CHAPTER

Candidate Materials: Basic Properties and Novel Architectures

During the 1940s the first nickel-based superalloys were introduced in aerospace industry. The first turbine blades were produced in the wrought form, then induction vacuum casting technologies were developed and in the 1970s casting with direct solidification enhanced their fatigue life. By the 1990s, single crystal alloys began to be produced (Sourmail, 2004; Reed, 2006). The application of ceramic TBCs allowed to increase the operating temperature. Turbine blades are air cooled by an internal cooling system which significantly increases the operating costs. The application of a TBC allows to reduce the requirements for the cooling system, in terms of lower cooling air flow and higher efficiency or, at constant air flow, lower metal surface temperature and thus higher lifetime (Strangman, 1986).

Generally speaking, a TBC system is composed of substrate, metal bond coat and ceramic top coat. The development of TBC systems should evolve at all levels: instead of independently developing these materials, a system approach should be adopted. Many investigators studied combinations of substrates, bond coats and top coats (Haynes et al., 1999; Cao et al., 2004, Zhu et al., 2004).

Regarding bond coats, MCrAlY (M = Ni and/or Co) coatings are usually applied by thermal spraying. They often contain reactive elements (Y, Re, Hf, Zr) to improve adhesion and oxidation resistance. In turn, β-NiAl intermetallic coatings are applied by thermal spraying or cementation pack method: they are less ductile than MCrAlY coatings. Finally, β-(Pt,Ni)Al coatings are also emerging as potential bond coats; electroplating of Pt is followed by aluminization, in order to allow diffusion of Pt into the coating. These coatings have exhibited improved oxidation and hot-corrosion resistance, by reducing inward diffusion of Al and outward diffusion of Ni and other elements, as well as by increasing the adhesion and Al diffusivity (Oguma et al., 2004; Kamal et al., 2009; Itoh & Saitoh, 2005; Darolia et al., 2001; Haynes et al., 2004).

However, looking at increasing temperature capability and efficiency of TBC systems, the attention is herein focused on ceramic materials. A ceramic TBC must meet some basic criteria such as:

- Low thermal conductivity (< 2 $Wm^{-1}K^{-1}$ or, preferably, less than 1 $Wm^{-1}K^{-1}$) (Clarke & Levi, 2003).
- High thermal expansion coefficient (in the range from room temperature to 1000 °C, the CTE of superalloy substrate is about 16×10^{-6} K^{-1}, while the CTEs of NiCrAlY and YSZ are approximately 17.5×10^{-6} K^{-1} and 10.7×10^{-6} K^{-1}, respectively).
- High phase stability from room to service temperature.
- Low elastic modulus.
- Strong resistance to high-temperature sintering.
- Environmental resistance against oxidative and corrosive media (molten salts, CMAS, oxygen) (Eliaz, 2002).

Thermal conductivity influences the heat transfer from the coating surface to the substrate. The thermal conductivity of a material depends on its intrinsic properties and on boundary convection, thermal radiation and gas pressure of the working atmosphere. It is also related to microstructure and porosity. A porous microstructure typically exhibits low Young's modulus and high strain tolerance, while the hardness affects wear and erosion resistance (Evans et al., 2006; Wellman & Nicholls, 2008).

The thermal expansion coefficient (CTE) is a basic thermo-physical property, because the TBC lifetime is strongly related to the strain level and to the mismatch in CTE values of coating and substrate.

TBCs have to provide thermal insulation to the substrate material for prolonged times and under thermal cycling conditions, as well as strain compliance by minimizing the stresses produced by expansion mismatch between metal and ceramic during heating and cooling cycles. These times are 1000 s of hours for jet engines being cycled numerous times between peak (~1300°C) and room temperature (take-off/landing and on-ground), 1000 s of hours for power-generation engines with fewer thermal cycles (maintenance shut-downs). TBCs have to maintain their mechanical integrity under cycling, as well as to be chemically compatible with the underlying metal. They must operate in an oxidizing environment at maximum pressure of ~ 10 atm and maximum gas velocity exceeding Mach 1.

Partially yttria stabilized zirconia (YSZ) has been the most used material for TBC fabrication: its properties have been studied for many years. The addition of one or more oxides to pure zirconia stabilizes the tetragonal or cubic phase, thus preventing detrimental phase changes during high-temperature exposure (Evans et al., 2008). Zirconia is stabilized in the tetragonal form by adding 6-8 wt. % yttria. After plasma spraying coatings with metastable tetragonal t' phase are generally obtained, by means of a diffusion-less shear transformation directly from the cubic state. This metastable phase does not directly transform to monoclinic phase on cooling to room temperature. YSZ coatings show very good thermo-mechanical properties and thermal conductivity in the range between 0.8 and 1.1 $Wm^{-1}K^{-1}$, which increases up to 1.5-1.7 $Wm^{-1}K^{-1}$ after 50 h at 1000°C (Trice et al., 2002; Schulz, 2000; Chen, 2006).

However, YSZ shows some limitations when used at temperature higher than 1200°C, because of phase changes as well as of the sintering of its porous microstructure. The phase changes, that can also occur at lower temperature in presence of molten salts, are typically accompanied by volume contraction (3-5%) and extended cracking, that reducing the barrier effect against penetration of corrosive agents and increasing the risk for coating delamination and spallation. The high-temperature exposure also assists the partial densification of the porous microstructure, by the closure of fine pores and microcracks and the formation of sintering necks, reducing the content of pores and increasing the stiffness, finally reducing the strain tolerance of the whole TBC system (Thompson & Clyne, 2001). In addition, oxygen penetration through the open pores and the microstructure of the ceramic layer leads to the oxidation of the underlying metallic surface, promoting the formation of TGO layer at the interface, that assists cracking and delamination.

Based on that, the researchers have focused their attention on alternative TBC materials. Many investigators have analysed combinations of bond coats and top coats in order to optimize TBC performance. The study of bond coat compositions is outside the objectives of this work, whose focus in on ceramic materials.

Other Stabilizer Oxides for Zirconia

The thermal conductivity and the high-temperature performance strongly depend on the nature and the amount of the dopants used as zirconia stabilizers. Tetravalent oxides as CeO_2 (ceria) can be successfully used to stabilize zirconia in the tetragonal or cubic phase. It has been reported that CYSZ (ceria-yttria co-stabilized zirconia) coatings exhibited lower thermal conductivity than YSZ ones at room temperature, higher phase stability and higher lifetime during long-term exposure. During exposure at temperatures higher than 1200-1300°C the metastable tetragonal t' phase of as-sprayed YSZ coatings decomposes to a mixture of high-yttria content cubic and low-yttria content tetragonal phases. On cooling to room temperature, the high-yttria cubic phase may be retained or transformed to high-yttria tetragonal phase. On the other hand, upon cooling the low-yttria tetragonal phase may transform to monoclinic phase depending on cooling rate and grain growth. In plasma sprayed CYSZ coatings the monoclinic transformation did not occur after isothermal exposure at 1315°C. CYSZ coatings showed higher CTE with respect to YSZ ones (12.6 x $10^{-6}K^{-1}$ and 10.7 x $10^{-6}K^{-1}$, respectively, between 50 and 900°C). The addition of CeO_2 improved toughness, thermal shock resistance and hardness. A thinner TGO was observed in CYSZ coatings cycled at high temperature (Sodeoka et al., 1997).

Plasma sprayed dysprosia (Dy_2O_3) stabilized zirconia (DySZ) coatings exhibited higher phase stability with respect to YSZ ones after thermal exposure at 1300°C for 50 h. However, the values of thermal conductivity and CTE were not quite different than those of YSZ, as shown in Table 1 (Di Girolamo et al., 2010; Di Girolamo et al., 2010; Ma et al., 2008; Subramanian, 2001; Jarligo et al., 2009; Dai et al., 2009; Gadow & Lischka, 2002; Paul et al., 2009; Cao et al., 2004).

Plasma sprayed lanthana-yttria (Y_2O_3-La_2O_3) co-doped zirconia (LaYSZ) TBCs partially decomposed after isothermal exposure at 1300°C for 50 h. They showed thermal conductivity and CTE values similar to those found for YSZ coatings, but better sintering resistance (Matsumoto et al., 2004; Matsumoto et al., 2006; Rauf et al., 2012).

In turn, plasma sprayed ytterbia (Yb_2O_3) partially stabilized zirconia (YbSZ) coatings have been applied as TBCs on NiCrAlY or NiCrAlYb bond coats, showing better thermal cycling resistance up to 1110°C. Their thermal conductivity and CTE were higher than those of YSZ TBCs.

Plasma sprayed scandia-yttria co-stabilized zirconia (ScYSZ) coatings showed better phase stability than YSZ ones up to temperatures of 1400°C (Jones, 1989; Jones, 1997; Liu et al., 2010).

The use of two or more rare-earth oxides is actually the best technological solution in order to produce coatings with enhanced thermophysical properties. Plasma sprayed zirconia coatings obtained by adding to ZrO_2 a primary dopant (Y_2O_3) and two rare-earth oxides co-dopants (the first is Nd_2O_3 or Gd_2O_3 or Sm_2O_3 and the second is Yb_2O_3 or Sc_2O_3) have exhibited lower thermal conductivity (0.5-0.7 $Wm^{-1}K^{-1}$), higher CTE (11.5-13.5 x 10^{-6} K^{-1}) and higher phase stability with respect to YSZ ones. Plasma sprayed ceria-neodymia co-doped zirconia coatings are also very promising, due to their high CTE (12.8 x 10^{-6} K^{-1}) from ambient temperature up to 1250°C.

However, in spite of their improved thermophysical properties, only a few of these TBCs have exhibited thermal cycle lifetime comparable or superior to that of YSZ ones, so that YSZ is actually difficult to be replaced. Looking at alternative zirconia-based systems, CYSZ and ScYSZ are much promising when severe operating conditions are expected.

Alternative TBC Materials

The search for alternative materials is aimed to develop TBC systems with increased temperature capability, enhanced phase stability, lower thermal conductivity and higher thermal gradient tolerance. In addition, provided that zirconia-based systems have exhibited any shortcomings as the temperature increased, due to sintering and detrimental attack of molten salts and CMAS, it is mandatory to develop materials having better resistance to sintering as well as to degradation promoted by infiltration of sulfates, vanadates and CMAS.

Some categories of materials have been identified in the last decade: rare-earth zirconates ($Sm_2Zr_2O_7$, $Gd_2Zr_2O_7$, $La_2Zr_2O_7$), perovskites ($SrZrO_3$, $Ba(Mg_{1/3}Ta_{2/3})O_3$, $La(Al_{1/4}Mg_{1/2}Ta_{1/4})O_3$), hexaaluminates

($LaMgAl_{11}O_{19}$) (Darolia, 2013; Vassen, 2010). However, there are many missing information on the long-term performance of these materials that are stopping their large diffusion.

Rare-earth zirconates ($RE_2Zr_2O_7$, where RE is La, Gd, Nd or Sm) are obtained by doping zirconia with very high wt. % of stabilizer oxides and are generally stable up to their melting point.

At temperatures between 700 and 1200°C rare-earth zirconates have relatively high CTE (between 8 and 11 x 10^{-6} K^{-1}), low thermal conductivity (1.1-1.5 $Wm^{-1}K^{-1}$) and can have pyrochlore or fluorite structure. Plasma sprayed lanthanum zirconate ($La_2Zr_2O_7$, LZ) coatings showed excellent thermal stability and thermal shock resistance at 1200°C. They also showed sluggish sintering kinetics. The major restrictions of LZ coatings are associated to its relatively low toughness and CTE (9.1-9.5 against 10.5-11.5 x 10^{-6} K^{-1} for YSZ), generally resulting in short thermal fatigue cyclic life (Wu et al., 2002; Vassen et al., 2000).

Sm, Gd and Nd zirconates exhibited lower thermal conductivity and higher CTE than lanthanum zirconate (Pan et al., 2007). The very low thermal conductivity of $Sm_2Zr_2O_7$ coatings (0.35-0.45$Wm^{-1}K^{-1}$) makes them very promising (Hu et al., 2010; Guo et al., 2015; Qu et al., 2007). It was found that, after exposure at 1200 and 1300 °C, both their mechanical properties and thermal conductivity increased, because of the formation of grain bridging at splat boundaries and reduction of porosity. The thermal conductivity of $Sm_{1-x}La_xZr_2O_7$ coatings was constant and near to 1 $Wm^{-1}K^{-1}$ (Hongsong et al., 2009; Chen et al., 2009)

The thermal conductivity of perovskites, such as $BaZrO_3$ and $SrZrO_3$, is higher than that of YSZ. Plasma sprayed $BaZrO_3$ coatings exhibited lower fracture toughness than YSZ and were subjected to BaO loss during thermal cycling at 1200°C. In turn, plasma sprayed $SrZrO_3$ coatings exhibited CTE and thermal cycling lifetime at 1250°C comparable to those observed for YSZ ones. $BaLa_2Ti_3O_{10}$ TBCs with

*Table 1. * Bulk samples or EB-PVD coatings*

	Thermal Conductivity [$Wm^{-1}K^{-1}$]	CTE [10^{-6} K^{-1}]
YSZ	0.8-1.2	10.5-11
CYSZ	0.5	12.6
DySZ (Dy_2O_3-ZrO_2)	0.6-1.2 1-1.4	~ 9.5
LaYSZ	0.6 1-1.3	9-10.5
YbSZ (Yb_2O_3-ZrO_2)	1.6	~ 12.5
(Nd/Gd/Sm)(Yb/Sc)YSZ	0.5-0.7	11.5-13.5
NdCSZ (Nd_2O_3-CeO_2-ZrO_2)	-	12.8
$La_2Zr_2O_7$	0.7-0.8 1.6 *	9.5 9.1-9.5
$Gd_2Zr_2O_7$	1-1.7 *	10.5
$Nd_2Zr_2O_7$	0.6-1 *	10.1
$Sm_2Zr_2O_7$	0.35-0.45	10.8
$BaZrO_3$	3.4 *	7.9
$SrZrO_3$	2.3 *	10.9
$La_2Ce_2O_7$	0.6-1.3	10.3
hexaaluminates	0.8-2	9.5-10.7

perovskite structure were stable up to 1400 °C, and resistant to sintering up to 1500 °C. Their CTE was close to that of YSZ, whereas their thermal conductivity was of about 0.7 at 1200 °C. They survived up to 1100 cycles (heating at 1100 °C and cooling to ambient temperature), while the appearance of the first cracks occurred after 1000 cycles in YSZ ones.

Some investigators studied specific compositions obtained by doping $SrZrO_3$ with Yb_2O_3 and Gd_2O_3: $Sr(Zr_{0.9}Yb_{0.1})O_{2.95}$ and $Sr(Zr_{0.8}Gd_{0.2})O_{2.9}$ exhibited lower thermal conductivity than SrZrO, 20% less than YSZ.

Plasma sprayed lanthanum cerate ($La_2Ce_2O_7$) coatings have exhibited lower thermal conductivity and longer thermal cycling lifetime with respect to YSZ ones. These TBC systems were stable after long-term exposures at 1400°C. However, lanthanum cerate is chemically incompatible with MCrAlY bond coat and their reaction typically assists the formation of $LaAlO_3$, making the TGO unstable and promoting delamination. Therefore, they are not suitable to be used as single layer TBC systems. In $La_2Ce_2O_7$ with fluorite structure thermal expansion drops rapidly in the range 200-400 °C: that is detrimental during thermal cycling (Ma et al., 2010; Wang et al., 2009).

In turn, plasma sprayed lanthanum hexa-aluminates ($LaMgAl_{11}O_{19}$) coatings were partially amorphous after deposition, so that during heating the crystallization induced shrinkage and cracks formation. However, by controlling the crystallization mechanism during processing advanced strain-tolerant microstructures with enhanced thermal cycling lifetime were obtained. Their plate-like structure exhibited excellent sintering resistance at 1300°C. The thermal conductivity of plasma sprayed lanthanum hexa-aluminates coatings is typically in the range between 0.8 and 2 $Wm^{-1}K^{-1}$, while the CTE is close to that of YSZ ones.

It has been suggested that $LaTi_2Al_9O_{19}$ can replace YSZ up to 1300 °C; the thermal conductivity was of 1.1-1.3 between 300 and 1500°C, whereas the CTE was in the range 8-11.2 x 10^{-6} K^{-1} between 200 and 1400°C (Dhineshkumar et al., 2016; Xie et al., 2011).

Multilayered Systems

Rare-earth zirconates, cerates and perovskites can be used as external top coats in multilayered TBCs capable for operating temperature above 1200-1300°C (Stover et al., 2004; Tamura et al., 1999; Cao et al., 2006; Di Girolamo et al., 2015).

Multilayered TBC systems are particularly suitable to reduce the thermal stresses induced during high-temperature exposure at the interface between metal and ceramic. The TBC is herein composed of two or more ceramic layers and YSZ is commonly applied as intermediate layer, due to its compatibility with TGO and other zirconia-based materials.

The intermediate layer generally provides the insulating property, while the upper layer, thanks to its higher temperature capability, density and resistance to oxygen infiltration, protects the inner layer from the environmental attack, acting as a sacrificial coating. Figure 4 shows a multilayered TBC composed of an inner CYSZ layer and an upper LZ layer, developed at ENEA.

Double APS TBCs, composed of an inner YSZ layer and an upper lanthanum zirconate layer, exhibited enhanced thermo-mechanical properties and very high durability during thermal cycling tests at different surface temperatures (1250, 1350 and 1450°C). At the lowest temperature the double system showed a thermal cyclic performance (more than 4000 cycles) better than that of single LZ layer (137-977 cycles) and comparable to that of single YSZ layer while a graded TBC, composed of a mixture of YSZ and LZ, survived 2000-2800 cycles. At about 1350°C, double-layer YSZ/$La_2Zr_2O_7$ system exhibited better cyclic lifetime (1000-2346 cycles) than those shown by single YSZ and double YSZ/$Nd_2Zr_2O_7$

Figure 4. Multilayered TBC system composed of plasma sprayed NiCrAlY bond coat (at the bottom), intermediate CYSZ layer and upper LZ layer
Courtesy of Alida Brentari.

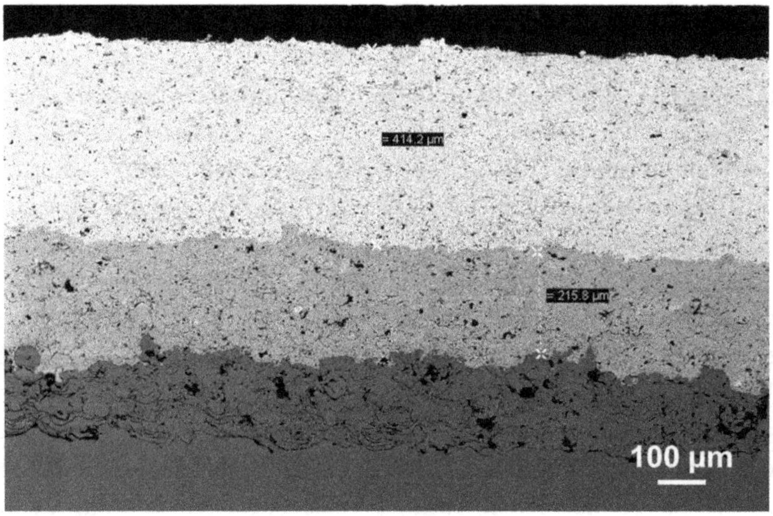

TBCs (220-1000 and 1748 cycles, respectively). In addition, YSZ/$La_2Zr_2O_7$ TBC survived more than 1200 cycles at 1450°C, showing better temperature capability than YSZ one (Vassen et al., 2004; Liu et al., 2015; Wang et al., 2012).

When exposed at 1300 °C for 50 h, single-layer LZ TBC is typically characterized by spallation from the TGO, because the relatively low toughness and CTE assist lateral crack formation and propagation at the interface.

Lanthanum zirconate exhibits lower oxygen ionic diffusivity than YSZ ($9.2 \pm 0.3 \times 10^{-4}\ \Omega^{-1}cm^{-1}$ against $0.1\ \Omega^{-1}cm^{-1}$) at 1000°C, so that can restrict the oxygen diffusion mechanism through the crystalline structure, whereas the most noticeable gas penetration mechanism is associated to the distribution of pores and microcracks within the microstructure. By reducing the oxygen partial pressure and the oxygen activity at the interface between ceramic TBC and metal bond coat the growth rate of TGO is then reduced, positively affecting the stress state at the interface. In the case of multilayered TBC the formation of a dense and continuous Al_2O_3-rich TGO improves both the oxidation resistance and the adhesive strength, thus prolonging TBC lifetime.

It has been also reported that the thermal cycling lifetime of plasma sprayed double YSZ/$SrZrO_3$ coatings at 1250°C was two times higher that of YSZ ones, while at temperatures higher than 1300°C double YSZ/$SrZrO_3$ and single YSZ TBCs approximately showed the same durability.

Gadolinium zirconate ($Gd_2Zr_2O_7$) was recently introduced as protective coating for parts of the aero-engine, but it exhibited low resistance against CMAS attack, erosion, and low toughness too, that reducing the lifetime of the coating system under thermal cycling (Zhong et al., 2014; Bakan et al, 2014; Bakan et al., 2015; Frommhertz et al., 2016).

However, functionally-graded EB-PVD $(GdYb)_2Zr_2O_7$/8YSZ TBCs were found to be more resistant to thermal cycling in comparison with double coatings (1346 against 942 cycles at 1350°C; substrate temperature ~ 1000°C), because of their enhanced sintering resistance. The failure of double TBC was dominated by layer-by-layer delamination within ceramic TBC, due to crack initiation and propagation generated by stress concentration: if the stress exceeded the adhesive-cohesive strength of the coating, delamination occurred due to cracking resulting from compressive stresses in the cooling stage. On the contrary, in the gradient coating, spallation at TBC/TGO interface was caused by TGO growth and CTE mismatch (Zhang et al., 2014).

Nanostructured TBCs

Nanostructured ceramic coatings are also emerging as potential candidates for TBC applications. Plasma sprayed nanostructured YSZ coatings typically exhibit a typical "bimodal" microstructure composed of well-melted splats and retained porous semi-molten nanozones, resembling the morphology of the powder feedstock (Di Girolamo & Serra, 2015).

During spraying, the hot gas mixture penetrates the agglomerated particles and melts their surface, whereas their core remains un-melted, so that the starting nanostructure is partially retained. The short dwell time of the particles in the plasma jet, the low heat transfer associated with their intrinsic porosity and the high cooling rate assist this mechanism. Some particles disintegrate into several smaller ones prior to impact upon the substrate, due to the heating of the gas entrapped in the agglomerates and the resulting pressure increase.

It has been demonstrated that nanostructured YSZ coatings have higher phase stability, lower thermal conductivity (0.7-0.8 against 0.8-1.2 Wm^{-1} K^{-1}) and higher thermal expansion coefficient (11 against 10.5-11*10^{-6} K^{-1}) than their conventional counterparts. Finer porosity and more uniform distribution of pores resulted in more interfaces and shorter phonon mean free path due to scattering. The retention of embedded porous nanozones increased the compliance and could counteract the sintering effects (Lima & Marple, 2008, Chen et al., 2002; Yu et al., 2010; Lamuta et al., 2015). In the case of nanostructured coatings, the formation of crack networks and the beginning of surface delamination required more cycles than in the conventional coatings, owing to higher inter-lamellar strength. Figure 5 shows the surface of nanostructured YSZ coating after thermal shock cycling: the coating is partially delaminated (Di Girolamo et al., 2011; Lima & Marple, 2007; Masoule et al., 2016).

SPS process can be also used to produce nanostructured TBCs. The thermal conductivity of YSZ coatings produced by SPS was in the range between 0.5 and 1.0 Wm^{-1} K^{-1}, close to that of APS coatings having the same porosity (Kassner et al., 2009). However, the SPS process allows to better control the final microstructure and to fabricate coatings with very high porosity. Their hardness was higher than that of APS coatings, while the Young's modulus, measured by three-point bending test, was found to be 3.3 GPa (for APS coatings it was 15-20 GPa at room temperature and decreased to 3 GPa at 1400°C) (Jadhav & Padture, 2008).

SPS TBCs survived 660 cycles at 1400°C (the temperature on the metal substrate surface was 1050°C), then the partial spallation of the top coat occurred, as a result of TGO formation and propagation of delamination cracks. The typical segmented cracks of SPS coatings survived, while the thermal cycling promoted sintering, pore coarsening and stiffening. The thermal cycling resistance was 50% lower than that of APS coatings because of the highest sintering rate of overspray particles with very small size and neck formation.

According to other investigators, SPPS offers the prospect of depositing highly durable nanostructured YSZ TBCs characterized by low thermal conductivity (0.55-0.65 Wm^{-1} K^{-1}). The presence of fine porosity and extended vertical segmentation cracks increased the strain tolerance and reduced the Young's modulus, thus improving the thermal cycling resistance by a factor of 2.5 with respect to APS TBCs.

Figure 5. Surface morphology of nanostructured YSZ TBC after thermal shock cycling, showing delamination
Courtesy of Francesco Marra.

Starting from these different opinions, further efforts are needed to understand the potential of SPPS/SPS coatings. It is reasonable to combine the advantages of these processes with the search for TBC materials with low thermal conductivity and better sintering resistance than YSZ (Padture et al., 2011).

Nanostructured CYSZ coatings have exhibited lower thermal diffusivity with respect to YSZ coatings ($0.33-0.55x10^{-6}$ m^2s^{-1} against $0.51-0.75x10^{-6}$ m^2s^{-1}), as well as higher thermal cycling lifetime at 1050°C (860 against 400-600 cycles for YSZ). This result was explained in terms of higher CTE and lower thermal conductivity (Jin et al., 2012; Latka et al., 2013).

In another work, it was reported that the thermal conductivity of Gd-doped YSZ coatings with porosity of 15% was 1.042 Wm^{-1} K^{-1} at room temperature and decreased to 0.894 $Wm^{-1}K^{-1}$ at 900 °C, while an apparent increase was observed at 1200 °C, because of the high-temperature radiative heat transfer through the material (the porosity values of conventional and nano YSZ TBCs were 10 and 24%, respectively.

Nanostructured lanthanum zirconate TBCs exhibited thermal conductivity lower than that of YSZ ones (between 0.73 and 0.79 Wm^{-1} K^{-1} versus 0.8-1.2 Wm^{-1} $K^{-1}$for YSZ coatings, in the range from 200°C to 1200°C). Their CTE was slightly lower (9.45×10^{-6} K^{-1}vs. 10.7×10^{-6} K^{-1}) (Wang et al., 2014). Thermal aging at 1300°C and 1400°C for 50 hours caused an increase in their thermal conductivity of 0.3 Wm^{-1} $K^{-1}$and 0.6 Wm^{-1} K^{-1}, respectively.

In turn, nanostructured Gd_2O_3-Y_2O_3-ZrO_2 coatings exhibited enhanced phase stability and lower thermal conductivity than conventional and nanostructured YSZ coatings: from room temperature to 1200°C, it was between 0.9 and 1.05 Wm^{-1} K^{-1}, against 1.1-1.2 for nano YSZ and 1.3-1.45 for conventional YSZ (Wang & Zhou., 2016).

Multilayered $Gd_2Zr_2O_7$/YSZ have been also developed by SPS (Mahade et al., 2016); they exhibited lower thermal conductivity than YSZ despite lower porosity, and longer thermal cyclic lifetime at 1300 °C, while maintaining the interface at 1050 °C. Single YSZ TBCs survived for 43 +/- 7 cycles, while the cycles to failure were found to be 395 for double GZ/YSZ and 521 for triple Gd dense/GD/YSZ TBCs.

In YSZ TBC the failure occurred within YSZ layer, at 60 μm from the TGO. Double- and triple-layered TBCs failed because of horizontal cracking within the GZ layer, near the interface with YSZ. The dense layer is also expected to increase resistance against CMAS and erosion. It should be noted that $Gd_2Zr_2O_7$ has a tendency to react with Al_2O_3 at high temperature, forming $GdAlO_3$, and its fracture toughness is lower than that of YSZ (1.02 against 2 $MPam^{1/2}$).

Mechanisms Affecting the Durability and Failure of TBCs

The performance and the durability of TBCs at operating conditions are affected by different damaging mechanisms or a combination thereof, such as:

- Formation and growth of the thermally grown oxide (TGO) at the interface between metal surface and ceramic TBC;
- High-temperature sintering of the porous microstructure;
- Thermal stresses arising from mismatch in CTE values of overlapped layers during repeated thermal cycling (and during cooling at higher extent, because above 600 °C the bond coating starts to creep and the system is stress-free, whereas during cooling and plastic deformation the stresses in the TBC system arise);
- Phase changes promoting cracking, delamination and TBC spallation;
- Environmental attack promoted by molten salts or CMAS;
- Impact of foreign particles on the surface, carried along with the hot combusted gas.

These effects can cause remarkable microstructural changes, affecting the thermal, mechanical and functional properties, and then the lifetime of the component (Evans et al, 2001). Failure of TBCs under thermal cycling is a complex process, influenced by high-temperature evolution of coating microstructure and by residual stress development due to TGO growth and thermal mismatch during cooling. Delamination, occurring normal to the spraying direction, is promoted by propagation of cracks formed because of thermal stresses associated to different processes, such as sintering, oxidation and TGO growth at the interface, degradation and phase changes promoted by high temperature and infiltration of contaminants. Once long enough, delamination can promote spallation of TBC segments, so that delamination cracking is a criterion for evaluating the failure of TBC systems.

The attention will be herein focused on: TGO formation promoted by high-temperature oxygen infiltration and oxidation; high-temperature sintering; environmental attack by CMAS and molten salts, highlighting the main effects on the microstructure and on the performance of TBCs with novel composition and architecture.

TGO Formation and High-Temperature Oxidation

From the observation of damaged components after service it has been ascertained that TBC failure occurs by cracking, delamination and spallation, typically on cooling to room temperature (Hutchinson et al., 2002; Sumner et al., 2013). The spallation is promoted by the swelling of the TGO, grown at the interface between ceramic TBC and metal bond coat during high-temperature exposure, because of oxygen infiltration through TBC and reaction with Al contained in the bond coat, as shown in Figure 6. Oxygen infiltrates by two transfer mechanisms. The former is the ionic diffusion through the microstructure of the ceramic layer, the latter is the penetration of gas through pores and microcracks.

Many failure mechanisms are associated to the formation and growth of the TGO. When it reaches the critical thickness (5-6 μm), the delamination takes place (Dong et al., 2014; Saremi et al., 2007). There are two contributions to the stress in the TGO layer, related to the growth strain as well as to the stresses generated by thermal expansion mismatch. The former essentially depends on thickening and lateral expansion of the TGO. The latter is produced by different thermal expansion of metal and ceramic.

By reducing the permeation of oxygen towards the bond coat, the growth rate of the TGO decreases and lesser tensile stresses arise at the interface. It is well accepted that high stress state can assist horizontal cracking and TBC spallation, that typically occurs when the length of lateral cracks reaches the critical value.

The gradual growth of the oxide scale promotes volumetric expansion and compressive stresses assisting TGO deformation, cracking and delamination. During cooling the thermal expansion mismatch drives residual in-plane compressive stresses, that reach a maximum at room temperature. During repeated cycles oxygen penetration advances and normal strains are induced in the TBC, promoting cracks that propagate and coalesce, causing failure at the interface (Fan et al., 2012; Liu et al., 2015).

Figure 6. Cross-sectional microstructure and EDS mapping of NiCrAlY bond coat after oxidation at 1100°C
Courtesy of Alida Brentari.

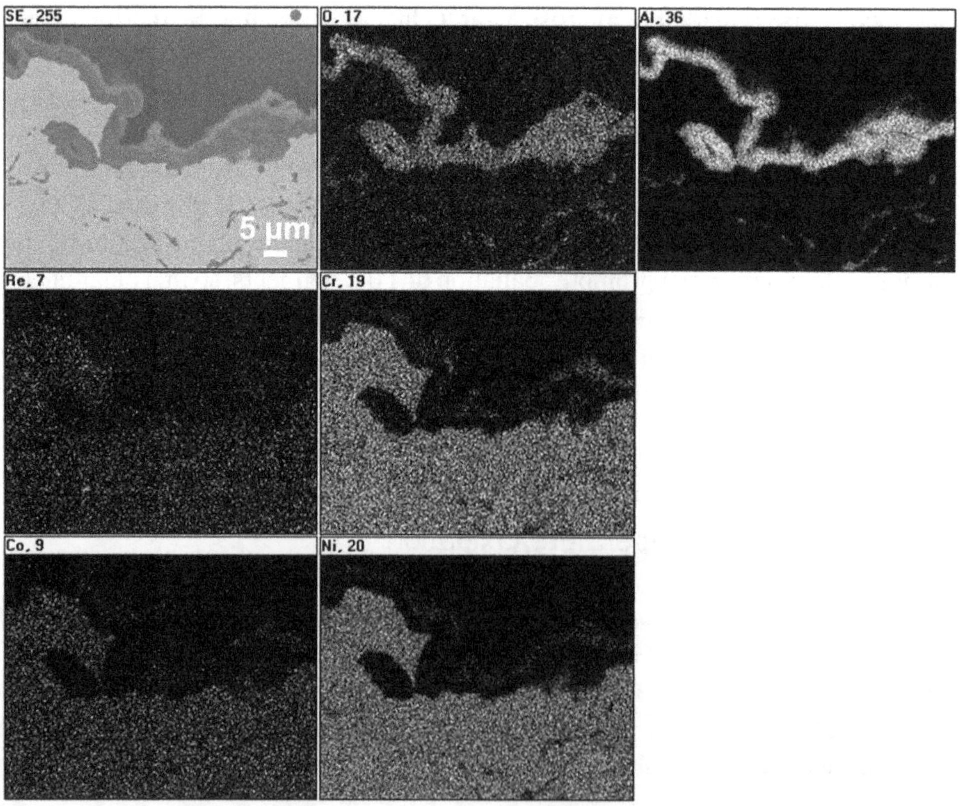

A TGO composed of α-Al_2O_3 with slow growth rate is recommended, because the formation of brittle mixed oxides like spinels promotes fast TGO growth, then crack formation and propagation. Al_2O_3 has low oxygen permeability and can retard further oxygen penetration.

TGO should remain elastic to the highest temperature and not creep to prevent rumpling or cavitation under thermal cycling: rumpling can cause local growing separations evolving in spallation (Ahrens et al., 2004). It has been reported that, at moderate heat fluxes and high temperature, failure is dominated by TGO formation and inelastic behavior, then fracture occurs at the interface between ceramic and metal. With increasing ceramic surface temperature and thermal gradient, thermal expansion mismatch plays a major role and fracture typically takes place within ceramic TBC, because of splat-by-splat delamination.

The maximum temperature for the bond coat is recognized to be 1150°C. The composition of the metal coating has to be optimized in order to minimize its deformation at high temperature and some effects of reaction and inter-diffusion with the substrate.

Thereby, the composition of the bond coat becomes important in order to reduce the growth rate of the TGO. Intermetallic β-NiAl and MCrAlY containing a high amount of Al are used as bond coats, because they enable the formation of Al_2O_3-rich scale on the surface, providing good oxidation resistance.

In a recent work, EB-PVD YSZ and $Gd_2Zr_2O_7$ TBCs were tested at 1100°C from 8 to 100h. The samples were exposed to vacuum annealing to promote the formation of a thin α-Al_2O_3 TGO (Doleker, 2016). No spallation was observed after 100 h. $Gd_2Zr_2O_7$ had better resistance in comparison with YSZ. However, $Gd_2Zr_2O_7$ was more prone to form cracks and its CTE was slightly lower than that of YSZ. Bobzin et al. (2013) also reported that $Gd_2Zr_2O_7$ layer tended to spall after 25 h at 1300°C. Substantially, double-layer TBCs were more resistant, due to the presence of stable Frenkel pair defects in $Gd_2Zr_2O_7$ and lower oxygen permeability.

In conclusion, aiming to reduce spallation promoted by TGO formation, it is necessary to reduce TGO growth rate by using multilayered systems and materials with better ionic conductivity able to reduce the infiltration of oxygen.

High-Temperature Sintering

Another mechanism of failure is associated to the sintering of the porous microstructure, typically occurring at temperatures above 1200°C, even if some effects of shrinkage have been also noticed at temperature of about 1000°C. During thermal exposure, sintering necks are formed at splat boundary, improving the interlamellar bonding, whereas the closure of microcracks and pores also occurs. Figure 7 shows the cross-sectional microstructure of plasma sprayed YSZ coatings exposed at 1315°C for 2, 10 and 50 h. Fine microcracks disappeared, while the large pores were affected by densification at lesser extent (Hutchinson et al., 2006; Cipitria et al., 2009; Alfano et al., 2010).

Sintering neck formation is responsible of temperature field perturbation, by providing paths to heat transfer. It increases the thermal conductivity and reduces the thermal insulation of the metal surface, accelerating the formation of the TGO at the interface (Takagi et al., 2011). The elastic modulus increases and the TBC becomes less strain tolerant to the thermal stresses arising from different thermal expansions of metal and ceramic during thermal cycling, so that extended crack formation and propagation promote delamination and spallation at the interface, leading the system to failure.

Strain relaxation can relieve the high thermal stresses arising during thermal exposure due to the mismatch in CTE values of coating and substrate, so that a high strain tolerance is needed (Malzbender et al., 2003).

Figure 7. Microstructure of annealed YSZ coatings exposed at 1315°C for (a) 2 h, (b) 10 h, (c-d) 50 h

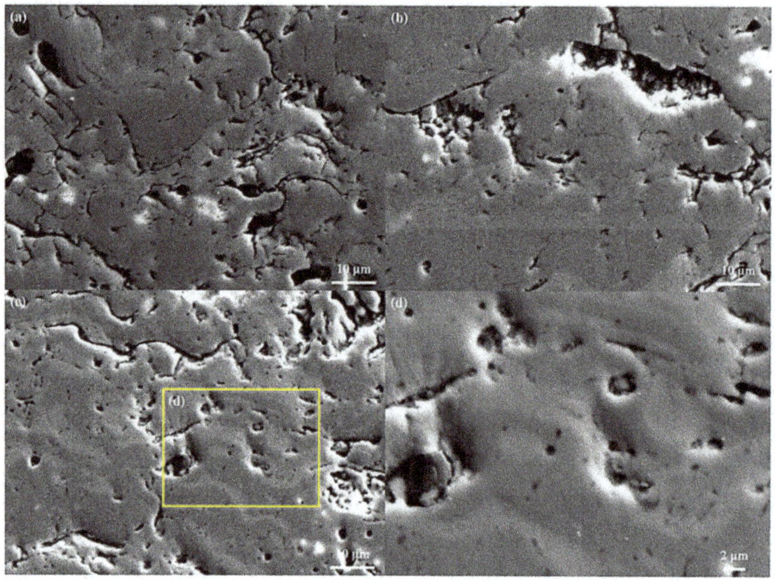

It has been reported that, after exposure at 1200 and 1350 °C, the thermal conductivity and the mechanical properties (hardness and Young's modulus) of LZ coatings increased, because grain bridging and sintering necks at splat boundary created more contact sites and reduced the interlamellar porosity, so that the lamellar structure tended to disappear. As detectable in Figures 8, bridging produced the fast healing of the interlamellar unbonded interfaces, increasing the lamellar bonding ratio. Both pore content and morphology affected sintering (Di Girolamo et al., 2015; Liu et al., 2015; Liu et al., 2016).

Figure 8. Cross sectional microstructure of LZ coating a) before and after thermal exposure at b) 1200°C and c) 1350°C for 50 h

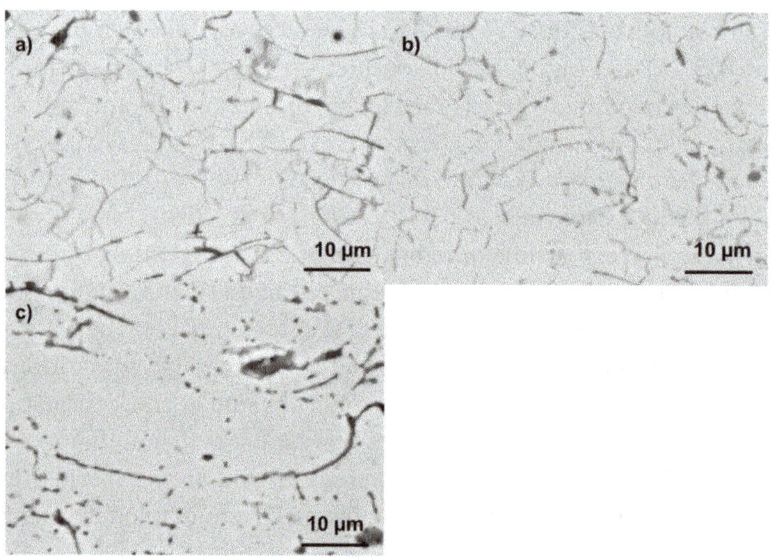

The formation of sintering necks increases the viscosity by raising the threshold load for the onset of splat sliding and promotes main crack propagation despite the healing of some finer cracks.

In the case of nanostructured coatings, sintering promotes grain coalescence, reducing the content of nanostructured areas (Lv et al., 2017). However, the partial retention of embedded porous nanozones increased the compliance and counteracted the sintering effects. The retention of partially melted particles and the presence of areas with different sintering rates were able to stop crack propagation as well as to prevent and retard the sintering effects occurring at high temperature (>1200°C) (Lima & Marple, 2007).

Hot-Corrosion From Molten Salts

Vanadium and sodium are common impurities found in low-grade petroleum fuels. The condensation of fuel combustion products generates extremely corrosive media. These deposits typically melt above 700 °C and penetrate into the TBC, destroying the microstructure, attacking bond coat surface and substrate. It is well known that in YSZ TBCs exposed to hot-corrosion of molten vanadates and sulfates, the failure occurs within the ceramic TBC, due to stresses related to the formation of YVO_4 crystals by chemical reaction and to monoclinic transformation, promoting extended crack growth and propagation and thus facilitating coating delamination (Mohan et al., 2007).

Some alternative stabilizers for zirconia have been studied to improve the hot-corrosion performance, in terms of lower phase decomposition and higher durability.

The addition of CeO_2 to YSZ allows to notably increase the resistance to molten salts. The formation of $CeVO_4$ crystals postponed the formation of YVO_4 ones, reducing their amount, as well as the transformation of tetragonal to monoclinic zirconia. YVO_4 and $CeVO_4$ crystals were formed by the reaction of $NaVO_3$ with Y_2O_3 and CeO_2, respectively (Ahmadi-Pidani et al., 2012).

In nanostructured YSZ coatings tested at 910 °C in presence of 45wt.% Na_2SO_4-55wt.%V_2O_5 molten salt, the reaction between Y_2O_3 and molten salt produced the formation of YVO_4, leached Y_2O_3 from YSZ and led to the destabilization of tetragonal zirconia and the following formation of monoclinic phase (Jamali et al., 2014).

Laser glazing treatment allowed to increase the lifetime by double, by reducing the reactive specific surface area of the dense glazed layer with the molten salt and improving the stress accommodation through network cracks produced by laser glazing treatment (Pidani, 2014; Ghasemi et al., 2010). Laser glazing typically promotes remelting and solidification of coating surface, reducing the surface roughness. As result of processing, the laser glazed coatings typically exhibit a columnar grain structure in the fracture section and an equiaxed grain structure on their surface.

Otherwise, in nanostructured scandia and yttria stabilized zirconia (3.6mol.% Sc_2O_3-0.4mol.% Y_2O_3) tested at 910°C in 45wt.% Na_2SO_4 -55wt.% V_2O_5, the failure of the TBC occurred in the bond coat, due to the formation of mixed oxides at the interface. Localized horizontal microcracks were visible because of the formation of hot-corrosion products within the ceramic layer, as found for coatings exposed for 100 and 300 h, respectively. Small amount of YVO_4 rod-like crystals was formed and monoclinic transformation took place, even if at lesser extent. After 300 h the coating detached from the substrate because of spallation occurring within the bond coat: that did not occur after 100 h. The thickness of TGO was smaller than that recognized as critical for spallation, so that the effect of TGO formation was considered as a minor factor affecting the performance of scandia-yttria co-stabilized zirconia under hot-corrosion (Liu et al., 2014).

In the case of nanostructured CYSZ TBCs tested at 1000 °C in presence of 45wt.%Na_2SO_4-55wt.%V_2O_5 molten salt for different times (from 6 to 300 h), YVO_4, $CeVO_4$ and $Ce_{0.75}O_2Zr_{0.25}$ crystals were formed by leaching Y_2O_3 and CeO_2. They showed better resistance than conventional YSZ and CYSZ TBCs, because of two factors: a) robust acidity and high content of the stabilizer in comparison with YSZ; b) better diffusional resistance related to the presence of nanostructured areas as a major factor in prohibiting further molten salt diffusion into the deeper layer. After hot-corrosion tests, restricted delamination at the external surface was observed, that being the major failure mechanism promoted by tensile stresses related to the presence of hot-corrosion products, such as monoclinic zirconia and crystals of YVO_4 and $CeVO_4$. The extraction of the stabilizer from the crystal lattice structure promoted the transformation of tetragonal to monoclinic phase, assisting crack formation within the ceramic layer (Hajizadeh-Oghaz et al., 2016).

Rare-earth zirconates were found to be more resistant to molten salt attack (Xu et al., 2012). It has been reported that lanthanum zirconate TBC was more resistant than YSZ one at the attack of V_2O_5 at 1000 °C: no spallation and meaningful changes in microstructure occurred. Very small amount of $LaVO_4$ crystals grew. Otherwise, YSZ TBC failed by cracking and spallation. However, when LZ TBC was attached by sulfates at 900 °C, it was rapidly degraded after a few hours, whereas after 360 h YSZ TBC preserved structural integrity and adhesion to the substrate. In the case of a mixture of salts, both the YSZ and LZ TBCs suffered fast degradation (Marple et al, 2006; Huang et al., 2011).

Recently it has been reported that LPO_4 can be added as toughening agent for $Gd_2Zr_2O_7$. Nanostructured 30mol.% $LaPO_4$ (lanthanum orthophosphate or monazite)-doped $Gd_2Zr_2O_7$ TBCs were tested at 900 °C in presence of Na_2SO_4-V_2O_5 and V_2O_5 molten salts. After 4 h in V_2O_5 a continuous and dense sealing layer formed on the coating surface, arresting further penetration of the molten salt. The coating kept its structural integrity. In the presence of the mixture, after 4 h at 900 °C a nearly continuous reaction layer grew on the surface (Guo et al., 2017, Wang et al., 2016). Na_2SO_4-V_2O_5 caused thinner reaction layer in comparison with V_2O_5, and some $NaVO_3$ traces were found in the inner nanostructured region of the coating, up to 100 μm. The penetration was not severe and the nanozones played a positive function on reducing the infiltration of the molten salts.

CMAS Attack

Low-melting temperature salts can represent a challenge in any applications, but they are not a barrier against increased performance, because the increase of gas temperature would be above the dew point of the molten salts. On the contrary, silicate deposits at higher temperature represent a major challenge. Siliceous particulates (airborne dusts, sand, ash, debris) are ingested with the intake of air and then adhere to the surfaces in the hot-gas path (combustors, airfoils, shrouds) and at peak temperatures of the engine cycles (take-off or landing) yield glassy melts based on CMAS (Borom et al., 1996). CMAS may not be a major concern for land-based turbines because of adequate in-take air filtration, but represent a major risk for aircraft engines.

The mechanism and rate of deposit build-up are under discussion. The siliceous particulates can melt in flight through the combustor and deposit as droplets. However, the examination of some airfoil deposits revealed agglomerates of solid particulates, implying that their melting might occur in-situ, during the hotter stages of the engine cycle. The rate of accumulation varies over the surface of the airfoil, concentrating on the hotter areas of the higher pressure side and, to some extent, on the leading edge, with little or no deposits on the suction side.

The melting point of CMAS is below the temperature of TGO/TBC interface. The glass wets and penetrate by capillarity action, chemically dissolving the material and reducing the strain tolerance. It promotes TBC stiffening and loss of thermal insulation. When the temperature drops the environmental deposits crystallize and cause thermal expansion mismatch between CMAS deposits and coating, leading to cracking and failure (Drexler et al., 2010). CMAS deposits attack the interface and promote delamination along the TGO or creep cavitation in the bond coat wherein the delamination path would propagate within the metal.

The melting point of CMAS deposits varies with their composition, even if many silicate deposits melt at 1200 °C. At temperature lower than their melting point these particles also bounce off or cause erosion. The deposits can contain oxides originating from the engine components by erosion of coating surface (NiO, TiO, ZrO_2). Anyway, the infiltrated CMAS were reported to have similar compositions for a wide variety of deposits, loadings and conditions of exposure.

The EB-PVD coatings with columnar microstructure are much susceptible to penetration of CMAS (Peng et al., 2012; Braue & Mechnich, 2011). Zhao et al. (2014) found that in YSZ coatings exposed at 1250 °C, CMAS glass penetrated into the ceramic layer along the inter-columnar gaps and interacted with coating material. After 4 h, an interaction area, composed of a mixture of YSZ and CMAS with equiaxal structure, with thickness of 20 μm, was formed in the YSZ surface layer. Floating of zirconia in CMAS melt occurred: Y_2O_3 was dissolved in CMAS glass and promoted fast monoclinic transformation. The mixture re-precipitated with different morphology and composition, as demonstrated by the presence of globular particles with composition depending on the local chemistry. When the infiltrated and precipitated layer solidified upon engine cooling, filled the pores, reducing porosity and accelerating sintering shrinkage. After 8 h delamination cracking of YSZ occurred: vertical cracks, wrinkles, spallation and delamination were observed. Partial spallation occurred. The delamination of YSZ was determined by CMAS infiltration. Monoclinic transformation also promoted cracking and delamination. Finally, the existence of vertical cracks produced by sintering accelerated corrosion mechanism.

As observed in the micrographs reported in the paper of Wu et al. (2011), in the upper region CMAS adhered to the surface, then it penetrated and deformed the column tips. Under the surface the authors detected an interaction zone, subjected to severe attack, followed by a region the columnar structure is faintly visible in. Yttria migrated out into the molten CMAS glass. In the deeper region the number of globular particles resulting from CMAS degradation decreased dramatically and the columnar structure was more clear. Kramer et al. (2006) also found that CMAS at 1240°C rapidly penetrated the interconnected pores of YSZ (coating thickness=100-200 μm).

Aiming to increase the resistance to CMAS attack, the use of alternative compositions based on rare-earth zirconates seems to be the most interesting solution, as in the case of $Gd_2Zr_2O_7$ (GZO) As reported in previous works, the glass melt rapidly wets the inner surface of the columns, with some features of the interaction suggesting intriguing topography effects on melt spreading; fluorite precipitates and apatite crystals filled the narrower gaps wherein supersaturation was achieved earlier; the column surface was covered by a dense layer of reaction product, that acts as a diffusion barrier. The column gaps were sealed, the reaction took place and eroded the column tips. The resistance of $Gd_2Zr_2O_7$ was higher than that of YSZ. After 4 h at 1300 °C the CMAS melt had infiltrated to a depth of 30 μm, owing to the rapid dissolution of GZO and precipitation of stable crystalline apatite as well as of fluorite and minor amount of spinels from the GZO/CMAS melt (Kramer et al., 2008; Habibi et al., 2012).

Samarium zirconate/YSZ double-layer TBCs were also tested at 1250 °C under CMAS attack (33CaO-9MgO-13Al_2O_3-45SiO_2). The penetration of CMAS was reduced by about a factor of five, but the double-layer coatings were completely penetrated after 4 hours of exposure. The interface between SZO (samarium zirconate) and YSZ was filled by CMAS. SZO dissolution and precipitation of fluorite and faceted apatite phases also occurred, with concomitant debonding at the interface between SZO and YSZ. The reaction layer contained faceted and globular particles, located in the region originally occupied by the column tips and co-existing with a high volume fraction of CMAS. After 4 h the columns were replaced by two-phase structure. The penetration of apatite was less effective with respect to gadolinium zirconate. The rate of CMAS penetration was slower than that found in YSZ. Substantially SZO showed excellent delamination resistance, lower thermal conductivity, enhanced thermal and sintering resistance (Zhao et al., 2009; Zhao et al., 2011).

SOLUTIONS AND RECOMMENDATIONS

Some possible solutions can be drawn in order to face current challenge as well as to explore future perspectives about the development of next-generation TBCs.

YSZ TBCs have been widely applied in energy and aerospace industries. They exhibited good service performance in several applications, even if their lifetime was drastically affected by severe operating conditions, such as temperatures higher than 1200°C, promoting phase changes and sintering, or by presence of molten salts and CMAS, that damaged YSZ TBC in a few hours.

Therefore, aiming to develop turbines with higher efficiency and to prolong service lifetime of the related components, it is mandatory to develop alternative compositions suitable to withstand higher temperatures and severe environmental attack.

Alternative compositions are generally based on oxides with low thermal conductivity and relatively high CTE: the TBCs should have thermal conductivity of about 0.5 $Wm^{-1}K^{-1}$ and a CTE higher than $10\text{-}11x10^{-6}K^{-1}$. New concepts suggest the employment of rare earth oxides (CeO_2 and Sc_2O_3) as zirconia stabilizers or the use of rare-earth zirconates ($Gd_2Zr_2O_7$ and $Sm_2Zr_2O_7$), perovskites ($BaLa_2Ti_3O_{10}$) and hexaluminates ($LaTi_2Al_9O_{19}$), in order to improve the resistance to sintering and hot-corrosion.

TBC materials with lower ionic conductivity than YSZ are able to reduce the oxidation of the bond coat and the growth rate of the TGO at the interface, that being recognized as one of the main factors for the increase of interfacial stresses responsible of delamination. Bond coat composition, microstructure and processing can also affect the formation of the TGO and then TBC lifetime.

Regarding sintering, it is worth noting that the loss of the strain tolerance can cause delamination or spallation of the coating, thus leaving the metallic component directly exposed to the extreme surrounding environment. High resistance to sintering and CMAS attack is a major request for novel TBCs: molten salts and CMAS are able to produce fast TBC damaging, cracking and failure.

The implementation of multilayered TBC systems seems to be the most effective solution for future investigations. These systems are generally composed of two or more ceramic layers with different functional properties: the most promising results have been obtained for functionally graded or double TBC systems, composed of an inner standard YSZ layer and an upper rare-earth zirconate layer. These systems are suitable to increase the temperature capability of YSZ TBCs. Multilayered TBCs are able to drastically reduce the thermal stresses during service and to prolong the lifetime of YSZ TBC systems. They also exhibit high resistance to erosion and CMAS attack. However, multilayered systems can involve higher manufacturing costs, so that their enhanced performance and durability should be validated at service conditions.

Moreover, nanostructured ceramic coatings are emerging as promising candidates for TBC applications but, at the moment, a wide market for plasma spray nanostructured powders does not exist. From this point of view, SPPS offers the prospect of depositing highly durable nanostructured TBCs. Nanostructured coatings with alternative compositions demonstrated to have enhanced resistance to high-temperature sintering and attack promoted by contaminants.

The application of new processes for fabrication of strain-tolerant nanostructured TBCs, in conjunction with the search for new materials with improved properties, is expected to provide high benefits in terms of performance, environmental impact and costs.

Substantially, a better knowledge of the operating behavior of materials and systems is mandatory for allowing large-scale production.

CONCLUSION

The development of next-generation turbines for aircrafts and power generation requires high efforts in increasing temperature capability and operating performance of TBC materials and systems.

In order to increase their efficiency, TBCs designers should achieve a better comprehension of TBC failure mechanisms related to different coating architectures and expected operating conditions, in order to predict their lifetime and to develop the best solutions. As herein discussed, TBC failure can be affected by several factors, often interconnected, so that it is not easy to identify common solutions for a lot of different applications.

Ceramic materials with low thermal conductivity are potential candidates for TBCs in conjunction with emerging processing technologies used for their manufacturing.

Processing affects the microstructure, that being subjected to high-temperature evolution during service, because of high-temperature phenomena such as oxidation, sintering and hot-corrosion.

TBC composition and microstructure have to be optimized in order to enhance the resistance against molten salts and CMAS: rare-earth doped zirconia, rare-earth zirconates, perovskites, hexaluminates, sometimes used in multilayered systems, are able to increase the temperature capability, as well as to reduce the thermal conductivity and the thermal stresses at the interface, that are responsible of delamination and spallation.

These materials are particularly promising when severe operating conditions are expected, provided that molten salts and CMAS can produce fast damage of the TBC, via phase changes and cracking mechanisms facilitating delamination and spallation.

SPS and SPPS processes are emerging for fabrication of nanostructured TBCs characterized by low thermal conductivity, high compliance and sintering resistance.

Further investigations are needed to well know the potentiality of new materials and architectures.

Last but not least, there are also high expectations for substitution and recycling of metals and rare-earth materials used in TBC. Metal superalloys and TBC materials usually contain small amounts of refractory metals, such as Ta, Nb, Zr, Hf, Cr, Ti, as well as of scarce metals and rare earths (Sm, La, Gd, Re, etx). Although the direct economic value of these metals is not much high, these materials are key to the development of low-carbon energy technologies and green aircraft engines, so that more attention should be paid on substitution and recycling issues. This challenge will be of higher interest in the coming years.

ACKNOWLEDGMENT

The author wishes to thank Francesco Marra and Alida Brentari for any SEM investigations aimed to previous studies.

REFERENCES

Ahmadi-Pidani, R., Shoja-Razavi, R., Mozafarinia, R., & Jamali, H. (2012). Evaluation of hot corrosion behavior of plasma sprayed ceria and yttria stabilized zirconia thermal barrier coatings in the presence of Na_2SO_4+V_2O_5 molten salt. *Ceramics International*, *38*(8), 6613–6620. doi:10.1016/j.ceramint.2012.05.047

Ahrens, M., Lampenscherf, S., Vassen, R., & Stover, D. (2012). Deposition and characteristics of submicrometer-structured thermal barrier coatings by suspension plasma spraying. *Journal of Thermal Spray Technology*, *21*(3-4), 416–424. doi:10.1007/s11666-012-9762-1

Ahrens, M., Vassen, R., Stoever, D., & Lampenscherf, S. (2004). Sintering and creep processes in plasma-sprayed thermal barrier coatings. *Journal of Thermal Spray Technology*, *13*(3), 432–442. doi:10.1361/10599630420434

Alfano, M., Di Girolamo, G., Pagnotta, L., Sun, D., Zekonyte, J., & Wood, R. J. K. (2010). The influence of high-temperature sintering on microstructure and mechanical properties of freestanding APS CeO_2-Y_2O_3-ZrO_2 coatings. *Journal of Materials Science*, *45*(10), 2662–2669. doi:10.1007/s10853-010-4245-6

Bakan, E., Mack, D. E., Mauer, G., Mucke, R., & Vassen, R. (2015). Porosity–Property Relationships of Plasma-Sprayed $Gd_2Zr_2O_7$/YSZ Thermal Barrier Coatings. *Journal of the American Ceramic Society*, *98*(8), 2647–2654. doi:10.1111/jace.13611

Bakan, E., Mack, D. E., Mauer, G., & Vassen, R. (2014). Gadolinium zirconate/YSZ thermal barrier coatings: Plasma spraying, microstructure and thermal cycling behavior. *Journal of the American Ceramic Society*, *97*(12), 4045–4051. doi:10.1111/jace.13204

Bast, U., & Schumann, E. (2002). Development of novel oxide materials for TBCs. *Ceramic Engineering and Science Proceedings*, *23*, 525–532. doi:10.1002/9780470294758.ch58

Bhatia, T., Ozturk, A., Xie, L., Jordan, E. H., Cetegen, B., Gell, M., & Padture, N. P. et al. (2002). Mechanisms of ceramic coating deposition in solution precursor plasma spray. *Journal of Materials Research*, *17*(09), 2363–2372. doi:10.1557/JMR.2002.0346

Bobzin, K., Bagcivan, N., Brogelmann, T., & Yildirim, B. (2013). Influence of temperature on phase stability and thermal conductivity of single- and double-ceramic-layer EB-PVD TBC top coats consisting of 7YSZ, $Gd_2Zr_2O_7$ and $La_2Zr_2O_7$. *Surface and Coatings Technology*, *237*, 56–64. doi:10.1016/j.surfcoat.2013.08.013

Borom, M. P., Johnson, C. A., & Peluso, L. A. (1996). Role of environmental deposits and operating surface temperature in spallation of air plasma sprayed thermal barrier coatings. *Surface and Coatings Technology*, *86-87*, 116–126. doi:10.1016/S0257-8972(96)02994-5

Boyce, M. P. (2002). *Gas Turbine Engineering Handbook* (2nd ed.). Portland, US: Gulf Professional Publishing.

Braue, W., & Mechnich, P. (2011). Recession of an EB-PVDYSZ Coated Turbine Blade by $CaSO_4$ and Fe, Ti-Rich CMAS-Type Deposits. *Journal of the American Ceramic Society*, *94*(12), 4483–4489. doi:10.1111/j.1551-2916.2011.04747.x

Cao, X. Q., Vassen, R., & Stover, D. (2004). Ceramic materials for thermal barrier coatings. *Journal of the European Ceramic Society*, *24*(1), 1–10. doi:10.1016/S0955-2219(03)00129-8

Cao, X. Q., Vassen, R., Tietz, F., & Stoever, D. (2006). New double-ceramic-layer thermal barrier coatings based on zirconia-rare earth composite oxides. *Journal of the European Ceramic Society*, *26*(3), 247–251. doi:10.1016/j.jeurceramsoc.2004.11.007

Chen, H., Gao, Y., Tao, S., Liu, Y., & Luo, H. (2009). Thermophysical properties of lanthanum zirconate coating prepared by plasma spraying and the influence of post-annealing. *Journal of Alloys and Compounds*, *486*(1-2), 391–399. doi:10.1016/j.jallcom.2009.06.162

Chen, H., Zhou, X., & Ding, C. (2002). Investigation of the thermomechanical properties of a plasma sprayed nanostructured zirconia coating. *Journal of the European Ceramic Society*, *23*(9), 1449–1455. doi:10.1016/S0955-2219(02)00345-X

Chen, L. B. (2006). Yttria-stabilized zirconia thermal barrier coatings - A review. *Surface Review and Letters*, *13*(05), 535–544. doi:10.1142/S0218625X06008670

Cipitria, A., Golosnoy, I. O., & Clyne, T. W. (2009). A sintering model for plasma-sprayed zirconia thermal barrier coatings. Part II: Coatings bonded to a rigid substrate. *Acta Materialia*, *57*(4), 993–1003. doi:10.1016/j.actamat.2008.10.058

Clarke, D. R., & Levi, C. G. (2003). Materials design for the next generation thermal barrier coatings. *Annual Review of Materials Research*, *33*(1), 383–417. doi:10.1146/annurev.matsci.33.011403.113718

Curry, N., Markocsan, N., Li, X. A., Tricoire, A., & Dorfman, M. (2011). Next generation thermal barrier coatings for the gas turbine industry. *Journal of Thermal Spray Technology*, *20*(1-2), 108–115. doi:10.1007/s11666-010-9593-x

Dai, H., Li, J., Cao, X., Meng, J. (2009). Thermal barrier coating material. *US20097597971.*

Darolia, R. (2013). Thermal barrier coatings technology: Critical review, progress update, remaining challenges and prospects. *International Materials Reviews*, *58*(6), 315–348. doi:10.1179/1743280413Y.0000000019

Darolia, R., Rigney, J.D., & Grylls, R., (2001). Nickel aluminide alloys with Cr and Zr for adherent coating or interlayer on superalloy turbine parts.

Davis, J. R. (2004). *Handbook of thermal spray technology*. Materials Park, OH, US: ASM International.

Dhineshkumar, S. R., Duraiselvam, M., Natarajan, S., Panwar, S. S., Jena, T., & Khan, M. A. (2016). Enhancement of strain tolerance of functionally graded $LaTi_2Al_9O_{19}$ thermal barrier coating through ultra-short pulse based laser texturing. *Surface and Coatings Technology*, *304*, 263–271. doi:10.1016/j.surfcoat.2016.07.018

Di Girolamo, G., Blasi, C., Brentari, A., & Schioppa, M. (2015). Microstructural, mechanical and thermal characteristics of zirconia-based thermal barrier coatings deposited by plasma spraying. *Ceramics International*, *41*(9), 11776–11785. doi:10.1016/j.ceramint.2015.05.145

Di Girolamo, G., Blasi, C., Pagnotta, L., & Schioppa, M. (2010). Phase evolution and thermophysical properties of plasma sprayed thick zirconia coatings after annealing. *Ceramics International*, *36*(8), 2273–2280. doi:10.1016/j.ceramint.2010.07.035

Di Girolamo, G., Blasi, C., Schioppa, M., & Tapfer, L. (2010). Structure and thermal properties of heat treated plasma sprayed ceria-yttria co-stabilized zirconia coatings. *Ceramics International*, *36*(3), 961–968. doi:10.1016/j.ceramint.2009.10.020

Di Girolamo, G., Marra, F., Blasi, C., Pilloni, L., & Valente, T. (2011). Microstructure, mechanical properties and thermal shock resistance of plasma sprayed nanostructured zirconia coatings. *Ceramics International*, *37*(7), 2711–2717. doi:10.1016/j.ceramint.2011.04.024

Di Girolamo, G., Marra, F., Schioppa, M., Blasi, C., Pulci, G., & Valente, T. (2015). Evolution of microstructural and mechanical properties of lanthanum zirconate thermal barrier coatings at high temperature. *Surface and Coatings Technology*, *268*, 298–302. doi:10.1016/j.surfcoat.2014.07.067

Di Girolamo, G., & Serra, E. (2015). Thermally sprayed nanostructured coatings for anti-wear and TBC applications: state-of-the-art and future perspectives. In M. Aliofkhazarei (Ed.), *Anti-abrasive nanocoatings* (pp. 513–545). Elsevier Ltd. doi:10.1016/B978-0-85709-211-3.00020-0

Doleker, K. M., & Karaoglanli, A. C. (2016). Comparison of oxidation behavior of YSZ and $Gd_2Zr_2O_7$ thermal barrier coatings (TBCs). *Surface and Coatings Technology*, *318*, 198–207. doi:10.1016/j.surfcoat.2016.12.078

Dong, H., Yang, G. J., Li, C. X., Luo, X. T., & Li, C. J. (2014). Effect of TGO thickness on thermal cyclic lifetime and failure mode of plasma-sprayed TBCs. *Journal of the American Ceramic Society*, *97*(4), 1226–1232. doi:10.1111/jace.12868

Drexler, J. M., Shinoda, K., Ortiz, A. L., & Padture, N. (2010). Air-plasma-sprayed thermal barrier coatings that are resistant to high-temperature attack by glass deposits. *Acta Materialia*, *58*(20), 6835–6844. doi:10.1016/j.actamat.2010.09.013

Eliaz, N., Shemesh, G., & Latanision, R. M. (2002). Hot corrosion in gas turbine components. *Engineering Failure Analysis*, *9*(1), 31–43. doi:10.1016/S1350-6307(00)00035-2

Evans, A. G., Clarke, D. R., & Levi, C. G. (2008). The influence of oxides on the performance of advanced gas turbines. *Journal of the European Ceramic Society*, *28*(7), 1405–1419. doi:10.1016/j.jeurceramsoc.2007.12.023

Evans, A. G., Fleck, N. A., Faulhalber, S., Vermaak, N., Maloney, M., & Darolia, R. (2006). Scaling laws governing the erosion and impact resistance of thermal barrier coatings. *Wear*, *260*(7-8), 886–894. doi:10.1016/j.wear.2005.05.005

Evans, A. G., Mumm, D. R., Hutchinson, J. W., Meier, G. H., & Petit, F. S. (2001). Mechanisms controlling the durability of thermal barrier coatings. *Progress in Materials Science*, *46*(5), 505–553. doi:10.1016/S0079-6425(00)00020-7

Fan, X. L., Xu, R., Zhang, W., & Wang, T. J. (2012). Effect of periodic surface cracks on the interfacial fracture of thermal barrier coating system. *Applied Surface Science*, *258*(24), 9816–9823. doi:10.1016/j.apsusc.2012.06.036

Frommhertz, M., Scholz, A., Oechsner, M., Bakan, E., & Vassen, R. (2016). Gadolinium zirconate/YSZ thermal barrier coatings: Mixed-mode interfacial fracture toughness and sintering behavior. *Surface and Coatings Technology*, *286*, 119–128. doi:10.1016/j.surfcoat.2015.12.012

Gadow, R., & Lischka, M. (2002). Lanthanum hexaaluminate – novel thermal barrier coatings for gas turbine applications – materials and process development. *Surface and Coatings Technology*, *151*, 392–399. doi:10.1016/S0257-8972(01)01642-5

Ghasemi, R., Shoja-Razavi, R., Mozafarinia, R., & Jamali, H. (2010). Laser glazing of plasma-sprayed nanostructured yttria stabilized zirconia thermal barrier coatings. *Ceramics International*, *39*(8), 9483–9490. doi:10.1016/j.ceramint.2013.05.066

Guignard, A., Mauer, G., Vassen, R., & Stover, D. (2012). Deposition and characteristics of submicrometer-structured thermal barrier coatings by suspension plasma spraying. *Journal of Thermal Spray Technology*, *21*(3-4), 416–424. doi:10.1007/s11666-012-9762-1

Guo, L., Li, M., He, S., Zhang, C., Wang, Q., & Ye, F. (2017). Preparation and hot corrosion behavior of plasma sprayed nanostructured $Gd_2Zr_2O_7$-$LaPO_4$ thermal barrier coatings. *Journal of Alloys and Compounds*, *698*, 13–19. doi:10.1016/j.jallcom.2016.12.241

Guo, L., Zhang, Y., & Ye, F. X. (2015). Phase Structure Evolution and Thermo-Physical Properties of Nonstoichiometry $Nd_{2-x}Zr_{2+x}O_{7+x/2}$ Pyrochlore Ceramics. *Journal of the American Ceramic Society*, *98*(3), 1013–1018. doi:10.1111/jace.13374

Habibi, M. H., Wang, L., & Guo, S. (2012). An investigation on hot corrosion resistance of plasma sprayed composite YSZ- $Gd_2Zr_2O_7$ and $Gd_2Zr_2O_7$ thermal barrier coatings in simulated turbine environment at 1050 °C. *Journal of the European Ceramic Society*, *32*, 1635–1642. doi:10.1016/j.jeurceramsoc.2012.01.006

Hajizadeh-Oghaz, M., Shoja-Razavi, R., Ghasemi, A., & Valefi, Z. (2016). Na_2SO_4 and V_2O_5 molten salts corrosion resistance of plasma-sprayed nanostructured ceria and yttria co-stabilized zirconia thermal barrier coatings. *Ceramics International*, *42*(4), 5433–5446. doi:10.1016/j.ceramint.2015.12.085

Hass, D. D., Slifka, A. J., & Wadley, H. N. G. (2001). Low thermal conductivity vapor deposited zirconia microstructures. *Acta Materialia*, *49*(6), 973–983. doi:10.1016/S1359-6454(00)00403-1

Haynes, J. A., Ferber, M. K., & Porter, W. D. (1999). Thermal cycling behavior of plasma-sprayed thermal barrier coatings with various MCrAlX bond coats. *Journal of Thermal Spray Technology*, *9*(1), 38–48. doi:10.1361/105996300770350041

Haynes, J. A., Pint, B. A., Porter, W. D., & Wright, I. G. (2004). Comparison of thermal expansion and oxidation behavior of various high temperature coating materials and superalloys. *Materials at High Temperatures*, *21*(2), 87–94. doi:10.1179/mht.2004.012

Hetmanczyk, M., Swadzba, L., & Mendala, B. (2007). Advanced materials and protective coatings in aero-engines applications. *Journal of Achievements of Materials and Manufacturing Engineering*, *24*(1), 372–381.

Hongsong, Z., Kun, P., Qiang, X., Fuchi, W., & Ling, L. (2009). Thermal conductivity of $(Sm_{1-x}La_x)Zr_2O_7$ (x=0, 0.25, 0.5, 0.75 and 1) oxides for advanced thermal barrier coatings. *Journal of Rare Earths*, *27*(2), 222–226. doi:10.1016/S1002-0721(08)60223-9

Hu, J., Zhao, H., Tao, S., Zhou, X., & Ding, C. (2010). Thermal conductivity of plasma sprayed $Sm_2Zr_2O_7$ coating. *Journal of the European Ceramic Society*, *30*(3), 799–804. doi:10.1016/j.jeurceramsoc.2009.09.010

Huang, H., Liu, C., Ni, I. Y., & Zhou, C. G. (2011). Evaluation of microstructural evolution of thermal barrier coatings exposed to Na_2SO_4 using impedance spectroscopy. *Corrosion Science*, *53*(4), 1369–1374. doi:10.1016/j.corsci.2010.12.023

Hutchinson, J. W., & Evans, A. G. (2002). On the delamination of thermal barrier coatings in a thermal gradient. *Surface and Coatings Technology*, *149*(2-3), 179–184. doi:10.1016/S0257-8972(01)01451-7

Hutchinson, R. G., Fleck, N. A., & Cocks, A. C. F. (2006). A sintering model for thermal barrier coatings. *Acta Materialia*, *54*(5), 1297–1306. doi:10.1016/j.actamat.2005.10.056

Itoh, Y., & Saitoh, M. (2005). Mechanical properties of overaluminized MCrAlY coatings at room temperature. *Journal of Engineering for Gas Turbines and Power*, *127*(4), 23–30. doi:10.1115/1.1688766

Jadhav, A. D., & Padture, N. P. (2008). Mechanical properties of solution-precursor plasma sprayed thermal barrier coatings. *Surface and Coatings Technology*, *202*(20), 4976–4979. doi:10.1016/j.surfcoat.2008.04.091

Jamali, H., Mozafarinia, R., Shoja-Razavi, R., & Ahmadi-Pidani, R. (2014). Comparison of hot corrosion behaviors of plasma-sprayed nanostructured and conventional YSZ thermal barrier coatings exposure to molten vanadium pentoxide and sodium-sulfate. *Journal of the European Ceramic Society*, *34*(2), 185–192. doi:10.1016/j.jeurceramsoc.2013.08.006

Jarligo, M. O., Mack, D. E., Mauer, G., Vassen, R., & Stöver, D. (2009). Atmospheric Plasma Spraying of High Melting Temperature Complex Perovskites for TBC Application. *Journal of Thermal Spray Technology*, *19*(1-2), 303–310. doi:10.1007/s11666-009-9377-3

Jin, L., Yu, Q., Ni, L., & Zhou, C. (2012). Microstructure and thermal properties of nanostructured 8 wt.% CeO_2 doped YSZ coatings prepared by atmospheric plasma spraying. *Journal of Thermal Spray Technology*, *21*(5), 928–934. doi:10.1007/s11666-012-9744-3

Jones, R. L. (1989). Scandia-stabilized zirconia for resistance to molten vanadate-sulfate corrosion. *Surface and Coatings Technology*, *39*, 89–96. doi:10.1016/0257-8972(89)90044-3

Jones, R. L. (1997). Vanadate-sulfate melt thermochemistry relating to hot corrosion of thermal barrier coatings. Naval Research Laboratory.

Kamal, S., Jayaganthan, R., & Prakash, S. (2009). Mechanical and microstructural characterisations of NiCrAoAlYTa coatings on superalloys deposited by detonation gun technique. *Surface and Coatings Technology*, *25*, 303–312.

Kassner, H., Stuke, A., Rodig, M., Vassen, R., & Stover, D. (2009). Influence of porosity on thermal conductivity and sintering in suspension plasma sprayed thermal barrier coatings. *Ceramic Engineering and Science Proceedings*, *29*(4), 147–158. doi:10.1002/9780470456323.ch13

Kramer, S., Yang, J., & Levi, C. G. (2008). Infiltration-Inhibiting Reaction of Gadolinium Zirconate Thermal Barrier Coatings with CMAS Melts. *Journal of the American Ceramic Society*, *91*(2), 576–583. doi:10.1111/j.1551-2916.2007.02175.x

Kramer, S., Yang, J., Levi, C. G., & Jonhson, C. A. (2006). Thermochemical interaction of thermal barrier coatings with molten CaO-MgO-Al_2O_3-SiO_2 (CMAS) deposits. *Journal of the American Ceramic Society*, *89*(10), 3167–3175. doi:10.1111/j.1551-2916.2006.01209.x

Lamuta, C., Di Girolamo, G., & Pagnotta, L. (2015). Microstructural, mechanical and tribological properties of nanostructured YSZ coatings produced with different APS process parameters. *Ceramics International*, *41*(7), 8904–8914. doi:10.1016/j.ceramint.2015.03.148

Latka, L., Cattini, A., Chicot, D., Pawlowski, L., Kozerski, S., Petit, F., & Denoirjean, A. (2013). Mechanical properties of yttria-and ceria-stabilized zirconia coatings obtained by suspension plasma spraying. *Journal of Thermal Spray Technology*, *22*(2-3), 125–130. doi:10.1007/s11666-012-9874-7

Lima, R. S., & Marple, B. R. (2007). Thermal Spray Coatings Engineered from Nanostructured Ceramic Agglomerated Powders for Structural, Thermal Barrier and Biomedical Applications: A Review. *Journal of Thermal Spray Technology*, *16*(1), 40–63. doi:10.1007/s11666-006-9010-7

Lima, R. S., & Marple, B. R. (2008). Nanostructured YSZ thermal barrier coatings engineered to counteract sintering effects. *Materials Science and Engineering A*, *485*(1-2), 182–193. doi:10.1016/j.msea.2007.07.082

Liu, H., Li, S., Li, Q., & Li, Y. (2010). Investigation on the phase stability, sintering, and thermal conductivity of Sc_2O_3-Y_2O_3-ZrO_2 for thermal barrier coating application. *Materials & Design*, *31*(6), 2972–2977. doi:10.1016/j.matdes.2009.12.019

Liu, H. F., Xiong, X., Li, X. B., & Wang, Y. I. (2014). Hot corrosion behavior of Sc_2O_3-Y_2O_3-ZrO_2 thermal barrier coatings in presence of Na_2SO_4+V_2O_5 molten salt. *Corrosion Science*, *85*, 87–93. doi:10.1016/j.corsci.2014.04.001

Liu, T., Chen, X., Yang, G. J., & Li, C. J. (2016). Properties evolution of plasma-sprayed $La_2Zr_2O_7$ coating induced by pore structure evolution during thermal exposure. *Ceramics International*, *42*(14), 15485–15492. doi:10.1016/j.ceramint.2016.06.201

Liu, T., Luo, X. T., Chen, X., Yang, C. X., Li, C. X., & Li, J. (2015). Morphology and size evolution of interlamellar two-dimensional pores in plasma-sprayed $La_2Zr_2O_7$ coatings during thermal exposure at 1300 °C. *Journal of Thermal Spray Technology*, *24*(5), 739–748. doi:10.1007/s11666-015-0236-0

Liu, Y., Vidal, V., Le Roux, S., Blas, F., Ansart, F., & Lours, P. (2015). Influence of isothermal and cyclic oxidation on the apparent interfacial toughness in thermal barrier coating systems. *Journal of the European Ceramic Society*, *35*(15), 4269–4275. doi:10.1016/j.jeurceramsoc.2015.07.018

Liu, Z. G., Zhang, W. H., Ouyang, J. H., & Zhou, Y. (2015). Novel thermal barrier coatings based on rare-earth zirconates/YSZ double-ceramic-layer system deposited by plasma spraying. *Journal of Alloys and Compounds*, *647*, 438–444. doi:10.1016/j.jallcom.2015.05.189

Loghman-Estarki, M. R., Nejati, M., Edris, H., Shoja Razavi, R., Jamali, H., & Pakseresht, A. H. (2015). Evaluation of hot corrosion behavior of plasma sprayed Scandia and yttria co-stabilized nanostructured thermal barrier coatings in the presence of molten sulfate and vanadate salt. *Journal of the European Ceramic Society*, *35*(2), 693–702. doi:10.1016/j.jeurceramsoc.2014.08.029

Lv, B., Fan, X., Xie, H., & Wang, T. J. (2017). Effect of neck formation on the sintering of air-plasma-sprayed thermal barrier coatings. *Journal of the European Ceramic Society*, *37*(2), 811–821. doi:10.1016/j.jeurceramsoc.2016.08.033

Ma, W., Dong, H., Guo, H., Gong, P., & Zheng, X. (2010). Thermal cycling behavior of $La_2Ce_2O_7$/8YSZ double-ceramic-layer thermal barrier coatings prepared by atmospheric plasma spraying. *Surface and Coatings Technology*, *204*(21-22), 3336–3370. doi:10.1016/j.surfcoat.2010.03.053

Ma, W., Mack, D. E., Vassen, R., & Stover, D. (2008). Perovskite-type strontium zirconate as a new material for thermal barrier coatings. *Journal of the American Ceramic Society*, *91*(8), 2630–2635. doi:10.1111/j.1551-2916.2008.02472.x

Mahade, S., Curry, N., Bjorklund, S., Markocsan, N., Nylen, P., & Vassen, R. (2016). Functional performance of $Gd_2Zr_2O_7$/YSZ multi-layered thermal barrier coatings deposited by suspension plasma spray. *Surface and Coatings Technology*, *318*, 208–216. doi:10.1016/j.surfcoat.2016.12.062

Malzbender, J., & Steinbrech, R. (2003). Determination of the stress-dependent stiffness of plasma-sprayed thermal barrier coatings using depth-sensitive indentation. *Journal of Materials Research*, *18*(08), 1975–1984. doi:10.1557/JMR.2003.0274

Marchand, O., Girardot, L., Planche, M. P., Bertrand, P., Bailly, Y., & Bertrand, G. (2011). An insight into suspension plasma spray: Injection of the suspension and its interaction with the plasma flow. *Journal of Thermal Spray Technology*, *20*(6), 1310–1320. doi:10.1007/s11666-011-9682-5

Marple, B. R., Voyer, J., Thibodeau, M., Nagy, D. R., & Vassen, R. (2006). Hot corrosion of lanthanum zirconate and partially stabilized zirconia thermal barrier coatings. *Transactions of ASME*, *128*, 144-152. doi:10.1115/1.1924534

Masoule, S. T., Valefi, Z., Ehsani, N., & Lavasani, H. Q. (2016). Thermal insulation and thermal shock behavior of conventional and nanostructured plasma-sprayed TBCs. *Journal of Thermal Spray Technology*, *25*(8), 1684–1691. doi:10.1007/s11666-016-0476-7

Matsumoto, M., Takayama, H., Yokoe, D., Mukai, K., Matsubaru, H., Kagyia, Y., & Sugita, Y. (2006). Thermal cycle behavior of plasma sprayed La_2O_3, Y_2O_3 stabilized ZrO_2 coatings. *Scripta Materialia, 54*(12), 2035–2039. doi:10.1016/j.scriptamat.2006.03.015

Matsumoto, M., Yamaguchi, N., & Matsubara, H. (2004). Low thermal conductivity and high temperature stability of ZrO_2–Y_2O_3–La_2O_3 coatings produced by electron beam PVD. *Scripta Materialia, 50*(6), 867–871. doi:10.1016/j.scriptamat.2003.12.008

Miller, R. A. (1997). Thermal barrier coatings for aircraft engines: History and directions. *Journal of Thermal Spray Technology, 6*(1), 35–42. doi:10.1007/BF02646310

Mohan, P., Yuan, B., Patetrson, T., Desai, V. H., & Sohn, Y. H. (2007). Degradation of yttria-stabilized zirconia thermal barrier coatings by vanadium pentoxide, phoshorous pentoxide, and sodium sulfate. *Journal of the American Ceramic Society, 90*(11), 3601–3607. doi:10.1111/j.1551-2916.2007.01941.x

Oguma, H., Okada, I., Torigoe, T., & Takahashi, K. (2004). High temperature corrosion resistant alloy, thermal barrier coating material, and gas turbine using high temperature corrosion resistant alloy.

Padture, N. P., Gell, M., & Jordan, E. H. (2002). Thermal barrier coatings for gas-turbine engine applications. *Science, 296*(5566), 280–284. doi:10.1126/science.1068609 PMID:11951028

Padture, N. P., Schlichting, K. W., Bhatia, T., Ozturk, A., Cetegen, B., Jordan, E. H., & Osendi, M. I. et al. (2001). Toward durable thermal barrier coatings with novel microstructure deposited by solution precursors plasma spray. *Acta Materialia, 49*(12), 2251–2257. doi:10.1016/S1359-6454(01)00130-6

Pan, W., Wan, C. L., Xu, Q., Wang, J. D., & Qu, Z. X. (2007). Thermal diffusivity of samarium–gadolinium zirconate solid solutions. *Thermochimica Acta, 455*(1-2), 16–20. doi:10.1016/j.tca.2006.12.001

Paul, S., Cipitria, A., Tsipas, S. A., & Clyne, T. W. (2009). Sintering characteristics of plasma sprayed zirconia coatings containing different stabilisers. *Surface and Coatings Technology, 203*(8), 1069–1074. doi:10.1016/j.surfcoat.2008.09.037

Pawlowski, L. (2008). Finely grained nanometric and submicrometric coatings by thermal spraying: A review. *Surface and Coatings Technology, 202*(18), 4318–4328. doi:10.1016/j.surfcoat.2008.04.004

Pawlowski, L. (2008). *The Science and Engineering of Thermal Barrier Coatings*. Chichester, UK: John Wiley & Sons Ltd. doi:10.1002/9780470754085

Peng, H., Wang, L., Guo, L., Miao, W., Guo, H., & Gong, S. (2012). Degradation of EB-PVD thermal barrier coatings caused by CMAS deposits. Progress in Natural Science. *Materials International, 22*, 461–467.

Pidani, R. (2014). The influence of laser treatment on hot corrosion behavior of plasma-sprayed nanostructured yttria stabilized zirconia thermal barrier coatings. *Journal of the European Ceramic Society, 34*(8), 2013–2021. doi:10.1016/j.jeurceramsoc.2014.01.031

Qu, Z., Wan, C., & Pan, W. (2007). Thermal expansion and defect chemistry of MgO doped $Sm_2Zr_2O_7$. *Chemistry of Materials, 19*(20), 4913–4918. doi:10.1021/cm071615z

Rauf, A., Yu, Q. H., Jin, L., & Zhou, C. G. (2012). Microstructure and thermal properties of nanostructured lanthana-doped yttria-stabilized zirconia thermal barrier coatings by air plasma spraying. *Scripta Materialia*, *66*(2), 109–112. doi:10.1016/j.scriptamat.2011.10.017

Reed, R. C. (2006). *The superalloys: fundamentals and applications*. Cambridge, UK: Cambridge University Press. doi:10.1017/CBO9780511541285

Richer, R., Yandouzi, M., Beauvais, L., & Jodoin, B. (2010). Oxidation behavior of CoNiCrAlY bond coats produced by plasma, HVOF and cold gas dynamic spraying. *Surface and Coatings Technology*, *204*(24), 3962–3974. doi:10.1016/j.surfcoat.2010.03.043

Saremi, M., Afrasiabi, A., & Kobayashi, A. (2007). Bond coat oxidation and hot corrosion behavior of plasma sprayed YSZ coating on Ni superalloy. *Transactions of JWRI (Joining and Welding Research Institute)*, *36*, 41-45.

Saremi, M., Afrasiabi, A., & Kobayashi, A. (2008). Microstructural analysis of YSZ and YSZ/Al_2O_3 plasma sprayed thermal barrier coatings after high temperature oxidation. *Surface and Coatings Technology*, *202*(14), 3233–3238. doi:10.1016/j.surfcoat.2007.11.029

Saremi, M., & Habibi, M. H. (2011). An investigation on hot corrosion resistance of plasma sprayed YSZ-ceria TBC in Na2SO4-V2O5 at 1050°C. In *TMS 2011- 140th annual meeting and exhibition, supplemental proceedings*. (pp. 429-437). San Diego, CA: John Wiley & Sons Inc.

Schulz, U. (2000). Phase transformation in EB-PVD yttria partially stabilized zirconia thermal barrier coatings during annealing. *Journal of the American Ceramic Society*, *83*(4), 904–910. doi:10.1111/j.1151-2916.2000.tb01292.x

Sodeoka, S., Suzuki, M., Ueno, K., Sakuramoto, H., Shibata, T., & Ando, M. (1997). Thermal and mechanical properties of ZrO_2-CeO_2 plasma-sprayed coatings. *Journal of Thermal Spray Technology*, *6*(3), 361–367. doi:10.1007/s11666-997-0071-z

Sourmail, T. (2004). *Coatings for high temperature applications*. University of Cambridge, UK. Retrieved from http://Thomas-sourmail.net/coatings/index.html

Stover, D., Pratch, G., Lehmann, H., Dietrich, M., Doring, J. E., & Vassen, R. (2004). New materials concepts for the next generation of plasma-sprayed thermal barrier coatings. *Journal of Thermal Spray Technology*, *13*(1), 76–83. doi:10.1007/s11666-004-0052-4

Strangman, T. E. (1986). Thermal barrier coatings for turbine airfoils. *Thin Solid Films*, *127*(1-2), 93–105. doi:10.1016/0040-6090(85)90215-9

Subramanian, R. (2001). Thermal barrier coatings having high phase stability.

Sumner, J., Encinas-Oropesa, A., Symms, N. J., & Oakey, E. J. (2013). High temperature oxidation and corrosion of gas turbine materials in burner rig exposures. *Materials Science and Technology*, *29*(7), 813–821. doi:10.1179/1743284712Y.0000000097

Takagi, K., Kudo, D., Kawasaki, A., & Harada, Y. (2011). Microstructural dependency of thermal expansion and sintering shrinkage in plasma-sprayed zirconia coatings. *Surface and Coatings Technology*, *205*(19), 4411–4417. doi:10.1016/j.surfcoat.2011.03.053

Tamura, M., Takahashi, M., Ishii, J., Suzuki, K., Sato, M., & Shimomura, K. (1999). Multilayered thermal barrier coating for land-based gas turbines. *Journal of Thermal Spray Technology*, *8*(1), 68–72. doi:10.1361/105996399770350584

Thompson, J., & Clyne, T. (2001). The effect of heat treatment on the stiffness of zirconia top coats in plasma-sprayed TBCs. *Acta Materialia*, *49*(9), 1565–1575. doi:10.1016/S1359-6454(01)00065-9

TMR-Transparency Market Research. (2017). Global thermal barrier coatings market. Retrieved from http://www.transparencymarketresearch.com

Trice, R. W., Su, Y. J., Mawdsley, J. R., Faber, K. T., Arellano-Lopez, A. R., Wang, H., & Porter, W. D. (2002). Effect of heat treatment on phase stability, microstructure, and thermal conductivity of plasma-sprayed YSZ. *Journal of Materials Science*, *37*(11), 2359–2365. doi:10.1023/A:1015310509520

Vassen, R., Cao, X., Tietz, F., Basu, D., & Stover, D. (2000). Zirconates as new materials for thermal barrier coatings. *Journal of the American Ceramic Society*, *83*(8), 2023–2028. doi:10.1111/j.1151-2916.2000.tb01506.x

Vassen, R., Jarligo, M. O., Steinke, T., Mack, D. E., & Stoever, D. (2010). Overview of advanced thermal barrier coatings. *Surface and Coatings Technology*, *205*(4), 938–942. doi:10.1016/j.surfcoat.2010.08.151

Vassen, R., Traeger, F., & Stoever, D. (2004). New thermal barrier coatings based on pyrochlore/YSZ double-layer systems. *International Journal of Applied Ceramic Technology*, *1*(4), 351–361. doi:10.1111/j.1744-7402.2004.tb00186.x

Wang, C., Wang, J., Wang, L., Hao, G., Sun, X., Shan, F., & Zou, Z. (2014). Nanocomposite lanthanum zirconate thermal barrier coating deposited by suspension plasma spray process. *Journal of Thermal Spray Technology*, *23*(7), 1030–1036. doi:10.1007/s11666-014-0068-3

Wang, C. M., Guo, L., & Ye, F. X. (2016). $LaPO_4$ as a toughening agent for rare earth zirconate ceramics. *Materials & Design*, *111*, 389–393. doi:10.1016/j.matdes.2016.09.014

Wang, L., Wang, Y., & Sun, X. G. (2011). Influence of pores on the thermal insulation behavior of thermal barrier coatings prepared by APS. *Materials & Design*, *32*, 36–47. doi:10.1016/j.matdes.2010.06.040

Wang, L., Wang, Y., Sun, X. G., He, J. Q., Pan, Z. Y., & Wang, C. H. (2012). Thermal shock behavior of 8YSZ and double-ceramic-layer $La_2Zr_2O_7$/8YSZ thermal barrier coatings fabricated by atmospheric plasma spraying. *Ceramics International*, *38*(5), 3595–3606. doi:10.1016/j.ceramint.2011.12.076

Wang, Y., Guo, H., & Gong, S. (2009). Thermal physical properties and thermal shock resistance of $La_2Ce_2O_7$ thermal barrier coatings with segmented structure. *Ceramics International*, *35*, 2639–2644. doi:10.1016/j.ceramint.2009.02.025

Wang, Y., & Zhou, C. (2016). Microstructure and thermal properties of nanostructured gadolinia doped yttria-stabilized zirconia thermal barrier coatings produced by air plasma spraying. *Ceramics International*, *42*(11), 13047–13052. doi:10.1016/j.ceramint.2016.05.084

Wellman, R. G., & Nicholls, J. R. (2008). Erosion, corrosion and erosion–corrosion of EB PVD thermal barrier coatings. *Tribology International*, *41*(7), 657–662. doi:10.1016/j.triboint.2007.10.004

Wu, J., Guo, H. B., Gao, Y. Z., & Gong, S. (2011). Microstructure and thermophysical properties of yttria stabilized zirconia coatings with CMAS deposits. *Journal of the European Ceramic Society*, *31*(10), 1881–1888. doi:10.1016/j.jeurceramsoc.2011.04.006

Wu, J., Wei, X., Padture, N. P., Klemens, P. G., Gell, M., Garcia, E., & Osendi, M. I. et al. (2002). Low-thermal conductivity rare-earth zirconates for potential thermal-barrier-coating applications. *Journal of the European Ceramic Society*, *85*(12), 3031–3035. doi:10.1111/j.1151-2916.2002.tb00574.x

Xie, X., Guo, H., Gonga, P., & Xua, H. (2011). Lanthanum-titanium-aluminum oxide: A novel thermal barrier coating material for applications at 1300 °C. *Journal of the European Ceramic Society*, *31*(9), 1677–1683. doi:10.1016/j.jeurceramsoc.2011.03.036

Xu, H., & Guo, H. (2011). *Thermal Barrier Coatings*. Cambridge, UK: Woodhead Publishing Limited. doi:10.1533/9780857090829

Xu, Z. H., He, L. M., Tang, Z. H., Mu, R. D., & Cao, X. Q. (2012). Evolution of high temperature corrosion behavior of $La_2(Zr_{0.7}Ce_{0.3})_2O_7$ with the addition of Y_2O_3 thermal barrier coatings in contacts with vanadate–sulfate salts. *Journal of Alloys and Compounds*, *536*, 106–112. doi:10.1016/j.jallcom.2012.04.074

Yu, Q. H., Rauf, A., & Zhou, C. G. (2010). Microstructure and thermal properties of nanostructured 4 wt.%Al_2O_3-YSZ coatings produced by atmospheric plasma spraying. *Journal of Thermal Spray Technology*, *19*(6), 1294–1300. doi:10.1007/s11666-010-9524-x

Zhang, H., Guo, L., Ma, Y., Peng, H., Guo, H., & Gong, S. (2014). Thermal cycling behavior of $(Gd_{0.9}Yb_{0.1})_2Zr_2O_7$/8YSZ gradient thermal barrier coatings deposited on Hf-doped NiAl bond coat by EB-PVD. *Surface and Coatings Technology*, *258*, 950–955. doi:10.1016/j.surfcoat.2014.07.051

Zhao, H., Begley, M. R., Heuer, A., Shargi-Mostaghimi, R., & Wadley, H. N. G. (2011). Reaction, transformation and delamination of samarium zirconate thermal barrier coatings. *Surface and Coatings Technology*, *205*(19), 4355–4365. doi:10.1016/j.surfcoat.2011.03.028

Zhao, H., Levi, C. G., & Wadley, H. N. G. (2009). Vapor deposited samarium zirconate thermal barrier coatings. *Surface and Coatings Technology*, *203*(20-21), 3157–3167. doi:10.1016/j.surfcoat.2009.03.048

Zhao, H., Levi, C. G., & Wadley, H. N. G. (2014). Molten silicate interactions with thermal barrier coatings. *Surface and Coatings Technology*, *251*, 74–86. doi:10.1016/j.surfcoat.2014.04.007

Zhong, X., Zhao, H., Zhou, X., Liu, C., Wang, L., Shao, F., & Ding, C. et al. (2014). Thermal shock behavior of toughened gadolinium zirconate/YSZ double-ceramic-layered thermal barrier coating. *Journal of Alloys and Compounds*, *593*, 50–55. doi:10.1016/j.jallcom.2014.01.060

Zhu, D., Nesbitt, J. A., Barrett, C. A., McCue, T. R., & Miller, R. A. (2004). Furnace cyclic oxidation behavior of multi-component low conductivity thermal barrier coatings. *Journal of Thermal Spray Technology*, *13*(1), 84–92. doi:10.1007/s11666-004-0053-3

Zhu, D. M., & Miller, R. A. (2000). Thermal barrier coatings for advanced gas-turbine engines. *Materials Research Society*, *25*(07), 43–47. doi:10.1557/mrs2000.123

KEY TERMS AND DEFINITIONS

CMAS: Siliceous particulates (airborne dusts, sand, ash, debris) ingested with the intake of air, that adhere to the surfaces in the hot-gas path (combustors, airfoils, shrouds) and at peak temperatures of the engine cycles (take-off or landing) yield glassy melts based on CMAS.

Molten Salts: Molten salts are solid deposits at standard temperature and pressure, but enters the liquid phase due to elevated temperature, so that they can infiltrate through TBC thickness and promote hot-corrosion. They are generally originated by contaminants present in low-grade fuels.

Sintering: Process of high-temperature partial densification occurring in porous materials at temperature generally above 1200°C.

TGO: Al_2O_3-based or mixed oxide scale growing during high-temperature exposure at the interface between ceramic TBC and metal bond coat, because of oxygen infiltration and its reaction with Al contained in bond coat material.

Thermal Barrier Coating: An advanced material system applied to metal components operating at high temperature, commonly used in gas turbines and aircraft engines. TBCs provide insulation from thermal loads and are able to sustain a thermal gradient through their thickness.

Thermal Spraying: A processing technology suitable to spray melted particles on a substrate. The feedstock is commonly heated by electrical (plasma or arc) or chemical (combustion flame) sources.

YSZ: Partially yttria stabilized zirconia is the most used material for TBC applications. The addition of yttria stabilizer allows to create oxygen vacancies in the crystal structure of zirconia, stabilizing itself in tetragonal and/or cubic phase.

Chapter 3
Investigation on Improved-Durability Thermal Barrier Coatings:
An Overview of Nanostructured, Multilayered, and Self-Healing TBCs

Mohammad Hassanzadeh
University of Tehran, Iran

Mohsen Saremi
University of Tehran, Iran

Zia Valefi
Malek Ashtar University of Technology, Iran

Amir Hossein Pakseresht
University of Tehran, Iran & Materials and Energy Research Center, Iran

ABSTRACT

Concurrent with the development of new generation of gas turbines, many attempts have been made to introduce advanced thermal barrier coatings with lower thermal conductivity and higher temperature stability. Most of the research to improve TBCs performance are based on two general approaches: structural modifications and chemical modifications. In most cases, the improvements in some properties are at the expense of loss of some other properties. Changing in the TBCs architecture and the application of multilayer coatings, consisting of layers with engineered properties based on the requirements, is a solution to achieve a combination of desired properties. In all of these development methods it is to be understood that the principle is reducing the possibility of formation of cracks, but, once formed, all such cracks can grow under and thermal cycles and eventually lead to coating delamination and spallation. Self-healing is the most precious phenomenon to overcome this problem.

DOI: 10.4018/978-1-5225-4194-3.ch003

INTRODUCTION

A large number of industrial processes operate under the most arduous conditions of temperature and stress. Some of the features of such environments include high temperature, increased temperature gradients, high pressure, large stresses on individual components, the presence of an oxidizing and corroding atmosphere, and also the possibility of foreign object damage (FOD). A few examples of such processes are steam turbines, aero, marine, and industrial gas turbines (Bose, 2011; X. Chen et al., 2003).

Turbine inlet temperature (TIT) is the critical parameter in improving gas turbine engine power and fuel efficiency. Even a small increase in TIT leads to less fuel consumption to perform the same function and thus will improve turbine efficiency (Rahman, Ibrahim, & Abdalla, 2011; Stöver & Funke, 1999; Vassen, Stuke, & Stöver, 2009). Investigation on Rolls-Royce aero gas turbine engines shown that via increasing turbine inlet temperature as much as approximately 700°C the efficiency increased from 12% to 50% during the period 1970–2000 (Nicholls, 2003). Today, for J-class turbine engines, the TIT and the efficiency of 1650°C and 61% were validated (Ai, Masada, & Ito, 2014). But despite all the advances in the gas turbine industry, this trend of increasing temperature is continuing such that it is estimated that over the next 20 years a 200°C increase in turbine inlet temperature will be required to meet the demand for improved performance (Singh, 2014). These improvements will be achieved by three principal routes: I) materials developments, II) advances in cooling systems and III) development of advanced protective coatings and deposition technologies (D. Clarke & Levi, 2003).

Material Development and Advances in Cooling System

In the past decades, there have been extensive efforts to development of advanced high-temperature materials for turbine engines. Nowadays, combustors and high-pressure turbine blades and vanes which work in the harshest parts of gas turbine engines are manufactured from nickel-based superalloys. The wrought superalloys were the primary forms that developed and applied to turbine blades. In the process of development, the cast forms of superalloys were preferred, because of the improvement in the temperature capability and mechanical properties. In the following, with the introduction of advanced casting technologies, directionally solidified (DS) and single-crystal (SC) superalloys are introduced and used in modern turbine engines. The investigations have shown that through improved alloy compositions and innovation in processing, the temperature capability of superalloys has been enhanced at the average rate of 6.7°C per year (Harada, 2003; Reed, 2006).

A further increase in the temperature tolerance of superalloys, such as nickel-based or cobalt-based superalloys, has become very difficult in recent years. The reason is that in the current generation of superalloys the temperature capability has been increased close to their maximum permissible (initial melting temperature: below 1500 K). On the other hand, according to published reports, each super-alloy generation can cost up to $ 10 million. This approximate amount only relates to the condition that the new generation is based on an adding material or a slight processing change from the previous generation. While if the change is fundamental, for example, a change in the casting process, will cost much more than this amount (Council, 1996; Stern, 1996).

In recent years, a significant part of the increase in turbine inlet temperature has been achieved due to the cooling of hot turbine components via internal flows of compressed air. Although compressing and pumping the air through the cooling system requires energy, and also the manufacture of turbine blades with internal cooling passages requires the use of sophisticated and costly casting processes, the science of air cooling is well developed, and now the role of cooling systems in increasing the turbine operating temperature is approximately 400°C (Council, 1996).

Development of Advanced Protective Coatings

Although the development of novel alloys and the use of advanced cooling systems have had a significant contribution to increasing the working temperature of alloys in recent years, their developments have always been accompanied by limitations and penalties. A part of the constraints of the development of cooling systems was mentioned earlier. Superalloys used in gas turbines are usually optimized to achieve a maximum mechanical strength at high temperatures and with less attention to the environmental conditions. For instance, the quantity of Al and Cr used in these alloys will never be sufficient to achieve resistance to oxidation and hot corrosion, because the presence of these elements more than a critical amount can cause lower melting point and creep strength of the alloy. In these circumstances, if the superalloys are directly exposed to the destructive environment, they will quickly deteriorate due to thermal damage, oxidation and high-temperature corrosion. Therefore, as the third means for improving the turbine efficiency, the use of protective coatings to provide barriers between the superalloys and the outer environment is essential (Bose, 2011). Diffusion, overlay, and thermal barrier coatings can be introduced as the three most important classes of high-temperature coatings that used in practice (Rhys-Jones, 1989).

Diffusion and overlay coatings have been discussed in detail in many pieces of research and are not covered in this chapter. The objective of this study is threefold: (a) to review the general concept of thermal barrier coatings (b) to summarize the recent advances in thermal barrier coatings to improve the coating performance and durability and (c) to introduce the novel concept of self-healing for thermal barrier coatings.

THERMAL BARRIER COATINGS STRUCTURE, MATERIALS, AND DEPOSITION PROCESSES

Thermal barrier coating (TBC) systems are commonly a multilayer coating structure which consists of a ceramic top coat (TC), a metallic bond coat (BC), and a thin layer of thermally grown oxide (TGO) forms at the interface between the TC and the BC during the service. The BC is a diffusion coating of platinum aluminide or a metallic overlay coating of the MCrAlY type (where M is Ni or/and Co). For this layer of TBC systems, two main tasks are considered, protecting the substrate against high-temperature oxidation and corrosion and also improving the adhesion of the ceramic top coat to the substrate.

The TC is a low thermal conductivity ceramic layer that can induce a significant drop in the surface temperature of the components (approximately 100-150°C). Thereby, it is possible to increase the lifetime of the component by reducing their surface temperature, and also it can improve the turbine efficiency by increasing the turbine entry temperature up to the melting point of the superalloys (~1300 °C) (Vaßen, Jarligo, Steinke, Mack, & Stöver, 2010). In addition to the thermal insulation ability (< 2 W/(mK), the materials used for the ceramic layer should also have other main features, including:

- High melting temperature
- Phase stability between room temperature and operating temperature

- Sufficient resistance to sintering at high temperatures
- The relatively compatible coefficient of thermal expansion (etc.) with a metallic substrate
- High fracture toughness at all temperatures (Cao, Vassen, & Stoever, 2004; David R Clarke & Phillpot, 2005; Sadowski & Golewski, 2016)

Since the 1960s, when the first attempts were made to deposition of a ceramic layer on commercial aero engine components by flame spraying, various materials for this layer were investigated for two decades. Eventually, in the 1980s, Yttria Stabilized Zirconia (YSZ) was identified as an exceptional ceramic topcoat material and has since been grounded as a standard for over the past 30 years (David R Clarke & Phillpot, 2005; Miller, 1997; Sadowski & Golewski, 2016).

The cross-sectional SEM image of a typical thermal barrier coating system along with the thickness range of each layer is shown in Figure 1. The inset shows the SEM image of TC and BC interface after 100 hr oxidation at 1100 °C in the air. As can be seen, due to oxidation of the bond coat, the TGO layer is gradually formed. The formed layer is usually α-Al_2O_3, and due to its thermodynamic stability, low oxygen diffusion coefficient, slow growth rate, and superior adherence provides excellent protection of the underlying metal surface (Darolia, 2013; Levi, 2004). However, nowadays, the overgrowth of the TGO has been considered as the primary factor of the spallation failure of TBCs (Bernard et al., 2017).

Although there are several methods for the deposition of thermal barrier coatings, two processes that are widely used today are (i) Electron Beam Physical Vapor Deposition (EB-PVD) and (ii) Atmospheric Plasma Spraying (APS). These two methods produce individual microstructures with particular desirable and undesirable characteristics. YSZ coatings deposited by EB-PVD have columnar microstructures

Figure 1. Cross-sectional SEM image of a TBC system deposited by atmospheric plasma spraying process on a NiCrAlY bond coat. Inset: SEM image of TBC after 100 h isothermal air oxidation at 1100 °C.

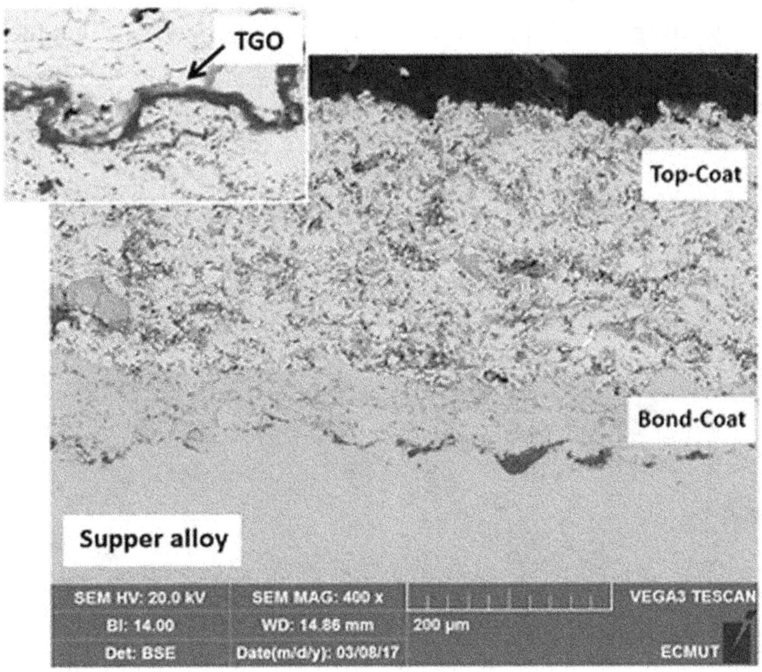

and often used for blades that are under severe thermomechanical loads at aero engines (Vaßen et al., 2010). Two main positive and negative aspects of this microstructure for thermal barrier coatings are:

- (+) Create high strain compliance and good erosion resistance
- (-) Set relatively high thermal conductivity: Due to the absence of microstructural characteristics to increase phonon and photon scattering (i.e., porosity or microcracks parallel to the surface) (Padture, Gell, & Jordan, 2002).

Versatility and lower production cost make atmospheric plasma spraying more attractive than EB-PVD commercially. In particular, the APS TBCs is widely used for static components such as combustion chambers and vane platforms. In APS TBCs, the structure is lamellar and resulting from the stacking of successive splats (plate-like structure of molten or semi-molten feedstock after impacting to the substrate). Figure 2 shows the SEM image (cross sections and top view) of APS YSZ top cost, as can be seen in this figure, pores and porosity, splat interface, intra-splat cracking, and inter-splat cracking are among the microstructural characteristics of these coatings. This porous structure (~10–15% porosity) is highly effective for the scatter of the photons. Therefore APS TBCs have lower thermal conductivity than EB-PVD TBCs. However, unfortunately, this particular microstructure is more sensitive to the CTE mismatch, and hence are less durable than EB-PVD coatings. Table 1 summarizes some of the typical properties of EB-PVD and APS YSZ (Gleeson, 2006; Huang, Eguchi, & Yoshida, 2003).

EMERGING APPROACHES FOR THERMAL BARRIER SYSTEMS

Concurrent with the development of new generation of gas turbines, many attempts have been made to introduce advanced thermal barrier coatings with lower thermal conductivity and higher temperature stability. As previously stated, 7-8 wt. % YSZ is the most popular selection for the ceramic top coat of TBCs. Despite its many advantages, it has a few limitations and drawbacks that the most significant of them are: (i) The probability of occurrence of a destructive phase transformation during thermal cycles, and (ii) The possibility of sintering at high temperature during the service (Padture et al., 2002).

Figure 2. (a) Lamellar microstructure of ceramic top coat manufactured by atmospheric plasma spray, (b) top views of APS TBC

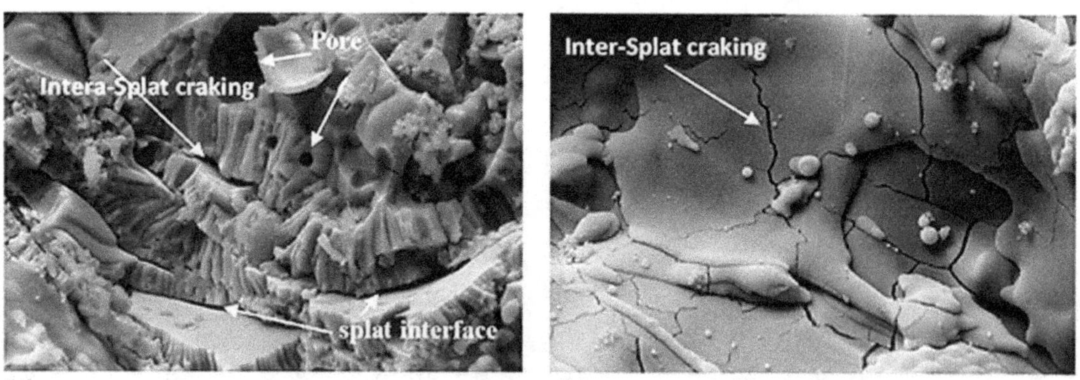

(a) (b)

Table 1. The main properties of thermal barrier coatings at ambient temperature

Deposition Method	Thermal Conductivity (W. m-1. K-1)	Surface Roughness (μm)	Adhesion Strength (MPa)	Young's Modulus (GPa)	Erosion Rate	TGO Growth Rate (mm/h)	Production Cost
EB-PVD	1.5–1.9	0.5–1	400	90	low	200–300	high
APS	0.8–1.1	4–10	20–40	200	high	10,000	low

In the past decade, extensive researches on advanced thermal barrier coatings have resulted in the presentation of several new TBC materials instead of YSZ which have higher melting points or improved thermophysical properties or both. Nevertheless, the studies have shown that the newly introduced materials have mainly lower coefficient of thermal expansion (CTE) and less toughness than YSZ. Therefore, in general, Thermal barrier coatings from new materials, despite lower thermal conductivity and higher temperature stability exhibit less durability than conventional coatings (Vassen et al., 2009). To overcome these limitations, structural modifications of thermal barrier coating systems have been considered as the second approach to the development of these coatings (Lee et al., 2017).

According to the above discussion, most of the research to improve TBCs performance are based on two general approaches: structural modifications and chemical modifications. The latter approach has been reported in several review articles in details (Cao et al., 2004; Clarke, 2003; Singh, 2014; Vaßen et al., 2010). In the following, firstly, some of the investigations about modification of the TBCs system architecture are reviewed in brief, and then a novel concept in designing that is a subset of chemical modification are discussed in particular.

NANOSTRUCTURED THERMAL BARRIER COATINGS

In the development process of the TBCs, there is a significant interest in achieving to the ceramic top-coats with lower thermal conductivity. By reducing thermal conductivity, not only the essential features to increase operating temperature and consequently gas turbines efficiency are provided, but also by decreasing the substrate temperature, the lifetime of the components is increased, and the destructive thermally activated processes responsible for TBC failure are delayed. To reduce thermal conductivity at elevated temperatures, it is necessary to increase phonon scattering centers, such as interfaces, grain boundaries, as well as micro or nano-porosities (Markocsan, Nylén, Wigren, & Li, 2007).

In previous researches, it has been illustrated that using fine (nanometer or submicron scale) YSZ powders instead of coarse ones (micrometer scale) as feedstock for plasma spraying in the APS process, eventually leads to coatings with lower thermal conductivity (Loghman-Estarki et al., 2015).

Figure *3* shows the microstructure of the coating produced by using the nanostructured agglomerated feedstock powders. As can be observed in Fig. 3a, the coating microstructure is formed by unmelted or partially melted particles that are spread throughout the coating microstructure and are surrounded by fully molted particles that operate as a binder, thereby retaining coating integrity; i.e., covering and anchoring the unmelted particles in the coating microstructure (R. Lima, Kucuk, & Berndt, 2002). The zones of consisting unmelted particles, which are known as "nanozone", have an utterly porous morphology (Fig.3b). These homogeneously distributed porosities that are in the nanometer dimension, along with

Figure 3. (a) Surface morphology of the as-sprayed nanostructured 7YSZ coating made from a nano-structured feedstock, (b) Higher magnification of nanozone

(a) (b)

fine grain structure, microcracks and existing voids act as the centers of phonon scattering and hence have a significant effect on decreasing the thermal conductivity of the TBCs (R. S. Lima & Marple, 2007).

In the past few years, several evaluations were carried out on the various properties of nanostructured TBCs, and the results were compared with conventional ones. Jing et al. evaluated the thermal conductivity of nano and conventional YSZ coatings. The porosity of nano-coating was 25% higher than the conventional one. Consequently, the thermal conductivity of nano-TBCs was about 0.8-1.1 W/mK, approximately 40% lower than that for the conventional coating. Also, thermal shock tests were carried out by heating the specimens in a furnace for 10 min at 1523 K (surface temperature), followed by subsequent cooling by compressed air from both sides for 2 min. The results showed that thermal cycling lifetime of The nano-TBC was more than 500 cycles versus 200 cycles for the conventional TBC (J. Wu, Guo, Zhou, Wang, & Gong, 2010). Valefi et al. have also investigated the insulation and thermal shock behavior of conventional and nanostructured plasma-sprayed TBCs. The coatings were heated to 950 °C for 10 min and then quenched with water. The best shock resistance was observed for the nanostructured coating and argued that nanopores in the structure lead to the excellent capability to absorb thermal shock stresses (Masoule, Valefi, Ehsani, & Lavasani, 2016).

If the nano-agglomerated powder is to be used as the feedstock for plasma spraying TBCs, it is necessary to avoid melting of all the feeding powder, such as YSZ nano-agglomerated powders, with engineering the spraying parameters. In fact, the proportion of fully melted to unmelted/partially melted zones must be such that the integrity and cohesion of the TBC to be acceptable, and on the other hand, the percentage of nanoparticles embedded in YSZ coating structure to be adequate for improving the coating lifetime and thermal conductivity. Recently, Solution Precursor Plasma Spraying (SPPS) as an alternative process has been developed. The SPPS is one of the efficient manufacturing techniques for obtaining inherently nanostructured (nanometer-scale grain sizes) TBCs (Pawlowski, 2009). In contrast to the conventional approach in which 10-100 μm flowable powders feed into the plasma torch, an aqueous or non-aqueous solution precursor is used as feedstock. The liquid precursor is injected into the spray torch via an atomizing system directly. The atomized droplets undergo a various thermophysical and chemical transformations within the plasma jet including droplets fragmentation, solvent evaporation, solute precipitation, pyrolysis, sintering and possibly melting and crystallization (Figure 4) (D. Chen, Jordan, & Gell, 2010; Fauchais et al., 2008).

Figure 5 shows the SEM micrograph cross-section and surface of the SPPS TBC. As can be observed, the microstructure of the SPPS TBCs have some exclusive features. Such as ultra-fine splats with enhanced splat-to-splat bonding, nanometer-sized grains, fine porosity and the presence of vertical cracks with a uniform spacing which are useful in improving strain tolerance (Gell et al., 2008; Valefi & Saremi, 2017). Based on comparisons made by some researchers, these coatings offer comparable or lower thermal conductivity, demonstrated higher thermal cycling durability, and costs comparable relative to the conventional APS TBCs. However, the relatively high energy needed for evaporating the liquid precursor and hence lower deposition rate are most important weaknesses of this process (Jordan, Jiang, & Gell, 2015; Ma, Wu, Roth, Gell, & Jordan, 2006).

Figure 4. Schematic illustration of the various stages of solution precursor plasma spraying process

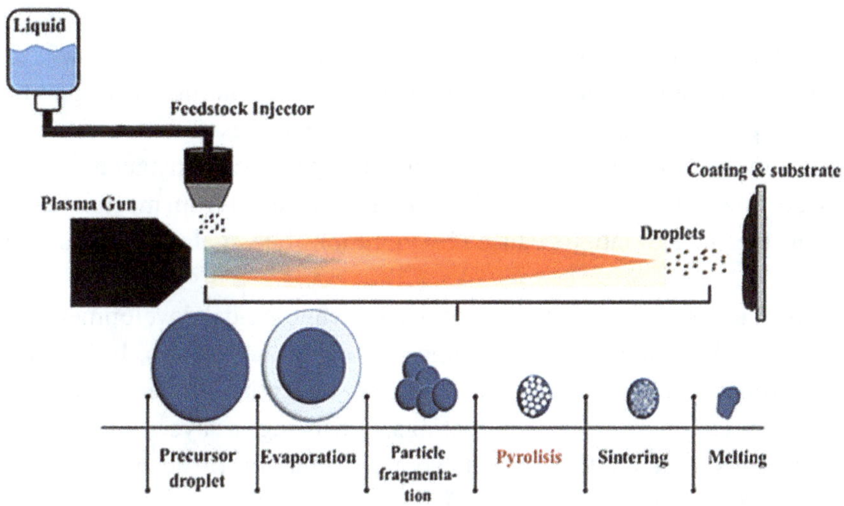

Figure 5. (a) Cross-section, and (b) surface SEM image of YSZ TBC system deposited by the solution precursor plasma spraying process

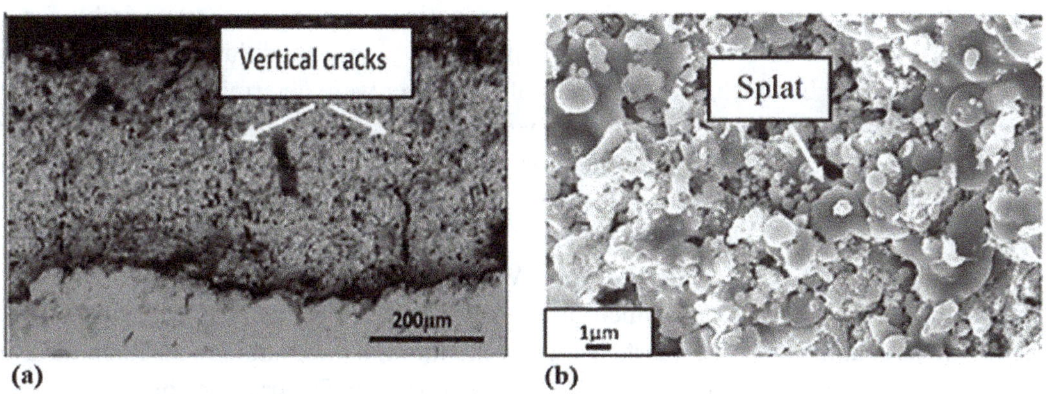

MULTILAYER THERMAL BARRIER COATING SYSTEMS

The improvements in thermal barrier coatings durability are based on the understanding of the failure mechanisms that these coatings face during the service. Numerous literature have reported the failure mechanisms of the TBCs. In general, three main failure mechanisms include:

1. In most cases, the failure location has been found in or close to the TGO-BC interface and attributed to stress concentration and elastic strain energy caused by TGO thickening and thermal expansion mismatch. During the thermal cycles, once the temperature cools down from operating temperature to ambient temperature, high compressive stresses develop in the ceramic layer because of the CTE mismatch between the oxide layer (TGO), the underlying coating, and the substrate. Concurrently, high compressive stresses develop in the TGO due to growth strain. Meanwhile, the system tries to release the strain energy by distorting the TGO and this result in developing the compressive and tensile stress gradient. Tensile stresses are the main driving force for TGO crack nucleation and propagation and ultimately spallation of the coating (Shi, Darzens, & Karlsson, 2005; Sloof, 2007).
2. In APS TBCs, in addition to the cracking that occurs at the BC/TGO interface, there is spallation at the ceramic top-coat splat boundaries. Toughening or eliminating these splat boundaries by the use of innovative plasma-spray process and change the coatings architecture and also suppression of sintering within the top-coat will increase top coat spallation resistance.
3. Besides damages due to thermal causes, TBC damage may result from mechanical causes, including FOD, as well as mechanical fatigue caused by vibration (Lee et al., 2017; Padture et al., 2002).

New proposals in all areas of chemical, structural and technological development of TBCs, are often incapable to creating suitable conditions for dealing with all failure factors. In fact, in most cases, improvements in some properties are at the expense of loss of some other properties. Changing in the TBCs architecture and the application of multilayer coatings, consisting of layers with engineered properties based on the requirements, is a solution to achieve a combination of desired properties (Pakseresht, Javadi, Ghasali, Shahbazkhan, & Shakhesi, 2016).

The durability of thermal barrier coatings is significantly influenced by the elastic modulus and fracture toughness of ceramic top coat. Recent studies have shown that the elastic modulus and toughness of conventional YSZ coatings are sensitive to microstructure. Coatings with higher levels of porosity have lower fracture toughness compared to the relatively dense structures. However, with increasing porosity, the elastic modulus and thermal conductivity of the coatings are increased, undesirable for a TBCs. Viswanathan et al. In order to achieve optimal properties in elastic modulus and fracture toughness, used a bilayer architecture in which, the properties of each layer were determined based on the location of that layer through providing a suitable control over the process parameters. They select the spraying parameters to introduce a high density and therefore high fracture toughness layer at regions prone to delamination failure (near BC/TC interface), while the majority of the overall coating was engineered to create a porous microstructure. The results indicated that this bilayer architecture showed significantly improved durability under thermal cycle exposure (Viswanathan, Dwivedi, & Sampath, 2014).

Segmented thermal barrier coatings, which can be defined as the coatings contain a network of cracks perpendicular to the coating/substrate interface that extend through the coating to the metallic layer, have been introduced to improve the durability of TBCs for several years. The presence of these segmentation cracks provide a mechanism for accommodating mismatch strains created by thermal expansion mismatch between the substrate and top coat and Therefore, lead to preserve the integrity and hence will improve the strain tolerance of TBC system (Duvall & Ruckle, 1982). Nevertheless, many investigators reported that the inward penetration of oxygen and the molten calcium–magnesium–alumina-silicate (CMAS) glass deposits, through the porous ceramic coating are the fundamental premature failure mechanisms for APS TBCs. Huang et al. in order to obtain good performance under these conditions, tried multilayered TBC architectures comprised of the initial layer of segmentation cracks, the middle porous layer and, the dense surface layer. The results showed that the durability of the multilayered TBCs under thermal cycles increased by a factor of 2 compared with that of the conventional one, due to the high strain tolerance and fracture toughness of the segmented microstructure. As well as, oxidation and CMAS corrosion resistance were also better for the multilayered TBCs, which could attribute to the suppression function of the dense top YSZ layer (Yu, Huang, Wang, Yu, & Wu, 2016).

In all these development methods it is to be understood that the principle is reducing the possibility of formation of cracks, but, once formed, all such cracks can grow under thermal cycles and eventually lead to coating delamination and spallation. Self-healing is the most precious phenomenon to overcome this problem. In the following, the use of self-healing concept to improve the durability of thermal barrier coatings is discussed in some detail.

SELF-HEALING CONCEPT

There are always proactive measures to prevent unplanned failure of components in engineering design, for example taking into account safety factors to cover the uncertainties. However, in many cases, premature failures do occur because of the extreme environmental condition (i.e., corrosive or/and high temperature, collisions and sudden blows, design faults, manufacturing defects, and maintenance mistakes) (Li & Meng, 2015).

Over the past two decades, many efforts have been made to find a solution to this unpredictable behavior of materials (especially, polymeric and ceramic materials). Particularly in cases where the cost of repairing and replacing is very high, as well as about the parts that applied to sensitive applications such as submarines and spacecraft. Accordingly, today's scientific and technological improvements have led to the establishment of a new branch in the scientific discipline of materials engineering called "self-healing materials." researchers have introduced this novel concept with inspiration from natural phenomena such as blood clotting or repairing of fractured bones. Conceptually, self-healing materials exhibit the ability to substantially repair (heal/recover) damages created in their structure and restore mechanical properties to near initial state autonomically (without any intervention) or non-autonomically (with the use of an external trigger) (Aissa, Haddad, & Jamroz, 2014; Amendola & Meneghetti, 2011).

From the thermodynamic point of view, self-healing is a non-equilibrium process. The self-healing material is driven away from the thermodynamic equilibrium because of either the deterioration event or by external triggers. Then, with the occurrence of the healing process, the material slowly restores thermodynamic equilibrium. In general, two main strategies are used in the literature to design self-healing materials: (i) applying the concept of reversible cross-link, and (ii) the use of a healing agent (Nosonovsky & Rohatgi, 2011).

Concept of Reversible Cross-Link

One of the common strategies to induce self-healing concept in polymeric materials is the incorporation of reversible cross-link groups in or on the polymer backbone. In these materials, chemical reactions or/and electrostatic attractive forces in combination with a modest temperature rise could result in the healing of the polymeric materials. These systems show a non-automatic healing response. Therefore, an external trigger such as thermal, photo or chemical driving force is required for activation of healing phenomenon. Two types of reversible cross-linking chemical reaction used in theses self-healing systems: the polymer systems based on the Diels-Alder (DA) and Retro Diels Alder (RDA) reaction and the polymers containing quadruple hydrogen bonds. As well as, for physical cross-linking systems, the presence of ionic groups give rise to the formation of electrically bound physical clusters and restoration of the mechanical properties at lower temperatures (Van der Zwaag, 2010). In this chapter, we focus mainly on "self-healing ceramic materials" and the healing mechanism of the polymeric materials can be found elsewhere in details (D. Y. Wu, Meure, & Solomon, 2008; Yang & Urban, 2013).

Release of Healing Agent

Most self-healing systems that have been introduced so far used some healing agent in forms of particles, microcapsules and hollow fibers into the matrix of the composite materials. When a healing phenomenon is activated either autonomically (by the deterioration process) or non-autonomically (by an external intervention such as heating), the healing agent is released and seals the void or crack by solidifying or through a chemical reaction. The mechanism is shown in figure 6. when a crack propagates, the healing agent container ruptures and the healing agent is released and exposed to the catalyst (i.e., oxygen) and Penetrates into the crack due to capillarity effect or because of the occurrence of a reaction with a high-volume expansion. Ultimately either the healing agent which mixes with the catalyst triggers the cross-linking phenomenon, or a strong bond between the healing agent and crack wall is created by a solid-state chemical reaction (Nosonovsky & Rohatgi, 2011; Zwaag, 2008).

Figure 6. Schematic diagram of the self-healing mechanism

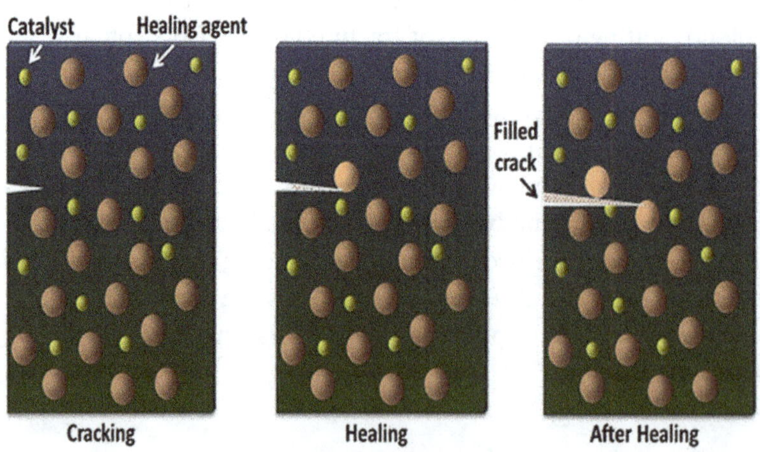

Based on the general mechanism self-healing employed, self-healing systems can be divided into the two categories of intrinsic and extrinsic systems. The concept of "extrinsic self-healing" is relied on an external healing agent (granular particles, fibers, etc.) dispersed in a nonresponsive matrix. The dormant healing agents become active when intersected by a microcrack in the matrix. Upon activation of healing mechanism, a chemical reaction is triggered which leads to filling of the crack and adhesion between the reaction product and the crack faces is in such a manner that the load bearing capabilities of the material are restored. These types of healing systems apply to materials such as concrete, asphalt, metals, and ceramics. In the case of "intrinsic self-healing," the healing action is based on the chemical nature of the material itself, and it has been demonstrated for polymers, metals, and metallo-ceramic MAX phases. Thereby, the second approach cannot be used for oxidic ceramics (i.e., zirconia), Because the ceramic matrix is already in minimum state of energy and cannot undergo any further reaction (Boatemaa, Kwakernaak, van der Zwaag, & Sloof, 2016; DiCarlo & Yun, 2005; Farle et al., 2016).

Self-Healing Thermal Barrier Coatings

As already mentioned, the ceramic materials have introduced as the appropriate selection for high-temperature environments because of their thermal/chemical stability, and also corrosion-and oxidation resistance. However, ceramic structures tend to have a brittle failure at all temperatures (and often before their end-use). This type of catastrophic failure that is usually caused by the stress concentration at the tip of the flaws leads to low mechanical reliability and hence as the primary limiting factor, inhibits their use in many applications. Many efforts have been made to overcome this shortcoming of ceramic materials in recent years by introducing the toughening mechanisms such as reinforcement with fibers, whiskers, ductile or brittle particles, and also by phase transformations (Ilhan et al., 2015; Nakao, Takahashi, & Ando, 2009). An alternative method that has recently been considered is to endow ceramic structures with the self-healing ability. The field of self-healing has a long history, but most of the investigations were related to the bulk material (often polymeric and metallic) and also polymeric coatings. However, we can find many reports on self-healing ceramic-matrix composites structures, but the published investigations of ceramic coatings are quite limited.

Today, most studies have been conducted into the self-healing ceramics, concentrated on the use of SiC as the healing agent. This selection is based on three main reasons: (a) SiC is relatively stable up to 2270 oC; (b) will decompose into SiO_2 which has a specific volume 113% larger than its parent phase when exposed to air at higher temperatures and (c) the reaction products SiO_2 bonds well to the ceramic matrix (Boatemaa et al., 2016).

Based on the literature, it was established that in SiC self-healing TBC, the healing occurred due to oxidation of the SiC particle and associated with volume expansion of healing produced into the cracks. Three important aspects of the healing mechanism are as follows. (a) after the formation of the initial microcracks and with their propagation in the ceramic matrix, SiC particles located on the crack walls can react with the oxygen in the atmosphere. (b) The oxidation reaction is accompanied by a significant volume expansion, thereby, the reaction product can fill the crack, assuming incorporation of adequate volume fraction of healing agent, (c) The healing reaction is exothermic, and the heat generated during reaction makes the formed oxide (SiO_2) and the matrix material (i.e., ZrO_2) to react or to melt once. Ultimately, the healing reaction might lead to formation of mechanically stable zircon ($ZrSiO_4$) and makes a strong bonding between the reaction product and crack walls (based on the following reactions) (Derelioglu, Carabat, Song, van der Zwaag, & Sloof, 2015; Nakao et al., 2009; Ouyang et al., 2016).

$$SiC + 2O_2 \rightarrow SiO_2 + CO_2 \quad (1)$$

$$ZrO_2 + SiO_2 \rightarrow ZrSiO_4 \quad (2)$$

In order to achieve full healing at lower temperatures, recently, some investigation have been carried out on the other various healing agent, and some advanced agents replacing SiC have been proposed. In general, there are three primary conditions for choosing a substance as a healing agent: (i) The appropriate selection of the healing agent should be a solid at the operating temperature, as the liquids have an unacceptably large CTE and thermal conductivity. (ii) After activation, the healing agent should turn into a fluid and fills the crack. (iii) The healing product formed inside the crack during healing treatment should turn into a solid by a subsequent solid-state chemical reaction with the matrix to create a load-bearing material. Also, the CTE, and the elastic modulus of healing agent after and prior to the healing reaction in relation to that of the parent phase, the diffusion kinetics and the effective shelf life of healing agent are the another important parameters (Derelioglu et al., 2015; Yoshioka, Boatemaa, van der Zwaag, Nakao, & Sloof, 2016).

The first step to the development of TBCs with self-healing capability is related to Sloof et al.'s researches. They used Boron alloyed $MoSi_2$ as the healing agent particle and indicated that adding B-$MoSi_2$ particles in YSZ coating will not generate extra stresses during thermal cycles. B-$MoSi_2$ particles exhibit compatible CTE with YSZ matrix and remain solid at 1100 °C in the absence of an oxidative atmosphere. Furthermore, alloying the $MoSi_2$ with B will increase the fluidity of the SiO_2 at the operating temperature that will facilitate the filling of cracks. It was demonstrated that, upon crack formation and with the exposure of $MoSi_2$(B) particles embedded into the TBC to high oxygen potential atmosphere, the oxidation reaction leads to the formation of amorphous SiO_2 which flows into cracks and establishes direct contact with the crack walls. Subsequently, the amorphous SiO_2 reacts with the ZrO_2 and leading to the formation of a well adhering solid reaction product such as $ZrSiO_4$. (Derelioglu et al., 2015).

As stated, one of the parameters affecting the performance of the self-healing systems is the shelf life of the healing agent. Considering that yttria partially stabilized zirconia is oxygen-transparent at high temperatures and therefore does not provide any oxidation resistance, premature activation of the healing particles in the absence of crack is to be expected. Carabao et al. encapsulated the $MoSi_2$(B) particles by a thin oxygen impermeable layer of alumina using a simple precipitation method to prevent the premature oxidation (Carabat, Zwaag, & Sloof, 2015).

With the occurrence of the healing reaction in a specific region of the coating, the reservoir of healing agents will decrease, and the self-healing capacity will drop in that particular zone gradually. Consequently, the probability of crack growth and ultimately deterioration of the coating under the subsequent thermal cycles will increase. Therefore, it is vital to consider appropriate strategies to optimize the self-healing capacity of the coatings to withstand repeated damages (Sloof, 2007). Multi-composite systems have been introduced to resolve this limitation. Self-healing systems containing healing agent with the forms of both whisker and particles called multi-composite systems. These systems not only enhance the fracture strength and toughness but also endow good self-crack healing capacity.

CONCLUSION

Most of the research to improve TBCs performance are based on two general approaches: structural modifications and chemical modifications. Nanostructured and multilayered TBCs and also the use of self-healing concept are three main methods which use to modify TBCs structure. The first step to the development of TBCs with self-healing capability is related to Sloof et al.'s researches. They used Boron alloyed MoSi2 as the healing agent particle and demonstrated that, upon crack formation and with the exposure of MoSi2(B) particles embedded into the TBC to high oxygen potential atmosphere, the oxidation reaction leads to the formation of amorphous SiO2 which flows into cracks and establishes direct contact with the crack walls. Subsequently, the amorphous SiO2 reacts with the ZrO2 and leading to the formation of a well adhering solid reaction product such as ZrSiO4.

ACKNOWLEDGMENT

The authors would like to acknowledge Iran National Science Foundation: INSF.

REFERENCES

Ai, T., Masada, J., & Ito, E. (2014). Development of the High Efficiency and Flexible Gas Turbine M701F5 by Applying "J" Class Gas Turbine Technologies. *Mitsubishi Heavy Industries Technical Review*, *51*(1), 1.

Aissa, B., Haddad, E. I., & Jamroz, W. (2014). *Self-Healing Materials: Innovative Materials for Terrestrial & Space Applications*. Smithers Rapra Technology.

Amendola, V., & Meneghetti, M. (2011). *Self-healing at the Nanoscale: Mechanisms and Key Concepts of Natural and Artificial Systems*. CRC Press. doi:10.1201/b11484

Bernard, B., Quet, A., Bianchi, L., Joulia, A., Malié, A., Schick, V., & Rémy, B. (2017). Thermal insulation properties of YSZ coatings: Suspension plasma spraying (SPS) versus electron beam physical vapor deposition (EB-PVD) and atmospheric plasma spraying (APS). *Surface and Coatings Technology*, *318*, 122–128. doi:10.1016/j.surfcoat.2016.06.010

Boatemaa, L., Kwakernaak, C., van der Zwaag, S., & Sloof, W. G. (2016). Selection of healing agents for autonomous healing of alumina at high temperatures. *Journal of the European Ceramic Society*, *36*(16), 4141–4145. doi:10.1016/j.jeurceramsoc.2016.05.038

Bose, S. (2011). *High temperature coatings*. Butterworth-Heinemann.

Cao, X., Vassen, R., & Stoever, D. (2004). Ceramic materials for thermal barrier coatings. *Journal of the European Ceramic Society*, *24*(1), 1–10. doi:10.1016/S0955-2219(03)00129-8

Carabat, A. L., Zwaag, S., & Sloof, W. G. (2015). Creating a Protective Shell for Reactive MoSi2 Particles in High-Temperature Ceramics. *Journal of the American Ceramic Society*, *98*(8), 2609–2616. doi:10.1111/jace.13625

Chen, D., Jordan, E. H., & Gell, M. (2010). The solution precursor plasma spray coatings: Influence of solvent type. *Plasma Chemistry and Plasma Processing*, *30*(1), 111–119. doi:10.1007/s11090-009-9200-4

Chen, X., Wang, R., Yao, N., Evans, A. G., Hutchinson, J. W., & Bruce, R. W. (2003). Foreign object damage in a thermal barrier system: Mechanisms and simulations. *Materials Science and Engineering A*, *352*(1), 221–231. doi:10.1016/S0921-5093(02)00905-X

Clarke, D., & Levi, C. (2003). Materials design for the next generation thermal barrier coatings. *Annual Review of Materials Research*, *33*(1), 383–417. doi:10.1146/annurev.matsci.33.011403.113718

Clarke, D. R. (2003). Materials selection guidelines for low thermal conductivity thermal barrier coatings. *Surface and Coatings Technology*, *163–164*, 67–74. doi:10.1016/S0257-8972(02)00593-5

Clarke, D. R., & Phillpot, S. R. (2005). Thermal barrier coating materials. *Materials Today*, *8*(6), 22–29. doi:10.1016/S1369-7021(05)70934-2

Council, N. R. (1996). *Coatings for high-temperature structural materials: trends and opportunities*. National Academies Press.

Darolia, R. (2013). Thermal barrier coatings technology: Critical review, progress update, remaining challenges and prospects. *International Materials Reviews*, *58*(6), 315–348. doi:10.1179/1743280413Y.0000000019

Derelioglu, Z., Carabat, A., Song, G., van der Zwaag, S., & Sloof, W. (2015). On the use of B-alloyed MoSi 2 particles as crack healing agents in yttria stabilized zirconia thermal barrier coatings. *Journal of the European Ceramic Society*, *35*(16), 4507–4511. doi:10.1016/j.jeurceramsoc.2015.08.035

DiCarlo, J. A., & Yun, H.-M. (2005). *Non-oxide (silicon carbide) fibers Handbook of ceramic composites* (pp. 33–52). Springer. doi:10.1007/0-387-23986-3_2

Duvall, D., & Ruckle, D. (1982). Ceramic thermal barrier coatings for turbine engine components. *ASME Paper*, *82*, 332.

Farle, A., Boatemaa, L., Shen, L., Gövert, S., Kok, J., Bosch, M., & Sloof, W. (2016). Demonstrating the self-healing behaviour of some selected ceramics under combustion chamber conditions. *Smart Materials and Structures*, *25*(8), 084019. doi:10.1088/0964-1726/25/8/084019

Fauchais, P., Etchart-Salas, R., Rat, V., Coudert, J.-F., Caron, N., & Wittmann-Ténèze, K. (2008). Parameters controlling liquid plasma spraying: Solutions, sols, or suspensions. *Journal of Thermal Spray Technology*, *17*(1), 31–59. doi:10.1007/s11666-007-9152-2

Gell, M., Jordan, E. H., Teicholz, M., Cetegen, B. M., Padture, N. P., Xie, L., & Roth, J. et al. (2008). Thermal barrier coatings made by the solution precursor plasma spray process. *Journal of Thermal Spray Technology*, *17*(1), 124–135. doi:10.1007/s11666-007-9141-5

Gleeson, B. (2006). Thermal barrier coatings for aeroengine applications. *Journal of Propulsion and Power*, *22*(2), 375–383. doi:10.2514/1.20734

Harada, H. (2003, November). High temperature materials for gas turbines: the present and future. *Paper presented at the International Gas Turbine Congress*, Tokyo, Japan.

Huang, H., Eguchi, K., & Yoshida, T. (2003). Novel structured yttria-stabilized zirconia coatings fabricated by hybrid thermal plasma spraying. *Science and Technology of Advanced Materials*, *4*(6), 617–622. doi:10.1016/j.stam.2004.01.004

Ilhan, Z., Hornes, A., Sata, N., Freitag, T. O., Ansar, A., Schiller, G., Krebs, A. (2015). Self-Healing Plasma Sprayed Ceramic Coatings.

Jordan, E. H., Jiang, C., & Gell, M. (2015). The solution precursor plasma spray (SPPS) process: A review with energy considerations. *Journal of Thermal Spray Technology*, *24*(7), 1153–1165. doi:10.1007/s11666-015-0272-9

Lee, D. H., Kang, N. K., Lee, K. S., Moon, H. S., Kim, H. T., & Kim, C. (2017). Evaluation of Thermal Durability of Thermal Barrier Coating and Change in Mechanical Behavior. *Journal of the Korean Ceramic Society*, *54*(4), 314–322. doi:10.4191/kcers.2017.54.4.05

Levi, C. G. (2004). Emerging materials and processes for thermal barrier systems. *Current Opinion in Solid State and Materials Science*, *8*(1), 77–91. doi:10.1016/j.cossms.2004.03.009

Li, G., & Meng, H. (2015). *Recent Advances in Smart Self-healing Polymers and Composites*. Elsevier Science.

Lima, R., Kucuk, A., & Berndt, C. (2002). Bimodal distribution of mechanical properties on plasma sprayed nanostructured partially stabilized zirconia. *Materials Science and Engineering A*, *327*(2), 224–232. doi:10.1016/S0921-5093(01)01530-1

Lima, R. S., & Marple, B. R. (2007). Thermal spray coatings engineered from nanostructured ceramic agglomerated powders for structural, thermal barrier and biomedical applications: A review. *Journal of Thermal Spray Technology*, *16*(1), 40–63. doi:10.1007/s11666-006-9010-7

Loghman-Estarki, M. R., Nejati, M., Edris, H., Razavi, R. S., Jamali, H., & Pakseresht, A. H. (2015). Evaluation of hot corrosion behavior of plasma sprayed scandia and yttria co-stabilized nanostructured thermal barrier coatings in the presence of molten sulfate and vanadate salt. *Journal of the European Ceramic Society*, *35*(2), 693–702. doi:10.1016/j.jeurceramsoc.2014.08.029

Ma, X., Wu, F., Roth, J., Gell, M., & Jordan, E. H. (2006). Low thermal conductivity thermal barrier coating deposited by the solution plasma spray process. *Surface and Coatings Technology*, *201*(7), 4447–4452. doi:10.1016/j.surfcoat.2006.08.095

Markocsan, N., Nylén, P., Wigren, J., & Li, X.-H. (2007). Low thermal conductivity coatings for gas turbine applications. *Journal of thermal spray technology, 16*(4), 498-505.

Masoule, S. T., Valefi, Z., Ehsani, N., & Lavasani, H. Q. (2016). Thermal Insulation and Thermal Shock Behavior of Conventional and Nanostructured Plasma-Sprayed TBCs. *Journal of Thermal Spray Technology*, *25*(8), 1684–1691. doi:10.1007/s11666-016-0476-7

Miller, R. A. (1997). Thermal barrier coatings for aircraft engines: history and directions. *Journal of thermal spray technology, 6*(1), 35-42.

Nakao, W., Takahashi, K., & Ando, K. (2009). Self-healing of Surface Cracks in Structural Ceramics. In *Self-healing materials: fundamentals, design strategies, and applications.*

Nicholls, J. (2003). Advances in coating design for high-performance gas turbines. *MRS Bulletin*, *28*(09), 659–670. doi:10.1557/mrs2003.194

Nosonovsky, M., & Rohatgi, P. K. (2011). *Biomimetics in materials science: self-healing, self-lubricating, and self-cleaning materials* (Vol. 152). Springer Science & Business Media.

Ouyang, T., Fang, X., Zhang, Y., Liu, D., Wang, Y., Feng, S., & Suo, J. (2016). Enhancement of high temperature oxidation resistance and spallation resistance of SiC-self-healing thermal barrier coatings. *Surface and Coatings Technology*, *286*, 365–375. doi:10.1016/j.surfcoat.2015.12.054

Padture, N. P., Gell, M., & Jordan, E. H. (2002). Thermal barrier coatings for gas-turbine engine applications. *Science*, *296*(5566), 280–284. doi:10.1126/science.1068609 PMID:11951028

Pakseresht, A., Javadi, A., Ghasali, E., Shahbazkhan, A., & Shakhesi, S. (2016). Evaluation of hot corrosion behavior of plasma sprayed thermal barrier coatings with graded intermediate layer and double ceramic top layer. *Surface and Coatings Technology*, *288*, 36–45. doi:10.1016/j.surfcoat.2016.01.012

Pawlowski, L. (2009). Suspension and solution thermal spray coatings. *Surface and Coatings Technology*, *203*(19), 2807–2829. doi:10.1016/j.surfcoat.2009.03.005

Rahman, M., Ibrahim, T. K., & Abdalla, A. N. (2011). Thermodynamic performance analysis of gas-turbine power-plant. *International Journal of Physical Sciences*, *6*(14), 3539–3550.

Reed, R. C. (2006). *The superalloys: fundamentals and applications*. Cambridge university press. doi:10.1017/CBO9780511541285

Rhys-Jones, T. (1989). Coatings for blade and vane applications in gas turbines. *Corrosion Science*, *29*(6), 623–646. doi:10.1016/0010-938X(89)90104-2

Sadowski, T., & Golewski, P. (2016). *Protective Thermal Barrier Coatings Loadings in Thermal Barrier Coatings of Jet Engine Turbine Blades* (pp. 5–11). Springer. doi:10.1007/978-981-10-0919-8_2

Shi, J., Darzens, S., & Karlsson, A. M. (2005). Aspects of the morphological evolution in thermal barrier coatings and the intrinsic thermal mismatch therein. *Materials Science and Engineering A*, *392*(1), 301–312. doi:10.1016/j.msea.2004.09.062

Singh, K. (2014). Advanced materials for land based gas turbines. *Transactions of the Indian Institute of Metals*, *67*(5), 601–615. doi:10.1007/s12666-014-0398-3

Sloof, W. G. (2007). Self Healing in Coatings at High Temperatures. In S. van der Zwaag (Ed.), *Self Healing Materials: An Alternative Approach to 20 Centuries of Materials Science* (pp. 309–321). Dordrecht: Springer Netherlands. doi:10.1007/978-1-4020-6250-6_14

Stern, K. H. (1996). *Metallurgical and ceramic protective coatings*. Springer Science & Business Media. doi:10.1007/978-94-009-1501-5

Stöver, D., & Funke, C. (1999). Directions of the development of thermal barrier coatings in energy applications. *Journal of Materials Processing Technology*, *92*, 195–202. doi:10.1016/S0924-0136(99)00244-7

Valefi, Z., & Saremi, M. (2017). The effects of plasma spray parameters on the microstructure and phase composition of thermal barrier coatings made by spps process. *Iranian Journal of Materials Science and Engineering*, *14*(2), 11–23.

Van der Zwaag, S. (2010). Routes and mechanisms towards self healing behaviour in engineering materials. *Bulletin of the Polish Academy of Sciences. Technical Sciences*, *58*(2), 227–236. doi:10.2478/v10175-010-0022-6

Vaßen, R., Jarligo, M. O., Steinke, T., Mack, D. E., & Stöver, D. (2010). Overview on advanced thermal barrier coatings. *Surface and Coatings Technology*, *205*(4), 938–942. doi:10.1016/j.surfcoat.2010.08.151

Vassen, R., Stuke, A., & Stöver, D. (2009). Recent developments in the field of thermal barrier coatings. *Journal of thermal spray technology, 18*(2), 181-186.

Viswanathan, V., Dwivedi, G., & Sampath, S. (2014). Engineered multilayer thermal barrier coatings for enhanced durability and functional performance. *Journal of the American Ceramic Society*, *97*(9), 2770–2778. doi:10.1111/jace.13033

Wu, D. Y., Meure, S., & Solomon, D. (2008). Self-healing polymeric materials: A review of recent developments. *Progress in Polymer Science*, *33*(5), 479–522. doi:10.1016/j.progpolymsci.2008.02.001

Wu, J., Guo, H., Zhou, L., Wang, L., & Gong, S. (2010). Microstructure and thermal properties of plasma sprayed thermal barrier coatings from nanostructured YSZ. *Journal of Thermal Spray Technology*, *19*(6), 1186–1194. doi:10.1007/s11666-010-9535-7

Yang, Y., & Urban, M. W. (2013). Self-healing polymeric materials. *Chemical Society Reviews*, *42*(17), 7446–7467. doi:10.1039/c3cs60109a PMID:23864042

Yoshioka, S., Boatemaa, L., van der Zwaag, S., Nakao, W., & Sloof, W. G. (2016). On the use of TiC as high-temperature healing particles in alumina based composites. *Journal of the European Ceramic Society*, *36*(16), 4155–4162. doi:10.1016/j.jeurceramsoc.2016.06.008

Yu, Z., Huang, J., Wang, W., Yu, J., & Wu, L. (2016). Deposition and properties of a multilayered thermal barrier coating. *Surface and Coatings Technology*, *288*, 126–134. doi:10.1016/j.surfcoat.2016.01.001

Zwaag, S. (2008). *Self healing materials: an alternative approach to 20 centuries of materials science*: Springer Science+ Business Media BV.

KEY TERMS AND DEFINITIONS

Self-Healing TBCs: Self-healing thermal barrier coatings are the novel coating systems which exhibit the ability to repair (heal/recover) damages created in their structure and restore mechanical properties to near initial state autonomically (without any intervention) or non-autonomically (with the use of an external trigger).

TBC: TBCs systems are commonly a multilayer coating structure which consists of a ceramic top coat (TC), a metallic bond coat (BC), and a thin layer of thermally grown oxide (TGO) forms at the interface between the TC and the BC during the service. These coatings usually applied to metallic surfaces, such as on gas turbine parts to increase their working temperature and protect them against environmental factors.

TIT: Turbine inlet temperature (TIT) is the critical parameter in improving gas turbine engine power and fuel efficiency.

YSZ: Yttria Stabilized Zirconia (YSZ) is the most common ceramic oxide that was identified as an exceptional ceramic topcoat material and has since been grounded as a standard for over the past 30 years.

Chapter 4
Thermal Barrier Coatings on the Base of Samarium Zirconates

Sebastian Oliwer Jucha
The Silesian University of Technology, Poland

Grzegorz Jan Moskal
The Silesian University of Technology, Poland

ABSTRACT

The zirconates of rare earth elements, such as Sm2Zr2O7, can be an alternative material for zirconia modified by yttria (8YSZ) usually used as an insulation layer in thermal barrier coatings (TBC) systems. This chapter describes the morphology of feedstock zirconate powder, internal morphology and selected properties of different samarium zirconate TBC systems. These included: composite TBC coatings of Sm2Zr2O7 + 8YSZ type with different ratio of both used to coatings deposition powders (25/75, 50/50 and 75/25) as well as the TBC of double ceramic layer (DCL) type with an 8YSZ internal layer and an outer layer of Sm2Zr2O7 type, and a monolayer TBC based on Sm2Zr2O7. Another presented subject is related to the oxidation resistance of TBC systems during static oxidation test at temperature 1100°C. In this case, the special emphasis was taken on the characterization of thermally grown oxides (TGO) zone thickness where the most important phenomena related with overall live-time of TBC systems usually take place.

INTRODUCTION

According to definition of The International Union of Pure and Applied Chemistry (IUPAC) the rare earth elements include 15 lanthanides (lanthanum, cerium, praseodymium, neodymium, promethium, samarium, europium, gadolinium, terbium, dysprosium, holmium, erbium, thulium, ytterbium, lutetium) scandium and yttrium, which tend to occur in the same ore deposits as the lanthanides and has similar chemical properties (Charewicz, 1990; Klupa, 2012; Hurst, 2010).

Rare earth elements are plentiful in Earth's crust; however, their name is associated with discovery them in rarely occurring minerals (Charewicz,1990). Their found widespread use in many industries. The best examples of their use are modern electronics, new hybrid or fully electric vehicles, materials

DOI: 10.4018/978-1-5225-4194-3.ch004

for permanent magnets and lasers (Klupa, 2012). Another area where for years it is observed the research on the use rare earth elements is energy storage systems such as SOFC (solid oxy-fuel cell) and aircraft engines industry with special attention to thermal barrier coating (TBC) systems, what is the main subject of the presented chapter.

TBC is a multilayer system, build with few types of materials with completely different physicochemical properties. It allows to achieve a level of properties that would not be obtainable for each of these materials separately.

TBC system is consist of external ceramic layer called as topcoat (TC), its most important mission is insulation properties and corrosion resistance, next is zone of thermally growth oxides (TGO) which forms the alumina (α-Al2O3) which is responsible for the oxidation properties, then is the aluminium reach metallic bond coat layer (BC) ((Ni, Co)CrAlY or aluminides of Pt and Ni) which has to bonding of the ceramic layer and protect the based material against oxidation. The aim of based material is a transfer of thermal and mechanical stresses (Bakan & Vaßen, 2017; Curry, 2014; Gupta, 2014).

Since over 40 years 6-8 mol% yttria-stabilized zirconia (8YSZ) has been the material of choice for ceramic top coats, as it has the exceptional combination of desired properties. In recent years mainly four different ceramic material groups have been suggested as new topcoat materials. Among them: hexaaluminates, perovskites, zirconia doped with different rare-earth (RE) cations (defect cluster zirconia) and pyrochlores (properties shown in Table 1). Other materials were also taken into account, like for example, silicates ($ZrSiO_4$), mullite, garnets (YAG and YAM) and $(Ca_{1-x}Mg_x)Zr_4(PO_4)_6$. However, their typically low coefficient of thermal expansion (CTE) excludes the possibility of their application (Bakan & Vaßen, 2017; Curry, 2014; Gupta, 2014).

The outer ceramic layer is obtained mainly by air plasma spraying (APS) suspension plasma spraying (SPS) or electron-beam physical vapor deposition (EB-PVD) method, the comparison of those method is set up in the Table 2 (Gupta, 2014; Bernard et al., 2017). Its primary function is to provide a thermal barrier, but the extremely aggressive thermochemical environment in which it must work, need to meet other relevant requirements (Clarke & Phillpot, 2005) those are among others:

1. *Ability to experience large deformations without failure*
2. *Compatibility of deformation*

Table 1. Properties of selected topcoat materials

	Example of Materials			
	Hexaaluminates: $LaMgAl_{11}O_{19}$	**Perovskites: $BaZrO_3$**	**Dopped Zirconia: 5 wt.%CaO+ZrO_2**	**Pyrochlore: $La_2Zr_2O_7$**
T_m (Melting point)	-	2963K	$T_{softening}$=2558K	2573K
λ (thermal conductivity)	1,7 W $m^{-1}K^{-1}$ (1273K)	3,42 W $m^{-1}K^{-1}$ (1273K)	-	1,56 W $m^{-1}K^{-1}$ (1273K)
α (thermal expansion coefficient)	10,1×$10^{-6}K^{-1}$ (293-1273K)	8,1×$10^{-6}K^{-1}$ (293-1273K)	9,91×$10^{-6}K^{-1}$ (293-1273K)	9,1×$10^{-6}K^{-1}$ (293-1273K)
C_p (heat capacity)	0,86 J $g^{-1}K^{-1}$ (1273K)	0,45 J $g^{-1}K^{-1}$ (1273K)	-	0,49 J $g^{-1}K^{-1}$ (1273K)
E (Young's modulus)	-	181 GPa (293K)	-	175 GPa (293K)

Cao, Vassen & Stoever, 2014.

3. *No phase transformations in cycles between room and operating temperature*
4. *High erosive resistance*
5. *High-temperature oxidation resistance*
6. *Liquid deposits, sands (CMAS – calcia-magnesia-alumina-silica) and volcanic ashes resistance*
7. *Thermodynamic compatibility the TBC material with alumina (Clarke & Phillpot, 2005; Cernushi, Bianchi, Leoni & Scardi, 1999; Vassen, Tietz, Kerkhoff & Stoever, 1998).*

TBC coatings are a great example of multifunctional materials. Failure to meet any of the above criteria by the material, makes it useless in this application. Suitable thermal conductivity properties are most important in choice the material on TBC (Clarke & Phillpot, 2005).

CHARACTERISTICS THE GROUPS OF MATERIALS DEDICATED FOR THE CERAMIC LAYERS IN TBC SYSTEMS

Initially, the choice of 8YSZ (yttria stabilized zirconia – 8%wt. $Y_2O_3 \times ZrO_2$) as a ceramic top-coat material in TBC system, was based on the simple fact that zirconia was one of the several refractory oxides, which could be deposited as thick films by using well-known technology like APS. The identification of yttria as an optimal stabilizer and its contribution required a number of tests, especially under thermal cycling conditions (Clarke & Phillpot, 2005;Stecura, 1985). Final mαaterial marked as 8YSZ shows desired properties such as (Moskal, Rozmysłowska, Gazda & Homa, 2009):

- *Melting point, approx. 2700^{0}C*
- *Very low coefficient of thermal conductivity, approx 2,3 W/mK,*
- *High coefficient of thermal expansion $11 \cdot 10^{-6}\,^0C^{-1}$,*
- *Low density approx. 6,4 g/cm^3*
- *Low Young modulus, (E=50GPa)*
- *High hardness, approx. 14 GPa, (*Moskal, Rozmysłowska, Gazda & Homa, 2009*).*

Despite the widespread use of YSZ layers, their durability is a major problem. These coatings undergo many destruction mechanisms, including: differences in thermal expansion coefficient between

Table 2. Comparison of the coatings obtained by the APS, EB-PVD and SPS method

Technology	APS	EB-PVD	SPS
1. Price and complexity	Cheap and easy technology	Very expensive and complicated technology	Relatively cheap and easy technology
2. Structure	Horizontal	Columnar	Columnar
3. Type of connection between BC and TC	Mechanical bonding	Chemical bonding	Mechanical bonding
4. Thermal Conductivity	Low	High	Very low
5. Oxidation protection	High	Low	Relatively high
6. Lifetime	Short	Very long	Relatively long

Gupta, 2014; Bernard et al., 2017.

the ceramic layer, a bond coat and substrate material, low hot corrosion, CMAS, AVA (artificial volcanic ash) and melted deposits resistance, degradation and phase transformations, changes the microstructure and chemical composition in different coating elements (Wang, 2006).

The consequence of unresolved exploitation problems is the strong emphasis on elaboration new ceramic materials and manufacture new TBC coating with greater durability and higher resistance to destructive processes. The critical point of these searches is to obtain a material with better insulating properties, thermodynamic compatibility with alumina, having no phase transformations in operation temperature, resistance in the external corrosive environment, which in the case of 8YSZ appeared by exhaustion of yttrium oxide with zirconia matrix. This results in displacement zirconium oxide to the area of phase transformations, with monoclinic into tetragonal (Wang, 2006).

Looking for potential new TBC materials, it should be choosing materials from the refractory group. However, there are numerous crystalline structures known in the mineralogical and crystallographic sciences, and each of them can be created with several other elements, there are thousands of possible compounds, which could be considered. Over the last 15 years, primarily four different ceramic material groups: (1) zirconia doped with different rare-earth (RE) cations (defect cluster TBC's) – e.g. ZrO_2-Y_2O_3-Gd_2O_3-Yb_2O_3, (2) perovskites – e.g. (Sr,Ba,Ca)/$SrZrO_3$, $Ba(Mg_{1/3}Ta_{2/3})O_3$ (3) hexaaluminates – e.g. $LaMgAl_{11}O_{19}$, and (4) pyrochlores – e.g. $La_2Zr_2O_7$, have been suggested as promising new top coat materials (Bakan & Vaßen, 2017; Bast & Schumann, 2002; Wu et al., pp. 3031-3035, 2002; Lehmann, Pitzer, Pracht, Vassen & Sover, 2003; Vassen, Cao, Tietz, Basu & Stover, 2000; Wu et al., pp. 3193-3200, 2002; Suresh, Seenivsan, Krishnaiah & Murti, 1997;Vaben, Traeger &Stover, 2004).

In the face of this complexity, initial attempts focused on the study of oxides with structures associated with zirconates (Clarke & Phillpot, 2005). Intensive researches focus on this area especially on oxides type of $RE_2Zr_2O_7$ (RE – rare earth elements) with pyrochlore/fluorite structure, which are characterized by many desirable properties. Belong to them (Clarke & Phillpot, 2005; Moskal, Rozmysłowska, Gazda & Homa, 2009):

Figure 1. Scheme of location operational damage on the rotating blade and distribution of thickness the ceramic coating of APS type

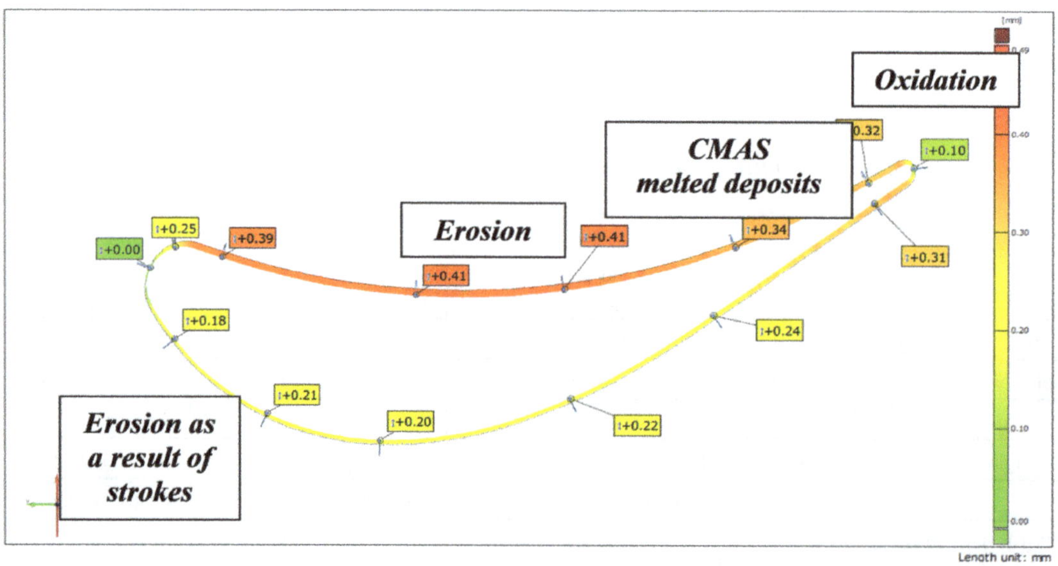

- *Very low coefficient of thermal conductivity, below 2,0 W/mK (from -1,1 W/mK to -1.7 W/mK in temperature between 700-1200⁰C in zirconates of Gd, Eu, Sm, Nd and La).*
- *High coefficient of thermal expansion* $>10\cdot10^{-6}\,{}^{0}C^{-1}$
- *Working temperature do 1600⁰C*
- *Young modulus <250GPa*
- *Hardness >6GPa*
- *Density<7g/cm*^{3} *(*Clarke & Phillpot, 2005; Moskal, Rozmysłowska, Gazda & Homa, 2009*).*

From a morphological point of view of the inner construction of TBC systems, much research has focused on recent years on double ceramic layer (DCL). They are promising both from a point of view thermal shock resistance, as well as, due to its compatibility with the TGO layer. In this issue seems to be interesting to use coatings on the base 8YSZ - $RE_2Zr_2O_7$. YSZ is characterized by good chemical stability in contact with Al_2O_3, whereas $RE_2Zr_2O_7$ provides very low thermal conductivity and much better operational damage resistance (Clarke & Phillpot, 2005).

CHARACTERIZATION OF SAMARIUM ZIRCONATE FEEDSTOCK POWDERS USED TO TBC SYSTEMS DEPOSITION BY APS METHOD

The powders of $Sm_2Zr_2O_7$ and 8YSZ are the basic materials of the analysis presented in this chapter. The powders are intended for plasma spraying with the APS method. The study scope included analysis of surface morphology and inner structure of powders by using the Hitachi 3400N scanning electron microscope. Analysis of chemical composition (EDX – the Noran System 7) and phase composition in the micro-areas (EBSD) was carried out at the same time. Phase composition was also determined with the standard XRD method with quantitative Ritveld calculations. The range of performed investigations covered also basic technological properties characterization that includes density measurement by pycnometer, bulk density measurement by Scott's volume indicator according to PN-EN 23923-2 and flow rate of powder determination using Hall's speed indicator by calculating the time of powder pouring according to PN-EN ISO 4490. Laser diffraction was used for calculation of average size and size distribution of the particles (Kippax, 2005).

The first stage of the study included the assessment of surface morphology of the powders and their inner structure. The analyzed powders were, in all cases, characterized by the structure similar to a spherical one in a predominant part (figure 2 a). The particles slightly deformed with the shape similar to a ball could also be noticed. The single particles, being a conglomerate of a spherical particle and much smaller "satellites" located on their outer surface were also noticed (Figure 2 b). In most cases, the tested powders based on rare earth zirconates were filled completely with a material. However, there were also powders, whose outer shell was filled with much smaller diameter particles.

$Sm_2Zr_2O_7$ type of powder granules was characterized by low surface smoothness. The morphology observed indicates the use of agglomeration process during their production. The initial fine dispersion particles, used for the production of powders, resulted in obtaining the porous and rough outer surface (Figure 2 b). The study of the outer structure showed generally a spherical shape of granules with single deformed particles. The inner structure is made up of an agglomerate of fine polyhedral particles of different shapes and sizes with voids occurring between them, which confirms the earlier observations concerning the production of powders with the agglomeration process. The initial particles of sizes not

Figure 2. a) SEM micro photography of Sm2Zr2O7 type powder with the spherical structure of particles; b) SEM micro photography of 8YSZ type powder with the spherical structure of particles with single conglomerates and with spherical outer shell filled with finer particles

exceeding 1µm were predominant. Analysis of the structure of 8YSZ powders revealed gas atomization as a rout of its production. The inner analysis also showed the presence of "hollow spheres" particles, where only external spherical particle shell occurred (Figure 3). The sieve analysis of zirconate powder is shown in Figure 4 a), where a relatively narrow range of granule's size can be observed.

Technological investigations of powders are related with parameters characterizing the ability of powders for appropriate transport through the inlet system and the gun (taped density and flow rate) are very similar and from APS process satisfying. The results of these measurements are presented in the Table 3.

The detailed chemical composition analysis in the micro-areas of investigated powders showed the high uniformity of chemical composition. No presence of the particles of initial materials for agglom-

Figure 3. Results of sieve analysis for $Sm_2Zr_2O_7$ feedstock powder

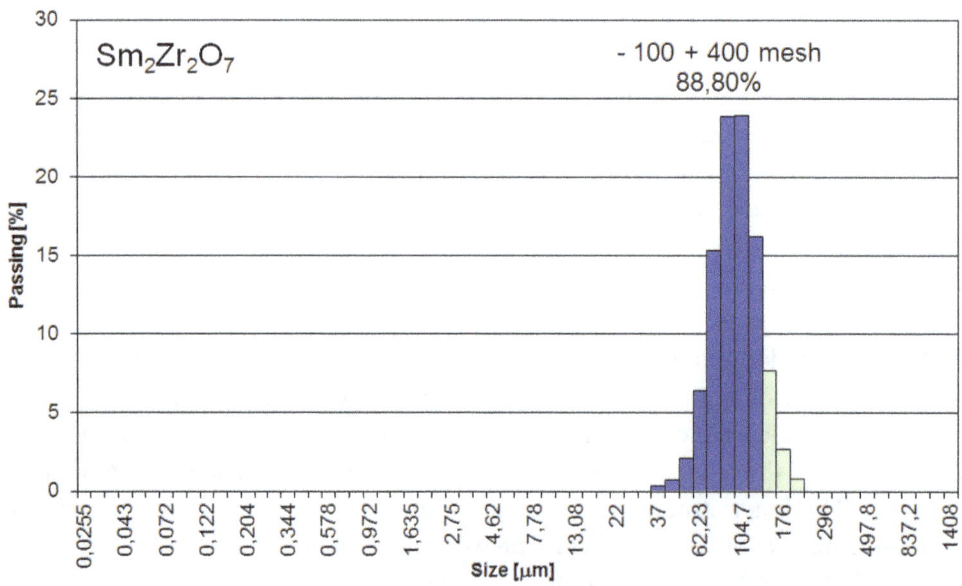

Figure 4. a) Sm2Zr2O7 type powder - results of chemical composition analysis in micro-areas; b) 8YSZ type powder - results of chemical composition analysis in micro-areas

a)

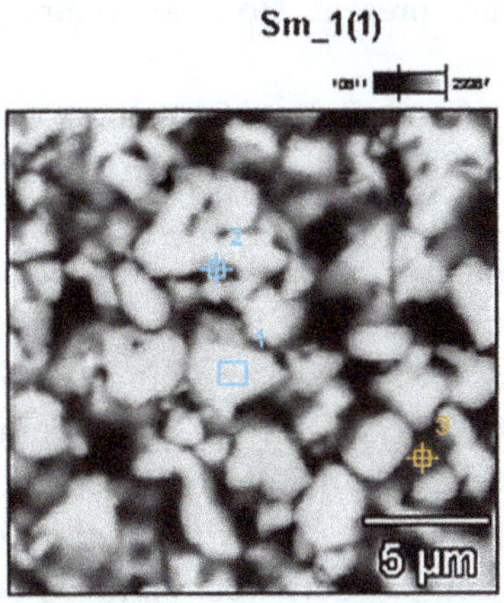

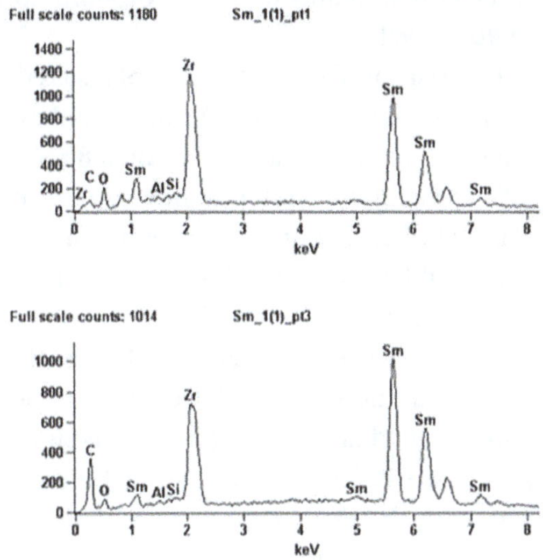

Wt %	Sm	Zr	Al	Si
Pt 1	55.51	44.21	0.15	0.13
Pt 3	64.49	35.13	0.16	0.22

b)

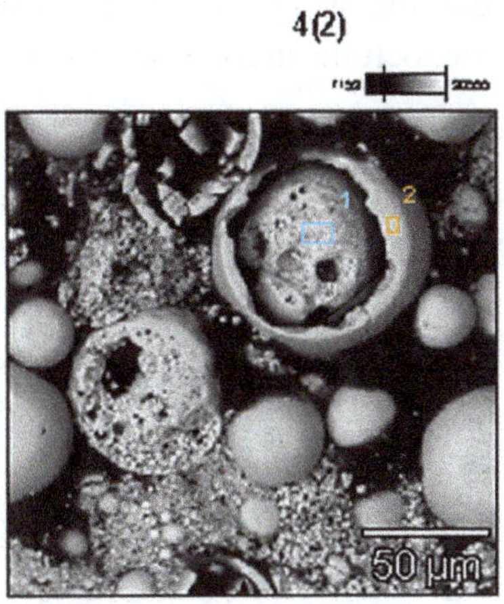

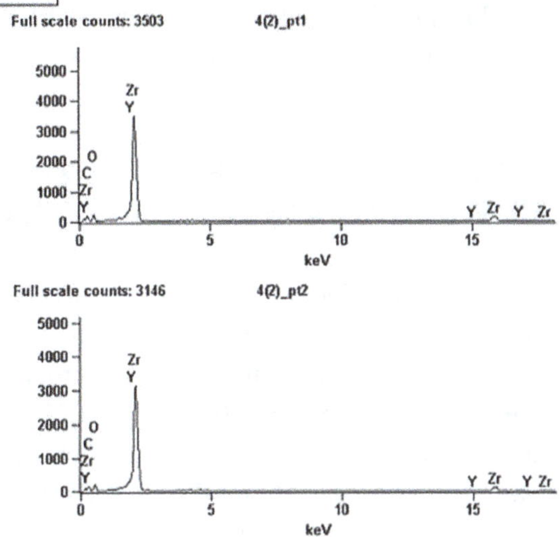

Wt %	Zr	Y
Pt 1	88.23	11.77
Pt 2	89.55	10.45

Table 3. Results of powders' technological properties measurements

Sample	Pycnometric Density (g/cm³)	Taped Density (g/cm³)	Flow Rate (s)
$Sm_2Zr_2O_7$	6,52	1,58	55,13
8YSZ	5,89	2,06	45,54

eration process was noticed. The results of the SEM analysis are presented in Figure 4 a) for $Sm_2Zr_2O_7$ and in Figure 4 b) for 8YSZ. The quantitative content of oxygen was not determined. Apart from main elements, contaminants in a form of aluminum and silicon are also present. However, their content is at very low level.

On the basis of X-ray diffraction of tested powders (presented at figure 5 a) and 5 b)), there was phases composition analyzed as well as the calculations of crystallite size (D) and lattice strain (e) was made. This analysis showed that conventional 8YSZ powder consists dominant tetragonal ZrO_2 phase with ca. 20% of monoclinic zirconia. In the case of samarium zirconate, the pyrochlore lattice was detected. To calculations of crystallite size three methods of approximation were used: Williamson–Hall (Cauchy/Cauchy), Halder–Wagner (Gauss/Gauss) and Cauchy/Gauss described elsewhere (Klug & Alexander, 1974; Mittemeijer & Scardi 2004; Bojarski & Łągiewka, 1995; Olszak & Olszyna, 2004; Tomiczek & Dobrzański, 2013; Kippax, 2005). The results of that analysis were shown in Table 4, where the similar value of crystallite size was detected. This fact is important for thermal conductivity point of view.

More detailed analysis of phase constituent of used to deposition powders were obtained by EBSD analysis. Those results were showed at Figure 6 and 7, where qualitative phase identifications were made.

That analysis revealed that main phase element constituent zirconate powder was $Sm_2Zr_2O_7$ with pyrochlore type of lattice. Additionally, single particles of Sm_2O_3, SmO and cubic ZrO_2 were detected not founded during XRD analysis. In the case of 8YSZ powder, the results of the EBSD analysis were in accordance with XRD results, where only tetragonal and monoclinic zirconia were found.

MICROSTRUCTURAL CHARACTERIZATION OF SAMARIUM ZIRCONATE BASED TBC SYSTEMS

Those studies were conducted on five types of TBC`s (with one reference based on 8YSZ). The first system was monolayer type of ceramic top coat based on $Sm_2Zr_2O_7$. Three of them were so called composite TBC deposited from a mixture of feedstock powders with a varying weight ratio of 8YSZ and $Sm_2Zr_2O_7$ feedstock powders: 50/50, 25/75 and 75/25. And the last one was TBC system of DCL (double ceramic layer) type with outer sublayer based on $Sm_2Zr_2O_7$ and inner sub layer deposited from 8YSZ.

The coatings were sprayed on 40 mm × 20 mm × 2 mm flat sheets of AMS 5599 Ni-based alloy commonly used for the combustion chambers. The surface preparation before the bond-coat spraying included blast-sanding (up to surface roughness Ra~ 4 μm) and degreasing. The Ni-22Cr-10Al-Y bond-coat was APS, its thickness was approximately 125 μm. The $Sm_2Zr_2O_7$ + 8YSZ top coat was obtained using air-plasma-spraying method with double powder feeder system and typical for ceramic powders parameters of process.

The scope of the research covered structural characterization of the ceramic coating via visual inspection, light and electron scanning microscopy (chemical composition, phase composition and crystallographic orientation were characterized using a Hitachi S-3400N microscope equipped with an EDS by Thermo NORAN (System Six) and an EBSD detector INCA HKL Nordlys II (Channel 5)), phase composition analysis using XRD from the surface, internal stress evaluation using $\sin^2\psi$ method, as well as qualitative and quantitative porosity analysis (spherical, horizontal and vertical). The surface morphology was analyzed using laser profilometry technique.

Figure 5. a) $Sm_2Zr_2O_7$ type of powder - results of phase analysis by XRD method; b) 8YSZ type of powder - results of phase analysis by XRD method

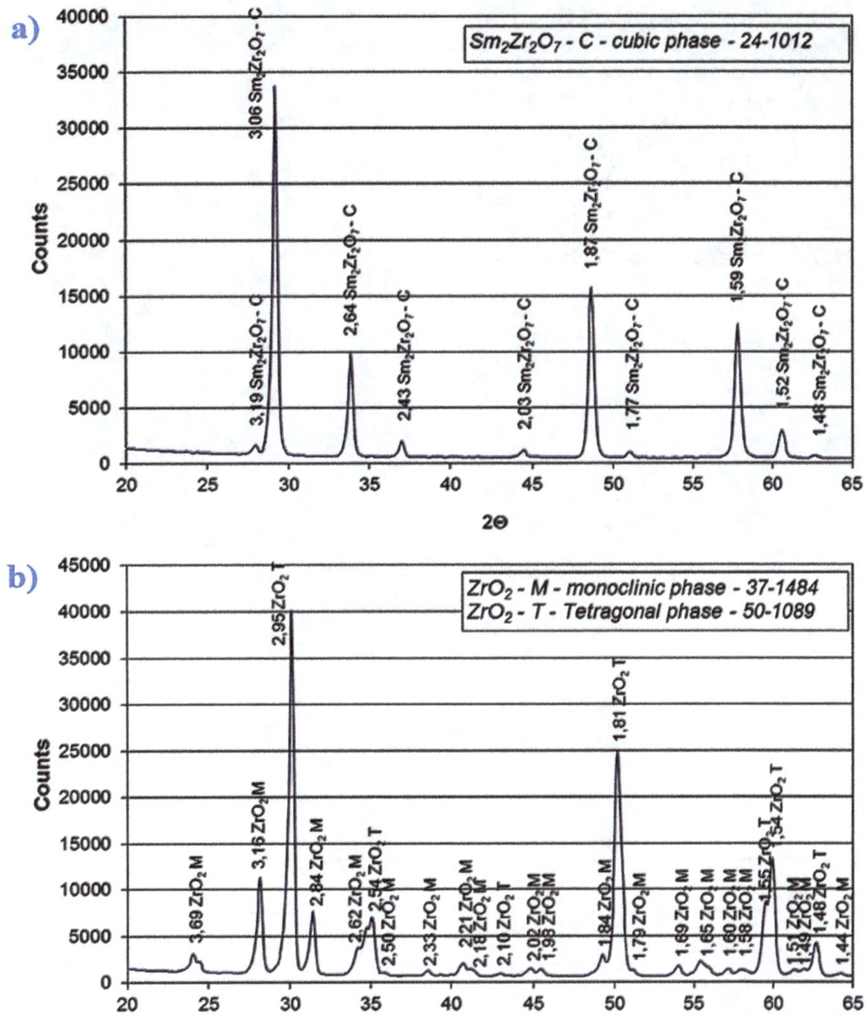

Table 4. The calculated value of crystallite size and lattice strain of analyzed powders

Method	Crystallite Size D (nm)	Lattice Strain e (%)	Pearson Coeff. R
	$Sm_2Zr_2O_7$		
Cauchy/Cauchy	70±11	0,168±0,025	0,95
Gauss/Gauss	42,1±2,2	0,204±0,013	0,96
Cauchy/Gauss	40,2±1,8	0,141±0,016	0,99
	8YSZ		
Cauchy/Cauchy	77±10	0,163±0,016	0,97
Gauss/Gauss	45,6±2,4	0,193±0,011	0,96
Cauchy/Gauss	45,7±2,4	0,146±0,018	0,99

Figure 6. $Sm_2Zr_2O_7$ type of powder: Results of phase analysis by EBSD method

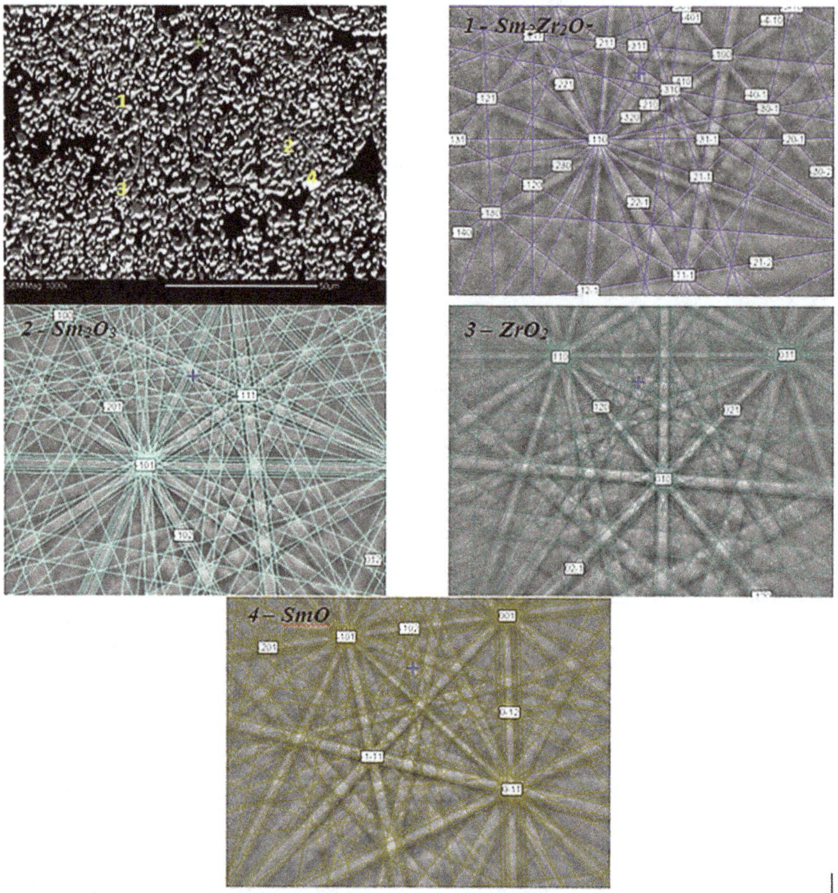

	$Sm_2Zr_2O_7$	*Sm_2O_3*	*ZrO_2*
Name	$Sm_{0,5}Zr_{0,5}O_{1,5}$	Sm_2O_3	O_2Zr
Composition	63,6\18,2\18,2	40\60	66,7\33,3
Laue and space group no.	11, m3m, 225	10, m3, 206	11, m3m, 225
Unit cell lengths	5,27, 5,27, 5,27 Å	10,91, 10,91, 10,91 [Å]	5,09, 5,09, 5,09 [Å]
Unit cell angles	90,0, 90,0 90,0 °	90,0, 90,0 90,0 °	90,0, 90,0 90,0 °
MAD	0,228°	0,275°	0,167°
Spec plane ~	(-632)	(-641)	(-6-41)
	SmO		
Name	SmO		
Composition	50\50		
Laue and space group no.	11, m3m, 225		
Unit cell lengths	4,99, 4,99, 4,99 Å		
Unit cell angles	90,0, 90,0 90,0 °		
MAD	0,228°		
Spec plane ~	(-632)		

The first stage of the study was a visual inspection of sprayed TBC`s. The obtained coatings were characterized by the even distribution of thickness throughout the whole specimens' surface with no thickening near the edges. No signs of macroscopic cracks and delamination in outer layer were observed. All of the coatings were similar in color to each other. The example macro graphs of deposited conventional and composite type of $Sm_2Zr_2O_7$+8YSZ thermal barrier coatings are presented at Figure 8.

Figure 7. 8YSZ type of powder: Results of phase analysis by EBSD method

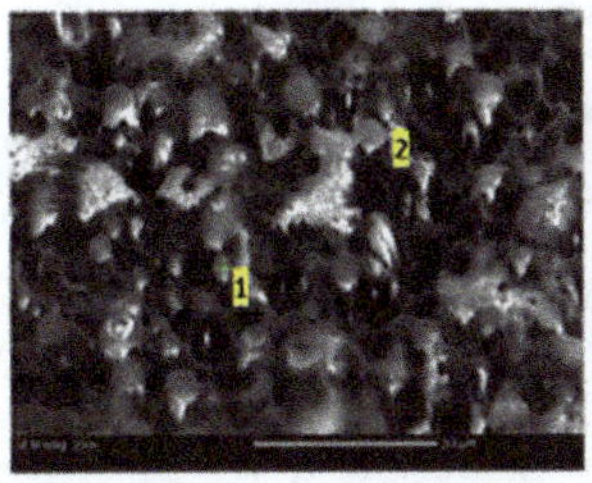

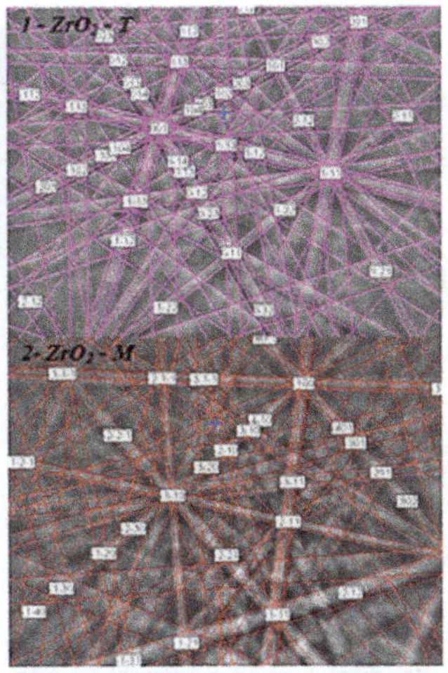

	ZrO_2	ZrO_2
Name	zirconia	Zirconia monoclinic
Composition	66,7/33,3	66,7/33,3
Laue space group no.	5, 4/mmm; 137	2, 2/m; 14
Unit cell lengths	3.51, 3.51, 4.99 Å	5.21, 5.21, 5.31 Å
Unit cell angles	90.0, 90.0, 90.0 °	90.0, 99.2, 90.0 °
MAD	0.264 °	0.433 °
Spac. plane ~	(425)	(-431)

Figure 8. Macro view of deposited TBC systems and surface topography detected by SEM method: Example pictures

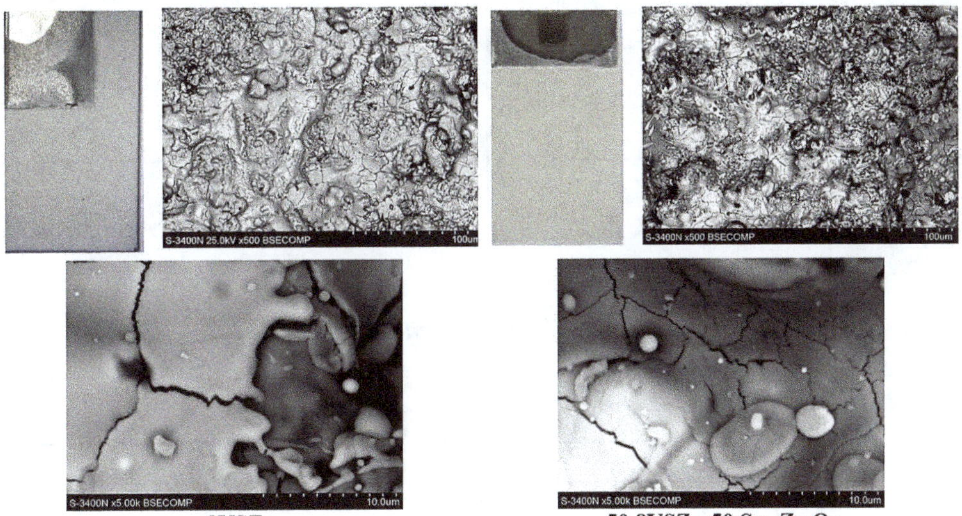

8YSZ **50 8YSZ - 50 $Sm_2Zr_2O_7$**

The roughness of the top surface is typical for the air-plasma sprayed coatings and results from the formation mechanism of the sprayed layer i.e. melting of powder followed by its settling as flattened disk-shaped liquid particles on the substrate surface. The outer layer, independently of the feedstock powder type, is covered with micro-cracks which are due to the rapid transition from liquid to solid state. The presence of these micro-cracks is favorable as it allows for the compensation of undesirable state of stress. The individual droplets of the solidified ceramic were also present on the examined surfaces (Figure 8). The surface topography of obtained coatings and the quantitative results of profilometric studies are presented in Figure 9 and Table 5 and Table 6. From these results, it can be concluded that the roughness parameters for different coatings are similar to each other and their values are typical for the APS process.

Figure 9. 2D isometric view of $RE_2Zr_2O_7$+8YSZ type coatings roughness with distributions of roughness peaks

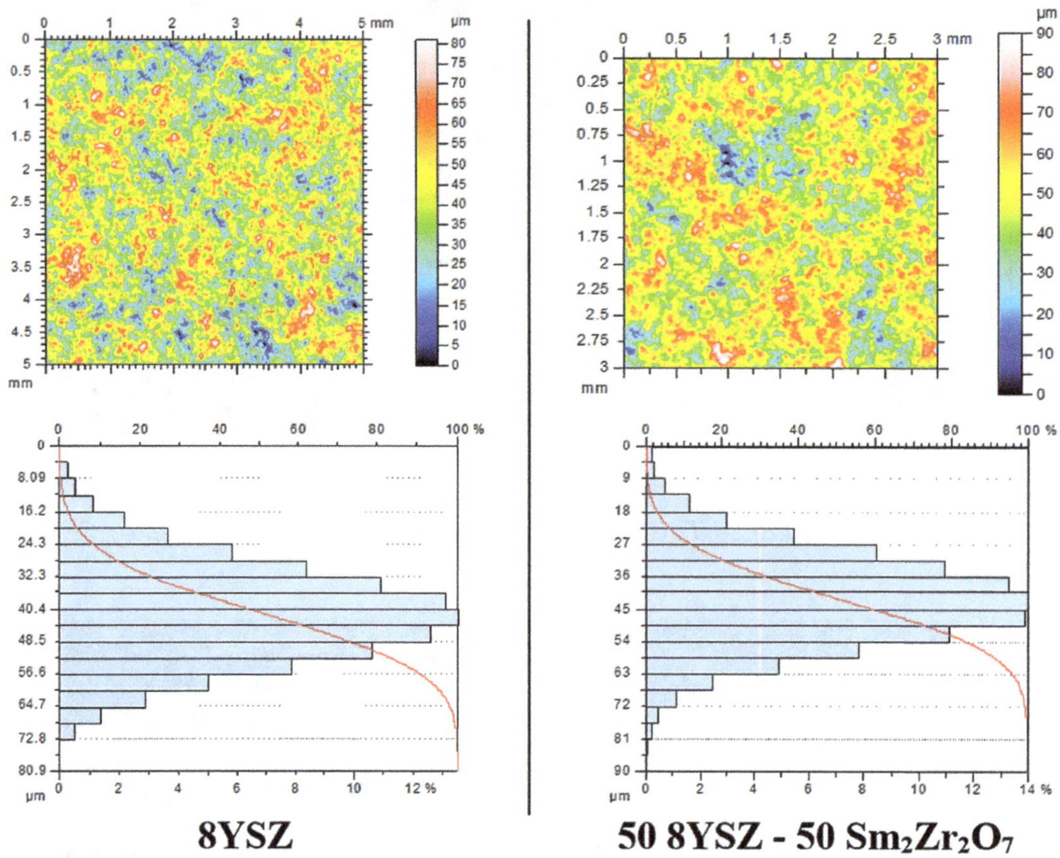

Table 5. Results of quantitative assessment of TBC roughness

	8YSZ	$Sm_2Zr_2O_7$	DCL	25 8YSZ – 75 $Sm_2Zr_2O_7$	50 8YSZ – 50 $Sm_2Zr_2O_7$	75 8YSZ – 25 $Sm_2Zr_2O_7$
R_z (μm)	52,1	34,7	38,1	34,7	35,4	42,7
R_a (μm)	9,5	7,9	9,0	7,0	6,8	6,9
R_q (μm)	11,8	9,5	10,5	8,5	8,4	8,9

Where:
R_z: Maximum peak height
R_a: Mean standard deviation from the centerline
R_q: Mean square deviation from the centerline

Exemplary results of microstructural and phase composition investigations of referential 8YSZ as well as $Sm_2Zr_2O_7$, $Sm_2Zr_2O_7$+8YSZ and DCL $Sm_2Zr_2O_7$/8YSZ type TBC`s are presented in Figure 10 and 11. In the standard quality control of TBC coatings, several factors need to be taken into consideration, including ceramic layer's thickness and its uniformity, porosity (in particular mass porosity, which decreases coating's adherence), the presence of non-melted ceramic particles and impurities.

Table 6. Values of parameters characterizing TBC surface topography

	8YSZ	$Sm_2Zr_2O_7$	DCL	25 8YSZ – 75 $Sm_2Zr_2O_7$	50 8YSZ – 50 $Sm_2Zr_2O_7$	75 8YSZ – 25 $Sm_2Zr_2O_7$
S_{pd} (1/mm^2)	83,3	72,3	59,9	77,6	90,7	61,8
S_{ha} (mm^2)	0,0697	1,101	0,067	0,0796	0,103	0,106
S_{hv} (mm^3)	5,64 x 10^{-6}	7,61 x 10^{-6}	1,45 x 10^{-5}	7,47 x 10^{-6}	1,45 x 10^{-6}	1,26 x 10^{-6}

Where:
S_{pd}: Number density of surface peaks
S_{ha}: Mean surface area of the peak
S_{hv}: Mean volume of the peak

The thickness of analyzed ceramic layers was uniform over the surface, as indicated by the results of measurements in 10 randomly chosen observation areas in each specimen. The morphology of the insulating layers in all cases is typical for APS process with visible pores and microcracks on the lamellar grains. The pores are uniformly distributed on the whole cross-section of the ceramic layer, regardless of the phase composition in particular areas (bright areas are a $Sm_2Zr_2O_7$ phase and darker ones are 8YSZ). No signs of mass porosity were found. The observed porosity can be divided by shape into three groups: spherical, vertical and horizontal (usually classified as cracks).

Due to the fact that two kinds of feedstock powders were used, the phenomena occurring on inter-phase boundaries also need to be investigated. Distinct boundaries between the samarium zirconate and yttria-stabilized zirconia areas indicate that no transient phases were formed (Figure 12).

Figure 10. The microstructure of conventional 8YSZ, new mono-layered Sm2Zr2O7 and DCL Sm2Zr2O7/8YSZ TBC systems

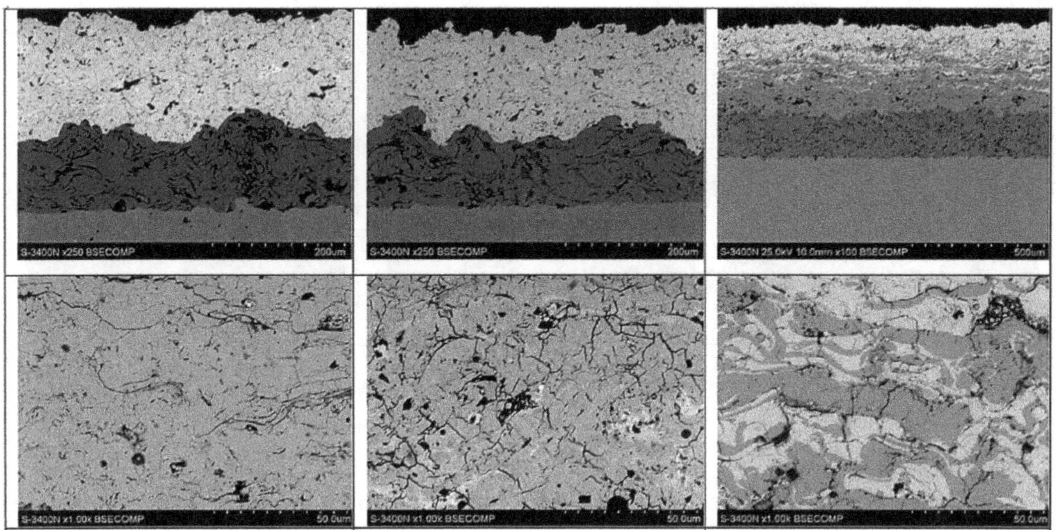

8YSZ	$Sm_2Zr_2O_7$	DCL- 8YSZ /$Sm_2Zr_2O_7$
~ 3 % of monoclinic ZrO_2	100 % of $Sm_2Zr_2O_7$	~ 55 % of 8YSZ

Figure 11. The microstructure of composite $Sm_2Zr_2O_7$ TBC systems

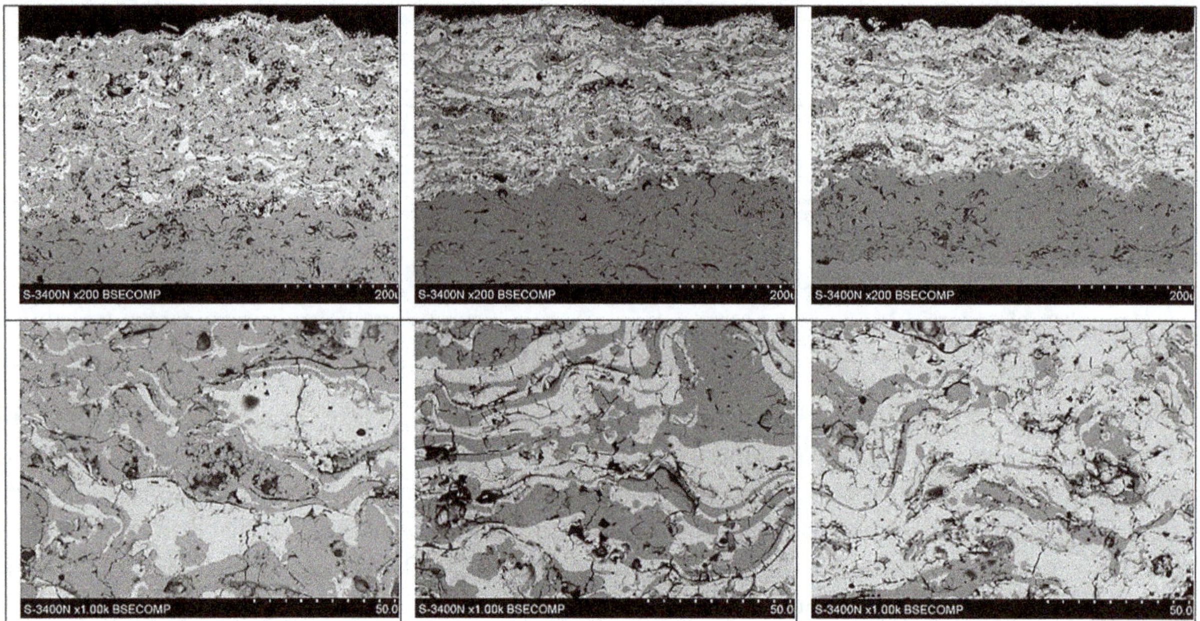

75 8YSZ –25 $Sm_2Zr_2O_7$	50 8YSZ –50 $Sm_2Zr_2O_7$	25 8YSZ –75 $Sm_2Zr_2O_7$
~ 76 % of 8YSZ	~ 55 % of 8YSZ	~ 35 % of 8YSZ

Figure 12. Element distribution in example composite 8YSZ + $Sm_2Zr_2O_7$ TBC

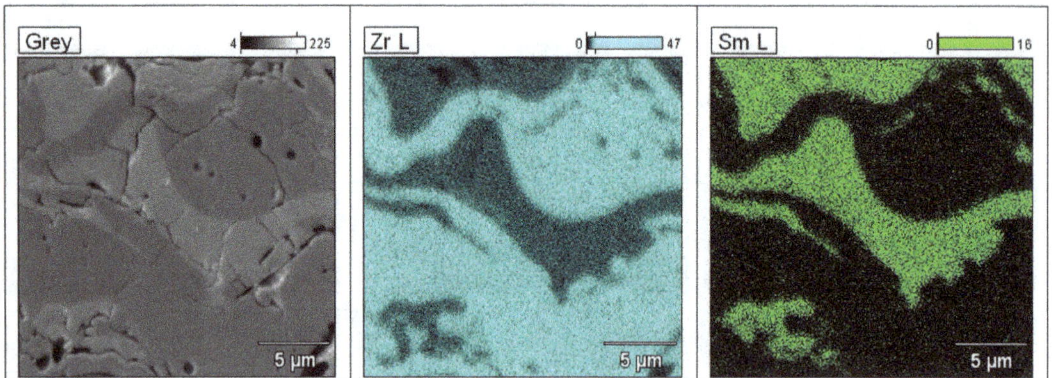

50 8YSZ - 50 $Sm_2Zr_2O_7$

The SEM results are in good agreement with XRD patterns, which proved the presence of tetragonal ZrO_2, a small amount of monoclinic ZrO_2 (the presence of cubic ZrO_2 is also possible) and $Sm_2Zr_2O_7$ with pyrochlore type of lattice. An example of diffraction pattern for composite TBC system of 50/50 type was showed at Figure 13, where those three phases were identified.

Figure 13. XRD pattern of composite 50% 8YSZ + 50% $Sm_2Zr_2O_7$ TBC system

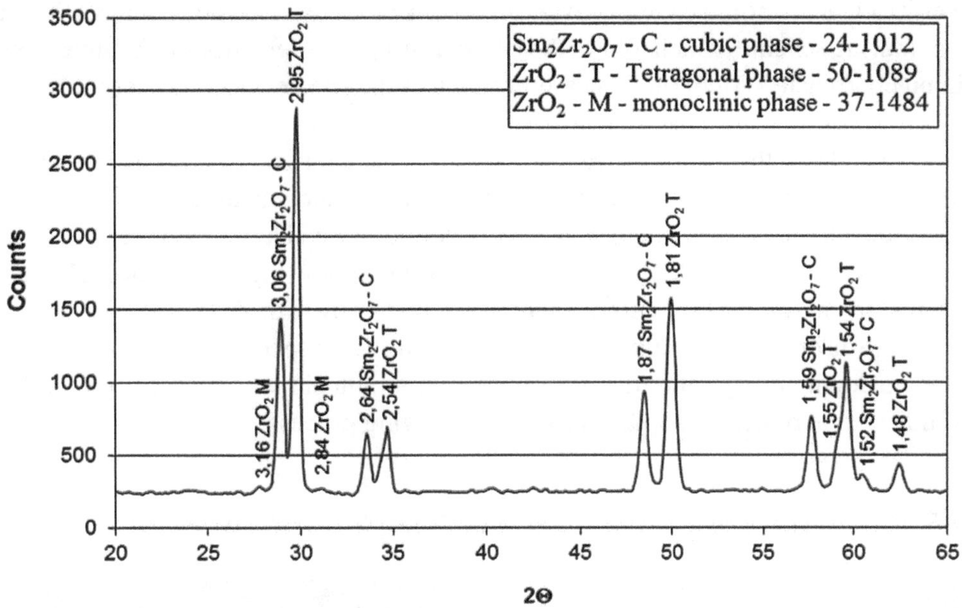

X-ray diffraction allowed also to evaluate the state of residual stresses, which in five cases were compressive and their values were 11,5MPa (8YSZ), 7,3 MPa (DCL - $Sm_2Zr_2O_7$/8YSZ),6,5MPa (50% 8YSZ+50% $Sm_2Zr_2O_7$), 3,5MPa (25% 8YSZ+75% $Sm_2Zr_2O_7$), and 1,8 MPa ($Sm_2Zr_2O_7$) respectively. Only in the case of 75% 8YSZ+25% $Sm_2Zr_2O_7$ coating, the internal stress was tensile and its value was18MPa.

Porosity measurements allowed to extract the spherical and horizontal constituents that have the most significant influence on thermal conductivity. Summary of porosity and TBCs thickness measurements is presented in Table 7. The thickness of the insulating layer in all cases are very similar. However, some differences can be noticed in case of porosity. The highest overall and spherical porosity is exhibited by the coating containing 75% wt. of 8YSZ, it is also the thickest one. The difference in porosity in comparison to the TBC containing 25% wt. of 8YSZ is almost 40%.

Table 7. Results of qualitative and quantitative analysis of porosity in TBC systems

Parameter	8YSZ	$Sm_2Zr_2O_7$	DCL $Sm_2Zr_2O_7$/8YSZ	50 8YSZ – 50 $Sm_2Zr_2O_7$	25 8YSZ – 75 $Sm_2Zr_2O_7$	75 8YSZ – 25 $Sm_2Zr_2O_7$
Total porosity (%)	13,16	11,25	11,03	9,85	7,88	12,69
Vol. of vert. pores (%)	3,10	2,95	2,21	2,09	1,34	2.37
Vol. of horiz. pores (%)	4,94	5,21	4,86	3,40	2,82	4.24
Vol. of spheroid. pores (%)	5,12	3,09	3,96	4,36	3,72	6.08
Av. leng. of horiz. pores (μm)	6,75	6,22	5,98	6,31	5,75	6.05
Av. thick. of horiz. pores (μm)	5,34	5,11	5,08	5,35	5,44	5.14
Lengt.to thick. ratio of horiz. p.	1,26	1,22	1,18	1,18	1.06	1.18
Thick.of ceramic layer (μm)	290	210	290	240	230	250

Detailed analysis of microstructure by EBSD method revealed the orientation and crystallite size in 8YSZ and $Sm_2Zr_2O_7$ areas of the ceramic layer. Figure 14 presents orientation maps with inverse pole figure (IPF) colouring scheme, which allows the crystallographic orientations to be interpreted in terms of the specimen coordinate system (in this case in terms of the specimen Z axis –perpendicular to their surface).

The maps revealed that the top coat ceramic layer was characterized by a homogeneity fine-grained microstructure in the case of $Sm_2Zr_2O_7$ areas and strong non-homogeneity crystallite grain size distribution in the case of 8YSZ areas, where average crystallite grain size was smaller or similar than for $Sm_2Zr_2O_7$ but there was two-modal type of microstructure with few big and much higher level of very small crystallites. In analyzed case of composite 50% 8YSZ+50% $Sm_2Zr_2O_7$ TBC system, the average crystallite size of $Sm_2Zr_2O_7$ was 2,2μm^2 and 2,0μm^2 for 8YSZ respectively. Analysis of inverse pole figure revealed no dominant crystallographic orientation for both phases, what suggest totally randomly process of solidification of ceramic splats during the spraying process.

Figure 14. EBSD results of composite 50% 8YSZ + 50% $Sm_2Zr_2O_7$ TBC system

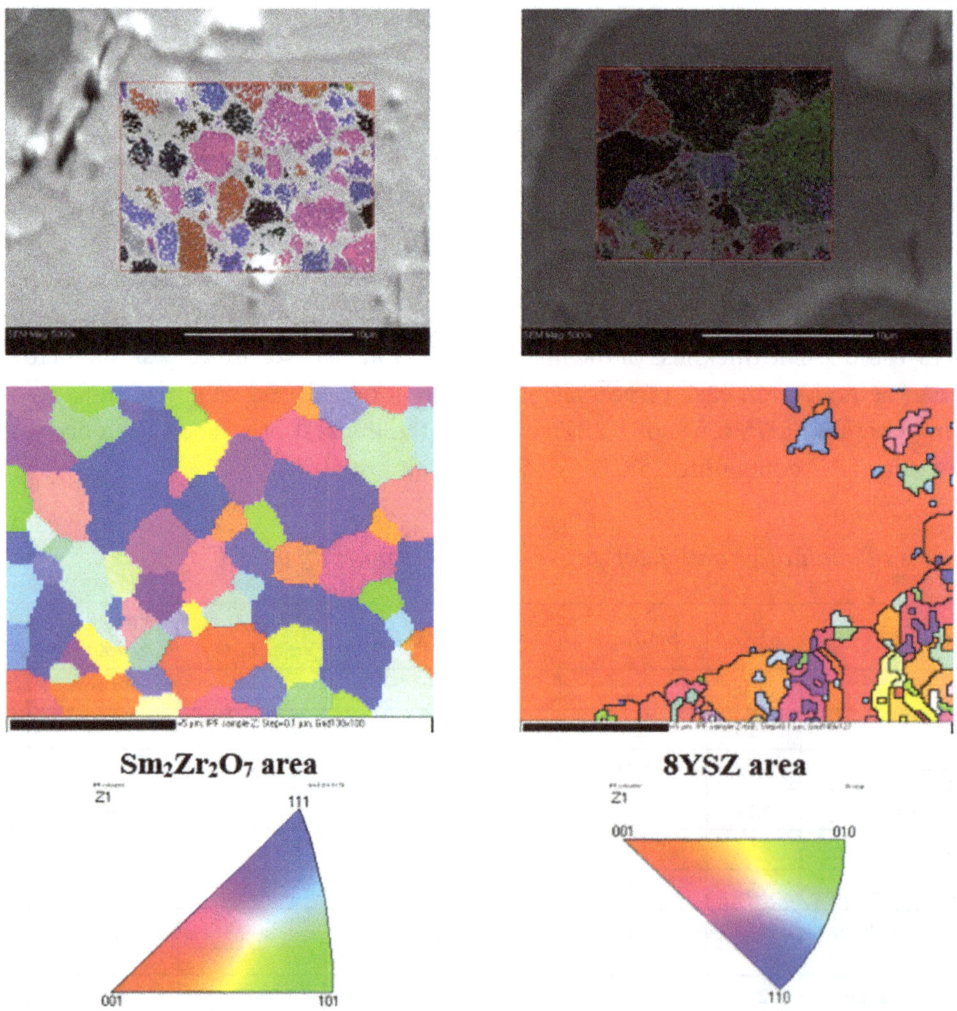

THERMAL DIFFUSIVITY CHARACTERIZATION OF SAMARIUM ZIRCONATE BASED TBC SYSTEMS

The thermal conductivity in multilayer TBC`s is governed by two main factors. The first one is the chemical composition of ceramic powder used for coating depositions (materials factor) and the second one is the microstructure of sprayed layer, including partially-melted ceramic particles, impurities and pores with different size and shape (technological factor). The size, shape, orientation and the share of individual microstructural elements depend on the spraying method, parameters and morphology of feedstock powders (Herman, 1998; Kulkarini et al., 2003; Wang, Kulkarini, Desphande, Nakamura & Herman, 2003; Kittle, 1996).

Considering material factors, it can be assumed that the heat transfer in solids is due to directional movement of phonon gas which carries energy (Kittle, 1996). The phonon mechanism involves the transfer of thermally excited atoms' energy via quantified elastic waves (phonons) between adjacent atoms. The velocity of wave propagation in solids is approximately several km/s and the rate of heat transfer is proportional to the temperature gradient. The efficiency of the heat flow is governed by the thermal resistance which involves dispersion of thermal waves in the crystal structure. One reason for this phenomenon is an enharmonic contribution to the atomic oscillation. The other one is the presence of crystal lattice defects which obstruct the propagation of the thermal wave, including linear defect (e.g. grain or subgrains boundaries) and foreign atoms which disturb the arrangement of atoms in crystal lattice and periodicity of atomic oscillation.

The thermal resistance and hence the heat conductivity of ceramic materials also depends on porosity. Similar to crystal lattice defects, pores are obstruction to the thermal wave propagation and decrease the thermal conductivity. However, in porous materials, besides the conduction, the heat is transferred by radiation and convection. The contribution of radiation in overall heat transfer is proportional to the size of the pores and to the third power of temperature. Thereby at high-temperature large pores promote the increase of the coefficient of heat conductivity, and with the increase of temperature the enhance heat transfer. On the other hand, small pores hinder the heat transfer independently of temperature (Herman, 1998; Kulkarini et al., 2003; Wang, Kulkarini, Desphande, Nakamura & Herman, 2003).

From the technological factors point of view, the heat conductivity of TBCs depend on porosity, grain size and the amount of impurities in the insulating layer. Considering the porosity of ceramic layer, it can be assumed that the decrease in heat conductivity is related mostly to the volume fraction and size of pores. The empirical formula for the correlation between porosity and heat transfer does not include the diversity of pores morphology which has an essential impact on insulating properties plasma-sprayed TBC`s. The influence of spherical and horizontal pores on the heat transfer (assuming that the influence of vertical porosity is negligible) is given by formula (3) (Klemens, 1991). It does not include the contribution of radiation heat transfer. The formula is based on the two component functions. The first one is Maxwell's equation considering mutually isolated pores with spherical shape (P_s) (Jadhow et al., 2006):

$$f_1\left[P_s\right]=1-\frac{3}{2}P_s \tag{1}$$

The second one takes into account the horizontal pores (P_h) and was formulated by Kachanov (Shafiro & Kachanov, 2000; Seviostanov & Kachanov, 2001):

$$f_2[P_h] = 1 - \left(\frac{2P_h}{\pi}\right)\left(\frac{l}{t}\right) \tag{2}$$

where:

l/t = Average length to thickness of the horizontal pores ratio.

Thereby, the effective value of heat conductance λ_{dens} can be given by the formula (Jadhow et al., 2006):

$$\frac{\lambda_{cot}}{\lambda_{dens}} = \frac{1}{2}\left\{f_2\left[\frac{P_h}{1-P_s}\right]xf_1\left[P_s\right] + f_1\left[\frac{P_s}{1-P_h}\right]xf_2\left[P_h\right]\right\} \tag{3}$$

where:

P_h = Horizontal pores fraction,
P_s = Spherical pores fraction,
f_1 = Function expressed by Eq. 2 formula,
f_2 = Function expressed by Eq. 2 formula.

The accuracy of the last formula has been experimentally confirmed in investigations concerning modeling of various shaped pores on the value of thermal conductivity of plasma sprayed ceramic layers (Jadhow et al., 2006; Maxwell, 1904; Cernuschi, Ahmaniemi, Vuoristo & Mantyla, 2004).

Thermal diffusivity and thermal conductivity investigation results for feedstock powders and TBC`s systems are presented in Figure 15. The results for TBC were obtained applying double layer model taking into account thermal resistance (Proteus® by Netzsch software) and were compared with results for 8YSZ. The basic measurement was made for thermal diffusivity. The value of thermal conductivity was calculated on the basis of specific heat, density and coefficient of linear expansion as a function of temperature. It should be noted that all composite coatings exhibit better insulating properties than a conventional 8YSZ coating. These results were adjusted using Eq. 3 and on that basis the values of effective thermal conductivity independent of porosity for insulating layers were determined.

Obtained data showed that the lowest thermal conductivity is the feature of $Sm_2Zr_2O_7$ included TBC systems of each type in comparison to conventional 8YSZ coating. The decreasing ranking of analyzed systems was as follow: -

- Monolayered 8YSZ and $Sm_2Zr_2O_7$,
- Composite systems of 25 8YSZ+75 $Sm_2Zr_2O_7$ and 50 8YSZ+50 $Sm_2Zr_2O_7$ type,
- DCL- $Sm_2Zr_2O_7$/ $Sm_2Zr_2O_7$+8YSZ/8YSZ coating
- Composite systems of 75 8YSZ+25 $Sm_2Zr_2O_7$.

The lowest thermal conductivity in the practically whole range of test temperature was exhibited by 75% wt. 8YSZ + 25% wt. $Sm_2Zr_2O_7$ coating system, what is unexpected results.

Figure 15. Thermal diffusivity, thermal conductivity and effective thermal conductivity of $Sm_2Zr_2O_7$ based TBC's system

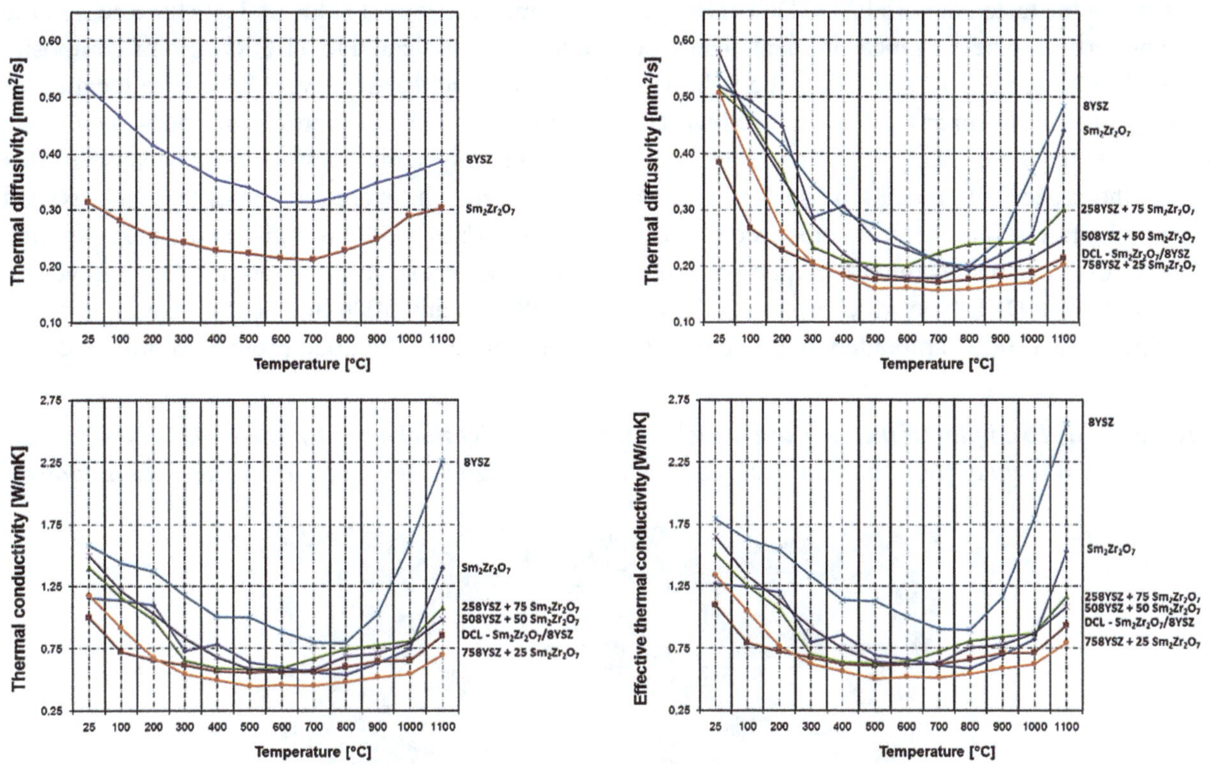

This difference can be due to some internal cracks in the ceramic layer not detected during microstructural investigations. The expected order was the lowest thermal conductivity should be the feature of DCL $Sm_2Zr_2O_7$/ $Sm_2Zr_2O_7$+8YSZ/8YSZ coating. It should be related with increasing content of samarium zirconate phase in the coatings as well as increasing the total length of interphase boundaries of $Sm_2Zr_2O_7$/8YSZ type. Due to a similar level of porosity, there is no changes in the observed order of effective thermal conductivity.

OXIDATION RESISTANCE AND TGO ZONE MORPHOLOGY OF SAMARIUM ZIRCONATE BASED TBC SYSTEMS

The last step of investigations was related with durability characterization during static oxidation test at a temperature of 1100°C. This analysis was related mainly with TGO (thermally growth oxides) thickness measurement on each sample after 2, 10, 48, 175 and 500 h of deposition. Based on the changes of TGO zone thickness, were created graphs of the kinetics of their growth in time. It was important to identify the best checking up material in oxidation environment. Because the surface area between BC and TBC is characterized by significant level roughness (technological requirement), and in consequence, the strong differences in TGO thickness were observed. The measurements of TGO thickness were performed in two places i.e. at the vertices of BC layer and the bottom. The places of measurement are presented in

Figure 16 a) and the final results of this analysis were showed at figure 16 b). The data obtained for 8YSZ coating are very similar to those for DCL coating, and only those were showed at figure 16 b).

Among the tested materials, the DCL TBC system exhibits the lowest value of TGO thickness in the case of vertices places of measurement. The second place was dedicated to TBC system of composite 75 8YSZ+25 $Sm_2Zr_2O_7$ type where the volume of 8YSZ ceramic was relatively high. The location of other system was similar tendency related with 8YSZ phase fraction. The vertices of bond-coat profile roughness are the critical place from final durability of TBC systems point of view because in this places the process of cracking nucleation and growth is started. These are the places to which oxygen arrives the fastest and therefore observed the most rapid growth of TGO layer and in the next step, the highest level of tensile strength can be observed. The best behavior from this point of view revealed TBC system of DCL type which average thickness after 48, 175 and 500h in oxidation environment at a temperature 1100°C amounted to 2,36 μm, 3,73 μm and 4,63 μm. The good protection provided also

Figure 16. a) Examples of the measurements places of the TGO thickness; b) The TGO zone thickness in the area of vertices (left), bottom (right) of bond-coat roughness as well as an average value (down).

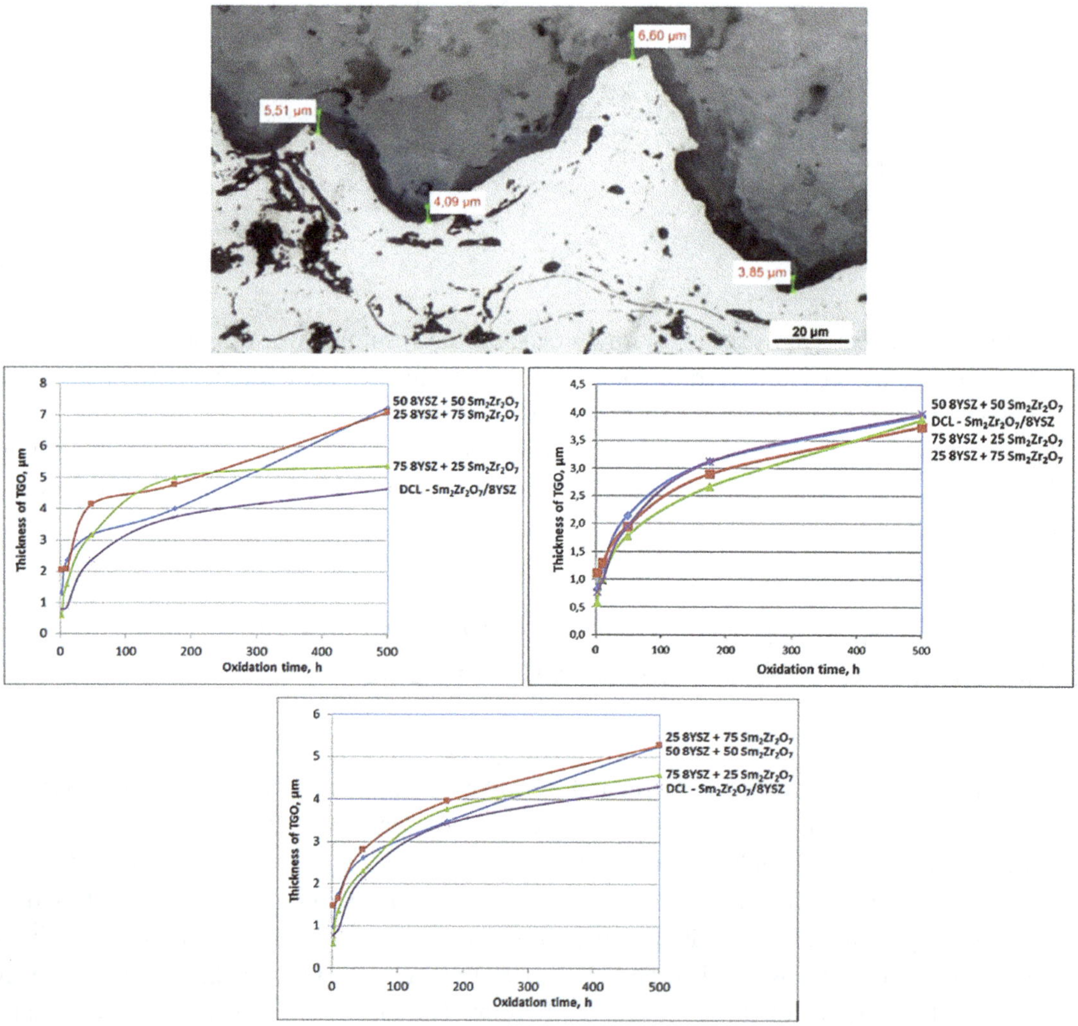

the coating marked as 75 8YSZ + 25 $Sm_2Zr_2O_7$, with an average thickness measured at the same time intervals and places (vertices) at 3,15 μm, 4,98 μm and finally 5,37 μm. Despite the initial high-speed growth TGO layer, that after 100h it were shown, like in the case double layer coating, the logarithmic course of the oxidation kinetics. In the case of coatings containing larger amounts of $Sm_2Zr_2O_7$ ceramic, that although initially the logarithmic character of the oxidation kinetics course, already after 48 hours it is able to observe change the kinetics course in the direction of linear course. TGO thickness measurements into the samples 50 8YSZ + 50 $Sm_2Zr_2O_7$ and 25 8YSZ + 75 $Sm_2Zr_2O_7$, had after 500 h of annealing in oxidation environment, respectively average thickness 7,23 μm and 7,08 μm respectively. Observed differences are related with phases constituent of TGO zone in this area, where the formation of more complex oxides than alumina was usually observed for example oxides of spinel type. The situation is looking somewhat different in the case of measurements carried out at the bottom of bond-coat profile roughness. The course of the TGO layer growth in all tested samples has similar to each other logarithmic character of the oxidation kinetics course. Differences between measurements of TGO thickness are within the range of the standard deviation. It can`t indicate the superiority any of the materials. Measurements carried out after 48, 175 and 500h are within the range of respectively 1,77-2,13 μm, 2,66-3,12 μm and 3,75-3,97 μm. This is the effect of similar phase constituent of TGO zone in this area, where the Al_2O_3 oxide is usually typical structural elements. In consequence, the local thickness of this zone is very similar. Presented analysis showed that application of $Sm_2Zr_2O_7$ as structural elements of deposited thermal barrier coatings has a strong influence on overall TGO thickness increasing in the area of vertices of bond-coat profile roughness. From the other site high level of zirconate ceramic in top-coat due to spallation effect of TBC system as was presented at Figure 17.

This negative effect of spallation is related primary with a much higher thickness of TGO zone and in the second step with the lower compatibility of this compound with Al_2O_3, and formation in consequence complexed oxides of rare elements with alumina. Additionally, in the ceramic, near the TGO layer occur yttrium diffusion from 8YSZ, therefore it is able to observe, with using EBSD locally occurring tetragonal

Figure 17. Spallation effect in $Sm_2Zr_2O_7$ TBC monolayer system

and monoclinic varieties of zirconium oxide. In unstabilized zirconium oxide, occur phase transformation which in the temperature range of operation, causes a significant decrease their durability. As given that this is a martensitic transformation it involves with change a volume of zirconium oxide, which results in formation a significant stress inside the material (Zhao et al., 2013; Kawasaki & Watanabe, 2002).

The cross-section micrographs of the DCL TBC system after annealing in 1100°C by 500h in the oxidizing environment are shown in Figure 18. Although after oxidation process in double ceramic layered TBC`s was observed growth of TGO, that there were no alarming signs of damage the coatings. The TBC layer has a compact character i.e. there were no there were no spallation areas as well as there was no notice the presence of large horizontal cracks. Both the $Sm_2Zr_2O_7$ layer and 8YSZ layer, characterized by the presence of porosity and tiny cracks Figure 18.

The DCL TBC sample covered with 8YSZ ceramic at the substrate and the $Sm_2Zr_2O_7$ as an outer layer demonstrated the smallest increase of TGO layer. The result was achieved thanks good thermochemical compatibility of 8YSZ with Al_2O_3 and better blockade of oxygen diffusion by yttria stabilized zirconia layer. In the case of monolayer TBC system with samarium zirconate this effect was much lower due to a high level of internal defects promoted the strong effect of oxygen diffusion (but lowering thermal conductivity). In consequence higher thickness of TGO zone was observed especially in the area of vertices of bond-coat profile roughness.

Composites systems of TBC samples didn't show such high resistant to working at 1100°C. Already at the samples annealed by 48h it can be observed clearly visible horizontal cracks, leading to final delamination (Figure 19). Among the coating prepared by spraying mixtures of two ceramics, the best high-temperature corrosion resistant in the oxidizing environment has demonstrated 75 8YSZ + 25 $Sm_2Zr_2O_7$ coating (Figure 20), where after 500 h of exposure no significant effects of destruction were observed.

Figure 18. The cross-section micrographs of the DCL type of TBC system after annealing in 1100°C by 500h a) an entire layer view, x80 b) areas with visible micro-cracks, pores and non melted particles zone, x500

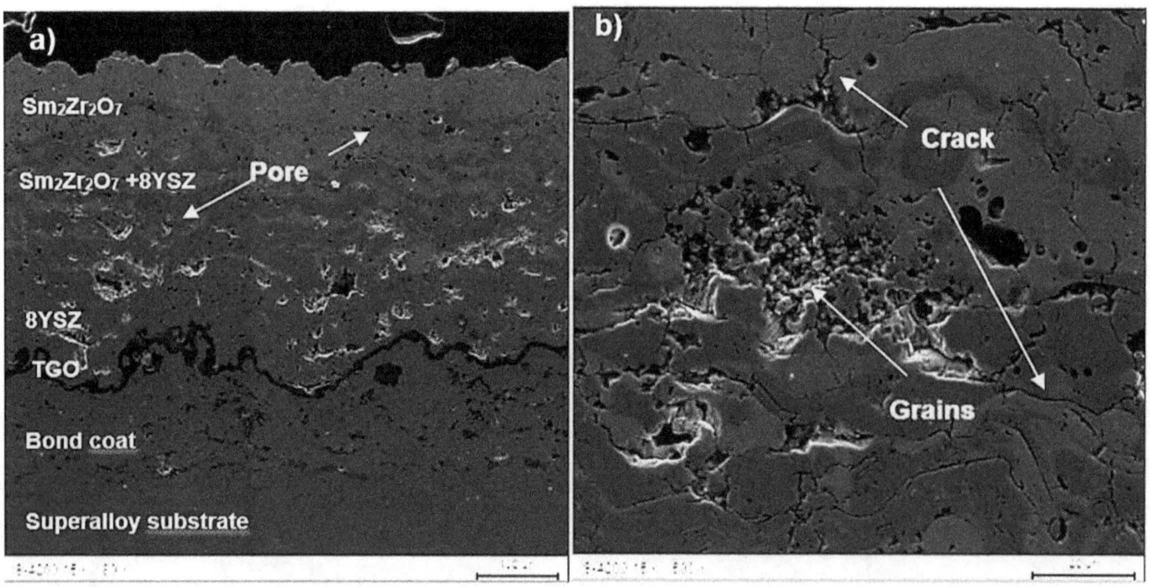

Figure 19. The cross-section micrographs of the 50 8YSZ + 50 $Sm_2Zr_2O_7$ TBC system after oxidation test in 1100°C by 48h

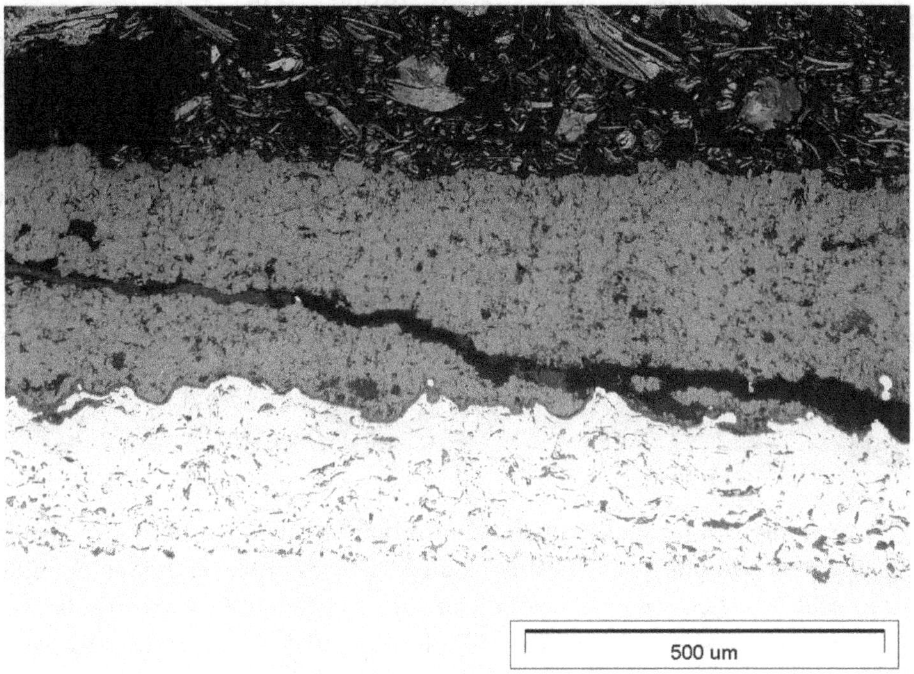

Figure 20. The cross-section micrographs of the 75 8YSZ + 25 $Sm_2Zr_2O_7$ TBC system after oxidation test in 1100°C by 500 h

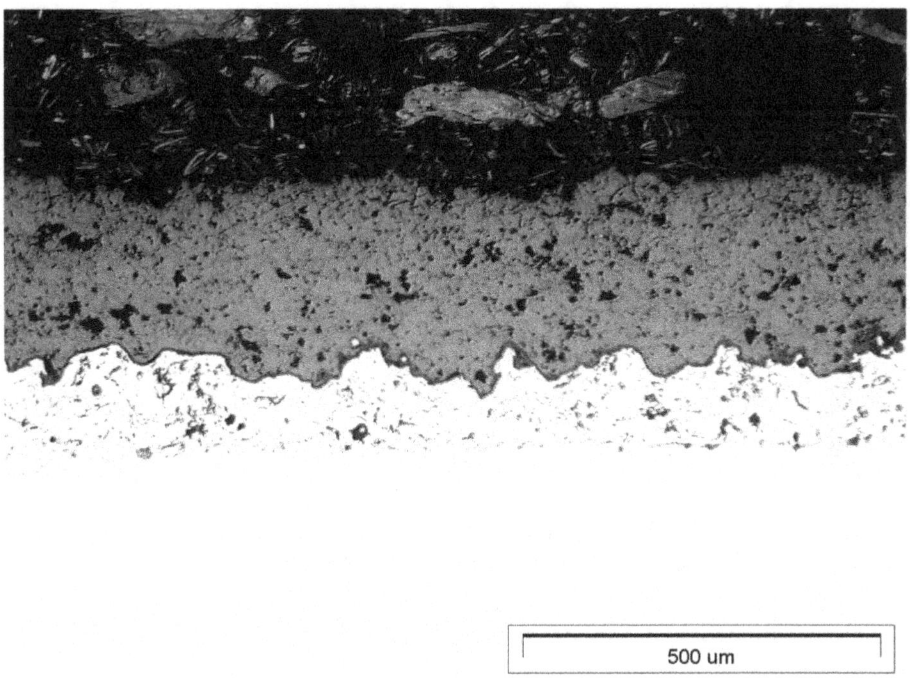

Results of the investigations on high-temperature corrosion resistant in oxidizing environment of different types of TBC systems based on samarium zirconates ceramic showed the superiority double layered TBC coating over the mixtures of two ceramics (8YSZ and $Sm_2Zr_2O_7$) in composites variant of sprayed TBC systems as well as the monolayer TBC systems based on $Sm_2Zr_2O_7$. The DCL gradient coatings due to its construction have a lot of advantages. They combine the advantages of two types ceramics: 8YSZ by excellent protection against high-temperature oxidation and thus reducing the speed of the TGO layer growth and in consequence it thickness, as well as keeping superior thermochemical resistance, and $Sm_2Zr_2O_7$ which has much lower thermal conductivity, and better resistance to AVA, CMAS or liquid salt deposits. These results confirm other studies, as well as literature data e.g. (Poerschke & Levi, 2015; Aygun, 2008; Drexler, 2011; Schulz & Braue, 2013; Drexler, Ortiz & Padture, 2012; Krämer, Yang & Levi, 2008; Levi, Hutchinson, Vidal-Sétif & Johnson, 2012; Drexler et al., 2011; Gledhill et al, 2011, Xu et al., 2010; Habibi, 2014).

CONCLUSION

DCL structure with gradient zone shows much better properties than currently use insulation layer of 8YSZ. As far as the protection of the TGO zone is on a comparable level, that applying the proposed solution allows to achieve much lower thermal diffusivity coefficient (approximately 30% lower) and lower thermal conductivity coefficient (approx. 40% lower, and 45% lower for effective thermal conductivity coefficient). The presence of 8YSZ zone next the bond coat ensure good oxidation protection, also slow and uniformity increase of TGO.

ACKNOWLEDGMENT

The research has been supported by National Science Centre within Sonata scheme, under contract UMO-2016/21/D/ST8/01687.

REFERENCES

Aygun, A. (2008). *Novel thermal barrier coatings (TBC) that are resistant to high temperature attack by CaO-Mgo-Al_2O_3-SiO_2 (CMAS) glassy deposits*. [Dissertation Doctor of Philosophy]. The Ohio State University, Columbus, OH.

Bakan, E., & Vaßen, R. (2017). Ceramic Top Coats of Plasma-Sprayed Thermal Barrier Coatings: Materials, Processes, and Properties. *Journal of Thermal Spray Technology*, *26*(6), 992–1000. doi:10.1007/s11666-017-0597-7

Bast, U., & Schumann, E. (2002). Development of Novel Oxide Materials for TBCs. *Ceramic Engineering and Science Proceedings*, *23*, 525–532. doi:10.1002/9780470294758.ch58

Bernard, B., Quet, A., Bianchi, L., Joulia, A., Malie, A., Schick, V., & Remy, B. (2017). Thermal insulation properties of YSZ coatings: Suspension Plasma Spraying (SPS) versus Electron Beam Physical Vapor Deposition (EB-PVD) and Atmospheric Plasma Spraying (APS). *Surface and Coatings Technology*, *318*, 122–128. doi:10.1016/j.surfcoat.2016.06.010

Bojarski, Z., & Łągiewka, E. (1995). *Rentgenowska analiza strukturalna*. Katowice: Wydawnictwo Uniwersytetu Śląskiego.

Cao, X. Q., Vassen, R., & Stoever, D. (2014). Ceramic materials for thermal barrier coatings. *Journal of the European Ceramic Society*, *24*(1), 1–10. doi:10.1016/S0955-2219(03)00129-8

Cernuschi, F., Ahmaniemi, S., Vuoristo, P., & Mantyla, T. J. (2004). Modelling of thermal conductivity of porous materials: application to thick thermal barrier coatings. *Journal of European Ceramic Society, 24*, 2657-2667.

Cernuschi, F., Bianchi, P., Leoni, M., & Scardi, P. (1999). Thermal diffusivity microstructure relationship in Y-PSZ thermal barrier coatings. *Journal of Thermal Spray Technology*, *8*, 102-109.

Charewicz, W. (1990). *Pierwiastki ziem rzadkich: surowce, technologie, zastosowanie*. Warszawa: Wydawnictwo Naukowo-Techniczne.

Clarke, D. R., & Phillpot, S. R. (2005, June). Thermal barrier coating materials. *Materials Today*, *8*(6), 22–29. doi:10.1016/S1369-7021(05)70934-2

Curry, N. (2014). Design of Thermal Barrier Coatings. *PhD Thesis Production Technology 2014 No.3*, Trollhattan Sweden: University West.

Drexler, J. M. (2011). *Thermal barrier coatings resistant to glassy deposits*. [Dissertation Doctor of Philosophy]. The Ohio State University, Columbus, OH.

Drexler, J. M., Gledhill, A. D., Shinoda, K., Vasiliev, A. L., Reddy, K. M., Sampath, S., & Padture, N. P. (2011). Jet Engine Coatings for Resisting Volcanic Ash Damage. *Advanced Materials*, *23*(21), 2419–2424. doi:10.1002/adma.201004783 PMID:21484889

Drexler, J. M., Ortiz, A. L., & Padture, N. P. (2012). Composition effects of thermal barrier coating ceramics on their interaction with molten Ca–Mg–Al-silicate (CMAS) glass. *Acta Materialia*, *60*(15), 5437–5447. doi:10.1016/j.actamat.2012.06.053

Gledhill, A. D., Reddy, K. M., Drexler, J. M., Shinoda, K., Sampath, S., & Padture, N. P. (2011). Mitigation of damage from molten fly ash to air-plasma-sprayed thermal barrier coatings. *Materials Science and Engineering A*, *528*(24), 7214–7221. doi:10.1016/j.msea.2011.06.041

Gupta, M. (2014). Design of Thermal Barrier Coatings. A modelling approach. *PhD Thesis Production Technology 2014 No.3*, Trollhattan Sweden: University West.

Habibi, M. H. (May 2014). *Hot corrosion behaviour of new candidates for thermal barrier coatings application in turbine simulated environments* [Dissertation]. Louisiana State University.

Herman, H. (1998). Plasma sprayed coatings. *Scientific American,* 259, 112-117.

Hurst, C. (March 2010). China's Rare Earth Elements Industry: What Can the West Learn? *Institute for the Analysis of Global Security (IAGS)*. Retrieved from http://fmso.leavenworth.army.mil/documents/rareearth.pdf

Jadhav, A.D., Padture, N.P., Jordan, E.H., Gell, M., Miranzo, P., & Fuller, Jr., E.R. (2006). Low thermal conductivity plasma-sprayed thermal barrier coatings with engineered microstructures. *Acta Materiallica, 54*, 3343-3349.

Kawasaki, A., & Watanabe, R. (2002). Thermal fracture behavior of metal/ceramic functionally graded materials. *Engineering Fracture Mechanics, 69*(14-16), 1713–1728. doi:10.1016/S0013-7944(02)00054-1

Kippax, P. (2005). Measuring particle size using modern diffraction techniques. *Paint & Coatings, 8.*

Kittle, C. (1996). *Introduction to Solid State Physics*. New York: Wiley.

Klemens, P. G. (1991). Thermal conductivity of inhomogeneous media. *High Temperatures-High Pressures, 23*, 241-248.

Klug, H. P., & Alexander, L. E. (1974). *X-ray Diffraction Procedures for Polycrystalline and Amorphous Materials* (2nd ed.). New York: Wiley.

Klupa, A. (2012). Cenniejsze niż złoto? Metale ziem rzadkich w światowej strategii gospodarczej. *Przegląd strategiczny, 1*, 239-250.

Krämer, S., Yang, J., & Levi, C. G. (2008). Infiltration-inhibiting reaction of gadolinium zirconate thermal barrier coatings with CMAS melts. *Journal of the American Ceramic Society, 91*(2), 576–583. doi:10.1111/j.1551-2916.2007.02175.x

Kulkarni, A., Wang, Z., Nakamura, T., Sampath, S., Goland, A., Herman, H., … Steinbrecht, R.W. (2003). Comprehensive microstructural characterization and predictive property modeling of plasma-sprayed zirconia coatings. *Acta Materiallia, 5*(1), 2457-2475.

Lehmann, H., Pitzer, D., Pracht, G., Vassen, R., & Stover, D. (2003). Thermal Conductivity and Thermal Expansion Coefficients of the Lanthanum Rare-Earth-Element Zirconate System. *Journal of the American Ceramic Society, 86*(8), 1338–1344. doi:10.1111/j.1151-2916.2003.tb03473.x

Levi, C. G., Hutchinson, J. W., Vidal-Sétif, M.-H., & Johnson, C. A. (2012). Environmental degradation of thermal-barrier coatings by molten deposits. *MRS Bulletin, 37*(10), 932–941. doi:10.1557/mrs.2012.230

Malony, M. J. (2001). U.S. Patent 6, 284,323.

Maxwell, J. C. (1904). *A treatise on electricity and magnetism*. Oxford: Clarendon Press.

Mittemeijer, E. J., & Scardi, P. (2004). *Diffraction Analysis of the Microstructure of Materials*. Berlin: Springer-Verlag. doi:10.1007/978-3-662-06723-9

Moskal, G., Rozmysłowska, A., Gazda, A., & Homa, M. (2009). Wybrane termofizyczne właściwości proszków cyrkonianowych na bazie pierwiastków ziem rzadkich typu RE2Zr2O7 (RE- Gd, La, Sm, Nd) przeznaczonych do natryskiwania cieplnego powłokowych warstw barierowych. *Prace Instytutu Odlewnictwa*, Tom XLIX. *Zeszyt, 4*, 15–26.

Olszak, D., & Olszyna, A. (2004). Określanie wielkości krystalitów i odkształceń sieciowych w nanokompozycie NiAl-Al_2O_3 na podstawie poszerzenia rentgenowskich linii dyfrakcyjnych. *Kompozyty, 4/11*, 284–288.

PN-EN 23923-2:1998: Proszki metaliczne -- Oznaczenie gęstości nasypowej -- Metoda wolumetru Scotta.

PN-EN ISO 4490:2014-11. Proszki metaliczne -- Oznaczanie czasu przesypywania przez kalibrowany lejek (sypkościomierz Halla).

Poerschke, D. L., & Levi, C. G. (2015). Effects of cation substitution and temperature on the interaction between thermal barrier oxides and molten CMAS. *Journal of the European Ceramic Society, 35*(2), 681–691. doi:10.1016/j.jeurceramsoc.2014.09.006

Schulz, U., & Braue, W. (2013). Degradation of La2Zr2O7 and other novel EB-PVD thermal barrier coatings by CMAS (CaO–MgO–Al2O3–SiO2) and volcanic ash deposits. *Surface and Coatings Technology, 235*, 165–173. doi:10.1016/j.surfcoat.2013.07.029

Seviostanov, I., & Kachanov, M. (2001). Plasma-sprayed ceramic coatings: Anisotropic elastic and conductive properties in relation to the microstructure; Cross-property correlations. *Materials Science Engineering* A, *297*, 235-243.

Shafiro, B., & Kachanov, M. (2000). Anisotropic effective conductivity of materials with nonrandomly oriented inclusions of diverse ellipsoidal shapes. *Journal of Applied Physic,* 87, 8561-8569.

Stecura, S. (1985). Optimization of the NiCrAl-Y/ZrO_2-Y_2O_3 Thermal Barrier System. *NASA Report Technical Memorandum* 86905, Cleveland.

Subramanian, R. (2002). U.S. Patent 6,387,539.

Suresh, G., Seenivsan, G., Krishnaiah, M. V., & Murti, P. S. (1997). Investigation of the thermal conductivity of selected compounds of gadolinium and lanthanum. *Journal of Nuclear Materials, 249*(2-3), 259–261. doi:10.1016/S0022-3115(97)00235-3

Tomiczek, B., & Dobrzański, L.A. (2013). Określanie struktury i wielkości krystalitów nanokompozytów wzmacnianych haloizytem, *XLI Szkoła Inżynierii Materiałowej Kraków – Krynica*, 24-27 IX 2013, 314-318.

Vaben, R., Traeger, F., & Stover, D. (2004). New thermal barrier coatings based on pyrochlore/YSZ double layer systems. *International Journal of Applied Ceramic Technology, 1*, 351–356.

Vassen, R., Cao, X., Tietz, F., Basu, D., & Stover, D. (2000). Zirconates as new materials for thermal barrier coatings. *Journal of American Ceramic Society*, 83, 2023-2028.

Vassen, R., Tietz, F., Kerkhoff, G., & Stoever, D. (November 1998). New materials for advanced thermal barrier coatings. In J. Lecomte-Beckers, F. Schuber, & P. J. Ennis, (Ed.), *Proceedings of the 6 th Liége Conference on Materials for Advanced Power Engineering*, Universite de Liége, Belgium (pp. 1627-1635).

Wang, C. (2006, September). Experimental and Computational phase studies of the ZrO_2-based systems for thermal barrier coatings [Dissertation an der Universität Stuttgart]. Universität Stuttgart, Stuttgart, Germany.

Wang, Z., Kulkarni, A., Deshpande, S., Nakamura, T., & Herman, H. (2003). Effects of pores and interfaces on effective properties of plasma sprayed zirconia coatings. *Acta Materiallia, 51*, 5319-5334.

Wu, J., Wei, X., Padture, N. P., Klemens, P. G., Gell, M., Garcia, E., & Osendi, M. I. et al. (2002). Low-Thermal-Conductivity Rare-Earth Zirconates for Potential Thermal-Barrier-Coating Applications. *Journal of the American Ceramic Society, 85*(12), 3031–3035. doi:10.1111/j.1151-2916.2002.tb00574.x

Wu, J., Wei, X. Zh., Padture, N. P., Klemens, P. G., Gell, M., & Garcia, E. (2002). Thermal conductivity of ceramics in the ZrO2-GdO1.5system. *Journal of Materials Research, 17*(12), 3193–3200. doi:10.1557/JMR.2002.0462

Xu, Z., He, L., Mu, R., He, S., Huang, G., & Cao, X. (2010). Hot corrosion behavior of rare earth zirconates and yttria partially stabilized zirconia thermal barrier coatings. *Surface and Coatings Technology, 204*(21-22), 3652–3661. doi:10.1016/j.surfcoat.2010.04.044

Zhao, S., Gu, L., Zhao, Y., Huang, W., Zhu, L., Fan, X., & Cao, X. et al. (2013). Thermal cycling behavior and failure mechanism of $La_2(Zr_{0.7}Ce_{0.3})_2O_7/Eu^{3+}$ - doped 8YSZ thermal barrier coating prepared by atmospheric plasma spraying. *Journal of Alloys and Compounds, 580*, 101–107. doi:10.1016/j.jallcom.2013.05.070

KEY TERMS AND DEFINITIONS

Laser Diffraction Analysis: Analysis used for calculation of average size and size distribution of the particles.

Laser Flesh Analysis: The method used to measure of thermal diffusivity. The energy pulse heats the parallel sample on one side and the opposite side is measured temperature increase by a time-dependent.

Pyrochlore Ceramic: Material used for insulation layers characterized by low thermal conductivity. The pyrochlore elemental cell can be seen as eight fluorite unit cell, each containing an average single oxygen vacancy.

Thermal Barrier Coating (TBC) System: Highly advanced materials applied to metallic surface, like gas turbine and combustion chamber in airplane engine. Presence the 8YSZ is the most widely used. TBC is a multilayer system, build with few types of materials with completely different physicochemical properties. It allows to achieve a level of properties that would not be obtainable for each of these materials separately. TBC system consists of external ceramic layer called as topcoat (TC), thermally growth oxides (TGO), aluminium reach metallic bond coat layer (BC) and based material.

Thermal Conductivity: Property of material to conduct heat. Under the same conditions, more heat will flow through the substance with higher thermal conductivity.

Thermal Diffusivity: Coefficient characterizing the material through which heat flows. Measures the heat transfer rate through the material from the hot side to the cold side.

Thermally Growth Oxides TGO: Part of TBC systems, which is formed by the oxidation of aluminum, component of BC layer. Forms the alumina (α-Al2O3) and is responsible for the oxidation properties.

Chapter 5
Oxidation Behavior of LTA/ YSZ Intermixed Layer Coating

Pritee Purohit
Army Institute of Technology, India

Shashikant T. Vagge
College of Engineering Pune, India

ABSTRACT

This chapter describes how for power generators like gas turbines and aero engines, the economic and environmental challenges are increasing day by day for producing electricity more efficiently. The efficiency of power generators can be increased by changing its operating conditions like inlet temperature and procedure. Currently, the inlet temperature to the industrial gas turbine is reaching up to 1400°C. Also, in aero engines, the ring temperature reaches around 1550°C. Therefore, the coatings used in aero engine applications undergo short duration thermal cycles at very high temperature. The mean metal temperatures reach around 950°C and can increase up to 1100°C. But in industrial gas turbines, it varies from 800 to 950°C. Operating temperature of industrial gas turbines slowly reaches to maximum and ideally remains constant for thousands of hours, unlike aero engines.

INTRODUCTION

Thermal barrier coatings TBCs are applied on hot section parts like blades and vanes of turbine and aero engines. In the last decade, a number of alternative ceramic materials were suggested for TBCs. It covers aluminates, doped zirconia, perovskites, pyrochlores, and fluorite (Xie, Guo, Gong, & Xu, 2011). Recently, lanthanum-titanium-aluminium oxide that is LaTi2Al9O19 (LTA) is proposed as TBC material, which is having excellent phase stability from ambient to 1600°C. It is the combination of rare earth element and ceramic that is Lanthanum oxide La2O3, titanium oxide TiO2 and aluminum oxide Al2O3. But due to its relative low fracture toughness, it is having a short life. This difficulty can be solved by using a lanthanum-titanium-aluminium oxide and yttria-stabilized zirconia (LTA/YSZ) double ceramic layer (DCL) (Xie et al., 2011).

DOI: 10.4018/978-1-5225-4194-3.ch005

In order to investigate the oxidation and hot corrosion behavior of the Al2O3 incorporated LTA/YSZ combined top coat different set of tests were performed. Inconel 718 was used as a substrate material for developing the coatings. Three combinations of LTA/Al2O3/YSZ coatings were developed using plasma spray method with varying topcoat thickness of LTA, 1. LTA 100 with 100 μm thickness, 2. LTA 150 with 150 μm thickness and five layered intermixed coatings. The conventional YSZ coatings were developed for comparison. An isothermal hot corrosion test was carried out at 900°C for 20, 45, 75 and 100 h in presence of sodium sulfate (Na2SO4) + 25wt. % sodium chloride (NaCl) salt. The cyclic hot corrosion test was carried out at 900°C. Each cycle comprises heating the coated sample for 45 min followed by 15 min air cooling. The test was continued till 20% failure of each sample. Isothermal oxidation of coatings was carried out for 240 h at 1000°C, 1100°C and 1200°. Cyclic oxidation of the samples was carried out at 1100°C. The cycle time was 1 h, consisting 45 min of heating at 1100°C and 15 min of air cooling. The test was continued till the 20% failure of each sample.

The coatings were evaluated using x-ray diffraction (XRD) technique by using BRUKER D8 X-Ray diffraction instrument. The x-ray diffractograms were obtained through Cu Kα radiation in the range of 2θ from 15 to 70°. Various phases obtained were identified using the X'Pert high Score plus. Surface morphology of samples was examined using field emission scanning electron microscope FESEM (model Carl Zeiss Sigma). The SEM images were taken in the range of 3 kV to 15 kV. Energy Dispersive X-Ray analysis (EDX) was carried out to determine the elemental percentage.

BACKGROUND

TBCs provide effective thermal resistance to the gas turbine engines components by maintaining higher operating and inlet temperatures and also optimum cooling requirements. Atmospheric plasma spray (APS), electron-beam physical vapor deposition (EBPVD), and solution precursor plasma spray process (SPPS) are used to apply metallic/ceramic spray on the superalloy substrates. In APS process compared to the other techniques the coating strain tolerance and TBC life is increased by the readily available cracks and splat morphology (Ghosh, 2015). The blades and vanes are air cooled. TBC is comprised of a ceramic topcoat which is made up of yttria stabilized zirconia and having lowest thermal conductivity, metallic bond coat applied on a superalloy (Seraffon, Simms, Sumner, & Nicholls, 2011). The TBC material composition should have the capability to sustain the temperature cycling, most extreme temperature and thermal and mechanical stresses. In commercial jet engines, TBC is expected to survive in thousands of take-offs and landings. In industrial gas-turbine engines, it should survive up to 30000 hours of continuous operation at high temperature up to 1000°C. TBC structure is much complex than other coatings due to its harsh and demanding operating conditions and multi-material nature.

TBC failure occurs due to many reasons. Few of them are, increase in its thermal conductivity with increase in temperature, stresses due thermal expansion coefficient mismatch of TBC materials, oxidation of metals due to availability of excess oxygen, variation in the material composition and hence continuously varying microstructure, properties and interfacial morphologies (Padture et al., 2002).

TBC undergoes oxidation, thermal fatigue, phase transformations, sintering, plastic deformation, diffusion, creep deformation, thermal conduction, radiation, thermal expansion and elastic deformation. These are the reasons behind the premature failure of TBC. These factors need to be considered and hence putting a limit on the maximum inlet temperature and thus the efficiency of power generation and maximum duration of heating. To achieve the maximum efficiency and longer TBC life it is essential to understand the basics of TBC materials and the reasons of TBC failure. (Padture et al., 2002). TBC of gas turbines have to sustain the higher thermal stress, mechanical stress, erosion and corrosive environment. To overcome these problems, a commonly used idea is the multilayer system was proposed by Takahashi (1999).

Cracking at the thermally grown oxide (TGO) layer between the top coat and bond coat is the most common failure mode for gas turbine blade coatings. TGO growth produces stress associated with the volume expansion which persists at all temperatures. When the system is cooled, even more mismatch is introduced from the mismatch in thermal expansion coefficients. The result is very high (2-6GPa) stresses which occur at low temperature and can produce cracking and ultimately fracture of the barrier coating. TGO formation also results in depletion of Al in the bond-coat. This can lead to the formation of undesirable phases which contribute to the mismatch stress. These processes are all accelerated by the thermal cycling which occurs in many thermal barrier coating applications. A key feature of all TBC components are well adapted thermal expansion coefficients between all layers. Thermal barrier coatings expand and contract at different rates by heating and cooling the environment so that materials when the different layers have poorly adapted thermal expansion coefficients introduce a deformation which can lead to cracking and ultimately failure of the coating.

COATING DEVELOPMENT

Substrate Material

Inconel 718 was used as a substrate material for developing the coatings. The chemical composition of Inconel 718 (ASTM B670 UNS N07718) provided by the supplier is shown in Table 1. Specimens of 10 mm x 10 mm x 4 mm were cut using wire cut electrical discharge machining (EDM). The edge effect of the specimen was eliminated by grinding. Using alumina powder the specimens were grit blasted before spraying.

Thermal Spray Powders

From the data available in previous literature the LTA powder was prepared using La_2O_3-99%, TiO_2-99%, and Al_2O_3-99% having size 325 μm. These powders were dried for removal of moisture at 200°C for 10 h, mixed together and then ball milled for 5 h. The mixed powder was dried at 200°C for 5 h and calcined at 1500°C for 24 h. It was compacted at 400 MPa and sintered at 1600°C for 72 h (Xie et al., 2011). Finally, sintered mass crushed and sieved with 90 μm sieve. The fluidity was in the range of 38 to 50 gm/min.

Table 1. Chemical composition of Inconel 718 (wt. %)

Ni	Cr	Fe	Co	Nb	Ti	Mn	V	Si	C	S	P
54.975	17.656	17.213	0.050	5.466	1.069	0.124	0.033	0.169	0.015	0.011	0.002

A 40 kW Metco thermal plasma spray unit having plasma spray F4 gun was used for developing coating. Thermal grade powders having fluidity in the range of 38 to 50 gm/min were selected for coating. The particle size of powders was kept 90 μm. The thickness of the LTA coatings was maintained in the range of 320-370 μm (Padture et al., 2002). LTA powder was preheated at 150°C before spraying. The coating parameters such as arc current, feed rate and spray distance were selected based on the data provided by the suppliers of these powders (Sigma-Aldrich India). Coating parameters generally depend on the melting temperatures and the particle size of the powders used for the coating. The large size particles and high melting temperature need a high arc current to melt the particles adequately before they are deposited onto the substrate (Sreedhar & Raja, 2010). Accordingly, it can be seen that LTA and Al_2O_3 need high arc current than NiCrAlY as they exhibit a much higher melting temperature. The coating parameters are described in Table 2.

While selecting the combination of coating layers in double ceramic layer coating, the material of top layer should have a low thermal conductivity with high phase stability, and act as a thermal insulator to protect the inner layer (Cao, Vassen, Tietz, & Stoever, 2006). Intermixed coating and conventional YSZ coatings were developed for comparison. In the above mentioned coatings bond coat of NiCrAlY of 90 μm thickness was sprayed on a substrate. In intermixed (INT) coatings the combination provides hybridized intermixed layers to develop a thermal barrier coating shown in Figure 1. The coatings developed by this combination are best as it is prepared like functionally graded coatings. In the case of functionally graded coatings, the change in coefficient of thermal expansion is gradual. This results in better cyclic oxidation performance. The weight ratios of the powders were taken in NiCrAlY/YSZ of 50/50 and YSZ/LTA of 50/50 proportion. The powders were ball milled for approximately 5 h using zirconia balls to prepare NiCrAlY 50 wt. % + YSZ 50 wt. %, YSZ 50 wt. % + LTA 50 wt. %. A bond coat of NiCrAlY 90 μm is applied to the substrate after shot blasting. The next layer of 50% YSZ + 50% bond coat having 50 μm thickness is applied followed by a layer of YSZ having thickness 100 μm. Further, a layer of thickness 50 μm with 50% YSZ + 50% LTA is applied. The top layer of LTA having the thickness of 100 μm is then applied.

The gradual change in the material composition of TBC provides a unique combination of both isothermal and cyclic oxidation and hot corrosion performance. The total thickness of the coating was maintained around 390 μm. Conventional coatings were developed with 90 μm bond coat thickness and 200 μm top coat thickness. The compositions of coatings are shown in Table 3 and Figure 1.

Annealing of intermixed coatings was done at 1050°C for 5 h in resistance heated muffle furnace for recrystallization of coating (Xie et al., 2011). Annealed LTA 100 and LTA 150 and intermixed samples were mentioned as LTA 100A, LTA 150A and INTA respectively. Two sets of coatings were tested to check the reproducibility.

Table 2. Parameters used for plasma spray coating method

S. N.	Powder	Parameters						
		Nozzle (mm)	Injector Diameter (mm)	Injector Distance (mm)	Current (A)	Voltage (V)	Feed Rate (kg s^{-1})	Spray Distance (mm)
1.	NiCrAlY	6	1.5	6	650	49	750	102
2.	YSZ	6	1.5	6	600	51	667	102
3.	LTA	6	1.5	6	700	52	500	102

Table 3. Compositions of coatings obtained by plasma spray method using F4 gun

S. N.	Coating Type	Composition				
		Layer 1	Layer 2	Layer 3	Layer 4	Layer 5
1.	INT	NiCrAlY (90 μm)	50%NiCrAlY+ 50%YSZ (50 μm)	YSZ (100 μm)	50%YSZ+ 50%LTA (50 μm)	LTA (100 μm)
2.	YSZ	NiCrAlY (90 μm)	YSZ (200 μm)	-	-	-

Figure 1. Diagram showing the types of coating layer combination along with thickness obtained by plasma spray method a. YSZ, b. intermixed

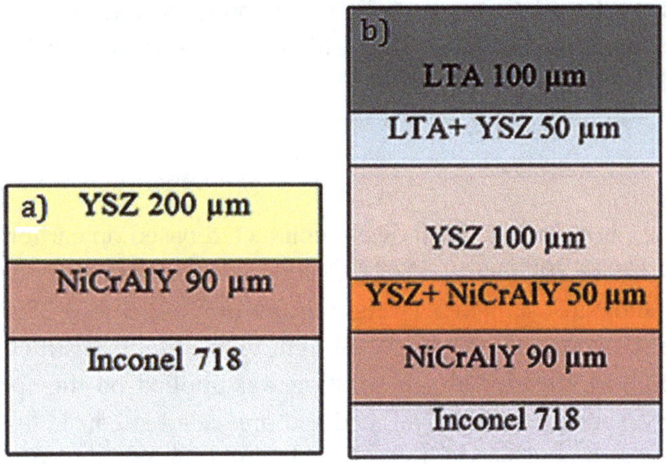

CHARACTERIZATION OF COATINGS

Field Emission Scanning Electron Microscope (FESEM) Analysis

A field emission scanning electron microscope (FESEM) is an electron microscope designed primarily for high magnification imaging, and may also be used for semi-quantitative analysis using an Energy-dispersive spectrometer. Due to the manner in which the image is created, SEM images have a characteristic three-dimensional appearance and are useful for judging the surface structure of the sample. Surface morphology and cross-section of as sprayed and surface morphology of hot corrosion tested specimens were examined using (FESEM) Carl Zeiss Sigma, available in the department of Metallurgy & Materials science of COEP. To obtain cross-section, the specimens were first mounted in an epoxy resin. The mounted specimens were then sectioned with an Isomet-slow speed abrasive diamond cutter and subjected to mirror polishing. The SEM images were taken in the range of 3 kV to 15 kV. Energy Dispersive X-Ray analysis (EDX) was carried out to determine the elemental percentage using point method and line scan method.

X-Ray Diffraction Analysis (XRD)

The X-ray powder diffraction technique allows one to identify the different phases present in the solid sample under study. Every crystalline material in its pure form always exhibits a characteristic X-ray diffraction pattern. Before and after subjecting to high-temperature oxidation test, the coatings were examined using x-ray diffraction technique by using BRUKER D8 X-Ray diffraction instrument available at Department of Physics of Pune University. The x-ray diffractograms were obtained using Cu Kα radiation in the range of 2θ from 15 to 70°. Various phases were identified using the X'Pert HighScore Plus and JCPDS-PCDFWIN software packages.

Oxidation

High-temperature isothermal oxidation and cyclic oxidation tests were performed at 1100°C to study the oxidation resistance of intermixed coatings. The high temperature muffle furnace Therelek TE-4673 was used for tests.

Hot Corrosion

In present work, the type I hot corrosion tests were conducted based on earlier published work and by referring standards ISO 17224:2015 and JIS Z2292-2004. Hot corrosion test was conducted at 900°C in presence of Na2SO4 and NaCl salts in 3:1weight proportion. Na_2SO_4 + 25% NaCl salt solution was prepared by mixing Na_2SO_4 and NaCl salts in 3:1 weight proportion in distilled water (Mudgal, Singh, & Prakash, 2015). The above-mentioned salt solution was applied on the specimen's surface in the following manner. Firstly, specimens were weighed and preheated at 250°C for a period of 1 h, as preheating provides good adhesion of the salt to the specimen. Then the salt solution was applied on the specimens using a small air spray gun on all sides of the specimens. These salt-coated specimens were kept in a furnace heated at 120°C for 2 h to remove the moisture from the salt coating. Thereafter, each salt-coated specimen was weighed individually to calculate the amount of the salt deposited per unit area. The above procedure provided 3-5 mg/cm^2 of the salt over the specimen. These salt coated specimens then arranged on a ceramic tray. A resistance heated tubular furnace, (make: Heat and control system) with ±1°C accuracy was used for the experiments.

Oxidation Behavior of Thermal Barrier Coatings

Isothermal oxidation of coatings was carried out for the duration of 240 h at 1100°C in a high temperature muffle furnace (make, model: Therelek TE-4673), to determine the oxidation kinetics and mechanism. Two sets of coatings were tested for isothermal oxidation under same conditions and the average of two reading was reported to check the reproducibility. Weights of the samples were measured after an interval of 24 h for each temperature, by using electronic balance Afcoset make having the least count 1 μg. Similarly, second, third, fourth, fifth, sixth, seventh, eighth, ninth and tenth set of samples were removed after 48 h, 72 h, 96 h, 120 h, 144 h, 168 h, 192 h, 216 h and 240 h respectively. Weight gain kinetics of the coatings was studied using this data. The activation energy was calculated for both samples using rate constant k_p and the gradient obtained by Arrhenius plots. To check the reproducibility, two sets of coatings were tested for oxidation under the same condition and the average of two reading was reported.

Oxidation Kinetics

The weight gain (mass change) per unit area vs. oxidation time plots at 1100°C for as-sprayed intermixed (INT), annealed intermixed (INTA) and YSZ coatings are depicted in Figure 2. It indicates that at all temperatures, the conventional YSZ samples have gained maximum weight of 0.05002 mg/cm^2 and annealed intermixed samples have gained the least weight of 0.03989 mg/cm^2.

As-sprayed intermixed sample shown a weight gain of 0.04223 mg/cm^2. All samples were survived till 240 h. Visible cracks were observed on all samples after isothermal oxidation at 1100°C for 240 h. These coatings were further examined to study isothermal oxidation kinetics of conventional YSZ, INT and INTA subjected to isothermal oxidation for 240 h at 1100°C, by plotting (weight gain/area)^2 vs. time as shown in Figure 3.

The Figures reveal a good fit between (weight gain/area)^2 and time. It can be predicted that the coatings follow the equation mentioned in equation 1 for isothermal oxidation kinetics. The parabolic rate constants (k_p) values for conventional YSZ, INT and INTA subjected to isothermal oxidation for 240 h at 1100°C was calculated from the above fit are summarized in Table 4 (Sreedhar & Raja, 2010).

$$\Delta w/A = (k_p \cdot t)^{1/2} + Co \quad (1)$$

where, $\Delta w/A$ is weight gain per unit surface area (mg/cm^2) at time t (s), Co is a constant, k_p is the parabolic rate constant (mg^2 cm^{-4} s^{-1}).

As-sprayed intermixed sample reveals a k_p values of 8.155x10^{-6} mg^2 cm^{-4} s^{-1} at 1100°C. Annealed intermixed sample reveals a k_p values of 7.295x10^{-6} mg^2 cm^{-4} s^{-1} at 1100°C. The rate constant was increased, from annealed to as-sprayed intermixed coating and from as-sprayed intermixed to YSZ sample. The similar trend is observed in weight gain data (Sreedhar, MasroorAlam, & Raja, 2009). Conventional YSZ samples have also shown k_p values of 1.11x10^{-5} mg^2 cm^{-4} s^{-1} at 1100°C. The k_p values found lower than YSZ for INT and INTA coatings, thereby indicating lower oxidation rates of INT and INT coatings.

Figure 2. Weight gain kinetics of intermixed coating during isothermal oxidation for 240 h at 1100°C

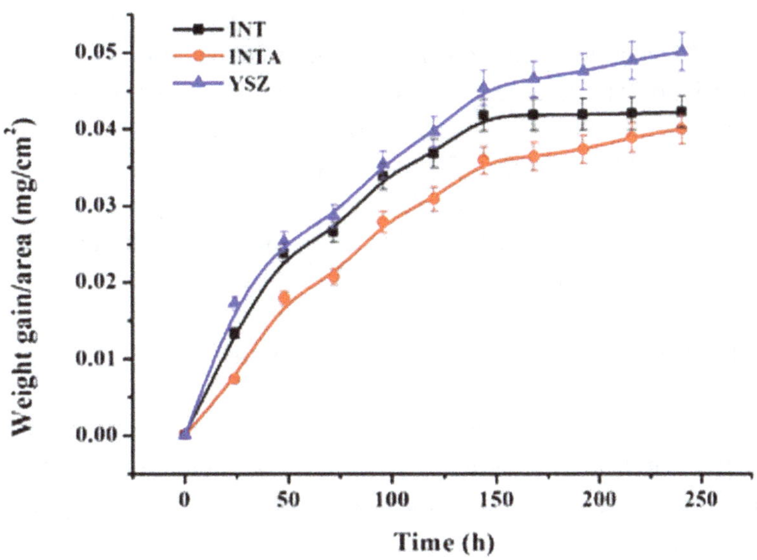

Figure 3. (Weight gain/area)^2 vs. time plots of intermixed coating after isothermal oxidation at 1100°C after 240 h

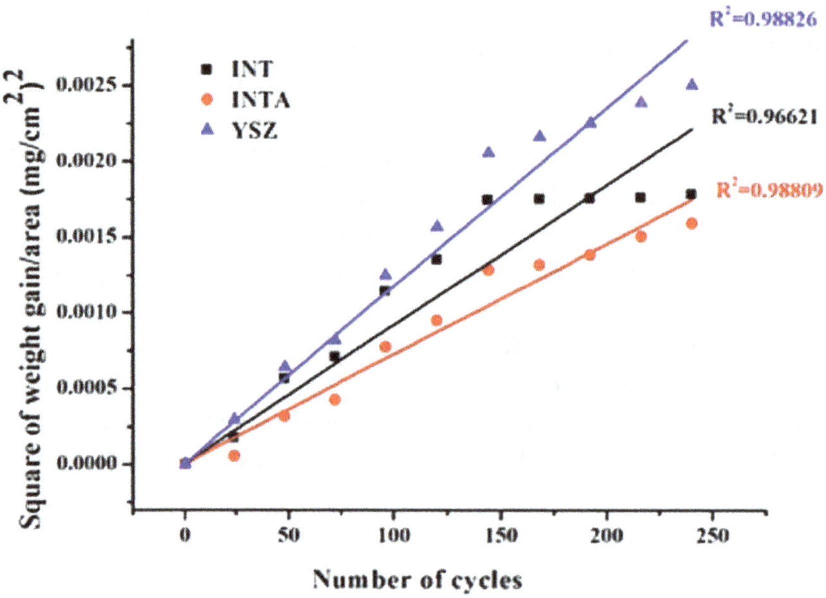

Table 4. Rate constant (k_p) of intermixed coating after isothermal oxidation at 1100°C after 240 h

S. N.	Specimen	k_p (mg^2 cm^{-4} s^{-1})
1	INT	8.155×10^{-6}
2	INTA	7.295×10^{-6}
3	YSZ	1.11×10^{-5}

XRD Analysis

The XRD patterns of INTA coatings before and after isothermal oxidation for 240 h at 1100°C are compiled in Figure 4 respectively. The surface phases of intermixed coatings before oxidation comprises of Al_2O_3, ZrO_2, TiO_2, and LTA only Phases of the isothermal oxidized intermixed coating after 24 h to 144 h at 1100°C are including of Al_2O_3, ZrO_2, TiO_2, and LTA.

After isothermal oxidation, LTA 100 coating for 240 h at 1100°C phases of LTA, Al_2O_3, TiO_2, and $LaAlO_3$ and spinels of Zr were observed. Peak diffraction intensities belonging to two new phases of $LaAlO_3$ and $Zr_{0.92}Y_{0.08}O_{1.96}$ are becoming strong at 1100°C after 240 h.

Surface Morphology of Coatings Before Oxidation

Surface morphology of intermixed as-sprayed and annealed coating is shown in Figure 5a and 5b respectively. The surface is very rough as it includes splats. Porosity, granular non-homogeneous and unmelted particles are more in a case of the as-sprayed sample. It can be seen that coating is not completely dense but it has some porosities on the surface.

Figure 4. X-Ray diffraction patterns of annealed intermixed coating after 24 h, 72 h, 144 h and 240 h isothermal oxidation at 1100°C

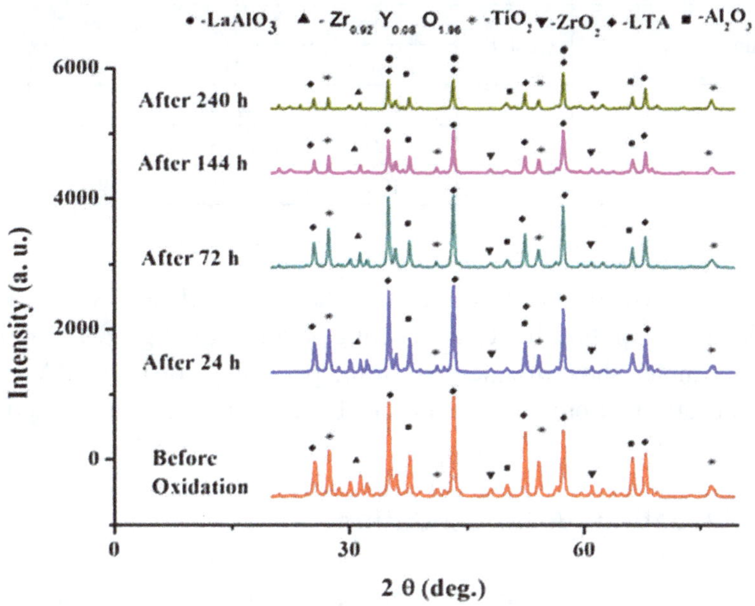

Figure 5. Surface morphology of a) as-sprayed intermixed coatings before oxidation, evaluated at point 1 and 2 and b) annealed intermixed coatings, evaluated at point 3 for elemental analysis using EDX method

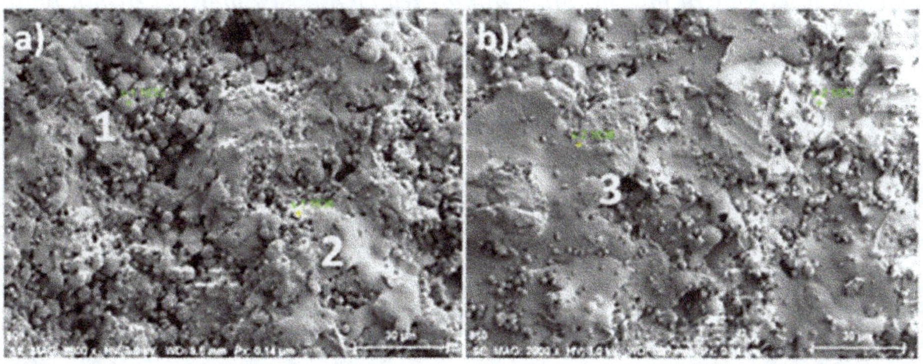

Annealed samples show the reduction in porosity and more uniformity compared to as-sprayed samples. The granular microstructure has been observed. After annealing, reduction in the porosity was observed due to re-crystallization (Xie, Guo, & Gong, 2009). A whisker-like structure has been observed. From the morphology of surface, melted top YSZ and LTA ceramic coating can be observed. It can be seen that coating is not completely dense but it has some porosities on the surface (Jiang et al., 2010).

The SEM microstructure of annealed surface is also non-uniform and compact. A certain quantity of pores can be observed (Wang et al., 2012). The coatings surface was further evaluated for elemental analysis using EDX method at locations shown in Figures 5a and 5b for as-sprayed and annealed coatings respectively.

Table 5. Elemental analysis of intermixed coatings obtained from EDX analysis in (at. %) before isothermal oxidation

Spectrum	Point	Elements				
		La	Ti	Al	O	Zr
INT As-sprayed Figure 5a	1	9.08	22.72	15.05	38.69	2.07
	2	3.27	10.45	36.2	36.76	0.76
INT annealed Figure 5b	3	10.9	24.31	29.22	34. 81	0.75

EDX analysis of as-sprayed and annealed intermixed coating surface has shown the presence of La, Ti, Al and O. The elemental composition of the coating obtained through EDX analysis is summarized in Table 5. Annealed samples show the increase in the percentage of La and Al and reduction in the percentage of O. It indicates that protective oxides of La and Al have been formed in the annealed intermixed samples.

Microstructure of Coatings After Oxidation

The surface SEM micrographs of INT and INTA samples after isothermal oxidation for 240 h at 1100°C are shown in Figure 6. EDX analysis of INT and INTA samples after isothermal oxidation for 240 h at 1100°C is summarized in Table 6.

Figure 6. Surface morphology of intermixed coating surface after isothermal oxidation at 1100°C for 240 h: area 1. INT, area 2. INTA.

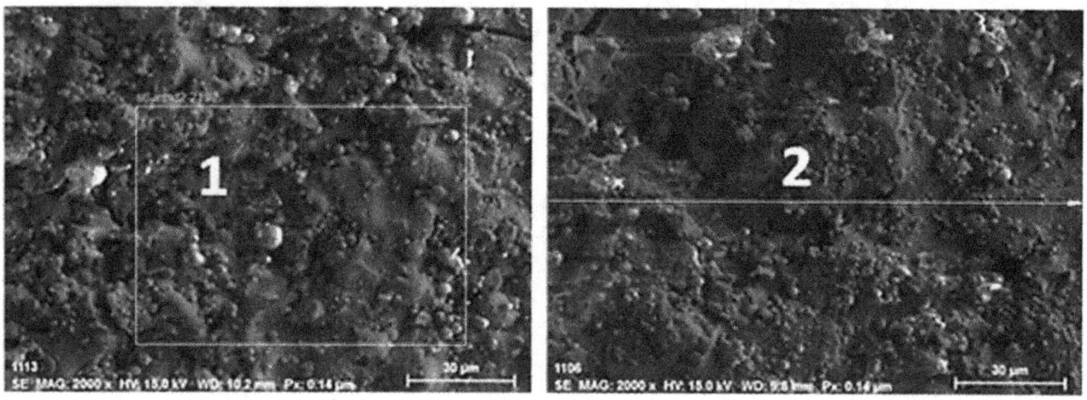

Table 6. Elemental composition of intermixed coatings obtained from EDX analysis in (at. %) after isothermal oxidation at 1100°C for 240 h

Spectrum	Point	Elements					
		O	Al	Ti	La	Zr	Ni
INT	1	68.13	25.62	5.13	0.88	0.17	0.08
INTA	2	63.78	25.11	9.22	1.68	0.21	-

The SEM microstructure of INT samples depicted in Figure 6 (1) and the SEM microstructure INTA shown in Figure 6 (2) reveals that, the porosity distribution is non-uniform across the entire coating surface after isothermal oxidation for 240 h at 1100°C. The porosity is the key parameter, which makes the coating strain tolerant to enhance the coating life. Cracks can be observed on the surface of both coatings.

The EDX analysis of area 1 and 2, shown in Figure 6 respectively reports the presence of elements like La, Ti, Al, Zr, Ti and O. It evidences the formation of $Zr_{0.92}Y_{0.08}O_{1.96}$, TiO_2, $LaAlO_3$, LTA and Al_2O_3 after isothermal oxidation for 240 h at 1100°C. XRD analysis confirmed the formation of phases observed in SEM and EDX analysis.

Cyclic Oxidation of Intermixed Coating

Cyclic oxidation of the samples was carried out at 1100°C in a high temperature muffle furnace of Therelek make having a maximum temperature of 1700°C. The cycle time was of 1 h, consisting 45 min of heating at 1100°C and 15 min of air cooling. Both the batches of annealed and as-sprayed samples were taken out of the furnace at 1100°C after 45 min. The samples were air cooled for 15 min and the weight of the samples was measured using the electronic balance of Afcoset having least count of 1μg. This 1 h cycle was repeated until a 20% of the coating got peeled off and the corresponding number of the cycle was recorded as the spallation life of the coating (Ramachandran, Balasubramanian, & Ananthapadmanabhan, 2012). Weight gain kinetics of the coatings was studied using this data. To check the reproducibility, two sets of coatings were tested for hot corrosion under the same condition and the average of two reading was reported

Oxidation Kinetics

Figure 7 shows the plot of weight gain observed in mg/cm^2 vs. the number of cycles of cyclic high-temperature oxidation at 1100°C. It shows that for all the types of coatings, the oxidation weight gain observed was more in the as-sprayed samples as compared to the annealed samples.

As-sprayed intermixed sample survived till 125 cycles and showed the oxidation weight gain of 2.3 mg/cm^2. The conventional sample showed the maximum oxidation weight gain of 2.274 mg/cm^2 and survived till 100 cycles. Whereas, annealed intermixed samples survived till 185 cycles and showed the minimum oxidation weight gain of 1.7 mg/cm^2.

X-Ray Diffraction Analysis

XRD analysis of annealed intermixed sample before and after cyclic oxidation at 1100°C is shown in Figure 8. LTA phases were retained in all samples before and after oxidation. It indicates that complete decomposition LTA phases don't occur in intermixed samples. After cyclic oxidation, the XRD pattern of annealed intermixed coatings revealed the presence of Al_2TiO_5, Al_2O_3, $Zr_{0.92}Y_{0.08}O_{1.96}$ and also the presence of LTA.

Controlled rate of oxidation was observed in annealed intermixed coating. After cyclic oxidation, the phase change of LTA from tetragonal to monoclinic (t' to m) and separation or dissolution of LTA phase into Al_2O_3 and other oxides were observed in both coatings.

The crystalline nature of LTA phase decreases with increase in oxidation exposure time. The presence of different elements observed in EDS analysis is in agreement with the phases formed in XRD analysis.

Figure 7. Weight gain plots of intermixed coatings after cyclic oxidation at 1100°C

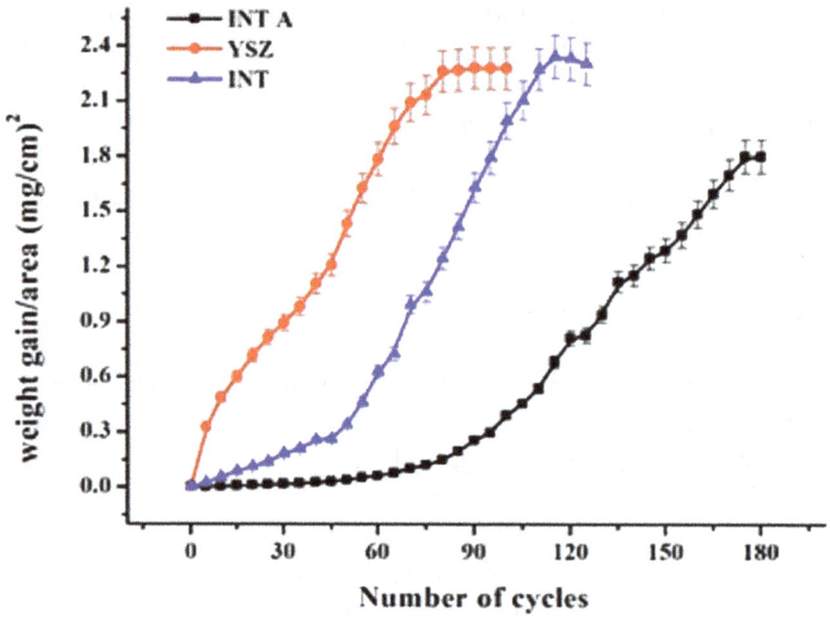

Figure 8. X-Ray diffraction patterns of intermixed coating after cyclic oxidation at 1100°C

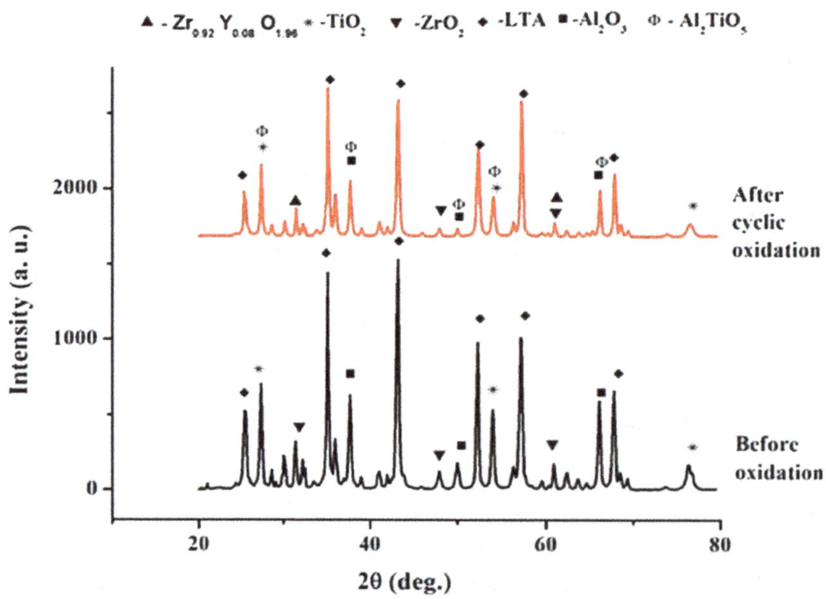

Surface Morphology

The surface SEM micrographs of as-sprayed and annealed intermixed samples after cyclic oxidation 1100°C is shown in Figure 9. The elemental composition obtained by EDX analysis (at. %) of both samples after cyclic oxidation for 1100°C is summarized in Table 7. The rough and dense surface was observed in both as-sprayed and annealed intermixed coating surfaces. Porosity was reduced even after cyclic oxidation at 1100°C. Micro-cracks were observed on the as-sprayed coating surface. The surface oxides are appeared much coarse and rough compared to the coatings. Annealed sample have shown better cyclic oxidation performance. The porosity was retained after cyclic oxidation. No sintering was observed in the microstructure. In a case of both samples oxides of $Zr_{0.92}Y_{0.08}O_{1.96}$, Al_2TiO_5 and Al_2O_3 was formed on the surface after subjecting to cyclic oxidation. XRD patterns showed peaks of these oxides. Also, EDX analysis was done at different locations shown in Figures 9a and 9b and the elemental composition is shown in Table 7. EDX analysis of as-sprayed and annealed coatings at various points showed the presence of elements like Al, O, La, Ti, Zr, Ni and Cr. Higher percentage of Al and O evidences the formation of oxides of Al. In Figure 9a point 1 may contain $Zr_{0.92}Y_{0.08}O_{1.96}$, Al_2TiO_5, point 2 may contain $Zr_{0.92}Y_{0.08}O_{1.96}$, Al_2TiO_5 and Al_2O_3. Also in Figure 9b, point 3 may contain Al_2TiO_5, Al_2O_3 and LTA. Point 4 shows the presence of $Zr_{0.92}Y_{0.08}O_{1.96}$, Al_2TiO_5, $LaAlO_3$ and TiO_2 as the EDX shows point 4 rich in La, Ti, Al and O. The same is validated through EDX and XRD. Annealed sample shows the higher amount of La, Ti and O and proves protection ability of coating.

Figure 9. Surface morphology of a) as-sprayed intermixed and b) annealed intermixed sample after cyclic oxidation

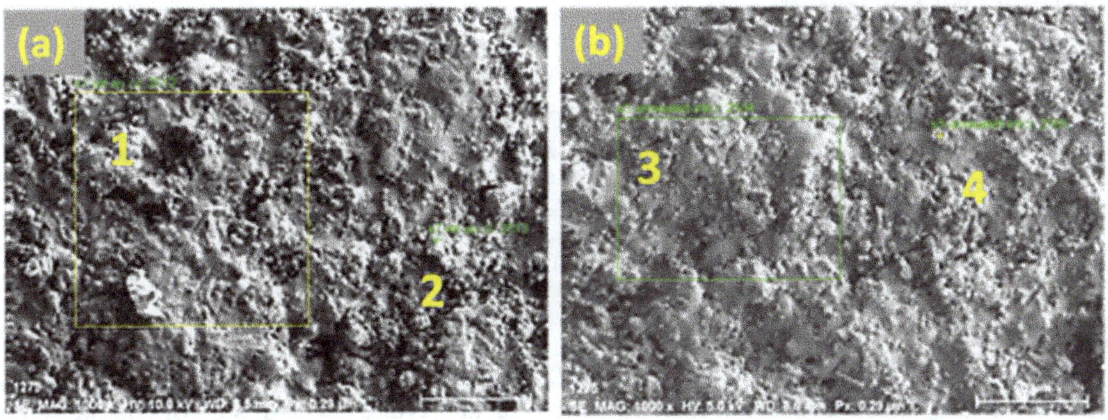

Table 7. Elemental composition obtained by EDX analysis in (at. %) after cyclic oxidation

Spectrum	Point	Elements						
		O	Al	Ti	La	Zr	Ni	Cr
INT Area 1 Figure 9a	1	67.67	27.81	3.49	0.66	0.28	0.08	-
INT Point 2 Figure 9a	2	67.16	29.54	2.56	0.36	0.32	0.05	-
INTA Area 3 Figure 9b	3	68.7	24.09	5.65	1.09	0.25	0.02	0.02
INTA Point 4 Figure 9b	4	70.37	17.53	9.63	1.77	0.15	0.21	0.01

Hot Corrosion Behavior of Thermal Barrier Coatings

In present work, the type I hot corrosion tests were conducted based on earlier published work and by referring standards ISO 17224:2015 and JIS Z2292-2004. A hot corrosion test was conducted at 900°C in presence of Na_2SO_4 and NaCl salts in 3:1weight proportion. Na_2SO_4 + 25% NaCl salt solution was prepared by mixing Na_2SO_4 and NaCl salts in 3:1 weight proportion in distilled water (Mudgal, Singh, & Prakash, 2015). The specimen weight was recorded, and was preheated for 1 h at 250°C, as it provides good adhesion of the salt to the specimen.

A small air spray gun was used to apply the salt solution on all sides of the specimen. The salt solution was applied on the specimens' surface. To remove the moisture, the salt coated specimens were kept in a furnace for 2 h at 120°C. Approximately 3-5 mg/cm^2 salt was coated on the specimen. The weight of the specimen was recorded as an initial weight of a sample. The salt coated specimens were kept in the furnace for hot corrosion test.

To check the reproducibility, two sets of coatings were tested for hot corrosion under the same condition and the average of two reading was reported. A resistance heated tubular furnace, (make: Heat and control system) with ±1°C accuracy was used for the experiments. Later the samples were tested for isothermal and cyclic hot corrosion.

Isothermal Hot Corrosion

An isothermal hot corrosion test was carried out at 900°C for 20, 45, 75 and 100 h in presence of Na_2SO_4 + 25wt. % NaCl salt. Specimens were coated with salt and kept in a furnace at 900°C in four different trays. The first tray was taken out after completion of 20 h, second after 45 h, and third after 75 h and the fourth tray after 100 h.

The weight of the specimen was measured after isothermal oxidation to study weight gain per unit surface area with respect to time. The hot corrosion kinetics was determined based on these results.

Hot Corrosion Kinetics

Weight gain data obtained from isothermal hot corrosion and cyclic hot corrosion of coatings are plotted to study the kinetics of hot corrosion. Isothermal hot corrosion was carried out for 20, 45, 75 and 100 h for as-sprayed and annealed intermixed coatings at 900°C. Figure 10 show the plot of weight gain observed in mg/cm^2 vs. time of hot corrosion for isothermal hot corrosion.

The weight gain of as-sprayed intermixed coatings at the end of 100 h isothermal hot corrosion was 0.076 mg/cm^2 and that of annealed coating was 0.066 mg/cm^2, which is negligible. The observed weight gain was due to a start of oxidation. After 100 h YSZ samples have shown a weight gain of 0.077 mg/cm^2. It shows that the hot corrosion weight gain increased with the decrease in LTA thickness. The hot corrosion weight gain followed the trend like,

YSZ with weight gain of 0.077 mg/cm^2 > INT with weight gain of 0.076 mg/cm^2 > INTA with weight gain of 0.066 mg/cm^2. These coatings were further examined for hot corrosion kinetics by plotting (weight gain/area)^2 vs. time as shown in Figure 11. The Figure reveals a good fit between (weight gain/area)^2 and time. From this, it can be predicted that the coatings follow the equation 1 for hot corrosion kinetics.

The parabolic rate constants (k_p) was calculated from the above fit are summarized in Table 8. INT and INTA coating revealed a k_p value of $6.277 x 10^{-5}$ mg^2 cm^{-4} s^{-1} and $4.59 x 10^{-5}$ mg^2 cm^{-4} s^{-1}. With anneal-

Figure 10. Weight gain plots of intermixed coating after exposed to Na_2SO_4 and NaCl salts in 3:1weight proportion at 900°C after 100 h of isothermal hot corrosion

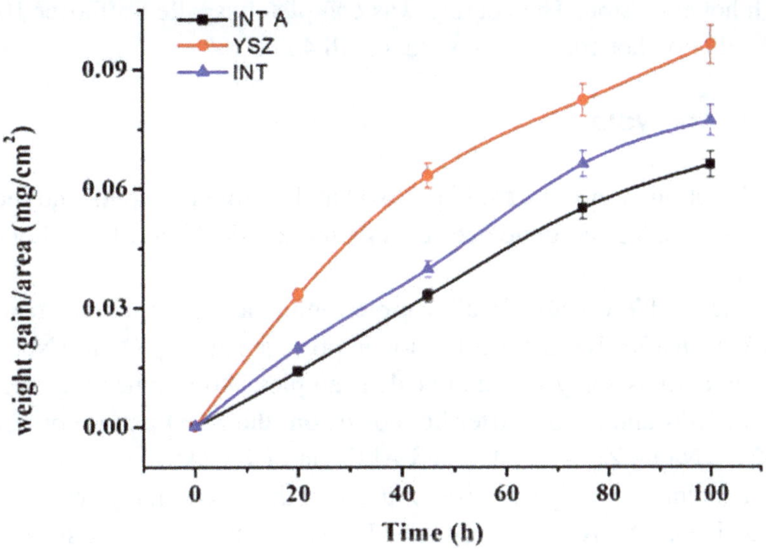

Figure 11. (Weight gain/area)^2 vs. Time plots of intermixed coating after exposed to Na_2SO_4 and NaCl salts in 3:1weight proportion at 900°C after 100 h of isothermal hot corrosion

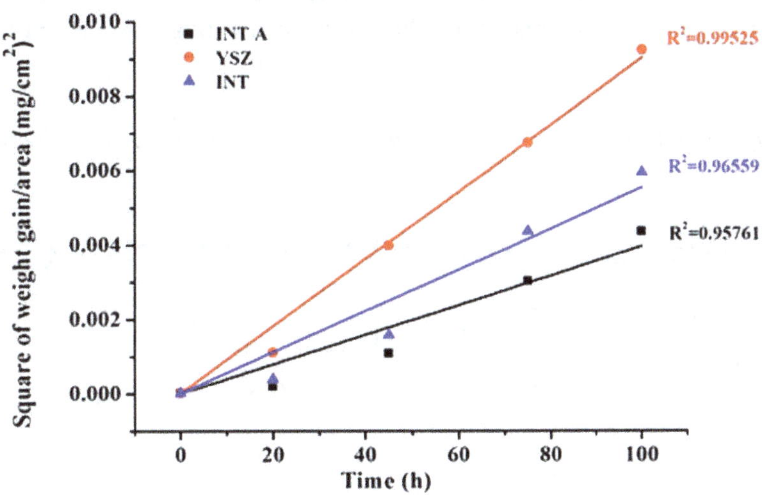

Table 8. Hot corrosion rate constant (k_p) of intermixed coating exposed to hot corrosion in presence of Na_2SO_4 and NaCl salts in 3:1weight proportion at 900°C for 100 h of isothermal hot corrosion

S. N.	Specimen	k_p ($mg^2\ cm^{-4}\ s^{-1}$)
1	INT	$6.277x10^{-5}$
2	INTA	$4.59x10^{-5}$
3	YSZ	$7.21x10^{-5}$

ing, the rate constant was decreased. The same order is observed in weight gain data (Sreedhar et al., 2009). Conventional samples have also shown the mass gain of 7.21×10^{-5} mg^2 cm^{-4} s^{-1} but cracks were developed after 45 h hot corrosion. The coating was completely spalled off after 100 h. Both INT and INTA coatings showed good hot corrosion resistance till 45 h.

X-Ray Diffraction Analysis

Annealed intermixed coatings were survived up to 100 h, therefore to understand the phases developed XRD analysis of INTA samples was done before and after 20, 45, 75 and 100 h hot corrosion at 900°C are shown in Figure 12.

It shows the presence of LTA phases in all samples before and after hot corrosion. It indicates that LTA phases remained even after 100 h hot corrosion in presence of Na_2SO_4 and NaCl salts in 3:1weight proportion at 900°C in both as-sprayed and annealed samples. After annealing, the XRD phases have shown the presence of LTA and Al_2O_3. After hot corrosion, the XRD pattern of coatings revealed the presence of TiO_2, ZrO_2, Na_2O, $Zr_{0.92}Y_{0.08}O_{1.96}$ and Al_2O_3 and LTA. After hot corrosion, the decrease in the intensity and broadening of LTA peak, observed due to the phase change of LTA from tetragonal to monoclinic (t' to m) and separation or dissolution of LTA phase into Al_2O_3 and other oxides. This indicates that the decrease in crystalline nature of LTA phase with the increase in hot corrosion exposure time.

Surface Morphology of Intermixed Coatings After 100 h of Hot Corrosion

It shows the presence of all element including La and Na. At some locations, the Al has been depleted completely and that can be a major reason of the failure of coatings. Annealed sample have shown

Figure 12. X-Ray diffraction patterns of intermixed coating after 20 h, 45 h, 75 h and 100 h isothermal hot corrosion at 900°C

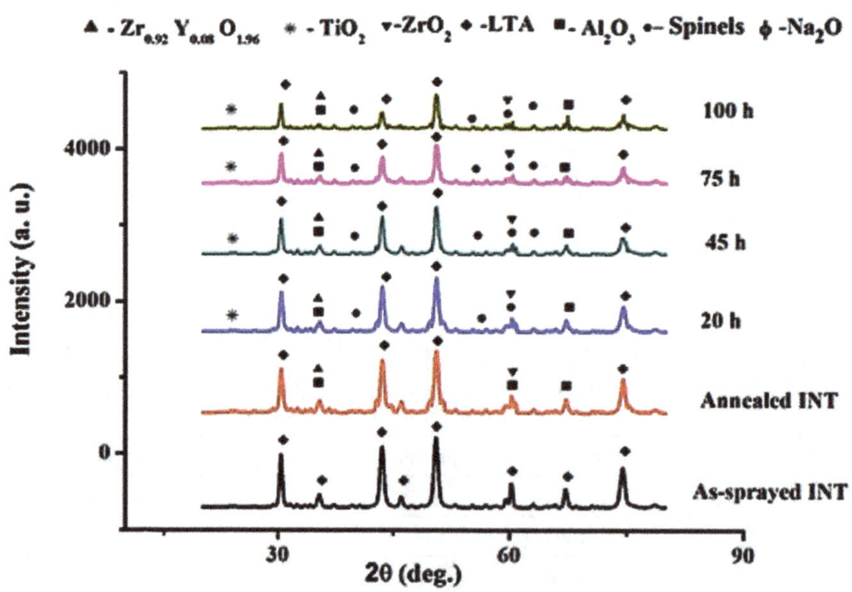

better results that the as-sprayed sample. Surface morphology of intermixed as-sprayed and annealed coatings after 100 h of hot corrosion is shown in Figures 13a and 13b respectively. An exposure to hot corrosion beyond 75 h has led to an increase in a number of micro-cracks on specimen surface in case of intermixed as-sprayed samples. After 100 h the as-sprayed coating surface has shown wide cracks with rough surface morphology. The granular micro-structure was observed on the surface. The cracks were propagated when the coating was removed from the furnace.

The surface oxides are appeared much coarse and rough compared to the coatings. Visible cracks appeared after 100 h of hot corrosion. Annealed sample have shown better hot corrosion performance. The porosity was retained till 100 h. No sintering was observed in the microstructure. In the case of both samples, Na_2O formed on the surface after subjecting to hot corrosion. XRD patterns showed peaks of theses oxides. EDX analysis of as-sprayed and annealed coatings after 100 h of hot corrosion was done at different locations as shown in Figures 13a and 13b and the elemental composition is shown in Tables 9.

EDX analysis of intermixed as-sprayed and annealed coatings at various points showed the presence of elements like Al, O, La, Ti, Na. It evidences the formation of oxides and spinels containing Al. As-sprayed coatings showed the presence of Cl and Ni along with other elements. In Figure 13a point 1 may contain Na_2O, Al_2O_3, LTA, TiO_2, point 2 may contain Al_2O_3, LTA and Na_2O. Both point 1 and 2 may contain spinel oxides of Ni, Al. In Figure 13b Point 1 may contain, TiO_2 and point 2 may contain Na_2O, Al_2O_3 and LTA. The same is validated through EDX and XRD. Annealed sample shows the higher amount of Al and O and proves protection ability of coating.

Figure 13. Surface morphology of a) as-sprayed intermixed coating surface and b) annealed intermixed coating surface after 100 h isothermal hot corrosion

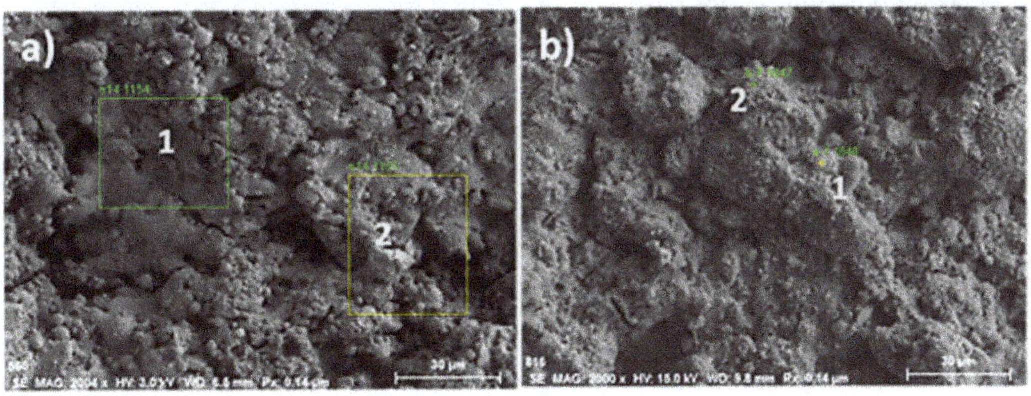

Table 9. Elemental composition of as-sprayed intermixed coating obtained from EDX analysis (wt. %) after 100 h isothermal hot corrosion

Spectrum	Point						
		O	Na	Al	S	Ti	La
As-sprayed intermixed coating after Figure 13a	1	35.74	1.31	28.20	0.13	13.99	5.21
	2	44.77	1.13	31.92	0	9.26	4.74
Annealed intermixed coating after Figure 13b	1	25.68	0.69	16.25	0	10.31	8.88
	2	29.14	1.10	17.51	0	3.50	1.40

Cyclic Hot Corrosion of Intermixed Coating

Cyclic hot corrosion of specimens was carried out in the same furnace as mentioned above at 900°C. The hot corrosion cycle was consisting of heating of the sample in the furnace for 45 min followed by 15 min cooling in air. This 1 h cycle was repeated until a 20% of the coating got peeled off and the corresponding number of the cycle was recorded as the spallation life of the coating (Ramachandran et al., 2012). The weight of the specimen was determined after each cycle and weight gain was determined. A microbalance with an accuracy of ± $1x10^{-5}$ g was used for measuring the weight. To check the reproducibility, two sets of coatings were tested for hot corrosion under the same condition and the average of two reading was reported

Hot Corrosion Kinetics

Figure 14 shows the plot of weight gain observed in mg/cm^2 vs. the number of cycles of cyclic hot corrosion at 900°C. It shows that for all the types of coatings, the hot corrosion weight gain observed was more in the as-sprayed samples as compared to the annealed samples. As-sprayed and annealed intermixed samples survived till 65 and 90 hot corrosion cycles and showed the weight gain of 0.1812 mg/cm^2 and 0.119 mg/cm^2 respectively. The conventional sample showed the maximum weight gain of 0.226 mg/cm^2 and survived till 40 hot corrosion cycles.

Figure 14. Weight gain plots of intermixed coating after cyclic oxidation in presence of Na_2SO_4 and NaCl salts in 3:1weight proportion at 900°C

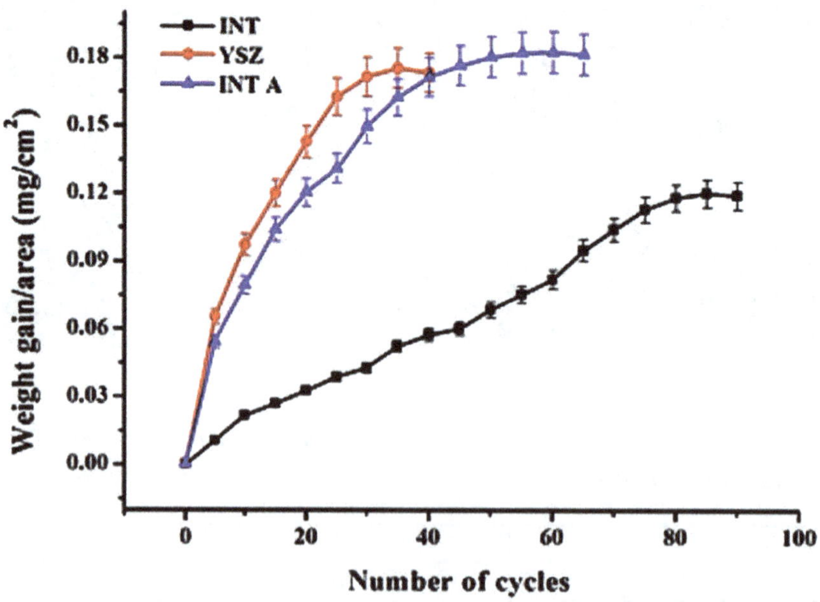

X-Ray Diffraction Analysis

XRD analysis of annealed sample before and after cyclic hot corrosion at 900°C is shown in Figure 15. LTA phases were retained before and after hot corrosion. It indicates that complete decomposition LTA phases don't occur in both LTA 100 and LTA 150 samples. After cyclic hot corrosion, the XRD pattern of coatings revealed the presence of Al_2O_3, and also the presence of LTA. The phases of Na_2O were observed after cyclic hot corrosion.

After cyclic hot corrosion, the crystalline nature of LTA phase decreases with increase in cyclic hot corrosion exposure time. The presence of different elements observed in EDX analysis was in good agreement with the phases obtained in XRD analysis.

Surface Morphology

The surface SEM micrographs of as-sprayed and annealed intermixed samples after cyclic hot corrosion in presence of Na_2SO_4 and NaCl salts in 3:1weight proportion at 900°C is shown in Figures 16a and 16b respectively. The elemental composition obtained by EDS analysis (at. %) of as-sprayed and annealed intermixed samples after cyclic hot corrosion in presence of Na_2SO_4 and NaCl salts in 3:1weight proportion at 900°C is summarized in Table 10.

The as-sprayed coating was survived till 65 hot corrosion cycles. To analyze the reason of earlier spallation, SEM and EDX analysis was carried out. Morphology of coating surface is shown in Figure 16a. It shows the reduction of porosity which resulted in the increase in thermal conductivity of coating surface. Elemental composition of the as-sprayed surface after cyclic hot corrosion obtained from EDX

Figure 15. X-Ray diffraction patterns of annealed intermixed coating after cyclic hot corrosion at 900°C

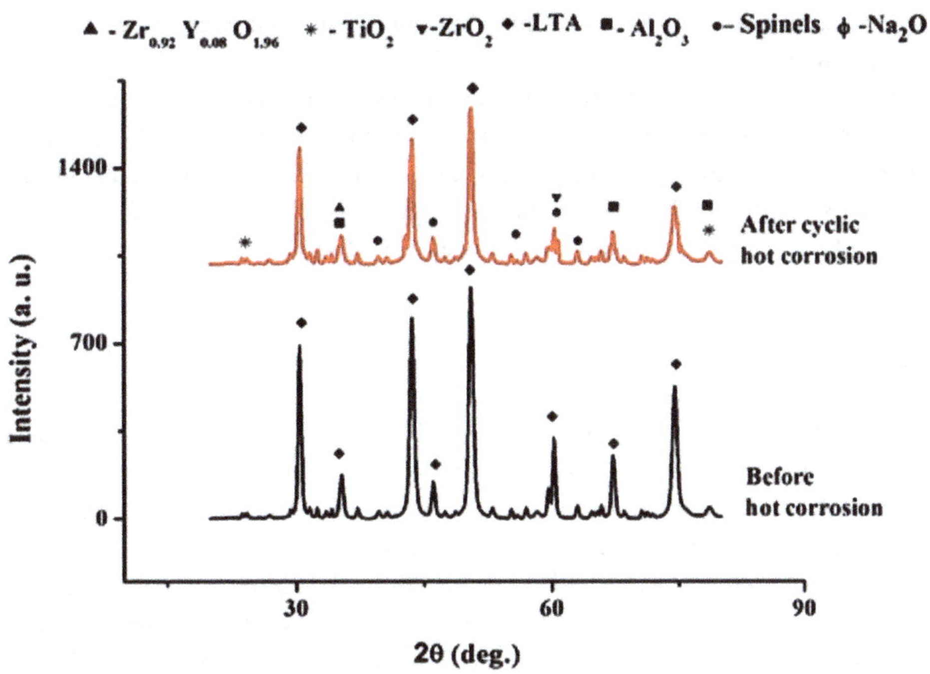

Figure 16. Surface morphology of a) as-sprayed and b) annealed intermixed coating surface after cyclic hot corrosion

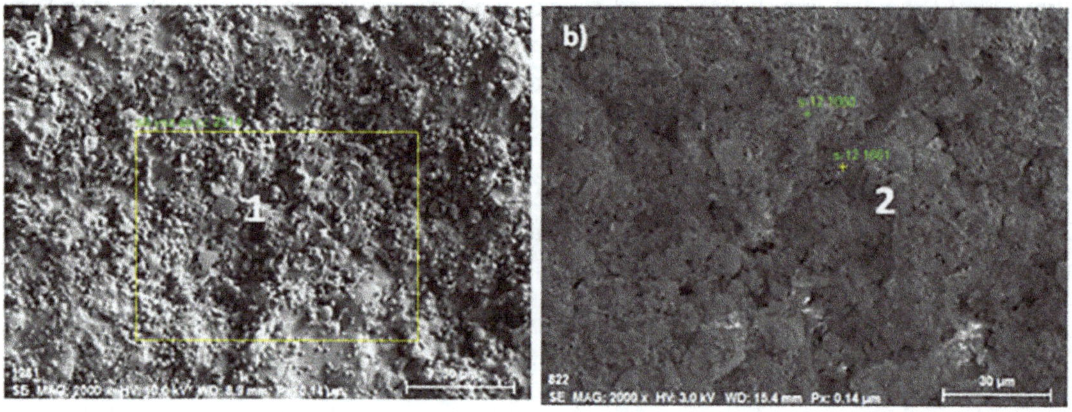

Table 10. Elemental composition of intermixed coating obtained from EDX analysis (wt. %) after cyclic hot corrosion

Spectrum	Point	Elements					
		La	Ti	Al	O	Zr	Na
As-sprayed intermixed coating Figure 16a	1	38.77	26.29	20.38	14.04	-	0.52
Annealed intermixed coating Figure16b	2	24.06	28.52	16.44	26.16	1.59	3.22

analysis (wt. %) is shown in Table 10. It showed the presence of O, Na, Al, Ti, and La. Presence of Na and Al evidences the formation of phases of Na_2O at point 1 in Figure16a. Various spinel oxides are contributing to weight gain.

Surface morphology of annealed coating after cyclic hot corrosion is shown in Figure 16b. The annealed samples showed good resistance to hot corrosion and were survived till 90 hot corrosion cycles. The microstructure shows a number of pores and fine cracks. EDX of annealed samples showed the presence of elements like O, Al, Na, Ti, Zr and La. Higher percentage of Al, Ti and O represents the formation of oxide layers of TiO_2 and Al_2O_3. No traces of Cl found in annealed samples.

DISCUSSION

At all temperatures studied intermixed annealed coatings shows higher oxidation resistance and conventional YSZ shows lower oxidation resistance. The lower oxidation resistance of conventional YSZ coating was mainly due to the porous top coat which acts as an oxygen diffusion path as well as lower adhesion of the top layer to NiCrAlY metallic bond coat (Keyvani, Saremi, & Sohi, 2011).

The oxidation of LTA coatings is resulted in the formation of different oxides and spinel oxides. LTA ($LaTi_2Al_9O_{19}$) is the only ternary compound reported in the Al_2O_3–La_2O_3–TiO_2 system. $LaAlO_3$ is among the 7 binary compounds of the ternary phase diagram (Skapin, Kolar, & Suvorov, 1993). According to the chemical composition ratio and the ternary Al_2O_3–La_2O_3–TiO_2 phase diagram the decomposed La and Ti-rich oxides on the coating surface during thermal cycling test is possible to be a solid solution of $La_{2/3}TiO_3$ and $LaAlO_3$ (Xie et al., 2011). As reported in previous literature the LTA is stable at higher temperatures. The formation of $LaAlO_3$ is relatively easy but the formation of higher aluminates is extremely slow below 1500°C that is more than 130 days. Hence, no other oxidation product is possible after isothermal oxidation of LTA. LTA can decompose to the only $LaAlO_3$, below 1500°C (Ropp & Libowitz 1978). Therefore, possible reaction of LTA is proposed here in equation (2).

$$LaTi_2Al_9O_{19} \rightarrow LaAlO_3 + 2\ TiO_2 + 4\ Al_2O_3 \tag{2}$$

The available α-Al_2O_3 and TiO_2 (anatase or rutile) together form Al_2TiO_5 (Nagano, Nagashima, Maeda, & Kato, 1999). The possible reaction is shown in equation (3),

$$Al_2O_3 + TiO_2 \rightarrow Al_2TiO_5 \tag{3}$$

A slight weight gain in case of as-sprayed intermixed coating with the increment of the oxidation time is probably attributed to the growth of non-protective spinel oxides of Zr, Ni and Cr, the formation of the oxide phases, $LaAlO_3$ and Al_2TiO_5. These are validated through SEM, XRD and EDX. Intermixed coatings have shown better oxidation resistance than conventional YSZ coatings. Annealing improved the oxidation resistance of the coating.

In intermixed samples $LaAlO_3$, $NaAlO_2$ and TiO_2 were formed after slight degradation of LTA. $LaAlO_3$ is among the seven binary compounds mentioned in phase equilibrium diagram of the system $A1_2O_3$-La_2O_3-TiO_2 (Skapin et al., 1993). The chemical reactions of LTA in the hot corrosion mechanism require detailed knowledge of the phase equilibrium in the system of salts and LTA. However, it is not available in the open literature. LTA is the only ternary compound of La_2O_3 and TiO_2 to Al_2O_3. According to the ternary Al_2O_3–La_2O_3–TiO_2 phase diagram (Skapin et al., 1993) and the chemical composition ratio, the decomposed La and Ti-rich compounds on the coating surface is possibly a solid solution of $La_{2/3}TiO_3$ and $LaAlO_3$ (Xie et al., 2011). Also, the phases observed in XRD and EDX of the present study were the combination of Al and Na. Therefore, the possible reaction is proposed here in equation (4) (Xie et al., 2011).

$$LaTi_2Al_9O_{19} + 4Na_2O \rightarrow LaAlO_3 + 2TiO_2 + 8NaAlO_2 \tag{4}$$

In a case of LTA 100 as-sprayed coatings, the thickness of LTA top layer was 100 μm, which was less than LTA 150 (150 μm) coating. With increase in the test duration along with slight degradation of LTA the molten salt infiltrated through the pores and micro-cracks of the 100 μm ceramic top layer of LTA 100 samples and seeped down by the edge effect, approached the bond coat and resulted in failure of coating (Xie et al., 2011).

Spallation of intermixed as-sprayed samples indicates that the molten salt has infiltrated to the edge of a specimen during thermal cycling. A protective alumina layer diminishes and oxygen reaches to the bond coat results in further bond coat oxidation. XRD phases and EDX analysis evidences the formation of oxides like Al_2TiO_5, Al_2O_3, Cr_2O_3, $LaAlO_3$ and Na_2O.

Intermixed annealed samples have shown the negligible weight gain of 0.119 mg/cm^2 till 90th cycle, after that the coatings were spalled off with weight loss. It is because the oxide layer formed and thermal expansion mismatch (Ramachandran, Balasubramanian, Ananthapadmanabhan, & Viswabaskaran, 2012). Major weight loss observed was due to the combined effect of spallation and volatile effect of chlorides as mentioned earlier.

CONCLUSION

Current research concludes that intermixed coatings proved better than conventional duplex YSZ coating. The oxidation and hot corrosion study showed the same result. Annealed intermixed coatings proved best among all coatings. Following conclusions are drawn from the present study. Intermixed annealed sample survived up to 185 oxidation cycles and as-sprayed samples survived till 125 oxidation cycles at 1100°C. Annealed intermixed samples show better performance in both isothermal and cyclic oxidation compared to YSZ coatings. Intermixed annealed sample survived up to 90 hot corrosion cycles and as-sprayed samples survived till 65 hot corrosion cycles.

REFERENCES

Cao, X. Q., Vassen, R., Tietz, F., & Stoever, D. (2006). New double-ceramic-layer thermal barrier coatings based on zirconia-rare earth composite oxides. *Journal of the European Ceramic Society*, *26*(3), 247–251. doi:10.1016/j.jeurceramsoc.2004.11.007

Ghosh, S. (2015). Thermal Barrier Ceramic Coatings - A Review. In A. Mohamed (Ed.), *Advanced ceramic processing* (pp. 11–138). Rijeka, Croatia: InTech; doi:10.5772/61346

Jiang, S. M., Li, H. Q., Ma, J., Xu, C. Z., Gong, J., & Sun, C. (2010). High temperature corrosion behavior of a gradient NiCoCrAlYSi coating II: Oxidation and hot corrosion. *Corrosion Science*, *52*(7), 2316–2322. doi:10.1016/j.corsci.2010.03.032

Jiang, S. M., Xu, C. Z., Li, H. Q., Ma, J., Gong, J., & Sun, C. (2010). High temperature corrosion behaviour of a gradient NiCoCrAlYSi coating I: Microstructure evolution. *Corrosion Science*, *52*(5), 1746–1752. doi:10.1016/j.corsci.2010.01.025

Keyvani, A., Saremi, M., & Sohi, M. H. (2011). Oxidation resistance of YSZ-alumina composites compared to normal YSZ TBC coatings at 1100°C. *Journal of Alloys and Compounds*, *509*(33), 8370–8377. doi:10.1016/j.jallcom.2011.05.029

Mudgal, D., Singh, S., & Prakash, S. (2015). Cyclic Hot corrosion behavior of Superni 718, Superni 600, and Superco 605 in sulfate and chloride containing environment at 900°C. *Metallography Microstructure and Analysis*, *4*(1), 13–25. doi:10.1007/s13632-014-0182-0

Nagano, M., Nagashima, S., Maeda, H., & Kato, A. (1999). Sintering behaviour of Al_2TiO_5 base ceramics and their thermal properties. *Ceramics International*, *25*(8), 681–687. doi:10.1016/S0272-8842(98)00083-2

Padture, N. P., Gell, M., & Jordan, E. H. (2002). Thermal barrier coatings for gas turbine engine applications. *Science*, *296*(5566), 280–284. doi:10.1126/science.1068609 PMID:11951028

Ramachandran, C. S., Balasubramanian, V., & Ananthapadmanabhan, P. V. (2012). Synthesis, spheroidization and spray deposition of lanthanum zirconate using thermal plasma process. *Surface and Coatings Technology*, *206*(13), 3017–3035. doi:10.1016/j.surfcoat.2011.10.052

Ramachandran, C. S., Balasubramanian, V., Ananthapadmanabhan, P. V., & Viswabaskaran, V. (2012). Influence of the intermixed interfacial layers on the thermal cycling behavior of atmospheric plasma sprayed lanthanum zirconate based coatings. *Ceramics International*, *38*(5), 4081–4096. doi:10.1016/j.ceramint.2012.01.066

Ropp, R. C., & Libowitz, G. G. (1978). The nature of the alumina-rich phase in the System La_2O_3-$A1_2O_3$. *Journal of the American Ceramic Society*, *61*(11-12), 473–475. doi:10.1111/j.1151-2916.1978.tb16119.x

Seraffon, M., Simms, N. J., Sumner, J., & Nicholls, J. R. (2011). The development of new bond coat compositions for thermal barrier coating systems operating under industrial gas turbine conditions. *Surface and Coatings Technology*, *206*(7), 1529–1537. doi:10.1016/j.surfcoat.2011.06.023

Skapin, S., Kolar, D., & Suvorov, D. (1993). X-ray diffraction and microstructural investigation of the Al_2O_3-La_2O_3-TiO_2 system. *Journal of the American Ceramic Society*, *76*(9), 2359–2362. doi:10.1111/j.1151-2916.1993.tb07777.x

Sreedhar, G., Alam, M. D. M., & Raja, V. S. (2009). Hot corrosion behavior of plasma sprayed YSZ/Al_2O_3 dispersed NiCrAlY coatings on Inconel-718 superalloy. *Surface and Coatings Technology*, *204*(3), 291–299. doi:10.1016/j.surfcoat.2009.07.026

Sreedhar, G., & Raja, V. S. (2010). Hot corrosion of YSZ/Al_2O_3 dispersed NiCrAlY plasma-sprayed coatings in Na_2SO_4-10 wt. % NaCl melt. *Corrosion Science*, *52*(8), 2592–2602. doi:10.1016/j.corsci.2010.04.007

Tamura, M., Takahashi, M., Ishii, J., Suzuki, K., Sato, M., & Shimomura, K. (1999). Multi-layered thermal barrier coating for land-based gas turbines. *Journal of Thermal Spray Technology*, *8*(1), 68–72. doi:10.1361/105996399770350584

Wang, L., Wang, Y., Sun, X. G., He, J. Q., Pan, Z. Y., & Wang, C. H. (2012). Thermal shock behavior of 8YSZ and double-ceramic-layer $La_2Zr_2O_7$/8YSZ thermal barrier coatings fabricated by atmospheric plasma spraying. *Ceramics International*, *38*(5), 3595–3606. doi:10.1016/j.ceramint.2011.12.076

Xie, X., Guo, H., & Gong, S. (2010). Mechanical properties of $LaTi_2Al_9O_{19}$ and thermal cycling behaviours of plasma-sprayed $LaTi_2Al_9O_{19}$/YSZ thermal barrier coatings. *Journal of Thermal Spray Technology*, *19*(6), 1179–1185. doi:10.1007/s11666-010-9529-5

Xie, X., Guo, H., Gong, S., & Xu, H. (2011). Lanthanum-titanium-aluminum oxide: A novel thermal barrier coating material for applications at 1300°C. *Journal of the European Ceramic Society*, *31*(9), 1677–1683. doi:10.1016/j.jeurceramsoc.2011.03.036

KEY TERMS AND DEFINITIONS

Al_2O_3: Aluminium oxide.
APS: Air plasma spray.
DCL: Double ceramic layer.
FESEM: Field emission scanning electron microscope.
FG-TBC: Functionally graded thermal barrier coating.
LTA: Lanthanum-titanium-aluminium oxide $LaTi_2Al_9O_{19}$.
XRD: X-ray diffraction.
YSZ: Yttria stabilized zirconia.

Chapter 6
Surface Engineering at High Temperature:
Thermal Cycling and Corrosion Resistance

John Henao
CIATEQ, Mexico

Oscar Sotelo
Universidad Nacional Autónoma de México, Mexico

ABSTRACT

Thermal and chemical properties are key aspects that determines whether a coating material is useful for high temperature applications. The progress in the application of coatings as high performance and functional materials is currently driven by their reliability under changing environmental conditions. Particularly, for high temperature applications, new gas turbines, propulsion systems, and cutting tools are outstanding examples for the development of enhanced coating systems. With the increase in the complexity of applications of coatings at high temperature, it became necessary that engineers and designers look inside into various aspects of material properties. In that order of ideas, this chapter focuses on reviewing the main physical and chemical concepts related with the performance of engineering coatings at high temperature. Especially, the subjects for reviewing here are thermal shock and hot corrosion.

INTRODUCTION

Coating technology is widely used in a number of applications associated with the protection of structural materials. Coating materials allow providing extra-protection, improving properties such as wear, corrosion and thermal resistance, phenomena that may induce catastrophic failures or damage to metallic materials. For example, in the automotive industry, transmission parts as synchronizing rings and shift forks are often coated with molybdenum using thermal spray techniques to prevent scuffing. Piston rings for diesel engines are also parts that are coated with carbides in order to improve wear resistance (Sundararajan et al., 2016).

DOI: 10.4018/978-1-5225-4194-3.ch006

In the aeronautic industry, research efforts have been devoted to the development and manufacturing of coating materials with high temperature resistance that can be able to protect the traditional turbine materials. Coating materials for high temperature protection are often deposited on combustion lines, first-stage blades and vanes, transition pieces, and other hot-path components of gas turbines (Hardwicke et al., 2013). Several ceramic coatings such as Al_2O_3, TiO_2, mullite, YSZ, and CeO_2+YSZ have been evaluated as thermal barrier coatings in gas turbine parts. Basic requirements for high temperature protection are: (1) low thermal conductivity (2) thermal expansion match with the substrate material, (3) high melting point, (4) no phase transformation between room temperature and operation temperature, (5) good adherence coating/substrate, (6) chemical inertness and corrosion resistance in hot environments. The work of engineers is being sure of selecting the coating materials according to basic requirements, however this task can usually result laborious since the number of materials satisfying all requirements are not much. In particular, two basic properties that compromise the structural stability of coatings at high temperature are the thermal expansion match and corrosion resistance (hot corrosion). Sometimes a material having an excellent thermal shock behavior may not have a good corrosion resistance and can work as thermal barrier. Thus, it is of great importance for engineers to know how these properties work together and to understand the key factors and mechanisms involved in the operation of coatings at high temperature. The present book chapter is focused on defining and reviewing the key aspects associated with thermal shock and hot corrosion of engineering coatings.

THERMAL SHOCK

Thermal shock is a concept associated with the thermal stresses that occurs in a material as a result of its exposure to a gradient of temperature between the surface and the interior and/or among various regions of the material. Thermal stresses $\left(\sigma_{th}\right)$ generated from temperature gradients during cooling are defined according to the following equation:

$$\sigma_{th} = \frac{E\alpha\Delta T}{1 - v} \tag{1}$$

where E is the elastic modulus, ΔT is the temperature difference, v is the poisson's ratio and α is the coefficient of thermal expansion. This equation is valid for bulk materials having spheres, cylinders and slabs shapes, and indicates that thermal stresses increase as the thermal expansion coefficient and elastic modulus of the material increase and also when the gradient of the temperature increases. In bulk materials, thermal stresses can give rise to crack initiation and propagation (brittle materials) and/or permanent deformations (ductile materials). With the aim of quantifying the thermal shock resistance, engineers have defined the thermal shock parameter R, which is commonly expressed in various forms as summarized in Table 1.

It is worth mentioning that E, σ, v present a different effect on the thermal shock resistance depending on their value. For crack initiation, the effects of these parameters are completely opposite than for crack propagation. For instance, high values of σ and low E and v offer high resistance to propagation of pre-exiting cracks. Another important observation is posed for ductile materials which are more resistance to thermal shock than brittle ones because they can withstand larger strains before rupture.

Table 1. Thermal shock parameters

Parameter	Definition	Interpretation	Units
$R = \frac{\sigma(1-v)}{E\alpha}$	Resistance to fracture initiation	Maximum gradient of temperature for steady state conditions	°C
$R' = \frac{\sigma(1-v)k}{E\alpha}$	Resistance to fracture initiation	Maximum heat flux for steady state conditions	Cal.cm/s
$R'' = \frac{E\gamma}{\sigma(1-v)}$	Resistance to propagation damage	Minimum in the extent of crack before initiating propagation	cm
$R''' = \frac{E}{\sigma^2(1-v)}$	Resistance to propagation damage	Minimum in elastic energy at fracture	$(psi)^{-1}$

σ ultimate tensile stress, E young modulus, v poison ratio, γ fracture surface energy, k thermal conductivity. (Hasselman, 1985).

However, ductile materials are also susceptible to damage by thermal shock since they can undergo yielding due to thermal cycling, leading to permanent deformations and thermal fatigue failure. In summary, both brittle and ductile materials are affected by temperature gradients, however thermal shock resistance is remarkably lower in brittle ones (ceramics). A bulk material with a large thermal diffusivity, a large ultimate strength, a small young's modulus, and a small thermal expansion coefficient should have a good thermal shock resistance (represented by a large value of R).

FACTORS AFFECTING THERMAL SHOCK

Overall, for high-temperature applications, coatings are used for providing thermal protection to the underneath metallic substrates. Ceramic materials are often employed to build-up coating systems due to their low thermal conductivity and high melting point. The degree of damage and strength degradation of ceramic coatings, subjected to severe fluctuating thermal environments, is a major limiting factor in relation to service requirements and lifetime performance. Such examples are found on coating systems exposed at high temperature and corrosive and wearing environments, corrosive and wearing environments such as rocket engines, gas turbines and cutting machines. Consequently, from an engineering point of view, it is essential to understand the thermal shock resistance behavior of the ceramic coating/substrate systems and comprehend factors affecting their thermal shock behavior.

Thermal shock resistance of coating systems depends on the thermal and residual stresses arising from the couple coating-substrate. Thermal stresses can rise from three sources: differential thermal contraction/expansion stresses, misfit strains, and/or thickness thermal gradients (Boley et al., 2012). Differential thermal contraction/expansion stresses occur when both coating and substrate change of temperature and their thermo-mechanical properties differ from one to another. Misfit strains are associated with the precipitation of second phases with very different thermal properties than the initial one. Thickness thermal gradient is another source of internal stresses in ceramic coatings. These types of stresses appear when the coating thickness is large enough to produce thermal contractions and expansions in the coating at the same time.

Alternatively, residual stresses are generated during the coating build-up process. These stresses are referred to those having mechanical, thermal, or chemical sources. Generally, three distinct processes have been singled out as the most conspicuous mechanisms active in inducing residual stresses in coatings. They are: 1) phase transformation of some ceramic during the cooling of the as-sprayed coating; 2) the rapid cooling of the molten spray particles from the impinging temperature to the elevated substrate temperature, namely the primary cooling or the solidification process of individual splats; and 3) the secondary cooling process of the whole coating-based system from an elevated deposition temperature to the room temperature, i.e., thermal mismatch effect (Tsui et al., 1997). Factors such as coating thickness, particle bonding, type of cooling and substrate material determine the amount of residual stresses. Of course, the degree of residual stresses will depend on the coating technique (Shen et al., 2000; Boley et al., 2012). In summary, the total internal stresses developed during a determined thermal load have the contribution from thermal and residual stresses and are associated with the driven force linked to the evolution of macro-defects and the final mechanical failure.

Because of the thermal and residual stresses, cracking and coating spalling can occur under thermal loading. The cracks may form patterns of a characteristic hierarchical structure, as can be seen from the results of a quenching experiment shown in Figure 1. It is shown that vertical surface cracking is one of the major failure mechanisms that appears in coatings and can affect coating life. This failure mechanism involves high heat flux thermal loading applied on the ceramic coating surface within a relatively short period of time. A high surface temperature and a large temperature gradient across the coating thickness are produced within seconds, which initially places the coating under high compression. As a result of the elevated temperature and compression in the hot zone, viscoplastic deformations of the coating material occur, and cause the compressive stresses to relax. Consequently, during cooling, the stresses become tensile and therefore initiate or propagate the surface cracks. The surface cracks can further induce delamination along the interface between the coating and the substrate, (Figure 2).

Due to cracking appearing, the presence of crystalline phases with a high fracture toughness is essential for keeping a good thermal shock resistance. This fact is observed in Figure 3 where two similar

Figure 1. a) Cross-section micrograph of a ceramic coating after thermal cycling (ΔT= 625 K). b) SEM image treated in the IMAGE J software to reveal the crack pattern

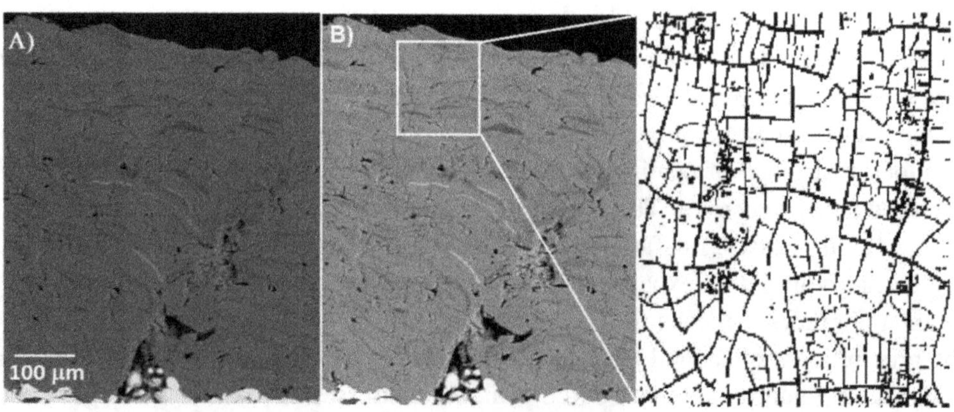

Figure 2. Thermal sprayed ceramic coating before and after 56 thermal cycles with a thermal gradient of 625 K

Figure 3. Thermal sprayed coatings a) Al_2O_3-%45 wt.TiO_2, b) Al_2O_3-%45 wt.TiO_2 with the presence of the Al_2TiO_5 phase. Coating in b) shows an improved thermal shock resistance due to the Al_2TiO_5 phase

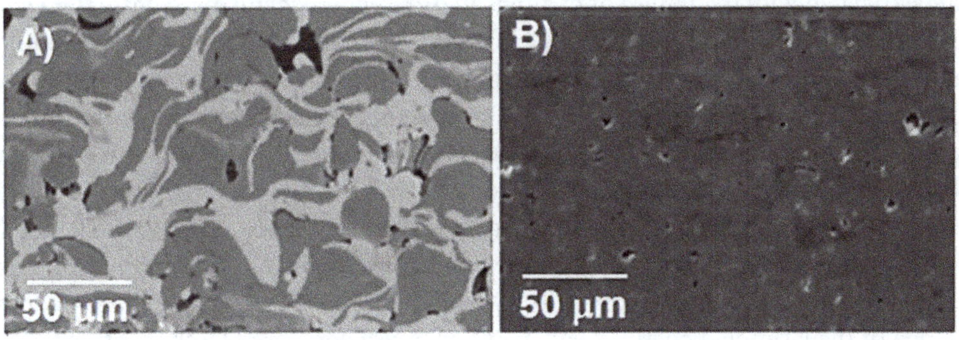

ceramic coatings were submitted to thermal load. Interestingly, the coating in Figure 3b presented a high number of thermal loads since it was composed for the Al_2TiO_5 phase; this phase is more resistance to thermal shock due to its low thermal expansion coefficient and high fracture toughness.

Operating efficiency of turbines and engines is usually dependent on the quality of their components. Coatings materials with good structural and high mechanical quality are often required to exploit the capabilities of aircraft components. Currently, one option to improve coating properties is the use of multilayer systems. However, a significant problem is that multilayer systems are not thermally stable at elevated temperatures. To extend lifetimes and increase service temperatures several modifications can be made to coating-substrate systems. Some solutions involve the utilization of the functionally graded material concept, Figure 4. A graded composition brings about beneficial changes in the thermo-mechanical properties and temperature distribution through the coating thickness and reduces the mismatch in properties between the outer barrier coating and the bond coat. A functionally graded coating is thicker compared to a single conventional coating layer. For a given thermal load, a graded coating has a higher bending (flexural) stiffness to resist the moment (created by the gradient in tensile stresses through the thickness) acting on the coating. A lower driving force for interface cracks (between the outer layer and the bond coat) is hence observed with increased compositional gradation. Further, the fracture toughness of the bond coat alloy is significantly higher than that of a ceramic one outer layer. The introduction of the bond coat alloy into the coating layers in a graded ceramic coating may enhance the fracture toughness of the whole coating system. Thus, it is expected to obtain a high thermal cycling resistance in a graded coating. For instance, in thermal barrier coatings (TBC), a graded TBCYSZ can offer higher thermal shock resistance compared to a monolithic TBCYSZ due to the enhanced crack propagation resistance across the system and the interface TBC-bond coat (Carpio et al., 2017).

On the other hand, during coating operation, a thermally grown oxide (TGO) is usually formed at the interface between the bond coat and the outer coating layer as a result of high temperatures. A number of investigations have shown that TGO also is a factor playing an important role in thermal shock resistance of coatings. For example, Evans (2001) have concluded that the thickness of TGO determines the stresses within thermal barrier coatings, and at the interface of TGO-bond coat. They found that the thickness of TGO increases with the thermal exposure time following a parabolic law. Moreover, according to Rabiei and Evans (2000) the propagation of crack in thermal barrier coatings moves toward the interface between the coating and TGO with increasing TGO thickness, existing a critical TGO thickness below delamination of the outer coating layer is exclusively within the thermal barrier coating. The composition of the TGO is another factor influencing thermal shock. Overall, four types of oxides are formed in TGO and are associated with the failure of coatings in different mechanisms. These TGO compositions are Al_2O_3, Cr_2O_3, (Ni, Co) $(Cr, Al)_2O_4$ and NiO. It is suggested that TGO must be limited in its growth by using thermal spraying techniques with low exposure to oxygen and high temperatures, for example the Cold-Spray technology. In fact, Li et al. (2010) found that a limited oxidation of a superalloy bond coat during cold spraying led to the increase of thermal cyclic life time by a factor of 10 with respect to a thermal barrier coating with a conventional bond coat. Recently, Soleimanipour et al. (2017) suggested laser cladding as an alternative method to improve thermal shock resistance of thermal spraying coatings. This technique uses a laser to deposit the bond coat. Again, as cold spray does, laser cladding is employed to avoid excessive formation of TGO which improves thermal shock resistance. Thermal shock resistance of the laser clad coatings had a 35% increase compared to conventional ones.

Figure 4. Schematic plot of a graded coating for high temperature applications

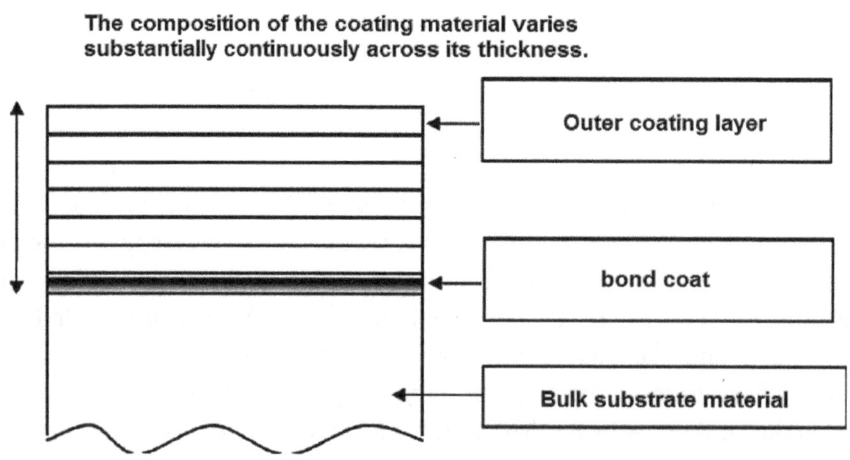

MODELING THERMAL SHOCK

A reliable and efficient numerical tool is often helpful to predict heat transfer and thermally-induced stresses in coatings. Modeling and simulation tools are commonly used to prevent and analyze failure of coatings under thermal cycling conditions. Currently, there are many researches using both analytical and numerical methods to evaluate stress state, fracture mechanics and failure mechanisms in coating materials exposed at high temperature. Some of these investigations incorporate the molecular dynamic (MD) model, finite difference method (FDM) and the finite element method (FEM) for multi purposes. Among these modeling approaches, FEM presents maximum similarity to the real operational conditions and has been used in a significant number of studies (Wang et al., 2016).

As stated in section 2, spontaneous microcracking during thermal shock is usually observed in ceramic coatings, which is attributed to misfit strains due to high temperature gradients, different coefficients of thermal expansion and the existence of various defects. One of the ways for modeling thermal shock conditions is the modeling of a bi-material structure containing an interfacial crack and subjected to a cooling shock based on a quasi-static uncoupled thermo-elasticity assumption. Alternatively, digital image-based modeling has been applied to the heterogeneity analysis of coatings in hot environments. For instance, the effect of pores and interfaces on effective properties of plasma sprayed zirconia coatings was studied based on a model constructed with object oriented finite method (Langer et al., 1997). Other studies have employed an extended FEM program to predict the effective Young' modulus and assess the fracture behavior of the coating (Michlik et al., 2006). However, up to the present time, modeling the micro-crack growth of thermal barriers by dynamic fracture criterion, based on the true microstructure, still needs further investigations.

Characterizing the fracture resistance of thermal spraying coatings, especially for thermal barriers, is of technological importance in terms of design. It can enable coating engineers to predict the expected thermal fracture response under various thermo-mechanical loading conditions. Usually, a methodology for characterizing fracture resistance of thermal spraying coatings consist on performing thermal shock experiments in conjunction with principles of crack arrest to estimate the crack growth resistance in functionally coatings. As stated by the Griffith theory (Lajtai et al., 1971), a crack propagates until the strain energy released because of the crack advance is enough to supply the surface energy required to build up the new crack surfaces shaped by its propagation. The fracture toughness of a material is often defined as the surface energy times the area of the crack surface. According to the fracture mechanics theory (Eftis et al., 1976), the energy balance for an infinitesimal crack extension (da) can be expressed as:

$$dW_E = dU + dW_P + Gd_a \tag{2}$$

where dU is the elastic strain energy in the cracked material, dW_E is the work done by the external forces, G is the strain energy released from crack extension and dW_P is the plastic work. Figure 5 shows a schematic of the strain energy release rate as a function of crack length for a crack initiating at the surface and growing through the thickness of a ceramic coating subjected to a thermal cycling. It is observed that higher the temperature, higher the energy released is, meanwhile the crack length increases.

A fracture resistance curve is also schematically illustrated in Figure 5. For a given maximum surface temperature (T_{max}) during various thermal cycles, this crack will be arrested at a location where its strain energy release rate falls below the fracture resistance of the coating at that location. Thus, for a ceramic coating with a certain architecture and submitted to thermal cycling, the final surface crack length (a_i)

Figure 5. Schematic representation of cracks growth and arrest in a ceramic coating subjected to a single heating–cooling cycle with maximum surface temperature (T_{max})

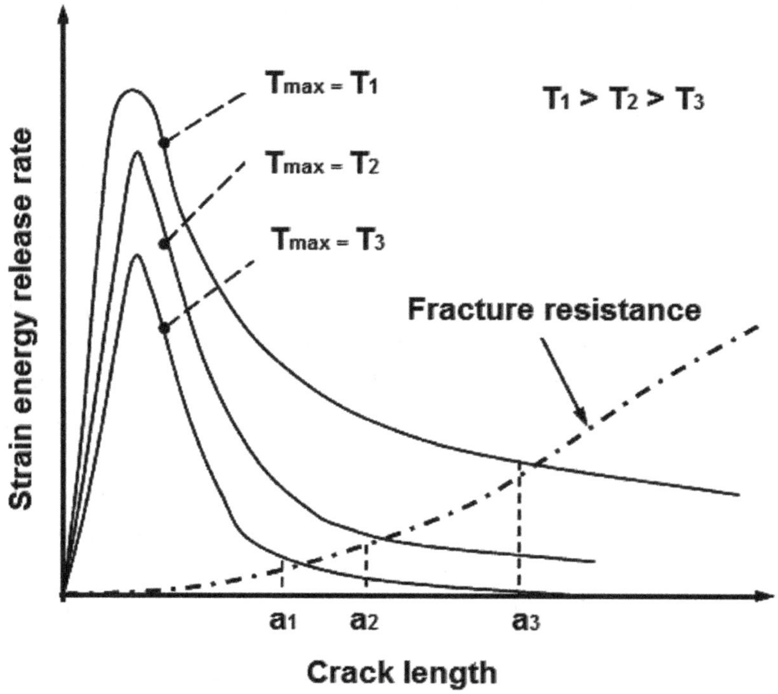

will be a function of the maximum surface temperature. Thus, for a given coating architecture, a longer surface crack would result a higher maximum surface temperature (T_{max}) during a given thermal cycle. Conversely, if the strain energy release rate for the surface crack were calculated with the measured final crack length (a_{im}), from a fracture mechanics model of I this coating subjected to a heating–cooling cycle with the corresponding maximum surface temperature T_i (also measured during the same experiment), the computed energy release rate should correspond to the fracture resistance of the coating at that location.

Using the above concepts, a surface thermal fracture analysis was carried out on a YSZ-coating prepared by High Velocity Oxy-fuel technique (HVOF) (one layer and three layers) in order to estimate the fracture resistance curves, Figure6. For the modeling part, the thermal shock experiments consisted of subjecting the surface of the specimen to a laser beam for 6 s with subsequent cooling in air. The laser beam was modeled by applying a heat flux with a Gaussian spatial distribution around the center of the top surface of the YSZ coating. The nodal temperatures were calculated by solving the transient heat transfer problem. All of these temperatures were entered into a transient structural analysis model to compute the resulting displacements and stresses. Time-dependent response of the various layers of the coating layers were considered in the analyses. The time dependent behavior of YSZ coating was described by means of the Norton–Hoff viscoplastic equation according to the equation 3 (Forestier et al., 2002).

$$\dot{\varepsilon} = A\exp\left(\frac{-\Delta H}{R * T}\right)\sigma^{n} \quad (3)$$

where ΔH , A, n, and σ are the creep activation energy, a constant, the stress-exponent and the stress value, respectively. Equation 3 describes the relation between the equivalent strain-rate and effective flow stress. The substrate material (steel 4130) was considered to be linear elastic. The modeling of the present problem was carried out with the ABAQUS/Explicit software. The elements used corresponded to the eight-node plane stress elements. The fracture resistance (J) was estimated from the simulation results using the integral method (Courtin et al., 2005). In order to compute the integrals accurately, re-meshing was employed in the vicinity of the crack tips. Thus, this analysis was useful for modeling the fracture mechanics of the system and to evaluate the effects of coating architecture on its resistance to crack growth. The inputs required for these fracture resistance calculations are shown in Table 2.

Figure 6. a) YSZ HVOF-coating experimental surface crack length determined by laser measurements after thermal cycling; b) Fracture resistance energy determined by modeling as a function of the crack length

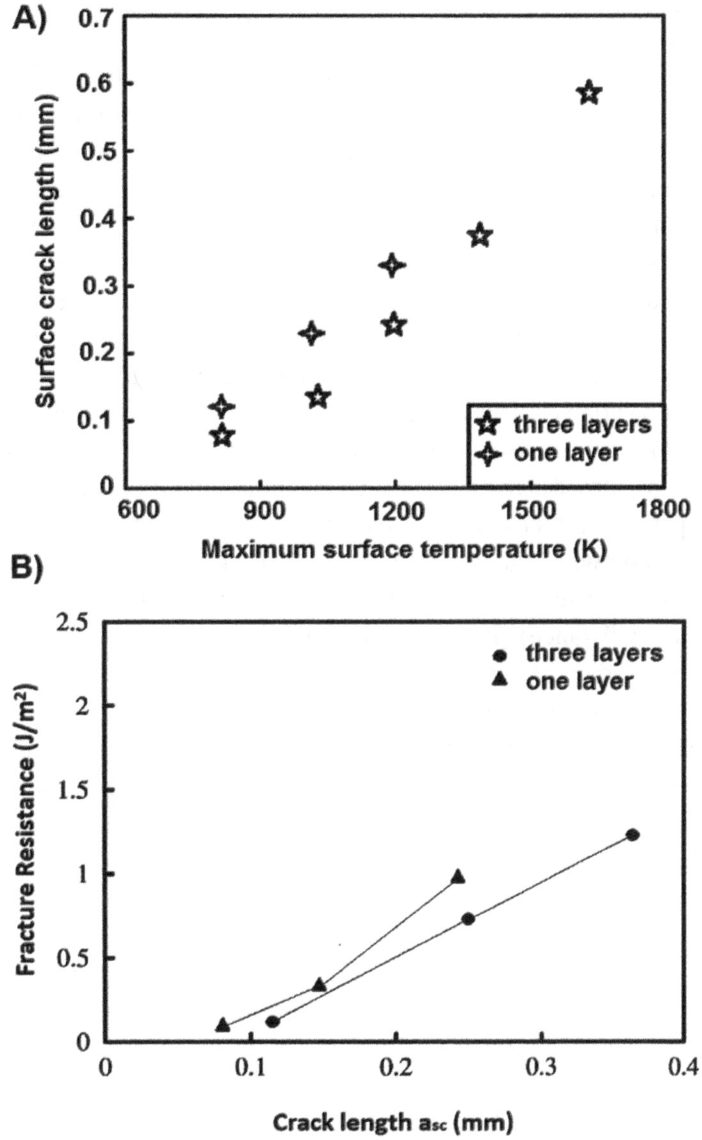

Table 2. Parameters and data required for fracture energy calculations

Parameter	YSZ- Ceramic Coating	NiCoCrAlY- Bond Coat	Steel 4130- Substrate	Units
density	6300	6290	7850	kg/m^3
Specific heat	580	460	477	J / kg K
Thermal conductivity	2.3	3.82	30.1	W/m K
Thermal expansion	7.5 E-6	10.3 E-6	11.2E-6	$^oC^{-1}$
Poisson ratio	0.25	0.3	0.29	dimensionless
Young modulus	210	64.5	205	GPa
Stress exponent, n	1.8	2.7	-	dimensionless
Activation energy, ΔH	217	150	-	kJ/mol
Constant, A	5E-7	9E-12	-	$N/m^2\ s^{-n}$

The results of the simulation show that the fracture resistance of the coating decreases when the coating thickness increased for a given crack length. Therefore, it is in agreement with the experiments where thick coatings presented smaller crack lengths at the same temperature, meaning that less energy was freed from crack formation. Although the present study takes into account the case of cracks initiating at the coating surface and growing through its thickness, it is possible to infer some results about the delamination and final failure of the coatings, since these are controlled by the growth of cracks. Similar results have been presented in recent investigations where the fracture resistance is sensitive to the coating architecture. In addition to the coating thickness, a graded composition has been also investigated in previous literature reports. In case of graded compositional ceramic coatings, the fracture resistance was lower than a non-graded one (Jian et al., 1995; Jin et al., 2008). Thus, fracture resistance increases with crack extension in the thick coatings as well as in the compositional graded ones (ceramic powder with the addition of the bond coat powder material). A graded composition brings beneficial changes in the thermo-mechanical properties and temperature distribution through the coating thickness and reduces the mismatch in properties between the ceramic outer layer and the bond coat. The effect of a graded composition is superior to the effect of having a thicker coating on the fracture resistance (more than three times higher fracture resistance) (Kokini et al., 2002).

Modeling has been also used to analyze micro-crack growth of ceramic coatings during thermal shock, employing digital image processing. This kind of study have been useful to determine the behavior of coatings based on its real microstructure. Characteristics such as pores, interfaces and Young' modulus has been investigated in order to evaluate the fracture behavior of coatings during thermal shock (Wang et al., 2003; Matejicek et al., 2003). Modeling in this way consists in using microstructural digital images of the coating followed by the definition of a finite element grid which is obtained by performing a segmentation of the real microstructure. Then, based on the finite element grid, the Tuler–Butcher failure criterion is employed to model the micro-crack growth of a ceramic coating undergoing thermal shock. This dynamic fracture criterion associates the fracture time with the maximum stress, suggesting that the fracture phenomenon is not instantaneous but it requires necessary time duration for the failure (the damage is cumulative). Mathematically, Tuler–Butcher failure criterion is defined according to the following equation (Tuler et al., 1968):

$$\sigma = \int_0^t \left(\max\left(0, \sigma_1 - \sigma_0\right)\right)^2 dt \geq K_{failure} \tag{4}$$

where σ_0 is a specified threshold stress, σ_1 is the maximum principal stress, and K_{failure} is the stress for failure. This general fracture criterion states that when the damage integral on some particular plane exceeds a critical value, fracture occurs on that plane. This model has shown to simulate crack formation with a good agreement against experimental observations of thermal barriers coatings after thermal shock.

Recently, one more technique has raised to simulate the crack propagation behavior under thermal shock conditions. This technique is the cohesive zone model (CZM). The CZM may be used to characterize the constitutive behavior of the interface under thermal cycling conditions. This method introduces failure mechanisms by using the hardening-softening relationships across the interface. In addition, the CZM method is also well suited for simulating the fracture process in a homogeneous medium. This method uses contact elements to allow the separation of the surfaces as well as a cohesive material model to describe the separation behavior of the surfaces. The CZM method applies to both the simulation of fracture in a homogeneous material and also for interfacial delamination along the interface between two materials. For instance, Hille (2009) have been one of the authors using the CZM method. They investigated the fracture patterns and evolutions of the overall crack growth of a thermal barrier coating. Their simulation results evidenced the importance of accounting for the effects of plasticity in the bond-coat layer and anisotropy in the top-coat layer which was in good agreement against experimental observations.

To summarize, at the present time, the research progress modeling has shown that computational tools are useful for analysis of coatings exposed at high temperature. Especially, the FEM models are commonly employed since they can account on factors such as heat conduction, heat convection and heat radiation at high temperature and geometry factors which make more realistic the simulation results. However, due to the irregular microstructure of the thermal spray coatings, micro-cracks, laminar interface, pores and layer interface co-existed at random, which make modeling a more difficult task. When two cracks in a given coating intersect, their behavior can be unpredictable. For example, these will become one crack and propagate continuously along the different direction or they can propagate separately along the different directions or they will stop. The question is how to predict the response of the material? When multiple cracks propagate toward to the interface, once they reach to the interface, these will propagate, but we cannot predict the way how these will propagate exactly. All of these aspects are important for the analysis of thermal barrier coatings under thermal cycling and have become an important research direction for the following years.

EVALUATION OF THERMAL SHOCK

One of the key points in the coating industry is the assessment of the coatings under environmental and operational conditions. For instance, aviation and industrial gas turbine manufacturers apply various thermal shock tests before release coating products to the market. The most common example are usually thermal barrier coatings which are used usually under Thermal Shock conditions in applications such as gas turbine blades and vanes. Currently, there is not an identified method that would provide advantage in comparing the results of the test with respect to the thermal shock behavior of coatings. This fact is mainly associated with the difficulty of finding a test method that can completely reflect the real operational conditions. At the present time, tests that are developed to simulate the real thermal shock conditions are the jet engine thermal shock test, burner heating thermal shock test and furnace cycle test.

In jet engine-thermal shock test, burner equipment is used to create a wide thermal gradient along the coating and thermo-mechanic stresses on the surface. This test is very suitable to provide data on the performance rating on the ceramic coating. By creating a large temperature gradient over the ceramic coating, surface temperature can rise up to temperatures superior to 1200°C and, therefore, sintering effect act on ceramic top coat can be tested during this method. In this test, the main stresses occur thermo-mechanically at the interface of the ceramic and bond coat. Due to the high temperature gradient within the ceramic layer, the coating/bond coat interface is oxidized at a small rate which allows to study fracture resistance of coatings as a function of the properties of the bulk electrode. Usually, the sample geometry is a disc shaped of around 25.4 mm in diameter. A typical test consists of 20 seconds of heating period, 20 seconds of cooling period (water immersion or nitrogen gas) and 40 seconds of waiting period in open atmosphere (Guo et al., 2016).

Alternatively, burner thermal shock test consists in creating high temperature gradients in the ceramic coating, introducing stresses that affect the integrity of ceramic coating. This test is based on cooling of the specimen after heating by a flame where propane and oxygen gases are used together. The specimen is often a disc shaped geometry. Since burner heating thermal shock test unit is an expensive system, jet engine thermal shock test results an economic alternative method for the thermal gradient tests (Kawasaki et al., 1996).

On the other hand, the furnace cycle test is another method widely used, primarily in aviation applications. This method is quite used since it better reflects the actual engine conditions. This process not only causes cyclic stresses in ceramic coating, but also promotes degradation of the bond coating as a result of severe oxidation. Consequently, complete failure and depletion of bond coating can be studied. The furnace thermal shock test is carried out in a furnace while air-cooling is implemented with the aid of a compressor. According to aviation standards, a cycle for aircraft engine component consists of 1 hour period, 45-50 minutes of which is in elevated temperatures and 10-15 minutes is spent for cooling. Nevertheless, for industrial thermal shock tests, the duration of exposure to high temperature is extended to cycles of 24 hours in elevated temperatures (1000 °C - 1200 °C). A period of 1 hour is spared for cooling at room temperature. The samples used in these tests are usually disc shaped and of 25.4 mm diameter and the criterion for failure is defined as a 10-20% spallation of coated surface (Bolcavage et al., 2004).

An alternative to the furnace cycle test for gas turbines have been recently proposed. The alternative method consists in heating coating samples at high temperatures with durations generally of 5 minutes heating and 2 minutes cooling (Wang et al., 2012). Cooling is often carried out in a liquid media (oil or water). Under these conditions, failure criterion is based on visual inspection of the coating surface for damages or loss of the coating. Overall, a total surface area of coating loss ranging from 10 to 20% is also considered as the criterion for failure. Figure 7 shows the images of Al_2O_3-TiO_2 coatings prepared by flame spray and submitted to furnace thermal shock.

As observed from Figure 7, this method is based, primarily, in the observation of macro-crack growth and delamination of coatings. The furnace thermal shock test with water cooling is considered a good method since it recreates operational conditions during heating and intensifies the cooling step, thus the failure is promoted by overestimated thermal shock conditions. It is supposed that in real conditions a given coating must have an extended life than that shown during the test.

Figure 7. Al_2O_3-TiO_2 flame spray coatings submitted to furnace thermal shock test at 650°C and water cooling

HOT CORROSION AND ITS PREVENTION BY MEANS OF THERMAL SPRAY COATINGS

"Hot corrosion" is the accelerated oxidation of a material at high temperatures which is often conducted by the presence of molten salts in the surface of metallic structures. While aqueous atmospheric corrosion is often controlled by the diffusion of dissolved oxygen in water, the hot corrosion process is controlled by the presence of soluble oxidants at the surface of metallic materials at high temperature (Arenas et al., 2005). Hot corrosion can appear in structures and equipment operated at high temperature such as internal combustion engines, gas turbines, boilers, fluidized bed combustion, and industrial waste incinerators (Sidhu et al., 2006). In hot corrosion, the environment plays an important role in terms of the level of degradation produced. For example, power plants, that use low-grade fuel oil, develop corrosive compounds such as V_2O_5, Na_2SO_4, $NaVO_3$. On the other hand, waste-to-energy power plants the main corrosive compounds are those associated with chloride-containing compounds such as $ZnCl_2$, KCl, NaCl, and PbCl (Porcayo-Calderón et al., 2012). The nature of the molten salt and its concentration will determine the severity of the corrosion phenomena. Anyway, it is crucial to understand when hot corrosion may appear and how it can be prevented. For instance, Figure 8 shows oil pipes where condensed films of molten salts were formed due to their operation at high temperature. In recent years, various studies have suggested that the operational life of these kind of pipes can be extended for several years by using thermal spray coatings (50%Ni-50%Cr) (Yamada et al., 2002).

At the present time, thermal spray coatings are a good option to prevent corrosion, especially, at high temperature. There are different thermal spray processes such as flame spraying, arc spraying, plasma spraying, cold spraying, and high velocity oxygen fuel (HVOF). The high velocity oxygen fuel (HVOF) process is one of the most popular thermal spraying technologies and has been widely adopted by many industries due to its flexibility, cost effectiveness and superior quality of the coatings produced. The HVOF process offers coatings with high bond strength, high hardness, and low porosity, having in

Figure 8. Pipes with condensed films of molten salts

most of cases superior mechanical properties than coatings prepared by means of other thermal spray techniques (Sidhu et al., 2006).

The architecture of thermal spray coatings is a key factor to prevent corrosion. For instance, low porosity is often needed to promote high corrosion resistance. Because of aggressiveness of species such as chlorine, vanadium, and sulfur, dense coatings are required in order to prevent permeation of these species. The penetration of molten salts into the coating may induce spallation and degradation of the metallic substrate. Recently, some studies have reported two possible mechanisms of molten salts permeation into thermal spray coatings. One way is penetrating the coating surface through a path leading directly to the substrate, see Figure 9, line A. Alternatively, the molten salts may permeate following a zigzag path, see Figure 9, line B. The second path is considered the most favorable to prevent corrosion since the corrosion process will occur along splat boundaries instead of taking place at the surface of the metallic substrate (Qin et al., 2017). Overall, low porosity is fundamental to avoid permeation of molten salts.

On the other hand, it is worth mentioning that inducing oxidation of the particles during the coating build-up process is beneficial for preventing hot corrosion. Although oxidation can affect the bonding force coating / metal substrate, oxide films are aimed to protect the outer layers against corrosion. For example, the formation of Cr_2O_3 oxide films in 80Ni20Cr coatings prepared by flame spray remarkably improves hot corrosion resistance. Figure 10 shows the cross-sectional microstructure of the 80Ni20Cr coating. It is possible to observe the presence of light and dark phases where the dark ones correspond to Cr oxides and partially oxidized particles. Overall, factors such as porosity, oxidation, residual stresses, and thickness must be always controlled in order to obtain dense coatings with high bonding strength and protective characteristics.

The composition of coatings is another key factor for high temperature applications. Ni-base, Co-base and Fe-base super alloys are often used for their good corrosion resistance and thermal shock resistance. These super alloys usually contain elements such as Cr, Al and Si since these elements allow to develop

Figure 9. Schematic showing parallel splat structure and possible permeation paths in the coating
Adapted from (Qin et al., 2017).

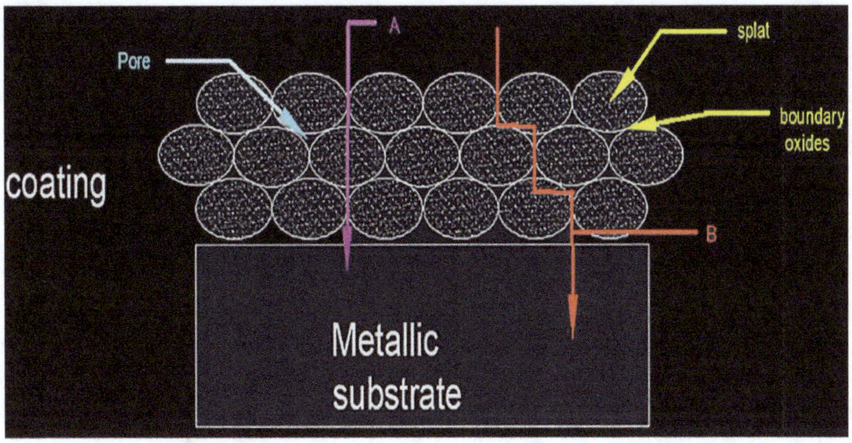

stable and protective oxide layers (Cr_2O_3, Al_2O_3 SiO_2) at high temperature. For example, Ni-Cr alloys are commonly used as protective coatings and/or bond coats since this type of alloys can develop stable oxide layers which are resistant to molten salts. Overall, the formation of Al_2O_3 and Cr_2O_3 is often wanted to obtain hot corrosion resistance in coatings. The formation of SiO_2 is specially to get hot corrosion resistance in acidic conditions. In fact, in the case of selecting a coating composition for hot corrosion is very important to know the operational conditions of the coating since basic or acidic conditions can affect the performance of the protective oxide layer (Sidhu et al., 2005; Rapp 2002). In the next section, a case study about hot corrosion of the 80Ni20Cr coating will be presented.

Electrochemical Evaluation in Molten Salts of a 80Ni20Cr Coating

Potentiodynamic polarization curves were carried out at 450 °C for HVOF coatings, Flame spray coatings and stainless steel 304 (SS304) substrates. The results of polarization tests are shown in Figure 11 and Table 3. The SS304 substrate shows the most noble behavior among the material studied with a corrosion potential of -141 mV. However, it does not show a passive zone, which is attributed to the

Figure 10. SEM micrograph of a 80Ni20Cr (%wt.) coating prepared by flame spray

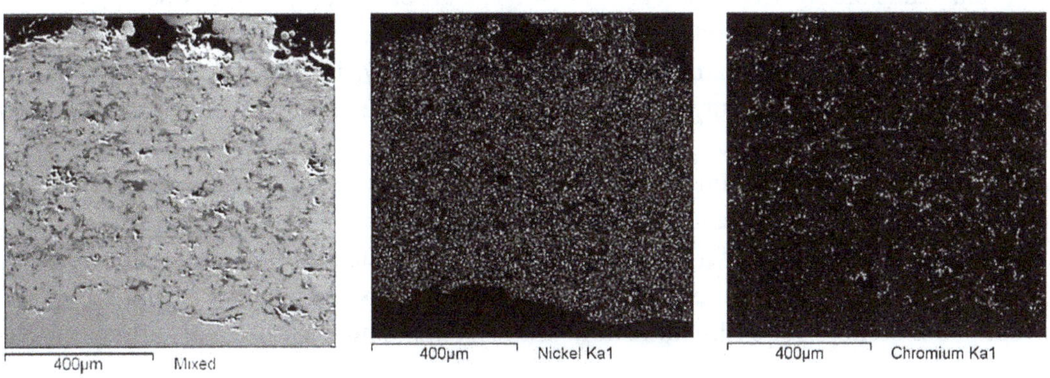

Figure 11. Polarization plots of the stainless steel 304 substrate and 80Ni20Cr coatings prepared by HVOF and flame spray. The tests were carried out in $ZnCl_2$-KCl molten salts at 450 °C

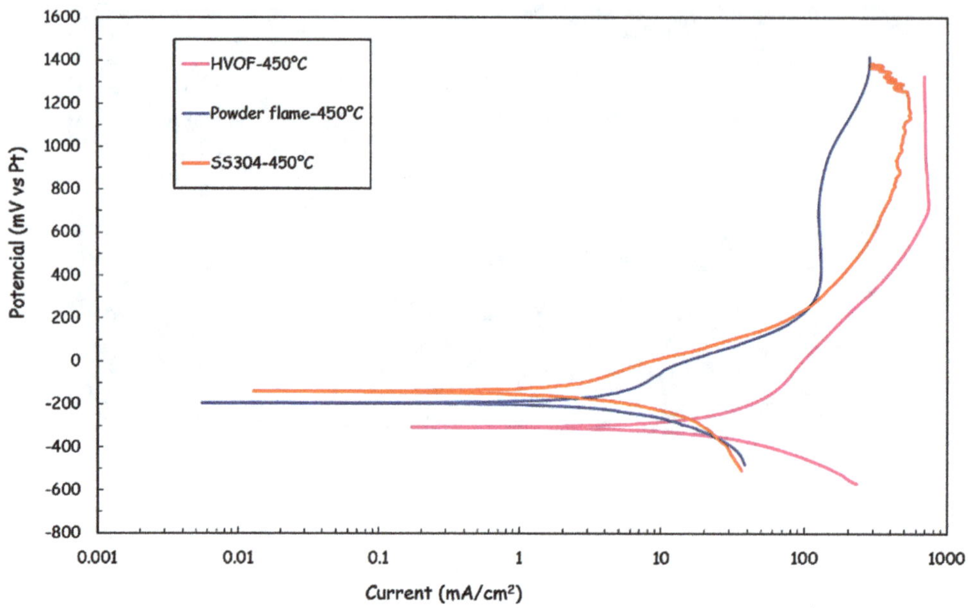

Table 3. Electrochemical parameters of potentiodynamic polarization tests

Materials	E_{corr} (mV)	ba	bc	I_{corr} (mA/cm^2)
HVOF	-306	698	300	36.19
Flame spray	-194	271	153	3.61
SS 304	-141	188	118	1.89

lack of a stable protective oxide layer. Alternatively, 80Ni20Cr coatings prepared by flame spray show a passive zone in the anodic branch. This fact is related to the presence of oxide films produced in the coatings during the spraying process. In the case of HVOF coatings, the corrosion potential value was -306 mV. The anodic polarization curves of these coatings reveal the occurrence of a continuous dissolution process. It also presented the highest corrosion current density in comparison with the other materials evaluated (Porcayo-Calderón et al., 2014). The absence of a passive zone and the lack of an exponential increase in the corrosion current is attributed to the smaller thickness of the oxide film at the top layer with respect to the coatings obtained by flame spray. The thickness variation of the oxide film between HVOF and flame spray coatings is the result of the short residence time of the particles in the HVOF process.

One simple way to study the film formation and passivation of materials immersed in molten salts is by monitoring the corrosion potential as a function of time. The increase of potential in the positive direction indicates the formation of a passive film, meanwhile a steady state potential indicates that the film remains intact and accomplish its protective role. A drop of potential in the negative direction indicates breaking, dissolution, or no film formation. The test temperature significantly affects the corrosion potential value of the alloys. Usually, increasing the temperature leads to boost the aggressiveness of the molten salts, and consequently, the corrosion resistance will only depend on the protective capacity of the oxide films formed on the alloys (Gurrappa, 2001).

According to Figure 12, the SS304 substrate, at 450°C within the first hour of testing, presents a sharp drop in the corrosion potential value from -500 mV to -950 mV and then slightly increases to -800 mV at the end of the test. This fact indicates a strong attack of the material and a subsequent attempt to form a protective oxide. Alternatively, the 80Ni20Cr coating prepared by flame spray presented a sharp drop in the corrosion potential values from -330 mV to -850 mV with a further increase up to -400 mV. This behavior indicates that these coatings showed a dissolution process of their oxide films at the beginning of the test and then experienced a passivation process (Porcayo-Calderón et al., 2012). On the other hand, the HVOF coatings presented a sharp drop in the corrosion potential value from -480 mV at -976 mV, and subsequently, it increased up to -572 mV. This behavior was similar to that observed in the flame spray coatings.

Linear polarization measurements at 450°C (see Figure 13) were also carried out in the present study.

Figure 13 shows a decrease in the values of current for the SS304 substrate during the first hour of immersion, however, at end the test, these values has the tendency to increase steadily. This behavior shows that, at the beginning, a protective oxide layer covered the material, however, this layer was not stable. These observations are consistent with the aspect of the SS304 substrate after the corrosion test (Figure 14a). According to Li (2003), the SS304 substrate shows intergranular corrosion. This kind of degradation does not occur in reducing environments. The behavior observed at 450° C is consistent with other studies (Lu et al., 2008), which indicate that at the beginning of the corrosion test, the corrosion rates will always be greater due to the presence of Cl^- anions. This fact happens because the molten salts dissolve the oxide layer and penetrate it rapidly.

Figure 12. E_{corr} values as a function of testing time for the SS304 substrate and 80Ni20Cr coatings prepared by HVOF and flame spray process. The tests were carried out in $ZnCl_2$-KCl molten salts at 450 °C

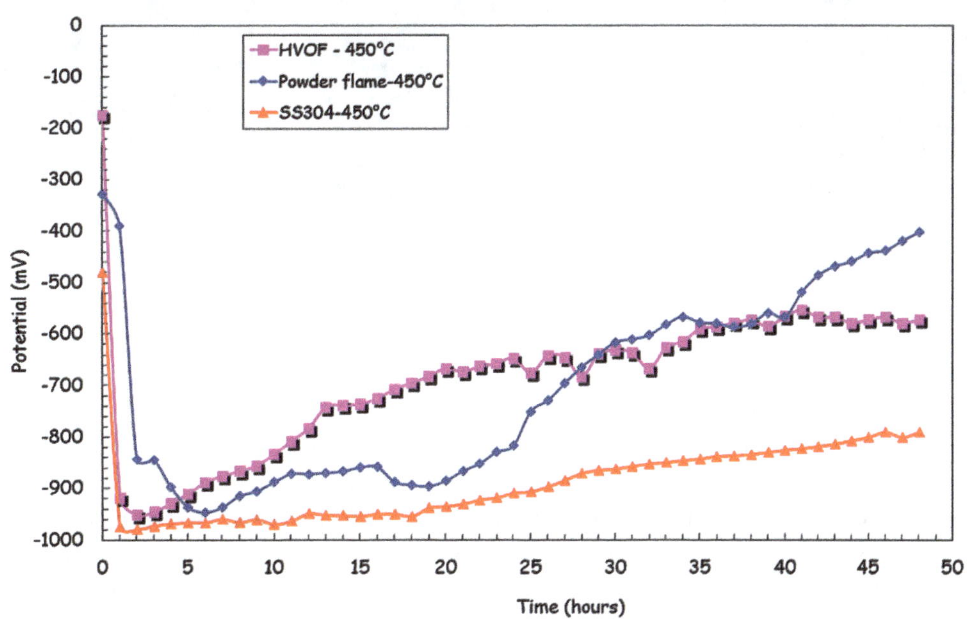

Figure 13. Current density (I_{corr}) plots of the SS 304 substrate and coatings in $ZnCl_2$-KCl molten salts at 450 °C

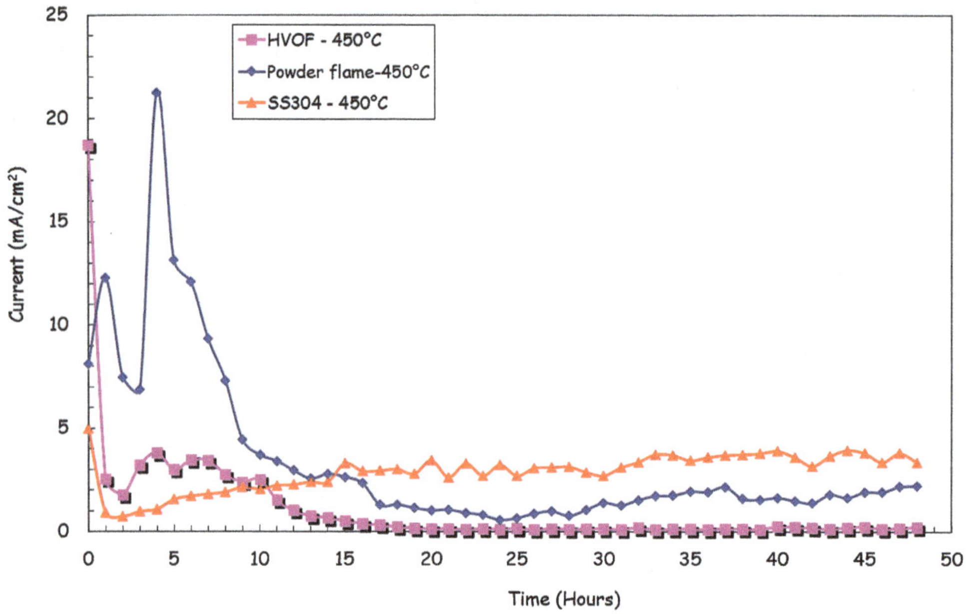

Figure 14. Cross-sectional micrograph of the SS 304, 80Ni20Cr coating prepared by HVOF and flame spray, after corrosion test in $ZnCl_2$-KCl at 450 °C

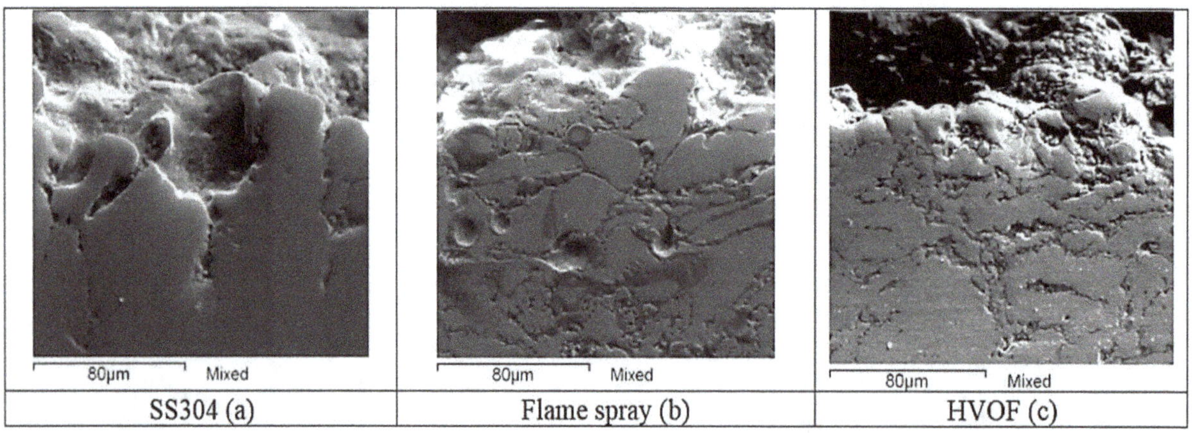

On the other hand, the flame spray coating presented a considerable increase in the value of current within the first 5 hours with a subsequent tendency to decrease. This fact is associated with the dissolution of protective oxides by the corrosive action of molten salts (Lu et al., 2008). Figure 14b shows the degradation of this coating as a result of the penetration of the molten salts (dark phases correspond to the chromium oxides formed during the spraying process). Alternatively, HVOF coatings showed the best corrosion behavior, presenting a decreased in the value of current density from 18 mA/cm^2 to 0.87 mA/cm^2. It presented the lowest rate of corrosion. This behavior is associated with the low porosity and formation of oxide films (mainly Cr_2O_3), see Figure 14c.

The impedance technique was also employed in the present study to evaluate the reaction kinetics and mechanisms involved in hot corrosion. Nyquist diagrams are often useful to get information about the corrosion process. For instance, the high frequency measurement of the Z 'axis delivers a measurement of the ionic solution or electronic resistance of the layer (molten salt), Rs, while the low frequency measurement delivers an estimate of the charge transfer process and resistance of the solution (Rs + Rct) (Stoff et al., 1991). On the other hand, three frequency regions referring to the high, intermediate, and low frequency values are obtained from Bode diagrams (Gonzalez et al., 1991; Tamilselvi et al., 2006). In the higher frequency region (f > 1000Hz), the Bode plot exhibits a plateau (horizontal line) of the $|Z|$ values with the phase angle approaching 0°. This fact is the response of the electrode ohmic or solution resistance, Rs, which includes the electrolyte resistance, cell geometry, impedance of the conductors, and the reference electrode. In the middle frequency region (1000 to 10Hz), the spectrum displays the maximum phase angle approaching −90° and a linear slope of about −1 in log$|Z|$ as log(f) decreases. This fact is the characteristic response of capacitive behavior of the electrode and describes the dielectric properties of the electronically conducting surface film. In the lower frequency region (f <10Hz) are detected the electron charge transfer process, the mass transfer processes, or other relaxation processes taking place at the film-electrolyte interface or within the pores of the surface film.

The Nyquist diagram, Figure 15, show two capacitive-like semicircles. In this case, the SS304 substrate has the highest charge transfer resistance (bigger loop), this behavior indicates an increase in the corrosion resistance and formation of a protective layer on the alloy surface. Of course, the plot Bode presents a higher resistance and phase angle. It is known that the magnitude of the maximum phase angle is associated with a more capacitive response of the protective oxide. Then, this behavior can be associated with the growth of a protective oxide and by the accumulation of corrosion products at the substrate surface which limited its contact with the corrosive agent.

Figure 15a. Results of EIS measurements for stainless steel 304 and coatings (HVOF and Flame sprayed) in molten salt $ZnCl_2$-KCl at 450 °C for 24 hours: Nyquist

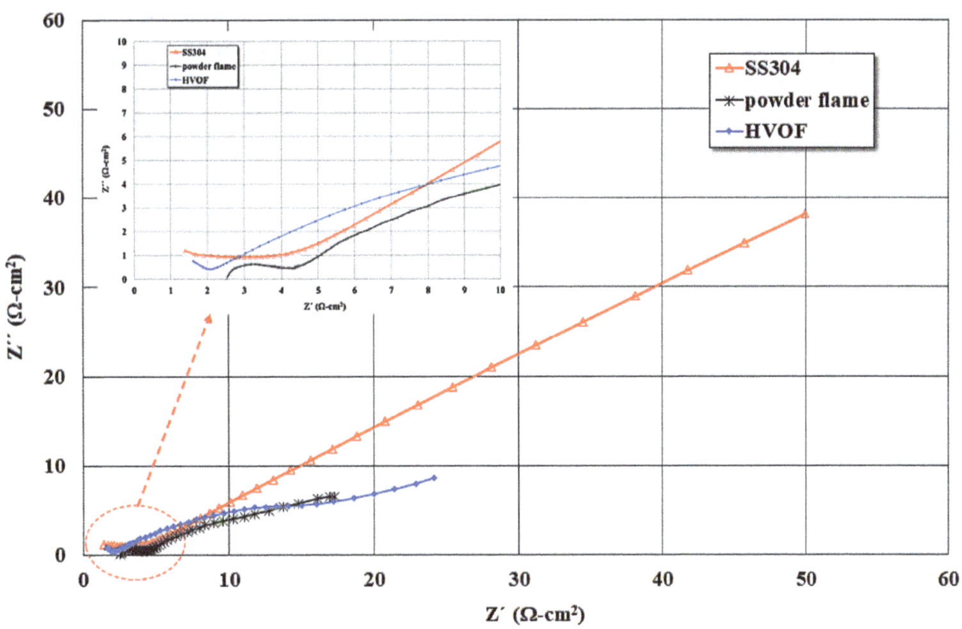

Figure 15b. Results of EIS measurements for stainless steel 304 and coatings (HVOF and Flame sprayed) in molten salt $ZnCl_2$-KCl at 450 °C for 24 hours: |Z|

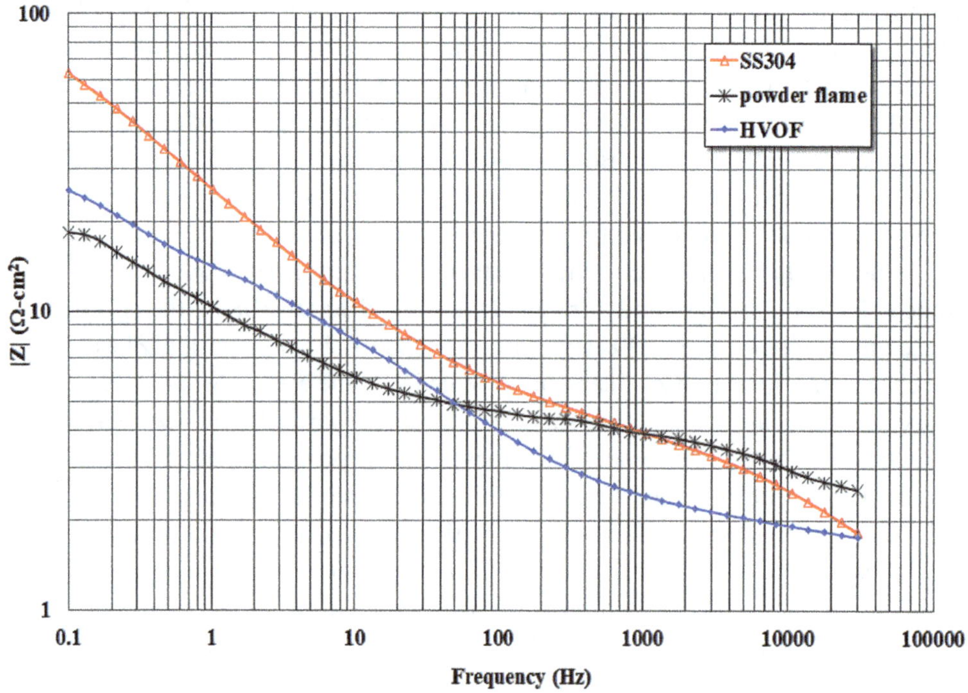

Figure 15c. Results of EIS measurements for stainless steel 304 and coatings (HVOF and Flame sprayed) in molten salt $ZnCl_2$-KCl at 450 °C for 24 hours: Angle phase

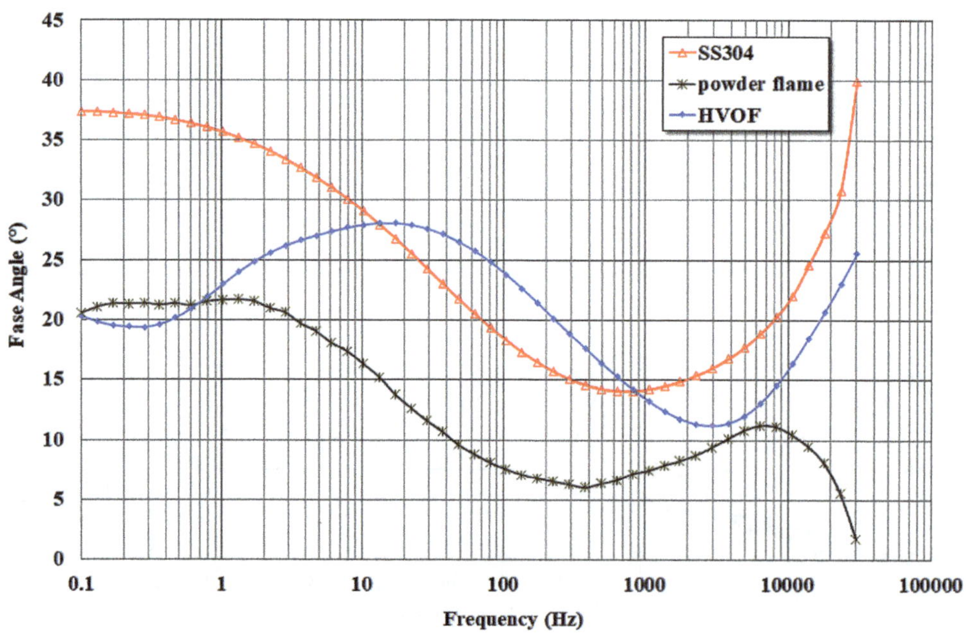

Both HVOF coatings and flame spray coatings also presented a high charger transfer resistance, but this time, the charge transfer resistance was lower in comparison to SS304 substrate. Also, those presented a two capacitive-like semicircle. This behavior indicates that the corrosion process was under charge transfer from the metal to the electrolyte through the double electrochemical layer.

MODELING HOT CORROSION

Equivalent circuit models are often useful for interpreting EIS data from hot corrosion experiments (Gonzales-Rodrigues et al., 2006). A wide variety of equivalent circuits are found in the literature. Those circuits are selected depending on the characteristics of the system. The Randles Circuit has been one of the most successful models in hot corrosion of thermal spray coatings since it allows to account for as-prepared and non-active coatings (Zeng et al., 2001). The Randles model can be described as the equivalent circuit for the impedance response which is characterized by a large semicircle (Nyquist diagram representation) (see Figure 16), where R_s represents the resistance of the molten salt, C_{dl} the capacitance of the double layer at the molten metal interface, and R_{ct} the electrochemical transfer resistance (charge transfer resistance).

Another equivalent circuit, derived from the Randles one, accounts for active coatings and their porosity, which are facts associated with reactions controlled by diffusion. This equivalent circuit is often represented with a semicircle at high frequency and a line at low frequency. Where R_{ct} represents the charge transfer resistance; and Z_w the resistance of Warburg i.e. diffusion of oxidant species. (see Figure 17).

Alternatively, a third equivalent circuit is often used. This alternative model represents materials with a tendency of building up a protective layer. In this case, the transport of ions is the limiting factor (see Figure 18). The corresponding Nyquist diagram is composed by double capacitive loops (two semicircles) where C_{ox} represents the capacitance of the oxide and R_{ox} represents the transfer resistance in the protective layer.

On the other hand, another equivalent circuit (see Figure 19) can be employed to model the corrosion behavior of thermal spray coatings. This model represents the response of an active metal undergoing localized corrosion. When localized corrosion occurs, the adjacent surface area of a pitting point is also susceptible to be corroded. In that case, the adjacent zone is characterized for undergoing corrosion at

Figure 16. Randles Circuit

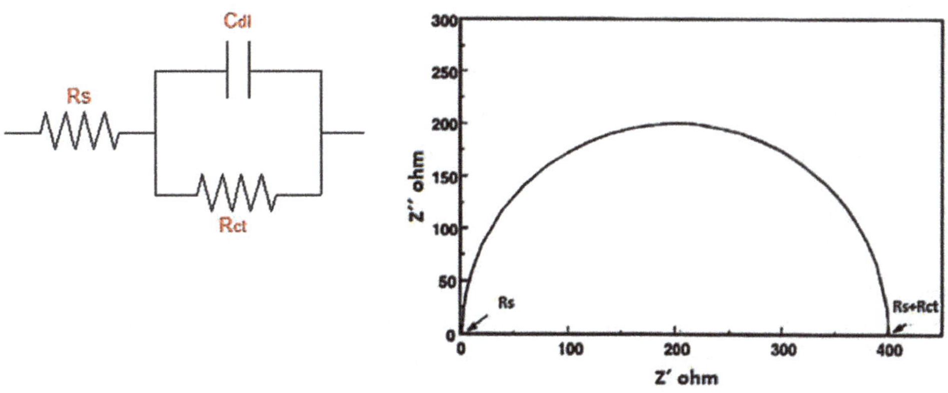

Figure 17. Equivalent circuit for modeling EIS data of metals forming a porous layer

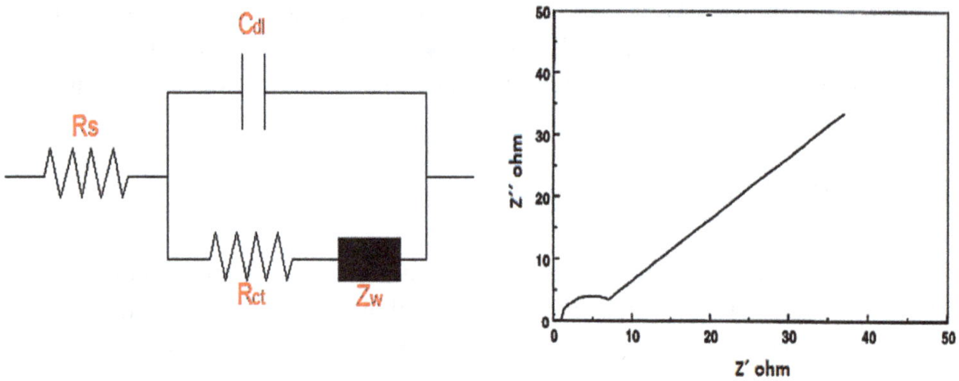

Figure 18. Equivalent circuit for modeling EIS data of metals forming a protective layer

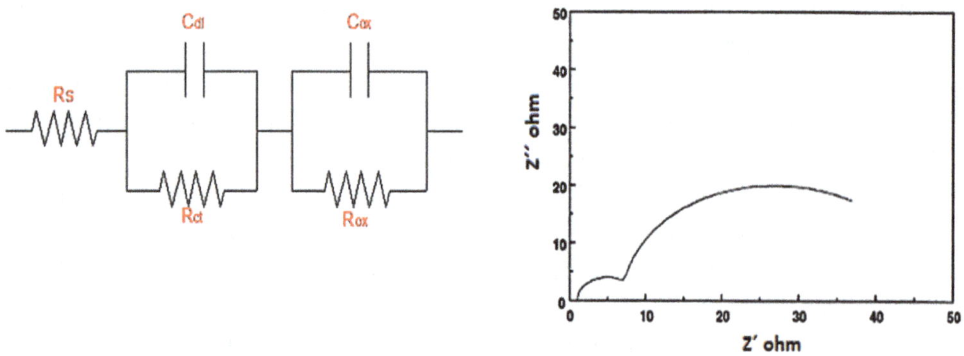

a slow rate. This fact is considered into the model represented in Figure 19, where C_{dl} and $R_{ct,}$ are the capacitance of the double layer and the charge transfer resistance within the surface area undergoing localized corrosion. C_{ox} and R_{ox} are the capacitance of the oxide and the charge transfer resistance within the corrosion zone, respectively. The Nyquist diagram of this model corresponds to an equivalent circuit consisting of two capacitive loops.

Overall, the choice of a model, representing the corrosion behavior of thermal spraying coatings at high temperature, will be based on the following criteria (Metikos-Hukovic et al., 2006; Valero-Vidal et al., 2011).

- The model must contain a small number of elements.
- The errors associated with each element must be up to 5%.
- It should be used a function defined as $Z_{CPE} = \left[Q\left(jw\right)^{n}\right]^{-1}$. This function is used for accounting defects influencing corrosion such as the roughness. Here, Q is the time constant, n is the phenomenological coefficient, w is the frequency (rad / sec) and j is $\sqrt{-1}$. The parameters "Q" and "n" are independent of the frequency. Moreover, the parameter "n" can be take values between 0 ≤ n

Figure 19. Equivalent circuit for modeling EIS data of metals forming a layer that experiment pitting corrosion

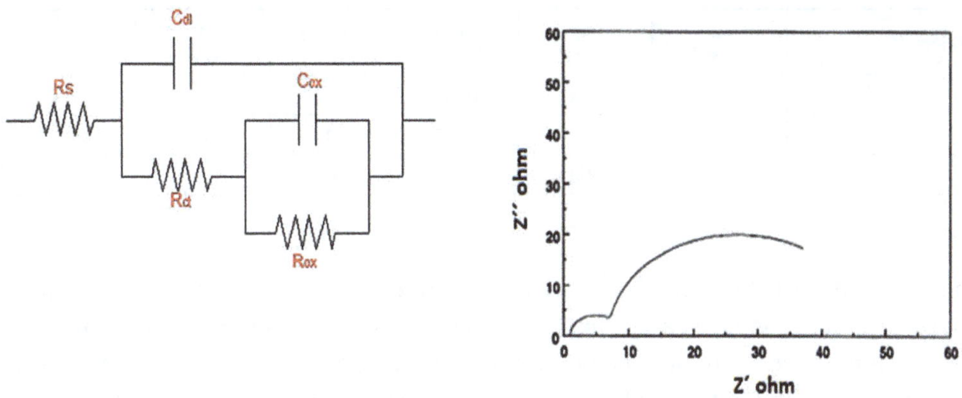

≤ 1. A value of "n" equal to 1 describes an ideal capacitor, meanwhile a value of "n" equal to 0 describes an ideal resistor. A value of "n" equal to -1 represents an inductor. A value of "n" between 0.5 <n <1.0 describes a distribution of dielectric relaxation times in the frequency space. Finally, when the value of "n" is equal to 0.5, it represents a Warburg impedance with diffusional character.

According to the description of the models in the present book chapter, an equivalent circuit model was employed for modeling the impedance data of the SS304 substrate, HVOF coatings and flame spray coatings, see Figure 20. Where R_s is the resistance of the molten salts, R_2 and C_2 are the resistance and capacitance of the layer corresponding to the corrosion products, respectively. R_1 and C_1 are the resistance and capacitance of the oxide layer. The modeling results are presented in Table 4.

Figure 20. Equivalent circuit used for modeling SS 304 substrate, and the 80Ni20Cr coatings prepared by means of HVOF and flame spray

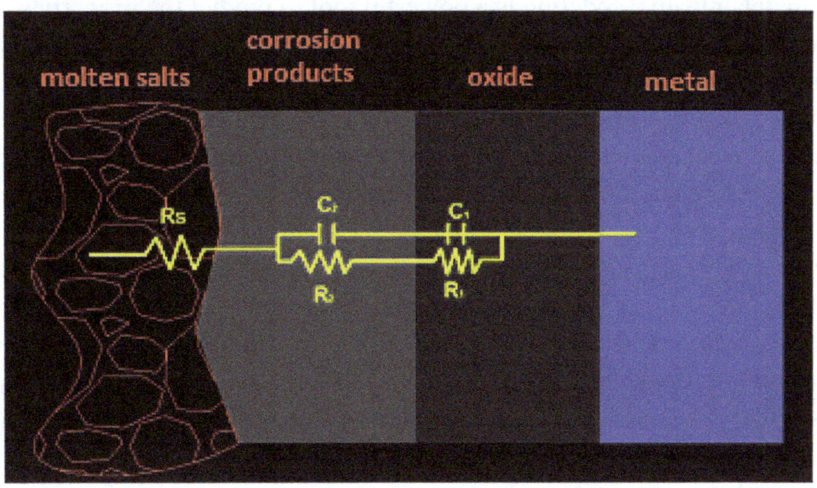

Table 4. Parameters calculated for the SS 304 and 80Ni20Cr coatings equivalent circuit tested for 24 h at 450 °C

Material	Rs ($\Omega.cm^2$)	R_1 ($\Omega.cm^2$)	Q_1 (F/cm^2)	n_{Q1}	R_2 ($\Omega.cm^2$)	Q_2 (F/cm^2)	n_{Q2}
SS 304	1	1000	0.018	0.45	2.12	0.00003	0.843
Powder flame	2	250	0.0476	0.5143	6	0.00109	0.5722
HVOF	1.458	500	0.7	0.9	36.7	0.024	0.402

Modeling results from Table 4 show a similar "R_s" value for all materials. However, the "R_1" values are considerably different for each case. The "R_2" value is lower than the"R_1"value which was expected since the corrosion products were in contact with molten salts, experiencing a continuous dissolution.

The "nQ" values are related to the charger transfer process obtained from the impedance diagrams, thus, if the Nyquist diagram gets a "nQ" value equal to one, it is possible to have a semicircle with high charge transfer resistance, but if "nQ = 0.8", the charger transfer resistance decreases, resulting in a small semicircle. Moreover, if "nQ = 0.5", the charger transfer resistance becomes even smaller. Some authors indicate that a "nQ" value of around 0 .5 indicates a diffusional behavior (Pan et al., 1998). The stainless-steel substrate and both HVOF and flame spray coatings turned out a "nQ" value of around 0.5. This fact is attributed to a possible penetration of aggressive species i.e. chlorine. The HVOF coatings presented the highest "nQ" value and the lowest porosity and penetration of aggressive species.

CONCLUSION

Thermal shock behavior, design and life prediction issues, modeling concepts and corrosion at high temperature have been described in this book chapter. Nature of coating materials determines their ability for thermal shock resistance under different thermal gradients. Under high heat flux operating conditions, significant creep-shrinkage strains will develop in a thermal barrier coating, resulting in the formation of vertical cracks in the that originate at the surface. The increased thermal cyclic stresses and accumulated creep strains provide the major driving force for the vertical crack growth. Eventually the coating can delaminate. Modeling is a powerful tool to predict the properties of coatings under thermal cycling and thus to prevent excessive cracking and delamination. At the present time, different experimental techniques involving high heat flux testing and rapid cooling have been developed to characterize important thermo-physical and thermomechanical properties of ceramic coating systems and to investigate the failure mechanisms under near-realistic environmental conditions. All of these methods have offered to engineers better elements to choose and propose new improved engineering coatings for high temperature applications.

Thermal spray coatings are a good choice for corrosion mitigation, especially, to prevent hot corrosion. Although, at the present time, there are different options for preventing metallic corrosion, using thermal spray coatings offers several characteristics from a technological point of view. For instance, they can act as thermal insulators and/or may accomplish a mechanical function. In addition, thermal spraying coatings can be designed for preventing hot corrosion. In the present book chapter, a case study was presented, giving the basis for carrying out the protection of metallic structures from hot corrosion.

It was discussed that electrochemical techniques are important tools for determining the rate of corrosion and mechanisms acting on the corrosion process. Moreover, equivalent circuits were presented as useful tools for modeling hot corrosion since they help to understand the mechanisms involved in the corrosion process.

ACKNOWLEDGMENT

The authors would like to thanks to the Universidad Nacional Autónoma de México, Campus Morelos, for the support during the preparation of the present book chapter. The authors thankfully acknowledge the support of the National Council of Science and Technology of Mexico (CONACYT). Many thanks to the Faculty of Engineering at the University of Antioquia (Colombia).

REFERENCES

Arenas, O. L., Porcayo-Calderon, J., Salinas-Bravo, V. M., Martinez-Villafane, A., & Gonzalez-Rodriguez, J. G. (2005). Effect of boron on the hot corrosion resistance of sprayed Fe40Al intermetallics. *High-Temperature Materials and Processes*, *24*(2), 93–100. doi:10.1515/HTMP.2005.24.2.93

Bolcavage, A., Feuerstein, A., Foster, J., & Moore, P. (2004). Thermal shock testing of thermal barrier coating/bondcoat systems. *Journal of Materials Engineering and Performance*, *13*(4), 389–397. doi:10.1361/10599490419883

Boley, B. A., & Weiner, J. H. (2012). *Theory of thermal stresses. New York.* US: Courier Corporation.

Carpio, P., Salvador, M. D., Borrell, A., & Sánchez, E. (2017). Thermal behaviour of multilayer and functionally-graded YSZ/Gd 2 Zr 2 O 7 coatings. *Ceramics International*, *43*(5), 4048–4054. doi:10.1016/j.ceramint.2016.11.178

Courtin, S., Gardin, C., Bezine, G., & Hamouda, H. B. H. (2005). Advantages of the J-integral approach for calculating stress intensity factors when using the commercial finite element software ABAQUS. *Engineering Fracture Mechanics*, *72*(14), 2174–2185. doi:10.1016/j.engfracmech.2005.02.003

Eftis, J., & Liebowitz, H. (1976). On surface energy and the continuum thermodynamics of brittle fracture. *Engineering Fracture Mechanics*, *8*(3), 459–485. doi:10.1016/0013-7944(76)90002-3

Evans, A. G., Mumm, D. R., Hutchinson, J. W., Meier, G. H., & Pettit, F. S. (2001). Mechanisms controlling the durability of thermal barrier coatings. *Progress in Materials Science*, *46*(5), 505–553. doi:10.1016/S0079-6425(00)00020-7

Forestier, R., Massoni, E., & Chastel, Y. (2002). Estimation of constitutive parameters using an inverse method coupled to a 3D finite element software. *Journal of Materials Processing Technology*, *125*, 594–601. doi:10.1016/S0924-0136(02)00406-5

Gonzales-Rodrigues, J. G., Haro, S., Martinez-Villafañe, A., Salinas-Bravo, V. M., & Porcayo-Calderon, J. (2006). Corrosion performance of heat resistant alloy in Na2SO4-V2O5 molten salts. *Materials Science and Engineering A*, 435, 258-265.

Gonzalez, J. E. G., & Mirza-Rosca, J. C. (1999). Study of the corrosion behavior of titanium and some of its alloys for biomedical and dental implant applications. *Journal of Electroanalytical Chemistry*, *471*(2), 109–115. doi:10.1016/S0022-0728(99)00260-0

Guo, X., Lu, Z., Jung, Y. G., Li, L., Knapp, J., & Zhang, J. (2016). Thermal properties, thermal shock, and thermal cycling behavior of lanthanum zirconate-based thermal barrier coatings. *Metallurgical and Materials Transactions E*, 3(2), 64-70.

Gurrappa, I. (2001). Identification of hot corrosion resistant MCrAlY based bond coatings for gas turbine engine applications. *Surface and Coatings Technology*, *139*(2), 272–283. doi:10.1016/S0257-8972(00)01156-7

Hardwicke, C., & Lau, Y. C. (2013). Advances in Thermal Spray Coatings for Gas Turbines and Energy Generation: A Review. *Journal of Thermal Spray Technology*, *22*(5), 564–576. doi:10.1007/s11666-013-9904-0

Hasselman, D. P. H. (1985). Thermal stress resistance of engineering ceramics. *Materials Science and Engineering*, *71*, 251–264. doi:10.1016/0025-5416(85)90235-6

Hille, T. S., Suiker, A. S. J., & Turteltaub, S. (2009). Microcrack nucleation in thermal barrier coating systems. *Engineering Fracture Mechanics*, *76*(6), 813–825. doi:10.1016/j.engfracmech.2008.12.010

Jian, C. Y., Hashida, T., Takahashi, H., & Saito, M. (1995). Thermal shock and fatigue resistance evaluation of functionally graded coating for gas turbine blades by laser heating method. *Composites Engineering*, *5*(7), 879–889. doi:10.1016/0961-9526(95)00041-K

Jin, Z. H., & Feng, Y. Z. (2008). Thermal fracture resistance of a functionally graded coating with periodic edge cracks. *Surface and Coatings Technology*, *202*(17), 4189–4197. doi:10.1016/j.surfcoat.2008.03.009

Kawasaki, A., Watanabe, R., Yuki, M., Nakanishi, Y., & Onabe, H. (1996). Effect of microstructure on thermal shock cracking of functionally graded thermal barrier coatings studied by burner heating test. *Materials Transactions*, 37(4), 788-795.

Kokini, K., DeJonge, J., Rangaraj, S., & Beardsley, B. (2002). Thermal shock of functionally graded thermal barrier coatings with similar thermal resistance. *Surface and Coatings Technology*, *154*(2), 223–231. doi:10.1016/S0257-8972(02)00031-2

Lajtai, E. Z. (1971). A theoretical and experimental evaluation of the Griffith theory of brittle fracture. *Tectonophysics*, *11*(2), 129–156. doi:10.1016/0040-1951(71)90060-6

Langer, S. A., Carter, W. C., & Fuller, E. R. (1997). Object Oriented Finite Element Analysis for Materials Science. *Center for Theoretical and computational Materials Science and the Information Lab.* NIST. Retrieved April 25, 2017 from www.ctcms.nist.gov/oof/

Li, Y., Li, C. J., Zhang, Q., Yang, G. J., Li, C. X. (2010). Influence of TGO composition on the thermal shock lifetime of thermal barrier coatings with cold-sprayed MCrAlY bond coat. *Journal of thermal spray technology*, *19*(1), 168-177.

Li, Y. S., Niu, Y., & Wu, W. T. (2003). Accelerated corrosion of pure Fe, Ni, Cr and several Fe-based alloys induced by ZnCl 2–KCl at 450 C in oxidizing environment. *Materials Science and Engineering A*, *345*(1), 64–71. doi:10.1016/S0921-5093(02)00386-6

Lu, W. M., Pan, T. J., Zhang, K., & Niu, Y. (2008). Accelerated corrosion of five commercial steels under a ZnCl 2–KCl deposit in a reducing environment typical of waste gasification at 673–773K. *Corrosion Science*, *50*(7), 1900–1906. doi:10.1016/j.corsci.2008.03.004

Matejicek, J., Sampath, S., Gilmore, D., & Neiser, R. (2003). In situ measurement of residual stresses and elastic moduli in thermal sprayed coatings: Part 2: processing effects on properties of Mo coatings. *Acta Materialia*, *51*(3), 873–885. doi:10.1016/S1359-6454(02)00477-9

Metikos-Hukovic, M., Pilic, Z., Babic, R., & Omanovic, D. (2006). Influence of alloying elements on the corrosion stability of CoCrMo implant alloy in Hank's solution. *Acta Biomaterialia*, *2*(6), 693–700. doi:10.1016/j.actbio.2006.06.002 PMID:16884967

Michlik, P., & Berndt, C. (2006). Image-based extended finite element modeling of thermal barrier coatings. *Surface and Coatings Technology*, *201*(6), 2369–2380. doi:10.1016/j.surfcoat.2006.04.003

Pan, J., Leygraf, C., Jargelius Pettersson, R. F. A., & Linden, J. (1998). Characterization of High-Temperature Oxide Films on Stainless Steels by Electrochemical-Impedance Spectroscopy. *Oxidation of Metals*, 50(5), 431-455.

Porcayo-Calderón, J., Sotelo-Mazón, O., Casales-Diaz, M., Ascencio-Gutierrez, J. A., Salinas-Bravo, V. M., & Martinez-Gomez, L. (2014). Electrochemical study of Ni20Cr coatings applied by HVOF process in ZnCl2-KCl at high temperatures. *Journal of Analytical Methods in Chemistry*. PMID:25210645

Porcayo-Calderón, J., Sotelo-Mazon, O., Salinas-Bravo, V. M., Arrieta-Gonzalez, C. D., Ramos-Hernandez, J. J., & Cuevas-Arteaga, C. (2012). Electrochemical Performance of Ni20Cr coatings applied by combustion powder spray in ZnCl2-KCl molten salts. *International Journal of Electrochemical Science*, *7*(2), 1134–1148.

Qin, E., Yin, S., Ji, H., Huang, Q., Liu, Z., Wu, S. (2017). Hot Corrosion Behavior of Arc-Sprayed Highly Dense NiCr-Based Coatings in Chloride Salt Deposit. *Journal of Thermal Spray Technology*.

Rabiei, A. G. E. A., & Evans, A. G. (2000). Failure mechanisms associated with the thermally grown oxide in plasma-sprayed thermal barrier coatings. *Acta Materialia*, *48*(15), 3963–3976. doi:10.1016/S1359-6454(00)00171-3

Rapp, R. A. (2002). Hot corrosion of materials: A fluxing mechanism? *Corrosion Science*, *44*(2), 209–221. doi:10.1016/S0010-938X(01)00057-9

Shen, Y. G., Mai, Y. W., Zhang, Q. C., McKenzie, D. R., McFall, W. D., & McBride, W. E. (2000). Residual stress, microstructure, and structure of tungsten thin films deposited by magnetron sputtering. *Journal of Applied Physics*, *87*(1), 177–187. doi:10.1063/1.371841

Sidhu, T. S., Agrawal, R. D., & Prakash, S. (2005). Hot corrosion of some superalloys and role of high-velocity oxy-fuel spray coatings—a review. *Surface and Coatings Technology*, *198*(1), 441–446. doi:10.1016/j.surfcoat.2004.10.056

Sidhu, T.S., Agrawal, R.D., Prakash, S. (2006) Hot corrosion of some superalloys and role of high-velocity oxy-fuel spray coatings—a review, *Surface & Coatings Technology, 198*, 441-446.

Sidhu, T. S., Prakash, S., & Agrawal, R. D. (2006). Evaluation of hot corrosion resistance of HVOF coatings on a Ni-based superalloy in molten salt environment. *Materials Science and Engineering A*, *430*(1-2), 64–78. doi:10.1016/j.msea.2006.05.099

Soleimanipour, Z., Baghshahi, S., & Shoja-razavi, R. (2017). Improving the Thermal Shock Resistance of Thermal Barrier Coatings Through Formation of an In Situ YSZ/Al2O3 Composite via Laser Cladding. *Journal of Materials Engineering and Performance*, *4*(26), 1890–1899. doi:10.1007/s11665-017-2591-0

Stott, F. H., & Rocchini, G. (1991). *Electrochemical Aspect of high temperatura corrosion reaction in combustión systems, Colombo, UK corrosion* (Vol. 2). Manchester: Europan Federation of Corrosion.

Sundararajan, G., Joshi, S. V., & Krishna, L. R. (2016). Engineered surfaces for automotive engine and power train components. *Current Opinion in Chemical Engineering*, *11*, 1–6. doi:10.1016/j.coche.2015.10.001

Tamilselvi, S., Raman, V., & Rajendran, N. (2006). Corrosion behaviour of Ti–6Al–7Nb and Ti–6Al–4V ELI alloys in the simulated body fluid solution by electrochemical impedance spectroscopy. *Electrochimica Acta*, *52*(3), 839–846. doi:10.1016/j.electacta.2006.06.018

Tsui, Y. C., & Clyne, T. W. (1997). An analytical model for predicting residual stresses in progressively deposited coatings Part 1: Planar geometry. *Thin Solid Films*, *306*(1), 23–33. doi:10.1016/S0040-6090(97)00199-5

Tuler, F. R., & Butcher, B. M. (1968). A criterion for the time dependence of dynamic fracture. *International Journal of Fracture*, *4*(4), 431–437.

Valero-Vidal, C., & Igual-Muñoz, A. (2011). Electrochemical Aspects in Biomedical Alloy Characterization: Electrochemical Impedance Spectroscopy. In A. N. Laskovski (Ed.), *Biomedical Engineering, Trends in Materials Science*. InTech. doi:10.5772/13039

Wang, L., Li, D. C., Yang, J. S., Shao, F., Zhong, X. H., Zhao, H. Y., & Wang, Y. (2016). Modeling of thermal properties and failure of thermal barrier coatings with the use of finite element methods: A review. *Journal of the European Ceramic Society*, *36*(6), 1313–1331. doi:10.1016/j.jeurceramsoc.2015.12.038

Wang, L., Wang, Y., Sun, X. G., He, J. Q., Pan, Z. Y., & Wang, C. H. (2012). Thermal shock behavior of 8YSZ and double-ceramic-layer La 2 Zr 2 O 7/8YSZ thermal barrier coatings fabricated by atmospheric plasma spraying. *Ceramics International*, *38*(5), 3595–3606. doi:10.1016/j.ceramint.2011.12.076

Wang, Z., Kulkarni, A., Deshpande, S., Nakamura, T., & Herman, H. (2003). Effects of pores and interfaces on effective properties of plasma sprayed zirconia coatings. *Acta Materialia*, *51*(18), 5319–5334. doi:10.1016/S1359-6454(03)00390-2

Yamada, K., Tomono, Y., Morimoto, J., Sasaki, Y., & Ohmori, A. (2002). Hot corrosion behavior of boiler tube materials in refuse incineration environment. *Vacuum*, *65*(3), 533–540. doi:10.1016/S0042-207X(01)00468-7

Zeng, L. C., Wang, W., & Wu, T. W. (2001). Electrochemical impedance models for molten salt corrosión. *Corrosion Science*, *43*(4), 787–801. doi:10.1016/S0010-938X(00)00108-6

KEY TERMS AND DEFINITIONS

Bond Coat: Metallic coating often used to ensure adhesion of thermal spraying coatings. It is deposited directly onto the substrate material and is responsible for allowing the formation of a second coating layer.

Delamination: Failure in a coating material which leads to separation of either some layers of the coating or the whole coating material. Factors inducing delamination are usually associated with mechanical solicitations, induced thermal stresses and poor processing.

Finite Element Modeling: A numerical method involving discretization of structures into elements and nodes.

Molten-Salt: Salt in liquid phase due to elevated temperature.

Thermal-Stresses: Internal stresses generated within a material because of temperature changes, inhomogeneous heating, or differential thermal expansion.

TGO: Thermally grown oxides that form at high temperatures in thermal spraying coatings due to the oxidation of the bond coat.

YSZ: Abbreviation of Yttria-stabilized zirconia, a ceramic material, often used as thermal barrier coating.

Chapter 7
Nanocrystalline, Enamel and Composite Coatings for Superalloys

Shenglong Zhu
Chinese Academy of Sciences, China

Fuhui Wang
Northeastern University, China

ABSTRACT

This chapter describes several innovations for improving the performance of high-temperature coatings. Nanocrystallization has been demonstrated to be a practical way to prolong the lifetime of high temperature coatings by decreasing the minimum Al concentration for sustaining the growth of thermally grown oxide (TGO) scales, and increasing the resistance against scale cracks and spallation. Enamel coatings with enhanced strength, toughness and thermophysical properties were developed for improving the hot corrosion resistance of superalloys. Low expansion nanocomposite coatings minimize the mismatch between coefficients of thermal expansion (CTEs) of the TGO scales and the underlying coatings, so allow growth of thicker TGO scales free of cracks and spallation and then prolong the lifetime.

INTRODUCTION

The major driving force for the development of high temperature protective coatings is the demand for advanced gas turbines with improved fuel efficiency and reduced carbon emissions, which may be achieved by raising the turbine inlet temperatures. For example, the turbine inlet temperatures of M501G and M501J (Mitsubishi Heavy Industries Ltd.) are 1500 °C and 1600 °C, while their combined-cycle efficiencies (gross, low heat value, ISO base) are ≥58% and ≥61.5%, respectively.

Earlier development of heat-resistant alloys was burdened by both oxidation resistance and mechanical properties at high temperatures. There are conflicts in chemical compositions, however, for simultaneously optimizing the oxidation resistance and the mechanical properties. For example, higher concentration of refractory elements, *e.g*. Mo, W, and Ta, is beneficial to longer high temperature creep-rapture lifetime

DOI: 10.4018/978-1-5225-4194-3.ch007

of an alloy while is detrimental to the oxidation resistance. The idea of developing protective coating technology releases half of the burdens, *i.e.* oxidation resistance, at least partially. By applying protective coatings on the surface of components, one can focus his goal on pursuing longer creep-rapture lifetimes by adding more refractory elements into his alloys while leave the task of optimizing the oxidation resistance to others who work on the development of the protective coatings.

Several protective coatings, *i.e.* diffusion coatings, overlay coatings and thermal barrier coatings (TBC) have been widely used in advanced gas turbines up to date. Diffusion coatings are mainly Cr-rich coatings, simple aluminide coatings and platinum modified aluminide (NiPtAl) coatings. Overlay coatings typically comprise β+γ' in γ matrix and are of composition MCrAlY (M is Ni, Co, or NiCo). TBC consists of a bondcoat, which is either an overlay coating or a diffusion coating, and a topcoat, which is a ceramic thermal insulator, *e.g.* yttria-stabilized zirconia (YSZ). In this chapter, the main achievement and the intrinsic issues of the above-mentioned traditional coatings will be briefly reviewed, and then several novel coatings will be introduced in purpose of exploring new pathways of protective technology for high temperature applications. The first one is the nanocrystalline coatings, which aim at improving the performance of coatings by nano-sized grain effects that decrease the minimum Al concentration for sustaining the growth of thermally grown oxide (TGO) scales and increase the resistance against cracks and spallation of TGO. The second one is the enamel coatings, which have quite low oxygen permeation rates and solubility in molten sulphates. The primary purpose of studying enamel coatings is to develop a coating that has hot corrosion rates same as oxidation rates, because the hot corrosion rates of the traditional coatings are much higher than their oxidation rates. The third one is low expansion nanocomposite coatings, which minimize the mismatch between the coefficients of thermal expansion (CTEs) of the TGO scales and the underlying coatings and thus allow growth of thicker TGO scales free of cracks and spallation than those on metallic coatings.

BACKGROUND

Theoretical Lifetimes of Protective Coatings

High temperature protective coatings are mainly based on aluminide, because the TGO formed on aluminide is slowly-growing α-Al_2O_3. The coating failure may be referred to as breakdown oxidation where the oxidation rates are dramatically accelerated because the growth of α-Al_2O_3 is replaced by less protective TGO due to insufficient Al supplies.

The growth of a protective α-Al_2O_3 scale is expressed by parabolic law:

$$\left(\Delta w\right)^2 = k_p t \tag{1}$$

where Δw is the mass gain per unit area, k_p the parabolic constant, and t the tested time. The scale thickness, h, can be calculated by

$$h = \frac{\Delta w}{\left(1-\beta\right)\gamma} \tag{2}$$

where β is the mass ratio of Al in Al_2O_3, *i.e.* ~47.075%, γ is the density of the Al_2O_3 scale, *i.e.* 3.97 g cm^{-3}. Thus, the Al loss per unit area due to Al_2O_3 growth is

$$m = \beta h \gamma \tag{3}$$

Theoretically, the maximum thickness of an Al_2O_3 scale, h^*, grown on a Ni-50Al (mol%) coating with thickness h_c should be expressed by

$$h^* = h_c \frac{50 - c_{\min}}{50} \frac{\gamma_{NiAl}\beta_{NiAl}}{\gamma\beta} \tag{4}$$

where c_{min} is the minimum molar percent of Al in the coating for sustaining the growth of the Al_2O_3 scale, β_{NiAl} the mass ratio of Al in NiAl, i.e. ~31.490%, γ_{NiAl} the density of NiAl, 5.86 g cm^{-3}. The lifetime t_c of the coating is then estimated to be

$$t_c = \frac{\left[\left(1-\beta\right)\gamma h^*\right]^2}{k_p} \tag{5}$$

The maximum thickness of TGO and the lifetime of coating calculated using Equations (4) and (5) are shown in Figure 1. Both of them are increased with the effective Al contents, *i.e.* 50-c_{min}, in the coating. The maximum thickness of TGO is independent on the k_p according to Equation (4) while the lifetime is reversely proportional to k_p according to Equation (5). Providing c_{min} = 10 mol%, $k_p = 10^{-13}$ $g^2cm^{-4}s^{-1}$, h_c=0.01 cm, then h^* is estimated to be ~79 μm, while t_c to be ~2.755 x 10^9 s, i.e. ~765 thousand hours. However, the practical lifetime of a real NiAl coating at 1100 °C is far less than 765 thousand hours.

Factors Shortening Lifetimes of Protective Coatings

There are mechanical and chemical factors may shorten the lifetime of the coatings:

- Spallation and cracks of TGO scale
- Surface rumpling
- Diffusion effects
- Accelerated oxidation due to oxidants other than oxygen

Spallation and Cracks of TGO Scale

The mechanical failures of TGO may be catalogued into several types: spallation and cracks of TGO, surface rumpling, and thermo-mechanical fatigue. These failures are associated with the residual strains of TGO due to combined effect of externally imposed stresses, isothermal growth stresses, thermal stresses, and stress releasing by creep and pseudoplastic deformation.

Figure 1. Calculated maximum thickness of TGO and lifetime of a β-NiAl coating with thickness of 0.01 cm. $k_p = 10^{-13}$ or 10^{-12} $g^2cm^{-4}s^{-1}$

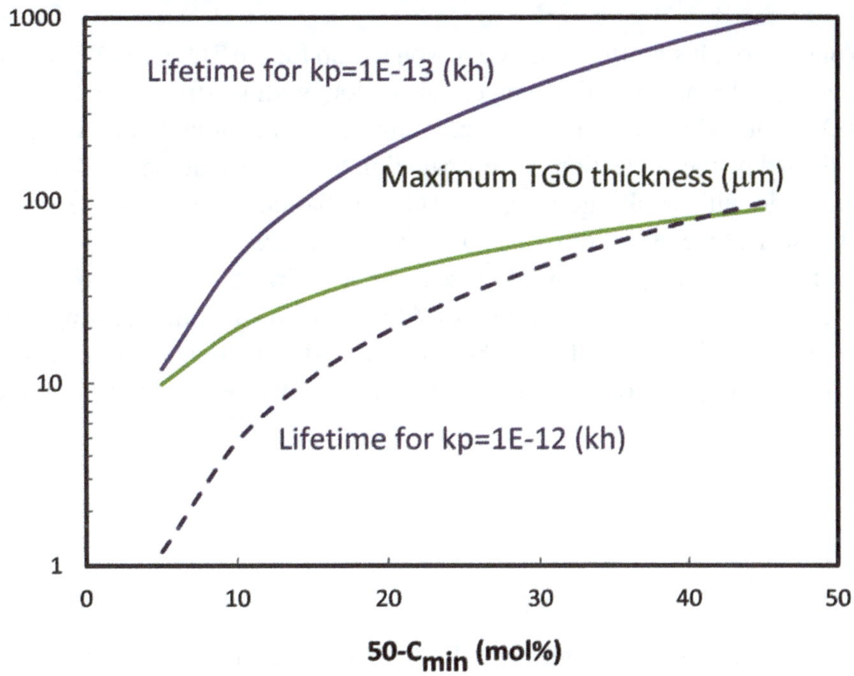

The growth stresses may be divided into compressive and tensile part. The compressive stresses may be associated with Pilling-Bedworth mechanism (1923), *i.e.* the volume differences associated with the metal-oxide conversion, and Rhines-Wolf mechanism (1970), *i.e.* diffusion-induced strains in metal (Suo, Kubair, Evans, Clarke, & Tolpygo, 2003) and lateral growth within TGO (Clarke, 2003). Two mechanisms for the initial tensile stresses in alumina have been observed: phase transformation induced volume contraction, *e.g.* ~ 5% from metastable cubic form to the most stable α form, and chemical compositional changes, *e.g.* from formation of mixtures of α-$(Al,Cr,Ni)_2O_3$ at transition stage to growth of pure α-Al_2O_3 at steady stage.

The thermal stresses are generated by temperature variations. The unconstrained mismatch strain, $\Delta\varepsilon$, associated with the CTE mismatch, $\Delta\alpha$, after a temperature drop, ΔT, is

$$\Delta\varepsilon = \Delta\alpha\Delta T \tag{6}$$

The residual stresses induced by a homogeneous mismatch strain depend on the configuration of the TGO/coating interface. The residual stresses for a TGO scale on a cylindrical substrate may be expressed by (Evans, Crumley, & Demaray, 1983)

$$\sigma_r \approx \left[\frac{E\Delta\varepsilon}{2\left(1-\nu\right)}\right]\left[1-\frac{r}{h}\right] \tag{7}$$

$$\sigma_\theta \approx \sigma_z = -\frac{E\Delta\varepsilon}{1-\nu} \tag{8}$$

where E is the Yong's modulus for the TGO, ν Poisson's ratio for the TGO, r the distance from the substrate center, and σ_r, σ_θ and σ_z are the radial, tangential and longitudinal stresses in the TGO, respectively.

For a weak TGO scale with strong TGO/substrate interface, the initial failure of the scale occurs by shear cracking followed by the nucleation and propagation of the crack along the interface. On the other hand, for a strong scale with weak interface, the TGO spallation follows decohesion at the interface and buckling of the scale. One may notice that strong and weak are words relative to the strength of the scale and the interface. For example, an α-Al_2O_3 scale on a NiCrAlY coating may be referred to as the case of weak scale with strong interface, while an α-Al_2O_3 scale on a β-NiAl coating may be referred to as the case of strong scale with weak interface because of void formation at the interface. In the former case, the critical strain for spallation, ε_c, can be expressed by (Evans, Mitchell, Lobb, &Owen, 1993; Schütze, 1995)

$$\varepsilon_c = \sqrt{\frac{2\gamma_0}{hE\left(1-\upsilon\right)}} \tag{9}$$

where $2\gamma_0$ is the fracture energy of the interface. In the latter case, the critical strain for buckling takes the form

$$\varepsilon_c = \frac{1.22}{1-\nu^2}\left(\frac{h}{R}\right)^2 \tag{10}$$

where R is the initial radius of the zone of decohesion.

For the α-Al_2O_3 scales on the NiCrAlY coatings, the maximum thickness free of spallation is usually about 6 to 7 μm, which is far less than the theoretic limit predicted by Equation (4). A computer software, named as COSP, that models the effects of the cyclic spallation on the oxidation kinetics was developed by Smiale and Auping (2002). For the growth of alumina free of spallation, the Al consumption with a cyclic time τ and the cycle number n is written as:

$$m_n^{spall\text{-}free} = \sqrt{nk_p\tau}\,\frac{\beta}{1-\beta} \tag{11}$$

In the case that the whole TGO scale spalls off after each cycle, however, the Al consumption is increased to

$$m_n^{spall} = n\sqrt{k_p\tau}\,\frac{\beta}{1-\beta} \tag{12}$$

That is, TGO spallation causes an increase in the Al consumption by $\sqrt{n}$ times. For a Ni-50Al (mol%) coating with thickness of 0.01 cm and parabolic constant of 10^{-13} $g^2cm^{-4}s^{-1}$, the lifetime is shortened to ~875 h if TGO spalls off totally after each hour.

Surface Rumpling

Surface rumpling, also termed interface undulation, surface roughening or ratcheting, is the phenomenon that the surface roughness of protective coatings increases gradually during oxidation tests. Rumpling has been observed in all the mainstream coatings including simple (Tolpygo & Clarke, 2004a) or platinum modified aluminide (Tolpygo & Clarke, 2000) and MCrAlY coatings (Pennefather & Boone, 1995). Rumpling yields shear stresses along interface and thus may cause instable growth of TGO scale, *e.g.* detachment, crack, and buckling. Rumpling is considered as one of the most important factors that cause TBC to fail. Besides, rumpling enlargers surface area, and then may increase the Al consumption and accelerate the coating degradation.

Stress releasing at high temperatures via creep of both the TGO scales and the underlying metals may cause rumpling (Bull, 1998; Tortorelli, Specht, More, & Hou, 2012; Veal, Paulikas, Gleeson, & Hou, 2007). Elastic mechanics analysis indicated that the principal factors influencing surface rumpling are the CTE mismatch between TGO and coating, the TGO modulus, the yield strength of the coating and the cyclic temperature range, while the thickness of the TGO scale has only a secondary importance (Evans, He, & Hutchinson, 1997).

Strains due to phase transformation in the coatings may cause rumpling too. The aspect ratio of undulations of thicker β-NiAl coatings on single crystal superalloy Rene N5 is generally smaller, probably because the thicker coating degrades slower (Tolpygo & Clarke, 2004b). Various diffusive processes (grain boundary, surface and bulk diffusion) in the coating driven by the coating stresses lead to the rumpling behavior were observed (Panat, Hsia, & Oldham, 2005). Adding Hf into single phase β-(Ni,Pt)Al coating decreased greatly the degradation rate from β to γ' and the extent of surface rumpling during oxidation tests at 1100 °C (Y. Yang, et. al., 2016).

Balint-Hutchinson model indicates that undulation growth might be reduced by increasing the coating creep strength or reducing the high temperature strength of the TGO (Balint & Hutchinson, 2005). With a dot over a symbol denoting its time rate of change, the total strain rate for the coating is given by

$$\dot{\varepsilon}_{ij} = \frac{1+\nu_c}{E_c}\dot{\sigma}_{ij} - \frac{\nu_c}{E_c}\dot{\sigma}_{kk}\delta_{ij} + \alpha_c \dot{T}\delta_{ij} + \frac{3}{2}\dot{\varepsilon}_R\left(\frac{\sigma_e}{\sigma_R}\right)^{r-1}\frac{s_{ij}}{\sigma_R}e^{-T_R/T} \tag{13}$$

where subscript c is with respect to the coating, T_R is a reference temperature defining the temperature dependence of the creep strength, $\dot{\varepsilon}_R$ is the reference creep-rate by power-law creep under reference stress σ_R, s_{ij} is the deviatory stress, σ_e is the effective stress, r is the creep exponent, typically between 3 and 5.

Interdiffusion

The interdiffusion in the coated superalloys is thermodynamically inevitable because the concentrations of Al in the protective coatings are higher than those in the substrates. Andersson and Ågren (1992) proposed a general formalism for multicomponent diffusion in simple phases. Assuming a vacancy exchange diffusion mechanism in a crystalline phase, diffusion coefficient defined in the lattice-fix frame of reference, D_{kj}^{L}, also termed intrinsic diffusion coefficient, can be defined as the product of a mobility term and a thermodynamic factor (Campbell, Boettinger, & Kattner, 2002):

$$D_{kj}^{L} = \sum_{i=1}^{n} M_{ki}^{L} \frac{\partial \mu_i}{\partial x_j} \tag{14}$$

The M_{ki}^{L} matrix, which is both composition and temperature dependent, defines the diffusion mobility in the lattice-fixed frame of reference. The partial derivative of the chemical potential, μ_i, with respect to the mole fraction, x_i, defines the thermodynamic factors, which can be calculated using an appropriate multicomponent thermodynamic database.

If the partial molar volumes are assumed to be constant, diffusivities in Equation (14) can be transformed to the volume-fixed frame of reference, D_{kj}^{V},

$$D_{kj}^{V} = \sum_{i=1}^{n} \left(\delta_{ik} - x_k\right) x_i M_i \frac{\partial \mu_i}{\partial x_j} \tag{15}$$

where the species j is the diffusing specie and k is the gradient specie, and δ_{ik} is the Kroncker delta. Following the work of Andersson and Ågren (1992), the off-diagonal terms of the diffusion mobility matrix are assumed to be zero, i.e., correlation effects are assumed to be negligible. Thus the number of parameters needed to describe diffusion in a multicomponent system is reduced because only the diagonal terms of the mobility matrix must be evaluated. The diagonal terms of the mobility matrix are assumed to have a form as follows:

$$M_{ki}^{L} = \delta_{ki} x_i M_i \tag{16}$$

$$M_i = \frac{1}{RT} \exp\left(\frac{\Delta Q_i^*}{RT}\right) \tag{17}$$

where R is the ideal gas constant, ΔQ_i^* can be expressed as functions of composition and temperature in terms of Redlich-Kister polynomial,

$$\Delta Q_i^* = \sum_j x_j Q_i^j + \sum_p \sum_{j>p} x_p x_j \sum_k A_i^{pj} \left(x_p - x_j\right)^k \tag{18}$$

where Q and A are linear functions of temperature.

Tracer or self diffusivity, D_i^*, is independent on the thermodynamic factor:

$$D_i^* = RTM_i \tag{19}$$

The chemical diffusivity, also termed interdiffusion coefficient, includes the effect of the mixing of atoms and is measured in the volume-fixed frame of reference from composition profiles, and can be expressed as

$$\tilde{D}_{kj}^n = D_{kj}^V - D_{kn}^V \tag{20}$$

where n is the solvent, e.g. Ni in Ni-base superalloys. The diffusion flux is written as functions of the n-1 independent concentration gradients,

$$J_k = -\sum_{j=1}^{n-1} \tilde{D}_{kj}^n \nabla c_j \tag{21}$$

The diffusion analysis is useful for assessing microstructural evolution of coated turbine blade systems (Karunaratne, Ogden, Kenny, & Thomson, 2009).

The interdiffusion between the coating and the substrate may lead to significant loss of Al in the coating (Dahl, Hald, & Horsewell, 2006), which may greatly shorten the lifetime of coating according to Equation (5). Over-aluminization has been developed in order to compensate the MCrAlY coatings for the Al loss due to interdiffusion (Lu, Zhu, & Wang, 2011).

Moreover, the interdiffusion problem of the coated single crystal (SC) superalloys caused large panics in recent decades (Chen, & Little, 1997), because the creep rupture life (CRF) of the coated single crystal superalloys may be significantly reduced as result of formation of SRZ (secondary reaction zone). For example, at 1100 ºC with stresses of 137 MPa, the CRFs of SC superalloys CMSX4, TMS-75 and TMS-138 coated by platinum aluminide were reduced to 35%, 23% and 14% of those without coatings, respectively (Narita, 2009). The SRZ, which comprises needle-shaped topologically close packed (TCP) phase and gamma phase in matrix of gamma prime, results from the phase zone transformation of the substrate from γ+γ' to γ+γ'+TCP and then to γ'+TCP.

Accelerated Oxidation Due to Oxidants Other Than Oxygen

In practical environments, alloys are usually exposed to complex atmospheres that contain a variety of gas species. Oxidants other than oxygen, e.g. S, Cl, C, and H_2O, may accelerate the oxidation rates.

In the presence of water vapor, the oxidation rates of protective coatings have been reported to be much higher because of formation of less-protective oxides (Zhou, Yu, Gong, & Xu, 2002). Water vapor might affect the rate of coating rumpling, too, and thus reduce the TBC lifetime (Lance, Unocic, Haynes, & Pint, 2013). A detailed review on the effects of water vapor on the oxidation behavior was presented by Sauders, Monteiro and Rizzo (2008).

Hot corrosion is an aggressive mode of accelerated oxidation associated with molten deposits, e.g. Na_2SO_4, NaCl, and vanadium oxides. There are two types of sulphur-induced hot corrosion. Type I hot corrosion, also termed high temperature hot corrosion (about 750 – 950 °C), involves basic fluxing (reactions between the molten sulphatic deposit with the preformed protective oxide layer) or acidic fluxing (elements such as molybdenum, tungsten, or vanadium in the alloy cause deposits to become acidic as oxides of these elements are incorporated in the deposits). Type II hot corrosion, also termed low temperature hot corrosion (about 600 – 850 °C), involves the formation of base metal (Ni or Co) sulphates, which react with alkali metal sulphates, transform into low melting point compounds, and then prevent the formation of protective oxide. The hot corrosion of virtually all susceptible alloys is observed to occur in two stages: an initiation stage during which the rate of corrosion is slow and similar to that in the absence of the deposit, and a propagation stage in which rapid, sometimes catastrophic, corrosion occurs (Birks, Meier, & Pettit, 2006). That is, the coatings are protective until the propagation stage starts.

Comparing the hot corrosion kinetics curves of two coatings may lead to false conclusions on their relative resistance against hot corrosion, because the kinetics curves contain both mass gain due to TGO growth and mass loss due to basic or acidic fluxing. To assess the corrosion rate for hot corrosion by mechanism of basic fluxing, one may apply a kinetics model of parabolic growth competing to linear declining (Zhu, Wei, & Liu, 2014). The growth rate d*h*/d*t* may be divided into two terms: parabolic thickening with a parabolic constant k_p in $g^2cm^{-4}s^{-1}$ and linear thinning with constant k_t in cm s^{-1}:

$$\frac{dh}{dt} = \frac{k_p}{2h\left[\left(1-\beta\right)\gamma\right]^2} - k_t \tag{22}$$

And the mass gain rate is

$$\frac{d\Delta W}{dt} = \frac{k_p}{2h\left(1-\beta\right)\gamma} - \gamma k_t \tag{23}$$

The thickness of TGO and the mass gain calculated using Equations (22) and (23) are shown in Figure 2. The horizontal broken thin lines in Figure 2A indicate both the thickness for maximum mass gain (ζ) and the maximum thickness of TGO (ξ), which will be explained later in this section. Figure 2B indicates that the loss rates of Al during hot corrosion are much higher than those during oxidation because the TGO grown during hot corrosion are much thinner than that grown during oxidation.

Equation (22) implies a trick: the thickness becomes constant when TGO grows to a special thickness ξ which meets the following relationship:

$$k_t = \frac{k_p}{2\xi\left[\left(1-\beta\right)\gamma\right]^2} \tag{24}$$

Figure 2. Calculated thickness of TGO, loss of Al in coating and mass gain of coating, during hot corrosion (A and B) or oxidation (B). $k_p = 10^{-13} g^2 cm^{-4} s^{-1}$, $k_t = 10^{-12}\ cm\ s^{-1}$

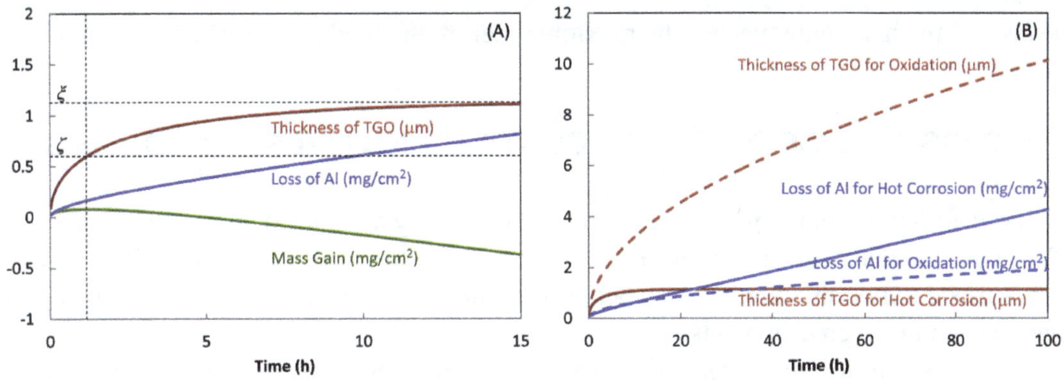

It means that the parabolic growth rate is equal to the linear thinning rate once the thickness is equal to ξ. Then the mass gain rate will also be kept being a constant, v_0, which is measurable and can be expressed as

$$v_0 = \frac{k_p}{2\xi(1-\beta)\gamma} - \gamma k_t \tag{25}$$

The v_0 is negative because it is a constant mass loss. By measuring v_0 and ξ, k_t and k_p are able to be estimated by

$$k_t = -\frac{v_0}{\beta\gamma} \tag{26}$$

and

$$k_p = 2\xi\left[(1-\beta)\gamma\right]^2 k_t \tag{27}$$

Similarly, Equation (23) implies that the mass gain increases to the maximum value when the TGO grows to a thickness ζ:

$$\zeta = \frac{k_p}{2(1-\beta)\gamma^2 k_t} \tag{28}$$

That is

$$\zeta = \left(1-\beta\right)\xi \quad (29)$$

It means that the mass gain reaches the maximum before the thickness reaches the maximum.

INTRINSIC ISSUSES OF DIFFUSION COATING AND OVERLAY COATING

In summary, the lifetimes of the protective coatings are not only associated with the chemical compositions of the coatings, but also can be shortened significantly by various factors, e.g. TGO cracking and spallation, surface rumpling, interdiffusion between coating and substrate, and acceleration by oxidants other than oxygen in the environments.

Extensive researches for improving the performance of the protective coatings have been carried out. Various engine manufacturers prefer nowadays overlay MCrAlY or platinum modified aluminide (NiPtAl) coatings to simple aluminide coating for several reasons. Both the simple aluminide coating itself and the TGO on the coating are prone to cracking during services. Replacing part Al by Cr or replacing part Ni by Pt increases the toughness of the coatings and therefore the resistance against thermomechanical fatigue (TMF) cracking, while replacing part Ni by Pt or adding small amount of reactive elements greatly enhances the TGO adherence. Besides, both NiPtAl and MCrAlY are much more resistant against high temperature (Type I) hot corrosion than the simple aluminide coating. As to the low temperature (Type II) hot corrosion, the best choice is NiPtAl. In general, the more expensive NiPtAl coatings work longer than the MCrAlY coatings.

Equations (1) to (5) indicate that the lifetimes of a coating may be prolonged by increasing the coating thicknesses, decreasing the growth rate of TGO, or decreasing the minimum molar percent of Al in the coating, c_{min}, for sustaining the growth of the Al_2O_3 scale. Practically, the coating thickness for any specific component has been limited for various considerations. It is difficult to decrease further in the growth rates of TGO after extensive investigations on reactive elements effects. So, works on decreasing the minimum molar percent of Al in the coating for sustaining the TGO growth may be helpful.

Equations (22) to (29) indicate that the lifetimes of a coating in hot corrosion environments is shorter than those in oxidation environments because the thickness of the TGO during hot corrosion is always thinner than that during oxidation and thus the loss rate of Al during hot corrosion is always faster than that during oxidation.

The mechanical failures of TGO are associated with the residual strain during thermal cycling. Equations (6) to (8) indicate that the residual stresses for a TGO scale on a cylindrical substrate are proportional to the CTE mismatch between the TGO and the coating. The CTE of alumina is about 50% to 60% of those of the NiPtAl coating and the MCrAlY coating. Practically, the maximum thickness of the TGO scales free of spallation is about 6 to 7 μm on the NiPtAl coating and the MCrAlY coating.

So, there still are three important intrinsic issues for the performance of both the overlay coatings and the NiPtAl coatings:

1. How to decrease c_{min}? This question leads to the development of the nanocrystalline coating.
2. How to decrease the loss rate of Al during hot corrosion? This question leads to the development of the enamel coating of which the performance does not rely on the TGO growth.

3. Are there other coatings which form alumina scale too while have lower CTE and thus allow the spallation-free alumina scale to grow thicker? This question will lead to the development of low expansion nanocomposite coating.

NANOCRYSTALLINE COATINGS

One of the greatest discoveries in materials science in 20th century is that nanocrystalline materials may exhibit various exotic physical and chemical properties. For example, superplasticity of nanocrystalline materials was reported (McFadden, Mishra, Valiev, Zhilyaev, & Mukherjee, 1999; Lu, Sui, & Lu, 2000). Nanocrystalline materials exhibit interesting oxidation behavior, too. In 1992, Lou, Wang, Xia and Zhang (1992) compared the high temperature oxidation behavior of a sputtered nanocrystalline coating and a cast superalloy which have the same chemical compositions, and revealed that the oxidation rates of the coating are much lower than those of the cast superalloy. They explained that nanocrystallization leads the materials from Cr_2O_3-forming alloy to Al_2O_3-forming one.

Preparation of Nanocrystalline Coatings

Nanocrystalline coatings can be prepared by various methods. Low pressure plasma spraying (LPPS) is able to produce nanocrystalline coatings by using nanocrystalline feedstocks. The most convenient method is magnetron sputtering. Magnetron sputtering is a physical vapor deposition method where atomic vapor is produced in high vacuum by positive ion sputtering from targets (the source materials) and then condenses onto the substrates to forming coatings. The energies of sputtered atoms are about several electron volts, about one order of magnitude of those of thermally evaporated atoms. So dense metallic coatings can be prepared using sputter deposition even at room temperatures. Low substrate temperature is essential to deposit nanocrystalline coatings. Another advantage of sputtering is that it copies the chemical composition of the target to the coatings.

Oxidation Mechanism of Nanocrystalline Coatings

The grain size effects on oxidation behavior of metals can be explained in two aspects: promoting selective oxidation and improving TGO adherence.

Formation of an external alumina scale on a Ni-Cr-Al coating requires a minimum Al concentration N_{Al} according to Wagner's selective oxidation theory based on diffusion kinetics analysis:

$$N_{Al} = \left(\frac{cN_O D_O V_m}{D_{Al} V_{ox}} \right)^{1/2} \tag{32}$$

where c is a constant, $N_O D_O$ is the oxygen permeability in the coating, V_m is the molar volume of the coating, D_{Al} is the diffusivity of Al in the coating and V_{ox} is the molar volume of alumina. In the oxide map plotted by Wallwork and Hed (1971) for the ternary system of Ni-Cr-Al at 1000 °C in air, the TGOs are divided into three zones: (1) Zone I: External NiO + internal oxides; (2) Zone II: External Cr_2O_3 +

internal oxides; (3) Zone III: External Al_2O_3 scale. There is a simpler rule to catalog Ni-base superalloys: Al_2O_3-forming ones are those whose mass ratios of Cr over Al are less than 4, and Cr_2O_3-forming ones are the others.

Greatly enhanced diffusivity in nanocrystalline materials is attributed to extremely high density of grain boundaries, which are short-cut diffusion paths as compared with bulk diffusion. The effective diffusivity D_{eff} of a metal stacked by simple cubic grains may be expressed as

$$D_{eff} = D_b \frac{1+\left(2r_d + r_d^2\right)r_D}{\left(1+r_d\right)^2} \tag{33}$$

where D_b is the bulk diffusivity, r_D is the ratio of grain boundary diffusivity to bulk diffusivity, and r_d is the ratio of grain boundary width to the grain size *d*. Using Equations (32) and (33), one can estimate the grain size effect on the effective diffusivity and the minimum Al concentration, as illustrated in Figure 3.

Therefore promoting selective oxidation by nanocrystallization of alloys can well explained by the grain size effects on diffusivity in the coatings (Lou, Wang, Zhu, Xia, & Zhang, 1994; Chen & Rühle, 2005). This inference is derived from two aspects of facts. Nanocrystallization does lead to change in oxidation mechanism, but it does not guarantee the formation of external scale of the most stable oxide on all alloys. For example, there are some Ni-Cr-Al alloys changed from Cr_2O_3-forming to Al_2O_3-forming ones after nanocrystallizing (Chen & Lou, 2000a), while there are also many alloys, *e.g.* IN738 (Geng, Wang, & Zhu, 2002), Ni-5Cr-5Al (at%) (Chen & Lou, 1999a), Ni–8.6Cr–7.2Al(at%) (Chen & Lou, 2000b) and Ni-3Cr-20Al (at%) (Chen & Lou, 1999b), form an outer mixture of Cr-rich oxide and a continuous inner Al_2O_3 layer. In other words, nanocrystallization results in the formation of the more stable TGO on all alloys during oxidation.

Homogeneity of chemical composition and single phase microstructure in sputter-deposited nanocrystalline coatings may also play important roles in TGO formation (Lou, Wang, Xia, & Zhang, 1992). The commercial alloys and coatings may consist of multi-phases at thermodynamically equivalent status.

Figure 3. Calculated ratio of effective diffusivity to bulk diffusivity (A), and size dependence of the ratio of minimum Al concentration (B). The grain boundary width is 0.5 nm, and the ratio of grain boundary diffusivity to bulk diffusivity is 1000. The effects are obvious when the grain sizes are less than 1 μm, and become more significant when the sizes are less than 100 nm.

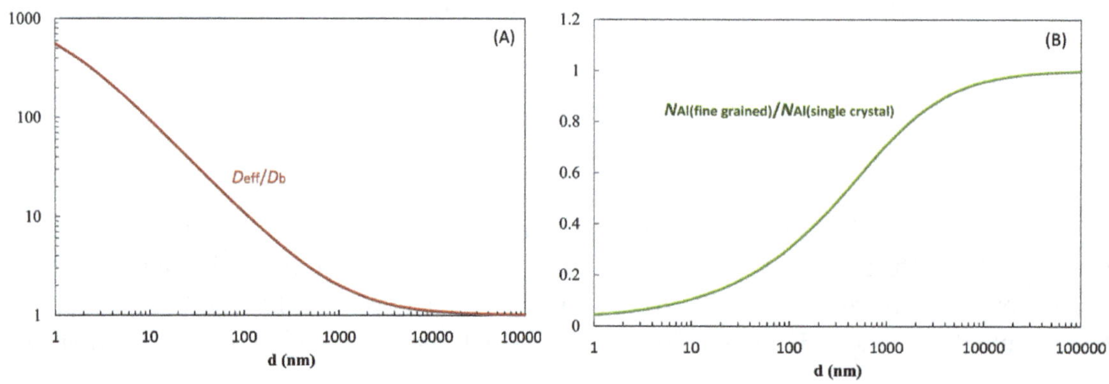

For example, the cast Ni-base superalloys consist of γ'-Ni_3(Al, Ti, Ta) cuboidal precipitates with γ-Ni matrix which contains refractory alloying elements Ta, Re, W and Mo. Some alloys may even contain large carbides. So, the different phases at the surfaces of cast alloys may form different oxide due to different chemical compositions.

It is interesting that nanocrystalline Y-free CoCrAl exhibited excellent resistance against TGO spallation same as CoCrAlY (Wang, Lou, Zhu, & Wu, 1996). Y and other reactive-elements are thought essential to prevent weak bonding of TGO to alloys due to sulfur effect (N'Gandu-Muamba & Streiff, 1993; Hou & Smialek, 1995). Evidence was found that nanocrystallization is able to prevent sulfur segregation at the TGO/metal interface of sulfur-bearing and RE-free alloys and hence enhance scale adhesion (Yang & Wang, 2006).

Nanocrystallization may improve TGO adherence also by fast releasing stresses due to accelerated creep rates, too. The elastic failure of TGO and the detachment of the TGO-metal interface may be avoid if the residual stresses have been decreased to a level low enough by ways of creep-induced plastic deformation of the TGO and the metals (Veal, Paulikas, Gleeson, & Hou, 2007). According to the diffusion-creep theory, the deformation rate, dε/dt, of a polycrystalline material is (Karch, Birringer, & Gleiter, 1987)

$$\frac{d\varepsilon}{dt} = \frac{\sigma\Omega}{d^2KT}\left(B_1D_v + \frac{B_2\delta D_b}{d}\right) \tag{34}$$

whereΩ is the atomic volume, B_1 and B_2 are numerical constants, D_v and D_b are the volume and boundary diffusivities, δ is the thickness of the grain boundary, k is Boltzmann's constant. Thus, the size effect on the diffusion-creep rate is reversal cubic of grain size if grain boundary diffusion predominates, i.e. the creep rate is increased to 3 orders of magnitude when the grain size is decreased to one order of magnitude. Evidence of enhanced deformation of TGO on nanocrystalline K38G superalloy was reported (Lou, Zhu, & Wang, 1995).

Applications for Suppressing SRZ Formation

The nanocrystalline coatings are useful for suppressing the SRZ problem of coated single crystal (SC) superalloys. A simple way is magnetron sputter deposition of nanocrystalline coatings with the chemical compositions same as the substrates (Wang, Chen, Zhu, & Wang, 2015; Wang, Chen, Yang, Zhu, & Wang, 2015; Chen, Zhu, & Wang, 2015; L. Yang, et. al., 2016). A more practical way is using the nanocrystalline coating as an interlayer. A system of duplex NiCrAlY/nano-N5 has been implemented on single crystal superalloy Rene N5 by Yang, Chen, Wang, Zhu and Wang (2016). This coating combines the advantages of traditional NiCrAlY and nanocrystalline coatings, *i.e.*, providing high resistance to oxidation and hot corrosion simultaneously, while avoids any disadvantages that the traditional coatings have suffered from, such as scale spallation, element interdiffusion (along with the formation of harmful TCP phases). In such a duplex system, interdiffusion between the outer layer NiCrAlY and the interlayer nanocrystalline N5 does occur though there is no SRZ formation by the following mechanism:

- The most important element of TCP is rhenium (Re). Re exists as solid-solution element in γ matrix of the γ+γ' SC superalloy substrate while is dispersed randomly in the coatings.

- The diffusivity of Re is about two magnitudes slower than Al and Ni at 1100 °C. So, Re stays where it was when interduffsion occurs.
- When part of γ transforms into γ'+TCP due to either outward diffusion of Ni or inward diffusion of Al, Re precipitates as TCP phases in the form of broken network in the case of SC superalloy. Therefore, SRZ forms.
- However, because Re is uniformly in the nanocrystalline layer, the phase zone transformation of γ to γ'+TCP leads to randomly dispersed TCP particles instead of TCP needles.

ENAMEL COATINGS

It has long been recognized that many glasses will, under suitable conditions, bond well to a number of metals and alloys. The strong, well-adherent bond has led to the extensive use of vitreous enamel as the coating for metals and alloys (Donald, 1993; X. Wang, Chen, Zhu, & Wang, 2015). An attractive chemical property of the SiO_2-based enamel coatings is that they are more resistant to hot corrosion than the MCrAlY and aluminide coatings (Xiong, Guan, Zhu, & Wang, 2006; Zheng, Zhu, & Wang, 2006; Xiong, Zhu, & Wang, 2008), because the solubility of SiO_2 in molten salts is extremely low, even much lower than Al_2O_3 and Cr_2O_3 (Rapp, 2002).

Oxidation Mechanism of Enameled Alloys

The oxidation kinetics of enameled alloys may be simplified as linear law when the oxidation behavior is predominated by oxygen ion diffusion through the coating, because the diffusion flux of oxygen, which is dependent on the coating thickness and the concentration gradient, is kept constant (Chen, Li, Shen, Zhu, & Wang, 2013).

The substrate components dissolved into the enamel coating may affect the oxidation mechanism. For example, as schematically illustrated in Figure 4, oxidation of enameled steel follows linear law at 800 °C by mechanism that the coating acts as a barrier from oxygen diffusion, while follows parabolic law at 700 or 600 °C by formation of a $(Cr,Fe)_2O_3$ particles dispersed in the coating (Chen, Li, Shen, Zhu, & Wang, 2014). For a Cr-free enamel coatings on a Cr-rich Ni-based superalloy K444, the increase in oxidation rate after certain time during exposures is believed to be associated with the formation of Cr-rich oxide particles (Wu, Chen, Zhu, & Wang, 2016).

The enamel coatings are more active than ceramic coatings. For example, Cr was not dissolved into the coating during oxidation of alumina-coated NiCr alloy at 1000 °C (Zhu, Wang, Lou, & Wu, 1995), while Al_2SiO_5 was observed at the coating/ substrate interface of enameled TiAlNb alloy at lower temperature of 800 °C (Zheng, Xiong, Zhu, Wang, & Wu, 2008). It has been reported that bubbling surface of the glassy coating results from gaseous products of Zn and Na due to the interfacial reaction between the SiO_2–Al_2O_3–ZnO–CaO-Na_2O based glass coating and the Ni-base superalloy K38G (Chen, Shen, Zhu, & Wang, 2013). The harmful interfacial reaction can be prohibited by deactivating the coating, *e.g.* alumina particle incorporation (Chen, Shen, Zhu, & Wang, 2013) or pre-crystallization of the enamel powders (Chen, Shen, Wang, Zhu, & Wang, 2013).

It looks like that the enameled alloy is oxidized at low oxygen partial pressure when no element of the enamel coating has been dissolved into the TGO. For instance, an alumina interlayer was observed in an enameled Ni-base superalloy K38G of which the surfaces had been sand blasted before enameling

Figure 4. Schematic illustration of the oxidation process of coated steel: (A) at 800°C and (B) at 700 °C

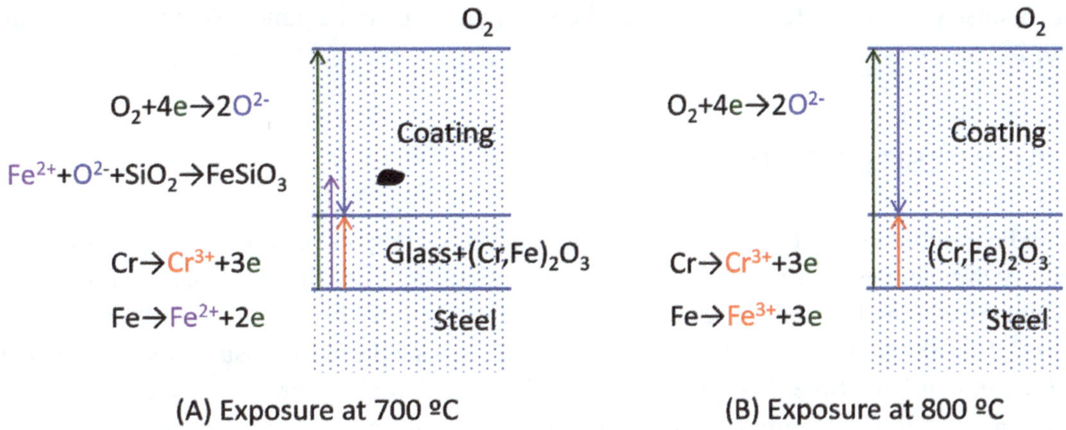

(Chen, Shen, Zhu, Wang, & Wang, 2013). Sand blasting nanocrystallizes the alloy surface, along with low oxygen partial pressure produced by the enamel coating, and then promotes the formation of an alumina layer at the coating/substrate interface.

Oxygen permeation of Ti-base alloys during exposure at high temperatures, *i.e.* formation of brittle oxygen-dissolved sublayer beneath TGO, is harmful because it decreases plasticity significantly (Xiong, Zhu, & Wang, 2005). The interfacial reactions of enameled Ti-based alloys have found to be useful in preventing from oxygen permeation by forming various lamellar interlayers beneath the enamel coating on Ti-based alloy Ti811 (Shen, Zhu, Chen, & Wang, 2011) and Ti6Al4V (Li, Zhu, Wang, Chen, Shen, & Wang, 2013), and intermetallics Ti–47Al–2Cr–2Nb (at%) (Li, Chen, Wang, Zhu, & Wang, 2013; Li, Zhu, Chen, Wang, & Wang, 2014; Li, Chen, Wu, Zhu, Wang, & Wang, 2014) and γ-TiAl (Shen, Zhu, & Wang, 2015).

Thermal Shocking Resistance of the Enamel Coatings

The most challenge work for developing enamel coatings is to improve the thermal shocking resistance because enamel is brittle and fragile. Particulate ceramic and/or metal inclusions are effective and economical for strengthening and toughening the coatings as well as tailoring the TEC (thermal expansion coefficients) to match the substrates. In general, metal inclusions are more effective in improving the fracture toughness (Chen, Zhu, Shen, Wang, & Niu, 2011), while ceramic ones are more effective in increasing the strength (Chen, Zhu, Shen, Wang, & Niu, 2013; Chen, Zhu, & Wang, 2013), and their combination provides greatly enhanced thermal shocking resistance (Chen, Shen, Zhu, Wang, & Niu, 2012).

One should pay enough attention to the microstructural stability of the enamel coatings for high temperature applications, because it may lead to large stresses and cause the coating to fail (Chen, Shen, Wang, Zhu, & Wang, 2012). The practical coatings with complex compositions may exhibit complicated crystallization behaviors at different temperatures (Chen, Zhu, & Wang, 2010), and the inclusions may react with the enamel glass by dissolving into glass or forming new phases (Wang, Chen, Zhu, & Wang, 2013). ZrO_2 inclusions in SiO_2-based enamel transformed into low expansion $ZrSiO_4$ and resulted in dramatic decrease in CTE. Replacing parts of the ZrO_2 inclusions with alumina particles is able to reduce the reaction rates between ZrO_2 and SiO_2-based enamel remarkably, probably by modifying the enamel chemistry due to preferential reaction between alumina and enamel (Wang, Shen, Chen, Zhu, & Wang, 2015).

NANOCOMPOSITE COATINGS

Table 1 indicates that the CTE mismatch between α-Al_2O_3 and typical coatings is large, over 41%. The CTE mismatch generates large residual strain during cooling stage, and then may cause spallation of TGO and shorten the coating lifetime. For example, though both the formation 1 μm thick alumina scale from scratch and the growth of 2 μm thick scale from 1 μm thick scale need to consume the same amounts of Al, *i.e.* 0.21 mg/cm^2, the time they takes is quite different, *i.e.* the former takes about 440 ks while the latter takes about 1760 ks, assuming that the parabolic rate constant is 10^{-13} g/cm^4 s.

Following the ideas of metal matrix composites, one can tailor the CTE of the composite coatings easily. The CTE of the composite coating might be estimated using the Turner equation:

$$\alpha_{eff} = \sum \alpha_i V_i + 4\frac{G_m}{K_m}\sum \frac{K_i - K_m}{4G\mu_m + 3kK_i}\alpha_i V_i \tag{35}$$

where K, G and V represent bulk modulus, shear modulus and volume fraction, respectively, while i and m refer to the inclusion and matrix components, respectively. The CTE of the coating is recommended to be lower than the substrate and higher than α-Al_2O_3. A good example is the composite of Ni-12CrN-26AlN (vol%), the CTE value is estimated to be ~12.5 ppm/°C at 1000 °C.

Preparation of Nanocomposite Coatings

Nanocomposite coatings can be prepared by various methods, *e. g.* physical vapor deposition and thermal spraying. The Ni+CrAlYN coatings, consisting of nanocrystalline grains of metallic γ-Ni and nitrides of Al and Cr, may be prepared by vacuum arc evaporation or reactive magnetron sputtering from NiCrAlY targets in gas mixture of Ar/N_2 (Zhu, Zhu, & Wang, 2012; Ren, Zhu, & Wang, 2015). The key mechanism for these methods is that Ni does not form stable nitride. For prolonging the lifetime of the coatings, adding more Al into the coatings is necessary. However, the total volume fractions of ceramic phases in the nanocomposite coatings should be less than that of the metallic phases. To overcome this difficulty, a multilayer coating with alternating Ni+CrAlYSiHfN and AlN layers was prepared by reactive magnetron sputtering from a pair of NiCrAlYSiHf and Al targets (Ren, Zhu, & Wang, 2016).

Table 1. Average coefficient of thermal expansion (ppm/°C)

Temperature (°C)	600	800	1000
Ni-20Cr-19Al-1Y(at%)	14.6	15.7	18.9
Ni-50Al (at%)	14.5	14.8	15.3
α-Al_2O_3	8.6	9.2	9.7

Haynes, Pint, Porter & Wright, 2004.

Oxidation Mechanism of Nanocomposite Coatings

Alpha-alumina scales were formed on the Ni+nitride nanocomposite coatings during high temperature exposures (Zhu, Zhu, Wang, & Zhang, 2014; Ren, Zhu, & Wang, 2016). TGO formed at the initial stage of oxidation consisted of mixed oxides of all metal elements, several nanometers in grain size. And then an inner Al_2O_3 layer of larger grains formed beneath. The growth of TGO was controlled by the oxygen inward diffusion because of the low activity of Al atoms in AlN, which resulted in the presence of bcc-NiCr particles in the alumina layer.

The Ni + CrAlYSiN nanocomposite coating exhibited slower TGO growth rates and better resistance against TGO spallation than metallic NiCrAlYSi coating during cyclic oxidation tests at 1000 °C. The excellent resistance against TGO spallation may be attributed to the low CTE mismatch between the TGO and the nanocomposite coating. A TGO crack and healing model is proposed for explaining why the TGO growth rate of the nanocomposite coating is lower than that of the metallic coating though α-Al_2O_3 scales were formed on both coatings. For the isothermal growth of a TGO, according to parabolic law, we have

$$h^2 = kt \tag{36}$$

The TGO scale thickness $h_{t+\Delta t}$ at time $t + \Delta t$ is

$$h_{t+\Delta t} = \left(k\Delta t\right)^{1/2}, t = 0 \tag{37a}$$

$$h_{t+\Delta t} = \left[k\left(\Gamma + \Delta t\right)\right]^{1/2}, \Gamma = \frac{h_t^2}{k}, t \neq 0 \tag{37b}$$

where Γ is the oxidation time for growing a scale with thickness h_t, calculated using Equation (37b).

In cyclic oxidation tests, cracking and detachment of the oxide scale might occur during cooling stages. Given a test cycle of Δt, at the cycle number of i ($i = 2, 3 \ldots$), healing the detachment η will take place firstly, and then the TGO scale grows from $h_t + \eta$ to $h_t + \eta + \chi$. Assume the healing time is τ, we have

$$\tau = \frac{\eta^2}{k} \tag{38}$$

Thus, the time for finishing an oxidation cycle is Δt-τ, assuming that the detachment is small enough. The scale keeps growing from an initial thickness of $h_i + \eta$ for time Δt-τ. According to Equation (37b), we have

$$\Gamma_i = \frac{\left(h_i + \eta\right)^2}{k} \tag{39}$$

At the end of a cycle, the scale thickness will be

$$h_{i+1} = \left[\left(h_i + \eta\right)^2 + k\left(\Delta t - \tau\right)\right]^{1/2} \tag{40}$$

Using Taylor's theorem for a series expansion of a function

$$\left(a + y\right)^m = a^m + ma^{m-1}y + \frac{1}{2}m\left(m-1\right)a^{m-2}y^2 + \varepsilon \tag{41}$$

By $a = (h_i + \eta)$, $y = k(\Delta t - \tau)$, $m = 1/2$, we have

$$h_{i+1} = \left(h_i + \eta\right)\left[1 + \frac{k\Delta t - \eta^2}{2h_i^2}\right] + \varepsilon_i \tag{42}$$

Given a 2 μm thick α-alumina scale that grew for 200 1-h cycles, $k = 0.02$ μm²/h. If the TGO scale detachment on average is 0.01 μm for each cycle, one can easily estimate that the thickness of the TGO scale is about 3.5 μm. For the same values k and TGO thickness, the growth of the TGO scale without cracking elapses over 600 cycles. So, the cracking and healing of TGO scales accelerates the growth rates and shortens the service lifetime of the coatings.

The MCrAlY coating and the nanocomposite coating exhibited different coating/substrate interdiffusion behavior. The most significant diffusion of the NiCrAlYSi coating was the inward diffusion of Al, while that of the nanocomposite coating of Ni+CrAlYSiN was the inward diffusion of N. The former led to γ→γ' phase transformation in the IDZ, accompanied by Cr-rich phase precipitating at the interface. The latter resulted in AlN and Ti(V)N formation in the IDZ, accompanied by a γ'→γ phase transformation in the substrate due to uphill diffusion of Al from the substrate to the coating. An hcp-AlN interlayer beneath the nanocomposite coating is helpful to eliminate the coating/substrate interdiffusion at 1000 and 1100 °C for 200 h (Ren, Zhu, & Wang, 2015).

FUTURE RESEARCH DIRECTIONS

The microstructural stability of the nanocrystalline coatings, the enamel coatings and the nanocomposite coatings during long-term exposure, *e.g.* 10 or 20 thousands hours, at high temperatures have not been well understood yet. Up-to-date, only several nanocrystalline coatings have been tested for 1000 h while most of these coatings were tested for several hundred hours only.

The nanocrystalline microstructures promote selective oxidation of the coatings due to enhanced element diffusivity. However, the diffusion and phase transformation in the nanocrystalline coatings and their effects on oxidation mechanism have not been studied quantitatively and far well understood yet. For example, it was revealed recently that the roles of Ta on oxidation of a Ta-containing nanocrystalline coating are strongly dependent on the contents of Y and Al in the coating (Yang, Chen, Wang, Bao, Zhu, & Wang, 2017).

CMAS (CaO-MgO-Al_2O_3-SiO_2) attack to TBC, which damages YSZ topcoat, has been received more and more concerns. It was revealed that the alumina inclusions in SiO_2–Al_2O_3–ZnO–CaO enamel greatly hindered the interfacial reaction between the enamel and YSZ (Wang, Shen, Chen, Zhu, & Wang, 2015). Thus, it is worth to explore the possibility of enamel coatings as environmental barrier coating against CMAS attack to TBC.

CONCLUSION

Recent progress on development of the nanocrystalline coating, the enamel coating and the low expansion nanocomposite coating were introduced briefly in this chapter.

The nanocrystalline microstructures of coatings promote selective oxidation and decrease minimum Al concentration for sustaining the growth of TGO scales due to enhanced element diffusivity, and increase the resistance against cracks and spallation of TGO by fast releasing thermal stresses due to accelerated creep rates. The idea of the nanocrystalline coating can be implemented in several ways: 1) Nanocrystalizing traditional coatings, *e.g.* MCrAlY or NiPtAl, so as to prolong their lifetimes; 2) Producing the nanocrystalline coatings with the chemical compositions same as the substrates so as to avoid the coating/substrate interdiffusion problem; 3) Combine the above ways by using the latter as interlayer and the former as surface coat.

The enamel coatings exhibit much slower oxidation and hot corrosion rates than metallic coatings because of their quite lower oxygen permeation rates and solubility in molten sulphates. The reactions between enamels and metals are much stronger than those between ceramics and metals. The thermal shocking resistance of the coating is associated mainly with the CTE mismatch between the coating and the substrate, the strength and toughness of the coating.

The low expansion nanocomposite coatings are designed to minimize the CTE mismatch between the TGO scales and the underlying coatings so as to allow growth of thicker TGO scales free of cracks and spallation. A simple way to produce such kind a coating is reactive physical vapor deposition using alloy targets in reactive gas, e.g. reactive sputtering using NiCrAl target in Ar/N_2 mixtrue. A diffusion barrier layer is required for the nitride particle reinforced Ni-base composite coatings on Ni-base superalloys.

REFERENCES

Andersson, J.-O., & Ågren, J. (1992). Models for numerical treatment of multicomponent diffusion in simple phases. *Journal of Applied Physics*, *72*(4), 1350–1355. doi:10.1063/1.351745

Balint, D. S., & Hutchinson, J. W. (2005). An analytical model of rumpling in thermal barrier coatings. *Journal of the Mechanics and Physics of Solids*, *53*(5), 949–973. doi:10.1016/j.jmps.2004.11.002

Birks, N., Meier, G. H., & Pettit, F. S. (2006). *High temperature oxidation of metals*. Cambridge, UK: Cambridge University Press. doi:10.1017/CBO9781139163903

Bull, S. J. (1998). Modeling of residual stress in oxide scales. *Oxidation of Metals*, *49*(1-2), 1–17. doi:10.1023/A:1018822222663

Campbell, C. E., Boettinger, W. J., & Kattner, U. R. (2002). Development of a diffusion mobility database for Ni-base superalloys. *Acta Materialia*, *50*(4), 775–792. doi:10.1016/S1359-6454(01)00383-4

Chen, G. F., & Lou, H. Y. (1999a). Oxidation behavior of sputtered Ni-3Cr-20Al nanocrystalline coating. *Materials Science and Engineering A*, *271*(1-2), 360–365. doi:10.1016/S0921-5093(99)00258-0

Chen, G. F., & Lou, H. Y. (1999b). The effect of nanocrystallization on the oxidation resistance of Ni-5Cr-5Al alloy. *Scripta Materialia*, *41*(8), 883–887. doi:10.1016/S1359-6462(99)00224-9

Chen, G. F., & Lou, H. Y. (2000a). Effect of nanocrystallization on the oxidation behavior of a Ni–8Cr–3.5Al alloy. *Oxidation of Metals*, *54*(1-2), 155–162. doi:10.1023/A:1004610931447

Chen, G. F., & Lou, H. Y. (2000b). The effect nanocrystallization on the oxide formation on Ni-11Cr-8Al alloy. *Scripta Materialia*, *43*(2), 119–122. doi:10.1016/S1359-6462(00)00379-1

Chen, G. F., & Rühle, M. (2005). Transient alumina transformation on a sputtered K38G nanocrystalline coating. *Surface and Coatings Technology*, *191*(2-3), 263–266. doi:10.1016/j.surfcoat.2004.02.025

Chen, J. H., & Little, J. A. (1997). Degradation of the platinum aluminide coating on CMSX4 at 1100 °C. *Surface and Coatings Technology*, *92*(1-2), 69–77. doi:10.1016/S0257-8972(96)03117-9

Chen, M. H., Li, W. B., Shen, M. L., Zhu, S. L., & Wang, F. H. (2013). Glass–ceramic coatings on titanium alloys for high temperature oxidation protection: Oxidation kinetics and microstructure. *Corrosion Science*, *74*, 178–186. doi:10.1016/j.corsci.2013.04.041

Chen, M. H., Li, W. B., Shen, M. L., Zhu, S. L., & Wang, F. H. (2014). Glass coatings on stainless steels for high-temperature oxidation protection: Mechanisms. *Corrosion Science*, *82*, 316–327. doi:10.1016/j.corsci.2014.01.033

Chen, M. H., Shen, M. L., Wang, X., Zhu, S. L., & Wang, F. H. (2012). Oxidation and thermal shock behavior of a glass-alumina composite coating on K38G superalloy at 1000 °C. *Journal of Materials Science and Technology*, *28*(5), 433–438. doi:10.1016/S1005-0302(12)60079-6

Chen, M. H., Shen, M. L., Wang, X., Zhu, S. L., & Wang, F. H. (2013). Interfacial reaction between SiO_2–Al_2O_3–ZnO–CaO based glass coatings and K38G superalloy substrates. *Surface and Coatings Technology*, *216*, 145–151. doi:10.1016/j.surfcoat.2012.11.045

Chen, M. H., Shen, M. L., Zhu, S. L., & Wang, F. H. (2013). Comparative study of interfacial reaction between superalloy substrate and glass coating with and without alumina particles incorporation. *Applied Surface Science*, *271*, 228–233. doi:10.1016/j.apsusc.2013.01.165

Chen, M. H., Shen, M. L., Zhu, S. L., Wang, F. H., & Niu, Y. (2012). Preparation and thermal shock behavior at 1000 °C of a glass-alumina-NiCrAlY tri-composite coating on K38G superalloy. *Surface and Coatings Technology*, *206*(8-9), 2566–2571. doi:10.1016/j.surfcoat.2011.11.015

Chen, M. H., Shen, M. L., Zhu, S. L., Wang, F. H., & Wang, X. L. (2013). Effect of sand blasting and glass matrix composite coating on oxidation resistance of a nickel-based superalloy at 1000 °C. *Corrosion Science*, *73*, 331–341. doi:10.1016/j.corsci.2013.04.022

Chen, M. H., Zhu, S. L., Shen, M. L., Wang, F. H., & Niu, Y. (2011). Effect of NiCrAlY platelets inclusion on the mechanical and thermal shock properties of glass matrix composites. *Materials Science and Engineering A*, *528*(3), 1360–1366. doi:10.1016/j.msea.2010.10.015

Chen, M. H., Zhu, S. L., Shen, M. L., Wang, F. H., & Niu, Y. (2013). Thermophysical properties of alumina particle reinforced glass matrix composites. *International Journal of Applied Ceramic Technology*, *10*(2), 224–233. doi:10.1111/j.1744-7402.2012.02806.x

Chen, M. H., Zhu, S. L., & Wang, F. H. (2010). Crystallization Behavior of SiO_2–Al_2O_3–ZnO–CaO Glass System at 1123–1273 K. *Journal of the American Ceramic Society*, *93*(10), 3230–3235. doi:10.1111/j.1551-2916.2010.03859.x

Chen, M. H., Zhu, S. L., & Wang, F. H. (2013). Strengthening mechanisms and fracture surface characteristics of silicate glass matrix composites with inclusion of alumina particles of different particle sizes. *Physica B, Condensed Matter*, *413*, 15–20. doi:10.1016/j.physb.2013.01.006

Chen, M. H., Zhu, S. L., & Wang, F. H. (2015). High temperature oxidation of NiCrAlY, nanocrystalline and enamel-metal nano-composite coatings under thermal shock. *Corrosion Science*, *100*, 556–565. doi:10.1016/j.corsci.2015.08.033

Clarke, D. R. (2003). The lateral growth strain accompanying the formation of a thermally grown oxide. *Acta Materialia*, *51*(5), 1393–1407. doi:10.1016/S1359-6454(02)00532-3

Dahl, K. V., Hald, J., & Horsewell, A. (2006). Interdiffusion between Ni-based superalloy and MCrAlY coating. *Defect and Diffusion Forum*, 258-260, 73-78.

Donald, I. W. (1993). Preparation, properties and chemistry of glass and glass-ceramic-to- metal seals and coatings. *Journal of Materials Science*, *28*(11), 2841–2886. doi:10.1007/BF00354689

Evans, A. G., Crumley, G. B., & Demaray, R. E. (1983). On the mechanical behavior of brittle coatings and layers. *Oxidation of Metals*, *20*(5-6), 193–216. doi:10.1007/BF00656841

Evans, A. G., He, M. Y., & Hutchinson, J. W. (1997). Effect of interface undulations on the thermal fatigue of thin films and scales on metal substrates. *Acta Materialia*, *45*(9), 3543–3554. doi:10.1016/S1359-6454(97)00062-1

Evans, H. E., Mitchell, G. P., Lobb, R. C., & Owen, D. R. J. (1993). A numerical-analysis of oxide spallation. *Proceedings of the Royal Society of London Series A*, 440(1908), 1-22.

Geng, S. J., Wang, F. H., & Zhu, S. L. (2002). High-temperature oxidation behavior of sputtered IN 738 nanocrystalline coating. *Oxidation of Metals*, *57*(3-4), 231–243. doi:10.1023/A:1014870101143

Haynes, J. A., Pint, B. A., Porter, W. D., & Wright, I. G. (2004). Comparison of thermal expansion and oxidation behavior of various high-temperature coating materials and superalloys. *Materials at High Temperatures*, *21*(2), 87–94. doi:10.1179/mht.2004.012

Hou, P. Y., & Smialek, J. L. (1995). Effect of sulfur removal by H_2-annealing on the oxidation behavior of a Ni-25Cr alloy. *Scripta Metallurgica et Materialia*, *33*(9), 1409–1416. doi:10.1016/0956-716X(95)00385-9

Karch, J., Birringer, R., & Gleiter, H. (1987). Ceramics ductile at low-temperature. *Nature*, *330*(6148), 556–558. doi:10.1038/330556a0

Karunaratne, M. S. A., Ogden, S. L., Kenny, S. D., & Thomson, R. C. (2009). A multicomponent diffusion model for prediction of microstructural evolution in coated Ni based superalloy systems. *Materials Science and Technology*, *25*(2), 287–299. doi:10.1179/174328408X355415

Lance, M. J., Unocic, K. A., Haynes, J. A., & Pint, B. A. (2013). Effect of water vapor on thermally-grown alumina scales on Pt-modified and simple aluminide bond coatings. *Surface and Coatings Technology*, *237*, 2–7. doi:10.1016/j.surfcoat.2013.06.041

Li, W. B., Chen, M. H., Wang, C., Zhu, S. L., & Wang, F. H. (2013). Preparation and oxidation behavior of SiO_2–Al_2O_3–glass composite coating on Ti–47Al–2Cr–2Nb alloy. *Surface and Coatings Technology*, *218*, 30–38. doi:10.1016/j.surfcoat.2012.12.022

Li, W. B., Chen, M. H., Wu, M. Y., Zhu, S. L., Wang, C., & Wang, F. H. (2014). Microstructure and oxidation behavior of a SiC–Al_2O_3–glass composite coating on Ti–47Al–2Cr–2Nb alloy. *Corrosion Science*, *87*, 179–186. doi:10.1016/j.corsci.2014.06.015

Li, W. B., Zhu, S. L., Chen, M. H., Wang, C., & Wang, F. H. (2014). Development of an oxidation resistant glass–ceramic composite coating on Ti–47Al–2Cr–2Nb alloy. *Applied Surface Science*, *292*, 583–590. doi:10.1016/j.apsusc.2013.12.012

Li, W. B., Zhu, S. L., Wang, C., Chen, M. H., Shen, M. L., & Wang, F. H. (2013). SiO_2–Al_2O_3–glass composite coating on Ti–6Al–4V alloy: Oxidation and interfacial reaction behavior. *Corrosion Science*, *74*, 367–378. doi:10.1016/j.corsci.2013.05.010

Lou, H. Y., Wang, F. H., Xia, B. J., & Zhang, L. X. (1992). High temperature oxidation resistance of sputtered micro-grain superalloy K38G. *Oxidation of Metals*, *38*(3-4), 299–307. doi:10.1007/BF00666917

Lou, H. Y., Wang, F. H., Zhu, S. L., Xia, B. J., & Zhang, L. X. (1994). Oxide formation of K38G superalloy and its sputtered micrograined coating. *Surface and Coatings Technology*, *63*(2), 105–114. doi:10.1016/S0257-8972(05)80014-3

Lou, H. Y., Zhu, S. L., & Wang, F. H. (1995). Rehealing ability of oxide scales formed on microcrystalline K38G coatings. *Oxidation of Metals*, *43*(3-4), 317–328. doi:10.1007/BF01047033

Lu, J. T., Zhu, S. L., & Wang, F. H. (2011). High temperature corrosion behavior of an AIP NiCoCrAlY coating modified by aluminizing. *Surface and Coatings Technology*, *205*(21-22), 5053–5058. doi:10.1016/j.surfcoat.2011.05.005

Lu, L., Sui, M. L., & Lu, K. (2000). Superplastic extensibility of nanocrystalline copper at room temperature. *Science*, *287*(5457), 1463–1466. doi:10.1126/science.287.5457.1463 PMID:10688789

McFadden, S. X., Mishra, R. S., Valiev, R. Z., Zhilyaev, A. P., & Mukherjee, A. K. (1999). Low-temperature superplasticity in nanostructured nickel and metal alloys. *Nature*, *398*(6729), 684–686. doi:10.1038/19486

N'Gandu-Muamba, J. M., & Streiff, R. (1993). The reactive element effect (R.E.E.): A tentative classification. *Journal de Physique. IV*, *03*(C9), 281–290.

Narita, T. (2009). A view of compatible heat-resistant alloy and coating systems at high-temperatures. *Paper presented at the International Workshop on Advanced Material for New and Renewable Energy*, Jakarta, Indonesia. doi:10.1063/1.3243272

Panat, R., Hsia, K. J., & Oldham, J. (2005). Rumpling instability in thermal barrier systems under isothermal conditions in vacuum. *Philosophical Magazine*, *85*(1), 45–64. doi:10.1080/14786430412331309344

Pennefather, R. C., & Boone, D. H. (1995). Mechanical degradation of coating systems in high-temperature cyclic oxidation. *Surface and Coatings Technology*, *76*(1-3), 47–52. doi:10.1016/0257-8972(95)02551-0

Pilling, N. B., & Bedworth, R. E. (1923). The oxidation of metals at high temperatures. *Journal of the Institute of Metals*, *29*, 529–582.

Rapp, R. A. (2002). Hot corrosion of materials: A fluxing mechanism? *Corrosion Science*, *44*(2), 209–221. doi:10.1016/S0010-938X(01)00057-9

Ren, P., Zhu, S. L., & Wang, F. H. (2015a). Characterization and oxidation behavior of a sputtered nanocomposite Ni+CrAlYSiHfN coating. *Corrosion*, *71*(4), 523–535. doi:10.5006/1492

Ren, P., Zhu, S. L., & Wang, F. H. (2015b). Microstructural stability of AlN diffusion barrier for nano-composite Ni + CrAlYSiHfN coating on single crystal superalloy at high temperatures. *Applied Surface Science*, *359*, 420–425. doi:10.1016/j.apsusc.2015.10.096

Ren, P., Zhu, S. L., & Wang, F. H. (2016a). Microstructure and oxidation behavior of a Ni+CrAlYSiHfN/AlN multilayer coating fabricated by reactive magnetron sputtering. *Corrosion Science*, *104*, 197–206. doi:10.1016/j.corsci.2015.12.012

Ren, P., Zhu, S. L., & Wang, F. H. (2016b). TEM study of the evolution of sputtered Ni+CrAlYSiHfN nanocomposite coating with an AlN diffusion barrier at high temperature. *Surface and Coatings Technology*, *286*, 262–267. doi:10.1016/j.surfcoat.2015.12.046

Rhines, F. N., & Wolf, J. S. (1970). The role of oxide microstructure and growth stresses in the high-temperature scaling of nickel. *Metallurgical Transactions*, *1*(6), 1701–1710. doi:10.1007/BF02642020

Saunders, S. R. J., Monteiro, M., & Rizzo, F. (2008). The oxidation behaviour of metals and alloys at high temperatures in atmospheres containing water vapour: A review. *Progress in Materials Science*, *53*(5), 775–837. doi:10.1016/j.pmatsci.2007.11.001

Schütze, M. (1995). Mechanical properties of oxide scales. *Oxidation of Metals*, *44*(1-2), 29–61. doi:10.1007/BF01046722

Shen, M. L., Zhu, S. L., Chen, M. H., & Wang, F. H. (2011). The oxidation and oxygen permeation resistance of quartz particle-reinforced aluminosilicate glass coating on titanium alloy. *Journal of the American Ceramic Society*, *94*(8), 2436–2441. doi:10.1111/j.1551-2916.2011.04587.x

Shen, M. L., Zhu, S. L., & Wang, F. H. (2015). Formation kinetics of multi-layered interfacial zone between γ-TiAl and glass–ceramic coatings via interfacial reactions at 1000 °C. *Corrosion Science*, *91*, 341–351. doi:10.1016/j.corsci.2014.11.036

Smialek, J. L., & Auping, J. V. (2002). COSP for windows—Strategies for rapid analyses of cyclic-oxidation behavior. *Oxidation of Metals*, *57*(5-6), 559–581. doi:10.1023/A:1015308606869

Suo, Z., Kubair, D. V., Evans, A. G., Clarke, D. R., & Tolpygo, V. K. (2003). Stresses induced in alloys by selective oxidation. *Acta Materialia*, *51*(4), 959–974. doi:10.1016/S1359-6454(02)00499-8

Tolpygo, V. K., & Clarke, D. R. (2000). Surface rumpling of a (Ni, Pt)Al bond coat induced by cyclic oxidation. *Acta Materialia*, *48*(13), 3283–3293. doi:10.1016/S1359-6454(00)00156-7

Tolpygo, V. K., & Clarke, D. R. (2004a). On the rumpling mechanism in nickel-aluminide coatings Part I: An experimental assessment. *Acta Materialia*, *52*(17), 5115–5127.

Tolpygo, V. K., & Clarke, D. R. (2004b). Rumpling induced by thermal cycling of an overlay coating: The effect of coating thickness. *Acta Materialia*, *52*(3), 615–621. doi:10.1016/j.actamat.2003.10.001

Tortorelli, P. F., Specht, E. D., More, K. L., & Hou, P. Y. (2012). Oxide growth stress measurements and relaxation mechanisms for alumina scales grown on FeCrAlY. *Materials and Corrosion*, *63*(10), 857–861. doi:10.1002/maco.201206760

Veal, B. W., Paulikas, A. P., Gleeson, B., & Hou, P. Y. (2007). Creep in α-Al_2O_3 thermally grown on β-NiAl and NiAlPt alloys. *Surface and Coatings Technology*, *202*(4-7), 608–612. doi:10.1016/j.surfcoat.2007.05.093

Wallwork, G. R., & Hed, A. Z. (1971). Some limiting factors in the use of alloys at high temperatures. *Oxidation of Metals*, *3*(2), 171–184. doi:10.1007/BF00603485

Wang, F. H., Lou, H. Y., Zhu, S. L., & Wu, W. T. (1996). The mechanism of scale adhesion on sputtered microcrystalline CoCrAl films. *Oxidation of Metals*, *45*(1-2), 39–50. doi:10.1007/BF01046819

Wang, J. L., Chen, M. H., Yang, L. L., Zhu, S. L., & Wang, F. H. (2015). Comparative study of oxidation and interdiffusion behavior of AIP NiCrAlY and sputtered nanocrystalline coatings on a nickel-based single-crystal superalloy. *Corrosion Science*, *98*, 530–540. doi:10.1016/j.corsci.2015.05.062

Wang, J. L., Chen, M. H., Zhu, S. L., & Wang, F. H. (2015). Ta effect on oxidation of a nickel-based single-crystal superalloy and its sputtered nanocrystalline coating at 900–1100oC. *Applied Surface Science*, *345*, 194–203. doi:10.1016/j.apsusc.2015.03.157

Wang, X., Chen, M. H., Zhu, S. L., & Wang, F. H. (2013). Interfacial reactions between a SiO_2-Al_2O_3-ZnO-CaO based glass and alpha alumina. *Surface and Coatings Technology*, *232*, 6–12. doi:10.1016/j.surfcoat.2013.04.044

Wang, X., Chen, M. H., Zhu, S. L., & Wang, F. H. (2015). Oxidation behavior of glass-based composite thermal barrier coating on K417G superalloy with a NiCoCrAlY bond coat at 1000 °C. *Surface and Coatings Technology*, *270*, 314–323. doi:10.1016/j.surfcoat.2015.02.014

Wang, X., Shen, M. L., Chen, M. H., Zhu, S. L., & Wang, F. H. (2015). Microstructure stabilization of a novel glass/YSZ composite coating material by adding alumina particles. *Ceramics International*, *41*(8), 9753–9762. doi:10.1016/j.ceramint.2015.04.047

Wu, M. Y., Chen, M. H., Zhu, S. L., & Wang, F. H. (2016). Protection mechanism of enamel–alumina composite coatings on a Cr-rich nickel-based superalloy against high-temperature oxidation. *Surface and Coatings Technology*, *285*, 57–67. doi:10.1016/j.surfcoat.2015.11.026

Xiong, Y. M., Guan, C. H., Zhu, S. L., & Wang, F. H. (2006). Effect of enamel coating on oxidation and hot corrosion behaviors of Ti-24Al-14Nb-3V alloy. *Journal of Materials Engineering and Performance*, *15*(5), 564–569. doi:10.1361/105994906X136151

Xiong, Y. M., Zhu, S. L., & Wang, F. H. (2005). The oxidation behavior and mechanical performance of Ti60 alloy with enamel coating. *Surface and Coatings Technology*, *190*(2-3), 195–199. doi:10.1016/j.surfcoat.2004.09.005

Xiong, Y. M., Zhu, S. L., & Wang, F. H. (2008). Synergistic corrosion behavior of coated Ti60 alloys with NaCl deposit in moist air at elevated temperature. *Corrosion Science*, *50*(1), 15–22. doi:10.1016/j.corsci.2007.06.007

Yang, L. L., Chen, M. H., Cheng, Y. X., Wang, J. L., Liu, L., Zhu, S. L., & Wang, F. H. (2016). Effects of surface finish of single crystal superalloy substrate on cyclic thermal oxidation of its nanocrystalline coating. *Corrosion Science*, *111*, 313–324. doi:10.1016/j.corsci.2016.04.023

Yang, L. L., Chen, M. H., Wang, J. L., Bao, Z. B., Zhu, S. L., & Wang, F. H. (2017). Diffusion of Ta and its influence on oxidation behavior of nanocrystalline coatings with different Ta, Y and Al contents. *Corrosion Science*, *126*, 344–355. doi:10.1016/j.corsci.2017.07.017

Yang, L. L., Chen, M. H., Wang, J. L., Zhu, S. L., & Wang, F. H. (2016). A duplex nanocrystalline coating for high-temperature applications on single-crystal superalloy. *Corrosion Science*, *102*, 72–83. doi:10.1016/j.corsci.2015.09.020

Yang, S. L., & Wang, F. H. (2006). Effect of nanocrystallization on sulfur segregation in Fe–Cr–Al alloy during oxidation at 1000 °C. *Oxidation of Metals*, *65*(3-4), 195–205. doi:10.1007/s11085-006-9014-1

Yang, Y. F., Jiang, C. Y., Yao, H. R., Bao, Z. B., Zhu, S. L., & Wang, F. H. (2016). Preparation and enhanced oxidation performance of a Hf-doped single-phase Pt-modified aluminide coating. *Corrosion Science*, *113*, 17–25. doi:10.1016/j.corsci.2016.09.014

Zheng, D. Y., Xiong, Y. M., Zhu, S. L., Wang, F. H., & Wu, W. T. (2008). Solid reactions between enamel and O-phase Ti-Al-Nb intermetallics at 800 °C. *Materials Scienc Forum*, 595-598, 233-238.

Zheng, D. Y., Zhu, S. L., & Wang, F. H. (2006). Oxidation and hot corrosion behavior of a novel enamel-Al_2O_3 composite coating on K38G superalloy. *Surface and Coatings Technology*, *200*(20-21), 5931–5936. doi:10.1016/j.surfcoat.2005.09.004

Zhou, C. G., Yu, J. S., Gong, S. K., & Xu, H. B. (2002). Influence of water vapor on the isothermal oxidation behavior of low pressure plasma sprayed NiCrAlY coating at high temperature. *Surface and Coatings Technology*, *161*(1), 86–91. doi:10.1016/S0257-8972(02)00398-5

Zhu, L. J., Zhu, S. L., & Wang, F. H. (2012). Preparation and oxidation behaviour of nanocrystalline Ni + CrAlYSiN composite coating with AlN diffusion barrier on Ni-based superalloy K417. *Corrosion Science*, *60*, 265–274. doi:10.1016/j.corsci.2012.03.027

Zhu, L. J., Zhu, S. L., Wang, F. H., & Zhang, J. Q. (2014). Comparison of the cyclic oxidation behavior of a low expansion Ni + CrAlYSiN nanocomposite and a NiCrAlYSi coating. *Corrosion Science*, *80*, 393–401. doi:10.1016/j.corsci.2013.11.052

Zhu, S. L., Wang, F. H., Lou, H. Y., & Wu, W. T. (1995). Reactive sputter deposition of alumina films on superalloys and their high-temperature corrosion resistance. *Surface and Coatings Technology*, *71*(1), 9–15. doi:10.1016/0257-8972(94)02289-3

Zhu, S. L., Wei, Z. W., & Liu, J. H. (2014). Kinetics modeling of parabolic growth competing to linear declining: A theoretical tool to interpret corrosion data in real world. *Advanced Materials Research*, *860-863*, 899–902. doi:10.4028/www.scientific.net/AMR.860-863.899

KEY TERMS AND DEFINITIONS

Enamel Coating: The vitreous coating prepared by fusing powdered glass to a substrate by firing at high temperatures. An enamel coating may be composed of glass matrix and dispersed metallic and/or ceramic inclusions.

Nanocomposite Coating: The multiphase coating where each phase has grain sizes less than 100 nanometers. A nanocomposite coating composed of metal matrix and dispersed ceramic particles may be referred to as cermet nano-coating.

Nanocrystalline Coating: The polycrystalline coating with grain sizes less than 100 nanometers (nm). The ultrafine-grained coatings are those with grain sizes ranged from 100 nm to half-micrometer.

Parabolic Constant: The constant for TGO growth that follows parabolic law, defined by the ratio of the square of mass gain per unit area to the tested time.

TGO: The abbreviation for thermally grown oxide, the corrosion products on the materials surfaces formed due to oxidation of materials during exposure in oxidizing environments at high temperatures. An alloy or a coating with excellent oxidation resistance is often referred to that forms continuous, slowly-growing and adherent TGO scales.

Thermal Expansion: The tendency of matter to change in shape, area, and volume in response to a change in temperature. The coefficient of thermal expansion (CTE) describes how the size of an object changes with a change in temperature. Specifically, it measures the fractional change in size per degree change in temperature at a constant pressure. Several types of coefficients have been developed: volumetric, area, and linear.

Thermal Shocking: A process that produces large thermal gradients in an object, *e.g.* a glass cube is dropped into ice water from a furnace at 600 °C. Thermal shocking causes different parts of the object to expand by different amounts. This differential expansion can be understood in terms of stress or of strain, equivalently. At some point, this stress can exceed the strength of the material, causing a crack to form. If nothing stops this crack from propagating through the material, it will cause the object's structure to fail.

Chapter 8
Slurry Sprayed Mullite Coatings and Their Corrosion Performances

Rajeev Verma
PEC University of Technology, India

Narendra Mohan Suri
PEC University of Technology, India

Suman Kant
PEC University of Technology, India

ABSTRACT

Slurry spray technique (SST) is a distinctive variant among the numerous and already established coating techniques. Functionally graded thermal barrier and environmental barrier coating have been the functionalities developed so far for the process. Among the choice of the various ceramic feedstocks available mullite and partially stabilized zircona have been found suitable and investigated for the coating deposition via SST. This chapter reports the findings of the corrosion studies in simulated industrial corrosive environments and characterization results of the six sets of slurry sprayed mulite-nickel based coatings. Decent protection against coating has been found during the immersion test performed on these coatings for evaluating their corrosion performance. The developed coatings are recommended for use in applications to endure the elevated temperature and inflict corrosion. Thermal cycling test was performed to support the acceptable thermal shock resistance and coating compliance of the developed coatings.

INTRODUCTION

Surface coatings are pursued as a protective agent applied to all or part of the substrate for the functional purpose intended. Environmental barrier coatings (EBCs) are pursued in the structurally critical components of the turbo-machineries to improve the stability and service life of the substrate material wherein the components are exposed to elevated temperature and corrosive environment. EBCs have

DOI: 10.4018/978-1-5225-4194-3.ch008

been used as "prime reliant" coatings since they serve as surface protection against combustion species containing water vapour to inhibit chemical reactions at the underlying substrate (Bezzi, et al., 2010).

In order to enhance the performance and efficiency, there have been increasing attempts over the past decades to raise the working temperatures of the turbomachinery (Walston, 2004). The areas in focus to the EBC application could include the components of combustion turbines, internal combustion engines, boilers, and waste incinerators etc., wherein the parts are exposed to the high temperature above 150°C and under oxidative environment. More EBC applications include aeronautical, nuclear, automotive industries and aerospace engine components like vanes, rotors, blades, combustor liners, shrouds, transition ducts and turbo airfoils.

Recent work by the authors has investigated mullite-nickel based slurry sprayed coatings fabricated with the aid of TiO_2 as liquid phase sintering additive and fly ash as a ceramic substitute (Verma et al., 2016a, 2016b, 2016c). The results confirmed that SST is a propitious alternative to the existing sophisticated techniques. Comparable performances have been reported in adhesion strength and coating thickness during these studies. It is therefore of interest to extend the scope of investigations to these developed SST coatings. The integrity and durability of the developed coatings shall be contingent on the evaluation of their use at elevated temperature. With these views, an attempt has been made to inclusively evaluate the longevity in context to the corrosion behavior, since no such information on these grounds could be found in the scientific literature. Small-scale laboratory studies which can provide a broad range of coating compliance data is a cost effective alternative to full-scale testing. The present contribution combines the observations described in previous reports by the authors (Verma et al., 2016a, 2016b, 2016c) and provides a new analysis based on the corrosion behaviour.

BACKGROUND

Over the past three decades, many fabricating methods have been developed for depositing ceramics and other coating materials on a metal substrate. All these fabricating techniques can be categorized into three main groups: deposition techniques, flame spray techniques, and bulk processes (Kieback et al, 2003). Each fabrication method differs from each other considerably, in the context of physical principal utilized, cost, and intricacy. However, the main hindrance in the pervasive utilization of these techniques is the sophisticated equipment and thus manufacturing cost involved. Besides, most of these already established fabrication methods have not been effective in large curved areas. Despite, the existence of pre-established and consolidated deposition techniques, an active research fraternity strives for novel and cost-effective alternatives, both for the method and feedstock materials. Moreover, the use of low fabrication cost stages and the possibility to utilize the industrial waste makes it easy to imagine the potential material and cost saving offered by SST. Slurry spray technique (SST) is conceived to be as an analogous to the thermal spray coating regime, because of the use of thermal energy required for sintering of the powder constituents. SST being relatively facile and unutilized deposition technique requires supplementary research efforts for it to make commercially viable.

ASTM 1018 low carbon steel have properties close to iron based super-alloys used for gas turbine parts like diffuser, compressor, mixer & exhaust which are subjected to relatively lower temperatures (Goward, 1998). Moreover, EBCs on economic materials like low carbon steel have the potential to replace the costlier materials like Ti-based alloys and Ni-based alloys used in steam turbine and other turbomachinery components. These parts are supposed to be used at relatively low cycle fatigue (LCF) and wherein the working temperature does not exceed 800°C (Fundamentals of Gas Turbine Engines, n.d.) By virtue of the predominant high temperature properties like coefficient of thermal expansion, thermal diffusivity and specific heat capacity (Anggono, 2005; Kokini, 1996), monolithic mullite $3Al_2O_3{\cdot}2SiO_2$ is a promising EBC material on low carbon steel.

In spite of desirable properties exhibited, extensive usage of mullite is not prevalent due to the difficulty of sintering mullite powder in its purest form owing to the poor interdiffusion rates of Si^{4+} and Al^{3+} within the mullite lattice (Lee & Rainforth, 1994; Tezuka, eta al., 2009). Since SST utilizes sintering of the powder mix, sintering problems pertaining to mullite at high temperature upto 1600°C might trigger defects in the substrate and coating. In response to this, it has been attempted to sinter mullite at lower temperatures by the virtue of liquid-phase sintering (LPS) additives like titanium dioxide (TiO_2) (Tripathi & Banerjee, 1998), ferrous oxide (Fe_2O_3) (Tripathi, et al., 2000), yttrium (III) oxide (Y_2O_3) (Chandran et al., 1996) and magnesium oxide (MgO) (Montanaro et al., 1997). While most commonly used sintering additives are oxide-based, however, some fluoride-based compounds have also been investigated as LPS aids.

The chapter elucidates the qualitative assessment of the corrosion behaviour, and characterization results of the developed coatings in molten salt (Na_2SO_4) environment, in view to correlate the functional behaviour relevant to EBC applications. The results have been compiled to provide the comparative performance of the coated and uncoated specimens in similar environments. Thus, the susceptibility to corrosion attacks for coated and uncoated ASTM 1018 low carbon steel substrate used in the study is investigated. Some of the common phenomena observed during the studies have also been mentioned. The corrosion products were analyzed with the help of various techniques such as visual and microscopic examination, X-ray diffractometry (XRD), and scanning electron microscopy/energy-dispersive analysis (SEM/EDS) for morphological examinations. The results of thermal cycling test conducted on the as-sprayed samples have also been presented in this article.

SLURRY SPRAY TECHNIQUE

Fabrication Materials

Functionally graded mullite ($3Al_2O_3{\cdot}2SiO_2$ or $Al_6Si_2O_{13}$) based deposits on ASTM 1018 low carbon steel substrate were coated by SST in this study. Nickel powder as bond coat material and reinforcement was chosen because of its closely related thermo-mechanical properties with the ASTM 1018 steel (Verma et al., 2016b). Fly-ash (FA), a principle ingredient of coal ash or coal combustion residuals (CCR), which has the rich percentage of metal oxides has been intriguing for its coating potential. The commercially available mullite with average nominal particle size distribution 66±14.6 μm; Class-F fly ash (251±24.1 μm) as feedstock ceramic materials and nickel powder (Plus Chem, 58.69 g/mol, 99.9%, 152±14.2 μm) were selected for utilization in SST as they inhold the requisite properties for the technique. As-received Class-F fly ash acquired from a thermal power plant in Bathinda, Punjab, India, has been utilized after baking. Its nominal chemical composition comprised 57.68% SiO_2, 26.24% Al_2O_3, 5.12% Fe_2O_3, 4.48% CaO, 1.92% K_2O, 1.69% TiO_2, 1.48% MgO, 0.86% Na_2O (wt. %), and traces of SO_3, P_2O_5, and MnO_2. In addition, TiO_2 (Plus Chem, 79.87 g/mol, 99.8%, 74±6.1 μm) as liquid phase sintering (LPS) supplement was investigated to lower the fabrication temperatures (Verma et al., 2016b).

Because of its favourable mechanical properties, ease of availability and low cost, ASTM 1018 mild/low carbon steel, was considered for the substrate material which is common to a wide range of domestic applications and industrial applications. ASTM 1018 steel offers a good balance of strength, toughness, and ductility besides it possesses excellent weldability. Moreover, ASTM 1018 hot rolled steel also includes enhanced Brinell hardness and machining characteristics (AZo Materials, 2012). The nominal chemical composition of the commercially procured ASTM 1018 steel substrate comprised of 0.17% C, 0.51% Mn, 0.17% Si, 0.07% Cr, 0.05% P, 0.04% S, 0.02% Ni, 0.06% Cu, 0.01% Mo, 0.02% Al (wt. %), and remaining Fe.

The substrate material is required to have a comparable value of the coefficient of thermal expansion (CTE) to that of the coating powders. This would avoid development of thermally induced residual stresses because of CTE mismatch, during sintering stage and in-service thermal fatigue due to repetitive temperature excursions. ASTM 1018 low carbon steel chosen as the substrate material, have low CTE similar to ceramics, it is thus likely to be resilient to cracking during sintering and in-service thermal excursions. The typical range of properties for selected materials is shown in Table 1.

Surface preparation prior to the spraying is vital for the coherence of the thermal spray coating. Since mechanical interlocking is considered as the adhesion mechanism, it requires a standard procedure for optimal coating performance. Sandblasting has been a prevailing surface preparation, for removal of any oil or grit present on the substrate surface. Sandblasting also makes the surface porous, which enable the coating powders to bond to the substrate during sintering, which is termed as 'diffusion bonding' (Bernard et al., 2008). ASTM 1018 mild/low carbon steel square specimens (25 x 25 x 5) mm for coating and subsequent characterization were sectioned. The coating coupons were ground to acquire the flatness of 0.05 mm and eventually grit blasted using alumina (60 grit) at 3 kg/cm^2 of pressure in order to achieve the average surface roughness (Ra) > 6.8μm. In order to avoid any structural changes in the specimens during substrate surface preparation, due care was taken while specimens were prepared manually.

Coating formulation via slurry spray technique (SST) encompasses the preparation of the slurry mixture of suspended sinterable coating ingredients in a carrier. Distilled water has been utilized as the carrier fluid as it offers limited electrostatic stabilization of the powder particles because of ionic dissociation. Moreover, it is advantageous of distilled water to leaves no salt residue after evaporation from the slurry.

Polyvinyl alcohol (PVA, sometimes referred to as PVOH) having the formula $[CH2CH(OH)]_n$ was chosen as the binder for the study. It is white (colourless in solution) and odourless, which melts and boils at 200°C and 228°C respectively. PVA is a water-soluble polymer binder also widely utilized in coatings, adhesives, sealants, paints, plastics, textiles etc. PVA as the binder used for slurry preparation significantly influences the viscosity of the mixture. It has been established experimentally that the percentage of the binder content affects the slurry viscosity almost linearly. Therefore, the variation of binder content in slurry mixture was determined to achieve right viscosity for it to be readily sprayable

Table 1. ASTM 1018 low carbon steel material properties

	Mean Coefficient of Thermal Expansion (x 10-6/K)	Thermal Conductivity (W/mK)	Melting Point (°C/K)
Low carbon steel	~6.7	~51.9	1449 / 1722.2
Nickel	12	90.0	1453 / 1726.2
Mullite	~5.4	3.5-6	1850 / 2123.2

which is the one of the prerequisites of SST. The optimum slurry viscosity for the ingredients used in the study was determined to be 25 mm^2/s, to minimize the porosity induction during drying stage (Verma et al., 2016b).

The dispersant in the mix serves a significant function to check the agglomeration of the powder constituents of different densities. Homogeneous dispersion of the slurry constituents like ceramic, metal powders and the binder is crucial for enabling them to be sprayed through a pressurised nozzle. The dispersant is a long chained polymeric compound which enables the intermolecular bonding within the powder constituents. These organic compounds graft themselves onto powder constituents through ionic exchange due to their non-uniform polarized regions of polymer molecules. The ionic charge of the trailing sections of these polymer molecules repulses the adjacent particles which is enclosed by the polarized dispersant molecules (Khan, et al., 2000). Tera-sodium pyrophosphate; $Na_4P_2O_7$, as a dispersant was added to the slurry mix for dispersing the differently dense ingredients apart.

Fabrication Through SST

Coating formulation by SST comprises of the following five principle stages as outlined in Figure 1.

Slurry Preparation

For preparing the slurry mix, initially precisely weighed tetra sodium pyrophosphate (TSPP), the dispersing agent was slowly added to a volume of distilled water (Greenwood et al., 1999) and allowed to mix with continuous stirring on a magnetic stirrer with a hot plate. Hydrosoluble PVA binder crystals were later added to this mixture at a slow, even rate with continuous stirring since the binder increases the

Figure 1. Principle stages of coating deposition by SST
Adapted from Verma et al., 2016a [SAGE Publications Ltd].

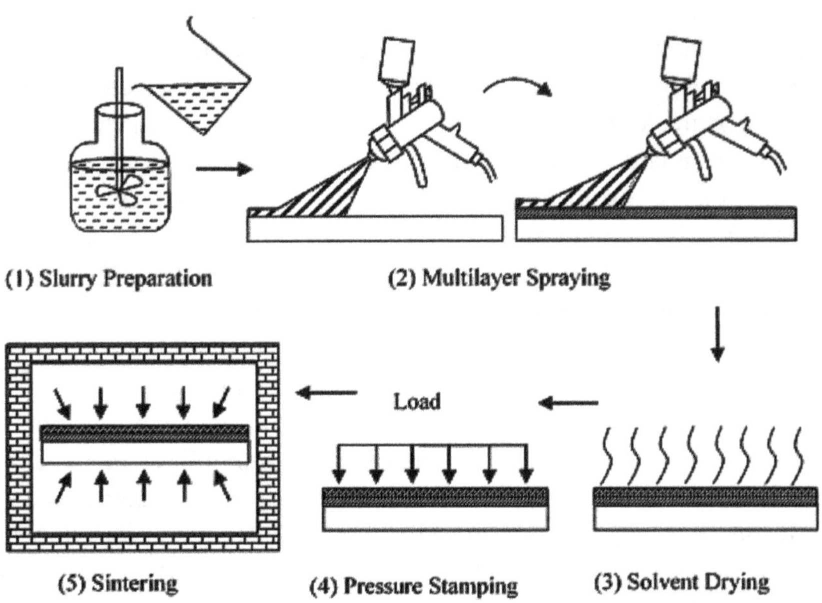

viscosity of the mixture and if introduced too quickly then it causes agglomeration (Baklouti et al., 1998). Although, PVA is soluble in hot as well as cold water. The solution becomes sticky and the tendency to form lumps increases as temperature rises. Lumps cannot readily be broken down by agitation and long mixing times are required to complete solubilization (Silverson, 2017). Moreover, vigorous agitation required for solubilizing the powders can lead to aeration problems. However, once the powder is fully dispersed increased rate of mixing while stirring is observed by heating the mixture up to 45°C at which the polymer becomes solubilized. Eventually, it needs to be ensured while mixing that the binder has been fully dissolved in the mixture after approximately 30 minutes continuous stirring.

Finally, the mix of ceramic and metal powders (mullite and nickel respectively) was gradually added to the solution. The addition of ceramic and metal powders was done with the mechanical stirring for nickel being ferromagnetic. Titanium dioxide (TiO_2) was later prepended to the slurry mixture as liquid phase sintering (LPS) additive. Sonication of the prepared slurry mixture was done with the ultrasonic bath sonicator to avoid flocculation by enhancing the dispersion of the prepared mixture. Appropriate selection of beaker and magnetic bead size used while stirring is important. It was noted that too brisk stirring and use of oversized magnetic bead in low volume resulted in the partial immersion of the magnetic bead in the solution, and leads to foam formation due to aeration which is undesirable for proper spraying. Moreover, a degree of shear is required to accelerate the process, since inefficient agitation takes longer to achieve full solubilization.

Multilayer Spraying

Multi-layered spraying allowed the production of functionally graded environmental barrier coatings (FG–EBCs), making SST wide open to a variety of applications. The coatings with SST could be orchestrated to cater to the specified needs, and not restricted by the material properties of the ceramic coating. The slurry thus prepared is then applied onto the substrate by a pressurized air stream of high velocity-low pressure (HVLP) system which is also commonly used in the paint spray technique. The slurry could be atomized using an air pressurized gravity fed or siphon fed type gun, offering the opportunity to coat irregular or complex geometries. Since, the gravity feed spray gun entails less air pressure than siphon fed type, to achieve sufficient atomization of the solution mixture with less overspray and greater control during spraying. Hence, gravity fed paint-spray gun was selected in the current study for spraying of the slurry mixture. The dispersed slurry mixture was fed into the paint hopper on top of the spray gun connected to a standard variable pressure air compressor. The high-velocity turbulent airflow atomizes the slurry into the high-velocity air stream through a small orifice. The disassociated slurry molecules allow a steady distribution of the sprayed coating onto the substrate held against the spray jet. This disassociation is greatly influenced by the velocity and pressure of the air stream although, at the cost of slurry losses by the spray disseminating to the surroundings.

The prepared slurry mixture was sprayed onto the coating coupons using air pressurized and gravity fed paint spray gun. Spraying is greatly influenced by the slurry properties, particle size, density and solid to liquid volume ratio. Their effect can be governed by the adequate selection of the deposition parameters like spraying pressure, nozzle-to-work distance and slurry flow rate. The cone-shaped spray stream was made uniformly directed onto the substrate surface in several quick successions, keeping a distance of 150 mm approximately across the nozzle tip and substrate to prevent localized deposition. The slurry ingredients used in the slurry preparation for right viscosity determined earlier are given in Table 2. To produce a functional coating, various slurry compositions were sprayed onto substrates in

Table 2. Composition used for the slurry mixture (wt.%)

Mullite + Ni Powder	LPS Additive (TiO_2)	Binder (PVA)	Dispersant (TSPP)	Mix Agent (Distilled Water)
44%	1%	2.5%	0.5%	52%

succession with intermediate drying. Schematic of the graded structure achieved by depositing successive layers of the varied composition as per Table 3 is shown in Figure 2.

Solvent Drying

The wet coating layer is fragile and susceptible to gravitational disturbance or other external force. The solvent drying or carrier removal stage starts immediately after the spraying of the slurry mixture. The duration of the drying stage depends on the thickness of deposited green coating and normally ranges from half an hour to two hours. During drying the wet coating is under stress by the structural changes within the coating due to carrier evaporation.

At the initial stage of evaporation mechanism of the drying in case of flat substrates, the slurry is rich in carrier (water) called the "water-supersaturated" condition, allowing the powder particle to move freely in the slurry media. In the next stage of drying termed as "fully saturated", water is in interstitial voids of the powder particles. In third and final stage referred as the dry powder stage, water evaporates completely and powder particles form a loosely packed structure. Evaporation of carrier is a lateral drying process, in which drying commences on coating edges having largest surface area and propagate towards the centre of the coating. Water is drawn toward the edges (saturated region) from the centre

Table 3. Composition of slurry mixture used for the experimentation, (125g): TiO_2: 1 wt.%; Fly-ash: 1 wt.%

Ceramic to Metal	Mullite (49%)	Fly-ash (1%)	Nickel (50%)	Additive: TiO_2	Binder	Dispersant	Mix. Agent
	44%			1%	2.5%	0.5%	52%
75-25	40.837	0.412	13.75	1.25	3.125	0.625	65
50-50	27.225	0.275	27.5	1.25	3.125	0.625	65
25-75	13.612	0.137	41.25	1.25	3.125	0.625	65

Figure 2. Functionally graded coating
Adapted from Verma et al., 2016a [SAGE Publications Ltd].

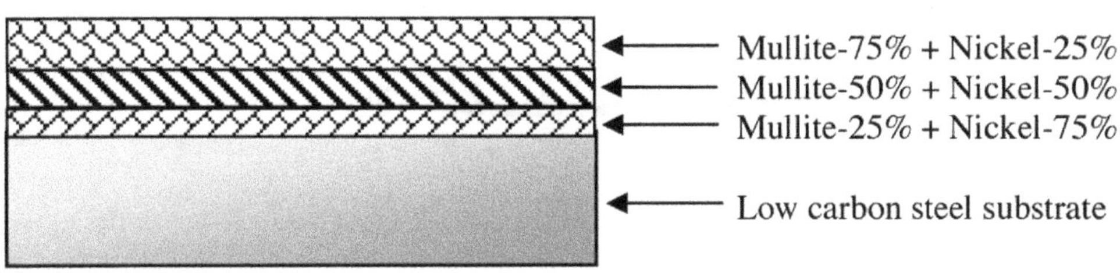

of the coating, which is a water-saturated region, by capillary action (Tanaka et al. 2006). As evaporation continues, the entire saturated region develops into the dry saturated region comprising of loosely packed powder particles also called as the green body. Upon further drying, the particles coalesce to form a composite, with the binder as a matrix phase to stabilize the powder particles as a particulate phase as shown in Figure 3. The binder forms a secondary bond among the binder monomer and the ceramic powder fragments.

The coated substrates were left to dry at ambient conditions (~40% humidity) for an hour after each consecutive layer to let the carrier fluid evaporate. The wet interlayer can be present if the base layer is not fully evaporated prior to application of subsequent slurry layers. The undesirable wet interlayer can spur dark indentation marks during the pressure stamping stage. These indentation marks contain the loosely packed powder with porosity created by evaporated water.

Pressure Stamping

The dried multi-layered green deposits were subsequently subjected to compressive loads on a universal stamping machine, to promote compaction prior to sintering. This is the pressure stamping stage which increases the overall proximity of the powder constituents, which in turn reduces the temperatures and time required for sintering (Dahl et al., 2007). The lower applied temperatures and reduced sintering times involved in coating formulation considerably reduce the thermal damage to the substrate and furthermore reducing the fabrication cost. The stamping pressures have shown grain boundary growth, however, beyond the pressure of 40 MPa defects like spallation and delamination were observed along the specimen during sintering stage of the pilot experiments. Thus, the stamping pressure was limited to 30 MPa for the current study. Surface grinding of the specimens during coupon preparation to achieve the required flatness was proved useful in attaining the uniform stamping pressure.

Figure 3. Powder-binder composite formed after drying

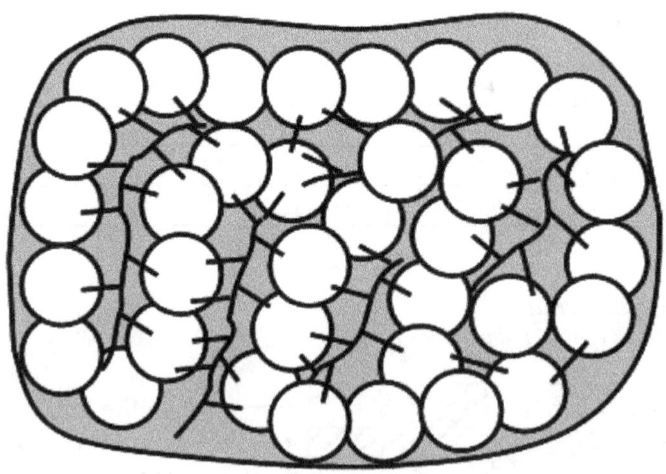

Debinding and Sintering

The binder maintains the integrity of the green deposit before sintering by acting as the matrix phase in the aqueous composite with the ceramic powder as a reinforcement. However, the binder's presence during the sintering stage is undesirable since it might form the contaminants and lead to cracking within the coating structure while sintering. Inclusions like dust, oil, or formation of some metastable phases due to these inclusions and binder present within the pores of the coating can cause the development of contamination. Moreover, these inclusions act as stress intensification areas and probable initiation points for coating failure during thermal cyclic loading (Konyashin, 2001). This would deteriorate the coating quality and cause plausible premature failure. Thus, the binder is ought to be vaporized prior to sintering stage, at a temperature lower than the sintering temperature of powder constituents. This process of extinguishing binder is termed as debinding. The debinding was performed at 300°C in a muffle furnace (Maximum temperature ~1400°C) and were soaked for about an hour in the furnace. During debinding, the PVA binder slowly vaporizes from the coating over this time and temperature to form water vapour and carbon dioxide, which are non-reactive to the remaining coating composition. These coating coupons after debinding were subsequently sintered in the same furnace to the appropriate sintering temperature and holding at this temperature for the sintering time at which the ceramic is fully sintered. Time for sintering is generally limited up-to fifteen hours since prolonged sintering escalates the possibility of coating cracks (Ruder et al., 1992).

During the sintering stage of SST, the dried weak form body comprising of powder particles bond together by diffusion and other atomic transport mechanisms to transform into coherent solid acquiring mechanical strength. The powder particles coalescence and transitions occur in the distinctive stages during the sintering process. During the initial stage of sintering, powder spheres are loosely packed with few contact regions after combustion of the binder. With the exertion of thermal energy, the grain boundaries begin to grow across the contact regions with an initiation of neck formation between the powder spheres. In the later stages of sintering, with the continuation of thermal energy, grain boundaries persist to expand forming a spherical pore within the coalesced powders, resulting in the porous microstructure. The sintering process is governed primarily by the reduction in the surface energy of the particle spheres, as a consequence of curtailment of the exposed surface area (Bernard-Granger et al., 2008).

Ceramic powders sinter roughly above 70% of their melting temperature, which corresponds to the temperature adequate to the promote grain growth amidst the powder spheres. Moreover, it has been stated that there persist an inverse relation among the sintering and sintering times (Choi et al., 2005). However, in case of sintering two dissimilar materials, the sintering temperature must be raised upto the commencement of the liquid phase sintering. After debinding, the furnace temperature was further raised to the sintering temperature, upto 1050°C at the constant heating rate of 10°C/minute. The coating coupons were allowed to sinter for respective sintering time. After sintering the furnace was shut off and the coupons were allowed to cool in the ambient conditions. The operating conditions under which coating formulation was carried out and the combination of parameters given by the design of experiment are presented in Table 4.

Table 4. Design spray parameters used for the coating deposition

Run No.	Process Parameters			
	Stamping Pressure (MPa)	**Sintering Additive: TiO_2 (wt.%)**	**Sintering Time (Minutes)**	**Sintering Temperature (°C)**
1	30	1	15	850
2	30	3	30	950
3	30	5	45	1050
10	40	1	45	1050
11	40	3	15	850
12	40	5	30	950

Spraying pressure: 4 bar; Fly-ash: 1wt.%;
The experiments were run in a randomized order

THERMAL CYCLING TEST

In order to ascertain the adequacy of EBCs, it is indispensable to affirm their thermal shock resistance. In view of this, low cycle fatigue (LCF) thermal behaviour of FG-EBC specimens fabricated by SST was investigated. The developed coatings were put through thermal cycling by a calibrated muffle furnace (maximum temperature ~1400°C). The coatings were scrutinized visually for any anomalies in respect of their structural integrity. The developed mullite coatings as per Table 4 were sectioned to the coupon size of (12 x 12 x 5) mm for the test.

For conducting the thermal cycling study, the test coupons were placed in alumina boat in a batch size of 6 and later moved to the furnace preheated to 750°C. The coupons were subjected to the heat at this temperature for 30 minutes. In the following stage, the coupons were withdrawn from the furnace and allowed to cool for an hour in ambient conditions. While the first batch of coupons was cooling, the second batch of specimens was placed in the furnace. The cyclic study was performed for 30 thermal cycles to test the coupons for failure. More than 20% region of coating spallation was considered the criterion for failure. Eventually, numbers of cycles were recorded for coating failure.

Thermal cycling test was conducted on mullite-nickel coatings to understand the thermal fatigue behavior by emphasizing the temperature gradient. It was observed during the 45 hours [(30+60)*30/60] long thermal cycling test that the specimens failing due to cracking and eventually spallation were those in which the failure commenced during the initial few cycles. Additionally, the specimens enduring these initial cycles (upto 6 cycles) did survive through the entire 30 cycle's thermal cycle test. On visual examination, if there were no signs of failure, the test was called off after 30 cycles.

Figure 4 manifests the probability of sustaining the thermal cycling test for the EBCs deposited by SST, where nearly 65% of the tested coupons survived the complete test. It is interesting to note that collapsing of the coatings due to thermal fatigue cracks took place during the initial cycles of thermal cycling test. It is worth mentioning that these initial collapses were ascribed to the processing defects encountered during the fabrication process. However, the test results manifest decent structural integrity of the slurry sprayed coatings during the thermal shock testing. This may be attributed to the thermal stress relieving caused by the strain tolerant structure across the coating thickness of the developed func-

Figure 4. Probability of survival against the number of thermal cycles

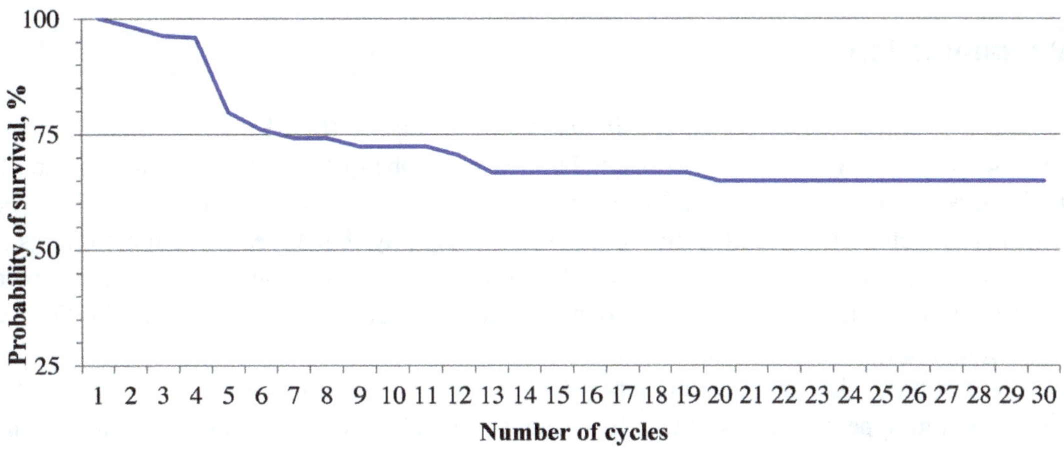

tionally graded coatings. Furthermore, favorable creep resistance and low strain rate of mullite support the conviction to good thermal shock resistance (Viswanath & Vijayarangan, 2012).

PREPARATION OF CORRODING MEDIA

Condensed alkali metal salts, especially Na_2SO_4 used for the current corrosion study is considered to be a prerequisite to hot corrosion, (Beltran et al., 1972). Moreover, it is well established that, Na_2SO_4 is a commonly found high-temperature corrosion promoter and is found to be the most common salt deposit in temperature excursions like gas turbines, power plants and boiler superheaters (Barbooti et al., 1988).

For corroding media to be used for the corrosion studies, a salt mixture of 0.5M Na_2SO_4 was prepared in distilled water. 16.25 mg of Na_2SO_4 salt in a dry paper and 100 ml distilled water in a washed and dried beaker was weighed on a digital electronic scale. All the reagents were analytical grade and used without further purification. The salt Na_2SO_4 was later added to a beaker containing some distilled water sufficient to dissolve the salt. This mixture after strong stirring was shifted into a 250 cm^3 flask. Distilled water was gradually added into the flask until it reached the meniscus level. In order to obtain similar conditions of reaction during corrosion studies, the coupons were polished upto 1μm levitated alumina powder suspension. Polished coupons were subsequently washed carefully with acetone and dried in hot air. Subsequently, to evaluate their surface areas, each physical dimensions of the coupons was recorded cautiously via a digital vernier caliper (Mitutoyo, Series 500, Resolution 0.01mm).

For experimentation, a 100ml solution was poured in the glass beakers and the coupons were immersed in each beaker individually to carry out the corrosion analysis in simulated industrial environment. The corrosion studies were carried out at room temperature. After 24 hours, to conduct the thermo gravimetric studies, the specimens were taken out, washed, degreased, dried and then weighed on an electronic balance (with sensitivity of 10^{-3} g).

CORROSION STUDIES

Visual Examination

The specimens exposed to the molten salt industrial environment (0.5M Na_2SO_4) had been visually examined each day for any signs of corrosion. The macrographs of these coupons after a month long exposure have been shown in Figure 5. The examination suggests iron oxides and iron sulfide (FeS) as the corrosion products were formed on the coated surface. By physical appearance, the products tended to be a combination of reddish-brown and greyish-black coloured compounds. These were ascertained as ferric oxide or hematite (Fe_2O_3; reddish-brown) and iron (II, III) oxide or magnetite (Fe_3O_4; greyish-black) were potentially the products of the mild steel corrosion.

In the event of bare mild steel coupon, a non-adherent corrosion product of rust colour was formed. The scale started to appear on the surfaces from the 4th day. However, this frangible scale did not last long and began flaking out of the surface by end of 9th day. By the end of the 10th day, the topmost layer of this scale separated from the coupon's surface. This spallation of the upper layer lasted till the end of

Figure 5. Macrographs of corrosion coupons after four weeks; (a) to (f) represents coatings for trial runs (as per Table 4), (g) uncoated coupon

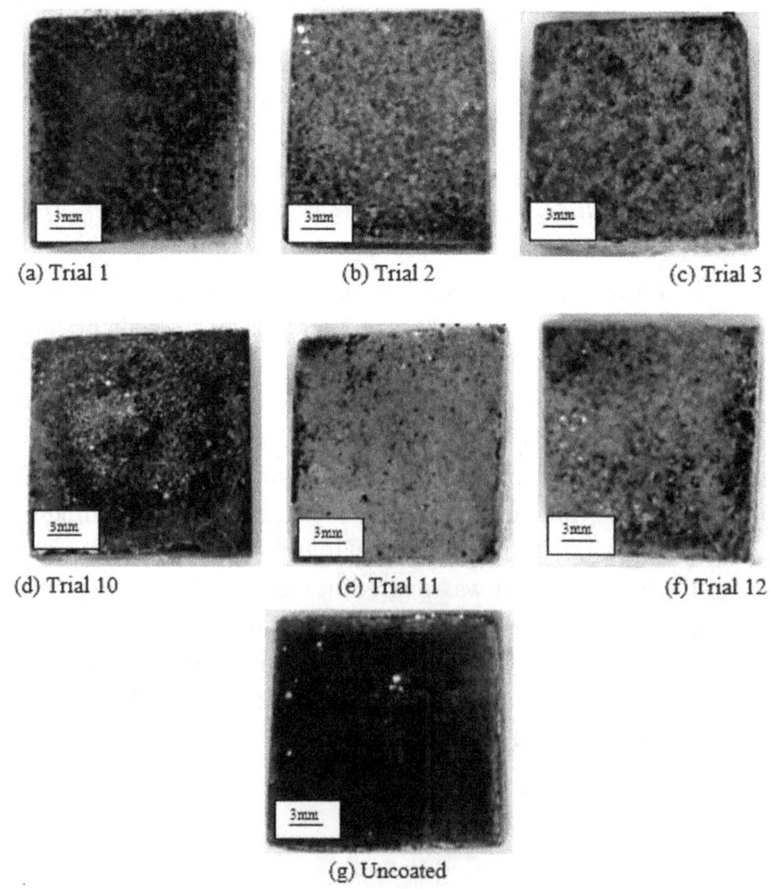

the 15th day. The colour of the coupons turned reddish-brown, resembling iron oxide in appearance by the end of 16th day. Moreover, the colour of the solution changed from colourless to brownish suspension after day 30 of immersion of specimens in 0.5M Na_2SO_4 concentration.

In general, the visual scrutiny of the mullite-nickel coatings after immersion test reveals the surface were free from any noticeable defects such as extensive rust, pitting attacks and scale in the coated area. No flaking, spallation and sputtering was observed from the coated surface. The coating was mainly noticed to be unscathed and without any disintegration along the substrate.

Gravimetric Studies

With a view to understanding the corrosion kinetics, weight change measurement was made for every corrosion coupons at the end of every 24 hours. For the gravimetric study, the spalled scale if any was included while measuring the weight change to determine the total rate of corrosion. The weight change data recorded for the corrosion coupons while the immersion test in the molten salt is demonstrated in Figure 6. After 30 days of immersion in simulated industrial environment, the cumulative weight-loss value for uncoated mild steel coupon was 687 mg (Table 5) which is fairly sizeable than the maximum weight-loss among the respective coated coupons (weight loss in trial 10 coupon is 169 mg).

A higher weight loss during the immersion test represents higher corrosion rates. It can be deduced from the plots that the requisite derogation in inhibiting corrosion of mild steel has been furnished by the developed mullite coating, as the weight loss for coated specimens were insignificant than those of their respective uncoated counterparts. Minimum weight loss among the coated coupons was for (Trial 2; weight loss: 150mg) corresponding to the sintering temperature of 950°C. However, in case of coupons

Figure 6. Plot for weight change/area vs. immersion time in days of exposure for slurry spray coated and uncoated mild steel subjected to corrosion for 30days in molten salt (0.5M Na2SO4)

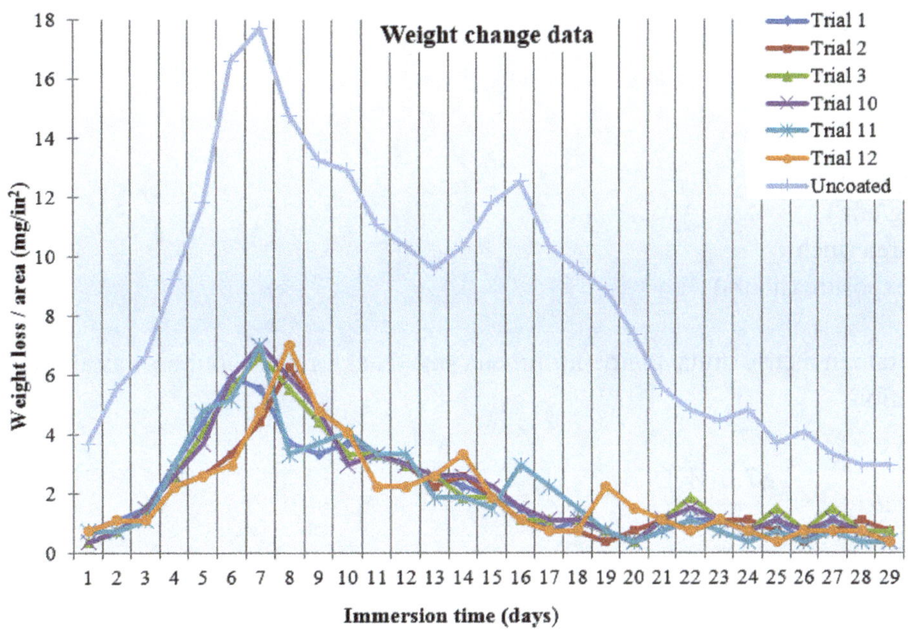

Table 5. Evaluated corrosion rate and inhibition efficiency for mullite-nickel coatings

Trial Run	Cumulative Weight Loss (mg)	Corrosion Rate (mpy)	Corrosion Rate (mm/y)	Protection Efficiency (%PE)
Uncoated	687	23.894	0.607	--
1	160	5.565	0.141	76.71
2	150	5.217	0.133	78.17
3	167	5.808	0.148	75.69
10	169	5.878	0.149	75.40
11	157	5.460	0.139	77.15
12	154	5.356	0.136	77.58

mg = milligram, mpy = mils per year, mm/y = millimetre per year

coated at the sintering temperature of 1050°C (Trial 10), the weight loss for the coated coupons were larger than those corresponding to 850°C and 950°C. This clearly suggests that the coupons coated with sintering temperature 950°C contribute more towards the corrosion inhibition than their counterparts coated at sintering temperature 1050°C & 850°C for the simulated industrial environment used in the study.

Corrosion Rate Evaluation (mpy)

In order to calculate the corrosion rate for the specimens after 30 days of immersion, the weight change data (mg), the surface area of the coupon (cm^2), and the density of the coating (g/cm^3) for each corrosion coupon were noted. Corrosion rate in *mpy* (mils per year) for each coupon was calculated using the following relation:

$$Corrosion\ Rate_{mpy} = \frac{534(W)}{\rho AT} \tag{1}$$

where,

W = Weight loss (mg)
ρ = Density (g/cm^3)
A = Surface area ($inch^2$)
T = Time of exposure (hours)

Corrosion rate in metric units' *mm/y* (millimeter per year) for each coupon was also calculated from following relation:

$$Corrosion\ Rate_{mm/y} = \frac{87.6(W)}{\rho AT} \tag{2}$$

where,

W = Weight loss (mg)
ρ = Density (g/cm^3)
A = Surface area (cm^2)
T = Time of exposure (hours)

The corrosion rate for the mullite coated and bare mild steel corrosion coupons as evaluated using equation 1 and 2, have been given in Table 5. The decreasing order of the corrosion rate for slurry sprayed specimens at distinct parameter conditions is:

Uncoated > Trial run 10 > Trial run 3 > Trial run 1 > Trial run 11 > Trial run 12 > Trial run 2

The trend of corrosion rate shows that the trial run specimens corresponding to sintering temperature 950°C are more corrosion resistant than those coated at sintering temperature 850°C and 1050°C. Furthermore, the specimen corresponding to 1050°C (Trial 2) is found to be the least corrosive among the slurry sprayed samples.

Protection Efficiency (%PE)

The protection efficiency *(%PE)* of the coated specimen was evaluated using the following relation:

$$\%PE = \left(\frac{CR_u - CR_c}{CR_u}\right)100 \quad (3)$$

where,

CR_u = Corrosion rate in absence of coating (mpy)
CR_c = Corrosion rate in presence of coating (mpy)

The values of percentage protection efficiency (%PE) obtained from weight loss measurements (as per the relation 3) for coated mild steel at different parameter conditions selected for slurry spray technique is summarized in Table 5. The variation of protection efficiency with trial runs is shown in Figure 7.

It was observed that the mullite coating inhibits the corrosion of mild steel in Na_2SO_4 solution, at the different sets of process parameters used in the study. This is due to the formation of more tenacious and protective metal-oxide compounds which passivate the metal surface. It is implicit that the coating shows the highest protection efficiency (≈78%) for the trial run 2 in 0.5M Na_2SO_4.

Effect of pH (Acidity)

A complex dependence of the corrosion rate of carbon steel lies on the hydrogen ion concentration which is the determinant of the pH of the solution. pH signifies the tendency of hydrogen ions to react with the metal in contact. Numerous commercial alloys tend to form an oxide layer comprising hydroxide-like species, in the presence of corrosive media. Therefore, the corrosion of these alloys depends on reaction governed by the hydrogen ion concentration (Lorbeer et al., 1980).

Figure 7. Results of immersion test for the mullite coatings: Immersion period 30 days

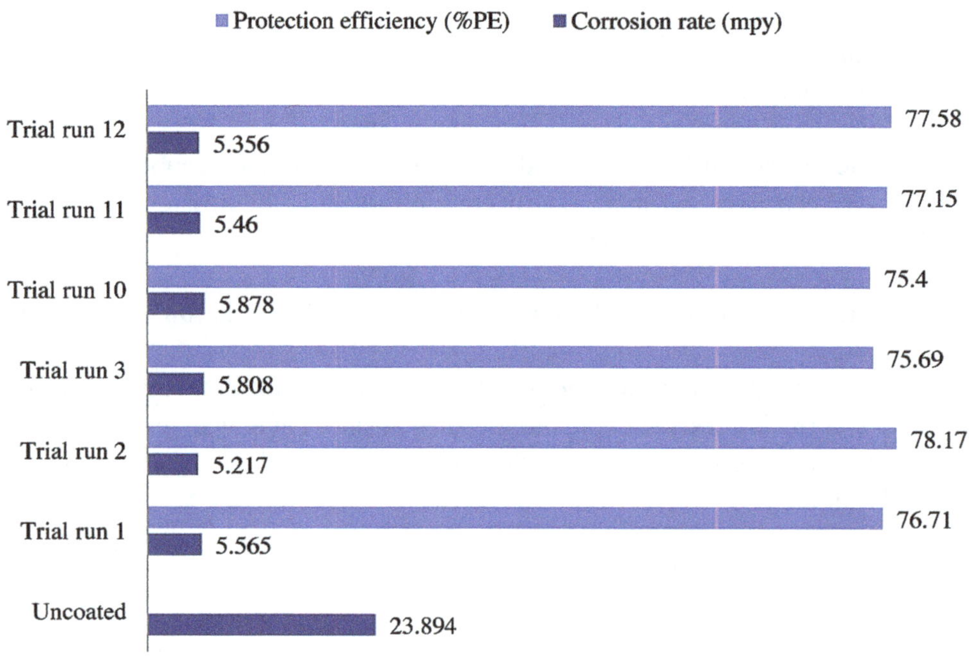

$$H_2O \rightarrow OH_{adsorbed} + H^+ + e^- \tag{4}$$

This reaction determines the corrosion rate which is dependent on hydrogen ion concentration corresponding to the pH concentration given as:

$$r = k\,(C_{H+})^n \tag{5}$$

where r is corrosion rate, k is rate constant, C_{H+} is the hydrogen ion concentration and n is an exponent which is dependent on the hydrogen ion concentration.

At strongly acidic condition (pH<5, low pH), the corrosion mechanism is dependent on both hydrogen ion concentration and the anions present in the solution. The corrosion rate in Na_2SO_4 is rapid at all acidic concentrations of pH<3, establishing the fact that the SO_4^{2-} seems to participate in corrosion reaction and accelerate the corrosion (Chin et al., 1972).

In the near-neutral pH ranges (5<pH<9) the corrosion behavior of mild steel differs notably from those under acidic conditions. Although, carbon steel has been one of the most extensively studied alloys in this range, it tends to be immune to corrosion in this range. At pH below 5, the oxide or hydroxide layers tend to dissolve. However, at pH above 5, the oxide or hydroxide layers tend to sustain on the metal surface, which is determined by the anions present in the solution (Fischer, 1978; ASM International, 1987). Formation of passivating film of magnetite (Fe_3O_4) is observed on steel in this pH range. Alloys such as the austenitic iron-base, ferritic alloys, nickel-base alloys, and duplex alloys also tend to have general corrosion rates that are independent of pH in this range (Shabnam, 2012).

In basic pH levels (pH>9) corrosion rate increase with pH or observed to be finite. Corrosion is observed to commence in some cases when at low levels of pH there was no corrosion. Fe corrosion persists even at high pH because of soluble species $[Fe(OH)_3]^-$ or $[Fe(OH)_4]^-$ being the most thermodynamically stable corrosion products (Silverman, 1982; Pourbaix, 1974). The phenomenon of stress corrosion cracking is observed in carbon steel at high pH and elevated temperature (Frank et al., 1985). However, austenitic and ferritic alloys are not significantly affected in this range, until very high pH is reached. This is ascribed to the evolution of passive oxide film of Cr present in these alloys. However, very high pH level and change in temperature do affect the corrosion rate.

Monitoring of pH

During the entire period of the immersion test, the pH of the Na_2SO_4 test solution was monitored on the daily basis. The pH was measured with laboratory digital pH meter (Li 120 Elico) and combined glass electrode. Standard buffers for pH meter calibration were used, where necessary the pH of each test Na_2SO_4 solution was adjusted to 6.5±0.1 by the addition of small quantities of 1M NaOH or 1M H_2SO_4. Figure 8 show pH vs. exposure period plot for bare and mullite-nickel coated low carbon steel specimens immersed in 0.5 M Na_2SO_4 (pH 6.5). During the initial days of the immersion in the molten salt environment, a slender rise in pH of the mullite coated steel coupons was noted. Although, in the later stages of the immersion, no considerable change in pH in the test solution was observed. Whereas in case of samples with sintering time 45 minutes and sintering temperature 1050°C condition (trial 3 and trail 10) the initial decrease in the pH valve was observed. However, no appreciable change in pH was observed during the entire one month of immersion. For the uncoated sample, the given trend is found to be amplified showing more dissociation of ions due to corrosion.

Figure 8. pH vs. time plot for mullite coated mild steel immersed in 0.5 M Na2SO4 at pH 6.5

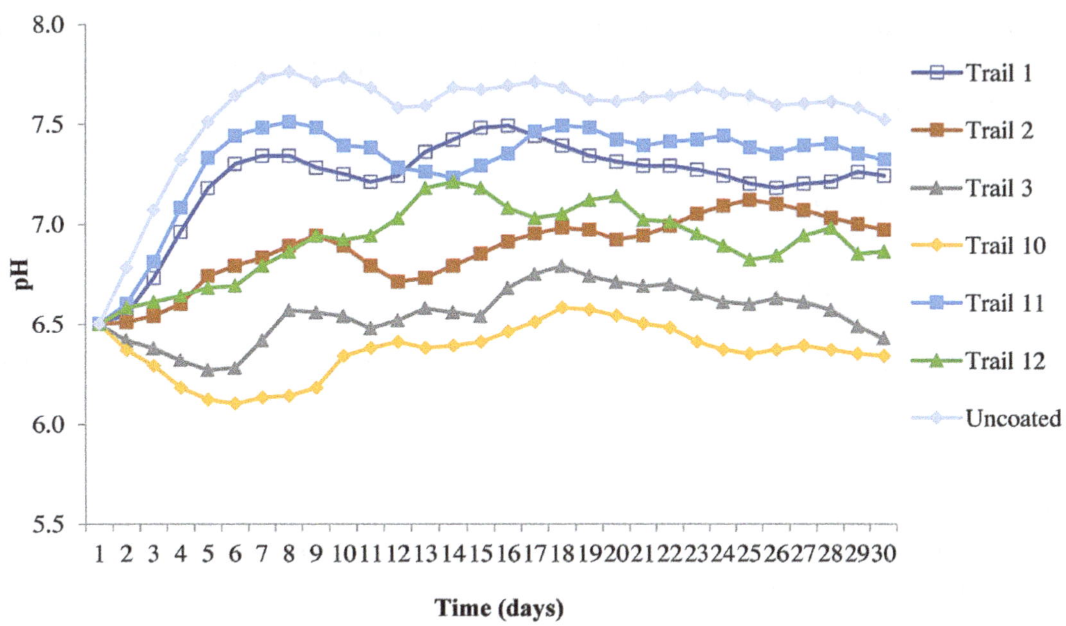

The effect of mullite coating with varying experimental conditions on corrosion behavior of low carbon steel in molten salt environment has been studied at room temperature using immersion test and weight loss method. The experimental conditions taken into consideration were immersion period, pH of the test solution and weight-loss of test coupons. The effect of the environmental variables like pH, immersion period, etc. on the corrosion behavior of alloy steels is fairly complicated and is conditional to the properties of the metal and the environment to which it is exposed (Fontana, 1978). Likewise, for the low carbon steel there is no anomaly of the complex reliance of corrosion rate on pH. In near neutral pH condition ($5< pH<9$) the direct contribution of pH on the corrosion rate of steel has not been recognized (ASM International, 1987).

In near neutral pH range, under static and deaerated conditions, the reduction of dissolves oxygen along with other oxidizers if any present in the solution are primarily responsible for the corrosion process. Besides, there is no influence of the activation polarization of the dynamic conditions (solution agitation) on the corrosion rate (Shabnam, 2012).

Considering the variation in the pH of the test solution recorded throughout the 30 days of immersion as shown in Figure 8. Except the pH of the solution of uncoated sample which increased to a value of approximately 7.75 on the initial days of immersion, in general, the pH of the solutions was found to vary between 6 and 7.5. The increase in the pH of the test solution is ascribed to the increase in OH^- ion concentration resulting from the H^+ ion reduction. The escalated corrosion rate of low carbon steel is accounted to the combined reduction reaction involving the reduction of both H^+ ions and oxygen. In consideration to the current examination, under static condition at room temperature, except for uncoated sample, the selected pH for the experimental study is implausible to influence substantially the corrosion rate of mullite coated low carbon steel.

The contrasting corrosion rate of coated low carbon steel in the presence of molten salt solution composition is expected to be caused by the variation in fabrication conditions for slurry spray technique. A decline in the corrosion rate for coated low carbon steel coupons as compared to the uncoated coupon is due to the mullite coating on the substrate acting as a protective impediment countering the oxidation of steel alloy. For uncoated coupon, the formation of a protective barrier is unlikely, which escalates the corrosion rate by stimulating the anodic reaction and emergence of the superficial micro galvanic cells resulting in an increased corrosion rate.

CHARACTERIZATION

X-Ray Diffraction Analysis

After the immersion tests, specimens were examined by X-ray diffraction (XRD) following the cleaning with distilled water. Literature reveals that, a real rust layer is very complex. Corrosion products of steel usually consist of an amorphous iron oxide magnetite (Fe_3O_4) and ferric oxy-hydroxides (α-, β- and/ or γ- FeOOH) (ASM International, 2003). Adding to the complexity of corrosion product, calcareous deposits ($CaCO_3$ and iron sulfide) are also known to form on steel surfaces in the molten salt environment (Jones, 1996). It was therefore decided to study by means of XRD the corrosion products formed after 30 days immersion in the solutions.

The XRD analysis for the coated mild steel coupons (Trial run-1; Table 4), subjected to the four weeks long molten salt exposure in simulated industrial environment (0.5M Na_2SO_4) has been shown in Figure 10. XRD analysis identified the phases similar to the composition of top-coat as were observed in the as-sprayed coating coupons (Figure 9). The surface indicated the presence of mullite ($3Al_2O_3{\cdot}2SiO_2$), nickel and anatase TiO_2 as the main phase. α-Al_2O_3 and quartz (SiO_2) have also been observed as minor phases. The XRD analysis also shows the formation of some reaction products and remains of Na_2SO_4 deposits in form of mainly hematite (α-Fe_2O_3), $NaFeO_2$ and FeS phases as corrosion scale in the simulated environment. The formation of these oxides impedes further penetration of the corrosive elements as they are compact and highly impervious. The confined substitution of the resilient Fe-oxide film by a non- resilient iron sulfide film is accounted for escalating the pitting corrosion (Hamdy et al., 2007).

SEM ANALYSIS

Surface Morphology

SEM micrographs specifying the surface morphology after 30 days immersion in 0.5M Na_2SO_4, of mullite-nickel the slurry sprayed low carbon steel substrate at different operating conditions as per Table 4 are shown in Figure 11(a-c). Microstructure reveals the similar feature as observed for the originally sprayed coating, except the presence of surface scale of irregular size globules developed on the top surface of the coatings in some trial runs (Trial run 1 & 3). The surface morphology in general shows surface free from any noticeable defects such as cracks, pits, and spallation. However, Trial run 3, at 45min sintering time and 1050°C sintering temperature, shows a surface scale of irregular shape.

Oxide scale of bare low carbon steel corroded after 30 days immersion in simulated industrial environment (0.5M Na_2SO_4) is shown in Figure 11(d), giving the appearance of randomly orientated flakes with irregular shapes. Further, the scale appears to be porous and fragile.

Figure 9. XRD spectra of as-slurry sprayed mullite coating (Trial Run 1: Table 4)

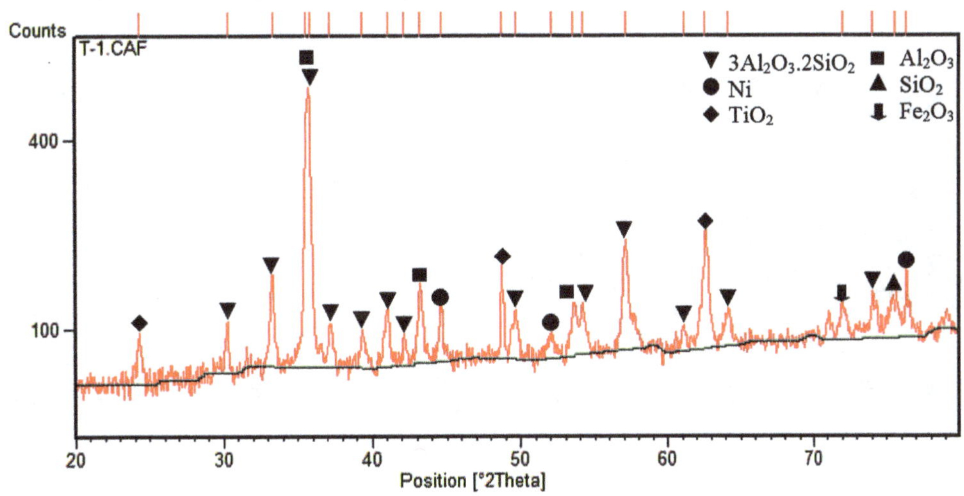

Figure 10. XRD spectra of slurry sprayed mullite coating exposed to molten salt (0.5M Na2SO4) environment (Trial Run 1: Table 4)

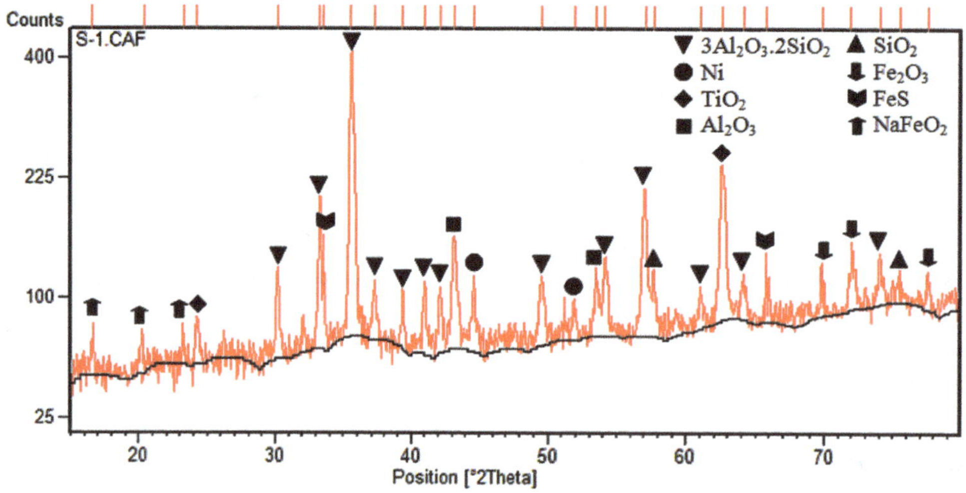

Figure 11. SEM Microstructure of slurry sprayed mullite coatings on low carbon steel, after 30 days of immersion in molten salt 0.5M Na2SO4: (a) to (c) Trial runs as per Table 4; (d) uncoated specimen

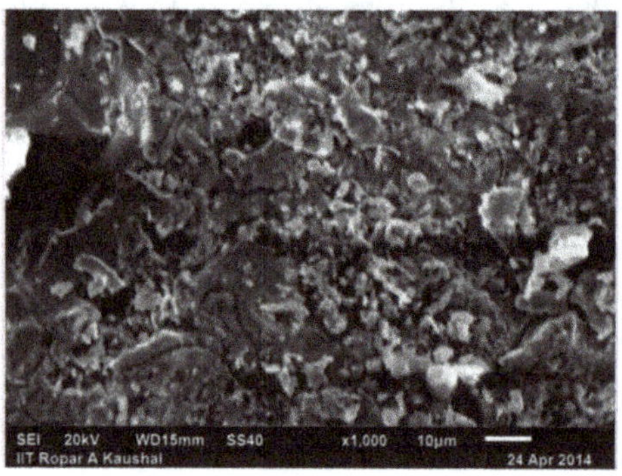

(a) Trial run 1

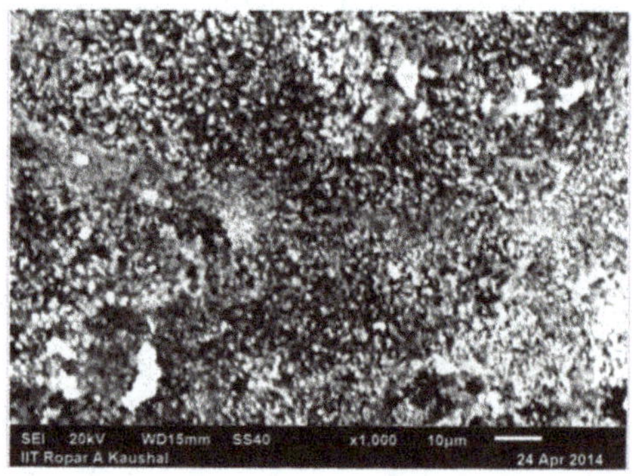

(b) Trial run 2

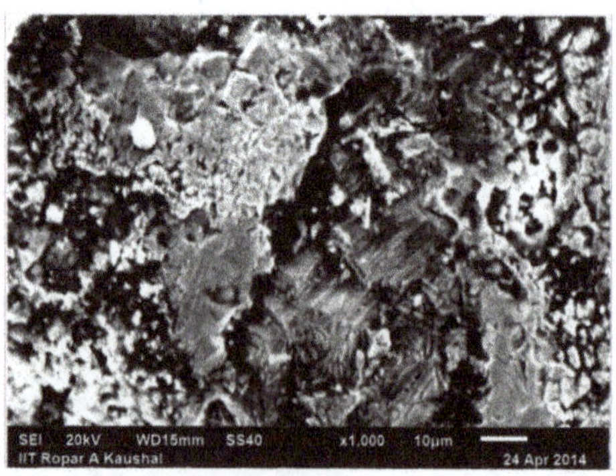

(c) Trial run 3

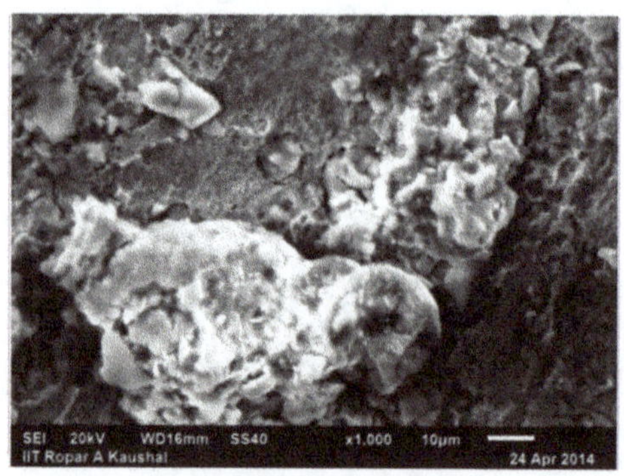

(d) Uncoated specimen

FE-SEM/EDS ANALYSIS

Surface Morphology

The SEM affixed with energy dispersive spectrometry shows the elemental analysis of mullite-nickel coating following the subjection to corrosion tests has been presented in Figure 12(a). The EDS analysis reveals the similar composition as observed for the originally sprayed top-coat, except the presence of additional Na and S. Presence of Na, S and Fe along with O indicates the possible evolution of the oxides and spinels of Na and S in the corrosion scale; like hematite (Fe_2O_3), $NaFeO_2$ and FeS phases. Similar affirmation is provided by the XRD evaluation of the coated specimens under the corrosive environment (Figure 10). The superior corrosion resistance of the mullite coatings was accredited to the formation of this dense tenacious and protective oxide layer of non-crystalline reaction products at inter-splat of the coating (Lee, 2009). The presence of Fe in the EDS elemental composition specifies the diffusion of Fe from substrate to the top surface.

The EDS spectra in Figure 12(b) shows the elemental composition on the surface of bare low carbon steel specimen corroded after thirty days of immersion period. EDS shows the oxide scale mainly composed of Fe and O, specifying the possible formation of an oxide of iron (Fe_2O_3). There is also a significant presence of C, Na, and S in the scale composition as the remains of Na_2SO_4. Moreover, EDS analysis revealed that the corroded sample comprised oxygen as the major elements. This culminates that oxygen might have been the key oxidizing agent in the simulated industrial environment, which instigated the corrosive effect on the metallic samples (Calderon et al, 1999). Furthermore, the oxide content in the EDS, propose that percentage weight content of oxygen on the corroded coupons is proportional to the corrosion rate (CR) of the mild steel sample (Table 5).

SEM ANALYSIS

Cross-Sectional Morphology

The SEM microstructures of the cross-sectional morphology of as-sprayed coating specimens following the subjection to the corrosion tests obtained at the different set of process parameters have been shown in Figure 13(b). The micrographs reveal analogous microstructural features like interlamellar pores, splat boundary and oxide stringers as observed in the originally sprayed specimen, Figure 13(a). The coating substrate interface is continuous, dense and adherent to the substrate free from cracks and any signs of spallation. The cross-sectional micrographs reveal that the interface is intact and suggest the coating is functional as an environmental barrier coating by acting as an effective diffusion barrier.

FE-SEM/EDS ANALYSIS

Cross-Sectional Morphology

The elemental characterization viewed by energy dispersive x-ray spectroscopy (EDS), along the cross-section of the slurry sprayed mullite-nickel coating, following the subjection to immersion tests is shown

Figure 12a. Part 1: FE-SEM with EDS analysis for slurry sprayed mullite based coating on low carbon steel substrate, after 30 days of immersion in molten salt 0.5M Na2SO4 (Trial 1)

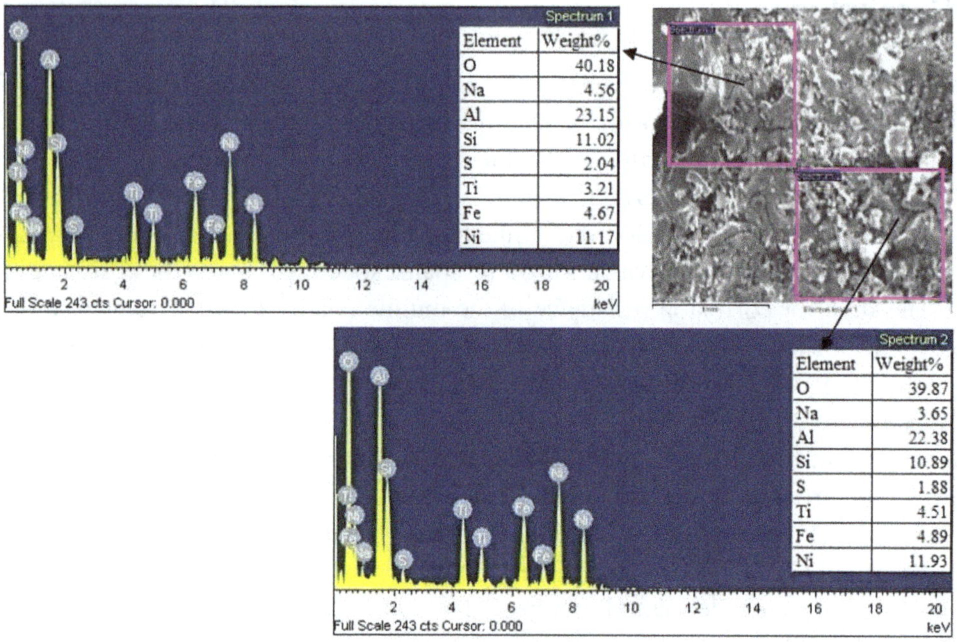

Trial 1: as per Table 4

Figure 12b. Part 2: FE-SEM with EDS analysis for slurry sprayed mullite based coating on low carbon steel substrate, after 30 days of immersion in molten salt 0.5M Na2SO4 (Uncoated specimen)

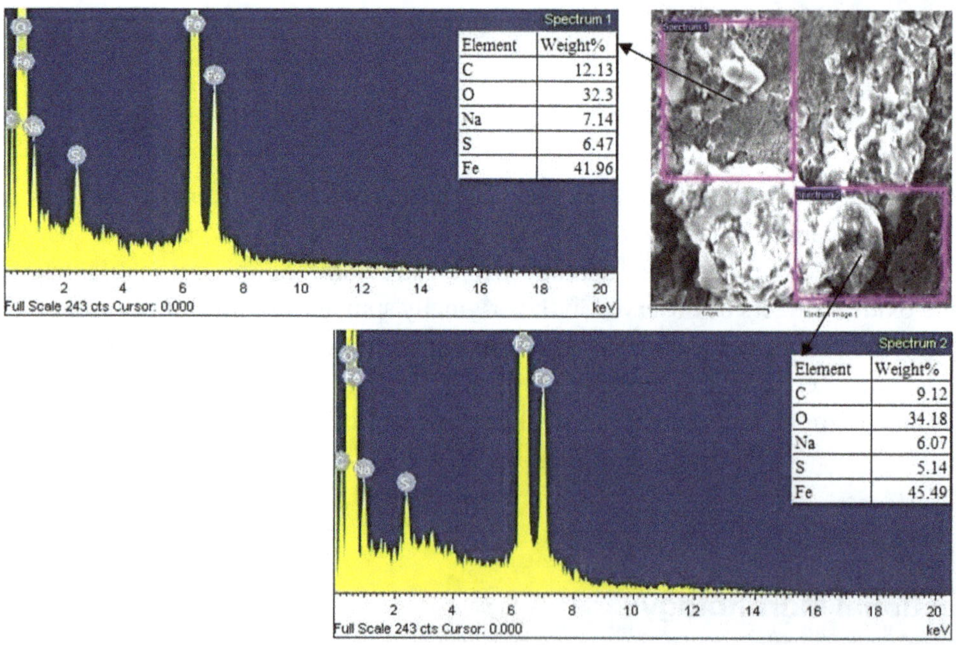

Figure 13. Cross-sectional morphology of Mullite based coatings on low-carbon steel specimens using SST: (a) Trial run 1 (as-sprayed); (b) Trial run 1 (after immersion test)

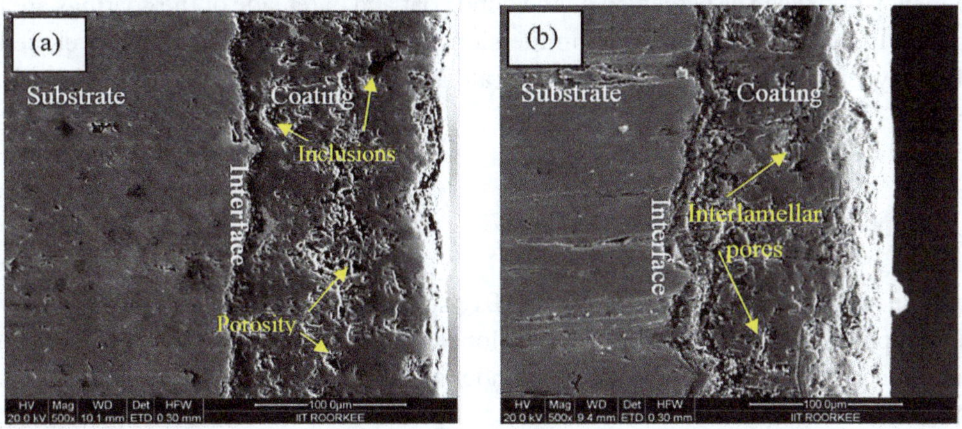

in Figure 14. Microstructure reveals similar feature as observed for the originally sprayed coating. The metal rich bond coat of (25% Mullite+75% Ni), could be clearly seen near the coating-substrate interface. The structure of (50% Mullite+50% Ni) interlayer remained same as the originally sprayed specimen.

The slurry sprayed mullite coatings exhibit no deterioration at the coating-substrate interface, manifesting the effectiveness of the coating to endure the diffusion of oxidizing species towards the substrate

Figure 14. FE-SEM with EDS analysis along the cross-section of mullite coating, after 30 days of immersion in molten salt 0.5M Na2SO4; Trial run 1, as per Table 4. (KV: 20.0, Mag.: 1600, Live Time(s): 30, Amp Time(μs): 7.68)

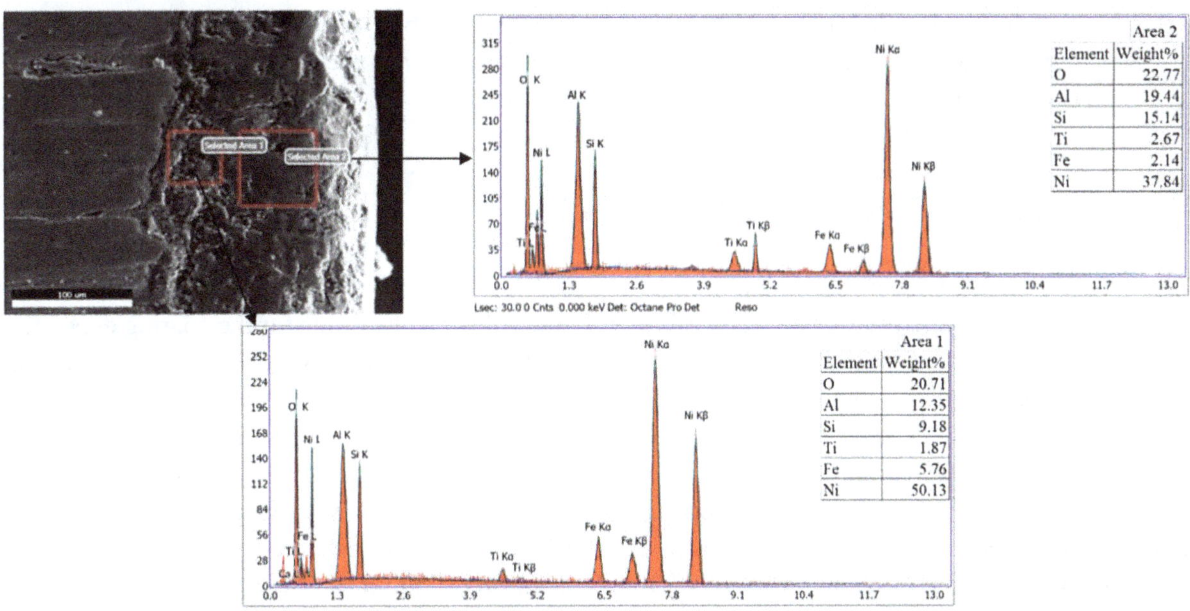

subjected to the simulated corrosive environment. Similar inferences can also be drawn from the gravimetric analysis and corrosion rates of the respective coated specimens.

The occupancy of oxygen as perceived on the bond coat and outer side of the coatings are in line with the EDS spectra that were obtained before immersion test. This suggests that the developed coating is inhibiting the oxygen to diffuse towards the substrate and is imparting the required corrosion aversion to the base alloy.

CONCLUSION

Coatings were successfully formulated on ASTM1018 steel using the slurry spray technique with the aid of TiO_2 as liquid phase sintering additive. The developed coatings through SST did possess good thermal shock behavior and protection against corrosion under simulated conditions. The salient conclusions of the present study are summarized as follows:

1. Despite the failure of the few samples during the initial cycles of the thermal cycling test, most of samples sustained up to 30 thermal cycles.
2. The corrosion studies conducted in the simulated industrial environment suggested that the specimens corresponding to sintering temperature 950°C were more resilient to corrosion among the slurry sprayed samples. In general slurry sprayed mullite coatings were found efficacious in furnishing the requisite corrosion aversion to low carbon steel against 0.5M Na_2SO_4.
3. The variation in the pH of the molten salt solution was observed during the 30 days of immersion period. In general, the pH of the solutions was found to vary between 6 and 7.5.
4. The XRD analysis shows the formation of some reaction products and remains of Na_2SO_4 deposits in the form of mainly hematite (α-Fe_2O_3), $NaFeO_2$ and FeS phases in the corrosion scale. The presence of these oxides and spinels contribute to the better corrosion resistance of mild steel in molten salt environment.
5. FE-SEM analysis of the mullite-nickel slurry sprayed low carbon steel substrate after 30 days immersion in 0.5M Na_2SO_4, in general shows surface free from any noticeable defects such as cracks, pits, and spallation.

CHALLENGES AND FUTURE DIRECTIONS

Slurry spray technique is a technique that embraces structural flexibility which is being thrust upon to be engineered to cater to the situational needs. While SST provides a new avenue for the coating deposition and the results documented in the present research are significant and useful in improving the life of coated steel in simulated industrial environment. Some open challenges are discussed based on the experimental study presented earlier. The challenges include the technological stages, in specific challenges such as achieving higher quality coatings in respect of coating density, adhesion, and microhardness. Lowering the sintering temperature for full ceramic top coat and effective stress relieving for materials with different physical properties. The end goal is to have an affordable yet robust technology for effective coating of a range of ceramic and metallic materials. Standardization of the process variables and automation and integration of fabrication stages play a pivotal role in accomplishing this goal.

In spite the challenges, it is suggested that the process development for production of FG ceramic-metal coating is an area wide open to be investigated. A roadmap for future directions and key development in SST research in context of thermal and environmental barrier coating (TEBCs) application could involve –the use of bond coat like NiCrAlY to eliminate thermally grown oxide (TGO) layer. TGO can adversely affect the creep resistance and thus in-service life of the TEBCs. In addition, the effect of inert gas environment on sintering stage can be studied to reduce the porosity and surface and sub-surface defects. Further, the behaviour of these coatings should further be studied in a real service environment for long term exposure to ascertain the integrity and longevity.

REFERENCES

Anggono, J. (2005). Mullite ceramics: its properties structure and synthesis. *Jurnal Teknik Mesin, 7*(1).

AZo Materials. (July 5, 2012). AISI 1018 Mild/Low Carbon Steel. Retrieved in May 2017 from http://www.azom.com/article.aspx?ArticleID=6115

Baklouti, S., Chartier, T., & Baumard, J. F. (1998). Binder distribution in spray-dried alumina agglomerates. *Journal of the European Ceramic Society, 18*(14), 2117–2121. doi:10.1016/S0955-2219(98)00107-1

Barbooti, M. M., Al-Madfai, S. H., & Nassouri, H. J. (1988). Thermochemical studies on hot ash corrosion of stainless steel 304 and inhibition by magnesium sulphate. *Thermochimica Acta, 126*, 43–49. doi:10.1016/0040-6031(88)87248-4

Beltran, A. M., & Shores, D. (1972). Hot corrosion. *Superalloys. Compilation of Monographs, 1972*, 317–339.

Bernard-Granger, G., Monchalin, N., & Guizard, C. (2008). Comparisons of grain size-density trajectory during spark plasma sintering and hot-pressing of zirconia. *Materials Letters, 62*(30), 4555–4558. doi:10.1016/j.matlet.2008.08.042

Bezzi, F., Fabbri, P., Brentari, A., Mingazzini, C., Burresi, E., Beaulardi, L., & Sangiorgi, S. (2010). Slurry coating of environmental barrier coating (EBC) on silicon carbide based material. In Advances in Science and Technology (Vol. 66, pp. 100-105). Trans Tech Publications.

Chandran, R. G., Chandrashekar, B. K., Ganguly, C., & Patil, K. C. (1996). Sintering and microstructural investigations on combustion processed mullite. *Journal of the European Ceramic Society, 16*(8), 843–849. doi:10.1016/0955-2219(96)00001-5

Chin, R. J., & Nobe, K. (1972). Electrodissolution kinetics of iron in chloride solutions III. Acidic solutions. *Journal of the Electrochemical Society, 119*(11), 1457–1461. doi:10.1149/1.2404023

Choi, S. R., Zhu, D., & Miller, R. A. (2005). Effect of Sintering on Mechanical Properties of Plasma-Sprayed Zirconia-Based Thermal Barrier Coatings. *Journal of the American Ceramic Society, 88*(10), 2859–2867. doi:10.1111/j.1551-2916.2005.00504.x

Dahl, P., Kaus, I., Zhao, Z., Johnsson, M., Nygren, M., Wiik, K., & Einarsrud, M. A. et al. (2007). Densification and properties of zirconia prepared by three different sintering techniques. *Ceramics International, 33*(8), 1603–1610. doi:10.1016/j.ceramint.2006.07.005

Fischer, M. (1978). Modellvorstellungen über den Reaktionsverlauf und Diskussion der Funktionen passivierender Zusätze bei der chemischen Passivierung von Eisen im pH-Bereich schwach sauer, neutral, basisch. *Materials and Corrosion*, *29*(3), 188–198. doi:10.1002/maco.19780290305

Fontana, M. G. (2005). *Corrosion engineering*. Tata McGraw-Hill Education.

Frank, W., & Graf, L. (1985). Mechanism of Intergranular Corrosion and Intergranular Stress-Corrosion Cracking in Mild Steels. *Res Mech.*, *13*(4), 251–264.

Fundamentals of Gas Turbine Engines. (n.d.) Retrieved in May 2017 from http://www.skybrary.aero/bookshelf/books/1621.pdf

Goward, G. W. (1998). Progress in coatings for gas turbine airfoils. *Surface and Coatings Technology*, *108*, 73–79. doi:10.1016/S0257-8972(98)00667-7

Greenwood, R., & Kendall, K. (1999). Selection of suitable dispersants for aqueous suspensions of zirconia and titania powders using acoustophoresis. *Journal of the European Ceramic Society*, *19*(4), 479–488. doi:10.1016/S0955-2219(98)00208-8

Günthner, M., Schütz, A., Glatzel, U., Wang, K., Bordia, R. K., Greißl, O., & Motz, G. et al. (2011). High performance environmental barrier coatings, Part I: Passive filler loaded SiCN system for steel. *Journal of the European Ceramic Society*, *31*(15), 3003–3010. doi:10.1016/j.jeurceramsoc.2011.05.027

Hamdy, A. S., Sa'eh, A. G., Shoeib, M. A., & Barakat, Y. (2007). Evaluation of corrosion and erosion–corrosion resistances of mild steel in sulfide-containing NaCl aerated solutions. *Electrochimica Acta*, *52*(24), 7068–7074. doi:10.1016/j.electacta.2007.05.034

International, A. S. M. (1987). Corrosion. [Metals park, Ohio.]. *ASM Metals Handbook.*, *13*, 38.

Jones, D. A. (1992). *Principles and prevention of corrosion*. Macmillan.

Khan, A. U., Briscoe, B. J., & Luckham, P. F. (2000). Interaction of binders with dispersant stabilised alumina suspensions. *Colloids and Surfaces. A, Physicochemical and Engineering Aspects*, *161*(2), 243–257. doi:10.1016/S0927-7757(99)00374-X

Kieback, B., Neubrand, A., & Riedel, H. (2003). Processing techniques for functionally graded materials. *Materials Science and Engineering A*, *362*(1), 81–106. doi:10.1016/S0921-5093(03)00578-1

Kokini, K., Takeuchi, Y. R., & Choules, B. D. (1996). Surface thermal cracking of thermal barrier coatings owing to stress relaxation: Zirconia vs. mullite. *Surface and Coatings Technology*, *82*(1-2), 77–82. doi:10.1016/0257-8972(95)02647-9

Konyashin, I. (2001). A technique for fabrication of coated TiCN-based cermets with functionally graded structure. *International Journal of Refractory Metals & Hard Materials*, *19*(4), 523–526. doi:10.1016/S0263-4368(01)00058-0

Lee, H. Y., & Baik, K. H. (2009). Comparison of corrosion resistance between Al 2 O 3 and YSZ coatings against high temperature LiCl-Li 2 O molten salt. *Metals and Materials International*, *15*(5), 783–787. doi:10.1007/s12540-009-0783-8

Lee, W. E., & Rainforth, M. (1994). *Ceramic microstructures: property control by processing*. Springer Science & Business Media.

Lorbeer, P., & Lorenz, W. J. (1980). The kinetics of iron dissolution and passivation in solutions containing oxygen. *Electrochimica Acta*, *25*(4), 375–381. doi:10.1016/0013-4686(80)87026-5

Montanaro, L., Tulliani, J. M., Perrot, C., & Negro, A. (1997). Sintering of industrial mullites. *Journal of the European Ceramic Society*, *17*(14), 1715–1723. doi:10.1016/S0955-2219(97)00043-5

Porcayo-Calderon, J., Brito-Figueroa, E., & González-Rodríguez, J. G. (1999). Oxidation behaviour of Fe–Si thermal spray coatings. *Materials Letters*, *38*(1), 45–53. doi:10.1016/S0167-577X(98)00130-X

Pourbaix, M. (1974). *Atlas of electrochemical equilibria in aqueous solutions*.

Ruder, A., Buchkremer, H. P., Jansen, H., Mallener, W., & Stöver, D. (1992). Wet powder spraying—a process for the production of coatings. *Surface and Coatings Technology*, *53*(1), 71–74. doi:10.1016/0257-8972(92)90105-J

Shabnam, H. (2012). Studies on the role of heavy metal ions on the corrosion behaviour of iron base alloys.

Silverman, D. C. (1982). Technical Note: Presence of Solid Fe (OH) 2 in EMF-pH Diagram for Iron. *Corrosion*, *38*(8), 453–455. doi:10.5006/1.3578222

Silverson. (2017). Welcome. Retrieved in May 2017 from http://www.silverson.com/images/uploads/documents/CPVASolutions.pdf

Tanaka, S., Pin, C. C., & Uematsu, K. (2006). Effect of Organic Binder Segregation on Sintered Strength of Dry-Pressed Alumina. *Journal of the American Ceramic Society*, *89*(6), 1903–1907. doi:10.1111/j.1551-2916.2006.01057.x

Tezuka, N., Low, I. M., Davies, I. J., Alecu, I. D., Stead, R. J., Mehrtens, E. G., & Latella, B. A. (2009). Effect of fluoride and oxide additives on the phase transformations in alumina/clay ceramics. *J. Aust. Ceram. Soc*, *45*(1), 19–27.

Tripathi, H. S., & Banerjee, G. (1998). Synthesis and mechanical properties of mullite from beach sand sillimanite: Effect of TiO 2. *Journal of the European Ceramic Society*, *18*(14), 2081–2087. doi:10.1016/S0955-2219(98)00149-6

Tripathi, H. S., Das, S. K., & Banerjee, G. (2000). Thermal shock behaviour of high alumina aggregates derived from sillimanite beach sand with and without Fe2O3 doping. *Ceramics International*, *26*(1), 1–6. doi:10.1016/S0272-8842(99)00010-3

Verma, R., Kant, S., & Suri, N. M. (2016a). Adhesion strength optimization of slurry sprayed mullite-based coating using Taguchi method. *Proceedings of the Institution of Mechanical Engineers. Part E, Journal of Process Mechanical Engineering*, *230*(2), 87–96. doi:10.1177/0954408915595948

Verma, R., Suri, N. M., & Kant, S. (2016b). Effect of parameters on adhesion strength for slurry spray coating technique. *Materials and Manufacturing Processes*.

Verma, R., Suri, N. M., & Kant, S. (2016c). Parametric Appraisal of Slurry-Sprayed Mullite Coatings for Coating Thickness. *Journal of Thermal Spray Technology*, *25*(7), 1289–1301. doi:10.1007/s11666-016-0437-1

Viswanath, B., & Vijayarangan, S. (2012). Characterization Studies of Mullite Coatings on Cast Aluminum. *Journal of thermal spray technology, 21*(2), 325-334.

Walston, W. S. (2004). Coating and surface technologies for turbine airfoils. *Superalloys*, 579-588.

KEY TERMS AND DEFINITIONS

Environmental Barrier Coatings: Coatings applied to engineering components to avoid their degradation in operating conditions.

Fly-Ash: Fly-Ash (FA), a principle ingredient of coal ash or coal combustion residuals (CCR), which is rich in metal oxides like silica and alumina.

Functionally Graded Coatings: Coatings with structural gradation in composition and properties, across their thickness to cater to the situational needs.

Liquid Phase Sintering: Liquid phase sintering (LPS) is a subclass of the sintering processes wherein during the thermal cycle, the particulate solid having solid solubility in the liquid coexists with a liquid.

Mullite: A ceramic compound of stable intermediate phase of silica and alumina, which rarely occur as natural mineral.

Protection Efficiency: The measure of relative effectiveness of corrosion protection offered by a coating against an uncoated substrate in a corroding media.

Thermally Grown Oxide: Thermally Grown Oxide (TGO) layer found at interface of bond and ceramic coat, is developed by volumetric expansion stresses associated with metal rich bond coat.

Section 2

Chapter 9
High Temperature Nanocomposite Coatings by Plasma Spraying for Friction and Wear Applications

Ehsan Sadri
Isfahan University of Technology, Iran

Fakhreddin Ashrafizadeh
Isfahan University of Technology, Iran

ABSTRACT

In conventional high temperature solid lubricants, during thermal cycling, a reduction of lubrication performance results in fluctuating friction coefficient and formation of abrasive wear debris. Moreover, the demand for high temperature solid lubrication in the range 300-1000 °C has led to development of self-lubricating wear resistant coatings of desired characteristics at high temperatures. This chapter introduces typical nanostructured ceramic matrix coatings produced by plasma spraying for self-lubricating wear resistance purposes in a wide temperature range. Production and properties of spray dried agglomerated nanocomposite granules are described and, then, structural, mechanical and tribological properties of Cr2O3-Ag nanocomposite coatings are presented. The lubrication mechanisms at moderate and high temperatures, by means of creation of a silver tribofilm on top surface, is discussed.

INTRODUCTION

There have been an increased interest in self-lubricating composite coatings in recent years for tribological applications at moderate to high temperatures. The most notably is adaptive and chameleon coatings produced by such processes as magnetron sputtering, chemical vapor deposition and hybrid vacuum arc, magnetron sputtering, pulsed laser deposition techniques (Donnet, & Erdemir, 2004; Muratore, Hu, Voevodin, 2009), PS series coatings produced by plasma spraying (Sliney, 1979) and laser surface modified adaptive coatings (Voevodin, & Zabinski, 2006). Solid lubricants include graphite, MoS_2, PTFE

DOI: 10.4018/978-1-5225-4194-3.ch009

and fluorides, and there is an increasing demand for self-lubricating wear resistant coatings that can accomplish the required lubrication at different environments as well as long wear endurance, mainly for bearings and other mechanical components in aerospace industries. Nevertheless, temperature range for conventional monolithic solid lubricants is particularly difficult to extend; their effective operation is limited to 250-400°C in air, above which the material is oxidized or decomposed, leading to loss of lubricious nature (Erdemir, 2001; Sliney, 2001). Self-lubricating wear resistant coatings contain several phases in a composite structure to accomplish the required lubrication. The example of these coatings are Ag/BaF_2-CaF_2 in Cr_2O_3 (Kim, Choi, Han, Uhm, & Lee, 2005), Ag in CrN (Mulligan, & Gall, 2005), MoS_2/Ag in Mo_2N (Aouadi et al., 2009) and adaptive coatings that can be operated over a wide range of working environments (Muratore, Hu, & Voevodin, 2009).

One successful approach to development of self-lubricating composite coatings with various environment operation is using solid lubricant interactions with environments to generate a low shear interface between the matrix and the wear counterpart, as self-lubricating contact surface, for various ambient conditions. Moreover, the high hardness of matrix combined with low friction of reinforcement gives very low wear rates and mechanical properties of the coatings; these characteristics coupled with good adhesion lead to high load-bearing capacity.

A number of hard coatings containing gold and silver, as solid lubricant particles, have been investigated (Aouadi et al., 2008; Mulligan, & Gall 2005). Soft metallic lubricants have multiple planes in crystal structures, thereby, do not work harden significantly during sliding. Silver is chemically inert and has a low shear strength even below 300 ^{0}C. During operation of such coatings, a thin lubricant Ag layer forms on the top surface; this soft-on-hard coating approach will avoid the high wear rate that would be observed in a pure silver layer. In addition, the hard coating avoids plastic deformation on consequent irregular surface morphologies that are common for pure soft coatings. A common disadvantage of the noble-metal based lubrication coatings produced is a lack of reversibility of the surface adaptations when subjected to temperature cycling (Mulligan Blanchet, & Gall, 2010).

Since crystalline solid lubricants are typically very soft and cause a significant reduction in hardness and elastic modulus of composite structure, amorphous and poorly crystalline nanoparticle inclusions with minimum detrimental mechanical effects were introduced in self-lubricating composite coatings. Such coatings have been developed generally in PVD processes but rarely considered in thermal spraying techniques.

Plasma spray process is an appropriate industrial method in order to fabricate self-lubricant wear resistant composite coatings. An essential point of the quality criteria is the homogenous distribution of reinforcements in the matrix. For this reason, the production of composite powders by spray-drying agglomeration procedure to fabricate thermal sprayed coatings is increasing.

Plasma sprayed nanocomposite ceramic matrix coatings may be employed for improved toughness, wear and friction properties over a wide range of temperatures; these properties are often provided by nanostructural semi-molten zones that reduce lamella thickness and increase crack propagation resistance. The key to good performance appears to be the avoidance of fully nanostructural features, rather, an essential point of the quality criteria is to produce a homogenous distribution of reinforcement and semi-molten nanostructured zones in the matrix of coating.

Several studies have focused on improvement of toughness and reduction of friction coefficient of Cr_2O_3 coatings, achieved by structural modification procedures. One of the recent methods is the double glow-discharge plasma technique combined chromizing and plasma oxidizing treatments (Liu, Tao, Xu, Chen, & Gao, 2009). Another one is production of plasma sprayed nanostructured Cr_2O_3 coatings (Cellard, Garnier, Fantozzi, Baret, & Fort, 2009). Moreover, addition of various particles, such as CaF_2 and Ag_2O (Ouyang, & Sasaki, 2001), Mo or MoO_3 (Ahn, Lyo, & Lim, 2000; Lyo, Ahn, & Lim, 2003), TiO_2 (Berger et al., 2009) and Ni 8.5Cr 7Al 5Mo 2Si 2B 2Fe $3TiO_2$ (Metco 442) (Cetinel, Celik, & Kusoglu, 2008), is an effective procedure for improvement of tribological behavior of Cr_2O_3 coatings. In all these studies, promotion of frictional and mechanical properties of Cr_2O_3 coatings has been reported to varying degrees. However, researchers rarely investigated the simultaneous influence of nanostructure and addition of solid lubricant nanoparticles on improvement of these coatings at various temperatures.

AGGLOMERATED NANOSTRUCTURED COMPOSITE POWDER

Thermally sprayed composite coatings generally consist of matrix and reinforcement phases. The matrix phase can be metal, alloys (for example nickel alloys), or ceramics (for example alumina and zirconia) and the reinforcements with three different morphologies are added to the matrix. These morphologies include whiskers, such as SiC and Si_3N_4, fibers, such as W and B, and particles, such as SiC and TiC. Regardless of the morphology, homogenous distribution of reinforcements in the matrix is a criterion determining the quality of sprayed composite coatings. Generally, thermal sprayed composite coatings are manufactured by following techniques (Pawlowski, 2007; Pawlowski, 2009):

- Blending the reinforcement and matrix powder and then thermal spraying of the blended feedstock powder on a substrate. In this technique, the coating will be inhomogeneous due to the different behavior of reinforcement and matrix in plasma jet, variant morphology and physical properties of materials.
- Simultaneous use of two thermal spraying systems for matrix and reinforcement particles.
- Production of agglomerated composite powder by spray drying of a slurry.
- Injection of liquid suspensions of fine powders and liquid precursors to produce suspension plasma spray coatings.

The third method, i.e. agglomeration of composite powder, is considered by researchers due to the relative homogeneity of the structure and uniform distribution of the reinforcement phase in the matrix.

In nanostructured materials, physical and mechanical properties are significantly improved with decreasing the grain size. For instance, the Hall-Petch experimental equation shows the improvement of the mechanical properties of materials with decreasing grain size. Indeed, nanostructured materials have characteristic improved properties such as hardness, toughness, friction coefficient in various environments, dry and wet, and wear resistance (Lu, & Liaw, 2001). Nanostructural characters in thermal sprayed oxide ceramic matrix composite coatings are obtained by regions containing nanoscale particles (Lima, & Marple, 2007). The surface area of particles is significantly increased with decreasing particle size. This larger surface area in nanoparticles may lead to the absorption of higher amounts of thermal energy during the particle dwell time in the spray jet, facilitating the melting process at the particle surface, which would also serve to promote the adhesion/cohesion of the impinging particles at the substrate.

The higher molten fraction of agglomerated nanoparticles in the spray jet causes a reduction of presence possibility of nanoparticles in the coatings. For this reason, the agglomerated nanostructured particles must be sprayed by controlling plasma spraying parameters for embedding the initial nanostructure of feedstock in the coating microstructure. Coatings produced by this technique contain a significant number of defects, and are inhomogeneous, anisotropic, and exhibit variability in properties (Sadri, & Ashrafizadeh, 2013; Lima, & Marple, 2007). Chwa et al. (2005) used the aluminum particles in nanostructured TiO_2 coatings for reducing the detrimental effect of low plasma power on microstructure of nanostructured coatings. The co-sprayed Al particles were considered to be well distributed and infiltrated in the splat boundary and play an important role, like an adhesive metal between partial/non-molten particles. The porosity was then decreased leading to improved mechanical properties. Although nanostructured TiO_2-Al composite coatings had lower porosity and higher hardness compared to nanostructured TiO_2 coatings, they had more porosity and lower hardness compared to conventional TiO_2 coatings due to the use of lower plasma power.

Improvement of wear properties has been reported in nanostructured coatings as compared to the conventional route. Lin et al. (2004) fabricated nanostructured alumina-titania coatings using agglomerated nanopowders by plasma spraying method and observed higher wear resistance and a decreased hardness. Other researchers have compared the hardness of other nanostructured coatings with their conventional coatings and have concluded that their hardness was almost the same, while having a different wear resistance behavior (Lima & Marple, 2007). However, it was reported that the nanostructured YSZ coating showed better wear and hardness properties than microstructured material (Tao, Liang, Ding, Lao, & Coddet, 2005). It has been debated that the improvement of wear properties of nanostructured coatings is not solely due to their higher hardness, but it is due to improved toughness, too. Indeed, hardness was increased in some materials while some others showed a decrease in hardness, compared to the conventional coatings (Lima, & Marple, 2007). The explanation was suggested by Ahn et al. (2006), who proposed that the higher toughness of nanostructured coatings was due to the improvement of fluidity of splats in the nanostructured coatings. Therefore, the adhesion between nanostructured splats is more and their inter-lamellar strength is increased consequently, which causes an improvement of toughness. The distribution of various phases is also more homogenous in nanostructured coatings than conventional ones. Such characteristics of nanostructured coatings are the main reasons for improvement of mechanical and tribological properties of these coatings even at moderate and high temperatures.

Production of Nanocomposite Powder

In plasma sprayed nanostructured ceramic matrix composite coatings, the feedstock must consist of nanoparticles of appropriate specifications. In this section, the procedure details for producing ceramic matrix agglomerated nanostructured composite powders such as Cr_2O_3-Ag, used to produce composite coatings has been described. For this, ceramic nanopowders are dispersed with metal reinforcement to obtain a homogeneous aqueous dispersion appropriate for spray drying process. In the second stage, Cr_2O_3-Ag composite powders are produced by agglomeration process.

Fragmentation and cold welding of particles are the two main processes reducing the size of particles during milling. Due to the brittle nature of oxide ceramics, the cold welding of particles is negligible; therefore, they are continuously fragmented during milling. But, very small particles behave as soft metals and their sizes cannot be decreased anymore (Suryanarayana, 2001). Figure 1 shows SEM images and laser dimensional distribution analysis of as-received and 20 and 30 hours milled Cr_2O_3 ceramic powders. After milling for 20 h nanoparticles, dia. $\leq$100 nm, were produced while particles in micron range were also present, Figure 1(b). However, after 30 hours of milling, uniform size particle distribution was observed and the particles size was 50-130 nm, Figure 1(c). During milling, Cr_2O_3 particles did

Figure 1a. SEM micrographs and DLS particle size distribution of commercial thermal spray Cr_2O_3 powder

Figure 1b. SEM micrographs and DLS particle size distribution of 20 h milling Cr_2O_3 powder

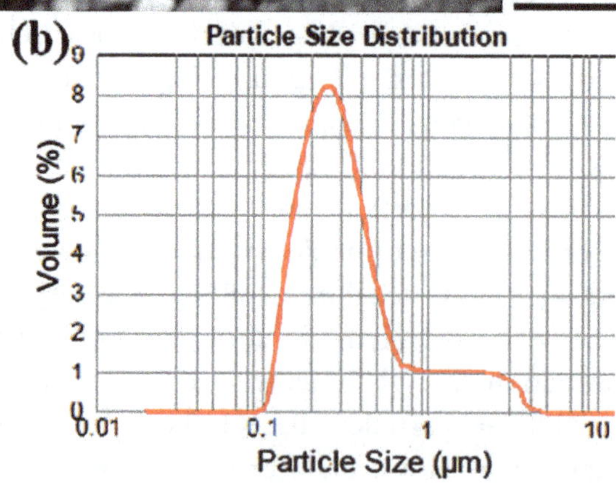

not undergo any noticeable plastic deformation, but were continuously fragmented. The near-spherical morphology of nanoparticles is an evidence of the mechanism occurring in most ceramic materials.

Nanoparticles cannot be directly used in plasma spraying for several reasons; 1- tiny nature and their low mass, 2- insufficient particle injection momentum to accelerate in the plasma flow, 3- low flowability, 4- high agglomeration tendency of precursors during spraying, 5- the need to increase cold carrier gas flow rate disrupting the high enthalpy flow, 6- low deposition rate due to low connect at impact of nanoparticles and, finally, 7- high adhesion of nanoparticles to powder feeder wall. Therefore, nanoparticles need to be agglomerated to a sprayable size granules composed of nanoparticles.

Agglomeration generally takes place by spray drying of the nanoparticle suspension and thermal treatment of the granules in a hot chamber. Although the mechanism of spray drying process is complex, for ceramic powders, it can be divided into four steps: droplet formation, droplet evaporation and ballooning, explosion and formation of the particle. The steps of agglomerated powder formation in

Figure 1c. SEM micrographs and DLS particle size distribution of 30 h milling Cr_2O_3 powder

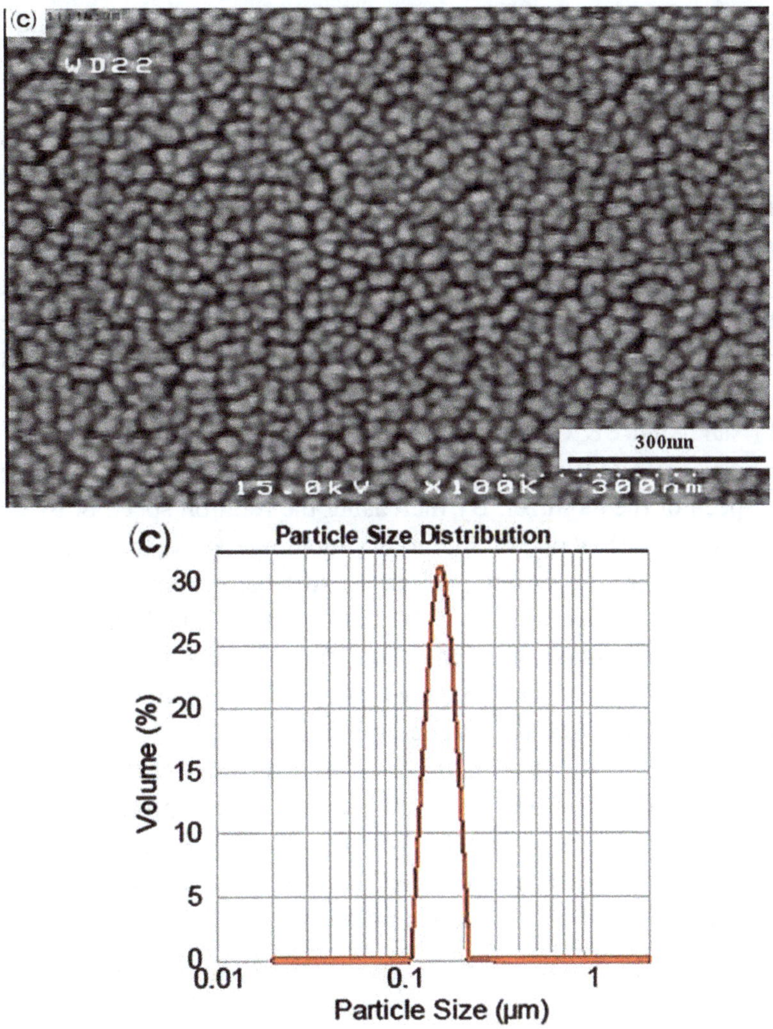

spray drying process are shown schematically by Cao et al. (2000). The metastable and homogeneous suspension in drying chamber is rapidly converted into fine droplets by hot feeding gas. Subsequently, the droplet is shrunk due to the fast evaporation of the solvent. When the liquid moves from the center to the outside of droplet, solid particles are also carried along. If the evaporation rate in the center of the droplet is higher than the diffusion rate of liquid into the droplet surface, a void is created inside the droplet, which is not suitable for most coatings, except for porous coatings such as TBCs. Therefore, the spray drying parameters should be adjusted in such a way that solvent evaporation rate is not higher than the diffusion rate into the droplet surface. In the fourth step, i.e. particle formation, the particle is formed by loss of solvent and addition of organic binder. The plasticized droplet shrinks continuously by the effect of surface tension until it is dried completely. Effect of operating parameters on final powder characteristics can be summarized as three factors, namely, drying temperature, feeding rate and suspension feeding pressure (Cao et al., 2000; Nandiyanto, & Okuyama, 2011).

The drying temperature has multiple effects. First, the droplet ballooning is intensified by increasing the drying temperature which causes a larger final particle size. Second, the surface tension is decreased with the temperature resulting in particle size increase. Third, the drying period is shortened by fast heating; the formation of crust starts earlier in the fourth step resulting in a decrease in the particle size. By increasing the feeding rate, powder density and final powder size are increased and the powder density and final powder size are decreased by decreasing suspension feeding pressure. It is worth noting that the effect of feeding rates on final powder characteristics is more than two other factors (Cao et al., 2000; Nandiyanto, & Okuyama, 2011). Figure 2 shows different particle morphologies which can be produced by spray drying under different conditions. Indeed, there are several parameters related to the mass and the heat transfer of the droplet affecting the morphology of the particles. Spherical shaped particles, Figure 2(a), are generated from solutions with maximum structural stability. Hollow particles, Figure 2(b), are generated by adjustment of the characteristic time for crust formation, heat transfer from surface to center of droplet, and drying residence time. Spherical particles with rough surface could be created by increasing the nanoparticle size in Figure 2(a); (c). Doughnut particles, Figure 2(d), are created from spherical, mushroom and convex morphologies by change in the hydrodynamics, structural stability and behavior of droplet (Nandiyanto, & Okuyama, 2011).

Figure 3 shows an agglomerated nanostructured chromium oxide powder produced by 10000 rpm and 17000 rpm rotation speed of the atomizer. By increasing the rotation speed of the atomizer, fixing the feeding rate of the suspension, flow rate of air and drying temperature, the conditions can be modified so that the diffusion rate of the solvent into the droplet surface is greater than evaporation rate inside the droplet. Doughnut (A) and unshaped hallow (B) Cr_2O_3 particles in 10000 rpm atomizer, Figure 3(a), are produced due to low diffusion rate and high evaporation rate of solvent by creation of micro pores. Indeed, if these voids rend skin of particle, doughnut shapes (A) are formed and if this does not happen, shapeless hollow particles (B) are formed. By increasing the atomizer rotation speed, atomization rate of the droplets, and fixing the drying temperature, solvent diffusion rate to skin of particle is increased up to the same evaporation rate. Therefore, the agglomerated granules produced by 17000 rpm atomizer speed are spherical with low porosity, without any micro voids and of a narrow size distribution, Figure 3(b). The spherical shape of the particles enhanced the flowability of powders in feeder and injector and consequently provided a more homogeneous microstructure of the coating as well as decreased coatings porosity.

Figure 2. Formation mechanisms of various particle morphologies: (a) spherical particles with smooth surface; (b) broken hollow particles; (c) Spherical particles with rough surface prepared from larger nanoparticle; (d) doughnut particles

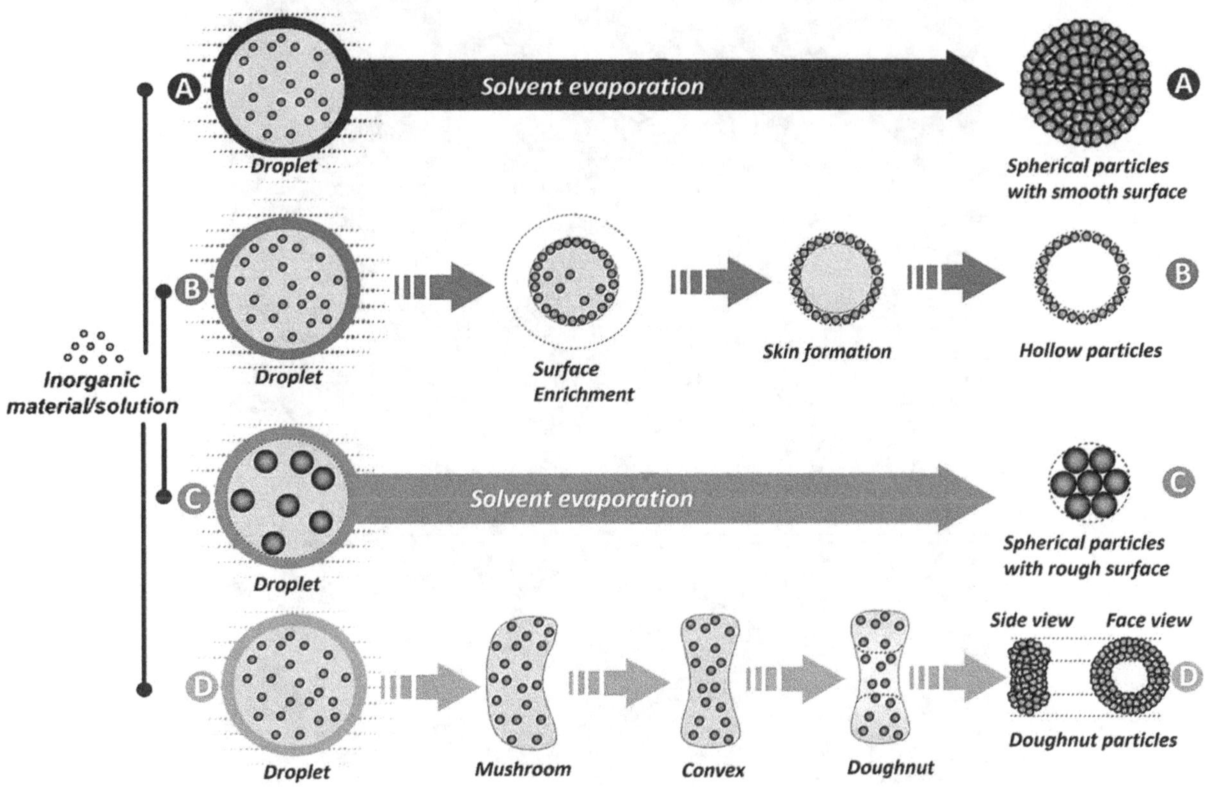

Evaluation of Nanocomposite Powder

Uniform distribution of reinforcement nanoparticle in composite powder lead to relative homogeneity in nanostructured coatings. Uniform distribution of nanoparticles in composite powder after spray drying process can be determined using typical EDS elemental distribution map performed on the cross section of mounted sample, which confirms the success of dispersion step, see Figure 4(c) for Cr_2O_3-5 wt. % Ag. Due to higher density of silver, approximately two times greater than that of Cr_2O_3, there is a great tendency for segregation in aqueous suspension or in spray drying process. This phenomenon can increase drastically in higher amounts of reinforcement. Reconstitution of individual nanoparticles in agglomerated powders is responsible for nanostructural characteristics in coatings, Figure 4(a-b) (Cellard, Zenati, Garnier, Fantozzi, & Baret, 2007).

In addition to the presence of nanoparticles, the crystallite size and internal lattice strain of thermally sprayed nanostructured coatings are evaluated for verification of nanostructural features (Flauchais, Heberlein, & Boulos, 2014). Crystallite size and internal lattice strain of the Cr_2O_3 powders versus milling time can be estimated using Williamson-Hall method, Figure 5. It is be observed that by increasing the milling time, the rate of decrease in crystallite size and increase in the lattice strain are both decreased. By comparing this with crystallite size and lattice strain variations of metals during mechanical milling,

Figure 3. Agglomerated nanostructured chromium oxide powder produced at (a) 10000 rpm and (b) 17000 rpm rotation speed of atomizer

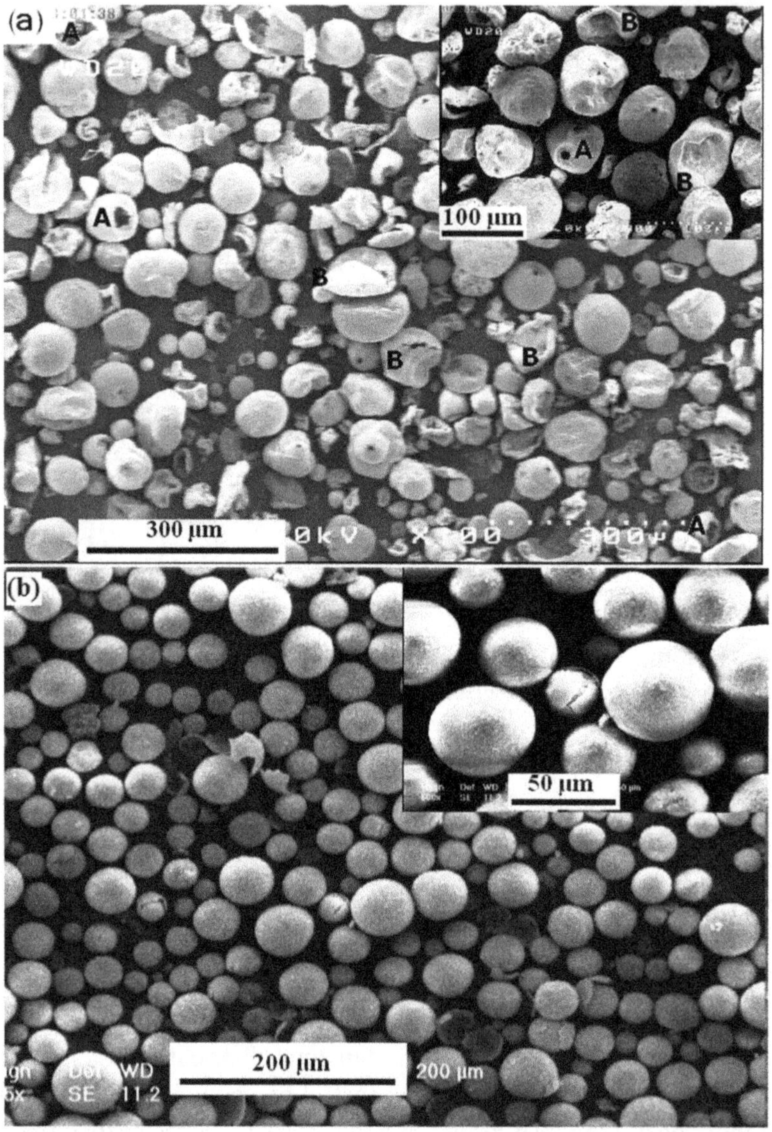

it can be noticed that the reduction rate of crystallite size for chromium oxide is higher than that of most metals. Accumulation of micro-strains in chromium oxide structure, and in most ceramics, was lower than that for metals. This can be due to the large amount of dislocations introduced by cold working during mechanical milling for metals which rarely occurs in ceramics. Also, deformation mechanisms of ceramics and metals during mechanical milling are substantially different. Although the precise mechanism of crystallite size reduction of ceramics during mechanical milling have not been yet established, researchers have found that decreasing of crystallite size occurs until nucleation and growth of defects within crystallite is possible (Delogu, & Cocco, 2006).

Figure 4. (a) Cr_2O_3-Ag composite powders produced by agglomeration (spray-drying), (b) individual nanosized Cr_2O_3-Ag particles in (a), (c) x-ray maps showing distribution of Cr and Ag in Cr_2O_3-5%Ag composite powders

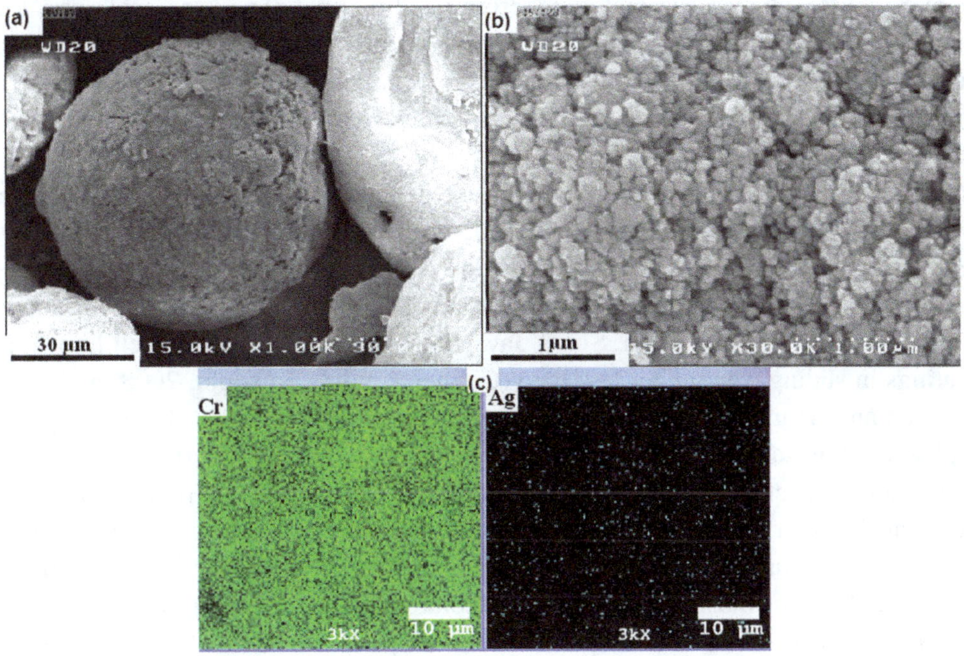

Figure 5. Variation of crystallite size and internal strain of Cr_2O_3 particles with increasing milling time

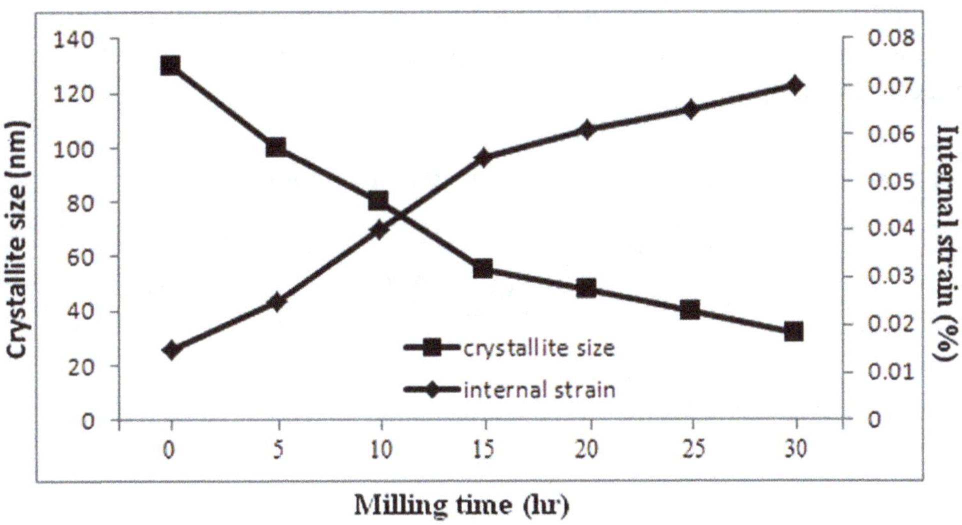

PHYSICAL PROPERTIES OF NANOCOMPOSITE Cr_2O_3-Ag COATING

Oxide ceramic coatings possess higher hardness and elastic modulus and higher melting point than metals, whereas their toughness, thermal conductivity and thermal expansion are lower. Furthermore these coatings have exceptional chemical, wear, corrosion and erosion resistance at high temperatures. Considering these characteristics, they are applied in such industrial applications that need simultaneously high corrosion and wear resistance of oxide films combined with the toughness of metal substrates (Bolelli, Cannillo, Lusvarghi & Manfredini, 2006).

Chromium oxide is one of the multi-purpose ceramic oxide coatings used in various industries. Cr_2O_3 is crystallized in a rhombohedral lattice with a c/a ratio of 2.74 leading to weak atomic bonding in c-axis. Thus, it has relatively lower shear strength and friction coefficient in comparison to other ceramic materials. In addition, Cr_2O_3 coatings can be operated in a wide temperature range from room temperature to 1000°C (Ahn, & Kwon, 1999; Singh, Sil, & Jayaganthan, 2011). They are well known as abrasion resistant coatings in sliding wear under low load (Cetinel, Celik & Kusoglu, 2008) and have indicated favorable friction and wear behavior under un-lubricated sliding conditions (Jin & Yang, 1996). High hardness and wear resistance, low electrical conductivity and chemical inertness, thus low corrosion rate, low coefficient of friction compared with other oxides, and high temperature oxidation resistance are other inherent characteristics of Cr_2O_3. Accordingly, it is accepted that the main application of Cr_2O_3 coatings is under conditions that corrosion and wear resistance are required at the same time (Liu, Tao, Xu, Chen & Gao, 2009). The most considerable shortcoming for Cr_2O_3 coatings is the rather high erosion rate due to the brittle nature of the oxide even at elevated temperatures (Cetinel, Celik, & Kusoglu, 2008).

The plasma spray parameters for deposition of three type of coatings including conventional micro-Cr_2O_3, nanostructured pure Cr_2O_3 and Cr_2O_3 with 2, 5 & 10 vol. %Ag as nanocomposite coupled with a Ni-5 wt. %Al bond coat are summarized in Table 1. The required conditions were adjusted in order to optimize bimodal nanostructured-microstructured coatings. As it is described extensively in the section "structure and morphology", different plasma spraying parameters were practiced to control the structure of the coatings.

Table 1. Process parameters and materials for plasma sprayed coatings

Coating	Powder Composition	Plasma Flame		Arc Current (A)	Powder Mass Rate (g/min)	Carrier Gas Flow Rate, Ar (L/min)	Spraying Distance (mm)	Number of Passes	As-Sprayed Thickness (μm)
		Primary Gas, Ar Flow Rate (L/min)	Secondary Gas, H_2 Flow Rate (L/min)						
NiAl bond coat	Ni- 5wt.% Al	45	12	420	20	2.5	120	2	50
Micro-Cr_2O_3	Pure Cr_2O_3	50	14	450	20	3	120	8	250
Nano-Cr_2O_3	Pure Cr_2O_3	50	8	400	20	3	80	10	260
Cr_2O_3-2%Ag	Cr_2O_3-2vol.% Ag	50	8	400	20	3	80	10	150
Cr_2O_3-5%Ag	Cr_2O_3-5vol.% Ag	50	8	400	20	3	80	10	170
Cr_2O_3-10%Ag	Cr_2O_3-10vol.% Ag	50	8	400	20	3	80	10	140

Porosity

Thermal sprayed coatings have relatively inhomogeneous structures including spherical and disc-shaped pores and cracks. This is due to various trajectories of particles in spray jet, different particle morphology and size, different thermo-physical properties of various materials in composite powder, the existence of temperature and velocity gradient by vertical deflection of plasma jet axis, and non-uniform powder injection by feeder. Different kinds of porosities are created during deformation at impact of splats with each other or substrate. Porosity is one of the most common characteristics in thermally sprayed coatings which change the mechanical properties, such as elastic modulus, fracture toughness and hardness, and physical properties, including thermal conductivity, thermal diffusion and resistance to thermal shock. Porosity of the thermally sprayed coatings link to various parameters such as physical and chemical properties of feedstock, morphology and size distribution of powder, particles velocity in thermal spraying, particle flattening at impact, amount of fully melted and semi-molten particles, thermal residual stress and substrate roughness.

In this section the porosity evaluations, values and types, of nanostructural plasma sprayed ceramic coatings with wear resistant performance are represented. Figure 6(a) shows two different kinds of porosity in the nanostructured Cr_2O_3 coating. The porosity values of coatings could be measured according to ASTM E-2109 standard by image analysis, Figure 6(b). The higher porosity of nanostructured than microstructured Cr_2O_3 coating could be due to the higher fraction of semi-molten and unmelted particles in the nanostructured coating. For this reason, the irregular micro pore values between semi-molten and fully melted zones, regions (A) in Figure 6(a), are greater in nanostructural coatings. Similar values of interlamellar pores, regions (B) in Figure 6(a), which present in most plasma sprayed coatings, were obtained for both microstructured and nanostructured coatings.

Structure and Morphology

Nanostructural characters in thermal sprayed ceramic matrix composite coatings are obtained by the presence of individual nanoparticles in the microstructure. Figure 7 shows the presence of nanostructured regions in nanostructured Cr_2O_3 and Cr_2O_3-5% Ag coatings as compared with microstructured Cr_2O_3. Zones of nanoparticles are originated from semi-molten splats in thermal spray process; such semi-molten particles, which were not fully melted during heating stage, are mainly responsible for obtaining nanostructural characteristics in the coatings. In fact, initial nanoparticles of the feedstock have been embedded in semi-molten regions, surrounded by fully molten particles that act as a binder in the structure of nano-coatings, Figure 7(c, e). If the fraction of semi-molten particles is higher than an optimum amount, the particles will not be fully melted in the plasma jet and, as a result, adhesion and cohesion strength and intersplat adhesion will not achieve an acceptable level required for the mechanical properties of the layer. This has been reported in particular for ceramic matrix coatings (Jin, & Yang, 1996; Liu, Tao, Xu, Chen, & Gao, 2009). Thus, the plasma spraying parameters must be adjusted so that a bimodal structure is obtained; this structure in thermal sprayed coatings is composed of both fully molten and semi-molten particles in such a ratio that gain sufficient adhesion/cohesion strength together with nanostructural features. These structures are observed in nano-Cr_2O_3, Figure 7(c), and Cr_2O_3-5% Ag, Figure 7(e), coatings; lighter and darker areas correspond to fully molten particles of micro-size zones and semi molten particles of nano-size zones, respectively. Individual nanoparticles embedded in the nanostructured zones were characterized at higher magnification of SEM images as in Figure 7(d, f).

Figure 6. (a) Two type of porosity within nanostructured Cr_2O_3 coating, (b) porosity content of coatings according to ASTM E-2109

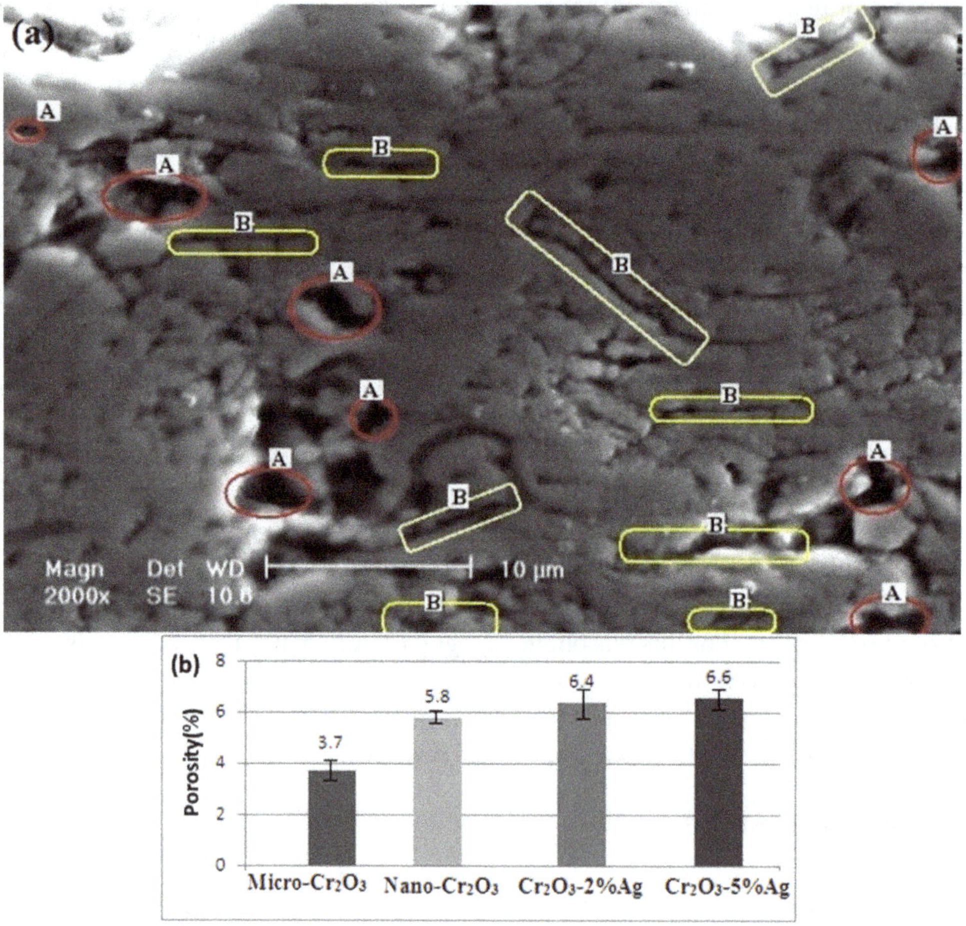

Similarities can be observed between these nanostructured zones of Figure 7(d, f) and the morphology of nanoparticles in the initial agglomerated feedstock of Figure 4(c). The dark regions in Figure 7(a) are interlamellar voids of micro-Cr_2O_3 coating.

Based on metallographic examination of samples prepared in Cr_2O_3 coating, the amount of nanostructured zones, obtained by image analysis, was 15-18%. Apparently, there is no agreement about what percentage of semi-molten nanostructured zones should be maintained from original feedstock in the coating microstructure in order to obtain optimum mechanical and tribological properties for various nanostructured coatings. Nevertheless, for APS nanostructured Al_2O_3-TiO_2 coatings, the optimized value of nano-size zones embedded in the composite coating has been reported in the range 11-20% by some researchers (Lima & Marple, 2007; Ibrahim, Hamid & Aal, 2010). Similar values have been reported as determined from XRD patterns (Berger-Keller, Bertrand, Filiatre, Meunier & Coddet, 2003) or image analysis (Sanchez et al., 2010) techniques.

The morphology of micro-Cr_2O_3 and nano-Cr_2O_3 coatings on the metallographic sections can be seen in Figure 8; the nano-Cr_2O_3 coating in Figure 8(a) has thinner lamellae than the micro-Cr_2O_3 coat-

Figure 7. FESEM micrographs of cross sections of (a- b) micro-Cr_2O_3, (c- d) nano-Cr_2O_3 and (e-f) Cr_2O_3-5% Ag coatings. The presence of nanoparticles within the semi-molten regions in nano-Cr_2O_3 (d) and Cr_2O_3-5% Ag (e) nanostructured coatings can be observed at higher magnification

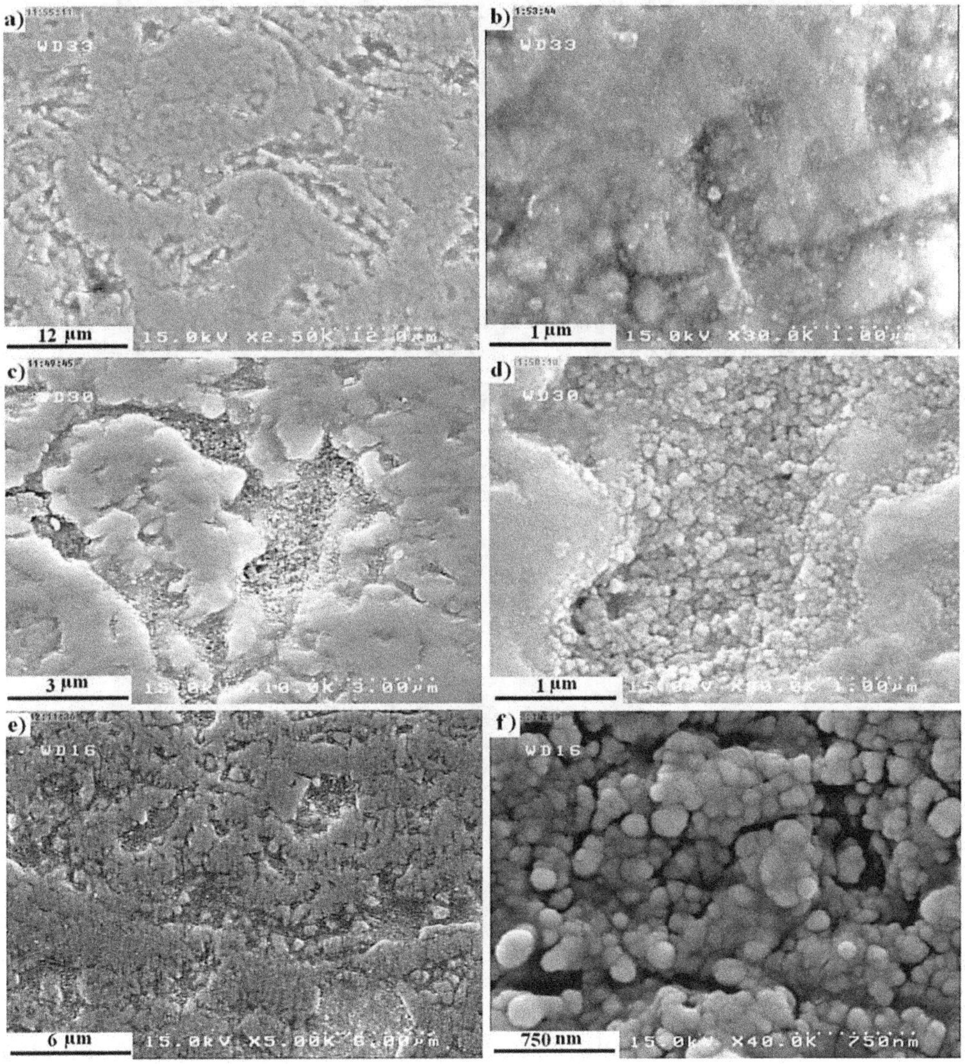

ing, Figure 8(b). This difference can be attributed to better rheological behavior of splats at impact to the substrate; when nanoparticles are present in the coating, splat flattening at impact is enhanced and this leads to formation of lamellae of smaller thickness in nano-Cr_2O_3 coatings. The existence of some micro-pores in nano-Cr_2O_3 coating, Figure 8(a), may imply that a considerable proportion of the feedstock remained as semi-molten or unmelted particles in nano- Cr_2O_3 coating. In fact, the semi-molten or solid particles in deposition process cannot be embedded onto the previous splats, therefore, such particles had a considerable contribution to the formation of porosity in the sprayed deposits.

Figure 8. SEM micrographs on metallographic sections of (a) nano-Cr_2O_3 and (b) micro-Cr_2O_3 coatings

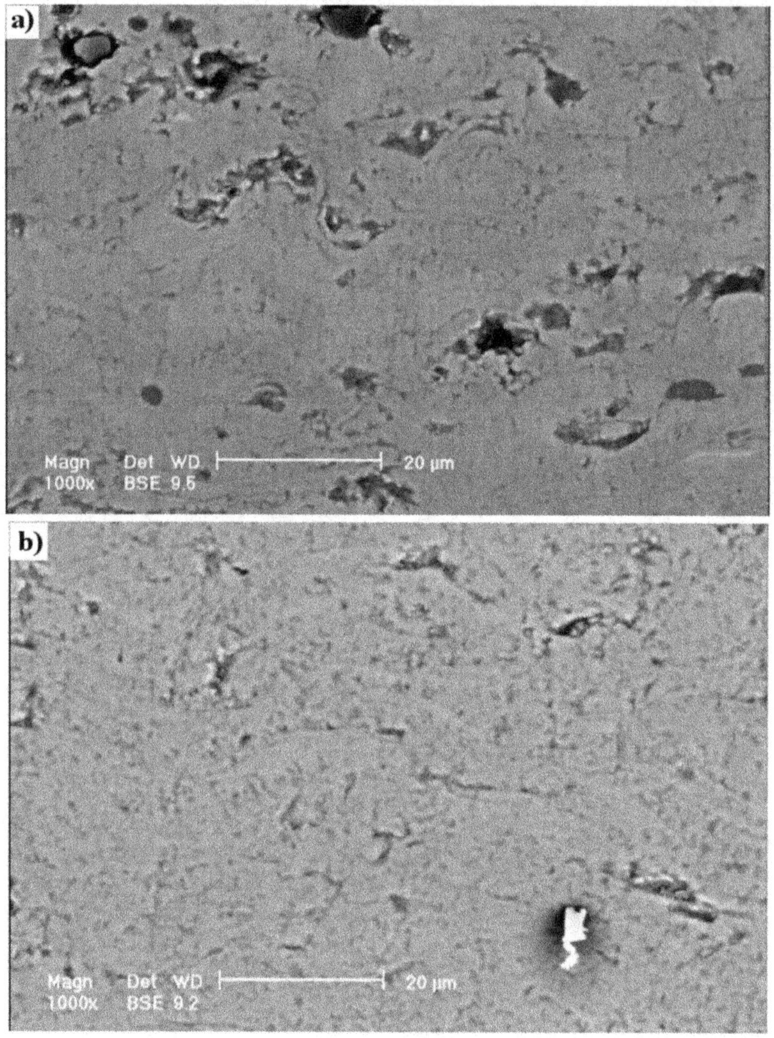

Crystallite size and internal lattice strain of the powders and coatings could be estimated using Williamson–Hall method, Figure 9 (Suryanarayana, 2001). Owing to the fact that all composite and pure nanostructured powders have been produced by mechanical milling, they have a decreased crystallite size and an increased lattice strain in comparison to microstructured powders. In general, the crystallite size of coatings was found to be higher than that of powders, Figure 9(a). The solidification of molten particles during plasma spraying process caused crystallite growth and, at the same time, the lattice strain decreased. This growth is not the case for all thermal spraying techniques as some researchers (He & Schoenung, 2002) reported refinement of crystallite size and enhancement of lattice strain in HVOF process. They suggested the main reason for these phenomena was the high velocity of particles at impact and rapid solidification of lamellae during this spraying technique. However, in plasma spraying, the particle velocity is substantially lower and the temperature of plasma jet is much higher than HVOF process; the crystallites mainly become columnar growth during solidification in APS. Notable point is that although growth of crystallite size and reduction of internal lattice strain of nanoparticles occurred during thermal spraying in pure and composite nanostructured coatings, the crystallite size in these coatings was still lower and the internal lattice strain was higher than conventional coatings. In the zones of nanostructured coatings formed from fully molten particles, the crystallites are coarser than those in the feedstock. On the other hand, since efforts were made to prevent complete melting of the particles during thermal spray process, crystallite size in semi-molten nano-particles which were embedded in the coating was equal to that of nanostructure powders at about 30 nm. As it is shown in Figure 9, the mean crystallite size of all nanostructured coatings was in the range 150-160 nm.

Figure 9. (a) Crystallite size and (b) internal lattice strain of micro-Cr_2O_3, nano-Cr_2O_3, Cr_2O_3-2% Ag, Cr_2O_3-5% Ag and Cr_2O_3-10% Ag powders and coatings

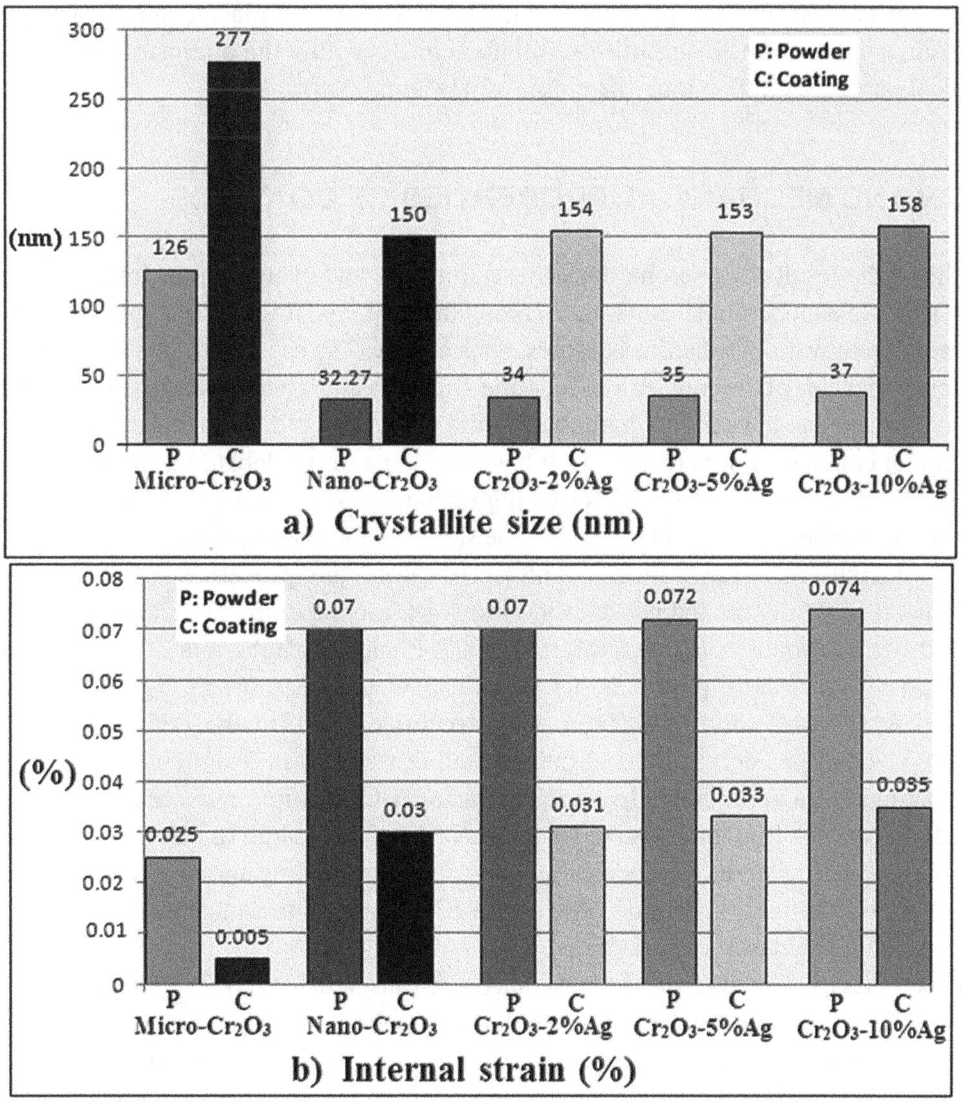

Composition and Reinforcement Distribution

In order to estimate dispersion uniformity of reinforcement nanoparticles in the microstructure, typical EDS elemental maps on coatings cross sections were prepared for typical elements. As presented in Figure 10(a, b), a rather uniform distribution of Ag nanoparticles was observed in Cr_2O_3-2% Ag and Cr_2O_3-5% Ag composite coatings. However, as the percentage of reinforcement was increased, distribution of reinforcement within the coating became nonuniform. As can be seen in Figure 10(c), silver nanoparticles have aggregated in the interlamellar spacing in Cr_2O_3-10% Ag coating, indicated by arrows. High concentration of silver in this specimen; 10 volume percentage of Ag equivalent to 22 weight percent in the feedstock, together with high surface energy of Ag nanoparticles caused their agglomeration in splats during plasma spray process. When molten droplets impact the previous lamellae, Ag particles of higher density, as compared with Cr_2O_3, penetrate the interface between the splat and the solidified lamella. Hence, as silver content was increased, the distribution of Ag in nanostructured composite coating became rather inhomogeneous. It is worth to mention that a key factor in utilization of reinforcement nanoparticles in producing composite coatings by plasma spraying depends on the ability to obtain a uniform fine distribution of reinforcements within the microstructure that is essential for tribological purposes (Cho, Xiong, Kim, Lee, & Hwang, 2008).

ADHESION AND MECHANICAL PROPERTIES OF COATINGS

Figure 11 shows the results of microhardness measurements and evaluation of fracture toughness performed on Cr_2O_3-Ag nanocomposite coatings; among the coatings, micro-Cr_2O_3 and Cr_2O_3-10% Ag had the highest and the lowest microhardness values, respectively. The crystallite size of the nanostructured coatings is lower than that of the conventional coating, Figure 9, and as such, based on Hall-Petch equation, they might be expected to have higher hardness. While the hardness difference between nanostructured and conventional coatings is not significant, it has been a matter of debate between some researchers. Lin et al. (2004) and Ahn et al. (2006) reported higher hardness for micro Al_2O_3-TiO_2 coatings. On the contrary, other researchers have observed an inverse situation for YSZ coatings (Chen, Ding, Zhang, La, & Lee, 2003). It appears that operational parameters of the coating process, such as particle velocity at impact and the ratio of fully melted particles in plasma jet, are the dominant factors affecting the hardness of nanostructured coatings (Lima & Marple, 2007; Flauchais, Heberlein & Boulos, 2014).

Comparison of hardness in pure nano-Cr_2O_3 and Cr_2O_3-Ag coatings in Figure 11a indicates that addition of silver as a soft solid lubricant to nanostructured Cr_2O_3 matrix decreased the hardness of the composite layers. Considering the fact that crystallite size and porosity content of nanostructured composite coatings were approximately similar to nano-Cr_2O_3 coating, reduction of hardness must be due to the addition of a soft metal phase in the hard matrix of chromium oxide. It is established that Ag particles of nanometric scale in a composite coating serve as a solid lubricant reservoir to reduce the coefficient of friction and exert minimal detrimental effect on hardness and elastic modulus in ceramic matrix coatings (Voevodin & Zabinski, 2005).

Fracture toughness, K_{IC}, of the coatings determined by Evans-Wilshaw equation (Sadri, & Ashrafizadeh, 2013), Figure 11b. As can be seen, K_{IC} of nanostructured coatings is higher than that of the conventional coating; a result of the presence of semi-molten nanostructured zones embedded in the microstructure. Some authors (Lima & Marple, 2007; Zois, Lekatou, Vardavoulias, 2009) have mentioned that semi-

Figure 10. SEM micrographs of the cross sections and corresponding x-ray maps of Ag and Cr of (a) Cr_2O_3-2% Ag, (b) Cr_2O_3-5% Ag and (c) Cr_2O_3-10% Ag nanostructured composite coatings

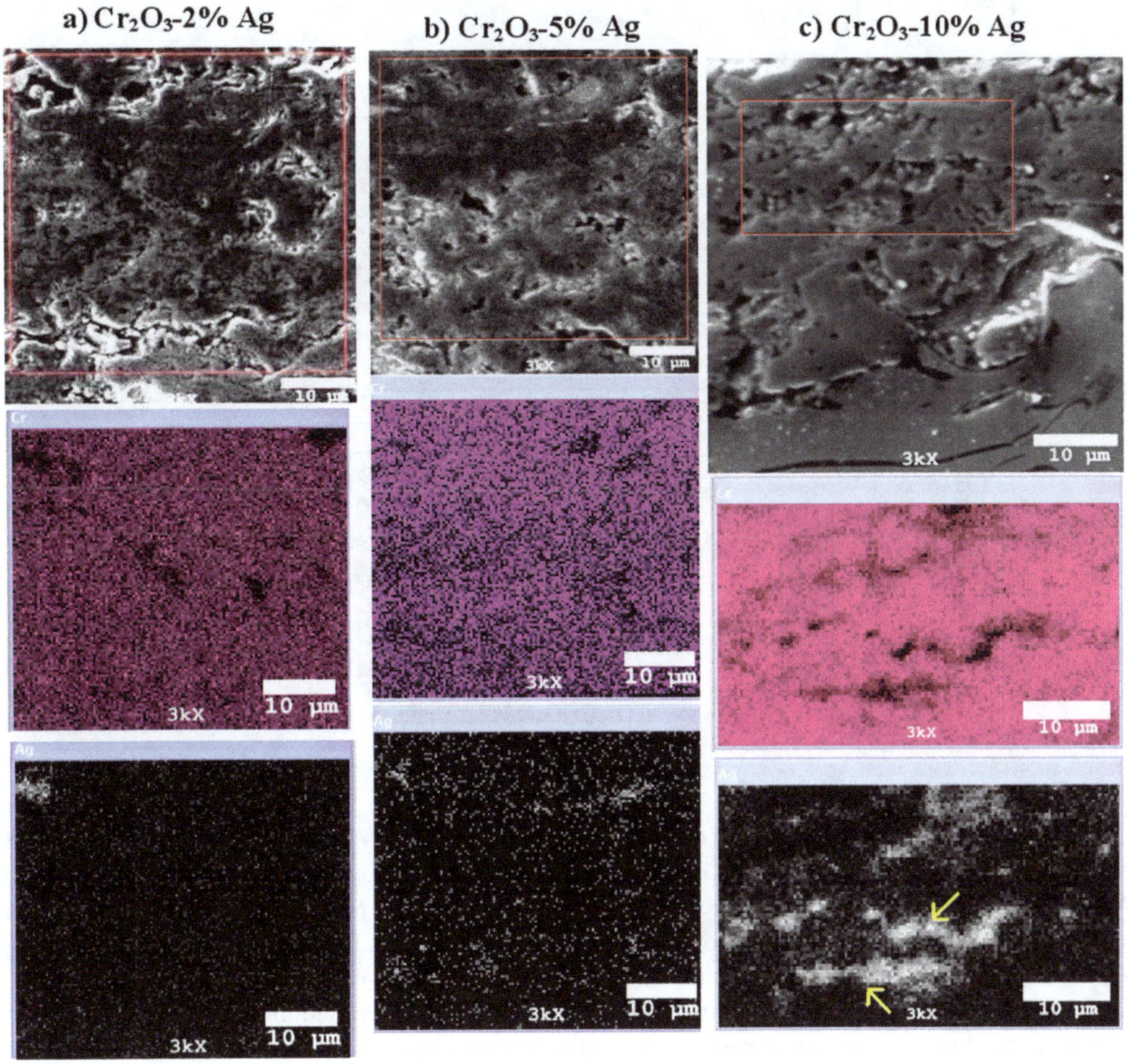

molten nanostructured particles embedded within the coating tend to act as crack arresters, leading to higher crack propagation resistance and enhancing the coating toughness. Indeed, in addition to the effect of nanoparticles on fracture toughness, the higher interlamellar strength in nanostructured coatings caused the fracture toughness of the coating to be improved (Ahn, Hwang, Song, Lee, & Kim, 2006) Under severe sliding conditions, wear of brittle materials, in particular ceramic matrix coatings, occurs by brittle fracture mechanisms. In such conditions, fracture toughness of a coating is the major factor in prevention of crack propagation that directly affects the tribological behavior of the surface (Tian, Wang, Yang, & Li, 2009).

Figure 11. Microhardness (a) and fracture toughness, K_{IC} (b) of coatings

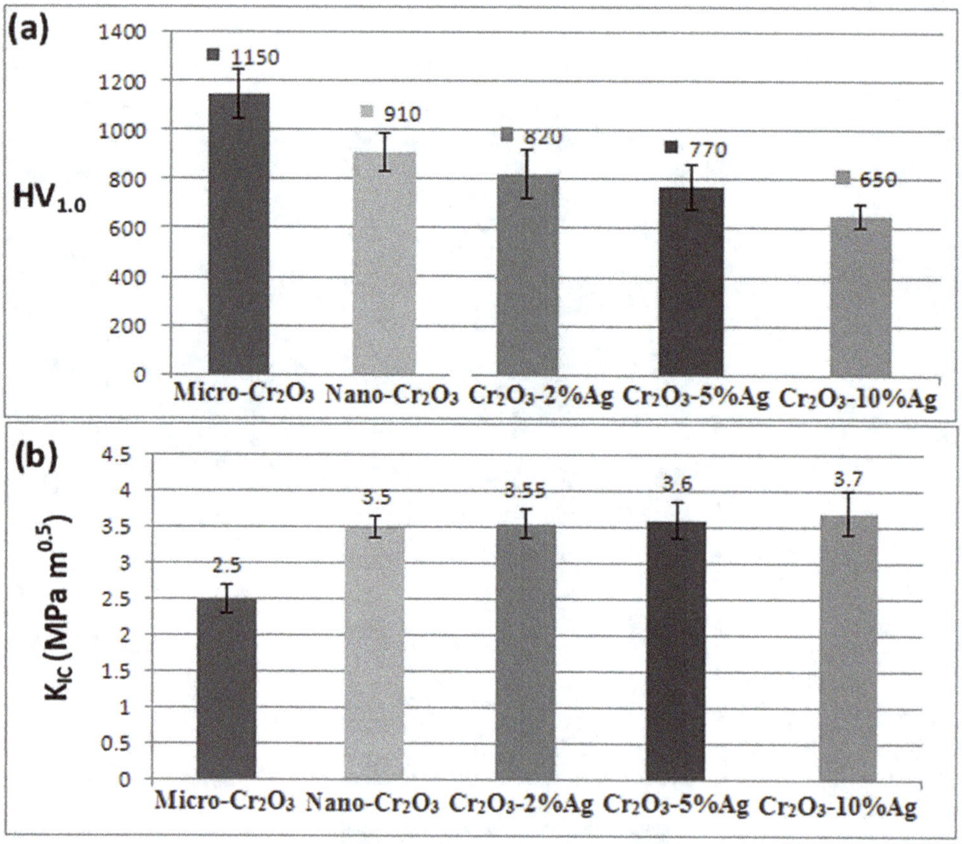

Typical crack propagation behavior of micro-Cr_2O_3 coating under Vickers indentation and influence of nanostructured zones on crack arresting conditions in nano-Cr_2O_3 coatings are indicated in Figure 12. As evident in SEM micrographs, in micro-Cr_2O_3 coating, cracks have propagated dominantly along the splat boundaries, Figure 12b. The nanostructured zones together with their crack arresting features are shown in Figure 12c-d. Arrows in Figure 12d illustrate that the crack is arrested when passing through the nanostructured, dark, zone and continues its propagation in fully melted, light, region.

From the above discussion, it may be concluded that two opposite factors affect the hardness of thermal sprayed nanostructured coatings. One effect is the finer crystallite size in nanostructured coatings that increases the hardness and other one is attributed to higher microstructural defects including unmelted particles, porosity and inclusions that cause a decrease in hardness. As these factors depend on control of process parameters, different results may be obtained in plasma sprayed coatings of the same composition. By optimizing the operational parameters, it would be possible to produce a composite coating of high hardness to resist abrasive wear and, at the same time, sufficient fracture toughness to withstand fracture mechanisms.

The results of tensile adhesion tests together with the fracture surface of Cr_2O_3-2% Ag coating are presented in Figure 13. More than 90% of failures were located within the ceramic top coat since the adhesion strength of AMDRY 962, Ni95 wt. %-Al wt. 5%, bond coat, being higher than 40 MPa, is much higher than that of the ceramic matrix coatings, Figure 13b. It is implied that cohesion strength

Figure 12. Crack propagation under Vickers indentation in micro-Cr_2O_3 coating (a-b) and crack tip arresting by nanostructured regions in nano-Cr_2O_3 coating (c -d)

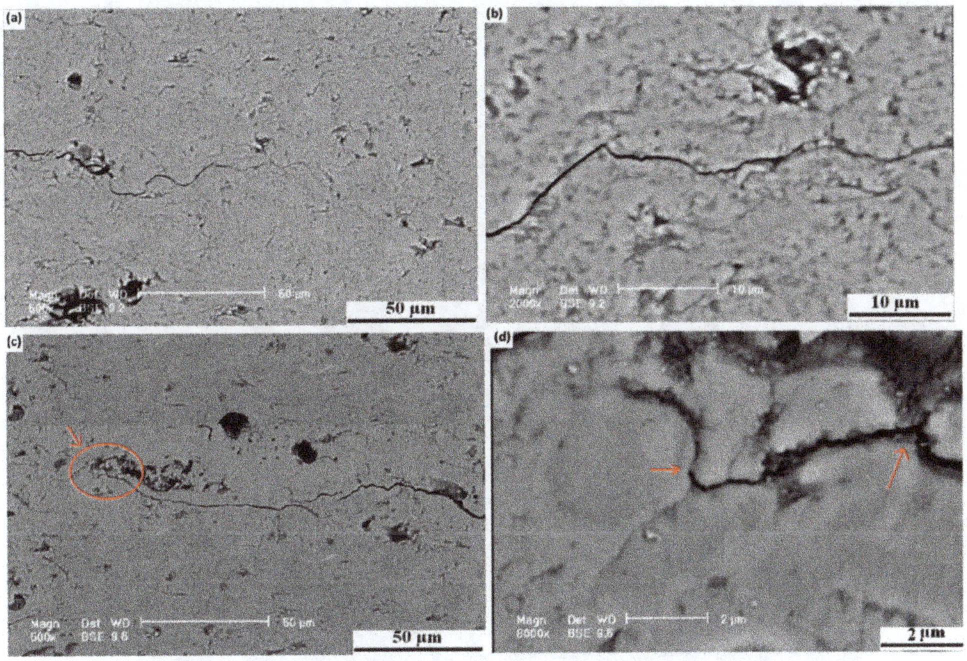

of the top coat is significantly lower than adhesion or cohesion strength of NiAl bond coat. Intrinsic embrittlement of the ceramic matrix, together with the lamellar morphology of thermal spray coatings are main factors for low values of cohesion strength of these duplex coatings. In effect, two main factors affect the adhesion of nanostructured coatings. First, improvement of fracture toughness that effectively promoted inter-splat adhesion and, secondly, existence of some micro-pores, Figure 8a, that decreased tensile test values since a pore of critical size can induce a macroscopic failure with tensile stress applied. Adhesion of nano-Cr_2O_3 was lower than micro-Cr_2O_3 coating, however, Cr_2O_3-2%Ag coating displayed the highest adhesion among the tested specimens, Figure 13b.

WEAR BEHAVIOR OF COUNTERPARTS AGAINST COATINGS

In order to have an accurate evaluation of frictional behavior, it is mandatory to avoid characteristic wear mechanisms in the counterpart material during tribological tests. Moreover, it would not be expected that counterpart reacts with the high temperature performance coatings. For these reasons, initially, 2 types of materials, namely AISI 52100 steel pin, alumina and WC balls were evaluated as counterpart in wear testing of micro-Cr_2O_3 coatings to examine their capability as appropriate counterface materials. Weight loss versus sliding distance during testing of counterparts against micro-Cr_2O_3 coating illustrates drastic decrease for steel pin and invariant weight for WC ball. Weight change of the coating was measured after each wear test; the weight gain of micro-Cr_2O_3 coating against steel pin, 2 mg, and its weight loss against alumina and WC balls, 4 and 0.3 mg, respectively, were observed after 1000 m sliding distance.

Figure 13. Adhesion strength of coatings (a) and fracture surface of Cr_2O_3-2% Ag coating after ASTM C633 test showing the failure within the ceramic matrix coating (b)

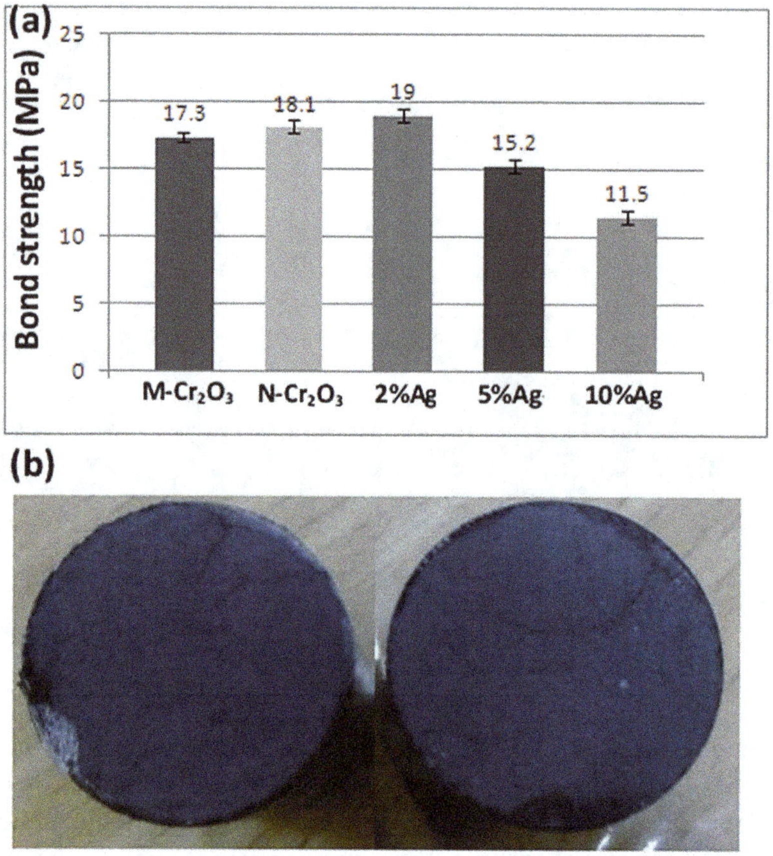

Sliding surface of micro-Cr_2O_3 coating against steel pin with the proposed schematic illustration of the wear mechanism are presented in Figure 14. Micro-analysis could be performed on two selected regions with different height on wear track. Considering that steel counterpart, hardness about 62 HRC, has a lower hardness than micro-Cr_2O_3 coating, microhardness about 1150 $HV_{1.0}$, during sliding, very small wear fragments were produced dominantly from steel pin. The particles then attached to the coating surface and, thus, transfer layers rich of iron element were generated, Figure 14(b); regions "E" with 74.4 and 7.3 wt. % Fe and Cr, respectively, indicate these transfer layers with much higher Fe than regions "L" that contain 32.2 wt. % Fe and 45.2 wt. % Cr. Moreover, the weight loss of steel pin during wear testing and lack of abrasion grooves on micro-Cr_2O_3 wear scar, Figure 14(a), confirm the proposed mechanism. The attached transfer layers, then, change the tribological behavior of the coating. However, it seems that steel pin is not an appropriate counterpart for wear testing of Cr_2O_3 coatings.

The WC ball, used as a counterpart in this study, is a standard typical material for ball-on-disc wear tests for which hardness is as high as 1400 HV and roughness (R_a) is lower than 0.1 μm. There was no considerable material loss neither from micro-Cr_2O_3 coating nor WC ball. Accordingly, it appeared that desirable counterpart features of WC ball lead to minimized surface interactions on the contact surface. Wear track of micro-Cr_2O_3 coating against WC ball revealed spalling of the hard asperity tips followed by

Figure 14. (a) Wear track and (b) the suggested wear mechanism of micro-Cr_2O_3 coating against steel pin at room temperature test

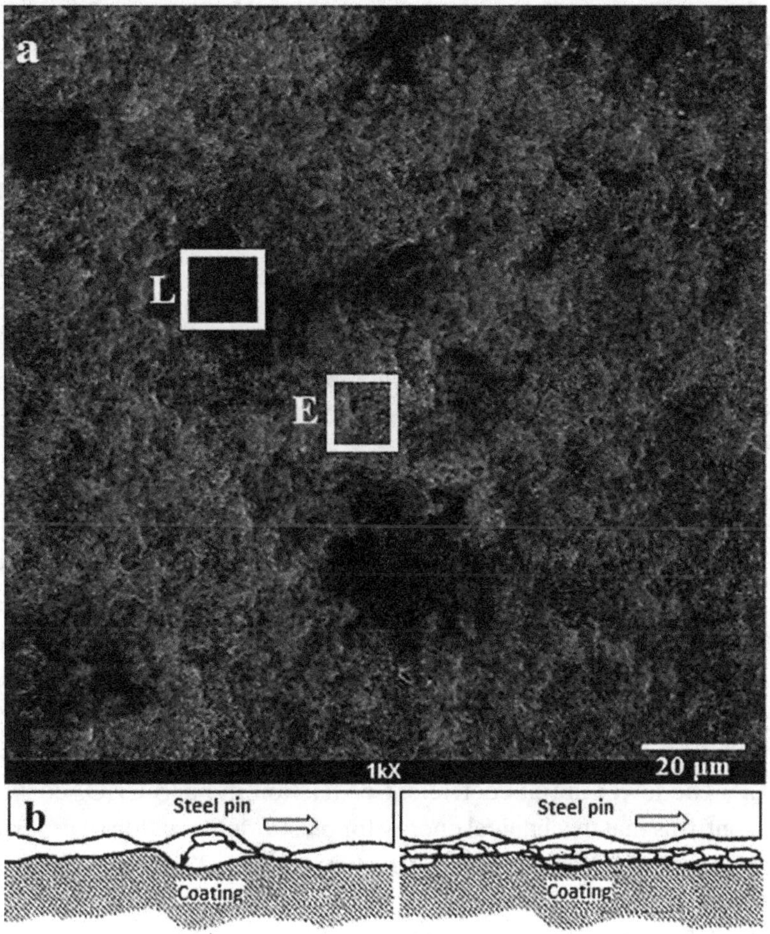

formation of tiny wear fragments and, consequently, smoothing of hills with unchanged valleys occurred on sliding surface of the coating (Sadri, 2012). After this stage, the contact surface remained almost unchanged since the applied load was rather low, 9.8 N, and the hardness of both contact surfaces was sufficiently high to avoid further considerable wear. As a result, WC ball could be employed as an appropriate counterface for examination of performance of Cr_2O_3-Ag self-lubricating nanocomposite coatings.

FRICTIONAL BEHAVIOR AND LUBRICATION MECHANISMS OF COATINGS

Due to high hardness of Cr_2O_3-Ag nanostructured composite coatings, no detectable weight loss (at 0.3 m/s linear velocity sliding tests) was observed, instead, surface asperities were smoothed to some extent. This means the tangential force that results in the coefficient of friction during sliding motion is originated only from interaction of asperities at the contact surface. The normal load employed was rather low to facilitate evaluation of solid lubrication behavior of the nanocomposite coatings; this was achieved by detecting coefficient of friction versus sliding distance over a wide range of temperatures as well as examination of corresponding wear tracks by SEM.

Room Temperature (25^{0}C) Lubrication Mechanisms

During running-in period, at a distance of less than 50 m, the significantly lower sliding friction of nano-Cr_2O_3 as compared to conventional micro-Cr_2O_3 coating, about 0.25, is a direct influence of nano-coating on friction. Indeed, semi-molten nanostructured zones distributed within the bimodal microstructure of plasma sprayed coatings, in this case nano-Cr_2O_3, act as crack arresters when a crack approaches these zones and, thereby, promote fracture toughness, K_{IC}, of the coated component (Sadri, 2012). Various researchers have shown that both ductile flow and brittle fracture can occur together in sliding surfaces of ceramic coatings (Liu, Zhang, & Deng, 2002; Malkin, & Huang, 1996). However, the greater extent of critical depth of cut causes a delay in brittle fracture and plastic deformation becomes dominant material removal mechanism in ceramic sliding surfaces. They also reported that the ratio of toughness to hardness (K_{IC}/HV) of materials directly affected critical depth of cut (Malkin, & Huang, 1996). The ratio K_{IC}/HV of coatings could be calculated from the data reported in Figure 11.

The critical depth of cut, dc, of nanostructured coatings is significantly higher than that of conventional micro-Cr_2O_3 coating. As mentioned before, this promotion of dc is the result of presence of semi-molten nanostructured zones. Moreover, wear scar of nano-Cr_2O_3 coating revealed that smearing and dulling of hill asperities as well as filling of valleys took place under sliding conditions, governing of plastic deformation. Whereas, in micro-Cr_2O_3 worn surfaces, spalling of hard asperity tips followed by formation of tiny fragments and, consequently, smoothing of hills with unchanged valleys as well as surface cracking and flaking at the micron scale are predominant mechanisms during sliding test of these coatings (sadri, 2012). The lower initial coefficient of friction in nano-Cr_2O_3 coating, compared with conventional Cr_2O_3, confirms that mechanical energy for plastic deformation of asperities is significantly lower than brittle fracture in Cr_2O_3 coatings. This feature produced a lower coefficient of friction and continued even after interaction of asperities.

Formation of solid lubricant tribofilm on top surface was found necessary to improve frictional behavior of composite coatings. Initially, transfer film was not present at sliding contact, instead, it was formed as a result of surface wear and subsurface loading (Aouadi et al., 2010). The reduction of friction coefficient and its fluctuations in Cr_2O_3-Ag nanocomposite coatings, compared to nano-Cr_2O_3, are indeed the direct influence of the presence of silver nanoparticles in these coatings (Sadri, 2012).

Figure 15 presents SEM micrographs of wear tracks for Cr_2O_3-5% Ag nanocomposite coating after testing at room temperature; other nanocomposite coatings, 2 & 10% Ag, had similar wear tracks. As can be seen, there are two distinct regions in wear scars that are different in both topography and chemical composition; "E" regions denote elevated areas that were exposed to real sliding contact, tribofilm, and "L" was used to identify lower worn regions. Results of EDS on worn surface of nanocomposite coatings revealed that elevated regions, "E", had higher Ag concentration, as solid lubricant material (Sadri, 2012).

Based on SEM examinations and microanalyses, a lubrication mechanism is schematically represented in Figure 16 that considers motion of a slider over hard matrix coatings with well-dispersed soft nanoparticles, consistent with solid lubrication characteristics of Cr_2O_3-Ag nanocomposite coatings. The "E" regions in Figure 15 are probably the islands of thin surface film on asperity tips, which were exposed to real sliding contacts. Initially, owing to localized surface heating at the interface contact of Cr_2O_3 and WC, silver nanoparticles penetrate near the surface toward asperity tips, Figure 16(a, b). The

Figure 15. Wear tracks of Cr_2O_3-5% Ag after wear testing at room temperature. "E" denotes elevated areas that were exposed to real sliding contact (tribofilm) and "L" denotes lower worn regions

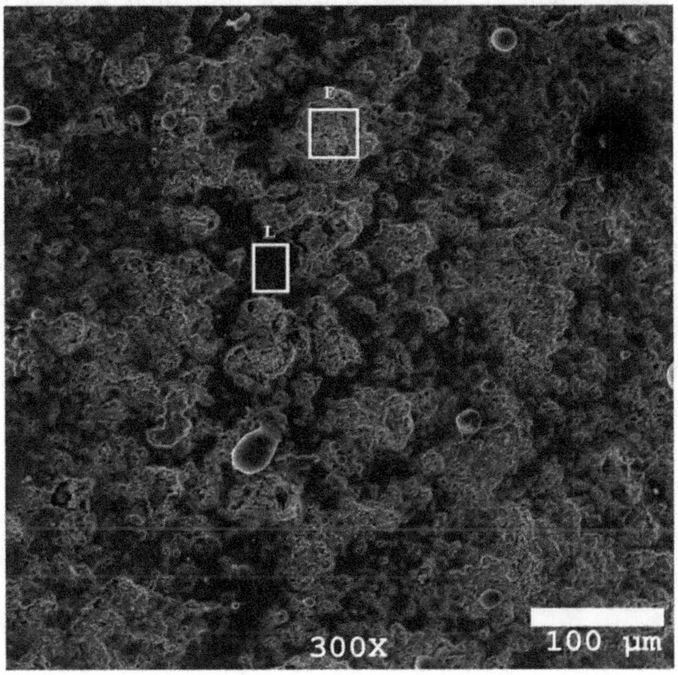

soft particles prevent contact of hard asperities of coating and counterface, leading to dissipation of frictional heat and, consequently, decreased coefficient of friction.

Some researchers (Scharf, Kotula & Prasad, 2010) reported that concentration of soft metal and, thereby, formation of large islands of soft particles on wear surfaces could be a result of gas-surface interactions; oxygen adsorption would enhance agglomeration of nanoparticles on the wear surface. This mechanism was suggested by comparing the location and size of gold islands clustered on the sliding surface of MoS_2/Sb_2O_3/Au self-lubricating nanocomposite coating in air and dry nitrogen. Coarsening of Au nanoparticles at subsurface and formation of large Au islands on the wear surface were observed in dry nitrogen and air, respectively. The combination of crystal structure with multiple slip planes and no appreciable work-hardening features under plastic loading, high thermal conductivity, low shear strength as well as chemical inertness and structural stability make silver an ideal metallic lubricant. In addition, the accumulated Ag flakes, Figure 16(b), form a very thin transfer layer of silver on contact surface which smears and shears plastically with minimized ploughing, inhibiting contact of hard counterbodies and, thereby, a reduced coefficient of friction, Figure 16(c).

Uniform distribution of Ag nanoparticles in the as-sprayed coating, Figure 10, and formation of Ag solid lubricant tribo-film on wear track as well as silver compositional gradient toward track surface in friction test confirm the proposed lubrication mechanism. It can be expressed that the regions comprising higher Ag, "E" in Figure 15, originated from attached silver islands, remained after plastic shearing at sliding contact. Fundamentally, strong adhesion of solid lubricant films to substrate is a main characteristic in composite or duplex tribological coatings that promises low friction as well as long wear life of the film. This feature is obtained easily in silver tribofilm which is diffused, created and compacted on contact surface during sliding tests.

Figure 16. Schematic illustration of lubrication mechanism for Cr_2O_3-Ag nanocomposite coatings; (a) as-sprayed; homogenous distribution of silver nanoparticles in Cr_2O_3 matrix, (b) the running-in step in friction test; the accumulation of Ag nanoparticles and formation of silver flakes on asperity tips, bulk diffusion of Ag nanoparticles toward contact surface, (c) the steady state period in friction test; formation of Ag solid lubricant tribofilm on wear track which smears plastically and reduced friction

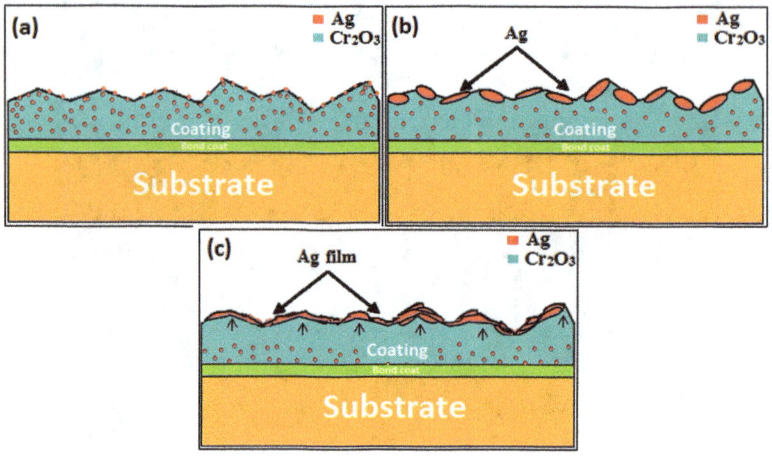

300°C Lubrication Mechanisms

Similar to the results of friction tests at room temperature, nanostructured coatings with dispersion of nano-silver particles and formation of thin solid lubricant layer on contact surface caused significant improvement of frictional behavior over the conventional coating. The significance of nanostructural characteristics in promotion of fracture toughness is illustrated by smoothed hills within wear tracks; plastic deformation in nano-Cr_2O_3 coating is attained by presence of semi molten nanostructured zones produced by plasma spraying. On the contrary, brittle fracture of hills, the presence of tiny debris, as well as flaking and spalling, were the feature observed in micro-Cr_2O_3 coating at 300°C tests (Sadri, 2012).

Frictional behavior of micro-Cr_2O_3 coating at 25°C and 300°C maybe compared versus sliding distance. The higher value of friction coefficient of and its fluctuations micro-Cr_2O_3 coating at 300°C, compared to 25°C, is in contradiction to the results reported by Ahn et al. (1999). They stated that the self-mated frictional behavior of Cr_2O_3 plasma spray coatings were promoted by formation of a compacted smooth film with major amount of CrO_2 at higher temperatures, 450°C, whereas this film had considerable amount of CrO_3 at room temperature. Then, the higher hardness of CrO_2, than CrO_3 and Cr_2O_3, and greater quantity of high temperature smooth layers compacted on the wear track, compared to room temperature, were attributed to the lower friction of plasma sprayed Cr_2O_3 coatings at high temperature wear tests. Moreover, since the friction of nanocomposite coatings strongly corresponds to the presence of thin silver surface layer (tribofilm) and most soft metals tend to have lower friction as the temperature increases, mainly due to softening and rapid recovery from strain hardening, a lower coefficient of friction would be expected at 300°C for nanocomposite coatings. However, conflicting results may imply the evidence of wear oxidation mechanisms of WC ball at high temperatures. More quantity of WC on wear scars at 300°C tests compared to room temperature tests, (Sadri, 2012), confirmed the oxidation wear of WC against Cr_2O_3 at high temperatures for all coatings.

The same as room temperature tests, two distinct regions inside the sliding tracks in Cr_2O_3-2, 5 and 10% Ag nanocomposite coatings after 300⁰C wear tests are observed, they are similar to wear scar in Figure 15; these regions show different topography and chemical composition, (Sadri, 2012). The elevated regions, "E", have higher Ag concentration, and reveal islands of low shear strength tribofilm formed by the mechanisms explained in Figure 16.

500ºC Lubrication Mechanisms

The friction traces of the coatings during testing at 500⁰C show significant improvement by nanostructural features and silver nanoparticle tribofilm comprising the solid lubrication mechanism described for room temperature tests. Similar to the other two temperatures, the elevated regions "E", which control the sliding contact, have higher silver content and confirm the tribofilm lubrication mechanism (Sadri, 2012). The wear mechanisms were intensified at high temperatures by oxidation degradation of WC ball. Thus, the coefficients of friction and their fluctuations of the coatings have increased at elevated temperature, compared to 25⁰C and 300⁰C (Sadri, 2012). Cr_2O_3-5% Ag nanocomposite coating shows optimum frictional behavior; Cr_2O_3-2% Ag and 10% Ag coatings had higher and lower thicknesses of smooth lubricant films, respectively.

CONCLUSION

Nanostructured plasma sprayed oxide ceramic matrix coatings for their high temperature self-lubricating wear resistance have been extensively studied. Since crystalline solid lubricants cause a significant reduction in hardness of composite structure, amorphous and poorly crystalline nanoparticle inclusions with minimum detrimental mechanical effects were introduced in self-lubricating composite coatings. This feature has rarely been considered in thermal spraying techniques. Nanostructured composite coatings can be successfully deposited by plasma spray process using composite granule feedstock with homogeneous distribution of solid lubricant nanoparticles. The initial individual nanoparticles of feedstock are embedded in semi-molten islands, surrounded by fully molten particles that act as a binder for the nanostructured coatings. The presence of semi-molten zones as crack arresters in the coatings caused a delay in brittle fracture and improved fracture toughness for tribological performance. As a result, the frictional behavior of the nanostructured coatings was significantly improved at various temperatures. Formation of low shear strength tribofilm of soft metal solid lubricant inclusions, silver, on top of the hard coatings, that smears and shears plastically with minimized ploughing, inhibited contact of hard counter bodies and, thereby, reduced coefficient of friction. Both low shear strength of tribofilm and high load bearing of ceramic oxide matrix contributed to self-lubricating characteristic of Cr_2O_3-Ag nanocomposite coating. The optimized thickness of soft metallic tribofilms developed at service temperature leads to low friction at conditions ranging from room temperature to 500 °C.

In tribological contacts, for low friction and extended wear endurance, particularly in aerospace industries, the current trend is towards adaptive self-lubricating thermal spray coatings with provision of lubrication over wide range of operating environments, such as ambient temperature, humidity, air content of atmosphere and contact loads, as well as during cycling of operating parameters. In this approach, superalloys are commonly selected as main matrix for their high temperature mechanical properties and oxidation resistance. Oxide phases are used as hard phase to enhance wear resistance and soft metals, alkaline fluorides and dichalcogenides are added to provide lubricant tribofilms by temperature and environment activated diffusion mechanisms. Chemical formation of binary lubricious oxide layers at contact interface, tribo-oxidation mechanism, appears to be applied as lubrication mechanism when temperature exceeds 500^{0}C in air, such as silver molybdenum binary oxide compounds. Plasma-transferred arc (PTA) coating process and suspension plasma spraying are incoming technologies for promotion of thermal spray self-lubricating composite coatings.

REFERENCES

Ahn, H. S., & Kwon, O. K. (1999). Tribological behaviour of plasma-sprayed chromium oxide coating. *Wear, 225*, 814–824. doi:10.1016/S0043-1648(98)00390-1

Ahn, H. S., Lyo, I., & Lim, D. S. (2000). Influence of molybdenum composition in chromium oxide-based coatings on their tribological behavior. *Surface and Coatings Technology, 133*, 351–361. doi:10.1016/S0257-8972(00)00892-6

Ahn, J., Hwang, B., Song, E. P., Lee, S., & Kim, N. J. (2006). Correlation of microstructure and wear resistance of Al_2O_3-TiO_2 coatings plasma sprayed with nanopowders. *Metallurgical and Materials Transactions. A, Physical Metallurgy and Materials Science, 37*(6), 1851–1861. doi:10.1007/s11661-006-0128-5

Aouadi, S. M., Luster, B., Kohli, B., Muratore, C., & Voevodin, A. A. (2009). Progress in the development of adaptive nitride-based coatings for high temperature tribological applications. *Surface and Coatings Technology, 204*(6-7), 962–968. doi:10.1016/j.surfcoat.2009.04.010

Aouadi, S. M., Paudel, Y., Luster, B., Stadler, S., Kohli, P., Muratore, C., & Voevodin, A. A. et al. (2008). Adaptive $Mo_2N/MoS_2/Ag$ tribological nanocomposite coatings for aerospace applications. *Tribology Letters, 29*(2), 95–103. doi:10.1007/s11249-007-9286-x

Aouadi, S. M., Paudel, Y., Simonson, W. J., Ge, Q., Kohli, P., Muratore, C., & Voevodin, A. A. (2009). Tribological investigation of adaptive $Mo_2N/MoS_2/Ag$ coatings with high sulfur content. *Surface and Coatings Technology, 203*(10-11), 1304–1309. doi:10.1016/j.surfcoat.2008.10.040

Aouadi, S. M., Singh, D. P., Stone, D. S., Polychronopoulou, K., Nahif, F., Rebholz, C., & Voevodin, A. A. et al. (2010). Adaptive VN/Ag nanocomposite coatings with lubricious behavior from 25 to 1000°C. *Acta Materialia, 58*(16), 5326–5331. doi:10.1016/j.actamat.2010.06.006

Berger, L. M., Stahr, C. C., Saaro, S., Thiele, S., Woydt, M., & Kelling, N. (2009). Dry sliding up to 7.5 m/s and 800°C of thermally sprayed coatings of the TiO_2–Cr_2O_3 system and (Ti, Mo) (C, N)–Ni (Co). *Wear, 267*(5-8), 954–964. doi:10.1016/j.wear.2008.12.105

Berger-Keller, N., Bertrand, G., Filiatre, C., Meunier, C., & Coddet, C. (2003). Microstructure of plasma-sprayed titania coatings deposited from spray dried powder. *Surface and Coatings Technology, 168*(2-3), 281–290. doi:10.1016/S0257-8972(03)00223-8

Bolelli, G., Cannillo, V., Lusvarghi, L., & Manfredini, T. (2006). Wear behavior of thermally sprayed ceramic oxide coatings. *Wear, 261*(11-12), 1298–1315. doi:10.1016/j.wear.2006.03.023

Cao, Q., Vessen, R., Schwarz, S., Jungen, W., Tietz, F., & Stoever, D. (2000). Spray-drying of ceramics for plasma-spray coating. *Journal of the European Ceramic Society*, *20*(14-15), 2433–2439. doi:10.1016/S0955-2219(00)00112-6

Cellard, A., Garnier, V., Fantozzi, G., Baret, G., & Fort, P. (2009). Wear resistance of chromium oxide nanostructured coatings. *Ceramics International*, *35*(2), 913–916. doi:10.1016/j.ceramint.2008.02.022

Cellard, A., Zenati, R., Garnier, V., Fantozzi, G., & Baret, G. (2007). Optimization of chromium oxide nanopowders dispersion for spray-drying. *Journal of the European Ceramic Society*, *27*(2-3), 1021–1027. doi:10.1016/j.jeurceramsoc.2006.04.132

Cetinel, H., Celik, E., & Kusoglu, M. I. (2008). Tribological behavior of Cr_2O_3 coatings as bearing materials. *Journal of Materials Processing Technology*, *196*(1-3), 259–265. doi:10.1016/j.jmatprotec.2007.05.048

Chen, H., Ding, C., Zhang, P., La, P., & Lee, S. W. (2003). Wear of plasma-sprayed nanostructured zirconia coatings against stainless steel under distilled-water conditions. *Surface and Coatings Technology*, *173*(2-3), 144–149. doi:10.1016/S0257-8972(03)00311-6

Cho, J., Xiong, Y., Kim, J., Lee, C., & Hwang, S. H. (2008). Tribological behavior of NiCr-base blended and nanostructured composite APS coatings by rig test. *Wear*, *265*(11-12), 1565–1571. doi:10.1016/j.wear.2008.03.008

Chwa, S. O., Klein, D., Toma, F. L., Bertrand, G., Liao, H., Coddet, C., & Ohmori, A. (2005). Microstructure and mechanical properties of plasma sprayed nanostructured TiO_2–Al composite coatings. *Surface and Coatings Technology*, *194*(2-3), 215–224. doi:10.1016/j.surfcoat.2004.07.080

Delogu, F., & Cocco, G. (2006). Microstructural refinement of ceramic powders under mechanical processing conditions. *Journal of Alloys and Compounds*, *420*(1-2), 246–250. doi:10.1016/j.jallcom.2005.10.037

Donnet, C., & Erdemir, A. (2004). Historical development and new trends in tribological and solid lubricant coatings. *Surface and Coatings Technology*, *180*, 76–84. doi:10.1016/j.surfcoat.2003.10.022

Fauchais, P. L., Heberlein, J. V. R., & Boulos, M. I. (2014). *Thermal Spray Fundamentals From Powder to Part*. Springer. doi:10.1007/978-0-387-68991-3

He, J., & Schoenung, J. M. (2002). Nanostructured coatings. *Materials Science and Engineering A*, *336*(1-2), 274–319. doi:10.1016/S0921-5093(01)01986-4

Ibrahim, A., Abdel Hamid, Z., & Abdel Aal, A. (2010). Investigation of nanostructured and conventional alumina–titania coatings prepared by air plasma spray process. *Materials Science and Engineering A*, *527*(3), 663–668. doi:10.1016/j.msea.2009.08.054

Jin, Y., & Yang, Y. (1996). Tribological behavior of various plasma sprayed ceramic coatings. *Surface and Coatings Technology*, *88*(1-3), 248–254. doi:10.1016/S0257-8972(96)02918-0

Kim, G., Choi, H., Han, C., Uhm, S., & Lee, C. (2005). Characterization of atmospheric plasma spray NiCr–Cr_2O_3–Ag–CaF_2/BaF_2 coatings. *Surface and Coatings Technology*, *195*(1), 107–115. doi:10.1016/j.surfcoat.2004.11.041

Lima, R. S., & Marple, B. R. (2007). Thermal Spray Coatings Engineered from Nanostructured Ceramic Agglomerated Powders for Structural, Thermal Barrier and Biomedical Applications: A Review. *Journal of Thermal Spray Technology*, *16*(1), 40–63. doi:10.1007/s11666-006-9010-7

Lin, X., Zeng, Y., Lee, S. W., & Ding, C. (2004). Characterization of Alumina-3wt. %Titania Coating Prepared by Plasma Spraying of Nanostructured Powders. *Journal of the European Ceramic Society*, *24*(4), 627–634. doi:10.1016/S0955-2219(03)00254-1

Liu, H., Tao, J., Xu, J., Chen, Z., & Gao, Q. (2009). Corrosion and tribological behaviors of chromium oxide coatings prepared by the glow-discharge plasma technique. *Surface and Coatings Technology*, *204*(1-2), 28–36. doi:10.1016/j.surfcoat.2009.06.020

Liu, X., Zhang, B., & Deng, Z. (2002). Grinding of nanostructured ceramic coatings: Surface observations and material removal mechanisms. *International Journal of Machine Tools & Manufacture*, *42*(15), 1665–1676. doi:10.1016/S0890-6955(02)00115-3

Lu, Y., & Liaw, P. K. (2001). The mechanical properties of nanostructured materials. *Journal of Metals*, *53*, 31–35.

Lyo, I. W., Ahn, H. S., & Lim, D. S. (2003). Microstructure and tribological properties of plasma-sprayed chromium oxide–molybdenum oxide composite coatings. *Surface and Coatings Technology*, *163*, 413–421. doi:10.1016/S0257-8972(02)00613-8

Malkin, S., & Hwang, T. W. (1996). Grinding mechanisms for ceramics. *Annals of the CIRP*, *45*(2), 569–580. doi:10.1016/S0007-8506(07)60511-3

Mulligan, C. P., Blanchet, T. A., & Gall, D. (2010). CrN–Ag nanocomposite coatings: Tribology at room temperature and during a temperature ramp. *Surface and Coatings Technology*, *204*(9-10), 1388–1394. doi:10.1016/j.surfcoat.2009.09.018

Mulligan, C. P., & Gall, D. (2005). CrN-Ag self-lubricating hard coatings. *Surface and Coatings Technology*, *200*(5-6), 1495–1500. doi:10.1016/j.surfcoat.2005.08.063

Muratore, C., Hu, J. J., & Voevodin, A. A. (2009). Tribological coatings for lubrication over multiple thermal cycles. *Surface and Coatings Technology*, *203*(8), 957–962. doi:10.1016/j.surfcoat.2008.08.073

Nandiyanto, A. B. D., & Okuyama, K. (2011). Progress in developing spray-drying methods for the production of controlled morphology particles: From the nanometer to submicrometer size ranges. *Advanced Powder Technology*, *22*(1), 1–19. doi:10.1016/j.apt.2010.09.011

Ouyang, J. H., & Sasaki, S. (2001). Effects of different additives on microstructure and high-temperature tribological properties of plasma-sprayed Cr_2O_3 ceramic coatings. *Wear*, *249*(1-2), 56–67. doi:10.1016/S0043-1648(01)00530-0

Ouyang, J. H., & Sasaki, S. (2001). Effects of different additives on microstructure and high temperature tribological properties of plasma-sprayed Cr_2O_3 ceramic coatings. *Wear*, *249*(1-2), 56–67. doi:10.1016/S0043-1648(01)00530-0

Pawlowski, L. (2007). *The science and engineering of thermal spray coatings* (2nd ed.). Wiley.

Pawlowski, L. (2009). Suspension and solution thermal spray coatings. *Surface and Coatings Technology, 203*(19), 2807–2829. doi:10.1016/j.surfcoat.2009.03.005

Sadri, E. (2012). Plasma spraying and tribological evaluation of Cr_2O_3-Ag nanostructured composite coating. Unpublished master's thesis, Isfahan University of Technology, Isfahan, Iran.

Sadri, E., & Ashrafizadeh, F. (2013). Structural characterization and mechanical properties of plasma sprayed nanostructured Cr_2O_3-Ag composite coatings. *Surface and Coatings Technology*, *236*, 91–101. doi:10.1016/j.surfcoat.2013.09.033

Scharf, T. W., Kotula, P. G., & Prasad, S. V. (2010). Friction and wear mechanisms in $MoS_2/Sb_2O_3/Au$ nanocomposite coatings. *Acta Materialia*, *58*(12), 4100–4109. doi:10.1016/j.actamat.2010.03.040

Singh, V. P., Sil, A., & Jayaganthan, R. (2011). Tribological behavior of plasma sprayed Cr_2O_3-3%TiO_2 coatings. *Wear*, *272*(1), 149–158. doi:10.1016/j.wear.2011.08.004

Sliney, H. E. (1979). Wide temperature spectrum self-lubricating coatings prepared by plasma spraying. *Thin Solid Films*, *64*(2), 211–217. doi:10.1016/0040-6090(79)90511-X

Sliney, H. E. (2001). Solid lubricants. In P.J. Blau (Ed.), ASM Handbook (Vol. 18, pp. 193-216) ASM International.

Suryanarayana, C. (2001). Mechanical Alloying and Milling. *Progress in Materials Science*, *46*(1-2), 1–184. doi:10.1016/S0079-6425(99)00010-9

Tao, S., Liang, B., Ding, C., Lao, H., & Coddet, C. (2005). Wear Characteristics of Plasma-Sprayed Nanostructured Yttria Stabilized Zirconia Coatings. *Journal of Thermal Spray Technology*, *14*(4), 518–523. doi:10.1361/105996305X76540

Tian, W., Wang, Y., Yang, Y., & Li, C. (2009). Toughening and strengthening mechanism of plasma sprayed nanostructured Al_2O_3–13 wt. %TiO_2 coatings. *Surface and Coatings Technology*, *204*(5), 642–649. doi:10.1016/j.surfcoat.2009.08.045

Voevodin, A. A., & Zabinski, J. S. (2006). Laser surface texturing for adaptive solid lubrication. *Wear*, *261*(11-12), 1285–1292. doi:10.1016/j.wear.2006.03.013

Zois, D., Lekatou, A., & Vardavoulias, M. (2009). A microstructure and mechanical property investigation on thermally sprayed nanostructured ceramic coatings before and after a sintering treatment. *Surface and Coatings Technology*, *204*(1-2), 15–27. doi:10.1016/j.surfcoat.2009.06.013

Chapter 10
Thermal Spray Coatings for Erosion-Corrosion Protection

Keshavamurthy R.
Dayananda Sagar College of Engineering, India

Naveena B. E.
Dayananda Sagar College of Engineering, India

Sekhar N.
Dayananda Sagar College of Engineering, India

ABSTRACT

This chapter discuss extensively on various thermal spray coatings for erosion-corrosion protection of industrial components, feedstock materials and their characteristics. Various coating removal mechanisms in thermal spray coatings have been presented. Influence of process variables, test parameters, coating thickness and working temperature on performance of thermal spray coatings for erosion-corrosion have been discussed comprehensively. Further, the concern of erodent's suspended in fluids, disintegrating and resulting in significant wear of the surfaces due to erosion-corrosion phenomenon also reported here.

INTRODUCTION

Friction, wear, corrosion, erosion, fatigue, oxidation are all surface phenomenon. Hence, it is of utmost importance that one focuses on improving the surface properties of the materials. This involves surface modification to tailor make the surface properties. The various industrial techniques of surface modifications mainly used are hot dipping, thermal spraying, heat treatment, electro less plating, electro plating, chemical/physical vapour depositions and weld overlaying. Among these techniques, thermal spray methods have been gaining widespread importance in order to produce products which will meet requirements of all surface phenomena and to ensure competitiveness in the market (Fauchais et al, 2014; Sidhu, 2005).

In this technique, the coating is formed in layers by accelerating fine molten, semi-molten or softened particles towards the component surface, where they rapidly flatten and solidify on impact. Upon impact, a mechanical interlocking takes place with the target surface. With succeeding striking particles,

DOI: 10.4018/978-1-5225-4194-3.ch010

depositing platelets called splats. With various layers of these splats forming the thickness of coating. Figure 1 presents the principle of a thermal spray process.

Feedstock material is usually introduced in powder, wire or rod form. The actual material can be anything between low melting plastics and refractory materials, e.g., oxide ceramics, as long as the material has a melting point, a stable molten state, and does not evaporate or excessively react with the surrounding atmosphere. The material is melted with the heat produced by an electric arc, fuel combustion or an ionized gas, i.e. plasma, and particles are accelerated towards target surface by the high velocity stream of process gases (Pawlowski, 2008; Saber-Samandari & Berndt, 2010).

The choice of the deposition process (flame spraying, high velocity oxy fuel flames (HVOF), high velocity air-fuel flames (HVAF), detonation guns, low pressure/vacuum plasma spraying, atmospheric plasma spraying, arc spaying, high/low pressure cold spraying) highly dependent on the desired application of coating and cost. Feedstock material, its form and set of deposition process parameters dictate directly the coating properties. Therefore, the correct choice of the spray process as well as respective parameters (type of fuel gas, velocity of particle, standoff distance and so on) is very significant in order to obtain low porosity and dense coatings with a high bond strength, as generally preferred for erosion-corrosion resistance and consequently, to enlarge the useful life in service of the components (Oska et al, 2011; Talib et al., 2003; Hermanek, 2001; Kembaiyan & Kashavan 1995). Table 1 shows the comparison of various thermal spray processes and coating characteristics.

Figure 1. Scheme of thermal spraying process

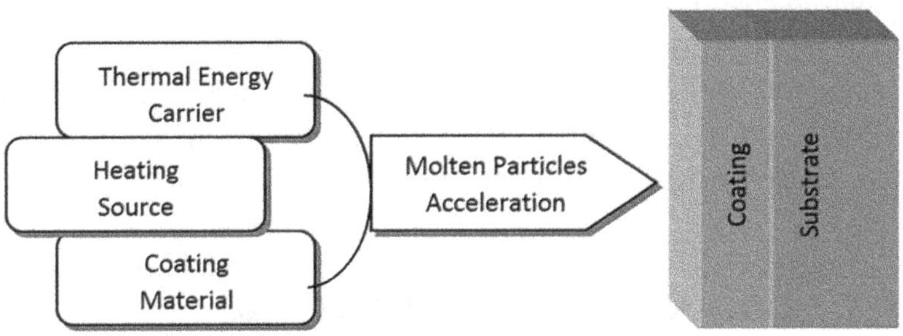

Table 1. Comparison of different thermal spray processes

Type of System	Temperature of the Heat Source (°C)	Particle Velocity (m/s)	Oxide Content (%)	Porosity (%)	Adhesion (MPa)
Flame Spraying	2500-3000	30-180	4-6	10-15	8
Electric Wire Arc	4000-6000	100-240	0.5-3	10-20	12
Detonation Gun	3000-4000	800-1200	1-5	1-2	>70
HVOF	2500-3000	500-800	0.2	1-2	>70
HVAF	1400-2000	600-1200	0.2	0-0.2	>70
Plasma Spraying	5500-12000	200-600	1-3	1.0 - 10	10-70
Cold Spray	200-1000	300-1000	0	<0.5	5->70

Talib et al., 2003; Vuoristo, 2014; Matikainen et al., 2015.

THERMAL SPRAY PROCESSES

Different processes of thermal spray, for example, detonation-gun, plasma and HVOF spraying techniques are generally utilized to combat erosion-corrosion actions during service at high temperature. One of the most promising methods among the available thermal spray processes is HVOF. Due to higher velocities of processing gas and flame temperatures lower than many other thermal spray techniques, makes it enable to produce desired high quality wear resistant cermet coatings (e.g., WC-based and Cr_3C_2-NiCr). These coatings usually posses very dense, low oxidization content and porosity, with excellent hardness and resistance to abrasion (Matikainen et al, 2015). Thus, problem of achieving erosion-corrosion resistance of coatings is fulfilled by HVOF process than the other available techniques of thermal sprays like plasma, arc or flame spraying.

HVOF process has the following additional advantages like high deposition efficiency, most convenient process for onsite applications with feedstock powder having low/high melting temperatures.

In HVOF, continuous combustion of oxygen with different gaseous fuel including hydrogen, propane, ethene and even liquid kerosene in a combustion chamber and expansion of gases are fed through a small orifice in a convergent/divergent nozzle to produce a supersonic gas jet with very high particle speeds. Powder feedstock to be coated via HVOF are injected axially into the expanding hot gases with carrier gas like nitrogen, helium etc.. Where they are impelled forward, heated and accelerated onto a surface at much higher rates to form an extremely dense and well bonded coating. Several torch passes creates a coating which will ensure low porosity, similar to the initial powder chemistry (Sidhu et al, 2005, Wank, n.d.; Gorlach, 2008).

Generally, the erosion-corrosion protection of thermal spray coatings is attributable to the chemistry and morphology of sprayed coatings. Coating resistance to erosion-corrosion is enhanced by the qualities like good adhesion, high density, small grain size and low porosity without cracks. This means that higher the value of surface roughness results in higher wear and corrosion attack due to higher contact surface area. (Pawlowski, 2008; El Rayes et al., 2013; Heimann & Lehmann, 2008; Fauchais & Vardelle, 2012; Chatha et al., 2012; Berget et al., 2007) .

EROSION, CORROSION, AND THEIR INTERACTIONS

Erosion

Erosion is defined as a degradation phenomenon affecting the target surface due to the impact of solid particles entrapped in a liquid medium. High strain rates are originated during the above impact leading to material loss and degradation.

The following three categories of influencing factors indicate the erosion rate and impact on target surface (Finnie, 1995).

- Operating parameters like velocity of erodent, angle of impingement, concentration of particles, viscosity, density and temperature of the carrier fluid are the some of influencing factors.
- Shape, size, composition, hardness of the erodent also play an important role.
- Mechanical properties of the target material like hardness, fracture toughness, yield and ultimate strength, ductility, microstructure and composition are the influencing parameters.

Other than the above mentioned factors, it has been reported that restitution coefficient plays an important role. This is a complex function of different mechanical properties of target and erodent materials. Here operating parameters such as impact velocity and angle of impingement are the most significant influencing factors.

The following two equations indicate the relationship between erosion rate E (in dimensionless units) impacting particles velocity, v, angle of impingent particles, α and their size, d (Taylor et al, 1987):

$$E = kv^{n}f(\alpha) \quad (1)$$

$$E = kd^{n} \quad (2)$$

where, k and n are constants related to materials and f(α) is a function dependent on the angle of impingent particles. “n” for bulk stainless steel (grade 304) and plasma-sprayed NiCrAl alloy is 0.63 as established by Taylor et. al (Taylor et al, 1987).

There are two types of erosion due to impingement of entrapped solids in liquid medium based on ductile and brittle mechanisms. Cutting or ploughing via sliding abrasion (e.g. along pipeline straights and risers) are the material removal mechanisms during ductile erosion. Here the angle of impingement is small, typically less than 30^0. Radial crack formations via impact abrasion (e.g. pipeline elbows, bends and protrusions) occur during brittle erosion at angles of impingement near normal to the surface.

Generally, ceramic hard coatings possess low fracture toughness and are capable of resisting ductile erosion mechanisms but fail due to brittle erosion at impingement angles above 60^0.

Whereas, the tough ductile phase of metal alloys is capable of elastic deformation and are therefore resilient to brittle erosion mechanisms. This is due to the fact it has low co-efficient of friction, resisting ductile erosion mechanisms (Bergmann & Vincenzi, 2011).

Figure 2. Ductile Erosion Mechanism and Brittle Erosion Mechanism demonstrate the mechanisms of erosion acting on a coating surface

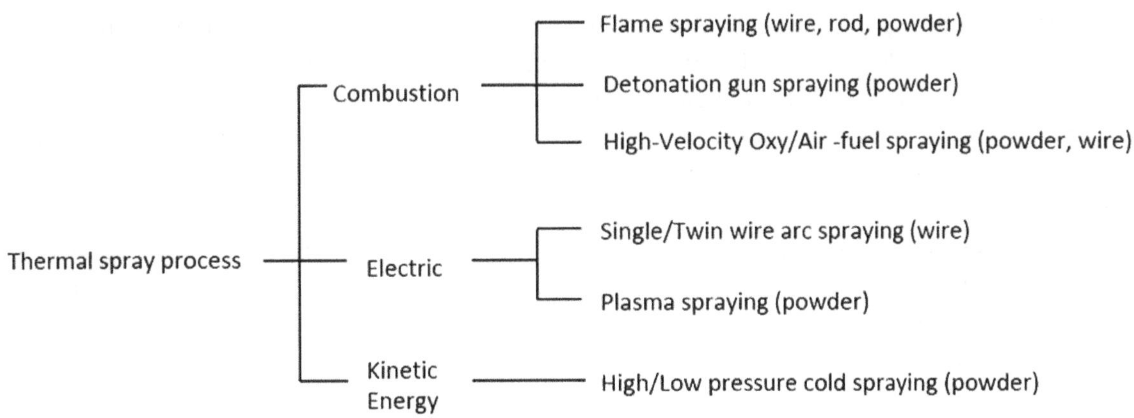

Corrosion

Corrosion is the deterioration (degradation) of a metal by an electrochemical reaction with its environment (Shreir, 1963b; Trethewey & Chamberlain, 1995). The phenomenon of corrosion produces corrosive attacks are two types namely (i) general corrosion (30% of failures) and (ii) localized corrosion (70% of failures).

During localized corrosion process the galvanic corrosion occurs when two different metals are in contact with each other in a conductive solution (electrolyte), the more cathodic metal is unaffected, whereas the more anodic metal is being corroded. Here, electrolyte and relative surface contact area play major roles. The smaller the ratio of anodic to cathodic area leads to sever anodic metal corrosion.

The other three types of corrosion are:

1. Inter-granular corrosion due to chemical element is depleted while bulk material manufacturing (heat treatment) or coating.
2. The pitting corrosion due to localized corrosion on the surface with pit formation
3. Transgranular corrosion due to high static tensile stress in the presence of a corrosive environment. This type of corrosion is influenced by the coating material and its microstructure (Fauchais & Vardelle, 2012)

For preventing corrosion of numerous materials, thermal spray coatings are extensively used with additional benefits of wear resistance properties. There is a very wide range of coating materials that can be sprayed.

Corrosion resistance can be improved by the following three types of coatings:

- **Anodic Coatings:** This is also known as sacrificial protection or cathodic protection. Here, the substrate is referred as cathode and the coating is the sacrificial corroding anode. For example Zn or Al, ZnAl and AlMg alloys on iron or steel, the thicker the coating around 250 μm will be the longer the protection. But above coating materials do not illustrate satisfactory wear resistance for most applications.
- **Cathodic Coatings:** In cathodic coatings the coating metal is cathode on the substrate. Nickel alloy or stainless steel is cathodic coating materials on steel can ensure good corrosion as well as wear resistances. But a major constraint in such coatings is that they must provide a complete protection to the substrate from the corrosive environment. If any defect in their microstructure like porosity may allow the electrolyte to reach the substrate. This will result in spalling of the coating due to accelerated corrosion.
- **Neutral Coatings:** Usually, even porous ceramics coatings of alumina or chromium oxide, will not accelerate the corrosion of the substrate. To prevent coating separation, corrosion of the substrate interface with the coating should be avoided. Resealing of the coatings is recommended along with dense and thick thermal spray coatings.

Increased rate of wearing may take place due to mechanical weakness on the surface caused by corrosion or oxidation and also due to the abrasive particles generated by the corrosion products and oxide particles. Stress and corrosion have a combined effect called the stress corrosion failure. Oxygen, sulfur, carbon, nitrogen or flux result in the formation of oxidized, sulfidized, carburized, nitride or slag layer on the surface at higher temperatures. Time and temperature are the significant factors controlling the severity and rate of high temperature corrosive attack.

Along with that other predominant influence factors on the corrosion rate are pH values of corrosion medium, concentration of chloride ion and time. As pH value increases, the corrosion rate decreases. Also, there is a rise in corrosion rate as increase in concentration of chloride ion.

The loss of material due to corrosion was determined by electrochemical measurements. The corrosion current density is used to determine the rate of metal dissolution in that particular environment by using Faraday's Law of Electrolysis (Equation 3). Thus monitoring of current density provides a powerful means of monitoring the corrosion rates of metallic components (Shreir, 1963a).

$$m/S = M.i.t/z.F \ (gcm^{-2}) \qquad (3)$$

where,

m = Mass of metal corroded (g)
M = Molecular mass ($gmole^{-1}$)
F = Faraday's constant 96500, coulombs (1 A~1 coulomb s^{-1})
I = Current (A)
i = Current density (Acm^{-2}) and i= I/S
t = Time (s)
z = Number of electrons involved in the corrosion reaction.
S = Area of metal involved (cm^2)

Erosion-Corrosion Interactions (Synergy)

The rate of deterioration of metal caused by the combined action of mechanical erosion and electrochemical attack is defined as erosion-corrosion acceleration. This is also known as synergy effect. This combined affect will cause greater damage and higher metal loss rate than erosion or corrosion alone. Here, not only the protective coating gets damaged by erosion but also substrate gets damaged due to corrosion.

Mathematical representation of a typical erosion-corrosion (T) attack is:

$$T = E + C + S \qquad (4)$$

E = Loss of material due to pure erosion mechanism,
C = Loss of material due to corrosion mechanism and
S = Synergy.

This leads to mechanisms of additive nature like spalling with intergranular corrosion or oxidative wear due to brittle fracture. Therefore, synergy can be represented by equations (5).

$$S = T - (E + C) \tag{5}$$

ΔE and ΔC are major two parts of synergy. Here, ΔE is the corrosion enhanced erosion (also called enhanced erosion loss due to corrosion: ΔE_C) and ΔC is the erosion-enhanced corrosion (also called enhanced corrosion loss due to erosion: ΔC_E), as in equation (6).

$$S = \Delta E + \Delta C \tag{6}$$

Protective corrosion film can be mechanically stripped by erosion resulting in surfaces of fresh reactive corrosion, i.e., creating ΔC, which depends on the repassivation rate and integrity of the film formed. Some more possible ΔC mechanisms are: (1) local acidification at erosion locations, accelerating corrosion rates and preventing oxide film formation; (2) high turbulence levels increased mass transport; (3) fatigue strength reduction by corrosion and (4) particle impact causing surface roughening of the specimen which enhanced mass transfer effects with increased corrosion rate. In addition Corrosion-enhanced erosion mechanisms (ΔE) are also possible, like: (1) corrosion processes removes the work hardened surfaces there by exposing the underlying base metal to erosion mechanisms; (2) corrosive attack at grain boundaries leading to grain loosening and eventual removal. (3) micropitting causes an increase in the number of stress concentration defects and (4) stress corrosion cracking causing detachment of plastically deformed flakes on the metal surface. The above-mentioned mechanisms are dominant then a positive synergy can be created but not always resulting in negative synergy. There are different mechanism reduces erosion rates (−ΔE) like (1) corrosion mechanisms increases the work hardening, (2) shot-peening effects by high velocity impinging particles (3) the presence adherent corrosion films (4) the speed of crack propagation is retarded by blunting of the crack tips due to lateral dissolution. A reduction in corrosion rate (−ΔC) occurs due to formation of a passive film, scaling and film growth would reduce corrosion rates dramatically.

Zero synergy occurs when E/T ratio is equal to 0.8. There is a negative effect, reducing overall mass loss when E/T> 0.8 due to a continuous oxide film formation protecting the surface. When E/T< 0.83 erosion and corrosion are highly interactive along with active corroding surface at denuded film sites (El Rayes et al., 2013; Liu et al., 2016; Tan et al, 2005; Saha et al., 2011; Souza & Neville, 2007).

FEEDSTOCK MATERIALS

Selection of feedstock material for coating against erosion-corrosion requires a careful study and consideration, as erosion-corrosion mechanisms are very complex in nature depending on chemical and physical interaction with its environments.

Single-Phase Materials

Metals

Erosion-corrosion protection structural components have been successfully sprayed with pure metals and metal alloys, including W, Mo, Zn, Co, Ni, Al, Fe, stainless steel, Ta, Ti, NiCr alloys, Fe-based alloys, cobalt-base Stellites, cobalt/nickel-base Tribaloys, and nickel aluminium bronze alloys. Sprayed alloys

Table 2. List of feedstock materials that are commonly used for Erosion-corrosion protection

Classification of Material	Examples
Metals and alloys	Al, Ni, Zn, W, Mo, Re, Nb, Co, Fe, stainless steel
Intermetallics	NiAl, TiAl, Ti_3Al, Ni_3Al, $MoSi_2$
Composites	WC–Co, Cr_3C_2–NiCr, TiC–NiCr, Al_2O_3-TiO_2
Ceramic oxides	Al_2O_3, ZrO_2, TiO_2, Cr_2O_3, MgO
Ceramic carbides	Cr_3C_2, WC, TiC, Mo_2C, SiC
Ceramic nitrides	TiN, Si_3N_4

Saber-Samandri & Berndt, 2010; El Rayes et al., 2013.

have advantages due to their similarity to many base metals requiring repair, their high strength, and their corrosion, wear, and/or oxidation resistance (Tan et al.,2005; Komaki et al, 2010).

Applications include turbine blades, combustors, bridges, marine, automotive, seawater handling systems, petrochemical industries, mining, agricultural equipments and other corrosion prone infrastructure.

Ceramics

Good thermal, electrical properties, resistant to corrosion, erosion and wear than metals in high-temperature environments make ceramic coatings more attractive.

Most forms of ceramics, including metallic oxides such as Al_2O_3, stabilized ZrO_2, TiO_2, Cr_2O_3, and MgO; carbides such as Cr_3C_2, WC, TiC, Mo_2C, and SiC; nitrides such as TiN and Si_3N_4 can be thermally sprayed for erosion-corrosion protection by providing suitable bond coats (Zavareh et al., 2014).

Applications include modern high performance machinery used in power plants, marine, mining, aerospace, petroleum etc.,

Composite Materials

Cermet Materials

Under aggressive erosion-corrosion condition metallic and ceramic materials fail and hence there is a strong necessity for developing alternate surface engineering solutions for implementation to resist the possible damages. Cermets are composites of ceramic and metal and are generally used to face severe erosion-corrosion conditions and there are a number of alternative thermally sprayed coatings. Ceramic hard face particles embedded in a ductile metal matrix is known as a cermet coating. High erosion resistance is offered by ceramic particles and more ductility is provided by a metallic binder phase than a pure ceramic coating.

Metallic components are protected by cermets in all respects including cost, bond strength, reduced friction as well as erosion-corrosion resistances. Generally, WC or CrC particles embedded in a metal (pure metal or alloys of Ni, Cr and Co) bound cermet coatings are used for erosion-corrosion protection. Normally, usage of Ni is avoided as a metal matrix due to its poor corrosion resistance but on the other hand Cr is added to the metallic binder as it improves the corrosive properties.

Combinations of WC-Co, Cr_3C_2-NiCr, CrC-NiCr, WC-CoCr, WC-Ni, TiC-NiCr and Fe-CrAlY–Cr_3C_2 cermets influence the erosion-corrosion resistance and decrease the coefficient of friction between the various sliding components and hence commonly preferred in thermal spraying process. Out of the above combinations, Cr_3C_2-NiCr is preferred for use at high temperature up to 850 °C to combat erosion-corrosion as cermet coatings. The size of ceramic particles, its phase and metallic binder composition determine the erosion-corrosion performance of thermally sprayed coatings (El Rayes et al., 2013; Berget et al., 2007; Toma et al., 2001)

Most of the oil and gas industries, manufacturing systems and machinery involving fluid/slurry flowing corrosive media which contain particles like sand etc. use cermet-based coatings as the material/ substrate damage is caused due to degradation mechanisms of erosion and corrosion (Kunioshi et al, 2005).

PROTECTIVE COATINGS AGAINST EROSION-CORROSION

There are many literature references available regarding the erosion-corrosion behavior of polymeric and metallic based coatings. Ductile coatings fail due to plastic deformation and brittle coatings fail due to effects of cracking and chipping mechanism. Coatings usually have both ductile and brittle constituents the erosion behavior is predominant major factor. Abrasive materials in a corrosive medium affect certain coatings and their mass transfer between coating/fluid and substrate/fluid can affect the erosion-corrosion behavior of the coating due to excessive oxide formation. This can also lead to the effects of permeability of liquid through porosity ending up in galvanic corrosion between coating and substrate. Here the accelerated degradation of material occurs due to process of erosion.

Bardal et al. (1995; Bjordal et al., 1995a; Bjordal et al. 1995b) have reported their findings during erosion-corrosion test on metallic and cermet coatings, deposited by chemical vapour deposition and HVOF. These tests have been conducted using special test rig. And they have concluded that corrosion of the metallic binder in the cermet coatings undermined the carbide particles, leading to enhanced material loss.

Espallargas et al.(2008a, 2008b) have reported in their trail on HVOF Cr_3C_2–NiCr and WC–Ni coatings that corrosion occurred at the metallic matrix, beginning at the interface between metal and ceramic thus leaving carbides unbounded as shown in Figure 3. This dislodges ceramic particles resulting in high material loss due to corrosion.

Yan Liu et al. (2016) have studied the erosion–corrosion mechanism of CeO_2-modified WC-12Co HVOF coating in saline water. The effect of corrosion on erosion and erosion on corrosion played an important role here. There was also an appreciable change in the roughness value on the polished coating surface due to the progress of erosion–corrosion process forming microelectrodes with an accelerated corrosion mass loss. Corrosion rate is accelerated due to presence of corrosive solution in corrosion pits in the Co binder phase resulting in WC particles lost support from the binder phase. This further increase the erosion rate and damages.

The following mechanisms occurred during the above erosion–corrosion process. (1) Microcracks developed by corrosion pits and abrasives action. (2) The density of corrosion pits increased leading to additional generation of microcracks at the edges of WC particles. (3) Expansion of adjacent corrosion pits making way to larger corrosion pit.

Figure 3. Scheme of the corrosion mechanism of a thermal spray coating during the erosion–corrosion process
Espallargas, 2008b.

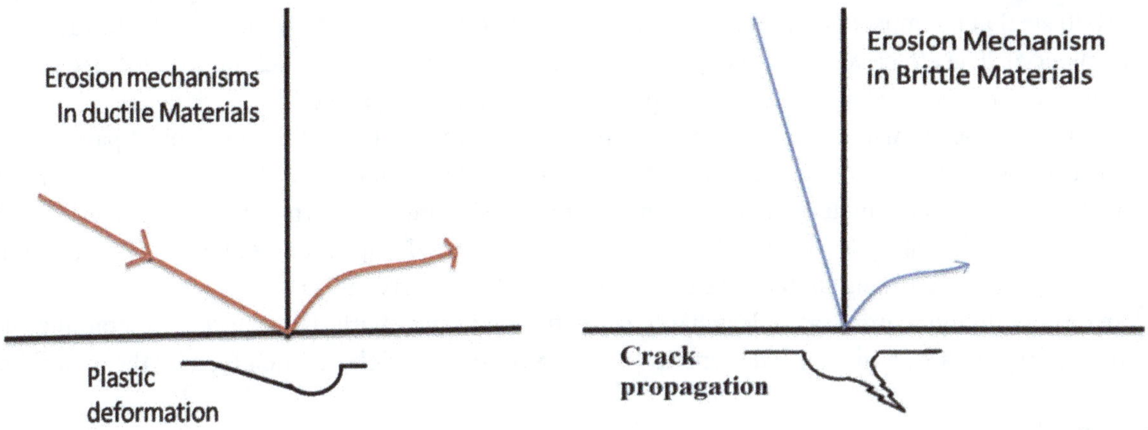

The propagation of microcracks over WC particles and their forward expansion over the coating surface or a deeper position resulting in falling of top layer of coating and exposing fresh surfaces beneath.

FACTORS INFLUENCING EROSION-CORROSION

Complex process of erosion-corrosion is influenced by a number of factors and has been suggested by many studies and their combination (Figure 4) suggests the following:

- **Particle Movement:** Impact velocity, impact angle, kinetic energy of the particles, interference between incoming and rebounding particles
- **Erosive Particle:** Hardness, size, shape, concentration and density
- **Coating Material:** Hardness, tensile strength, ductility and metallurgical structure
- **Chemistry of Water in Use:** pH, salinity, temperature, dissolved oxygen and corrosivity
- **The Variables of Flow:** Velocity of fluid, flow diameter, fitting geometry, upstream influences, and void fraction (if two-phase flow). (Humphrey, 1990; Heitz, 1991; Levy, 1995; Neville, 1995).

Erosion rate is most drastically affected by velocity of impact of the erodent particles. The erosion rate (E_R) of a material is often expressed in the form

$$E_R \propto u^n \tag{7}$$

where, 'u' is impact velocity and 'n' is the velocity exponent.

As the impact velocity of erodent particle increases, the mass loss also increases as stated in the above equation. This is due to the fact that, increased impact velocity exhibits higher energy and can cause more damage to the target surface when compared with lower velocity particles (Ramesh & Keshavamurthy, 2011).

The velocity exponent 'n' is around 2.4 for ductile metals at oblique impact conditions (Hutchings, 1992), around 2.55 for erosion under normal impact (Sundarrajan & Roy, 1997) and values for brittle materials range from 2 to 4 (Humphrey, 1990). Finnie (1960) has found the velocity exponent of 2 for SAE 1020 steel at an impact angle of 20° and remarked that the case might be absolutely different for a brittle material. The effective velocity exponent is lower and in the range of 1-2 if corrosion takes place. The velocity exponent is also influenced by particle size, impact angle etc.

Another parameter influencing the erosion rate is angle between the velocity vector of particles and the target surface. Erosion rate on the impact angle is defined by the nature of the target material. Figure 5 represents how a ductile material exhibits a maximum erosion rate at intermediate angle (e.g 15°-30°), whereas brittle material exhibits its highest erosion rate at normal impact angle, i.e. at 90° and the rate falls steadily with decreasing impact angle (Humphrey, 1990, Clark, 1992).

Particles with high hardness are less likely to shatter on impact with the surface and retaining their full kinetic energy during impact. The kinetic energy (K.E) of a particle can be given by the expression:

$$K.E. = mv^2/2 \quad (8)$$

where, 'm' - mass of a single particle and 'v' - particle velocity

For 'N' number of particles striking the surface where, $N = m_s.a/m$ where, m_s mass of sand, a- surface area, v- particles velocity

$$K.E. = m_s.a.v^2/2 \quad (9)$$

It is an established fact that, as kinetic energy of erodent increases, erosion wear rate would also increase.

In the various studies of erosion by solid particles, attention to damage occurring to the particle itself on impact has been neglected. Generally, at higher concentration of sand particles in slurry would lead to the inter collision of particles among themselves resulting in losing their surface strength and kinetic energy. This may lower the erosion rate. This fact is also observed by Naveena et al. (2017) during their study on slurry erosive wear behaviour of plasma-sprayed flyash–Al_2O_3 coatings and can be seen SEM photograph of erodent after the slurry erosion test and is shown in Figure 6 from this it can be inferred

Figure 4. Schematic diagram showing factors influencing erosion-corrosion

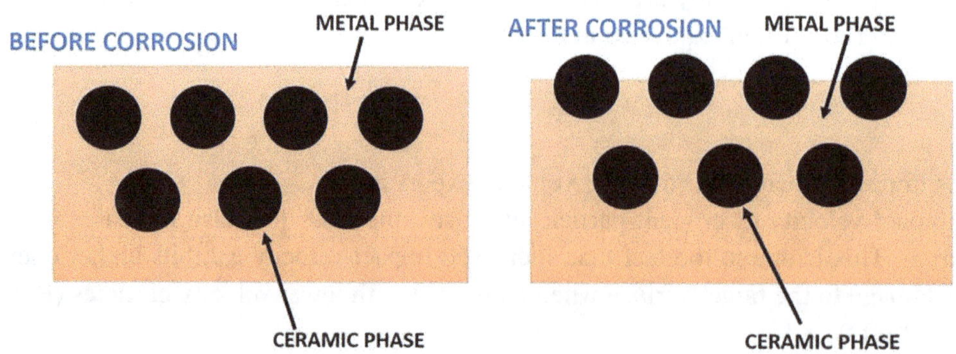

Figure 5. Graph showing the influence of impact angle on erosion rate in the case of ductile and brittle material
Humphrey, 1990.

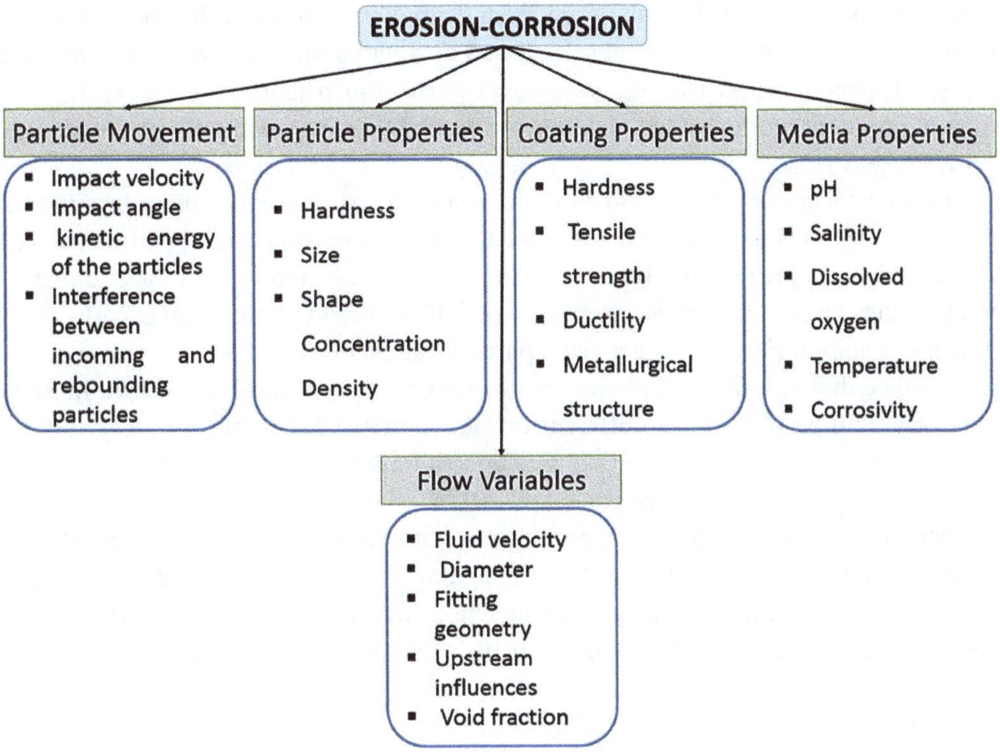

Figure 6. SEM photography of sand particles after slurry erosion test (higher concentration)
Naveena et al, 2017.

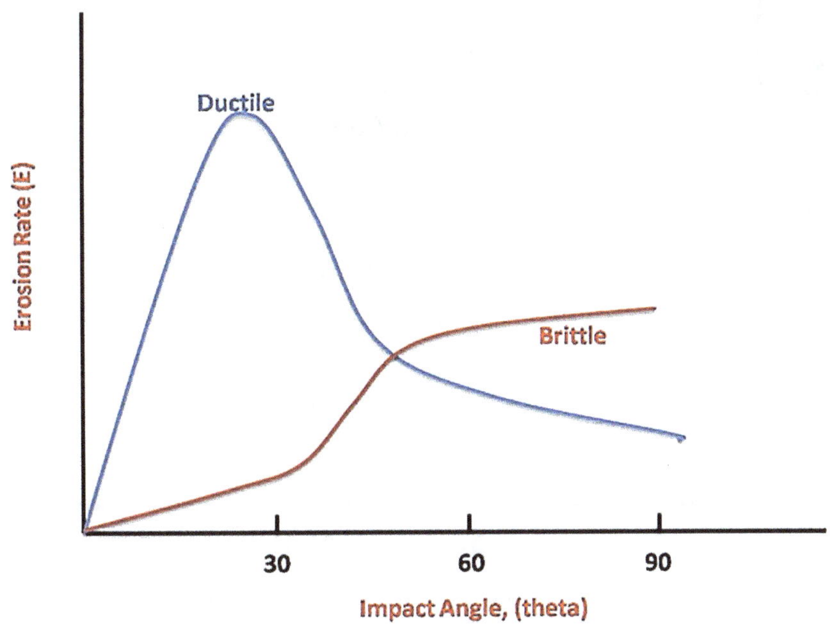

that due to inter collision most of the erodents are severely damaged and result in large-sized sand particles getting broken into smaller ones resulting in drop of weight loss.

The erosion rates of a target material largely depend on particle shape, size and hardness. Angular and larger particles are more erosive than rounder and smaller particles. It has been illustrated (Levy, 1995) that for AISI 1020 steel, the erosion rate caused by angular steel grit was four times greater than that caused by the spherical shot. The angular erodent caused much sharper, deeper craters to form with a more efficient production of extruded platelets or lips and their removal with repeated bombardment of the stream.

Particle size plays a crucial role in erosion rate of a material. Increased mass loss with increasing impinging particle size in slurry can be mainly due to the fact that small sized particles tend to deviate near the surface of the target material because of their lower mass and lower impact energy. However, bigger sized particles has higher impact energy thereby hit the target surface efficiently thus removing higher volume of material. Further, bigger sized particles are also able to produce more abrasive action on the target surface due to large area of contact and higher mass. It has been shown in Figure 7, how erosion rates vary with particle size at different impact velocity (Sundarrajan & Roy, 1997). The erosion rate increases with an increase in particle size up to a limiting size only beyond which the effect of particle size becomes unimportant (Naveena et al., 2017).

Particle shape often plays a key role in erosion rate and has been studied by several investigators. Wear rates strongly depend on particle shape, with angular particles causing greater wear than rounded particles. As the angularity of the erodent particle increases, the erosion rate increases and in addition, the erosion rate-impact angle behavior shifts more towards a ductile response (Figure 8) (Hutchings, 1992).

Figure 7. Effect of erosion rate with particle size at different impact velocities
Sundarrajan & Roy, 1997.

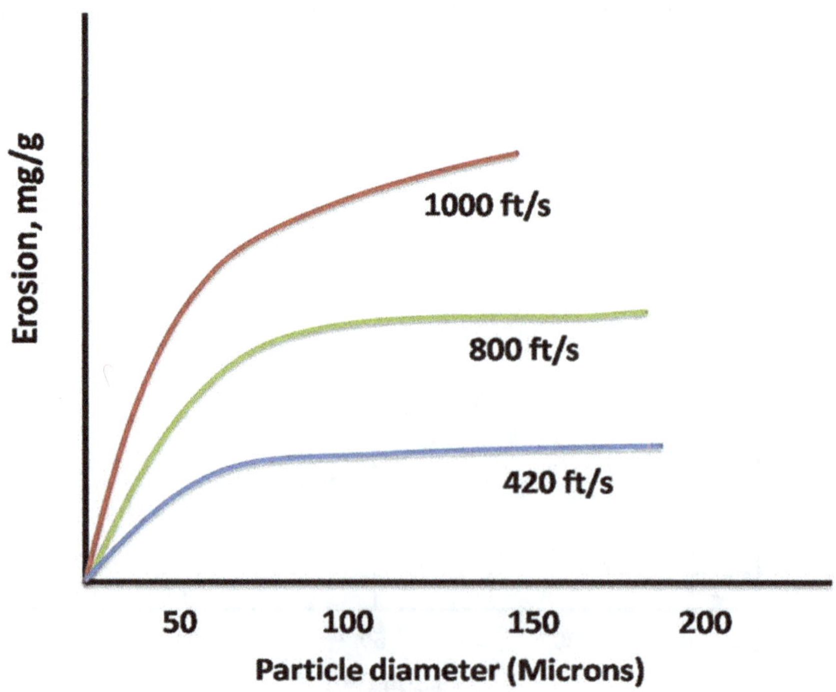

Figure 8. Influence of impact angle on erosion rate in the case of spherical and angular particles
Hutchings, 1992.

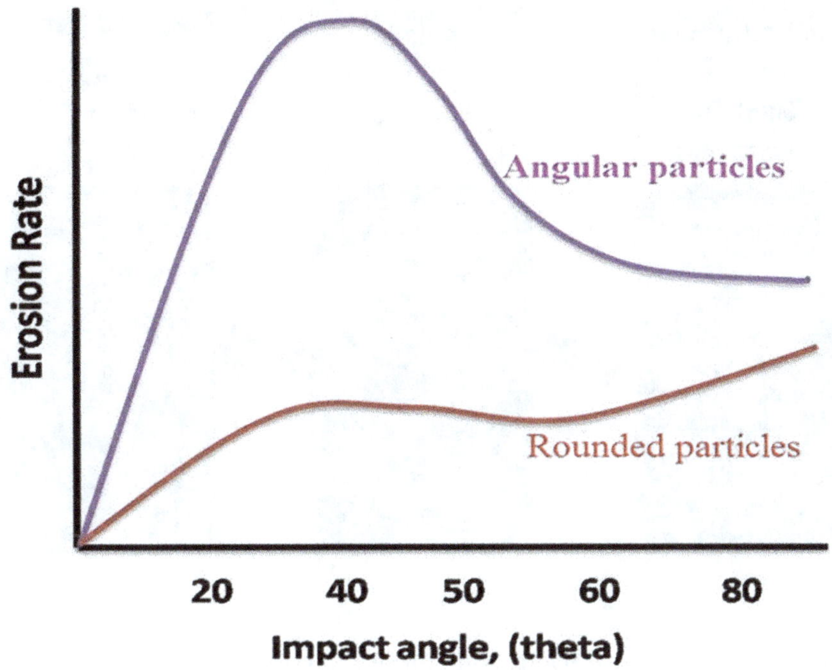

The ratio of erodent particle hardness to target surface hardness (Hp/Ht) is to be considered for evaluating the erosion behaviour of the coatings. The following interpretations have been arrived for the following ratios of Hp/Ht:

1. **When Hp/Ht< 1.2:** Less erosion damages.
2. **When Hp/Ht> 1.2:** Brittle materials where material removal by the intersection of radial and lateral
3. **When Hp/Ht< 1:** Particles cannot penetrate into the target to generate lateral cracks. (Sundarrajan & Roy, 1997; Wierderhorn & Hockey, 1983)

Increased concentration of erodent particles in the slurry leads to higher mass loss. This is due to the fact that, the higher the concentration of sand particles in the slurry leading to higher probability of particle impingement on the target material irrespective of the particle size. This will in turn resulting in increased mass loss (Ramesh & Keshavamurthy, 2011; Naveena et al, 2017).

Figure 9 illustrates the number of influencing factors and interdependencies. These influence the intrinsic coating properties. The performance of thermal spray coatings during service is dependent on several characteristics such as coating thickness, surface finish (roughness), microstructure (grain morphology and orientation, defects such as density and porosity, heterogeneity, micro-defects such as dislocations etc.) and mechanical properties (micro-hardness, tensile strength, fracture strength, toughness, wear resistance, elastic moduli), service conditions etc. Premature failure of the coatings can occur due to interactions between service conditions (such as abrasion, thermal cycling, corrosion,

Figure 9. Illustration of the number of influencing factors and their interdependencies
Oksa et al, 2011.

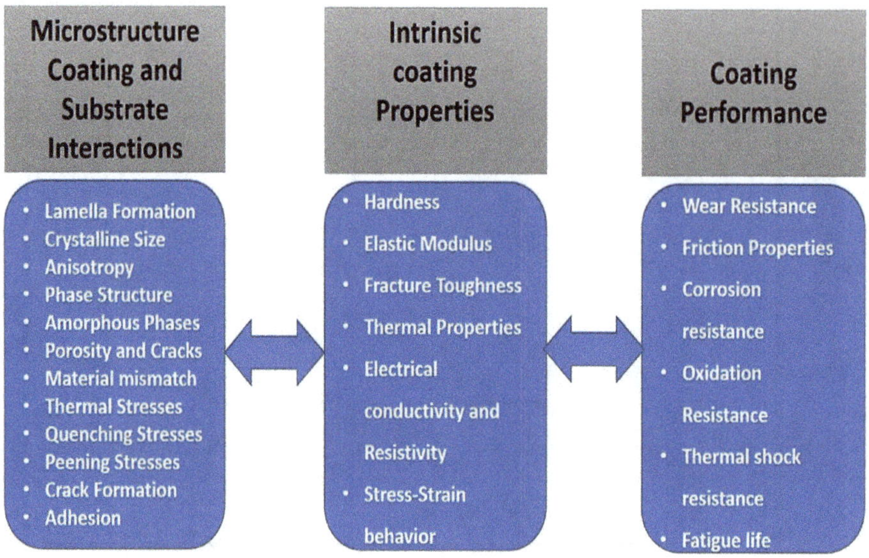

the environment) and coating defects. Defects that result in coating failure include: i) disbonds at the coating/substrate interface and inter-splat boundaries; ii) the occurrence of pores within the coating; iii) formation of oxides within the coating; iv) residual stress build-up (Souza & Neville, 2007).

Cermets are potentially attractive options for erosion-corrosion resistance. The dominant mechanism of erosion depends on the scale of individual particle contacts. If the erosive particles are small enough, wear can occur by preferential erosion of the metallic binder phase, leading to undercutting and eventual removal of whole carbide grains. If the binder regions are too small to allow this type of localised attack, then overall erosion of the carbide and binder occurs, usually by ductile mechanisms if the erodent particles are small (Hutchings, 1992). Levy (1991) has separated cermets into two different categories such as low carbide content and high carbide content and discussed the mechanisms of material losses.

The general concept of erosion-corrosion involves two interdependent phenomena: the fluid mechanics of the solid-fluid two-phase flow (or momentum exchange between the solid and liquid phase) and the wear of materials by collisions of abrasive particles on the metallic surface (Humphrey, 1990). In a fluid stream the material loss of a target surface is dependent on number of striking particles, their direction relative to the surface and their velocity. The quantity is largely determined by the flow conditions, in particular pump components and fluid handling units, where the flow direction changes fairly rapidly and that the amount of material removed is related to the work done or energy dissipated by the impacting particles (Finnie, 1960). Moreover, several factors affecting the rate of material loss are viscosity of liquid, density, size of particle, target shape etc.

The velocity of particle largely depends on carrier fluid viscosity and particle size. It has been reported that the impact velocities of impinging particles are much less than the flow velocities.

The ratio of the number of particles striking unit area of the surface in unit time and the number of particles contained within the volume of suspension swept by that area in unit time is defined as striking or collision efficiency.

Low values of collision efficiency will result in a low value of erosion rate. Also a liquid of high viscosity will exert a greater drag force on a particle than a liquid of low viscosity, which in turn has more effect in moving the particles laterally. As a consequence, at high viscosities, relatively few particles will be able to collide with the contacting surface. The impact velocity as reported depends upon interactions between particles, liquid and the specimen and the prevailing flow regime (Humphrey, 1990; Clark, 1992).

Particle erosion can influence the corrosion rate by different mechanisms, such as: increased mass transport by high turbulence levels caused by surface roughening (Poulson, 1990), removal of protective oxide or corrosion product films exposing fresh reactive material and local acidification in erosion pits that accelerates corrosion rates and prohibits film formation. Similarly, corrosion can affect the erosion rate by removing work hardened surface, by preferentially attacking at grain boundaries resulting in loosened and eventual lost grains, by lowering the fatigue strength of the material and by introducing micro-pits that act as stress-concentrating defects (Wood & Fry, 1989). However, in situation where hard oxides can reduce erosion at elevated temperatures, corrosion can also be considered to "inhibit" erosion (Stack & Bray, 1995).

Mechanical behavior of ductile-to-brittle transitions can affect the mechanisms of mass removal from the surface under the influence of variations of temperature are to be considered as significant factor in wear and corrosion applications. At higher temperature, ductile wear mechanisms like cutting and plastic deformation are more sever and the corrosion processes at the surface are aggressive. During low temperatures, micro-cracks at the surface are favored and

leading to material loss in the form of flakes. (Wellmann & Nicholls, 2004; Mesa et al., 2003; Uusitalo et al., 2002).

There is a linear variation of mass loss to duration or exposure time up to certain period. Beyond this, continuous impinging action of sand particles is the reason for the work hardening of target surface resulting in the stability. Also, due to continuous mechanical action the suspended sand particles lose their strength and sharpness.

SOLUTIONS TO INDUSTRIAL CONCERNS

Corrosion due to electrochemical actions and mechanical wear processes are the main root causes of failure in industrial application like petroleum industries, power plants, marine, mineral processing and mining and chemical industries, due to an interaction of erosion and corrosion. Therefore there arises an urgent need of materials developed with superior erosion-corrosion resistance in specific environmental conditions.

Power Plant Boilers

Power station boiler-walls, pipes, reheater, economizer, ash removal system, and other auxiliary parts of thermal plants undergo erosion–corrosion issues which have relevance to the reliability and running cost of the installation. The environment of the furnaces is highly aggressive with corrosive deposits and erosive process due to the presence of ash particles.

During recent years, several researchers are continuously working on development of suitable structural materials and coatings to withstand these sever working conditions.

One of the possible approaches to the above problems is to provide wear and corrosion resistant coatings with good thermal properties, like detonation gun, plasma or HVOF sprayed nickel based or cermet alloys. Generally carbide type of cermet coatings such as Cr_3C_2-NiCr, have shown an exceptional performance in various industrial areas such as aeronautics, spaceflight, oil, metallurgy, and machinery industry (Szymański et al., 2015; Wang, 1996; Mishra et al., 2013).

Marine Pump

A case study of failure due to erosion-corrosion has been reported on marine pump impeller made out of 304 stainless steel after two years of exposure to high-velocity sand and saline water at Qinshan, China (Wang et al, 2011).

HVOF WC–Co–Cr, nickel 625 alloy coatings, Al bronze, Ni–Al alloy coatings and Ni–Cr–Mo–Si–B thermal sprayed coatings are suggested in marine applications where erosion-corrosion resistances are required. This involves high cost. Therefore, recently new sort of HVOF-sprayed Fe-based amorphous metallic coatings (FeCrMoMnWBCSi) have been suggested to bridge the gap between cost and functional surface requirements (Wang et al., 2011; Zhao et al., 2010).

Petroleum Industry

There are many instances of petroleum industrial application where in aggressive conditions of erosion and corrosion has to be combated. Therefore, material selection is very crucial. The critical elements involved are system operation and equipment including piping systems and their accessories, like valves, piping items etc. Also, included are devices that propel the fluid under saline conditions.

In this industry, carbon steel is the most popular and economical material. Used in structural, upstream and downstream piping etc. This is an important material and has excellent mechanical properties. But this has a drawback of having poor corrosion resistance. Therefore, the surface has to be protected with the proper suitable protective coatings to combat aggressive environments.

Thermally sprayed Plasma/HVOF coatings of metallic materials ranging from low grade cast irons and carbon steels to the higher grades of austenitic and duplex stainless steels, oxide ceramics such as alumina, zirconia, titania, chromia, silica, yttria. To improve wear, erosion, corrosion, cavitation and fretting cermet materials are considered as good surface coating materials. Ceramics made from Al_2O_3-TiO_2 composite powders are prospectively attractive (Zavaraeh, 2014; Shrestha, 2005).

CONCLUSION

Thermal spraying as an attractive option to provide the desired properties of the surface material for use in severe erosion-corrosive environments should be based from a starting designing stage to final manufacturing process. On the manufacturing stage for a coated component, the selection of proper feedstock and spraying parameters could be of most importance.

Many thermal spray processes, like plasma, detonation-gun and HVOF spraying techniques are usually used to combat erosion-corrosion wear. From the existing thermal spray processes, the HVOF sprayed cermet coatings are showing excellent erosion-corrosion resistance abilities in petroleum industries, power plants, marine, mineral processing, mining and chemical industries.

The erosion-corrosion behavior of thermal spray coatings is greatly affected by resulting microstructure and morphology, defects such as porosity and nature of splat boundaries, anisotropy of mechanical properties, properties of interfaces between binder and reinforcement particles. Erosion and corrosion conditions such as impact velocity, impact angle, erodent size and shape, chemistry of fluid in use and fluid variables greatly influences erosion-corrosion rate.

REFERENCES

Matikainen, V., Koivuluoto, H., Milanti, A., & Vuoristo, P. (2015). Advanced coatings by novel high-kinetic thermal spray processes. Tiede & Tekniikka.

Bardal, E., Eggen, T. G., Rogne, T., & Solem, T. (1995). The erosion and corrosion properties of thermal spray and other coatings. In *Proceedings of the Int. Therm. Spray. Conf.*, Kobe, Japan.

Berget, J., Rogne, T., & Bardal, E. (2007). Erosion–corrosion properties of different WC–Co–Cr coatings deposited by the HVOF process—influence of metallic matrix composition and spray powder size distribution. *Surface and Coatings Technology*, *201*(18), 7619–7625. doi:10.1016/j.surfcoat.2007.02.032

Bjordal, M., Bardal, E., Rogne, T., & Eggen, T. G. (1995a). Erosion and corrosion properties of WC coatings and duplex stainless steel in sand-containing synthetic sea water. *Wear*, *186–187*, 508–514. doi:10.1016/0043-1648(95)07148-2

Bjordal, M., Bardal, E., Rogne, T., & Eggen, T. G. (1995b). Combined erosion and corrosion of thermal sprayed WC and CRC coatings. *Surface and Coatings Technology*, *70*(2-3), 215–220. doi:10.1016/0257-8972(94)02266-S

Carlos, P. (2011). *Bergmann, Juliane Vicenzi, Protection against Erosive Wear Using Thermal Sprayed Cermet-A Review*. Springer-Verlag Berlin Heidelberg; doi:10.1007/978-3-642-21987-0

Chatha, S. S., Sidhu, H. S., & Sidhu, B. S. (2012). Characterisation and corrosion-erosion behaviour of carbide based thermal spray coatings. *Journal of Minerals & Materials Characterization & Engineering*, *11*(6), 569–586. doi:10.4236/jmmce.2012.116041

Clark, H. (1992). The influence of the flow field in slurry erosion. *Wear*, *152*(2), 223–240. doi:10.1016/0043-1648(92)90122-O

El Rayes, M. M., Abdo, H. S., & Khalil, K. A. (2013). Erosion - Corrosion of Cermet Coating. *International Journal of Electrochemical Science*, *8(1)*, 1117–1137.

Espallargas, N., Berget, J., Guilemany, J. M., Benedetti, A. V., & Suegama, P. H. (2008a). Cr3C2–NiCr and WC–Ni thermal spray coatings as alternatives to hard chromium for erosion–corrosion resistance. *Surface and Coatings Technology*, *202*(8), 1405–1417. doi:10.1016/j.surfcoat.2007.06.048

Espallargas, N., Berget, J., Guilemany, J. M., Benedetti, A. V., & Suegama, P. H. (2008b). Cr_3C_2–NiCr and WC–Ni thermal spray coatings as alternatives to hard chromium for erosion–corrosion resistance. *Surface and Coatings Technology*, *202*(8), 1405–1417. doi:10.1016/j.surfcoat.2007.06.048

Fauchais, P., & Vardelle, A. (2012). Thermal Sprayed Coatings Used Against Corrosion and Corrosive Wear. In H. Jazi (Ed.), Advanced Plasma Spray Applications. doi:10.5772/34448

Fauchais P.L., Heberlein, J.V.R., & Boulos, M. (2014). *Thermal Spray Fundamentals: From Powder to Part*. New York: Springer Science+Business Media. doi:10.1007/978-0-387-68991-3_2

Finnie, I. (1960). Erosion of surfaces by solid particles. *Wear*, *3*(2), 87–103. doi:10.1016/0043-1648(60)90055-7

Finnie, I. (1995). Some reflections on the past and future of erosion. *Wear*, *186-187*(Part 1), 1–10. doi:10.1016/0043-1648(95)07188-1

Frank, J. (2001). *Hermanek, Thermal Spray Terminology and Company Origins, First Printing*. Materials Park, OH: ASM International.

Gorlach, I. A. (2008). Development of the Thermal Spraying Process for Anticorrosion Surface Protection. In *Proceedings of the International MultiConference of Engineers and Computer Scientists IMECS '08*, Hong Kong, March 19-21.

Heitz, E. (1991). Chemo-Mechanical Effects of Flow on Corrosion. *Corrosion*, *47*(2), 135–145. doi:10.5006/1.3585229

Humphrey, J. A. C. (1990). Fundamentals of fluid motion in erosion by solid particle impact. *Int. J. of Heat and Flow*, *11*(3), 170–195. doi:10.1016/0142-727X(90)90036-B

Hutchings, I. M. (1992). *Friction and Wear of Engineering Materials*. In Tribology.

Kembaiyan, K.T., & Keshavan, K. (1995). *Combating severe fluid erosion and corrosion of drill bits using thermal spray coatings*.

Komaki, M., Mimura, T., Kusumoto, Y., Kurahasi, R., Kouzaki, M., & Yamasaki, T. (2010). Formation of Fe-Based Amorphous Coating Films by Thermal Spraying Technique. Materials Transactions, 51(9), 1581-1585.

Kunioshi, C. T., Correa, O. V., & Ramanathan, L. V. (2005). Erosion-Oxidation Behavior of Thermal Sprayed Ni20Cr Alloy and WC and Cr_3C_2 Cermet Coatings. *Materials Research*, *8*(2), 125–129. doi:10.1590/S1516-14392005000200005

Levy, A. V. (1995). *Solid Particle Erosion and Erosion-corrosion of Materials*. USA: ASM International.

Liu, Y., Hang, Z., Chen, H., Ceng, S., Gou, G., Wang, X., ... & Wu, X. (2016). Erosion–Corrosion Property of CeO2-Modified HVOF WC-Co Coating. *Journal of Thermal Spray Technology*, *25*(4), 815-822. doi:.10.1007/s11666-016-0391-y

Mesa, D. H., Toro, A., Sinatora, A., & Tschiptschin, A. P. (2003). The effect of testing temperature on corrosion–erosion resistance of martensitic stainless steels. *Wear*, *255*(1-6), 139–145. doi:10.1016/S0043-1648(03)00096-6

Mishra, S. B., Chandra, K., & Prakash, S. (2013). Erosion–corrosion performance of NiCrAlY coating produced by plasma spray process in a coal-fired thermal power plant. *Surface and Coatings Technology, 216*, 23–34. doi:10.1016/j.surfcoat.2012.09.044

Naveena, B. E., Keshavamurthy, R., & Sekhar, N. (2017). Slurry erosive wear behaviour of plasma-sprayed flyash–Al_2O_3 coatings. In *Surface Engineering*. Taylor & Francis. doi:10.1080/02670844.2017.1288341

Neville, A. (1995). An Investigation of the Corrosion Behaviour of a Range of Engineering Materials in Marine Environments [PhD Thesis]. University of Glasgow, UK.

Neville, A., Hodgkiess, T. & Dallas, J. T, (1995). A study of the erosion-corrosion behaviour of engineering steels for marine pumping applications. *Wear,* 186-187, 497-507.

Oksa, M., Turunen, E., Suhonen, T., Varis, T., & Hannula, S.-P. (2011). Optimization and Characterization of High Velocity Oxy-fuel Sprayed Coatings: Techniques, Materials, and Applications. *Coatings, 1*(2), 17–52. doi:10.3390/coatings1010017

Pawlowski, L. (2008). *The Science and Engineering of Thermal Spray Coatings* (2nd ed.). John Wiley & Sons, Ltd. doi:10.1002/9780470754085

Poulson, B. (1990). Mass transfer from rough surfaces. *Corrosion Science, 30*(6/7), 743–746. doi:10.1016/0010-938X(90)90037-6

Ramesh, C. S., & Keshavamurthy, R. (2011). Slurry erosive wear behavior of Ni-P coated Si3N4 reinforced Al6061 composites. *Materials & Design, 32*(4), 1833–1843. doi:10.1016/j.matdes.2010.12.024

Heimann, R. B., & Lehmann, H. D. (2008). Recently Patented Work on Thermally Sprayed Coatings for Protection Against Wear and Corrosion of Engineered Structures. *Recent Patents on Materials Science, 1*(1), 41–55. doi:10.2174/1874464810801010041

Saber-Samandari, S., & Berndt, C.C. (2010). *Thermal spray coatings – A technology review,* International Heat Treatment and Surface Engineering, 4(1). doi:.10.1179/174951410X12572442577381

Saha, G. C., Khan, T. I., & Zhang, G. A. (2011). Erosion–corrosion resistance of microcrystalline and near-nanocrystalline WC–17Co high velocity oxy-fuel thermal spray coatings. *Corrosion Science, 53*(6), 2106–2114. doi:10.1016/j.corsci.2011.02.028

Shreir, L. L. (1963b). [Newnes-Butterworths, London.]. *Corrosion, 2.*

Shrestha, S., Hodgkiess, T., & Neville, A. (2005). Erosion–corrosion behaviour of high-velocity oxy-fuel Ni–Cr–Mo–Si–B coatings under high-velocity seawater jet impingement. *Wear, 259*(1-6), 208–218. doi:10.1016/j.wear.2005.01.038

Sidhu, T. S., Prakash, S., & Agrawa, R. D. (2005). Studies on the properties of high-velocity oxy–fuel thermal spray coatings for higher temperature applications. *Materials Science, 41*(6), 805–823. doi:10.1007/s11003-006-0047-z

Souza, V. A. D., & Neville, A. (2007). Aspects of microstructure on the synergy and overall material loss of thermal spray coatings in erosion–corrosion environments. *Wear*, *263*(1-6), 339–346. doi:10.1016/j.wear.2007.01.071

Stack, M., & Bray, L. (1995). M and Bray L, *Interpretation of wastage mechanisms of materials exposed to elevated temperature erosion-corrosion using erosion-corrosion maps and computer graphics. Wear*, *186-187*, 273–283. doi:10.1016/0043-1648(95)07174-1

Sundarrajan, G., & Roy, M. (1997). Solid particle erosion of metallic materials at room and elevated temperatures. *Tribology International*, *30*(5), 339–359. doi:10.1016/S0301-679X(96)00064-3

Szymański, K., Hernas, A., Moskal, G., & Myalska, H. (2015). Thermally sprayed coatings resistant to erosion and corrosion for power plant boilers - A review. *Surface and Coatings Technology*, *268*, 153–164. doi:10.1016/j.surfcoat.2014.10.046

Talib, R. J., Saad, S., Toff, M. R. M., & Hashim, H. (2003). Thermal spray coating technology – A review. *Solid State Science and Technology*, *11*(1), 109–117.

Tan, K. S., Wharton, J. A., & Wood, R. J. K. (2005). Solid particle erosion–corrosion behaviour of a novel HVOF nickel aluminium bronze coating for marine applications—correlation between mass loss and electrochemical measurements. *Wear*, *258*(1-4), 629–640. doi:10.1016/j.wear.2004.02.019

Taylor, M. L., Murphy, J. G., & King, H. W. (1987). Brittle/ductile erosion of plasma sprayed ceramic and metallic coatings. In D. L. Houck (Ed.), *Thermal Spray: Advances in Coatings Technology* (pp. 143–149). Materials Park, OH: ASM International.

Toma, D., Brandl, W., & Marginean, G. (2001). Wear and corrosion behaviour of thermally sprayed cermet coatings. *Surface and Coatings Technology*, *138*(2-3), 149–158. doi:10.1016/S0257-8972(00)01141-5

Trethewey, K. R., & Chamberlain, J. (1995). *Corrosion for Science and Engineering* (2nd ed.). England: Longman Group Ltd.

Uusitalo, M. A., Vuoristo, P. M. J., & Mäntylä, T. A. (2002). Elevated temperature erosion–corrosion of thermal sprayed coatings in chlorine containing environments. *Wear*, *252*(7-8), 586–594. doi:10.1016/S0043-1648(02)00014-5

Vuoristo, P. (2014). Thermal spray coating processes. In Comprehensive materials processing (1st ed., Vol. 4, pp. 229-276). doi:10.1016/B978-0-08-096532-1.00407-6

Wang, B. (1996). Erosion-corrosion of thermal sprayed coatings in FBC boilers. *Wear*, *199*(1), 24–32. doi:10.1016/0043-1648(96)06972-4

Wang, Y., Zheng, Y. G., Ke, W., Sun, W. H., Hou, W. L., Chang, X. C., & Wang, J. Q. (2011). Slurry erosion–corrosion behaviour of high-velocity oxy-fuel (HVOF) sprayed Fe-based amorphous metallic coatings for marine pump in sand-containing NaCl solutions. *Corrosion Science*, *53*(10), 3177–3185. doi:10.1016/j.corsci.2011.05.062

Wank, A. (n.d.). *Basics of thermal spray technology*. GmbH, Luckenbach, Germany: GTV Verschleiss-Schutz.

Wellman, R. G., & Nicholls, J. R. (2004, May). High Temperature Erosion-Oxidation Mechanisms, Maps and Models. *Wear*, *256*(9-10), 907–917. doi:10.1016/j.wear.2003.04.003

Wierderhorn, S. M., & Hockey, B. J. (1983). Effect of material parameters on the erosion resistance of brittle materials. *Materials Science*, *18*(3), 766–780. doi:10.1007/BF00745575

Wood, R. J. K. & Fry, S.A. (1989). The Synergistic Effect of Cavitation Erosion and Corrosion for Copper and Cupro-Nickel in Seawater. *Journal of Fluids Engineering*, *111*, 271-276.

Zavareh, M. A., Sarhan, A. A. D. M., Bushroa, B. A. R., & Basirun, W. J. (2014). Plasma thermal spray of ceramic oxide coating on carbon steel with enhanced wear and corrosion resistance for oil and gas applications. *Ceramics International*, *40*(9), 14267–14277. doi:10.1016/j.ceramint.2014.06.017

Zhao, W., Wang, Y., Liu, C., Dong, L., Yu, H., & Ai, H. (2010). Erosion–corrosion of thermally sprayed coatings in simulated splash zone. *Surface and Coatings Technology*, *205*(7), 2267–2272. doi:10.1016/j.surfcoat.2010.09.011

Chapter 11
Microstructural and Tribological Characteristics of Air Plasma Sprayed Alumina–Titania Coatings

Venkateshwarlu Bolleddu
VIT University, India

Vikranth Racherla
Indian Institute of Technology Kharagpur, India

Partha Pratim Bandyopadhyay
Indian Institute of Technology Kharagpur, India

ABSTRACT

Thermally sprayed coatings from nanostructured agglomerated ceramic powders have received considerable attention in recent past due to their improved performance over corresponding coatings from conventional powders. When complete melting of a fraction of agglomerated particles is prevented, unmelted and partially melted (PM) particles appear as inclusions within the fully melted matrix resulting in a bi-modal coating microstructure. In addition to uniform mixing of constituents, the enhanced performance of nanostructured coatings has also been attributed to above described bi-modal nature of the microstructure. Interestingly, even though nitrogen is cheaper and more widely available than argon, essentially all earlier works on plasma spraying of nanostructured coatings use argon as the primary plasma gas. In this chapter, the effect of critical plasma spray parameter (CPSP), which controls the specific power input to the plasma gas, on microstructural and wear characteristics of conventional and nanostructured alumina-titania (Al2O3-13wt%TiO2) coatings is studied systematically.

DOI: 10.4018/978-1-5225-4194-3.ch011

INTRODUCTION

Air plasma spraying is one of the most adaptable surface modification technologies that can be used to enhance surface properties of engineering components like wear resistance, corrosion resistance, etc. In plasma spraying, the material which to be deposited is supplied in powder form. In addition to other plasma spraying process parameters, powder characteristics also strongly influence the performance of the deposited coatings. Plasma sprayed ceramic coatings deposited using nanostructured agglomerated powders, whose synthesis and use started around a decade back, have been reported to possess superior properties over coatings obtained using corresponding conventional powders. Despite the fact that Nitrogen (N_2) is more readily available and less expensive plasma gas, Argon (Ar) was used as the primary plasma gas in all earlier studies on nanostructured coatings deposited using plasma spraying.

Air plasma sprayed alumina and alumina-titania coatings are used in several applications because of their high wear resistance, heat and thermal shock resistance, oxidation and hot corrosion resistance and electrical insulation. They are also attractive as they bond well with metallic bond coats, resist wear, solid particle erosion and have high hardness and chemical stability even at elevated temperatures. Coatings obtained from nanostructured agglomerated alumina-titania powders have been shown to provide better wear resistance over corresponding conventional coatings. *Bi-modal microstructures* in these coatings with uniformly dispersed partially melted regions in fully molten matrix are believed to be responsible for improved properties of nanostructured coatings. Nearly all earlier works, analyze the performance of nanostructured alumina-titania coatings that are deposited using argon as the primary plasma gas. In this work, we study the microstructural characteristics, mechanical properties, phase distribution, and wear performance of alumina-13wt% titania coatings deposited using conventional and nanostructured agglomerated powders. Air plasma spraying is used to deposit the coatings. While the main emphasis has been on comparing the properties of conventional and nanostructured coatings deposited using nitrogen as the primary plasma gas, for comparison, nanostructured coatings deposited using argon as the primary plasma gas were also analyzed. In nanostructured alumina-titania coatings, as critical plasma spray parameter (CPSP) - which is defined as the ratio of gun or arc power to the primary gas flow rate - controls the volume fraction of unmelted/partially melted regions, which in turn control the properties of coatings. In this study, coatings deposited with nitrogen as well as argon as primary plasma gases were obtained as a function of CPSP.

In this chapter, effect of process parameters on microstructural and tribological properties of nanostructured and conventional Al_2O_3-13wt%TiO_2coatings, deposited using air plasma spraying (APS) with nitrogen as the primary plasma gas, is discussed in detail.

BACKGROUND

Plasma sprayed alumina coatings are used in many applications, e.g. valves, plungers, pump seals, shields for guided missiles, rocket nozzles, engine components, integrated circuits, vacuum tube envelops, railroad components, etc., (Ono et al., 1993), because of their high wear resistance, heat and thermal shock resistance, oxidation and hot corrosion resistance and electrical insulation (Gadow et al., 1998; Krishnan et al., 2006). They also bond well with metallic bond coats, resist wear and solid particle erosion and have high hardness and chemical stability even at elevated temperatures (Pantelis et al., 2000; Erickson et al., 2001). Alumina exists in nature in a number of hydrated states, e.g. Diaspore (α-AlO(OH)),

Boehmite (γ-AlO(OH)), and gibbsite (γ-Al$(OH)_3$) (Carim et al., 1997). Typically, the Bayer process is used to refine bauxite, the most important ore of aluminium, to obtain alumina.

The effect of spraying distance, substrate pre-heat temperature (25-500°C), and coating thickness on coating properties such as hardness, porosity, and surface roughness, for alumina coatings, was studied by Sarikaya et al. (2005). It was found that an increase in coating thickness leads to decrease in coating hardness and increase in porosity and surface roughness. Highest coating hardness, lowest porosity and lowest surface roughness were obtained at a spraying distance of 120 mm. The effect of arc current, spraying distance, nozzle diameter and substrate cooling on in-flight particle characteristics (velocity and temperature) and coating properties such as coating thickness, hardness, and surface roughness were investigated by Tekmen et al. (2008). They observed that particle velocities increased with arc current and decreased with increase in spraying distance, at least for spraying distance in the range of 100 to 150 mm. The effect of arc current on particle velocities was more noticeable at higher arc currents.

Sabiruddin et al. (2011) studied the effect of substrate preheating temperature (100-300°C), primary gas flow rate (50-90 slpm), secondary gas flow rate (5, 10 slpm), and nozzle size (5, 7 mm) on splat morphologies for agglomerated and crushed alumina, nanostructured agglomerated Al_2O_3-13wt%TiO_2, Ni-5wt%Al, and Ni-20wt%Cr powders. Splats were taken on polished low carbon steel substrates, glass slides, as well as on polished bond coats. While substrate pre-heating temperature and roughness were seen to affect the splat shape, substrate chemistry was seen to play an important role in determining the wetting of splats. Nozzle diameter and primary gas flow rate which influence plasma gas velocity and temperature were seen to control the degree of splashing of splats. Rapid quenching of splats, from heat conduction to the substrate, was seen to result in formation of amorphous phases in splats (Mizuta et al., 1992; Gualtieri et al., 2006; Guilemany et al., 1997; Rosenflanz et al., 2004).

Mcpherson (1973, 1980) did a detailed study of phase composition of plasma sprayed alumina splats. It was shown that rapid cooling of plasma sprayed alumina splats (~ 10^6-10^7 K/s), from heat conduction to substrate, results in the formation of dominant γ-Al_2O_3 phase. However, if process parameters are chosen such that powder particles melt only partially, then splats are seen to be composed of unmelted α-Al_2O_3 crystals in γ-Al_2O_3 matrix (Shaw et al., 2000; Goberman et al., 2002). Presence of α-Al_2O_3 crystals in γ-Al_2O_3 matrix is seen to improve the toughness as well as wear resistance of coatings. Interestingly, while alumina coatings are mostly composed of crystalline phases, coatings deposited from agglomerated powders containing alumina alloyed with certain oxides, e.g. CaO, La_2O_3 and Gd_2O_3, in eutectic compositions, are expected to be predominantly amorphous in nature (Rosenflanz et al. (2004)). The percentage of amorphous and crystalline phases in coatings can be determined by Rietveld analyses as e.g. demonstrated by Gualtieri et al. (2006). It is noteworthy that none of the earlier works have studied either the effect of process parameters on content of various crystalline and amorphous phases in coatings *or* on correlation between percentage of crystalline and amorphous phases and the macro properties of coatings, such as wear resistance, fracture toughness, and micro hardness.

Erickson et al. (2001) demonstrated that addition of titanium oxide to alumina leads to considerably high coating toughness while maintaining enough coating hardness. This, in addition to their excellent wear, corrosion, and erosion resistance, have made plasma sprayed alumina-titania coatings the preferred choice for several industrial applications (Tian et al. (2009)). Differences in coating microstructures have been shown to significantly influence coating properties such as wear resistance, adhesive fracture toughness etc., for alumina-titania coatings. In particular, several works have shown that plasma sprayed coatings deposited using nanostructured agglomerated powders result in *bi-modal* microstructures which greatly enhance their fracture toughness and wear resistance (Wang et al., 2000; Jordan et al., 2001; Luo et al., 2003).

Several earlier works (Shaw et al., 2000, Gell et al., 2001; Goberman et al., 2002; Lin et al., 2003; Wang et al., 2009) studied the effect of critical plasma spray parameter (CPSP) on microstructure of conventional and nanostructured alumina-titania coatings. Shaw et al. (2000) studied the effect of heat treatment of reconstituted nanostructured alumina-titania powders on microstructure of nanostructured coatings. Heat treatment of agglomerated powders before thermal spraying was found to be very useful in burning out binder materials used in spray drying as well as in leading to high-density sprayable granules with good flow characteristics. Phase transformation of nano-sized Al_2O_3 and TiO_2 during heat treating, sintering, and thermal spraying was seen to be in general identical to that of micrometer sized counterparts. The ratio of the electrical power to primary argon gas flow rate (CPSP defined as amp x volt/SCFH) was identified as the key plasma spray parameter that controls temperatures experienced by powder particles during plasma spraying. When CPSP was less than 240, particle temperatures were low and most particles did not experience any melting. When CPSP was between 250 and 300, particle temperatures were intermediate and partial melting occurred. For CPSPs greater than 310 plasma temperatures were high and nearly complete melting took place. The phase content, microhardness, density, and grain sizes in coatings were all seen to strongly depend on CPSP.

Gell et al. (2001) and Goberman et al. (2002) investigated the effect of CPSP on volume fraction of partially melted (PM) and fully melted (FM) regions in coatings along with the effect of CPSP on volume percentage of γ-Al_2O_3 phase in conventional and nanostructured coatings. Volume fraction of PM regions and γ-Al_2O_3 phase in nanostructured coatings were seen to decrease with increasing CPSP. Nanostructured coatings deposited using argon as the primary plasma gas at lowest CPSP (of 300) contained about 50% PM region and about 70% γ-Al_2O_3 phase. On the other hand, conventional coatings were seen to be composed of nearly completely fully melted splats. Crack growth resistance and abrasive wear resistance of coatings was seen to be highest for nanostructured coatings deposited at 350 CPSP.

Wang et al. (2009) did a detailed study of microstructure of coatings deposited using nanostructured agglomerated alumina-13wt% titania powders. They showed that nanostructured alumina-titania coatings consist of fully melted and partially melted regions. The fully melted regions had lamellar-structure as in conventional coatings. The partially melted regions consisted of liquid-phase sintered (a three-dimensional net or skeleton-like structure with Al_2O_3-rich submicron particles embedded in TiO_2-rich matrix) and solid-phase sintered regions (composed of unmelted nano-particles). It was noted that while particles that reach temperature of 2045°C or above undergo complete melting, those that reach in between 1840°C (melting point of TiO_2) and 2045°C (melting point of Al_2O_3), experience liquid phase sintering, while those that reach below 1840°C experience solid phase sintering. Formation of solid-phase and liquid-phase sintered regions in partially melted regions is attributed to the melting point difference of Al_2O_3 and TiO_2.

Guessasma et al. (2004) obtained sliding wear characteristics of plasma sprayed alumina-titania coatings using pin-on-disc (POD) tests. It was found that the friction coefficient increases with increase in applied load and sliding velocity. Separately, Bounazef et al. (2004) and Guessasma et al. (2006) found that the friction coefficient increases linearly with distance between the plasma flame and the powder injection point. They also correlated plasma spraying process parameters such as primary and secondary gas flow rates, carrier gas flow rate, powder injection distance etc., with wear characteristics of plasma sprayed alumina-titania coatings. It was found that there is a combined effect of spray temperature and in-flight particle characteristics on the wear resistance of the alumina-titania coatings.

Tian et al. (2008) conducted fretting wear tests for alumina-titania coatings using 52100 steel balls. Nanostructured coatings were found to exhibit much better fretting wear resistance than conventional coatings due to presence of nano-sized grains, reduced lamellar structures and amorphous phases. While no crack propagation pattern was seen in nanostructured coatings, fretting cracks were seen to propagate along splat boundaries in conventional coatings. In fretting wear tests for nanostructured coatings, a transfer of iron oxide layer was observed on the worn coating surface and the coefficient of friction ranged from 0.7 to 0.9.

Rico et al. (2009) investigated the dry sliding performance of nanostructured and conventional Al_2O_3-13wt%TiO_2 coatings. It was observed that nanostructured coatings exhibited enhanced wear resistance at all conditions considered. Transition from mild to severe wear was observed for nanostructured as well as conventional coatings. However, the critical pressure at which the transition occurred was higher for nanostructured coatings as compared to conventional coatings. Crack propagation controlled the wear of coatings. At low loads the cracks propagated along splat boundaries, but at high contact loads they passed through the splats. The composite hierarchical structure of nanostructured coatings improved their wear resistance. At low loads, the cracks are intercepted and deflected by first level of microstructural features, i.e., the PM particles. As normal load increases, the second level of microstructure begins to be operative. The cracks are then deflected by the nanoparticles located inside the PM zones. This deflection process leads to a toughening effect and subsequent wear resistance improvement.

EXPERIMENTAL METHOD

Substrate Preparation

Coatings in all cases of air plasma spraying were deposited on AISI 1020 steel substrates of around 65 x 55 x 5 mm size. A power saw was used to cut the coupons from bars of 55 mm width and 5 mm thickness. An Alex NH 500 surface grinder was used to grind the top and bottom faces of substrates to remove the oxide layers and generate flat and smooth surfaces with a surface roughness of about 0.1 μm. Before thermal spraying, the substrates were grit blasted inside a suction type grit blasting cabinet (Sandstorm, Bangalore, India) using alumina grits of mesh size 24 at 100 psi air pressure and at 125 mm stand-off distance to increase surface roughness to around 5 μm R_a for providing better mechanical anchorage between the bond coat and substrate. Grit blasting also helps in removing the oxide layer that forms during time gap between surface grinding and grit blasting. Next, the grit blasted substrates were cleaned ultrasonically for about 10 minutes in iso-propanol bath to remove dust and any other foreign matter present on substrate surfaces. The substrates, prepared using the said procedure, were then preheated to around 200°C using the plasma gun, with nitrogen as the only plasma gas, before depositing the bond coat. Pre-heating was done to remove grease, oily substances, water vapour, etc., and produce a clean, nascent surface for depositing the bond coat.

Plasma Spraying

Coatings were deposited using Sulzer Metco 3MB plasma gun mounted on a CNC X-Y table. Nitrogen and hydrogen were used as the primary and secondary plasma gases for all the coatings. Nitrogen was also used as the carrier gas for carrying the powders from powder feeder to the gun. For a given coating system, the process parameters used will influence the resultant voltages and consequently the arc power per unit primary gas flow rate. For all coatings spraying was done at a spray angle of 90°. Two auxiliary air jets, parallel to the plasma flame, are used to cool the substrate during the deposition process and to remove any unmelted and weakly bonded powder particles from the coating surface. Typically, bond coats of about 100-150 μm were deposited prior to depositing ceramic or cermet top coatings of about 250-350 μm thickness. Bond coats were deposited to improve the adhesion of the top coats and to reduce the effect of mismatch in thermal expansion coefficients between ceramic/cermet top coats and metal substrates.

Microstructural and Mechanical Characterization

Cross-sectional specimens of 10 x 5 x 5 mm size were cut from coated samples for analyzing the microstructural characteristics and mechanical properties of coatings. A low speed diamond saw (150 low speed diamond saw, MTI Corporation) was used to cut the specimen from the coated coupons. Thermoplastic mounts with the specimen inserts were then prepared using a cold mounting technique with Geosyn cold mounting compound powder and liquid acquired from Geologists syndicate private limited, Kolkata, India. The cold mounted inserts, with exposed coatings cross-section, were then polished using SiC (Silicon Carbide) abrasive papers of grades 220, 400, 600, 800 and 1000 and diamond pastes of grades 8, 6, 3, 1 and ¼ spread on selvyt cloths on an auto polisher (Struers, LaboPol-21). The polishing was done for 20 minutes for each abrasive paper and for 30 minutes for each grade of diamond paste. During polishing, the abrasive paper was kept wet at all times.

Optical micrographs of coatings cross-sections were obtained using an inverted Olympus microscope (GX41, DP 12). Six micrographs taken at six randomly selected locations, at 50x magnification, were used for estimation of coatings porosity. Porosity was measured from the micrographs using "ImageJ" software at a fixed grey scale threshold value of 188, which was seen to give consistent results. Mean and standard deviation of porosity were estimated based on area fraction of pores obtained from six images. Distribution of fully and partially melted regions in coatings was observed using a ZEISS EVO 60 scanning electron microscope (SEM) and a Zeiss Supra 40 field emission scanning electron microscope (FE-SEM). For estimating the volume percentage of partially melted/unmelted regions in nanostructured coatings, for each coating, ten SEM/FE-SEM images at randomly selected locations were taken at a fixed magnification of 1000x. Each image was printed on an A4 sheet and via visual inspection the partially melted/unmelted regions were identified and cut out. The initial weight of the sheet and the cumulative weight of partially melted/unmelted regions were then used to calculate the area/volume fraction of partially melted/unmelted regions in the image. Mean and standard deviation of volume fraction of partially melted/unmelted regions in the coating were then calculated based on readings obtained from all images obtained for the coating.

Hardness measurements were made using a LECO LM 700 microhardness tester with a Vickers indenter at 100 g load and 15 s dwell time. Ten readings were taken for each coating and the average and standard deviation were computed. X-ray diffraction studies were carried out using a PANalytical X'pertPRO PW1070 instrument with CuKα radiation with tension and intensity of generator at 40 kV and 30 mA, respectively, scanning step of 0.016711° and step time of 0.13 s. XRD data was analyzed using "X'Pert High Score" software to obtain the phase distributions in coatings. The phase distributions have been quantified using relative phase intensity ratio method as implemented by Khor *et al.* (2001). Surface roughness of as-deposited coatings was measured using a Taylor-Hobson precision, surtronic

3+ profilometer and TalyProfile software (version 3.1.9). Surface roughness profiles at ten randomly selected locations, obtained using the profilometer, were used to estimate the mean and standard deviation of average surface roughness R_a of the coatings.

Tribological Characterization

Wear tests were conducted on as-deposited coatings using Ducom tribometer (wear and friction monitor, TR-20-M24) under different loads (0.5-1.5 kgf) and sliding speeds (0.17-0.5 m/s) for a constant sliding distance of 300 m. Tungsten carbide-cobalt balls of 6 mm diameter were used as the counter bodies for all tests. Weight loss of coatings for the wear tests done was estimated by measuring the sample weight before and after the wear test using an electronic balance (Mettler Toledo Classic Plus, AB265-s/Fact). Wear debris is removed, before measuring the sample weight after the wear test, by manually blowing air over the sample. Three wear tests were conducted for each load and speed condition to confirm consistency of results. Mean and standard deviation of the obtained readings for the weight loss were also calculated. Surface morphology of wear tracks was observed using ZEISS EVO 60 scanning electron microscope (SEM) to identify the wear mechanism.

RESULTS AND DISCUSSION

Powder Morphology and Phase Distribution

Powder morphologies for conventional and nanostructured agglomerated Al_2O_3-13wt%TiO_2 feedstock powders are shown in Figure 1 and Figure 2. It is clear from the figures that, while conventional alumina-titania powders prepared using the fusing and crushing technique are blocky, dense and angular with sharp edges, the corresponding nanostructured agglomerated powders prepared using spray drying and

Figure 1. (a, b) SEM images showing powder morphology of conventional aluimina-13wt% titania powders at different magnifications

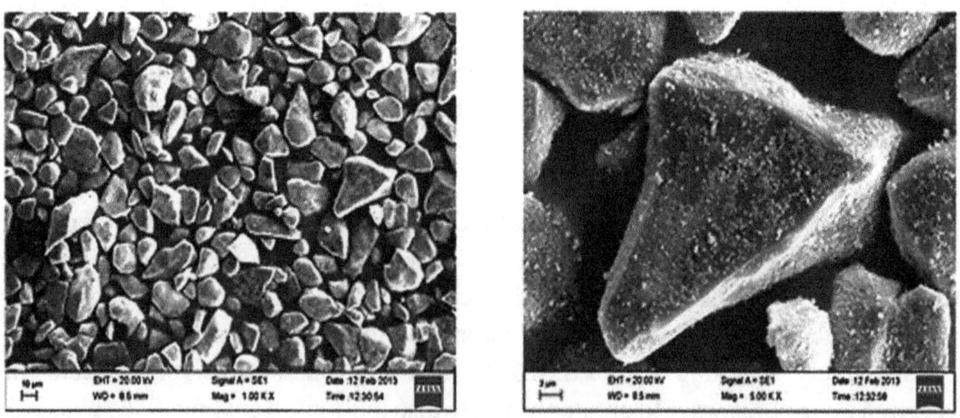

(a) (b)

Figure 2. (a, b) SEM images showing powder morphology of nanostructured agglomerated alumina-13wt% titania powders at different magnifications

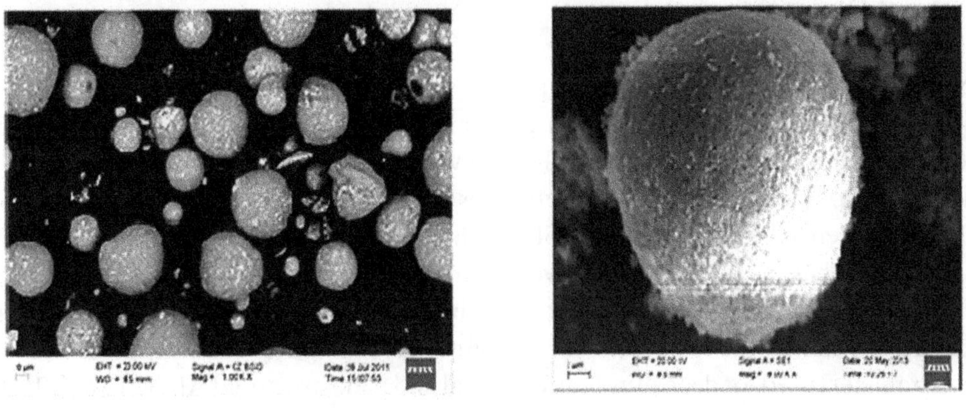

(a) (b)

sintering technique are porous and nearly spherical. It has been observed in experiments that, the spherical morphology of agglomerated powders results in better powder flow characteristics. The porous nature of agglomerated powders results in lower thermal diffusivities for the powders and is expected to result in partial melting of some of the powder particles. It may be noted that, nanostructured agglomerated Al_2O_3-13wt%TiO_2 powders have a tap density of 1.4 g/cm^3 (information on tap density for conventional powders is not reported as the same was not provided by the manufacturer).

X-ray diffraction (XRD) patterns for conventional and nanostructured Al_2O_3-13wt%TiO_2 feedstock powders are shown in Figure 3. It can be seen that, while alumina occurs in α-phase in both conventional and nanostructured powders, titania occurs in anatase phase in conventional and in rutile phase in nano-structured powders. Peak broadening can be noted in alumina as well as titania phases in nanostructured powders signifying the presence of micro-stresses, varying from point to point in both the phases.

Figure 3. XRD patterns for (a) conventional and (b) nanostructured alumina-13wt%titania powders

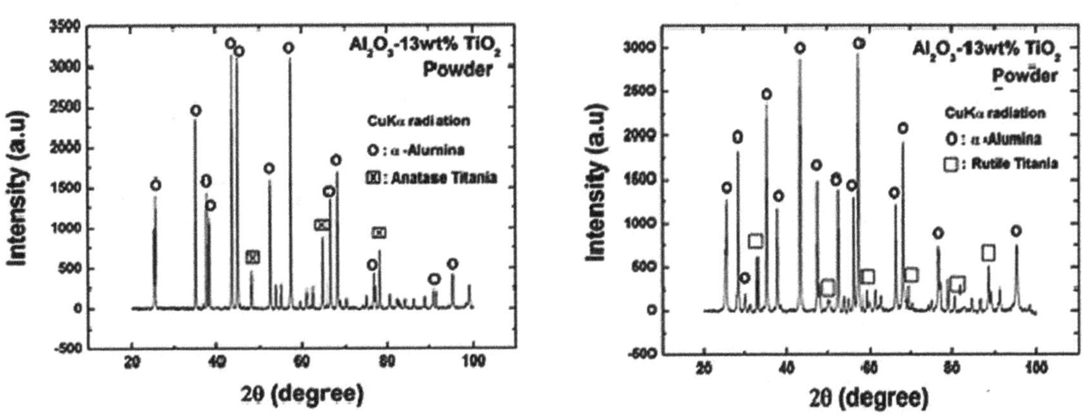

(a) (b)

Characterization of Air Plasma Sprayed Coatings

Microstructural Characteristics

A typical coating architecture obtained for air plasma sprayed (APS) alumina-titania coatings is shown in Figure 4. It is evident from the figure that, a good interface bonding exists between the top and bond coat. Apart from chemical bonding, a rough interface with a surface roughness of around 9.6 μm resulting from air plasma spraying of the bond coat provides a strong mechanical interlocking of bond and top coats which can be seen to be about 100 and 250 μm thick. Note that since Ni, which is the predominant element in bond coat, has the highest atomic weight among Ni, Cr, Fe, Al, O, and Ti (elements in bond coat, substrate, and top coat), the bond coat appears in a light shade in the back scattered SEM image shown in Figure 4, whereas the top coat, which is largely composed of Al_2O_3 appears in a dark shade. The process parameters used for depositing conventional and nanostructured Al_2O_3-13wt%TiO_2 coatings, with nitrogen and argon as the primary plasma gases, respectively, are given in Table 1. Note that the conventional coatings were deposited *only* using nitrogen as the primary plasma gas at the highest CPSP.

Figure 4. Typical coating architecture of air plasma sprayed alumina-13wt% titania coatings. Note that AISI 1020 steel is used as the substrate and Ni-20wt%Cr is used as the bond coat material.

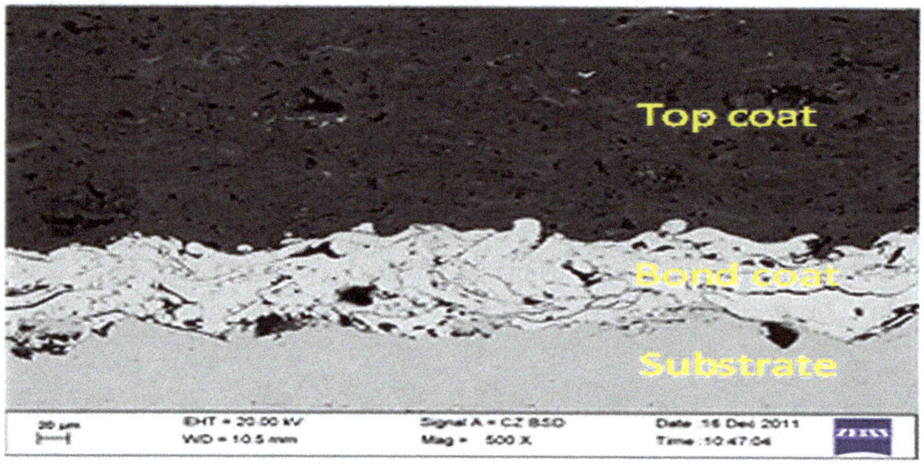

Table 1a. Process parameters used for depositing nanostructured alumina-13wt% titania coatings using nitrogen as the primary plasma gas

Serial No.	Current (A)	Voltage (V)	N_2 Flow Rate (SCFH)	H_2 Flow Rate (SCFH)	CPSP
1	400	84	50	15	672
2	400	88	65	20	542
3	450	90	50	15	810
4	450	91	65	20	630
5	500	90	50	15	900
6	500	92	65	20	707

Table 1b. Process parameters used for depositing nanostructured alumina-13wt% titania coatings using argon as the primary plasma gas

Serial No.	Current (A)	Voltage (V)	Ar Flow Rate (SCFH)	H_2 Flow Rate (SCFH)	CPSP
1	500	67	60	10	279
2	500	67	50	10	335
3	500	67	40	10	419

Cross-sectional SEM micrographs of conventional and nanostructured Al_2O_3-13wt%TiO_2 coatings deposited using nitrogen as primary plasma gas are shown in Figure 5. The following are evident from the Figure 5: (i) Four distinct features can be seen in coatings deposited from nanostructured alumina-titania powders: Partially melted (PM) regions composed of *nano* alumina particles in titania rich matrix embedded in fully melted alumina-titania matrix, titania rich spots that appear as white patches in micrographs observed using a backscattered electron detector, uniform distribution of micro-cracks nearly aligned with the coating direction, and preferential distribution of nearly spherical pores in PM regions, (ii) PM regions in nanostructured coatings tend to arrest micro-cracks as seen from insert in Figure 5 b, (iii) Segregation of titania in fully melted regions decreases with increasing CPSP as confirmed by minimal presence of *white* titania rich splats in Figure 5d, (iv) Porosity of nanostructured coatings is higher than that for conventional ones, particularly at lower CPSPs, (v) Inter splat bonding for conventional as well nanostructured coatings, particularly those at higher CPSPs, is "good", and (vi) Nearly uniform distribution of micro-cracks, in the coating direction, occurs in conventional as well as nanostructured coatings. It may be noted that *white* titania rich regions in conventional and nanostructured coatings, PM regions in nanostructured coatings, and nearly spherical pores appearing in nanostructured coatings act as crack arrestors. It may also be noted from Figures 5 b-d that, the percentage of PM regions in coatings obtained with nitrogen as the primary plasma gas, even at lowest CPSP considered here, is far smaller than those typically obtained for corresponding coatings deposited using argon as the primary plasma gas (Luo et al. (2003), Gell et al. (2001), Jordan et al. (2001)). This is attributed to the fact that, nitrogen in the ionised state has a much higher specific enthalpy at a given temperature as compared to argon (Ingham et al. (1975)). In addition, presence of hydrogen increases the thermal conductivity of the gas followed by more heat transfer to the particles (Heiman (1996), Escure et al. (2003), Tekmen et al. (2009)). Hence, the nitrogen-hydrogen combination brings about a more intensive particle melting. As stated earlier, titania has a lower melting point (1855°C) than that for alumina (2072°C). Hence, at a lower CPSP, i.e., at lower particle temperatures, melting and segregation of titania is promoted while alumina is still in solid state (Wang et al. (2009)).

SEM micrographs of coating cross-sections for nanostructured coatings deposited with argon as the primary plasma gas are shown in Figure 6. It can be clearly seen that, the CPSPs for coatings deposited with argon are much smaller than those deposited with nitrogen. It may be noted that, what is imposed during the coating process is the current and primary and secondary plasma gas flow rates. Voltage difference between the electrodes, which in turn controls the arc power and resultant CPSP, is decided by the plasma gas composition. Argon, with a much lower dielectric breakdown strength than nitrogen, is seen to result in lower arc powers and lower CPSPs for the imposed current and gas flow rates (see Table 1). It is also clear that, percentage of partially melted regions in nanostructured coatings is much higher for coatings deposited with argon. However, the percentage decreases rapidly with increasing CPSP.

Figure 5. Microstructures of alumina-titania coatings deposited using nitrogen as the primary plasma gas with (a) conventional powders at 900 CPSP and (b-d) nanostructured powders at 542, 707 and 900 CPSPs, respectively

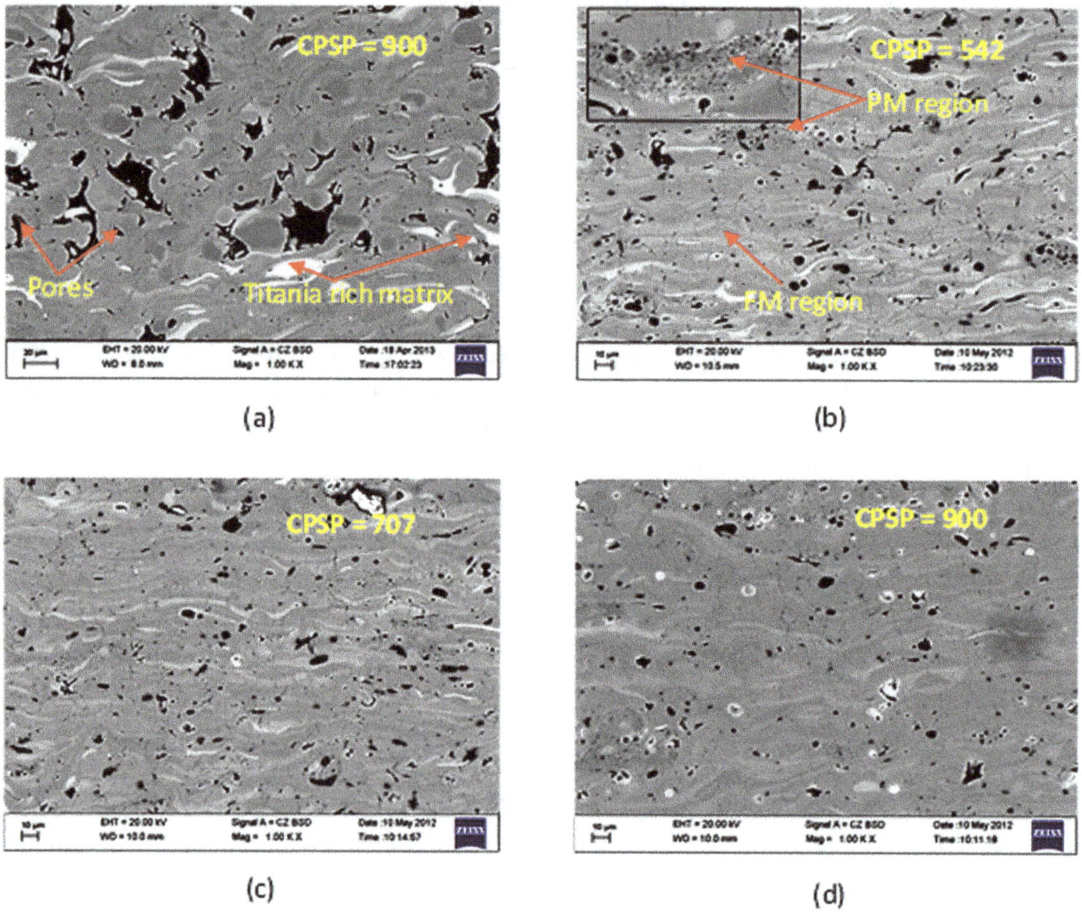

X-ray diffraction (XRD) patterns for conventional and nanostructured alumina-titania coatings deposited using nitrogen are shown in Figure 7. Contrary to observations in coatings deposited with argon (Tian et al. (2009), Ibrahim et al. (2010), Yugeswaran et al. (2010)), in coatings deposited here with nitrogen, conventional coatings have greater α-alumina content than the corresponding nanostructured coatings. Surprisingly, the volume percentage of α-alumina in nanostructured coatings, deposited at 542 as well as 900 CPSPs, as estimated from the relative phase intensity ratio method (Khor et al. (2001), is less than 4 vol. %, while that in the conventional coating for a CPSP 900 it is estimated to be around 7 vol. %.

XRD patterns for nanostructured alumina-titania coatings deposited using argon as the primary plasma gas at different CPSPs are shown in Figure 8. Note that in contrast to nanostructured coatings deposited with nitrogen as primary plasma gas, corresponding coatings deposited with argon show a significant fraction of alpha alumina phase. A trace amount of rutile titania phase can also be seen in the coatings. Volume percentage of alpha alumina phase in nanostructured coatings deposited with nitrogen and argon as primary plasma gases, determined using the relative phase intensity ratio method, is shown in Figure 9. While the volume percentage of alpha alumina phase in nanostructured coatings deposited

Figure 6. Microstructures of nanostructured alumina-titania coatings deposited using argon as the primary plasma gas at (a) 279, (b) 335 and (c) 419 CPSPs, respectively

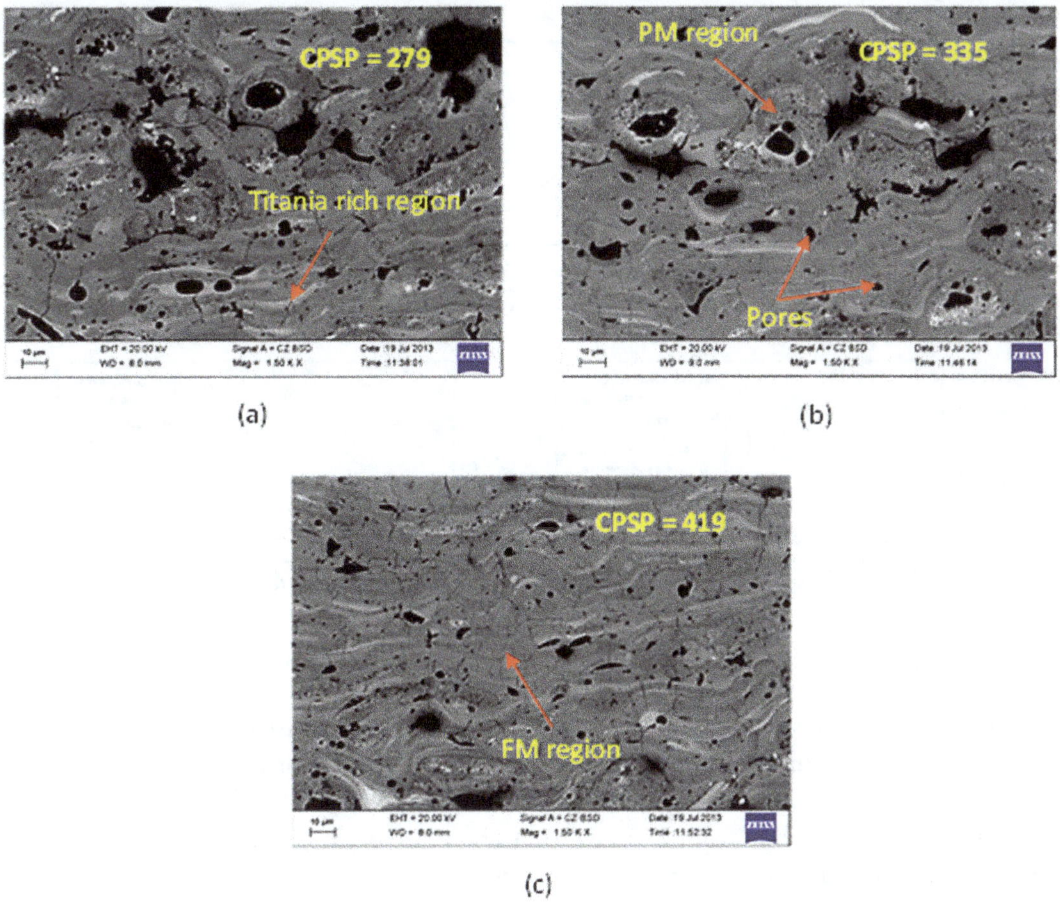

with nitrogen is varying from around 3 to 4%, it varies from about 27 to 29% in nanostructured coatings deposited with argon. It may be noted that percentage of alpha alumina phase in nanostructured coatings is primarily affected by the choice of primary plasma gas. The effect of CPSP on percentage of alpha alumina phase is minimal. Volume percentage of alpha alumina phase in conventional coatings deposited with nitrogen at 900 CPSP is about 7%.

Characteristics of air plasma sprayed alumina-titania coatings deposited at different CPSPs using nitrogen and argon as primary plasma gases, respectively, are shown in Table 2. Porosity as a function of CPSP for coatings deposited with nitrogen and argon are shown in Figure 10. Note that porosity decreases nearly linearly with CPSP for both the primary gases and that the average porosity in nanostructured coatings is around 9%. Percentage of partially melted regions in nanostructured coatings deposited with nitrogen and argon as primary plasma gases, respectively, as a function of CPSP is shown in Figure 11. It may be noted from Figure 11 that, the resultant CPSPs for argon are much smaller than that for nitrogen. More specifically, note from Table 1 that the voltage across the electrodes in the plasma gun, which governs the resultant CPSP, is primarily governed by the type of primary gas used: nitrogen or argon. Voltage across the electrodes is relatively unaffected by the magnitude of plasma current or by the volume flow rates of primary and secondary plasma gases.

Figure 7. XRD patterns for alumina-13wt% titania coatings (a) conventional coatings deposited at 900 CPSP and (b, c) nanostructured coatings deposited at 542, 900 CPSPs, respectively. Note that these coatings were deposited using nitrogen as the primary plasma gas

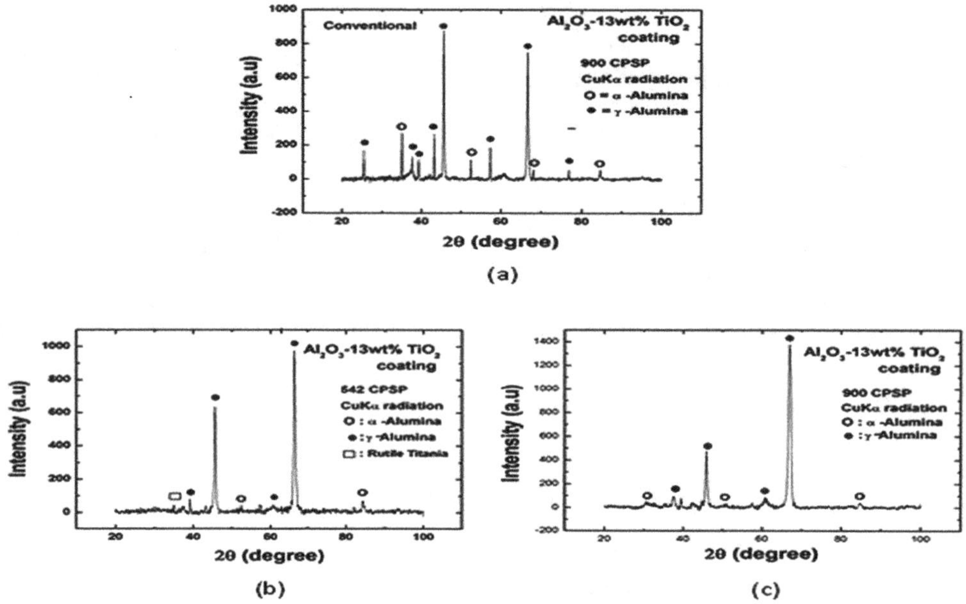

For both the primary gases, percentage of PM regions is seen to decrease nearly linearly with increasing CPSP. However, while the variation in percentage of PM region for nitrogen as primary gas, for CPSP varying from 540 to 900, is about 17 to 10%, the variation in percentage of PM region for argon as primary gas, for CPSP varying from 280 to 420, is about 48 to 13%, respectively. Thus, mechanical properties that are affected by presence of PM regions, e.g. wear resistance, are expected to be much more sensitive to the CPSP at which the nanostructured coatings are deposited for argon as compared to nitrogen as the primary plasma gas. Higher specific enthalpy of nitrogen at a given plasma gas temperature is believed to be responsible for lower percentage of PM regions in nanostructured coatings deposited with nitrogen. Microhardness of nanostructured coatings deposited with nitrogen and argon, as a function of CPSP, are shown in Figure 12. It is interesting to note that, the hardness increases nearly linearly with increasing CPSP for both the primary gases. However, hardness of coatings deposited with argon is seen to be much smaller than hardness of corresponding coatings deposited with nitrogen.

The surface roughness of coatings deposited with nitrogen and argon, as a function of CPSP, are shown in Figure 13. Surprisingly, for coatings deposited with argon, there seems to be no effect of CPSP on surface roughness of coatings. On the other hand, coatings deposited with nitrogen at lower CPSPs had a lower average surface roughness. The only exception to this trend is for coatings deposited at 810 CPSP.

Wear Characteristics

Test parameters used for conducting the wear tests using tungsten carbide-cobalt ball in a ball-on-disc tribometer are listed in Table 3. Weight loss of coated specimen in ball-on-disc wear tests, with 6 mm diameter tungsten carbide-cobalt ball, in conventional and nanostructured alumina-titania coatings de-

Figure 8. XRD patterns for nanostructured alumina-13wt% titania coatings deposited using argon as primary plasma gas at (a) 279, (b) 335 and (c) 419 CPSPs, respectively

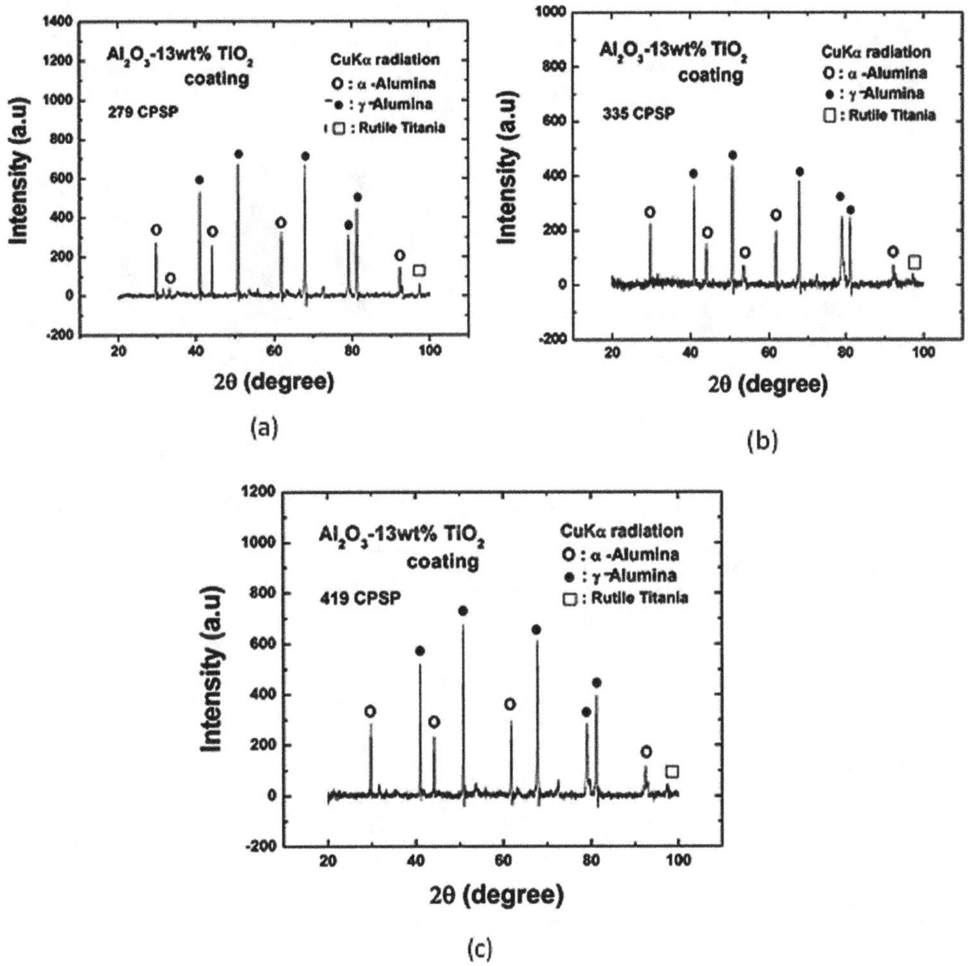

posited with nitrogen are shown in Figure 14. It may be noted that the wear tests are done on as-sprayed coatings at different normal loads and sliding speeds for a constant sliding distance[1] of 300 m. Figure 14 a-c shows the effect of normal load on weight loss at different sliding speeds. For conventional coatings, deposited at 900 CPSP, it can be seen that the weight loss increases nearly linearly with normal load. The trends for nanostructured coatings are more complex but the weight loss in general increases with increasing normal load. The effect of CPSP on weight loss in nanostructured coatings is very limited. In particular, at the highest sliding speed of 0.5 m/s, there is effectively no effect of CPSP on weight loss in nanostructured coatings. The effect of normal load on weight loss at three different sliding speeds, for coatings deposited at 900 CPSP is shown in Figure 14 d. The weight loss in wear increases steadily with increase of sliding speed for the nanostructured coatings. The results for corresponding conventional coatings at a sliding speed of 0.33 m/s are also shown in the Figure 14 d. It can be noted that, conventional coatings give better wear performance at all loads except for the highest load of 1.5 kgf. In particular, it performs far better than nanostructured coatings at 0.5 kgf normal load.

Figure 9. Percentage α- alumina as a function of CPSP in nanostructured alumina-titania coatings deposited using (a) nitrogen and (b) argon. Note that percentage α- alumina in conventional alumina-titania coatings deposited at 900 CPSP using nitrogen is found to be 7.3%.

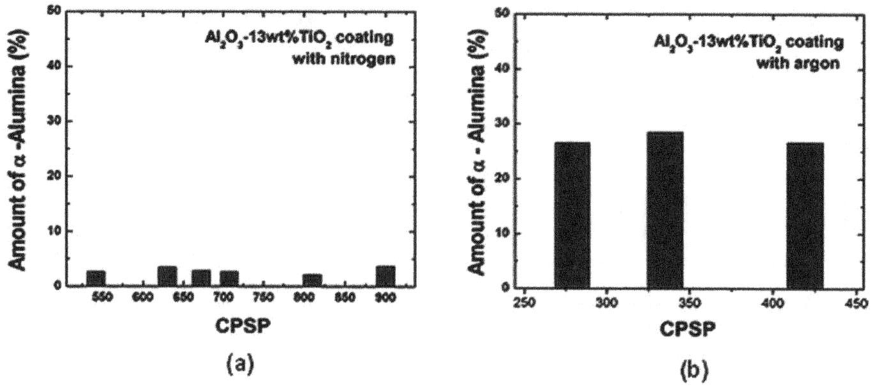

Table 2a. Characteristics of alumina-13wt% titania coatings deposited at different CPSPs using nitrogen

Coating System	CPSP	Porosity (%)	PM Region (%)	Microhardness HV_{100} (GPa)	Surface Roughness, R_a (μm)
Nanostructured alumina-titania	542	11±0.2	16.7±0.7	7.9±0.6	5.0±0.7
	630	10±0.2	14.8±0.8	9.0±0.6	5.9±1.0
	672	9.7±0.3	13.4±0.6	9.4±0.6	6.9±1.0
	707	9.4±0.2	12.6±0.5	9.6±0.5	10.9±0.9
	810	8.6±0.2	11.3±0.9	11.1±0.5	4.7±0.5
	900	8.0±0.5	10.2±1.2	12.2±0.8	11.2±1.7
Conventional alumina-Titania	900	6.0±0.4	NIL	9.9±0.8	9.6±0.9

Table 2b. Characteristics of alumina-13wt% titania coatings deposited at different CPSPs using argon

Coating System	CPSP	Porosity (%)	PM Region (%)	Microhardness HV_{100} (GPa)	Surface Roughness, R_a (μm)
Nanostructured alumina-titania	279	10.3±0.6	47.5±0.6	5.8±0.5	10.8±1.3
	335	9.5±0.3	37.0±0.3	6.3±0.6	10.6±1.6
	419	8.9±0.4	13.2±0.3	7.4±0.7	10.6±1.6

Weight loss results in nanostructured alumina-titania coatings deposited with argon, at different CPSPs, are shown in Figure 15. Note from the figure that, there is a strong effect of CPSP in coatings deposited with argon. It can be seen from Figure 11 b that volume percentages of PM regions in coatings deposited at 335 and 419 CPSPs are around 37% and 13%, respectively. It has been observed by Luo *et al.* (2003) that such coatings containing around 15-20 vol% PM regions exhibit optimum crack growth resistance. Thus, in the current scenario coatings deposited at CPSP 419 are expected to possess

Figure 10. Percentage porosity as a function of CPSP in nanostructured alumina-titania coatings deposited using (a) nitrogen and (b) argon as the primary plasma gases

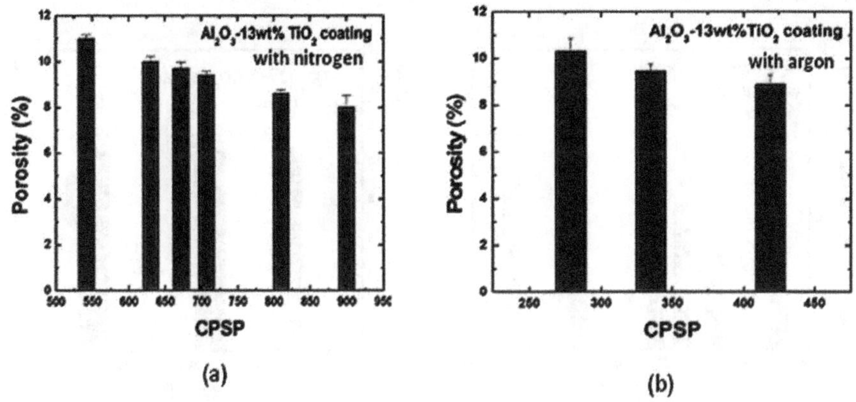

Figure 11. Percentage PM region as a function of CPSP in nanostructured coatings deposited using (a) nitrogen and (b) argon

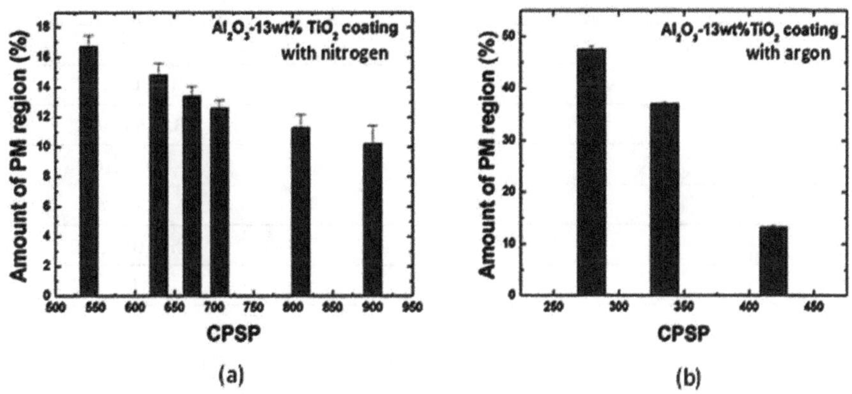

Figure 12. Microhardness as a function of CPSP in nanostructured coatings deposited using (a) nitrogen and (b) argon

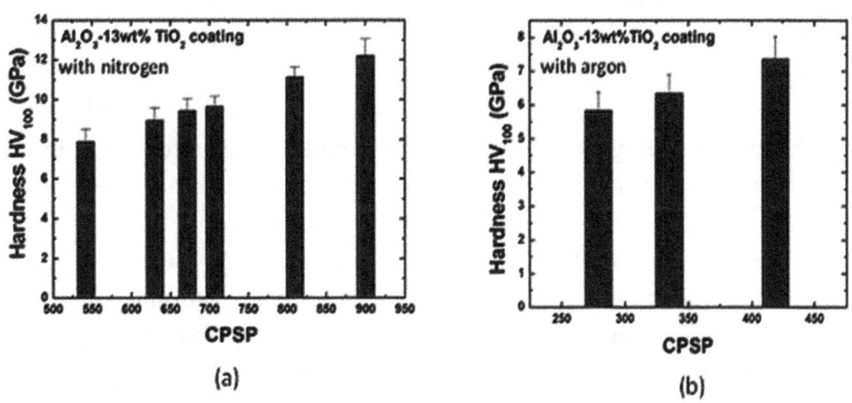

Figure 13. Surface roughness as a function of CPSP in nanostructured coatings deposited using (a) nitrogen and (b) argon. Surface roughness of conventional alumina-titania coatings deposited at 900 CPSP using nitrogen is 9.6 μm.

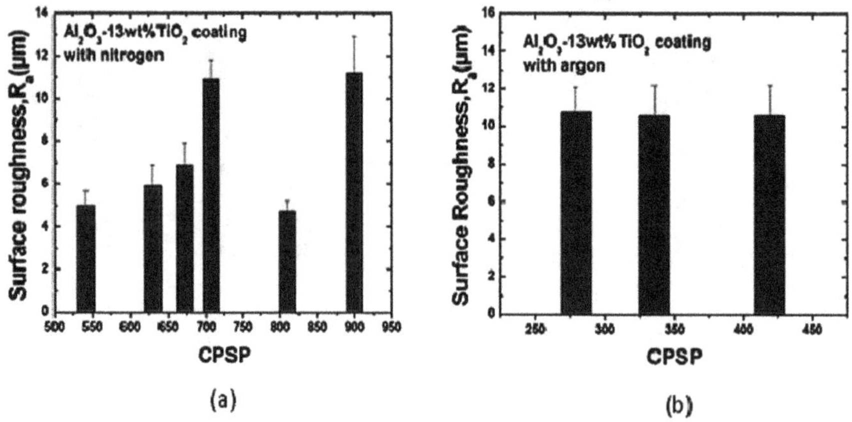

Table 3. Test parameters used for conducting wear tests

Test Parameter	Value
Load (kgf)	0.5,1.0,1.5
Speed (rpm)	319,637,955
Test duration (min)	30,15,10
Sliding distance (m)	300
Sliding velocity (m/s)	0.17,0.33,0.5
Track diameter (mm)	10

higher crack growth resistance and higher wear resistance. Further, interestingly, the effect of normal load on weight loss for coatings deposited at 419 CPSP is seen to be minimal. This is in contrast to the other coatings deposited with argon where the weight loss has increased substantially under the action of a normal load of 1.5 kgf. Also note that, for sliding speed of 0.33 m/s at normal load of 1.5 kgf, for nanostructured coatings deposited with argon at 419 CPSP, weight loss is lowest and it is about 1.65 mg in comparison to weight loss of about 3 and 2.25-2.5 mg for conventional and nanostructured coatings deposited with nitrogen (see Figure 14b).

Morphologies of wear tracks for coatings deposited with nitrogen at 900 CPSP, in wear tests done at a normal load of 1.5 kgf and sliding speeds of 0.17, 0.33, and 0.5 m/s, are shown in Figure 16. Figure 16 b shows that the wear track surface is relatively "smooth" at lowest sliding speed. Shipway and Hogg (2005) show that wear in WC-Co ball, in ball-on-disc wear tests, occurs through preferential removal of binder phase followed by pull out of WC particles especially when the ball rubs against a material harder than the binder but softer than WC. Thus, in this case where alumina-titania coating is harder than cobalt but softer than WC, three body abrasive wear could occur from WC grits removed from the ball. Thus, based on "smooth" wear track surfaces and possibility of WC grits between the ball and the coating, abrasive wear (both two-body and three-body) is believed to be *the dominant wear mechanism* at

Figure 14. Weight loss as function of normal load for nanostructured alumina-titania coatings deposited at 707 and 900 CPSPs and corresponding conventional coatings deposited at 900 CPSP at different sliding speeds of (a) 0.17, (b) 0.33 and (c) 0.5 m/s respectively, and (d) for conventional and nanostructured coatings deposited at 900 CPSP at all sliding speeds. The conventional coating has been tested at a sliding speed of 0.33 m/s only.

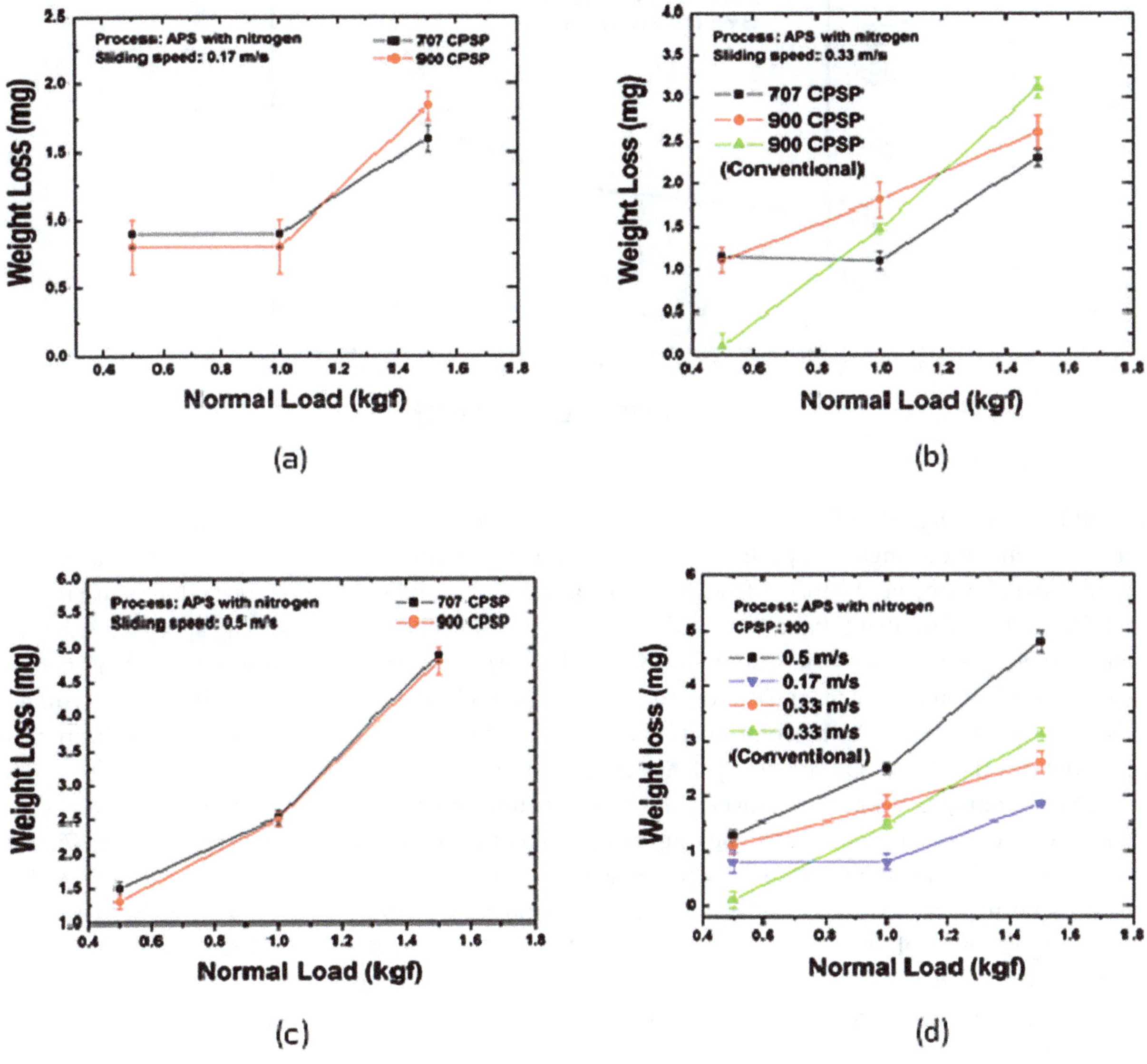

lower sliding speeds. Separately, it can also be noted from the figure that wear track width increases and wear track surface becomes more "rough" and "pitted" as sliding speed increases to 0.33 m/s or higher.

Miguel et al. (2003) show that higher crack density in plasma sprayed coatings can result in fatigue wear leading to "rough" wear track surfaces. Figures 5 b-d show that nanostructured coatings deposited with nitrogen have much higher crack density than corresponding conventional coatings. Consequently, it is expected that fatigue wear would be important for nanostructured coatings particularly at higher

Figure 15. Weight loss as a function of normal load at a sliding speed of 0.33 m/s in nanostructured alumina-titania coatings deposited using argon at 335 and 419 CPSPs

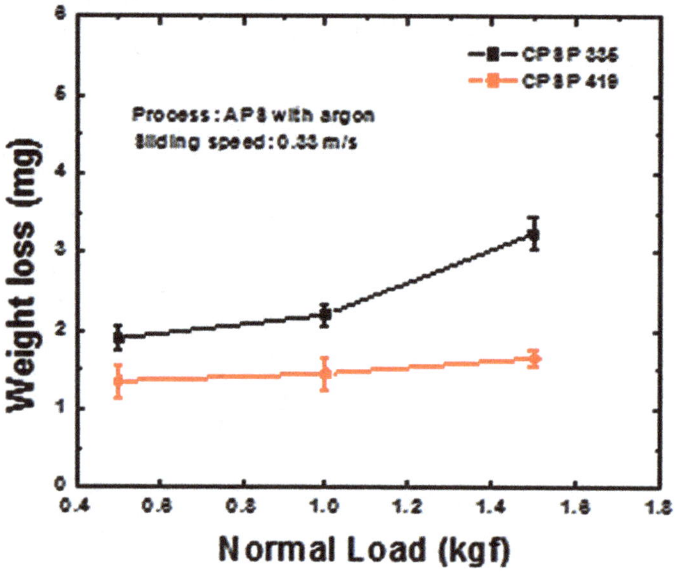

normal loads and higher sliding speeds. In addition, presence of unmelted and partially melted regions in nanostructured coatings is expected to result in wear fragments that can undergo further pulverization to give nanostructured debris. "Rough" wear track surfaces and sub-micron wear debris in Figures 16 d-f indeed confirm that fatigue wear and pulverization do occur at higher sliding speeds at least for nanostructured coatings deposited at highest CPSP. Pulverization resulting in sub-micron wear debris is clearly evident in higher magnification micrograph shown in the inset of Figure 16 f. Furthermore, pulverised debris appears to be consolidating and redepositing on wear track surface in the form of a discontinuous film between "smooth" patches as seen in Figure 16 f.

SEM micrographs of wear track surfaces for nanostructured coatings deposited at a lower CPSP (707), from wear tests done at lowest and highest normal loads and 0.33 m/s sliding speed, are shown in Figure 17. Groove marks indicating abrasive wear are clearly visible in Figure 17 b showing track surfaces for the lowest normal load. Extensive pulverization and redisposition of pulverized grains, suggesting fatigue wear at highest normal load of 1.5 kgf, are evident from Figures 17 d, e. Note that these observations are in line with weight loss plots shown in Figure 14 b where a sudden increase in weight loss occurs as normal load is increased from 1.0 to 1.5 kgf, indicating operation of a new wear mechanism in addition to abrasive wear at highest load. Lastly, note that with increase in normal load, there is an increase in wear track width and the wear track surface changes from being discontinuous to continuous. In order to understand as to why coatings deposited at highest CPSP result in lower wear resistance note that while hardness, volume percentage of porosity and PM regions of nanostructured coatings deposited at 707 CPSP are 9.6 GPa, 9.4% and 12.6%, respectively, the same for coatings deposited at 900 CPSP are 12.2 GPa, 8% and 10.2%, respectively. Thus, even though coatings deposited at highest CPSP are harder, they have slightly lower percentage of PM regions making them less tough and hence more vulnerable to brittle fracture. Further, as pores are known to arrest cracks, lower porosity in coatings deposited at highest CPSP could make them more prone to brittle fracture.

Effect of CPSP on wear track surfaces for nanostructured agglomerated coatings, for wear tests done at 0.33 m/s sliding speed and 1.5 kgf normal load, is shown in Figure 18. It is clear that both coatings undergo fatigue wear. While consolidation of pulverized grains for coatings deposited at 707 CPSP is evident in Figure 18 b, pitting of wear track surfaces is seen for coatings deposited at 900 CPSP in Figure 18 d. Higher magnification images shown in insets for coatings obtained at 707 and 900 CPSPs (Figures 18 b and 16 f) show that while average size of pulverised debris is larger for coatings deposited at 707 CPSP, the amount of pulverization is greater for coatings deposited at 900 CPSP. Once again, this is in line with weight loss results for these coatings shown in Figure 14 b.

Scanning electron micrographs of wear track surfaces for *conventional* alumina-titania coatings deposited with nitrogen at 900 CPSP, for wear tests done at a sliding speed of 0.33 m/s and normal loads of 0.5, 1.0 and 1.5 kgf, are shown in Figure 19. It is clear that wear track surfaces for conventional coatings are much smoother than that for nanostructured coatings. Plastic deformation appears to be dominant in conventional coatings. Cracks appearing on track surfaces in Figures 19 b-e suggest that delamination wear (which is a special kind of fatigue wear) could *also* be occurring in addition to abrasive wear in conventional coatings. However, as suggested by linear weight loss plot in Figure 14 d, magnitude of weight loss from delamination wear appears to be small. Note that this is in contrast to what happens in nanostructured coatings where fatigue wear and pulverisation are predominant at higher loads. A detailed discussion on delamination wear in materials that exhibit plasticity can be found in Suh (1973).

Note from Figures 14 a-c, 17, and 19 that super linear behaviour is observed for weight loss as a function of normal load for nanostructured coatings, where there is a change in wear mechanism as normal load is increased from 0.5 to 1.5 kgf, whereas nearly linear behaviour is observed for conventional coatings. Linear increase in wear with load (Archard (1953)) could suggest adhesive wear, where plastic deformation of coating material creates the wear debris. On the other hand, WC grits pulled out from WC-Co ball during wear tests could also result in abrasive wear. The presence of fine *nano* scale particles on wear tracks after the tests is far lower for conventional coatings than for nanostructured coatings. Absence of any significant amount of partially melted or unmelted particles is thought to be the reason for the lack of wear debris, whereas presence of titania rich regions are thought to make the coating ductile and wear track surfaces smooth.

Wear track surface morphologies for nanostructured coatings deposited with argon at two different CPSPs, in wear tests done at 0.33 m/s sliding speed and 1.5 kgf normal load, are shown in Figure 20. Note from the figure that while coatings deposited at 419 CPSP have microscopically "smooth" wear track surfaces, those deposited at 335 CPSP appear to undergo greater pulverization and result in re-deposition of pulverized wear debris. As shown by Luo et al. (2003), micro-cracks initiated in stressed regions in coatings having *high* percentage of PM regions, could be resulting in higher fatigue wear in coatings deposited at 335 CPSP which have about 37% of PM regions. Note that this observation is in line with super linear weight loss behaviour observed for these coatings (see Figure 15). Figure 21 shows morphologies of worn track surfaces for nanostructured coatings deposited with argon at the highest CPSP (i.e., 419) for wear tests conducted at 0.5 and 1.5 kgf normal loads. For both normal loads, the coating surfaces are smooth and wear fragments appear to have been pulverized. Note from Figure 15 that there is very little effect of normal load on weight loss for this coating. This is in clear contrast to tribological behaviour of nanostructured coatings deposited with nitrogen. Surface morphologies of the wear tracks for nanostructured coatings deposited with argon at CPSP of 335 for the wear tests conducted at sliding speed of 0.33 m/s and normal load of 0.5 and 1.5 kgf, respectively, are shown in Figure 22.

Figure 16. Morphology of wear tracks at lower and higher magnifications corresponding to the wear tests done on nanostructured alumina-titania coatings at 1.5 kgf normal load and a sliding speed of (a, b) 0.17, (c, d) 0.33 and (e,f) 0.5 m/s, respectively. Note that these coatings were deposited at 900 CPSP using nitrogen as primary plasma gas.

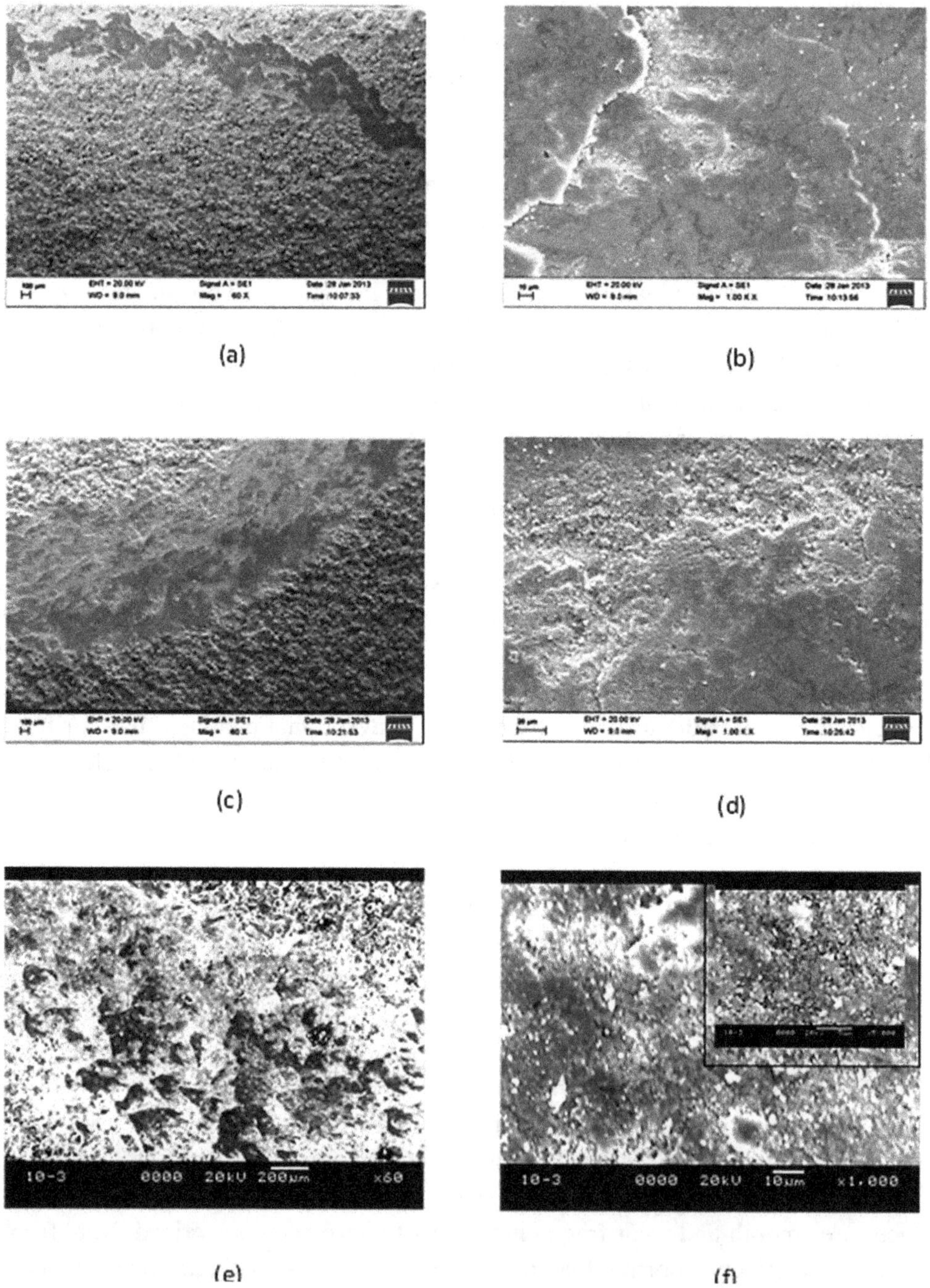

Figure 17. Morphology of wear tracks at lower and higher magnifications in wear tests done on nano-structured alumina-titania coatings deposited at 707 CPSP using nitrogen at a sliding speed of 0.33 m/s and a normal load of (a, b) 0.5 and (c-e) 1.5kgf, respectively.

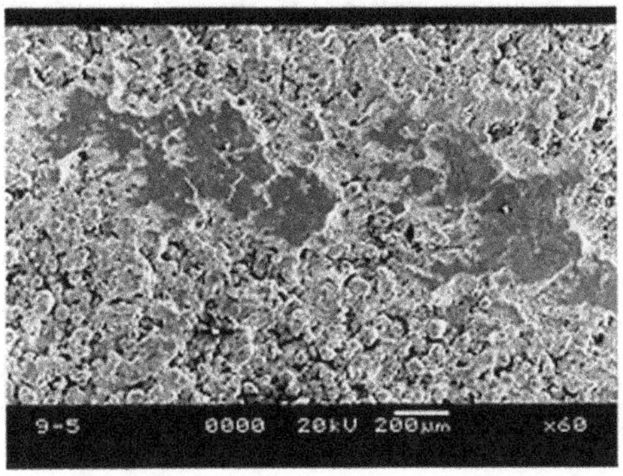

(a)

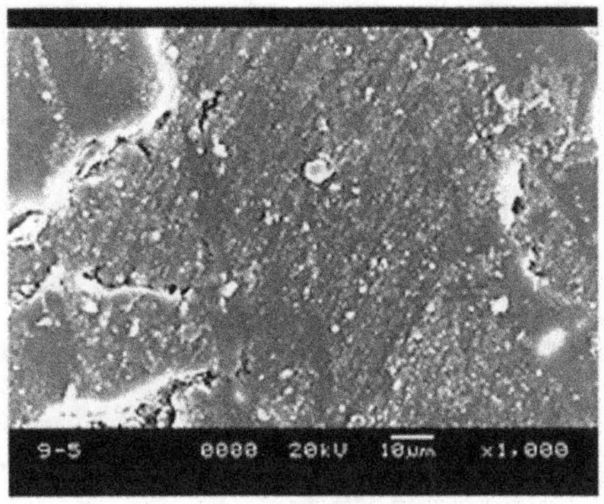

(b)

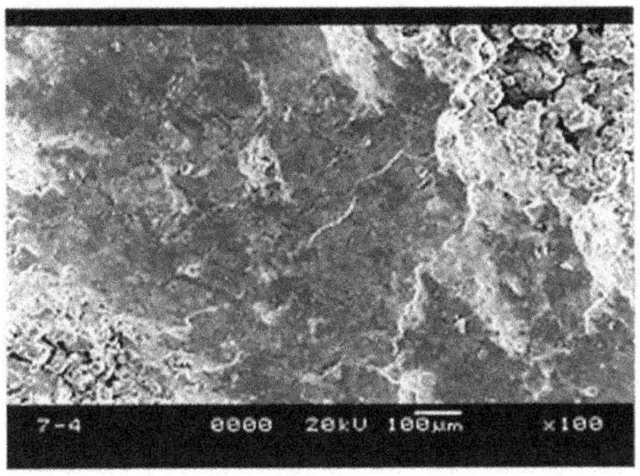

(c)

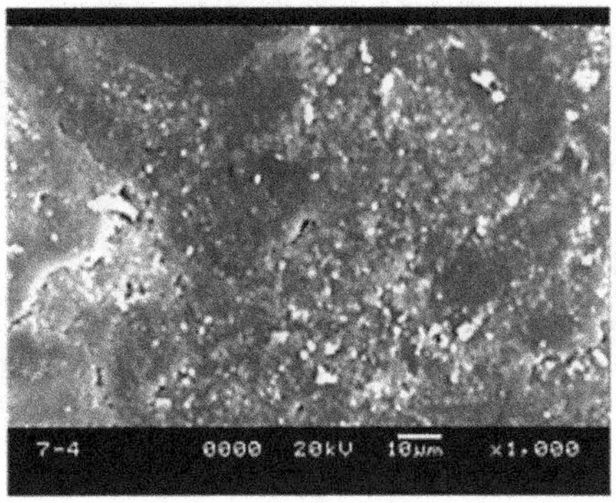

(d)

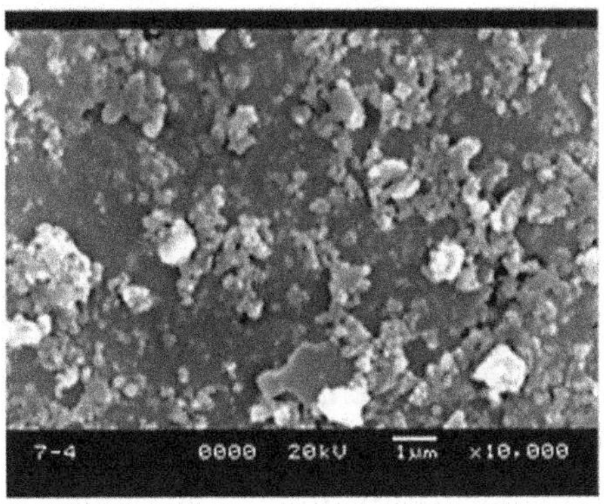

(e)

Figure 18. Morphology of wear tracks at lower and higher magnifications corresponding to the wear tests done on nanostructured alumina-titania coatings at a sliding speed of 0.33 m/s, normal load of 1.5 kgf and (a, b) 707 and (c, d) 900 CPSPs, respectively.

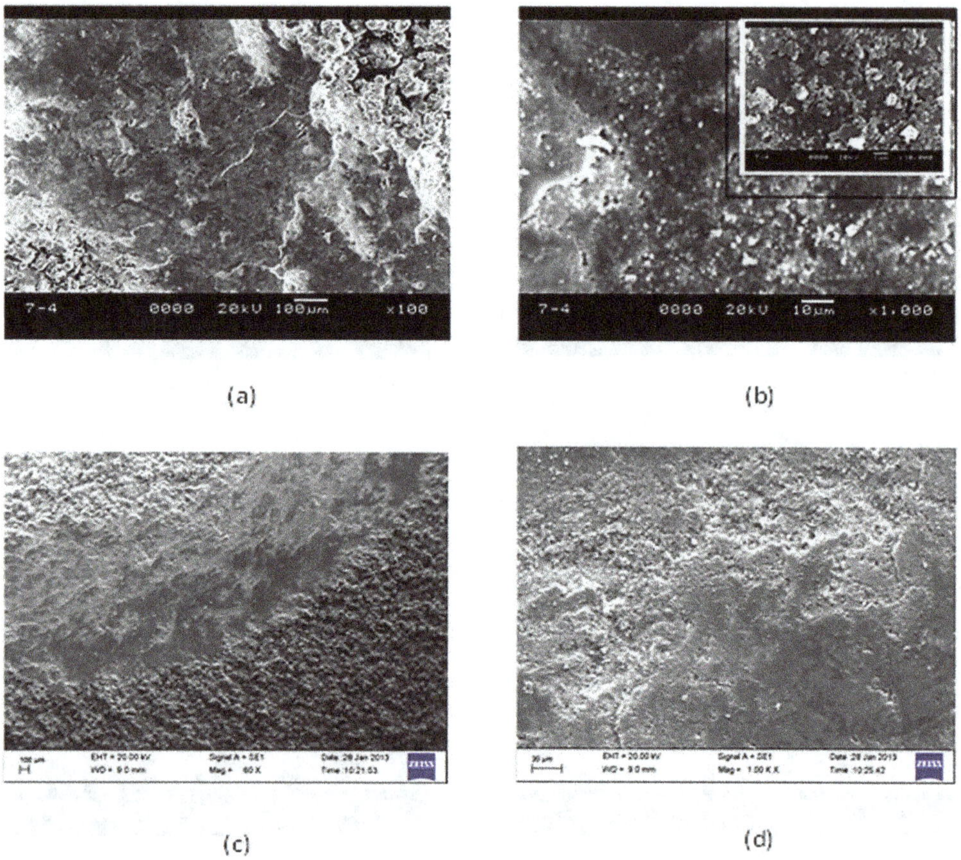

Note that for this CPSP, weight loss for the coatings increases with increase of normal load (see Figure 15). However, no major difference in wear mechanism can be observed.

FUTURE RESEARCH DIRECTIONS

This work can be taken forward in the following directions. Wear performance of nanostructured and conventional coatings in this work has been obtained in as- sprayed conditions. For air plasma sprayed alumina-titania coatings deposited with nitrogen, weight loss for nanostructured coatings at lowest load was considerably high as compared to that for conventional counterparts. However, the increase in weight loss with increase in normal load was much lower for nanostructured coatings deposited at identical CPSP. Uneven, rough top surfaces of coatings could be playing an important role in wear tests particularly at lower normal loads. Thus, the wear performance of coatings, particularly the alumina-titania coatings, could be evaluated after surface grinding the coatings to eliminate the effect of uneven top surfaces.

Figure 19. Morphology of wear tracks at lower and higher magnifications corresponding to the wear tests done on conventional alumina-titania coatings at a sliding speed of 0.33 m/s and a normal load of (a, b) 0.5, (c) 1.0 and (d, e) 1.5 kgf, respectively. Note that these coatings were deposited at 900 CPSP using nitrogen as primary plasma gas.

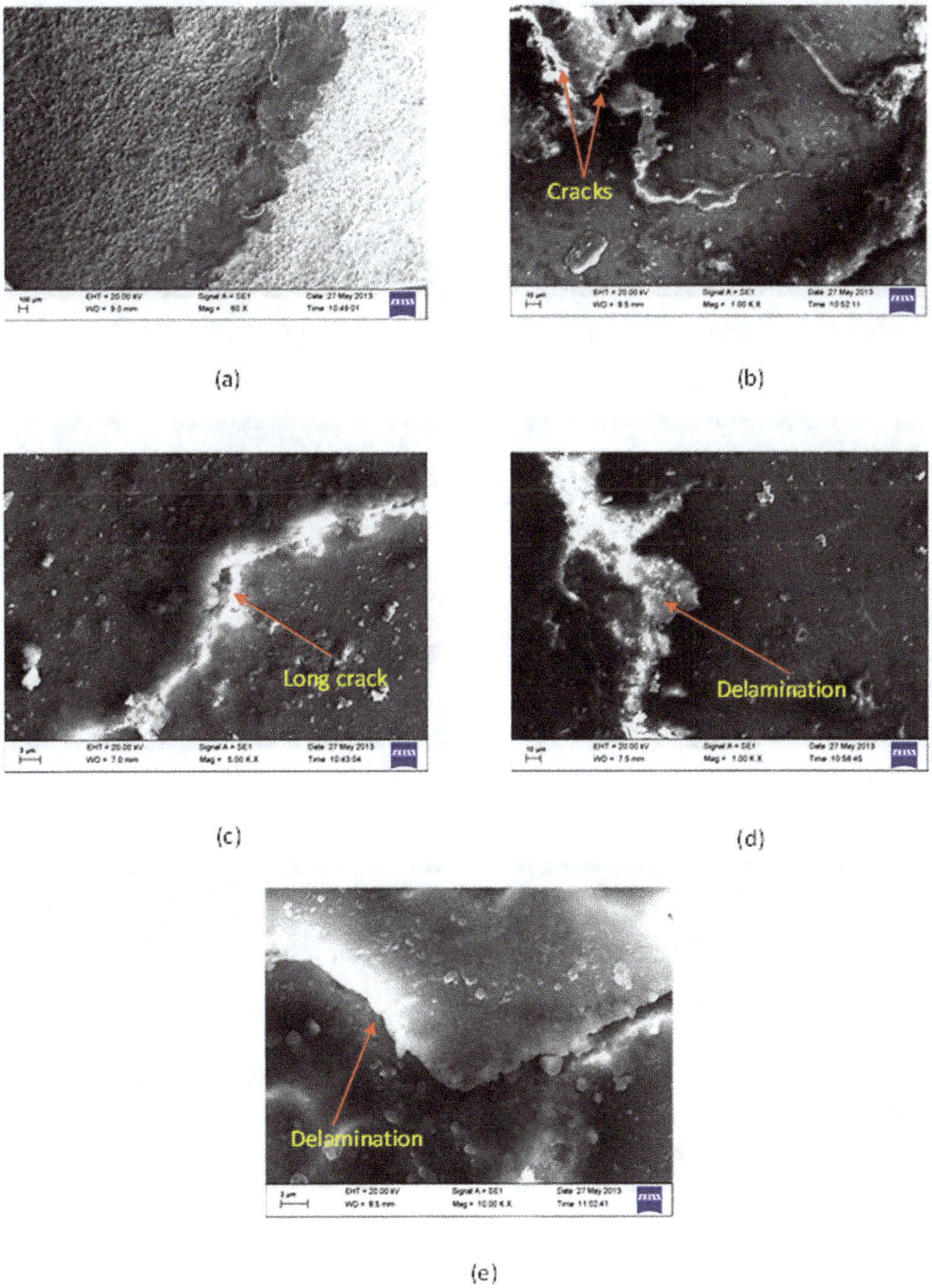

Figure 20. Morphology of wear tracks at lower and higher magnifications corresponding to the wear tests done on nanostructured alumina-titania coatings deposited using argon as primary plasma gas at a sliding speed of 0.33 m/s, normal load of 1.5 kgf and (a-c) 335 and (d-f) 419 CPSPs, respectively.

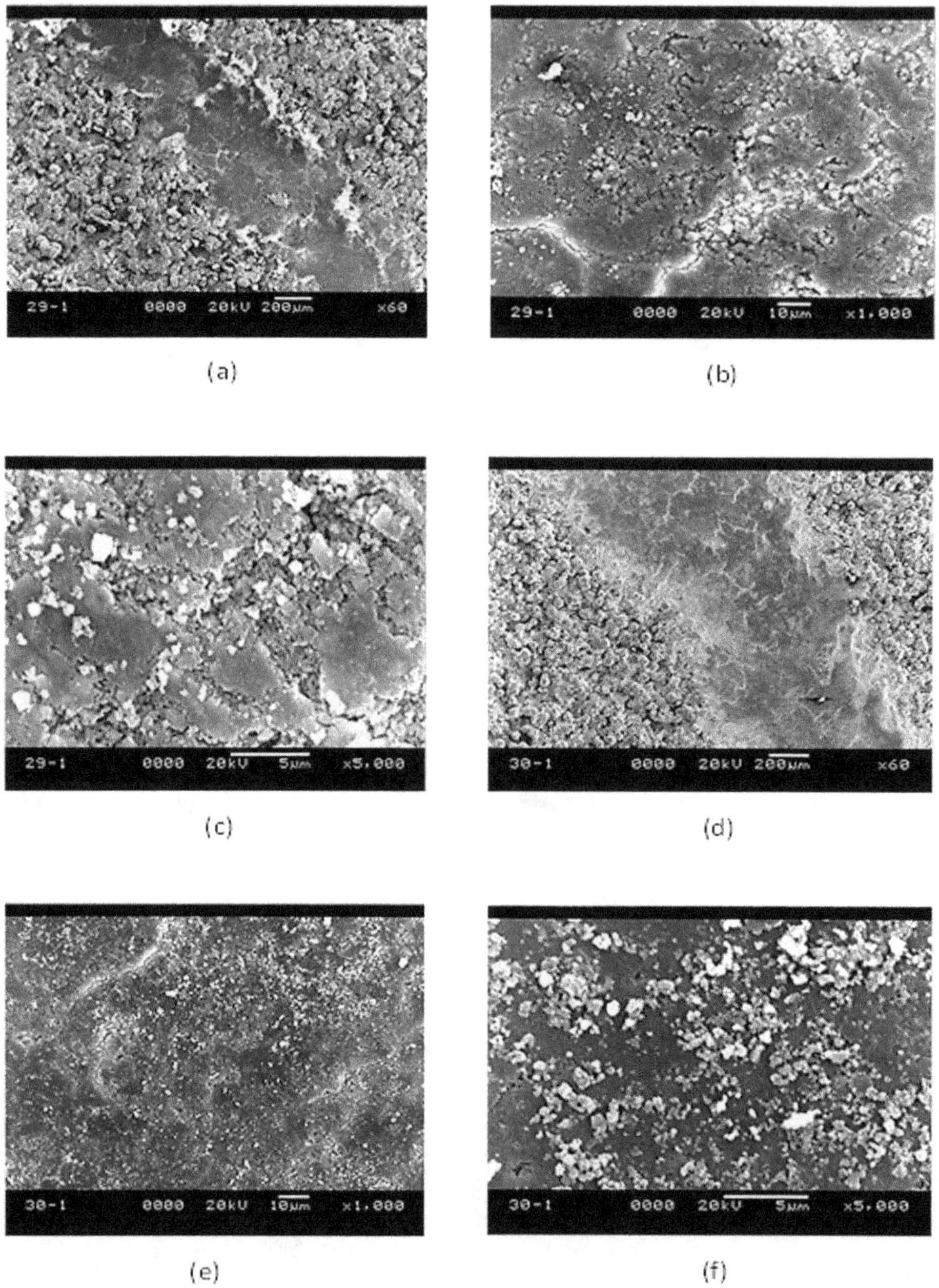

Figure 21. Morphology of wear tracks at lower and higher magnifications corresponding to the wear tests done on nanostructured alumina-titania coatings deposited using argon at a sliding speed of 0.33 m/s and a normal load of (a-c) 0.5 and (d-f) 1.5kgf, respectively. Note that these coatings were deposited at highest CPSP 419.

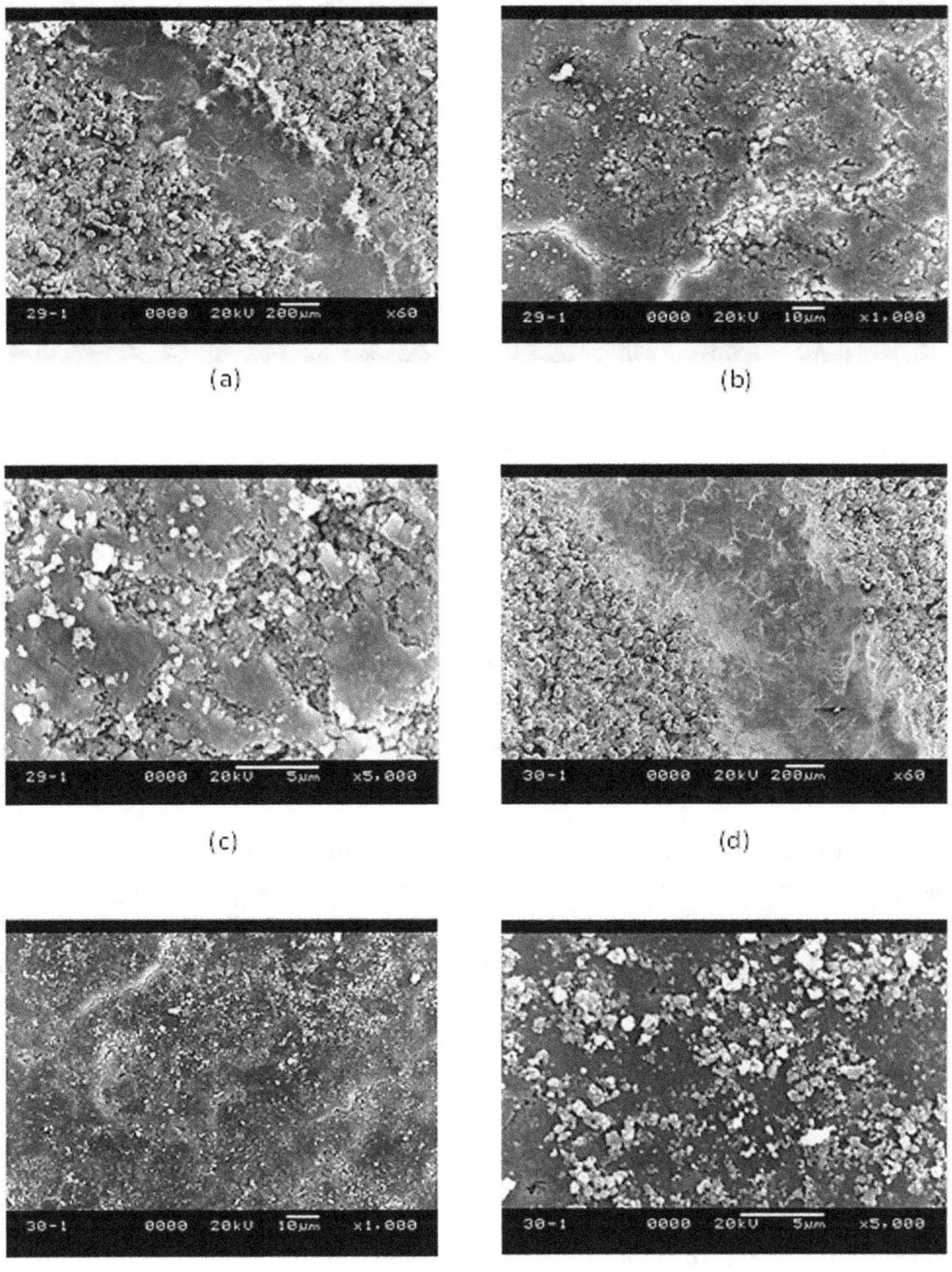

Figure 22. Morphology of wear tracks at higher magnifications corresponding to the wear tests done on nanostructured alumina-titania coatings deposited using argon at a sliding speed of 0.33 m/s and a normal load of (a) 0.5 and (b) 1.5 kgf, respectively. Note that these coatings were deposited at CPSP 335.

CONCLUSION

This chapter presents a detailed study of microstructural characteristics and wear performance of air plasma sprayed conventional and nanostructured alumina-13wt% titania coatings deposited using nitrogen and argon as primary plasma gases. While conventional coatings that exhibit a fully melted splat structure were obtained at a single critical plasma spray parameter (CPSP), nanostructured coatings for both the primary gases were obtained as a function of critical plasma spray parameter (CPSP). The following are the conclusions:

- Nanostructured powders were seen to result in a much more continuous, uniform powder flow than conventional powders that are blocky and angular. Spherical morphology of nanostructured powders was believed to be responsible for their enhanced flow ability.
- Nanostructured coatings deposited with argon *as well as* nitrogen were seen to exhibit bi-modal microstructures with homogenous distribution of partially melted regions in fully melted matrix. On the other hand, conventional coatings were seen to be composed of only of the fully melted matrix.
- Percentage of partially melted (PM) regions in nanostructured coatings decreased nearly linearly with increasing CPSP. However, while the percentage change of PM regions for coatings deposited with nitrogen was about 10-17%, the same for coatings deposited with argon was about 15-45%. Higher specific enthalpy for nitrogen at a specified gas temperature, which can bring about more intensive melting, is thought to be responsible for this.
- Percentage of α-alumina in nanostructured coatings was seen to be primarily dependent on choice of primary plasma gas. While the average volume percentage of α-alumina in nanostructured coatings was about 3% for coatings deposited with nitrogen, the same for coatings deposited with argon was about 27%. The volume percentage of α-alumina in conventional coatings deposited with nitrogen was found to be about 7%. Effect of CPSP on percentage of α-alumina content in

nanostructured coatings deposited with nitrogen was seen to be minimal. Percentage porosity for nanostructured coatings decreased almost linearly with increasing CPSP and was of the same order for coatings deposited with nitrogen as well as argon.

- Microhardness of nanostructured coatings deposited with nitrogen was significantly higher than that for corresponding coatings deposited with argon. While average microhardness of nanostructured coatings deposited with nitrogen was about 10 GPa that for coatings done with argon was about 6.5 GPa.
- Average surface roughness of as-sprayed nanostructured coatings deposited with nitrogen was found to be much smaller than that for coatings deposited with argon. However, while surface roughness of coatings deposited with nitrogen was found to be dependent on CPSP, there was nearly no effect of CPSP on surface roughness of coatings deposited with argon.
- Effect of CPSP on weight loss in ball-on-disc wear tests, for nanostructured coatings deposited with nitrogen, was found to be small. In addition, while conventional coatings were seen to result in lower weight loss at smaller normal loads, nanostructured coatings deposited at an intermediary CPSP performed better at higher loads, albeit by a small amount.
- In nearly all cases, weight loss was seen to increase super linearly with normal load for nanostructured coatings deposited with nitrogen. Scanning electron micrographs of wear track surfaces indicate that fatigue wear followed by pulverization was responsible for super linear increase of weight loss.
- Effect of CPSP on weight loss in wear tests, for nanostructured coatings deposited with argon, was found to be significant. Minimal weight loss was observed for coatings deposited at the highest CPSP.
- Weight loss in wear tests done at the highest normal load and intermediate sliding speed considered, for best performing nanostructured coatings deposited with nitrogen was about 50% higher than that for best performing coatings deposited with argon. Further, weight loss for coatings deposited at 419 CPSP with argon was found to be nearly unaffected by the normal load at least in the load range considered.
- In contrast to wear track morphology for nanostructured coatings deposited with nitrogen, track morphologies for coatings deposited with argon, particularly at highest CPSP, were *smooth*. In addition, weight loss for coatings deposited with argon increased nearly linearly with normal load. These observations suggest that abrasive wear mechanism was dominant for the coatings deposited with argon.

REFERENCES

Archard, J. F. (1953). Contact and rubbing of flat surfaces. *Journal of Applied Physics*, *24*(8), 981–988. doi:10.1063/1.1721448

Bounazef, M., Guessasma, S., Montavon, G., & Coddet, C. (2004). Effect of APS process parameters on wear behaviour of alumina-titania coatings. *Materials Letters*, *58*(20), 2451–2455. doi:10.1016/j.matlet.2004.02.026

Carim, A. H., Rohrer, G. S., Dando, N. R., Tzeng, S. Y., Rohrer, C. L., & Perotta, A. J. (1997). Conversion of Diaspore to corundum: A new α-alumina transformation sequence. *Communications of the American Ceramic Society*, *80*(10), 2677–2680. doi:10.1111/j.1151-2916.1997.tb03171.x

Erickson, L. C., Hawthorne, H. M., & Troczynski, T. (2001). Correlations between microstructural parameters, micromechanical properties and wear resistance of plasma sprayed ceramic coatings. *Wear*, *250*(1-12), 569–575. doi:10.1016/S0043-1648(01)00608-1

Escure, C., Vardelle, M., & Fauchais, P. (2003). Experimental and theoretical study of the impact of alumina droplets on cold and hot substrates. *Plasma Chemistry and Plasma Processing*, *23*(2), 185–221. doi:10.1023/A:1022976914185

Gadow, R., Killinger, A., Friedrich, C., & Voss, A. (1998). Combined metallurgical and ceramic coating in the development of tubular ozone generators. In *Proceedings of 15th International Thermal Spray Conference*, Nice, France (pp. 1081-1090).

Gell, M., Jordan, E. H., Sohn, Y. H., Goberman, D., Shaw, L., & Xiao, T. D. (2001). Development and implementation of plasma sprayed nanostructured ceramic coatings. *Surface and Coatings Technology*, *146*, 48–54. doi:10.1016/S0257-8972(01)01470-0

Goberman, D., Sohn, Y. H., Shaw, L., Jordan, E., & Gell, M. (2002). Microstructure development of Al_2O_3-13wt.% TiO_2 plasma sprayed coatings derived from nanocrystalline powders. *Journal of Acta Materialia*, *50*(5), 1141–1152. doi:10.1016/S1359-6454(01)00414-1

Gualtieri, M. L., Prudenziati, M., & Gualtieri, A. F. (2006). Quantitative determination of the amorphous phase in plasma sprayed alumina coatings using the Rietveld method. *Surface and Coatings Technology*, *201*(6), 2984–2989. doi:10.1016/j.surfcoat.2006.06.009

Guessasma, S., Bounazef, M., & Coddet, C. (2004). Note on POD test parameters to study wear behaviour of alumina–titania coatings. *Materials Characterization*, *52*(3), 237–241. doi:10.1016/j.matchar.2004.05.002

Guessasma, S., Bounazef, M., Nardin, P., & Sahraoui, T. (2006). Wear behavior of alumina–titania coatings: Analysis of process and parameters. *Ceramics International*, *32*(1), 13–19. doi:10.1016/j.ceramint.2004.11.014

Guilemany, J. M., Nutting, J., & Dougan, M. J. (1997). A Transmission Electron Microscopy Study of the Microstructures Present in Alumina Coatings Produced by Plasma Spraying. *Journal of Thermal Spray Technology*, *6*(4), 425–429. doi:10.1007/s11666-997-0025-5

Heiman, R. B. (1996). *Plasma Spray Coating: Principle and Application* (1st ed.). Weinheim, Germany: VCH. doi:10.1002/9783527614851

Ingham, H. S., & Fabel, A. J. (1975). Comparison of plasma flame spray gases. *Welding Journal*, 101–105.

Jordan, E. H., Gell, M., Sohn, Y. H., Goberman, D., Shaw, L., Jiang, S., & Strutt, P. et al. (2001). Fabrication and evaluation of plasma sprayed nanostructured alumina-titania coatings with superior properties. *Materials Science and Engineering A*, *301*(1), 80–89. doi:10.1016/S0921-5093(00)01382-4

Khor, K. A., & Li, Y. (2001). Crystallization bahaviors of in the plasma-spheroidized alumina/zirconia mixtures. *Materials Letters*, *48*(2), 57–63. doi:10.1016/S0167-577X(00)00280-9

Krishnan, R., Dash, S., Kesavamoorthy, R., Babu Rao, C., Tyagi, A. K., & Raj, B. (2006). Laser surface modification and characterization of air plasma sprayed alumina coatings. *Surface and Coatings Technology*, *200*(8), 2791–2799. doi:10.1016/j.surfcoat.2005.05.002

Lin, X., Zeng, Y., Zhou, X., & Ding, C. (2003). Microstructure of alumina-3wt.% titania coatings by plasma spraying with nanostructured powders. *Materials Science and Engineering A*, *357*(1-2), 228–357. doi:10.1016/S0921-5093(03)00164-3

Luo, H., Goberman, D., Shaw, L., & Gell, M. (2003). Indentation fracture behaviour of plasma-sprayed nanostructured Al_2O_3-13wt%TiO_2 coatings. *Materials Science and Engineering A*, *346*(1-2), 237–245. doi:10.1016/S0921-5093(02)00523-3

McPherson, R. (1973). The structure of Al_2O_3-Cr_2O_3 powders condensed from a plasma. *Journal of Materials Science*, *8*(6), 859–862. doi:10.1007/BF02397915

Miguel, J. M., Guilemany, J. M., & Vizcaino, S. (2003). Tribological study of NiCrBSi coatings obtained by different processes. *Tribology International*, *36*(3), 181–187. doi:10.1016/S0301-679X(02)00144-5

Mizuta, H., Oda, K., Shibasaki, Y., Maeda, M., Michida, M., & Ohshima, K. (1992). Preparation of high strength and Translucent Alumina by Hot Isostatic Pressing. *Journal of the American Ceramic Society*, *75*(2), 469–473. doi:10.1111/j.1151-2916.1992.tb08203.x

Ono, H., Teramoto, T., & Shinoda, T. (1993). Development of plasma sprayed lubricant free slide plates with ceramic coat for railway use. *Materials and Manufacturing Processes*, *8*(4-5), 451–463. doi:10.1080/10426919308934850

Pantelis, D. I., Psyllaki, P., & Alexo-poulos, N. (2000). Tribological behaviour of plasma sprayed Al_2O_3 coatings under severe wear conditions. *Wear*, *237*(2), 197–204. doi:10.1016/S0043-1648(99)00324-5

Rico, A., Rodriguez, J., Otero, E., Zeng, P., & Rainforth, W. M. (2009). Wear behaviour of nanostructured alumina-titania coatings deposited by atmospheric plasma spray. *Wear*, *267*(5-8), 1191–1197. doi:10.1016/j.wear.2009.01.022

Rosenflanz, A., Frey, M., Endres, B., Anderson, T., Richards, E., & Schardt, C. (2004). Bulk glasses and ultra hard nanoceramic based on alumina and rare-earth oxides. *Nature*, *430*(7001), 761–764. doi:10.1038/nature02729 PMID:15306804

Rosenflanz, A., Frey, M., Endres, B., Anderson, T., Richards, E., & Schardt, C. (2004). Bulk glasses and ultra-hard nano ceramic based on alumina and rare-earth oxides. *Nature*, *430*(7001), 761–764. doi:10.1038/nature02729 PMID:15306804

Sabiruddin, K. (2011). Properties of splats and coatings obtained from plasma sprayed alumina based powders [Ph.D Thesis]. Department of Mechanical Engineering, Indian Institute of Technology Kharagpur, India.

Sarikaya, O. (2005). Effect of some parameters on microstructure and hardness of alumina coatings prepared by the air plasma spraying process. *Surface and Coatings Technology*, *190*(2-3), 388–393. doi:10.1016/j.surfcoat.2004.02.007

Shaw, L. L., Goberman, D., Ren, R., Gell, M., Jiang, S., Wang, Y., & Strutt, P. R. et al. (2000). The dependency of microstructure and properties of nanostructured coatings on plasma spray conditions. *Surface and Coatings Technology*, *130*(1), 1–8. doi:10.1016/S0257-8972(00)00673-3

Shipway, P. H., & Hogg, J. J. (2005). Dependence of microscale mechanisms of WC-Co hard metals. *Wear*, *259*, 44–51. doi:10.1016/j.wear.2004.12.021

Suh, N. P. (1973). The delamination theory of wear. *Wear*, *25*, 111–124. doi:10.1016/0043-1648(73)90125-7

Tekmen, C., Tsunekawa, Y., & Okumiya, M. (2008). Effect of plasma spray parameters on in-flight particle characteristics and in-situ alumina formation. *Surface and Coatings Technology*, *203*(3-4), 223–228. doi:10.1016/j.surfcoat.2008.08.057

Tekmen, C., Tsunekawa, Y., & Okumiya, M. (2009). In-situ TiB_2-Al_2O_3 formed composite coatings by atmospheric plasma spraying: Influence of process parameters and in-flight particle characteristics. *Surface and Coatings Technology*, *203*(12), 1649–1655. doi:10.1016/j.surfcoat.2008.12.016

Tian, W., Wang, Y., & Yang, Y. (2008). Fretting wear behaviour of conventional and nanostructured Al_2O_3-13wt%TiO_2 coatings fabricated by plasma spray. *Wear*, *256*(11-12), 1700–1707. doi:10.1016/j.wear.2008.04.012

Tian, W., Wang, Y., Zhang, T., & Yang, Y. (2009). Microstructure, spallation and corrossion of plasma sprayed Al_2O_3-13% TiO_2 coatings. *Corrosion Science*, *51*(12), 2924–2931. doi:10.1016/j.corsci.2009.08.005

Wang, D., Tian, Z., Shen, L., Liu, Z., & Huang, Y. (2009). Microstructural characteristics and formation mechanism of Al_2O_3-13wt.% TiO_2 coatings plasma-sprayed with nanostructured agglomerated powders, *Journal of Surface and coatings. Technology (Elmsford, N.Y.)*, *203*, 1298–1303.

Wang, Y., Jiang, S., Wang, M., Wang, S., Xiao, T. D., & Strutt, P. R. (2000). Abrasive wear characteristics of plasma sprayed nanostructured alumina/titania coatings. *Wear*, *237*(2), 176–185. doi:10.1016/S0043-1648(99)00323-3

Yugeswaran, S., Selvarajan, V., Vijay, M., Ananthapadmanababhan, P. V., & Sreekumar, K. P. (2010). Influence of critical plasma spraying parameter (CPSP) on plasma sprayed alumina-titania coatings. *Ceramics International*, *36*(1), 141–149. doi:10.1016/j.ceramint.2009.07.012

ENDNOTE

1 In all the wear tests done in this work sliding distance is taken as 300 m and this information will not be mentioned in the remainder of the chapter during discussion of wear test results.

Chapter 12
Reactive Plasma Spray: A Method for Nitride Coatings Deposition in Thermal Spray

Mohammed Shahien
Central Metallurgical R&D Institute (CMRDI), Egypt & Toyohashi University of Technology (TUT), Japan

ABSTRACT

Thermal spraying is a well-known coating technology with many variations in spraying techniques, feedstock materials and substrate materials. These unique variations increased its industrial applicability in different fields, including aerospace, automotive, chemical process, corrosion protection, and medical applications. However, one of the main limitations of thermal spray is the difficulty of depositing several nitride ceramics directly using conventional techniques. This is due to the decomposition of nitride particles under high temperature without a stable melting phase. This chapter presents reactive plasma spraying (RPS) technology as a promising solution for the in situ fabrication of several nitride ceramic coatings. The main attractive prospects of RPS for fabricating nitride coatings are specifically highlighted. Successful development of various high-temperature nitride coatings, such as AlN, Fe4N and Si3N4, are presented. Process optimization, the relationship between reaction and process parameters and the influence on coatings formation are comprehensively discussed.

INTRODUCTION

Surface coating is a suitable solution to improve the performance of industrial parts by improving the surface function ability, enhancing the component lifetime and reducing the manufacturing cost. The coating techniques are divided into thin film and thick film deposition techniques. Thin film deposition techniques are mainly based on chemical/physical vapor phase deposition phenomena. Thick films can be deposited using several techniques, including wet chemical or electrochemical process, overlay welding, brazing, and thermal spraying. Each technique has certain advantages and certain limitations. Among the thick film deposition techniques, thermal spray deposition is well known industrially due to the broad variations of the following:

DOI: 10.4018/978-1-5225-4194-3.ch012

- Materials (metallic or nonmetallic materials)
- Material forms (powder, ceramic rod, wire, molten materials, suspension, solution-precursor)
- Spraying systems (high velocity oxy-fuel, flame, wire arc, detonation, plasma, laser-plasma, atmospheric vacuum, and high frequency)
- Substrate materials, forms, shapes and dimensions (length, width and thickness)

These unique variations in thermal spray increase its industrial applicability in different fields, including aerospace, automotive, power, chemical process, corrosion protection, and medical applications. In general, thermal spray technology is based on molten or semi-molten phase deposition to form the coatings. Thus, the solid feedstock (in the form of a powder, wire, rod, suspension, or solution precursor) is introduced into a hot environment (gas, flame, or plasma jet) to form liquid droplets, which are accelerated and propelled onto a substrate material. The molten or semi-molten particles impact on the substrate, rapidly solidify and coalesce to form the thermal-sprayed coating. However, one of the limitations of thermal spray technology is the difficulty of depositing several nitride ceramics directly using conventional thermal spray techniques. This is attributed to the decomposition of nitride particles under high temperature without a stable melting/semi-melting phase.

One suitable solution to fabricate nitride ceramics is to use the reactivity of plasma and to encourage the plasma nitriding reaction of the molten feedstock materials. Reactive plasma spraying (RPS) is a promising solution for *in situ* fabrication of nitride ceramic coatings and other compositions, which are difficult to deposit using conventional thermal spray techniques. The process is based on the chemical reactions between the feedstock materials (elemental metal or non-metal powder) and the surrounding active species in the plasma (such as atoms, ions, and radicals) to develop the required materials. Several nitride ceramics, including titanium nitride (TiN), aluminum nitride (AlN), iron nitride (Fe_4N) and silicon nitride (Si_3N_4) have been successfully fabricated using the RPS process. In the last decade, development of the reactive plasma spray technology to fabricate thick nitride ceramic coatings has received significant attention in the field of thermal spraying.

This chapter presents the main concepts of reactive plasma spraying technology and the applicability of reactive spraying in different spraying systems and feedstock materials. Furthermore, it summarizes the state of current knowledge about plasma nitriding as well as the main aspects and challenges for implementing RPS to fabricate different nitride ceramic coatings using the RPS process.

BACKGROUND

Definition and History of Thermal Spray

Thermal spray is a generic term for a group of commonly used processes for depositing metallic, nonmetallic (ceramic or plastic) or any desired combination of coatings (Fauchais et al., 2001; Schiefler Filho et al., 2004). It is defined as the method of coating application by the deposition of molten or semi-molten (softened) particles propelling onto the substrate. Thermal spraying represents a group of processes which employ heat and velocity to coat the surface of one material with another, using wire or feedstock (Fauchais et al., 2001). Thermal spray coating involves the use of a torch to heat a material in a powder or wire form to a molten or near-molten state. The heat source is typically a flame or an electric arc, which uses gas to propel the material to the targeted substrate, creating a completely new surface. The major advantage of thermal spraying is the wide variety of spray materials and techniques (Schiefler Filho et al., 2004).

The origins of thermal spraying date back to the early 1900s, approximately 1911, when Max Ulrich Schoop in Zurich, Switzerland patented a process that incorporate an electric arc as the heat production source for melting two wires of lead (Pb) and tin (Sn) (Ishikawa et al., 1993). This process used an oxy-acetylene flame as the heat source to spray low melting point materials such as Sn and Pb. The invention of the metallizing process by Schoop is considered the birth date of the technology. Metallizing is defined as the thermal spraying of metallic coatings. Several years later, their efforts produced the first instrument for the spraying of solid metal in wire form. This simple device was based on the principle that if a wire rod were fed into an intense, concentrated flame, it would melt. Furthermore, if the flame were surrounded by a stream of compressed gas, the molten metal would become atomized and readily propelled onto a surface to create a coating. Previously, the process was called flame spraying since only flame combustion spray guns were available. Other methods were later invented, which used electrical and plasma energies instead of flames to generate the required heat. Hence, the more generic term "thermal spray" was adopted.

Initial applications of this technology were repairing tanks and aircraft components. Later, the arc spraying process was first started in Germany by the OSU company (Zerfass, 2006), developing power sources as well as arc spray pistols ready to be used in industry (Ruijin et al., 2004; Pawlowski, 2008). High Velocity Oxy-Fuel (HVOF) thermal spray was developed at the end of 1970s and the beginning of 1980s. Today, there are many commercially available spray system. They have different designs, but all are based on the same fundamental principles. Thermal spraying has many advantages (Pawlowski, 2008; Bunshah 2001):

- Most materials can be used as substrates, i.e., metals, ceramics and organic materials, including wood, cloth and paper.
- The feedstock for the thermal spray material can be widely selected, such as metals, inter-metallic compounds, ceramics, plastics and their compounds.
- There is no restriction in the size of the substrates.
- High deposition rate compared with other coating processes.
- It is easy to undertake construction on-site.
- These processes are characterized by zero dilution of the substrate as a result of mechanical bonding.
- The ability to apply thin coatings (control the coating thickness) and a high rate of area coverage compared with arc welding processes.
- The melting point of the coating material can be higher than of the substrate.
- Other advantages of spraying include little or no pre- and post-heat treatment and minimum distortion of the substrate (Harvey, 1996). Therefore, usually no influence on heat treatments, chemical compositions, etc., is observed due to moderate heat input by thermal spray coating processes, and substrate temperatures seldom exceed 150°C.

Thermal Spray Materials

The major advantage of thermal spraying is the wide variety of materials that can be used. The coating material may be a single element, an alloy or a compound in a powder form with unique physical properties, or in a rod or wire form. In most cases, this is achievable only using the thermal spray process. A thermal spray gun uses plasma arc, combustible gases or an electric arc to generate the necessary heat to melt the coating material. The material changes to a plastic or molten state when heated and is accelerated by process gases. The acceleration of the molten material produces a confined stream of particles that the process gases transfer to the substrate. The particles strike the substrate, flatten upon impact, and form thin platelets that bond to the substrate and to each other. The sprayed particles build up and cool down into a lamellar structure that forms a coating.

Since the first demonstration by Schoop, there have been significant developments in the technology of thermal spray with respect to range of materials that can be deposited, heat source employed to heat the material, velocity attained by the particles and properties of the coatings. Thermal spraying processes vary according to the heat source and feedstock material as summarized in Table 1 (Pawlowski, 2008; Bunshah 2001; Al-Sabouni, 1999).

Thermal Spray Basic Operation and Techniques

Thermal spraying processes vary according to the heat source, spray velocity and feedstock material. They can be categorized into two groups, which are of the lower and higher energy processes. The lower (traditional) processes include electric arc spraying (EAS) and flame spraying, which is widely used for general atmospheric corrosion resistance, such as zinc and aluminum coatings on bridges and other structures.

High-velocity oxy-fuel (HVOF) and plasma arc spraying (PAS) are higher (advanced) energy processes. The coatings produced are of lower porosity and increased adhesion to the substrate. This process also allows spraying of consumables that have higher melting points, which increases the range of deposited materials and achieves higher resistance to more severe thermal and mechanical environments. The above-mentioned processes can also be further divided into combustion or plasma/electric depending on the type of energy used to heat the sprayed material or wires to form coatings. For example, both HVOF and flame spraying processes use combustion of oxygen and fuel gas.

Currently, there are many commercially available spray systems that have different designs. However, all of them are based on the same fundamental principles. Figures 1 provides an overview of different thermal spray processes and the trend of thermal spraying development processes. Flame spraying, especially HVOF (High-Velocity Oxy-Fuel) and HVAF (High-Velocity Air Fuel), has been developed for the fabrication of metals, cermets and oxide ceramic coatings. However, for the fabrication of ceramics

Table 1. Thermal spray process, its material form and energy source

Process	Material Form	Energy Source
Flame spraying	Powder, Rod, Wire	Oxyacetylene flame.
Electric arc spraying	Wire (motor driven)	Electric arc.
Detonation flame spraying	Powder	O_2-acetylene-N_2 gas detonation.
High velocity oxy-fuel (HVOF)	Powder	Oxygen, fuel.
Atmospheric Plasma Spray (APS)	Powder	Plasma arc.
Vacuum Plasma Spraying (VPS)	Powder	Plasma arc.

Pawlowski, 2008; Bunshah 2001; Al-Sabouni, 1999.

Figure 1. Present thermal spraying development process trend

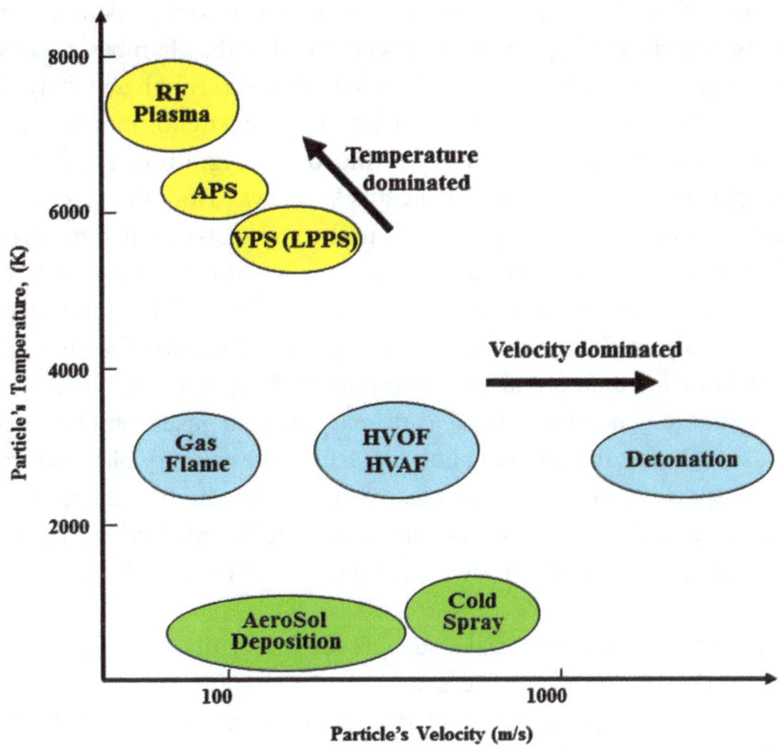

and high melting point metals, plasma spraying processes have been developed, such as RF (Radio Frequency) plasma spraying and DC (Direct Current) plasma spraying which includes APS (Atmospheric Plasma Spray) and LPPS (Low Pressure Plasma Spray) (Al-Sabouni, 1999; Yamada, 2006). Recently, a coating process without a heat source also been developed as cold spray and aerosol deposition.

Plasma Spraying Techniques

The most widely used electrical methods to produce thermal plasma are based on generating plasmas using either Direct Current (DC) or Radio Frequency (RF) sources. In the case of DC sources, generally, the generator is essentially a DC electric arc and two electrodes, front (anode) and rear (cathode), which are contained in a chamber in addition to the arc through which the operating gas passes. Plasma generators work on the concept that if a sufficient voltage is applied to two electrodes separated by a small gap, the difference in potential at that moment causes electrons to be extracted from the cathode. The electrons accelerate and speed towards the anode. If gas is inserted in the gap between the two electrodes, its atoms will collide with the ensuing electrons and themselves will causing more electrons to detach and travel towards the anode. Meanwhile, the nuclei, which are stripped of their electrons and are positively charged, move to the cathode. Thereby, the gas in the gap has been ionized and becomes electrically conductive (plasma arc); it exits through an orifice in the anode as a plasma stream, containing only electrons and ionized gas.

Most commercial plasma guns are fundamentally simple in design, consisting of a chamber and a front nozzle (anode) with an orifice. The chamber and nozzle are water-cooled. At the rear of the chamber is a cathode, which is also water-cooled. A port, somewhere within the chamber, allows the high-pressure plasma forming gas (or gases) to enter. The powder is introduced through an external port at the nozzle orifice; hardware is also available to inject powder internally upstream into the nozzle bore. A high-frequency spark initiates the operation and is discontinued upon ignition. The high-pressure gas cools the outer layer of the plasma arc so that extreme heat is kept away from the nozzle bore.

The plasma spray process is based on spraying of molten or heat-softened material onto a surface to provide coatings. The material, in the form of powder, is injected into a very high temperature plasma flame, where it is rapidly heated and accelerated to a high velocity. The hot materials impact on the substrate surface and rapidly cool, forming the coatings. Typical plasma forming gases include argon, nitrogen, hydrogen, helium and their combinations (argon-hydrogen, argon-helium, nitrogen-hydrogen, etc.). Argon and nitrogen are generally utilized as primary plasma gases, and hydrogen is used in most cases as a secondary, since it aids in producing hotter plasma. The primary plasma forming gas is usually used as a carrier to transport the powder to the plasma stream; the high temperatures and high thermal energy generated in the secondary plasma make it a very suitable method to coat materials with high melting points (i.e., ceramic materials). Plasma spray has the following advantages:

- Spraying of high melting point materials, such as refractory metals (e.g., tungsten) and ceramics (e.g., zirconia), unlike combustion processes.
- Plasma-sprayed coatings are generally much denser, stronger and cleaner than other thermal spray processes, with the exception of HVOF and detonation processes.
- Plasma spray coatings probably account for the widest range of thermal spray coatings and applications and make this process the most versatile.

Plasma spraying is considered as the most versatile of all thermal spray processes. Thus, the coating quality is in general higher than that obtained using flame spraying. Plasma spraying is most commonly performed in normal atmospheric conditions and referred to as Atmospheric Plasma Spray (APS). Some other spraying is conducted in protective environments using vacuum chambers that are normally back filled with a protective gas at low pressure. This is referred to as Low Pressure Plasma Spray (LPPS) or Vacuum Plasma Spraying (VPS).

Atmospheric Plasma Spraying (APS)

A DC atmospheric plasma spray gun consists of a circular anode, usually copper, and a cathode made of thoriated tungsten. The electric arc discharge produced by a generator heats up the working gases, which expand to form a plasma jet. The powder, which is suspended in a carrier gas, is injected into the plasma flame. The powder particles are melted and accelerated in the plasma flame, and upon impact with the substrate, the coatings are formed. Bond strengths for ceramic materials are typically in the range of 15-25 MPa, and for metallic alloys, bond strengths are in excess of 70 MPa. Porosity is usually in the range of 1-7% but may be much higher. Coating thicknesses are typically 50-500 μm.

In general, an APS system is characterized by its high particle velocity, and it will easily enable the fabrication of dense coatings. In addition, the APS process is characterized by its rapid cooling and solidification rates, which makes it an attractive process for the formation of non-crystalline materials (such as metallic glass) and non-stable phases. Moreover, there is a flexibility of substrate materials, shapes and dimensions (length, width and thickness) in the APS process. Thus, any shapes and dimensions are acceptable according to the desired applications. These characteristics make it a preferable process in industrial applications compared with CVD and PVD techniques.

Vacuum Plasma Spraying (VPS)

In addition, this process is sometimes called Low Pressure Plasma Spraying (LPPS). It is an effort to improve the properties of APS coatings, often for special applications. For example, for electronic applications, a process was required that minimized undesirable reactions with the environment. A vacuum plasma spraying installation is composed of a plasma torch (similar to the type used in APS) enclosed in a vacuum chamber. The chamber is initially evacuated to a certain pressure (~ 10-2 mbar) and then backfilled with argon to a certain value usually approximately 40 mbar. Reducing the amount of oxygen in the environment prevents oxidation of both particles and substrate. Consequently, excellent coatings result with superior densities and adhesion than those obtained using atmospheric plasma spraying. However, the requirement for a vacuum chamber means that the instillation is fixed, which leads to high capital cost and necessary automation. Furthermore, only small components can be coated, and the higher heat input to the work piece may cause some metallurgical or distortion problems.

Radio Frequency (RF) Plasma Spraying

A variant of thermal plasma generation is to use the high frequency or radio frequency (RF) plasma spray process, in which the electrical discharge is a radio frequency (>400 kHz) instead of a direct current arc discharge. In the RF plasma spraying, an RF discharge is maintained either by capacitive or inductive coupling using the power source. For capacitive coupling, the frequency electric field is in charge of maintaining the discharge. However, the time varying magnetic field is responsible for maintaining the inductively coupled discharge. Generation of thermal plasma via inductively coupled discharges is far more common. The plasma torch consists of a confinement tube surrounded by several turns of an induction coil connected to the RF power supply. Different gas streams are supplied into the discharge chamber from the head of the torch, and the required power to sustain the discharge is based on the frequency and gas composition (ex. lower power level achieved using monatomic gases such as argon). The most commonly used RF torches in plasma spraying have the following parameters: RF power in the range of 30-100 kW, plasma gas flow rates of 50–100 slm, temperature of the resulting plasma jet is in the range of 8,000 K and 10,000 K, with exit plasma velocity below 100 ms^{-1} and pressure of 30-100 kPa (Fauchais et al., 2014).

Basically, compared with most DC torches, the main difference is in the RF torch internal diameter, which leads to lower flow velocities below 100ms^{-1}, higher temperatures and the possibility of using larger particles. Furthermore, the axial injection function of feedstock powders is possible, where the feedstock powders are injected almost at the center of the RF coil. Additionally, RF plasma offers a larger plasma volume and longer residence times of the particles in the plasma suitable for melting larger size particles. I practical applications, radio frequency plasma has been employed for a wide range of powder processing (metals and ceramics). Table 2 summarizes the main differences between DC and RF plasma spraying (Fauchais et al., 1997).

Table 2. Summary of DC and RF plasma spraying characteristics

	RF Plasma Spraying	DC Plasma Spraying
Particle diameter range, μm	50-120	10-70
Particle velocity range, m/s	20-80	120-300
Particle flight time, ms	5-15	1-3
Particle flattening time, μs	2-3	1
Cooling time to 500 K, μm (a)	0.23-03.5	0.1-0.2
Splat solidification time, μm (a)	1-12	0.5-4
Time between two successive impacts, ms	5-20	1-5
Time between two successive passes, s	0.1-30	0.1-30

Fauchais et al., 1997.
(a) The values closely linked to the thermal contact resistance between splat and substrate or previously deposited layer

NITRIDE COATINGS DEPOSITION IN THERMAL SPRAY

Nitride ceramics [such as titanium nitride (TiN), silicon nitride (Si_3N_4), aluminum nitride (AlN), and iron nitride (Fe_4N)] are promising materials for several applications due to their excellent properties. Moreover, it is possible to improve many characteristics by fabricating nitride ceramic coatings on the material surface. Usually, nitride ceramic films are produced via vapor phase deposition techniques such as Chemical Vapor Deposition (CVD) and/or Physical Vapor Deposition (PVD). While these processes enable the fabrication of high-quality nitride ceramic thin films, it is difficult to produce thick coatings (more than 100 μm). This is attributed to the relatively low deposition rate of these vapor phase techniques (Sundgren, 1985). Nitride ceramic thin films are useful for a wide variety of electrical and electronic applications. However, it is difficult to apply thin films to structural parts (e.g., as protective layers in reaction vessels of CVD, PVD or etching, insulation for heat sinks and power device applications), which require thick coatings.

However, thermal spray has been widely used to produce thick coatings with a wide variety of spraying materials (metals, inter-metallic compounds, ceramics, plastics, or any desired combinations). However, it is difficult to deposit most nitride ceramics via the thermal spray process. This is due to the decomposition of nitride particles under high temperature without a stable melting phase. Thus, the thermal spray process requires melting or semi-melting for the feedstock powder to deposit on the substrate surface. Figure 2 presents a trial of direct spraying of aluminum nitride (AlN) powder using an atmospheric DC plasma spraying system. It is clear that few AlN particles impact on the substrate and are easy to remove even via scratching. This strongly reveals that it has been impossible to fabricate nitride coatings using conventional thermal spraying techniques.

Figure 2. Direct spraying of AlN feedstock powder by conventional plasma spray process: (a) spraying gun using Ar/H2, APS: 9MB, Sulzer Metco, Switzerland, (b) schematic expression of spraying experiment and (c) the real image of the sample after spraying

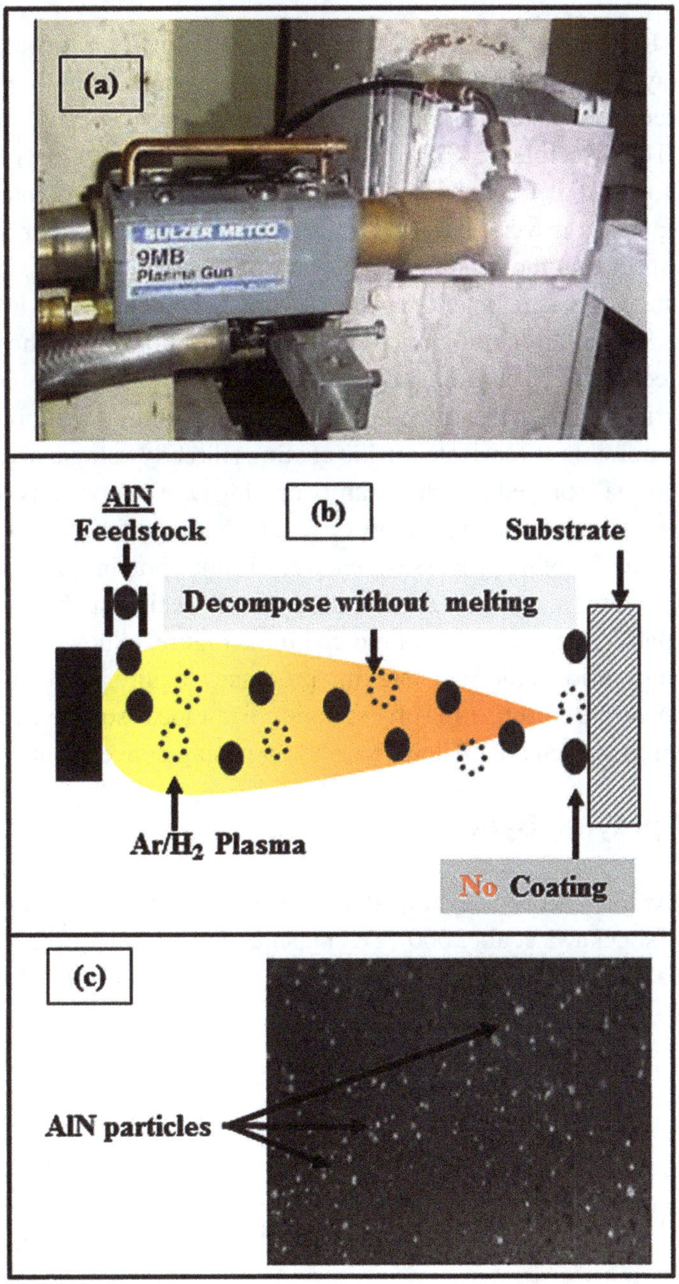

SOLUTION: REACTIVE PLASMA SPRAY A METHOD FOR NITRIDE COATINGS

Main Concept and Principle

Reactive plasma spray (RPS) is not a new spray system. It is a coating process that depends on using commercial plasma spray techniques and adjusting plasma environment to be reactive. This enables a reaction between the feedstock materials and the surrounding active species such as atoms, ions and radicals in the plasma. The principle of RPS is to encourage reactions within plasma under controlled conditions. The goal is to tailor the composition, which is difficult or impossible to achieve using other production routes. The process is based on the reaction of elemental metal or non-metal powder with the surrounding active species in the plasma such as atoms, ions and radicals. In the RPS process, the reactants are injected into or downstream of the plasma flame, where they react in-flight with the powder being sprayed. Reactants may be any combination of solids, liquids, or gases. The RPS process is a recent development, and its success has been limited due to the materials and spraying system environment.

For example most nitride ceramics decompose under high temperature without a stable melting phase. Therefore, it has been impossible to fabricate nitride coatings directly using conventional thermal spraying techniques. Thus, thermal-sprayed nitride ceramic coatings are not produced as industrial products (Yamada, 2006). Reactive plasma spraying is a promising technology for the *in situ* formation of nitride coatings. Among the plasma spraying processes, reactive plasma spraying (in which elemental metal or non-metal powder [rod or wire] reacts with surrounding active species, such as atoms, ions and radicals in plasma) is useful for fabricating nitride ceramic coatings. Figure 3 shows the schematic explanation of RPS process. Thus, conventional thermal spraying processes require molten or semi-molten droplets to form the coatings. Therefore, reactive plasma spraying is an ideal solution for *in situ* fabrication of non-oxide ceramic coatings, which are difficult to fabricate using conventional thermal spraying.

Reactive Plasma Spraying Systems

The reactive spraying process has been developed using Radio Frequency (RF) plasma (Yamada, 2006; Fukumoto et al., 1997; Tsunekawa et al., 2000), DC plasma (Tsunekawa et al., 1996, 1997), laser (Lugscheider et al., 1996; Bacci et al., 2000), or their hybrids (Ohmori et al., 1991, 1993) as reactive heat

Figure 3. Schematic expression of reactive plasma spraying process

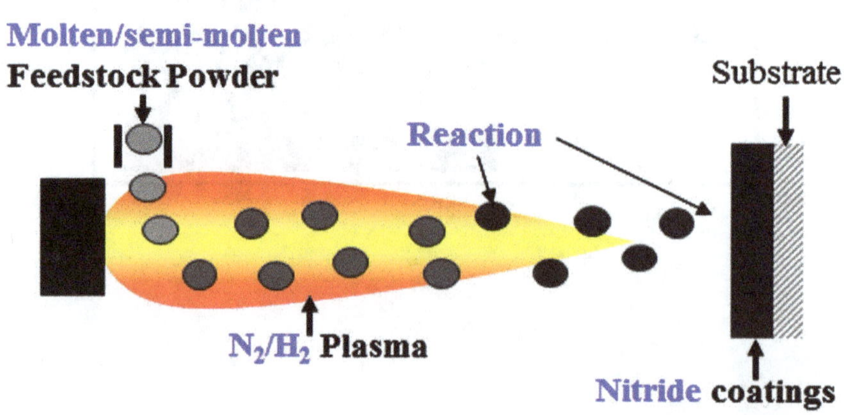

sources. Among these, reactive RF plasma spraying is considered the most attractive process (Yamada, 2006). The RF plasma spraying system provides the highest temperature and lowest particle velocity among the thermal spraying processes (Fauchais et al., 1997; Okada et al., 1989; Selezneva et al., 2001; Bianchi et al., 1995). This means that, RF plasma spraying provides more opportunities for the reaction between the sprayed particles and surrounding plasma compared with other reactive spraying systems. The characteristic times of RF plasma spraying and DC plasma spraying are summarized in Table 2 (Fauchais et al., 1997). Furthermore, the RF plasma spraying apparatus does not include an electrode, which is an advantage for reactive plasma spraying. Thus, electrode materials never diffuse into the plasma and deposited coatings, and the electrode is never corroded by active plasmas. Thus, a wide choice of reactants is possible. A commonly used RF plasma spraying apparatus is based on a hybrid plasma spraying apparatus made by Nihon Koshuha Co. Ltd., (Fukumoto et al., 1997; Yamada et al., 2004; Fukumoto et al., 2004; Yamada et al., 2006a, 2006b, 2007).

In addition, a DC plasma jet has unique properties, such as a high-energy content and high velocity, and has been used successfully for melting, spraying and fabrication of dense coatings. Unfortunately, the residence time of the injected particles is very short. Among the DC plasma spray systems, atmospheric plasma spraying (APS) is an industrially well-established process and is widely used to produce various coatings, especially for ceramic coatings, which are used for wear and corrosion protection and thermal barrier function. The APS process is based on the creation of a plasma jet to melt a feedstock powder. The powder particles are injected with the aid of a carrier gas and gain their velocity and temperature via thermal and kinematic transfers from the plasma jet. The atmospheric plasma process is an atmospheric system, and it is possible to adjust it as a reactive plasma technique because its design enables the use of N_2 gas as a plasma gas. Shahien et al., 2010a, 2010b, 2010c, 2011a, 2011b, 2011c, 2011d, 2011e, 2012, 2013a, 2013b, 2013c) used an atmospheric DC plasma spray system (APS: 9 MB, Sulzer Metco, Switzerland) in RPS mode using N_2 as the primary gas and H_2 as the secondary gas. Basically, the atmospheric plasma spray process is characterized by the following:

- Its high deposition rate (several μm per min), which makes it a promising process for fabrication of thick coatings.
- No limitations for the substrate shape or dimensions (length, width and thickness) in APS process. Thus, any shape and dimensions are acceptable according to the desired applications, which makes it a preferable process for industrial applications compared with the RF system.
- The APS process has a high particle velocity (Fauchais et al., 1997); therefore, it enables the fabrication of dense coatings.
- It is characterized by rapid cooling and solidification rates, which makes it an attractive process for the formation of non-crystalline materials (such as metallic glass) and non-stable phases.
- The APS process is already commercially used, and its design enables the use of N_2 as plasma gas. Therefore, by adjusting and optimizing it as a plasma nitriding technique, it will be a versatile process for several nitride coating fabrications.

However, because it is an atmospheric process and characterized by high particle velocity and high deposition rate, controlling and completing the nitriding process of the feedstock powder is expected to be difficult.

Reactive Spraying of Si_3N_4 Coatings

Several attempts have been made to fabricate thermal-sprayed silicon nitride (Si_3N_4), such as using Si_3N_4 and a metallic binder phase like Ni as a starting feedstock powder (Lugscheider et al., 1990; Thiel et al., 1996; Berger et al., 1998; Heimann et al., 1998; Gawne et al., 2001; Thiele et al., 2002). Even when it was possible to fabricate the coatings, the chemical reaction between Ni and Si_3N_4 formed a brittle silicide (Lugscheider et al., 1990; Thiel et al., 1996). In addition, several trials were performed using different binder materials to fabricate Si_3N_4 coatings via conventional thermal spray techniques such as plasma spraying, high velocity oxy-fuel (HVOF), and detonation gun spray. The binder materials used were silicon, alumina, yttria, and their mixed powder. Although thick and dense Si_3N_4-based coatings were fabricated, the coatings included impurities originating from the binder materials. In addition, the mechanical properties of the fabricated coatings were not good due to their undesirable microstructure. Therefore, development of a thermal spray coating technology to fabricate thick nirtide ceramic coatings is strongly required.

The reactive plasma spray nitriding of pure Si particles with the surrounding active species in the plasma (e.g., N atoms or ions) during spraying showed a promising way to fabricate Si_3N_4 plasma-sprayed coatings (Eckardt et al., 1994, 1995, 1996; Tahara et al., 2004; Yamada et al., 2004; Yamada 2006; Moriyama et al., 2003, 2004). The developed coatings contain a low concentration of a Si_3N_4 phase. Then, Yamada et al. successfully improved the Si_3N_4 content through spraying with $Ar/N_2/H_2$ plasma (Yamada et al., 2004). A thick and dense coating with a lamellar structure was fabricated with a deposition rate of approximately 20 μm/min. The hardness of this coating (fabricated with $Ar/N_2/H_2$ plasma) was approximately 1390 Hv, which is higher than the hardness of the coating fabricated with Ar/N_2 plasma (using the same starting powder and the same RF plasma spraying system). Furthermore, increase of the N_2 or H_2 fraction on the plasma improves the intensity ratio of β-Si_3N_4 compared with other phases.

Table 3 summarize the relationship between the intensity ratio of the XRD spectra and plasma gas fraction. The N_2 flow rate was substituted with a H_2 flow rate of 0.4 L/min, and the H_2 fraction was substituted with a N_2 flow rate of 4.0 L/min (Yamada et al., 2004, Yamada 2006). Therefore, N_2 and H_2 fractions in the plasma gas are the controlling factors for nitriding Si and for the fabrication of the Si_3N_4 phase during the RF plasma spray process. Increasing the N_2 fraction in the plasma gas increases the active nitriding species, such as N atoms (Tamaki et al., 1997), which contribute to nitridation. The H_2 addition to plasma gas is necessary to synthesize the Si_3N_4 phase and to enhance the nitriding reaction through the removal of the oxide layer which exists on the surface of the particles. However, adding H_2 to plasma improves the thermal conductivity of plasma, increases RF power, and increases the particle and substrate temperature, which contributes to nitriding (Bourdin et al., 1983). The nitriding reaction of Si in reactive RF plasma spraying occurred after deposition onto the substrate. The liquid phase of Si on the substrate was formed via H_2 addition into plasma gas and reacted with gaseous nitrogen.

Reactive Spraying of AlN Coatings

Among the nitride ceramics, the development of thermal-sprayed AlN ceramic coatings has received significant attention. Thus, the fabrication of thermal-sprayed AlN coatings will be a cost-effective solution for a number of applications (Krishna et al., 2002). Several attempts have been undertaken to fabricate AlN coatings using thermal spraying of AlN and Al_2O_3 powder via detonation spray (Krishna et al., 2002) and using a low-pressure plasma spray system (LPPS) (Cao et al., 1999). Furthermore, fabrication

Table 3. Relation between intensity ratio of $I(\beta\text{-}Si_3N_4)/I(Si)+I(\beta\text{-}Si_3N_4)$ and H_2 and N_2 flow rates

N_2 Gas Flow Rate Effect			H_2 Gas Flow Rate Effect		
Gas Flow, Q/l/min		$I(\beta\text{-}Si_3N_4)/I(Si)+I(\beta\text{-}Si_3N_4)$	Gas Flow, Q/l/min		$I(\beta\text{-}Si_3N_4)/I(Si)+I(\beta\text{-}Si_3N_4)$
N_2	H_2		H_2	N_2	
1	0.4	0.0530	0.5	4.0	0.14
2	0.4	0.2155	0.6	4.0	0.3
3	0.4	0.3162	0.7	4.0	0.315
4	0.4	0.6078	0.8	4.0	0.2
			0.9	4.0	0.355
			1	4.0	0.608

Yamada et al., 2004, Yamada 2006.

of AlN coatings with a carbon reduction of Al_2O_3 via plasma spraying was investigated (Ohmori et al., 1993; Kassabji et al., 1983; Khor et al., 2001). In addition, reactive plasma spraying showed a promising impact for the fabrication of AlN coatings through the nitriding of Al particles using an RF plasma spray system (Yamada 2006; Fukumoto et al., 2004; Yamada et al., 2006a, 2007).

Reactive spraying of Al particles using the N_2/Ar plasma from the RF plasma spraying system leads to the formation of AlN/Al-based coatings (Yamada 2006; Fukumoto et al., 2004). Dense coating can be achieved with low AlN and high Al content. Furthermore, the AlN content in the coating was improved by increasing the plasma power. However, the coating shifted to a porous structure. The increase in the AlN content is related to the change in the plasma temperature and reactivity with increase of the RF power and N_2 flow rate (Yamada et al., 2005). More clear explanations about this process will be addressed later during the discussion of the spectroscopic measurement of the plasma. Improvement of the AlN content is in good agreement with the solubility of nitrogen in liquid Al, as summarized in Table 4 (Wriedt 1986). Conversely, the coating microstructure is significantly influenced by the reactivity and plasma temperature and is shifted to a porous structure. The porosity increase is due to the formation of agglomerates during the nitriding of Al particles and/or thermal decomposition of AlN-reacted particles. In addition, it was difficult to fabricate sound coatings using a higher RF power and N_2 flow rate. Therefore, the fabrication of a thick and dense Al-based Al/AlN composite coating was possible. However, it was difficult to fabricate a dense AlN-based Al/AlN composite coating via reactive RF plasma spraying using a large Al powder.

However, AlN-based coatings were successfully fabricated at ambient atmosphere through the reactive spraying and nitriding of Al feedstock particles in the N_2-H_2 plasma of the atmospheric plasma spray system (Shahien et al., 2010a). When using an APS system, one of the main challenges was the formation of a meta-stable cubic-AlN (c-AlN) phase (Shahien et al., 2010a). It is attributed to the rapid

Table 4. Nitrogen solubility in liquid Al at N_2 pressure of 0.1 Mpa

N Content, at % $\times10^3$	25	28	36	44	53	61	70
Temperature, (K)	933.5	973	1073	1173	1273	1373	1473

Wriedt 1986.

solidification and high quenching rates of the APS system. Normally, during the AlN nuclei growth in the N_2 atmosphere, the cubic phase gradually crystallizes to form a single-phase hexagonal AlN phase structure. However, due to the high quenching rate of the APS system, the time was not sufficient for the crystal growth and formation of a hexagonal AlN phase. This led to the formation of cubic phase in the APS process. Compared with conventional solidification of relatively heavy sections, the rapid solidification provides large departures from an equilibrium constitution and results in the substantial extension of solid solubility and, therefore, formation of new equilibrium crystalline phases and metallic glasses. In other words, during rapid solidification, a two-phase mixture of equilibrium phases in a eutectic alloy can be replaced by an extended solid solution, such as a new crystalline phase, a non-crystalline phase (i.e., metallic glass), or non-stable phases simply by quenching with a sufficient rapidity from the liquid state (Chraska et al., 2001; Duwez et al., 1960; Rodriguez-Diaz et al., 2009; Tilly et al., 1889; Kramer et al., 2007; Kong et al., 2009).

Furthermore, during the APS process, the high particle velocity does not provide enough time for complete crystal growth before impacting on the substrate and rapid cooling. One important issue in the APS spraying is to prevent the oxidation of Al particles by keeping the Al particles in the center of the plasma stream (where there is no oxygen) using a low carrier gas flow rate of 1 L/min. Such low carrier gas flow is enough to supply the Al particles to the beginning of the plasma stream. Then, the particles are simultaneously accelerated by the plasma stream and, thus gain velocity and temperature via thermal and kinetic transfers toward the substrate material. Additionally, in APS reactive spraying, the present oxygen content (Al_2O_3: from the oxide layer on the surface of Al particles or even if some air oxygen contaminated and reacted with Al) interacts with the Al particles in the surrounding N_2 plasma and forms an aluminum oxide nitride phase (Al_5O_6N). Aluminum oxide nitride phases are formed through the reaction between Al, Al_2O_3 and N_2 (Borovinskaya et al., 2015; Gromov et al., 2005). The formed aluminum oxide nitride phase interacts with the surrounding N_2/H_2 plasma to form c-AlN during spraying. The nitriding of aluminum oxide nitride led to the formation of a cubic AlN phase due to the same cubic symmetry as reported by J. C. Kuang et al (Kuang et al., 2010). The crystal structure of c-AlN is cubic and closely packed, and the non-metal arrangement in aluminum oxide nitride is also cubic and closely packed.

Reactive Spraying of Fe_4N Coatings

Among the nitride ceramics, iron nitride (Fe_4N) has also received interest as an attractive nitride ceramic, which has excellent mechanical, electrical and magnetic properties. Fabrication of Fe_4N thin films has been reported (Grachev et al., 2002; Hayashi et al., 1995; Adachi et al., 2003), as well as fabrication of its sintered compacts (Tanaka et al., 2000; Salak et al., 1999; Takada et al., 1988; Nakajima et al., 1991; Hirose et al.,1994; Ito et al.,1996). However, it has been difficult to fabricate sintered compacts with dense microstructure. This is attributed to the thermal decomposition of Fe_4N above 974 K (Jack 1951), and the decomposed nitride easily transforms into the oxide. Furthermore, the decomposition temperature of Fe_4N is lower than the melting point of Fe. It is also known that the Fe_4N decomposition temperature is much lower than of other nitride ceramics, as shown in Table 5 (Toth 1971; Pierson 1996; Ichinose et al.,1998). In addition, Fe_4N is an unstable material, as illustrated by the generation free energy of various nitrides (Sun et al.,1997). Therefore, it was impossible to fabricate iron nitride coatings via conventional atmospheric reactive plasma spraying due to the oxidation of iron powder during spraying (Niizuma 2000; Niizuma et al., 2001). In addition, it was also considered that, using reactive plasma spray nitriding, it is difficult to fabricate Fe_4N coatings. Thus, the decomposition temperature of Fe_4N (974 K) is lower than the melting point of Fe (1809 K), as shown in Table 5.

To fabricate a thermal spray Fe_4N coating, a water-cooled substrate holder was used to prevent the thermal decomposition of Fe_4N formed on the substrate by keeping the substrate temperature below its decomposition temperature. Thick and dense Fe_4N-based coatings were successfully fabricated via reactive RF plasma spraying of fine Fe in Ar-N_2-H_2 plasma (Yamada 2006; Yamada et al., 2006b). The coatings showed a lamellar structure, which is the typical characteristic of thermal-sprayed coatings. In addition, the concentration of the Fe_4N phase gradually improved with the N_2 flow rate, and the metallic Fe phase was suppressed and disappeared with the N_2 flow rate.

Reactive Spraying of Other Coatings

Reactive plasma spraying is promising for fabricating different materials, which is difficult to do using conventional techniques. The fabrication of titanium nitride (TiN) coatings had been widely investigated due to their excellent mechanical properties (Fukumoto et al., 1997; Smith 1993; Grenier et al., 1996; Smith et al., 1993; Ohmori et al., 1991, p. 3; Deneirjean et al., 2003; Lugscheider et al., 1996; Bacci et al., 2000; Steffens et al., 1995). In addition, utilization of the reactive spraying mode to fabricate non-oxide ceramics, inter-metallic compounds and ceramics-dispersed coatings is receiving a lot of interest (Smith 1993; Fauchais et al., 1997; Hoshiyama et al., 2002).

FACTORS OF REACTIVE PLASMA SPRAY PROCESS

While reactive plasma spraying is showing a promising impact on fabrication of several non-oxide ceramics, intermetallic and ceramics-dispersed coatings, the optimization of the process is significantly related to several controlling factors. The parameters, which control the RPS process, can be divided into process-related parameters (e.g., spray system, electric power, shielding nozzle and process related issues) and reaction-related parameters (e.g., feedstock materials, active and reacting species, reaction time and the interaction between particle-plasma). In this part, some of the main controlling factors during the RPS process will be discussed.

Table 5. Melting point and decomposition temperature of various materials

Elements	Melting Point, (K)	Nitrides	Decomposition Temperature, (K)
Ti	1933	TiN	3223 (melt)
Si	1700	Si_3N_4	2173
Al	933	AlN	2273
Cr	2163	CrN	1773
Fe	1809	Fe_4N	974

Toth 1971; Pierson 1996; Ichinose et al., 1998.

Starting Feedstock Particle Size

The powder features (size, shape, morphology, specific mass, and injection rate) must be adapted according to the spraying process and spray parameters (Kolman et al., 1994; Fauchais et al., 2001, 2009, 2010; Allen et al., 2001; Wang et al., 2007; Vardelle et al., 2001; Ettouil et al., 2008). Thus, even the powders have the same chemical composition and size distribution. The coatings significantly vary due to the difference in the powder morphology according to their manufacturing method. During reactive spraying, the particle size is one of the main factors. The feedstock particle size strongly affects the nitriding process, in-flight diagnostics, melting, flowability behaviors of the particle in the plasma and vaporization behaviors of the particles. The nitriding reaction starts from the particle surface in the plasma stream. Decreasing the particle size of the feedstock powder increases the specific surface area of the particles, which increases the possibility of surface nitriding and promotes the nitriding conversion (AlN formation). In addition, using the smaller size of the particles improves their melting behavior. Thus, the heat capacity of large particles is larger than of smaller particles, which makes the increase of their temperature slower than for small particles. In addition, the total heat transfer to large particles is smaller than that to smaller particles at a constant weight. Therefore, the heat conduction inside the large particles is slower, and its decomposition rates are lower than those of the small particles.

However, using the fine particle size lowers the deposition efficiency and suppresses the coating thickness (especially with increasing the flight time). This is attributed to the loss of fine powders during the injection and excessive vaporization. Furthermore, the completion of nitriding of some fine particles during flight in the plasma suppressed the coating thickness (Shahien et al., 2011b). The particle size significantly influences the nitriding and the coating microstructure of the coatings.

In addition, the feedstock particle size significantly affects the flowability of the powder during spraying, and achieving a stable supply and spraying of fine particles to the plasma stream is difficult (this phenomenon was more pronounced in the case of radial injection, such as in the case of the APS system). Thus, fine particles have a low momentum and have difficulty penetrating into the plasma jet and, therefore, are carried around the jet envelope. In addition, Vardelle et al. reported that fine particles tend to bypass the plasma jet and are sucked downstream by it, which creates defects in the coatings when the particles stick to it (Vardelle et al., 2001). Furthermore, during the injection, the fine particles have a tendency to tilt toward the wall of the injector tube due to the turbulent dispersion, which leads to low velocity at the injector exit. These fine particles may be entrained by the gas flow further downstream and create defects in the coatings. However, the spraying of large particles leads to a much better flowability with no pipe clogging, less pipe wear and feeding systems with smoother particles and correlatively sprayed denser coatings. Moreover, be being further downstream in the plasma jet, the trajectory of large particle is generally unaffected by the turbulence of the plasma flow; smaller and less massive particles follow the fluid motion and may be entrained in an eddy.

Upon spraying of the fine particles, the coating peeled from the substrate due to the complete nitriding (AlN formation) and thus provide the thermal expansion difference between the fabricated ceramic coating and the metallic substrate. It is a good indication of the complete nitriding of fine Al particles and the AlN coating formation. The thermal mismatch between the AlN phase in the coating and the substrate can be avoided using a substrate material with a low thermal expansion coefficient material such as AlN or quartz. High content AlN based coating was fabricated when using an Al powder with an average particle size of 3 μm and a quartz substrate in the RF system (Yamada et al., 2006a, 2007). Although, the coating showed a high content of the AlN phase, it is worth mentioning that, in the case of the RF system, most of the coatings fabricated onto quartz substrate include a great deal of agglomerates. These agglomerates are quite brittle, and the coatings were easily removed just by scratching (Yamada et al., 2007). Therefore, the use of fine particles requires an optimization of the spraying parameters to fabricate thick coatings.

In-Flight Time

The in-flight time or spraying distance is an important factor during the RPS process. It has a significant influence on the nitriding reaction and the coating microstructure (Shahien et al., 2010a, 2010b). Figure 4 represents the tendency of increasing the nitride content and decreasing the thickness of the coatings with an increase of the spraying distance. An increase of the spraying distance enhances the in-flight nitriding, which increases reaction intervals and yields a higher nitride content in the coatings. This leads to a longer time for the feedstock particles to start the nitriding reaction with active species in the plasma during the flight. However, an increase of the spraying distance reduces the particle deposition efficiency and lowers the coating thickness. Thus, the sprayed particles lose their kinetic energy and viscosity with the increase spraying distance. This is due to lowering both the particle velocity and temperature with the spraying distance. In addition, the spraying distance results in complete nitriding of some particles (formation of AlN) during flight. This prevents their deposition and suppresses the coating thickness. In addition, the spraying distance affects vaporization rates, especially when using low melting point materials such as Al.

To achieve a thick coating with a higher nitride content, a shorter spraying distance and improvement of nitriding reaction are recommended. To improve the nitriding reaction at short spray distance, control of materials and use of a chemical additive significantly affect the enhancement of the chemical reaction. For example, the use of a NH_4Cl additive promotes the in-flight nitriding conversion of Al particles (Shahien et al., 2011a). The addition of NH_4Cl enhances the formation of AlN by changing the Al reaction pathway to vapor-phase chlorination-nitridation sequences. This makes the nitriding reaction milder.

Figure 4. The coatings thickness and the AlN content in the fabricated coatings at different spray distance, (Al 30 μm in N_2/H_2 plasma, N_2:330.9 kPa, 100 L/min and H_2:344.7 kPa, 5L/min)

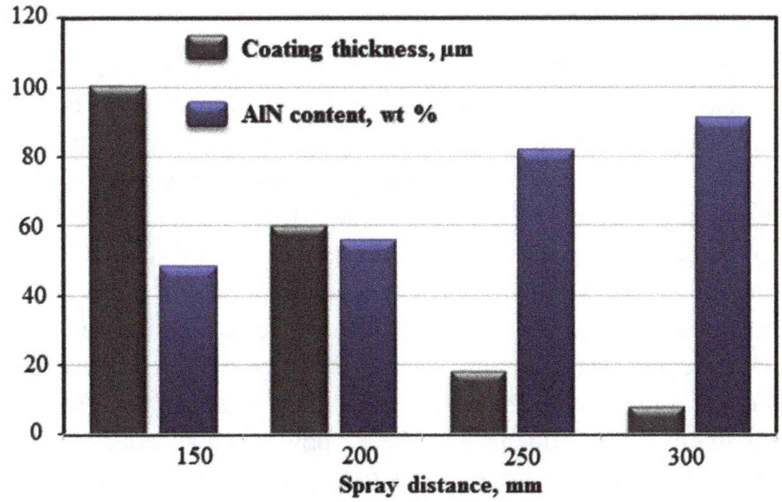

Thus, it is more controlled with no explosive mode, and this prevents Al particles from coalescing after melting. Another way to improve the AlN content in the fabricated coating during RPS is to use AlN additives (Shahien et al., 2011e). The AlN addition improves the content and prevents the coalescence of Al particles after deposition on the substrate. This opens an interesting way to improve the nitriding conversion through adjusting and controlling the reacting materials and the ambient environment.

Reactive Plasma Characteristics

One of the main controlling factors in the RPS process is the plasma characteristics, including the plasma temperature and active species. This is directly influences the reaction between the feedstock material and the plasma and, thus the formation of the required coating. The plasma temperature during reactive spraying and the influence of plasma power and gases has s significant impact on understanding the RPS process (Yamada et al., 2005; Yamada 2006). Thus, plasma temperature and active species are significantly influenced by the flow rates and plasma power. Plasma temperature was measured from the emission spectral of the excited Ar species in the visible region ranging from 772 to 852 nm (due to the strong emission), (Wiese et al., 1996; Yamada et al., 2005). The measurement was investigated using a fixed point at the center of the plasma jet, and the plasma temperature measurement was based on the Boltzmann plot method of the corrected emission intensities, as follows.

$$I_{mn} = \frac{1}{4\pi} A_{mn} n_a \frac{g_a}{Z_a} \exp\left(-\frac{E_m}{kT}\right) h v_{mn}$$

where A_{mn} is the transition probability of m to n transition, g_a is the known statistical weight or degeneracy of the excited state, Z_a is the partition function, E_m is the excitation energy, k is the Boltzmann constant, T is the absolute temperature and $h_{\upsilon mn}$ is the energy of transition from m to n. Then resolve and taking the natural logarithm yields:

$$\ln\left(\frac{I_{mn}}{g_m A_{mn} v_{mn}}\right) = -\frac{E_m}{kT} + \ln\left(\frac{N_a h}{4\pi Z_a}\right)$$

According to the above-mentioned relation, the plot of the left-hand side versus the excitation energy, E_m, results in a linear relationship with a slope of $(-1/kT)$ since all other terms are constant. From the slope, the plasma temperature can be calculated (referred to as the Boltzmann plot method) (Fauchais et al., 1996; Hiraoka et al., 1996; Semenov et al., 2001). The relationship between the plasma temperature and plasma gas flow rates is shown in Table 6 (Yamada et al., 2005; Yamada 2006). The addition of N_2 and H_2 gases significantly increased the temperature. In other words, N_2 and H_2 and their flow rates improved the quantity of diatomic molecule in unit volume. The further increase of the H_2 gas above the 0.3 L/min flow lowered the temperature. Thus, the total gas flow rate affected the temperature (excessive plasma gas flow rate decreased the plasma temperature). Table 7 summarize the influence of the RF power on the plasma temperature and the plasma gas fraction (Yamada et al., 2005; Yamada 2006). Although, plasma temperature is significantly influenced by the RF power, it does not change with the plasma gas fraction. This reveals that plasma temperature and active species are very important factors in the RPS

Table 6. Relationship between plasma temperature and plasma gases flow rate (RF power: 10 kW, Ar flow rate: 10 l/min)

N_2 Plasma Gas Effect			H_2 Plasma Gas Effect		
Gas Flow, l/min		Temperature, (K)	Gas Flow, l/min		Temperature, (K)
N_2	H_2		H_2	N_2	
0	0	6109.11	0	1	7080.15
0.1	0	5689.18	0.1	1	9091.63
0.3	0	5853.93	0.3	1	11286.5
0.5	0	6154.8	0.5	1	10689.68
0.7	0	6432.54	0.7	1	10493.44
1	0	7080.15	1	1	8325.19

Yamada et al., 2005, Yamada 2006.

Table 7. Summarize Relationship between plasma temperature and plasma gas fraction

Plasma Gas Fraction N_2/H_2 (l/min)	Temperature, (K)	
	5 kW	10 kW
0.8/0.2	36318.16	36375.08
0.6/0.4	36329.53	36708.8
0.4/0.6	36375.08	36732.04
0.2/0.8	36375.08	36720.41

Yamada et al., 2005, Yamada 2006.

process and are significantly influenced by the plasma gas flow rates and the RF plasma power. This affects the chemical reaction as well as the coatings microstructure, as has been previously explained during reactive spraying of Al particles using the N_2/Ar plasma from the RF plasma spraying system.

In addition, the spectroscopic measurements of RF plasma revealed that the H_2 addition is necessary to generate NH radicals in the plasma (Yamada et al., 2005; Yamada 2006). A distinct peak was observed at 336 nm in the spectrum. Furthermore, the amount of generated NH radical is significantly influenced by the plasma gas fraction, as shown in Figure 15(b) (Yamada et al., 2005; Yamada 2006). The highest amount of NH radical is achieved using the same flow rate of N_2 and H_2. Thus, the NH radical was generated from the same number of N atoms and H atoms (Hudis 1973). The generation of the NH radical is one of the most important factors for the nitriding reaction in reactive RF plasma spraying. The amount of NH radicals was controlled by the plasma gas fraction, and it can be used to control nitriding reaction in reactive plasma spraying.

Nitriding of the Feedstock Particle

Plasma nitriding of the particle is the main challenge of the RPS process, and it directly determines the resulting coating composition. Nitriding is significantly related to the decomposition/melting temperature of the nitrides as well as the generation free energy of the reaction. During spraying, plasma nitriding

Table 8. The emission intensity of NH radicals excited in plasma as a function of plasma gas fraction

Plasma Gas Fraction N_2/H_2, (l/min)	Emission Intensity, (Arbit. Unit.)	
	5 kW	10 kW
0.8/0.2	470.682495	481.698456
0.6/0.4	582.845093	699.01355
0.4/0.6	549.79718	750.087585
0.2/0.8	453.657806	613.890137

Yamada et al., 2005, Yamada 2006.

of Ti particles occurs after deposition on the substrate (under the solid phase) (Fukumoto et al., 1997; Tsunekawa et al., 1998). Plasma nitriding of Si particles occurs under the liquid phase on the substrate (Tilly et al., 1889; Yamada et al., 2004), and plasma nitriding of Al particles starts during flight and completes the reaction after the particles are deposited on the substrate while in the liquid phase (Fukumoto et al., 2004; Yamada et al., 2007; Shahien et al., 2011b). Conversely, plasma nitriding of Fe particles occurs on the substrate during the solid phase (Yamada et al., 2006b). Thus, Fe particles never form the nitride during the liquid phase. This is due to the low decomposition temperature of Fe_4N, which is much lower than the Fe melting point, as shown in Table 5.

On the other hand, the bonding characteristic is an important issue for the nitride formation in reactive plasma spraying. From this view, the nitrides are classified into five main categories, as summarized Table 9 (Pierson 1996). Both AlN and Si_3N_4 are covalent nitride which react during the liquid phase. The plasma nitriding of Al particles during the RPS process is explained as follows (Shahien et al., 2011b, 2012).

- Nitriding of the particles occurs at the surface, and a crystalline nitride shell is formed. Then, Al starts to break the AlN shell to leave it (it is attributed to the thermal stress of large volume of molten Al). In addition, due to the high temperature of the plasma, the generation rate of reaction heat is enough to vaporize the molten Al present inside the shell, and this is the beginning of the next step.
- The molten Al inside the shell expands enough to break the shell and flow out in the form of molten or vaporized Al core via the capillary-like phenomenon.
- Finally, diffused Al vapors react with the surrounding N_2 via volume nitriding outside the shell with a remaining hole or an empty core.

Basically, the surface of Al particle is coated with an initial protective oxide film, (Rosenband et al., 2004). This film prevents good contact with nitrogen gas and therefore inhibits the nitridation reaction. The particles start the reaction after melting, and even the nitridation is insignificant below the melting point of Al itself. Once the particle melts, the initial protective oxide layer is broken due to the tensile stress of increased volume. Then, nitrogen gases react intensely with liquid Al particles to form aluminum nitride and/or aluminum oxynitride, and the nitriding reaction occurs via the abovementioned three steps.

The egg-shell microstructure formation is a well-known feature of direct nitriding of Al particles and formation of AlN (Lin et al., 2001; Juang et al., 2003; Kimura et al., 1989; Fujii et al., 2001; Radwan et al., 2007). Plasma nitriding start on the particle surface during the flight, and once the partially reacted particle or droplet hits the substrate, its surface collapsess and the inside content flows out and bonds

Table 9. The classification of the elements forming nitrides

Saltike Nitrides				Interstitial Nitrides		Intermediate Nitrides	Covalent Nitrides		Volatile (Molecule Forming) Nitrides	
Be	Sc	Cu	Ge	Ti	Cr	Mn	B	P	C	Po
Mg	Y	Ag	Sn	Zr	Mo	Te	Al		N	F
Ca	La	Au	Pb	Hf	W	Re	Ga		O	Cl
Sr	Ac	Zn	As	V		Fe	In		S	Br
Ba		Cd	Sb	Nb		Co	Ti		Se	I
Ra		Hg	Bi	Ta		Ni	Si		Te	At

Pierson 1996.

with the substrate. Then, the partially reacted particle completes the nitriding reaction via plasma irradiation during deposition onto the substrate. However, the completely reacted particles during flight (AlN particle) will not deposit onto the substrate because of its poor deposition efficiency. The egg-shell formation was clearly observed during reactive plasma nitriding of Al particles of APS and RF spraying (Fukumoto et al., 2004; Yamada et al., 2007; Shahien et al., 2011b, 2012).

Particle Diagnostics During RPS

Generally, chemical reactions are time and temperature-dependent phenomena. For the particles during plasma spraying, the time can be controlled by controlling it velocity, which decides its residence time in the plasma. Therefore, the in-flight particle diagnostics (temperature and velocity) are important factors to clarify the in-flight interaction between the plasma and feedstock particles. The in-flight particle diagnostics are influenced by the spraying parameters such as spray distance, particle size, plasma power as well as plasma nozzle design and diameter. Therefore, to understand and optimize the entire RPS process, understanding the particle in-flight diagnostics and their relations with the processing parameters is an important factor.

The influence of spray distance and particle sizes during the plasma nitriding of Al particles were clarified by understanding particle diagnostics during spraying. Figure 5 presents the average particle velocities as a function of the spraying distance and feedstock particle sizes (Shahien et al., 2012). The particle velocity decreases gradually with the spray distance for both powders. Therefore, an increase of the spraying distance provides more time for the particles to start the reaction with the nitriding species in the plasma during flight. However, the increase in the spray distance results in decreasing the deposition efficiency, excessive vaporization and the complete nitriding of some particles (formation of AlN phase) during flight. Therefore, it is difficult to fabricate thick and dense AlN coatings at a longer spray distance.

From the particle velocity observations, it is revealed that the particle velocity is not influenced by the particle size. This indicates that reaction time is the same for both sizes. Although the average velocities were the same for different particle sizes, velocity distributions showed that the number of detected particles significantly decreased with decreasing the particle size (Shahien et al., 2012). This is related to the low intensity of the emitted radiation from the fine particles. Thus, decreasing the particle size (diameter) reduces the intensity of light, as expressed in the following relation (Salhi et al., 2009):

$$I(\lambda,T) = \propto \frac{D^2 d\lambda}{\lambda^5 \left(e^{\beta/\lambda T-1}\right)}$$

where I is the intensity of emitted radiation, λ is the wave length, T is the particle surface temperature, and D is its diameter, α is a factor depending on the particle emissivity and β is a constant.

In addition, complete nitriding of the fine particles during flight and formation of AlN led to a decrease in the number of the measured particles.

However, the in-flight particle temperature is an important factor for clarifying the influence of spraying parameters during the RPS process. Figure 5 presents average particle temperatures at a different spray distance of the Al powder with a 30-µm particle size during RPS in an APS system (detected using DPV-2000 measurements) (Shahien et al., 2010c). The measurements confirmed the decrease in particle temperature with increasing spraying distance. However, it was difficult to detect the appropriate temperatures for the fine Al particles using the DPV-2000 system. Thus, the fine Al particles easily evaporate under the high temperature environment of the spraying conditions due to their low melting point. This occurs due to the higher evaporation rates of fine particles compared with larger diameter particles (Fauchais et al., 2010). Thus, high emissions from the intense Al vapor led to higher values of the measured temperatures using the DPV2000 measurement, especially for fine particles due to higher evaporation rates. Furthermore, different sources of radiation may affect the temperature measurement of fine Al particles under this spray condition. Reflected emission from the plasma on the particle surface and direct emission from the plasma species around the measured wavelength range have a significant effect on the temperature measurement value.

It is well known that the in-flight particle temperature of fine particles is much higher than that of the large particles (Vardelle et al., 1993). Because of the greater heat absorbed per unit volume of fine particles, they will attain a higher temperature than a large particle under the same plasma temperature condition. A particle requires eight times the amount of heat for attaining the same temperature increase

Figure 5. The average particles velocities of Al 30 and 15µm and average particles temperature of Al 30 µm at different spray distances
Shahien et al., 2012.

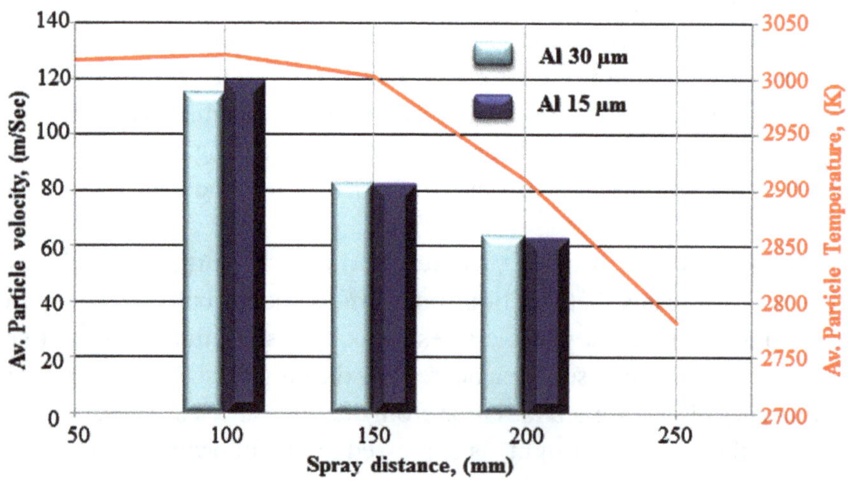

as compared with another particle half its diameter (Remesh et al., 2003). In addition, the surface area of the particle through which heat transfer is increased fourfold when compared with a particle of half the diameter. Hence, the heat absorbed per unit volume is the inverse of the particle diameter.

Herein, the in-flight particle temperature and velocity can clarify the particle nitriding. The nitriding reaction of Al particles was improved using a smaller particle size. However, the coating thickness was significantly suppressed. This is related to increasing the particle temperature which led to an excessive vaporization of Al particles during the flight. Thus, a smaller particle, which has penetrated the flow, may evaporate after a short time of traveling in the plasma. Furthermore, mixing of these vapors with the plasma gas can alter the heating of other particles and evaporation, which lowers the deposition efficiency and coating thickness. In addition, the complete nitriding of some fine particles (the formation of AlN) during flight with decreasing particle size led to a lower deposition efficiency and decrease of the coating thickness. This is attributed to the poor deposition efficiency of AlN particles. In addition, decreasing the particle velocity by increasing the spray distance suppresses particle deposition efficiency and decreases the coating thickness. This is due to decreasing the particle momentum energy when the particles leave the plasma jet. Moreover, the particle velocity declines due to the characteristic spreading of a multi-phase spray cone with increase of the standoff distance. In addition, with the increase of the standoff distance (flight time), some sprayed particles deviate from the substrate.

Nitriding Species

One of the main important factors in RPS is the reactive species, such as N_2, during the plasma nitriding. It significantly influences the reaction process as well as the coating microstructure. The nitriding reaction of the Al particle gradually improves during flight with the increase of the N_2 gas flow rate. This is due to the increase of the enthalpy of plasma and improvement of the reactive temperature with the N_2 plasma gas flow rate (Krishna et al., 2002). In addition, the N_2 gas flow rate enhanced the amount of nitriding species surrounding the sprayed particle. The influence of the N_2 plasma gas should be optimized with other spraying parameters, such as the spray distance. This is due to the completion of the nitriding reaction at a shorter spray distance with increase of the N_2 gas flow rate (Shahien et al., 2010c).

Furthermore, the N_2 plasma gas showed a strong influence on the coating microstructure as well as on deposition efficiency and thickness. The influence of N_2 gas should be optimized with other spraying and reaction parameters. Figure 6 shows the coating thickness as a function of N_2 gas flow rate and spraying distance (during RPS of Al 30 µm in an APS system) (Shahien et al., 2010c). An increase of the N_2 gas flow rate to 120 L/min at a short spray distance improved the coating thickness. The further increase of the N_2 gas low rate as well as the spray distance decreased the deposition efficiency and thus the coating thickness, and it is difficult to fabricate homogenous coatings at higher flow rates of more than 160 L/min. The initial improvement of the coating thickness is related to increasing the driving force for the particles toward the substrate due to the increase of particle velocities during the increase of the N_2 gas flow rate, as shown in Table 10 (Shahien et al., 2010c).

The decrease of the thickness with a further increase of the N_2 gas flow rate is attributed to improving the in-flight nitriding reaction. This led to the completion of nitriding of some particles (AlN formation during the flight) and therefore prevented their deposition on the substrate surface (AlN particle did not have a molten phase). In addition, an increase of the flow rate led to an excessive vaporization and loss of some Al particles during the flight. The influence of N_2 gas on the coating thickness became more significant at longer spraying distances. Thus, increase of the flight time under high N_2 gas amount led

Figure 6. The coatings thickness tendency as function of N_2 plasma gas and spray distance, (in N_2/H_2 plasma, N_2:330.9 kPa and H_2:344.7 kPa, 5L/min)
Shahien et al., 2010c.

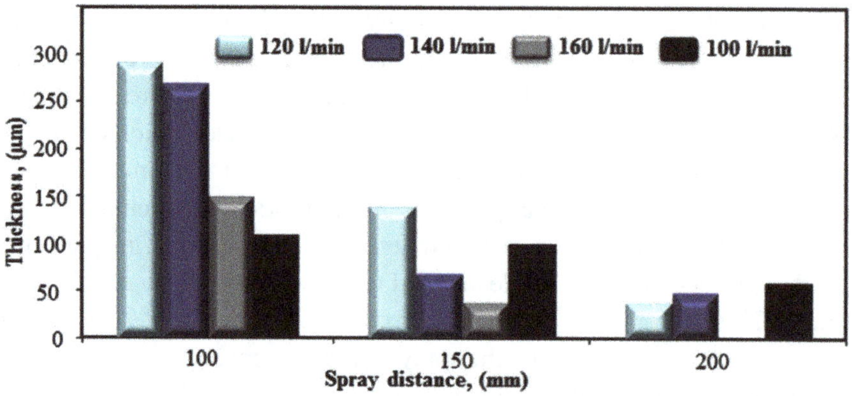

to the further improvement of particle nitriding, and this assists the complete nitriding of a larger number of particles during the flight and in the formation of AlN. In addition, the plasma gas velocity decreased and caused a deviation in the particle trajectory, and it did not hit the substrate at a longer spraying distance. It is also considered that decrease of the particle temperature with the spraying distance increase reduced the deposition efficiency and suppressed the coating thickness.

FUTURE RESEARCH DIRECTIONS

Implementation of the reactive spraying mode for the industrial applications of the high temperature coatings is a promising solution to solve many industrial challenges. This can be achieved through the following:

- Improving the level of reactive species in the plasma jet.
- Utilization of reactive metals and closely controlling the feedstock materials synthesis to enhance interaction between the feedstock and plasma species.
- Computational design of the reactive mode is an essential aspect to implement the RPS for industrial application. Computational fluid dynamics modeling is considered a reliable method to investigate the reaction phenomenon during spraying and deposition.

Table 10. The particle velocity at different flow rate of N_2 plasma gas

N_2 Gas Flow Rate, (l/min)	100	120	140	160	180
Av. Particle Velocity, (m/s)	114.7114	127.3568	133.2452	138.0977	157.7956

Shahien et al., 2010c.

- Investigation of the coating performance and evaluation to meet the industrial requirements of improved reliability, lower manufacturing cost, optimized material/process and low maintenance cost.
- Coating formation and microstructure design to enable and predict the coating microstructure, performance and life time, by emphasizing its design approaches, 3D model design, design tools, and simulations.

CONCLUSION

The use of the reactive plasma spraying (RPS) process is a pioneer solution to overcome the limitations of conventional plasma technology and to fabricate nitride ceramic coatings. The technology is based on tailoring the required compositions (which are difficult to fabricate using conventional spraying techniques) through utilization and enhancement of the reactivity of the plasma to assist chemical reactions between the feedstock materials and the surrounding active species in the plasma. The RPS process has shown a promising impact for the *in situ* fabrication of several nitride ceramics including aluminum nitride (AlN), iron nitride (Fe_4N), titanium nitride (TiN) and silicon nitride (Si_3N_4), using the plasma nitriding of metallic elements of these materials. Utilization and optimization of the RPS process strongly depends on adjusting both the reaction-related and process-related parameters. These parameters significantly influence the chemical reaction, formation process, coating microstructure and properties. The main key aspects in RPS are feedstock materials and their characteristics, particle size, reactive species, plasma power, in-flight time, plasma characteristics, nitriding mechanism, in-flight particle diagnostics, N_2 plasma gas, splat morphology and reactive spraying system.

REFERENCES

Adachi, N., Endo, M., Menjo, N., & Okuda, T. (2003). Annealing effects of iron nitride thin film grown by plasma-assisted evaporation technique. *Journal of the Ceramic Society of Japan, 111*(6), 369–371. doi:10.2109/jcersj.111.369

Al-Sabouni, O. (1999). *Reactive plasma spraying.* (Unpublished doctrol dissertation). Cranfield University, London, UK.

Allen, A. J., Long, G. G., Boukari, H., Ilavsky, J., Kulkarni, A., Sampath, S., & Goland, A. N. et al. (2001). Microstructure characterization studies to relate the properties of thermal-spray coatings to feedstock and spray conditions. *Surface and Coatings Technology, 146-147*, 544–552. doi:10.1016/S0257-8972(01)01433-5

Bacci, T., Bertamini, L., Ferrari, F., Galliano, F. P., & Galvanetto, E. (2000). Reactive plasma spraying of titanium in nitrogen containing plasma gas. *Materials Science and Engineering A, 283*(1-2), 189–195. doi:10.1016/S0921-5093(00)00704-8

Berger, L. M., Herrmann, M., Nebelung, M., Thiele, S., Heimann, R. B., Schnickt, T., & Vuoristo, P. et al. (1998). Investigations on thermal spraying of silicon nitride-based powders. In *Proceeding of 15th International Thermal Spray Conference,* Nice, France (pp. 1149-1154).

Bianchi, L., Grimaud, A., Blein, F., Lucchese, P., & Fauchais, P. (1995). Comparison of plasma-sprayed alumina coatings by RF and DC plasma spraying. *Journal of Thermal Spray Technology, 4*(1), 59–66. doi:10.1007/BF02648529

Borovinskaya, I. P., Ignat'eva, T. I., Semenova, V. N., & Chemagina, E. A. (2015). Aluminum oxynitride by SHS in chemical furnace. *International Journal of Self-Propagating High-Temperature Synthesis, 24*(3), 142-147.

Bourdin, E., Boulos, M., & Fauchais, P. (1983). Transient conduction to a single sphere under plasma conditions. *International Journal of Heat and Mass Transfer, 26*, 567–582. doi:10.1016/0017-9310(83)90008-X

Bunshah, R. F. (2001). *Handbook of hard coatings, deposition technologies, properties and applications*. New York, USA: Noyes Publication.

Cao, L. H., Khor, K. A., Fu, L., & Boey, F. (1999). Plasma spray processing of Al_2O_3/AlN composite powders. *Journal of Materials Processing Technology, 89-90*, 392–398. doi:10.1016/S0924-0136(99)00063-1

Chraska, T., & King, A. H. (2001). Transmission electron microscopy study of rapid solidification of plasma sprayed zirconia part I. First splat solidification. *Thin Solid Films, 397*(1-2), 30–39. doi:10.1016/S0040-6090(01)01360-8

Deneirjean, A., Lefort, P., & Fauchais, P. (2003). Nitridation process and mechanism of Ti-$_6$Al-$_4$V particles by dc plasma spraying. *Physical Chemistry Chemical Physics, 5*(22), 5133–5138. doi:10.1039/B311186P

Duwez, P., Willen, R. H., & Klement, W. (1960). Continuous series of metastable solid solutions in silver-copper alloys. *Journal of Applied Physics, 31*(6), 1136–1137. doi:10.1063/1.1735777

Eckardt, T., Mallener, W., & Stover, D. (1994). Reactive plasma spraying of silicon in controlled nitrogen atmosphere. In *Proceeding 7th National Thermal Spray Conference,* Boston, MA (pp. 515-519).

Eckardt, T., Mallener, W., & Stover, D. (1995). Influence of a reactive nitrogen atmosphere on plasma sprayed silicon coatings. *VDI-Berichte, 1166*, 341–348.

Eckardt, T., Mallener, W., & Stover, D. (1996). Development of plasma sprayed silicon/silicon nitride coatings by in situ nitriding. In Proceeding of Thermal Spray '96 (pp. 309-312), German.

Ettouil, F. B., Mazhorova, O., Pateyron, B., Ageorges, H., El Ganaoui, M., & Fauchais, P. (2008). Predicting dynamic and thermal histories of agglomerated particles injected within a d. C. plasma jet. *Surface and Coatings Technology, 202*(18), 4491–4495. doi:10.1016/j.surfcoat.2008.04.032

Fauchais, P., & Coudert, J. F. (1996). Measurement of temperatures in thermal plasmas. In R. Benocci, G. Bonizzoni, & E. Sindoni (Eds.), *Thermal Plasmas for Hazardous Waste Treatment* (pp. 39–66). PA: World Scientific. doi:10.1142/9789814447171_0002

Fauchais, P., Heberlein, J., & Boulos, M. (2014). *Thermal Spray Fundamentals: From Powder to Part.* New York: Springer. doi:10.1007/978-0-387-68991-3

Fauchais, P., Montavan, G., & Bertrand, G. (2009). Influence of powders on thermal spray coating structures: Recent developments in nano or finely structured coatings and some safety issues. In B. R. Marple, M. M. Hyland, Y.-C. Lau, C.-J. Li, R. S. Lima, & G. Montavon (Eds.), *Thermal Spray 2009: Expanding Thermal Spray Performance to New Markets and Applications* (pp. 799–817). Materials Park, OH, USA: ASM International.

Fauchais, P., Montavan, G., & Bertrand, G. (2010). From powders to thermally sprayed coatings. *Journal of Thermal Spray Technology*, *19*(1-2), 56–80. doi:10.1007/s11666-009-9435-x

Fauchais, P., Vardelle, A., & Denoirjean, A. (1997). Reactive thermal plasmas: Ultrafine particles synthesis and coating deposition. *Surface and Coatings Technology*, *97*(1-3), 66–78. doi:10.1016/S0257-8972(97)00294-6

Fauchais, P., Vardelle, A., & Dussoubs, B. (2001). Quo vadis thermal spraying? *Journal of Thermal Spray Technology*, *10*(1), 44–66. doi:10.1361/105996301770349510

Fujii, T., Yoshida, K., Suzuki, K., & Ito, S. (2001). Direct nitriding of large grains of aluminum with 2 mm size. *Solid State Ionics*, *141-142*, 593–598. doi:10.1016/S0167-2738(01)00773-1

Fukumoto, M., Itoh, S., & Itoh, S. (1997). Fabrication of functionally graded TiN coating by reactive plasma spraying. *Surface Engineering*, *13*(4), 315–319. doi:10.1179/sur.1997.13.4.315

Fukumoto, M., Yamada, M., Yasui, T., & Takahashi, K. (2004). Fabrication of aluminum nitride coating by reactive RF plasma spray process. *In Proceeding of the International Thermal Spray Conference* (CD). Osaka, Japan.

Gawne, D. T., Bao, Y., & Zhang, T. (2001). Plasma-spray deposition of silicon nitride-based coatings. In *Proceeding of International Thermal Spray Conference*, Singapore.

Grachev, S. Y., Borsa, D. M., & Boerma, D. O. (2002). On the growth of magnetic Fe_4N films. *Surface Science*, *516*(1-2), 159–168. doi:10.1016/S0039-6028(02)02029-0

Grenier, S., Shanker, K., Tsantrizos, P., & Ajersch, F. (1996). Deposition of TiN films using a thermal plasma reactive forming technology. *Surface and Coatings Technology*, *82*(3), 311–316. doi:10.1016/0257-8972(95)02744-0

Gromov, A., Ilyin, A., Ditts, A., & Vereshchagin, V. (2005). Combustion of Al–Al_2O_3 mixtures in air. *Journal of the European Ceramic Society*, *25*(9), 1575–1579. doi:10.1016/j.jeurceramsoc.2004.05.017

Harvey, D. (1996). The tough truth - Wear-resistant coatings using high velocity oxyfuel. *Industrial Lubrication and Tribology*, *48*(2), 11–16. doi:10.1108/00368799610781230

Hayashi, Y., & Morinaga, S. (1995). Evaluation of iron nitride film resistors deposited by reactive sputtering method [in Japanese]. *Journal of the Vacuum Society of Japan*, *38*(3), 350–353. doi:10.3131/jvsj.38.350

Heimann, R. B., Thiele, S., Berger, L. M., Herrmann, M., Nebelung, M., Wielage, B., & Vuoristo, P. et al. (1998). Thermally sprayed silicon nitride-based coatings on steel for application in severe operation environments: Preliminary results. *Microstructural Science*, *26*, 389–394.

Hiraoka, K., Shiwaku, T., & Ohji, T. (1996). Temperature distributions of gas tungsten arc plasma by spectroscopic methods. *Quarterly Journal of the Japan welding* [in Japanese]. *Society, 14*(4), 641–648.

Hirose, J., Ito, S., Akashi, K., & Koura, N. (1994). Preparation of Fe_4N sintered compact by HIP. [in Japanese]. *Journal of the Japan Society of Powder and Powder Metallurgy, 41*(8), 989–993. doi:10.2497/jjspm.41.989

Hoshiyama, Y., Miyake, H., Murakami, K., & Nakajima, H. (2002). Titanium aluminide-based composite depositions with dispersed nitride particles produced by reactive thermal spraying. *Journal of Japan Institute of Metals, 66*(7), 784–791. doi:10.2320/jinstmet1952.66.7_784

Hudis, M. (1973). Study of ion-nitriding. *Journal of Applied Physics, 44*(4), 1489–1496. doi:10.1063/1.1662398

Ichinose, N., & Kuwabara, H. (1998). *Nitride Ceramics.* Tokyo, Japan: Nikkan-Kogyo-Shinbunsha.

Ishikawa, K., & Tobe, M. S. S. (1993). Application of thermal sprayed coatings to prevent corrosion of construction in Japan. In *Proceeding of 5th National Thermal Spray Conference,* Anaheim, CA (pp. 679-684). .

Ito, S., Shinmura, S., Hosokoshi, T., & Akashi, K. (1996). HIP sintering and saturation magnetization of Fe_4N [in Japanese]. *Journal of the Japan Society of Powder and Powder Metallurgy, 43*(12), 1415–1419. doi:10.2497/jjspm.43.1415

Jack, K. H. (1951). The iron-nitrogen system: The preparation and the crystal structures of nitrogen-austenite (γ) and nitrogen-martensite (α'). *Proceedings of the Royal Society of London. Series A, Mathematical and Physical Sciences, 208*(1093), 200–215. doi:10.1098/rspa.1951.0154

Juang, R. C., Lee, C. J., & Chen, C. C. (2003). Combustion synthesis of hexagonal aluminum nitride powders under low nitrogen pressure. *Materials Science and Engineering A, 357*(1-2), 219–227. doi:10.1016/S0921-5093(03)00163-1

Kassabji, F., Tourenne, F., Derradji, A., & Fauchais, P. (1983). Aluminium and aluminium nitride deposition by low pressure nitrogen arc plasma spraying. In *Proceeding of 10th International Thermal Spray Conference*, Düsseldorf, Germany (Vol. 80, pp. 82-84).

Khor, K. A., Boey, F. Y. C., Zhao, X. L., & Cao, L. H. (2001). Aluminum nitride by plasma spraying of an Al_2O_3-C-Sm_2O_3 system. *Materials Science and Engineering A, 300*(1-2), 203–210. doi:10.1016/S0921-5093(00)01658-0

Kimura, I., Ichiya, K., Ishii, M., Hotta, N., & Kitamura, T. (1989). Synthesis of fine AlN powder by a floating nitridation technique using an N_2/NH_3 Gas Mixture. *Journal of Materials Science Letters, 8*(3), 303–304. doi:10.1007/BF00725505

Kolman, B., Forman, J., Dubsky, J., & Chraska, P. (1994). Homogeneity studies of powders and plasma sprayed deposits. *Mikrochimica Acta, 114*(1), 335–342. doi:10.1007/BF01244560

Kong, J., Ye, Z., & Lv, F. (2009). Non-equilibrium solidification character of $Zr_{56.2}Ti_{13.8}Nb_{5.0}Cu_{6.9}Ni_{5.6}Be_{12.5}$ bulk metallic glass composites containing ductile dendrite phase. *Journal of Alloys and Compounds, 478*(1-2), 202–205. doi:10.1016/j.jallcom.2008.11.130

Kramer, M. J., Mecco, H., Dennis, K. W., Vargonova, E., McCallum, R. W., & Napolitano, R. E. (2007). Rapid solidification and metallic glass formation – Experimental and theoretical limits. *Journal of Non-Crystalline Solids, 353*(32-40), 3633–3639. doi:10.1016/j.jnoncrysol.2007.05.172

Krishna, L. R., Sen, D., Rao, Y. S., Rao, G. V. N., & Sundararajan, G. (2002). Thermal spray coating of aluminum nitride utilizing the detonation spray technique. *Journal of Materials Research, 17*(10), 2514–2523. doi:10.1557/JMR.2002.0366

Kuang, J. C., Zhang, C., Zhou, X., Liu, Q., & Ye, C. (2005). Formation and characterization of cubic AlN crystalline in a carbothermal reduction reaction. *Materials Letters, 59*(16), 2006–2010. doi:10.1016/j.matlet.2005.02.039

Lin, C., & Chung, S. (2001). Combustion synthesis of aluminum nitride powder using additives. *Journal of Materials Research, 16*(8), 2200–2208. doi:10.1557/JMR.2001.0302

Lugscheider, E., Limbach, R., Liden, A., & Lodin, J. (1990). Plasma spraying of silicon nitride (Si_3N_4). *High Temperature Materials Powder Engineering, 1*, 877–880.

Lugscheider, E., Remer, P., & Zhao, L. (1996). Reactive plasma spraying of wear resistant coatings of Ti composites. In C. C. Berndt (Ed.), *Thermal spray: practical solutions for engineering problems* (pp. 927–932). PA: ASM International, Materials Park, OH.

Moriyama, M., Tahara, H., & Yoshikawa, T. (2003). Silicon nitride coating using electromagnetic acceleration plasma generators [in Japanese]. *Koon Gakkaishi, 29*(4), 13.

Moriyama, M., Tahara, H., & Yoshikawa, T. (2004). Silicon nitride coatings by magneto-plasma-dynamic plasma spraying [in Japanese]. *Koon Gakkaishi, 30*(1), 40–41.

Nakajima, K., Taki, K., Takata, M., & Okamoto, S. (1991). An effort on preparing sintered bodies of Fe_4N [in Japanese]. *Journal of the Ceramic Society of Japan, 99*(4), 311–314. doi:10.2109/jcersj.99.311

Niizuma, K. (2001). Synthesis of iron nitride by high energy plasma [in Japanese]. *Collaboration Report in Joining and Welding Research Institute Osaka University, 2000*, 194–195.

Niizuma, K., Utsushikawa, Y., & Kobayashi, A. (2001). Magnetic properties of Fe coating prepared by gas tunnel plasma spraying [in Japanese]. *Journal of IAPS, 9*, 136–138.

Ohmori, A., Hirano, S., & Kamada, K. (1993). Spraying TiN by a combined laser and low-pressure plasma spray system. *Journal of Thermal Spray Technology, 2*(2), 137–144. doi:10.1007/BF02652021

Ohmori, A., Wakamatsu, M., & Kamada, K. (1993). Synthesis of Al2O3-AlN coatings by low pressure plasma spraying and nitriding. *Transactions of JWRI, 22*(2), 227-232.

Ohmori, A., Zhang, Y. C., Hirano, S., & Arata, Y. (1991). Reaction spray by laser-plasma jet synthesis of TiN coatings. *Transactions of JWRI, 20*(2), 27-33.

Okada, T., Hamatani, H., & Yoshida, T. (1989). Radio-Frequency plasma spraying of ceramics. *Journal of the American Ceramic Society, 72*(11), 2111–2116. doi:10.1111/j.1151-2916.1989.tb06040.x

Pawlowski, L. (2008). *The science and engineering of thermal spray coatings*. England: John Wiley& Sons Ltd. doi:10.1002/9780470754085

Pierson, H. O. (1996). *Handbook of refractory carbides and nitrides –Properties, characteristics, processing and applications*. New Jersey, USA: Noyes Publications.

Radwan, M., & Miyamoto, Y. (2007). Growth of quasi-aligned AlN nanofibers by nitriding combustion synthesis. *Journal of the American Ceramic Society*, *90*(8), 2347–2351. doi:10.1111/j.1551-2916.2007.01747.x

Remesh, K., Ng, H. W., & Yu, S. C. M. (2003). Influence of process parameters on the deposition footprint in plasma-spray coating. *Journal of Thermal Spray Technology*, *12*(3), 377–392. doi:10.1361/105996303770348258

Rodriguez-Diaz, R. A., Banos, L., Novelo, O., Flores, C., Colin, J., Arenas-Alatorre, J., & Juarez-Islas, J. A. (2009). Microstructural characterization of $Fe_{40}Al_{15}Cr$ (% at.) intermetallic alloy produced by rapid solidification. *Acta Microscopica*, *18*(2), 169–173.

Rosenband, V., & Gany, A. (2004). Activation of combustion synthesis of aluminium nitride powder. *Journal of Materials Processing Technology*, *147*(2), 179–203. doi:10.1016/j.jmatprotec.2003.12.017

Ruijin, W., Tianjian, X. L. Z., & Xiaoou, H. (2004). The properties of high productive high velocity arc sprayed coatings and its applications. In *Proceedings of International Thermal Spray Conference*. Osaka, Japan.

Salak, A., Rundnayova, E., & Selecka, M. (1999). Internal nitriding of sintered iron in nitrogen-low ammonia atmospheres. *Powder Metallurgy*, *42*(1), 45–50. doi:10.1179/pom.1999.42.1.45

Salhi, Z., Guessasma, S., & Fenineche, N. (2009). Yttria-Stabilized zirconia in-flight particle characteristics under vacuum plasma spray conditions. *Vacuum*, *83*(11), 1382–1387. doi:10.1016/j.vacuum.2009.04.051

Schiefler Filho, M. F. O., Buschinelli, A. J. A. B., Gärtner, F., Kirsten, A., Voyer, J., & Kreye, H. (2004). Influence of process parameters on the quality of thermally sprayed X46Cr13 stainless steel coatings. *Journal of the Brazilian Society of Mechanical Sciences and Engineering*, *26*(1), 98–106. doi:10.1590/S1678-58782004000100016

Selezneva, S. E., & Boulos, M. (2001). Supersonic induction plasma jet modeling. *Nuclear Instruments & Methods in Physics Research. Section B, Beam Interactions with Materials and Atoms*, *180*(1-4), 306–311. doi:10.1016/S0168-583X(01)00436-0

Semenov, S., & Cetegen, B. (2001). Spectroscopic temperature measurements in direct current arc plasma jets used in thermal spray processing of materials. *Journal of Thermal Spray Technology*, *10*(2), 326–336. doi:10.1361/105996301770349411

Shahien, M., Yamada, M., Yasui, T., & Fukumoto, M. (2010a). Cubic Aluminum Nitride Coating through Atmospheric Reactive Plasma Nitriding. *Journal of Thermal Spray Technology*, *19*(3), 635–641. doi:10.1007/s11666-010-9469-0

Shahien, M., Yamada, M., Yasui, T., & Fukumoto, M. (2010b). Fabrication of AlN coatings by reactive atmospheric plasma spray nitriding of Al powders. *Materials Transactions*, *51*(5), 957–961. doi:10.2320/matertrans.T-M2010804

Shahien, M., Yamada, M., Yasui, T., & Fukumoto, M. (2010c). Influence of Plasma Gases on Fabrication of AlN Coatings through Atmospheric Plasma Nitriding Process. In S. Ikezawa (Ed.), *Industrial Application of Plasma Process* (Vol. 3, pp. 1–10). India: Transworld Research Network.

Shahien, M., Yamada, M., Yasui, T., & Fukumoto, M. (2011a). Influence of NH_4Cl Powder Addition for Fabrication of Aluminum Nitride Coating in Reactive Atmospheric Plasma Spray Process. *Journal of Thermal Spray Technology*, *20*(1-2), 205–212. doi:10.1007/s11666-010-9584-y

Shahien, M., Yamada, M., Yasui, T., & Fukumoto, M. (2011b). Reactive Atmospheric Plasma Spraying of AlN coatings: Influence of Aluminum Feedstock Particle Size. *Journal of Thermal Spray Technology*, *20*(3), 580–589. doi:10.1007/s11666-010-9582-0

Shahien, M., Yamada, M., Yasui, T., & Fukumoto, M. (2011c). Fabrication of AlN/Al_2O_3 coatings via atmospheric reactive plasma nitriding of A l_2O_3 powders. *Key Engineering Materials*, *484*, 166–171. doi:10.4028/www.scientific.net/KEM.484.166

Shahien, M., Yamada, M., Yasui, T., & Fukumoto, M. (2011d). Controlling of Nitriding Process on Reactive Plasma Spraying of Al Particles. In *IOP Conference Series: Mater. Sci. Eng.* doi:10.1088/1757-899X/18/20/202006

Shahien, M., Yamada, M., Yasui, T., & Fukumoto, M. (2011e). In situ fabrication of AlN coating by reactive plasma spraying of Al/AlN powder. *Coatings*, *1-2*(2), 88–107. doi:10.3390/coatings1020088

Shahien, M., Yamada, M., Yasui, T., & Fukumoto, M. (2012). Reactive Plasma Nitriding of Al_2O_3 Powder in Thermal Spray. *International Journal of Modern Physics*, *6*, 546–551.

Shahien, M., Yamada, M., Yasui, T., & Fukumoto, M. (2013a). Synthesis of Cubic Aluminum Nitride Coating from Al_2O_3 Powder in Reactive Plasma Spray process. *Materials Transactions*, *54*(2), 207–214. doi:10.2320/matertrans.T-M2012831

Shahien, M., Yamada, M., Yasui, T., & Fukumoto, M. (2013b). N_2 and H_2 Plasma Gases' Effects in Reactive Plasma Spraying of Al_2O_3 Powder. *Surface and Coatings Technology*, *216*, 308–317. doi:10.1016/j.surfcoat.2012.11.062

Shahien, M., Yamada, M., Yasui, T., & Fukumoto, M. (2013c). Reactive Plasma Spraying of Fine Al_2O_3/AlN Feedstock Powder. *Journal of Thermal Spray Technology*, *22*(8), 1283–1293. doi:10.1007/s11666-013-9992-x

Smith, R. W. (1993). Reactive plasma spray forming for advanced materials synthesis. *Powder Metallurgy International*, *25*(1), 9–16.

Smith, R. W., Lugcheider, E., Jokiel, P., Mueller, U., Merz, J., & Wilbert, M. (1993). Synthesis of composite materials by reactive plasma spray processing. In C. C. Berndt & T. F. Bernecki (Eds.), *Thermal Spray: Research, Design and Applications* (pp. 439–444). Materials Park, OH, USA: ASM International.

Steffens, H.-D., Wilden, J., & Buchmann, C. (1995). Laser beam nitriding of thermally sprayed titanium coatings. In C.C. Berndt & S. Sampath (Ed.), Thermal Spray Science & Technology (pp. 981-986). Materials Park, OH: ASM International.

Sun, Y., & Bell, T. (1997). A numerical model of plasma nitriding of low alloy steels. *Materials Science and Engineering A*, *224*(1-2), 33–47. doi:10.1016/S0921-5093(96)10561-X

Sundgren, J.-E. (1985). Structure and properties of TiN coatings. *Thin Solid Films*, *128*(1-2), 21–44. doi:10.1016/0040-6090(85)90333-5

Tahara, H., Moriyama, M., & Fujiuchi, K. (2004). Ceramic spraying using electromagnetically accelerated plasma. In *Proceeding of the International Thermal Spray Conference* (CD), Osaka, Japan.

Takada, S., Okumura, Y., & Sasaki, T. (1988). Preparation of iron-nitride by reduction sintering method. *Research reports of the Faculty of Engineering* [in Japanese]. *Gifu University*, *38*, 43–48.

Tamaki, M., Tomii, Y., Yamamoto, N., Kuwabara, H., & Nishikawa, I. (1997). Behavior and role of hydrogen during plasma nitriding [in Japanese]. *Journal of the Japan Society of Powder and Powder Metallurgy*, *44*(7), 712–715. doi:10.2497/jjspm.44.712

Tanaka, K., & Hayashi, K. (2000). Preparation of Fe_4N sintered compact by heating green and sintered compacts of Fe powder in flowing NH_3+H_2 mixed gas. *Journal of the Japan Society of Powder and Powder Metallurgy*, *47*(6), 600–606. doi:10.2497/jjspm.47.600

Thiel, S., Heimann, R. B., Herrmann, M., Berger, L. M., Nebelung, M., Zschunke, M., & Wielage, B. (1996). Thermal spraying of silicon nitride – based powders. In C. C. Berndt (Ed.), *Thermal spray: practical solutions for engineering problems* (pp. 325–331). Materials Park, OH: ASM International.

Thiele, S., Heimann, R. B., Berger, L. M., Herrmann, M., Nebelung, M., Schnick, T., & Vuoristo, P. et al. (2002). Microstructure and porperties of thermally sprayed silicon nitride-based coatings. *Journal of Thermal Spray Technology*, *11*(2), 218–225. doi:10.1361/105996302770348871

Tilly, D. J., Lofvander, J. P. A., & Levi, C. G. (1996). Solidification paths and carbide morphologies in melt processed $MoSi_2$-SiC in situ composites. *Metallurgical and Materials Transactions*, *28A*, 1889–1900.

Toth, L. E. (1971). *Transition metal carbides and nitrides*. New York, London: Academic Press.

Tsunekawa, Y., Hiromura, M., & Okumiya, M. (1998). Nitriding of Metal Droplets in Synthesis of Intermetallic Matrix Composite Coatings by Reactive Plasma Spraying. In C. Coddet (Ed.), *Thermal Spray: Meeting the Challenges of the 21st Century* (pp. 1143–1148). PA: ASM International, Materials Park, OH, USA.

Tsunekawa, Y., Hiromura, M., & Okumiya, M. (2000). Nitride formation in synthesis of titanium aluminide matrix composite coatings by reactive RF plasma spraying. *Journal of Thermal Spray Technology*, *9*(1), 83–89. doi:10.1361/105996300770350096

Tsunekawa, Y., Okumiya, M., Kobayashi, T., Okuda, M., & Fukumoto, M. (1996). Chromium-nitride in situ composites with a compositional gradient formed by reactive DC plasma spraying. *Journal of Thermal Spray Technology*, *5*(2), 139–144. doi:10.1007/BF02646427

Tsunekawa, Y., Okumiya, M., Okuda, M., & Fukumoto, M. (1997). Reaction of metal droplets during flight and on substrate in reactive low pressure DC plasma spraying. In *Proceeding of the Eleventh International Conference on Surface Modification Technologies,* Paris, France (pp. 234-245).

Vardelle, M., Vardelle, A., & Fauchais, P. (1993). Spray parameter and particle behavior relationships during plasma spraying. *Journal of Thermal Spray Technology, 2*(1), 79–91. doi:10.1007/BF02647426

Vardelle, M., Vardelle, A., Fauchais, P., Li, K.-I., Dussoubs, B., & Themelis, N. J. (2001). Controlling particle injection in plasma spraying. *Journal of Thermal Spray Technology, 10*(2), 267–284. doi:10.1361/105996301770349367

Wang, M., & Shaw, L. L. (2007). Effect of the powder manufacturing method on microstructure and wear performance of plasma sprayed Alumina-Titania coatings. *Surface and Coatings Technology, 202*(1), 34–44. doi:10.1016/j.surfcoat.2007.04.057

Wiese, W. L., Smith, M. W., & Miles, B. M. (1996). Atomic Transition Probabilities, Vol. II: Sodium through Calcium. Washington, DC: U.S. Government.

Wriedt, H. A. (1986). The Al-N (aluminum-nitrogen) system. *Bulletin of Alloy Phase Diagrams, 7*(4), 329–333. doi:10.1007/BF02873001

Yamada, M. (2006). *Fabrication of nitride ceramic coatings by reactive plasma spraying*. Unpublished doctoral dissertation, Toyohashi University of Technology, Toyohashi, Japan.

Yamada, M., Inamoto, T., Fukumoto, M., & Yasui, T. (2004). Fabrication of silicon nitride thick coatings by reactive RF plasma spraying. *Materials Transactions, 45*(12), 3304–3308. doi:10.2320/matertrans.45.3304

Yamada, M., Inamoto, T., Fukumoto, M., & Yasui, T. (2005). Spectroscopic measurement of plasma in the reactive plasma spray process [in Japanese]. *Journal of the Vacuum Society of Japan, 48*(3), 241–243. doi:10.3131/jvsj.48.241

Yamada, M., Kouzaki, Y., Yasui, T., & Fukumoto, M. (2006). Fabrication of iron nitride coatings by reactive RF plasma spraying. *Surface and Coatings Technology, 201*(3-4), 1745–1751. doi:10.1016/j.surfcoat.2006.02.046

Yamada, M., Nakamura, H., Yasui, T., Fukumoto, M., & Takahashi, K. (2006). Influence of substrate materials upon fabrication of aluminum nitride coatings by reactive RF plasma spraying. *Materials Transactions, 47*(7), 1671–1676. doi:10.2320/matertrans.47.1671

Yamada, M., Yasui, T., Fukumoto, M., & Takahashi, K. (2007). Nitridation of aluminum particles and formation process of aluminum nitride coatings by reactive RF plasma spraying. *Thin Solid Films, 515*(9), 4166–4171. doi:10.1016/j.tsf.2006.02.054

Zerfass, H. (2006). *ArC Spraying*. Osu Gunter Hessler GmbH & Co.KG. Retrieved from http://www.osu.de/e/arcspraying.html

KEY TERMS AND DEFINITIONS

Feedstock Powder: The starting material for spraying, and it can be metallic, non-metallic: ceramic or plastic, or any desired combination.

In-Flight Time: The time of the particles inside the plasma during spraying until deposition.

Nitriding Species: The active species in the plasma which are responsible for particles nitriding and formation of the nitride phases, like the N_2/H_2 gases.

Particles Diagnostics: The particle in-flight temperature, velocity and size during spraying in the plasma.

Plasma Characteristics: The plasma temperature, plasma velocity and active species during spraying.

Plasma Spray Deposition: A coating technology based on deposition of molten/semi-molten (softened) particles propelling onto a substrate by using plasma flame.

Reactive Plasma Spraying: A coating process based on enabling the reaction between the feedstock materials and surrounding active species in the plasma.

Thermal Spraying: A generic term for a group of commonly used processes employ the heat and velocity to coat the surface of substrate material with another feedstock material.

Section 3

Chapter 13
Hybrid Sol-Gel Coatings:
Erosion-Corrosion Protection

Rita Bacelar Figueira
Engiprojects Lda, Portugal

ABSTRACT

The properties and wide application range of organic-inorganic hybrid (O-IH) sol-gel materials have attracted significant attention over the past decades. The combination of organic polymers and inorganic materials in a single-phase provides exceptional possibilities to tailor electrical, optical and mechanical properties concerning diverse applications. This unlimited design concept has led to the development of diverse coatings for several applications such as glasses, and metals to mitigate mechanical abrasion, erosion and corrosion. This class of materials could be easily obtained by sol-gel method at mild synthesis conditions. Furthermore, the large variety of available chemical precursors allows producing a diversity of coatings with tuned mechanical and thermal properties. This chapter will introduce the fundamentals of the sol-gel method to produce O-IH protective thin coatings and discuss the methodologies used to apply these materials onto different metallic substrates for erosion and corrosion protection.

INTRODUCTION

Corrosion degradation is a major cause of metallic structural deterioration (Figure 1). It severely affects their service life and may result in structural failure, leakage, product loss and environmental pollution, causing large financial outlays for recovery/prevention. The World Corrosion Organization (WCO) estimates that the annual cost of corrosion worldwide is around 3% of the world's Gross domestic product (Bhandari, Khan, Abbassi, Garaniya, & Ojeda, 2015), not considering the environmental costs due to corrosion. A passive approach to corrosion protection involves depositing a barrier layer that prevents contact of a material with the corrosive environment. Actives approaches reduce the corrosion rate when the protective barrier is damaged and corrosive agents come into contact with the metal substrate. Only the combination of both methodologies can provide reliable protection against corrosion of metallic structures.

DOI: 10.4018/978-1-5225-4194-3.ch013

The technological advances require more and more materials with high resistance to aggressive environments. Coatings to protect materials against erosion and corrosion are of extreme importance for both service life and maintenance cost of installations operating in aggressive environments (Szymański, Hernas, Moskal, & Myalska, 2015). The combination of degradation mechanisms, corrosion with erosion, can be several times higher than the effect of each, acting alone. Therefore, materials selection must be considered at every stage of design, construction and operation of equipment and systems. This will minimize premature failures that ultimately lead to severe economic losses and compromise the structure safety.

Sol-gel process has been considered a powerful and versatile strategy to produce functional inorganic and organic-inorganic hybrid (O-IH) materials. The interest in this method started with Ebelman and Graham's studies in the mid-1880s on silica gels (Hench & West, 1990). However, it was only in the 1980s that this method was adopted by Schmidt (1985) and Wilkes (Wilkes, Orler, & Huang, 1985) in the synthesis of their O-IH materials. The understanding of the mechanisms underlying the sol–gel methods have been subject of several text books and reviews (Aegerter & Mennig, 2004; Atkinson, 1991; Brinker, Hurd, Schunk, & Frye, 1992; Brinker & Scherer, 1990; Ciriminna et al., 2013; Dimitriev, Ivanova, & Iordanova, 2008; Livage, 1997, 1999; Pierre, 1998; Sanchez et al., 2001, 2010; Sanchez & Ribot, 1994; Clément Sanchez, Belleville, Popall, & Nicole, 2011). Numerous definitions of sol-gel process have also been proposed and discussed (Brinker & Scherer, 1990; Dislich, 1986; Pierre, 1998; Segal, 1984). In this chapter a broader definition, given by Pierre (1998), will be used, in which a sol-gel process is "every process that starts from precursors solutions with intermediate stages including a gel and/or a sol". Sol-gel method is a low temperature versatile route for the preparation of inorganic and O-IH materials in the form of monoliths, fibres, films or nanoparticles. O-IH materials are extremely interesting once they allow an interface between two worlds of chemistry (organic and inorganic) each with distinctive

Figure 1. Corrosion of a metallic support exposed to marine environment

properties. This enables obtaining materials with characteristics that cannot be found on the individual components. Moreover, it is a useful method, particularly, in the production of coatings since enables coating deposition on substrates with a large surface area involving simple and inexpensive equipment. Besides the excellent control of precursor stoichiometry, the existence of a large number of precursor reagents, with tuned functional groups, allows obtaining a diversity of O-IH gels following the same trends. It is also a method that eases the incorporation of additional components, introducing complementary functions to the materials (e.g. UV protection, antifouling, antireflection, moisture resistance, corrosion inhibition, and adhesion protection). Overall, the reagents involved to produce such matrices have very low environmental impacts. The tested corrosion protection processes and materials based on sol–gel technology demonstrate positive performances when several properties are considered, such as corrosion resistance, adhesion, fatigue resistance, reliability, and quality control. Although some of the possible alternatives show potential for corrosion protection, continuous efforts are being made in order to improve the production of O-IH materials with enhanced properties, to understand he mechanisms involved in coating production and therefore their corrosion protection performance (Figueira, Silva, & Pereira, 2014).

Moreover, O-IH materials represent an important part of the R&D activities and technology portfolio in companies such as Bayer, Boeing GE, Honeywell, BASF, Akzo-Nobel, Sumitomo, ASAHI Kasei, ASAHI Glass, Dow Corning, amongo others. They are also important in more specialized enterprises such as Henkel, Procter & Gamble, Wacker, Nissan Chemicals, Shin-Etsu, Toppan, Mitsubishi Chemicals, NEPES, Merck, MicroResistTechnologies, among others (Figueira et al., 2014).

In the following sections the different strategies with a focus on highlighting the influence that sol-gel processing imparts on the properties of O-IH sol-gel coatings will be discussed. Illustrative examples from the recent literature will be described, and the advantages and current limitations of each approach will be debated. This chapter will briefly introduce the corrosion and environment, the fundamentals of sol-gel method to produce O-IH protective thin coatings and discuss the methodologies used to apply these materials onto different metallic substrates (e.g. galvanized steel, aluminium, copper, magnesium, nickel-titanium and their alloys) for erosion-corrosion protection. Finally, the future developments in the area that are required to further strengthen the utility of sol-gel techniques for generating improved O-IH coatings systems, with even higher functionality, will be discussed.

CORROSION

The major degradation mechanism of metallic materials is corrosion. Both the physical and chemical properties of the products are important since they may influence the subsequent rate of reaction. In the case of steel, Cl^-, O_2, and H_2O species, as well as electron transport, all play extremely important roles in the corrosion process. The deterioration of structures leads to defects that may compromise the performance of a structural element. In extreme cases may conduct to its failure compromising the function and/or safety of the entire structure. Therefore, the understanding of the deterioration processes and mechanisms together with the causes and related effects is essential for the durability and performance of the infrastructures. The main deterioration processes may be categorized as biological, chemical, physical or other damage processes (Figure 2). Fatigue and corrosion are the main mechanisms of deterioration in steel structures. Therefore, these two mechanisms are the most important when designing new structures or maintaining the existing ones.

Figure 2. Classification of deterioration processes and defect types
Adapted from Bardal, 2004.

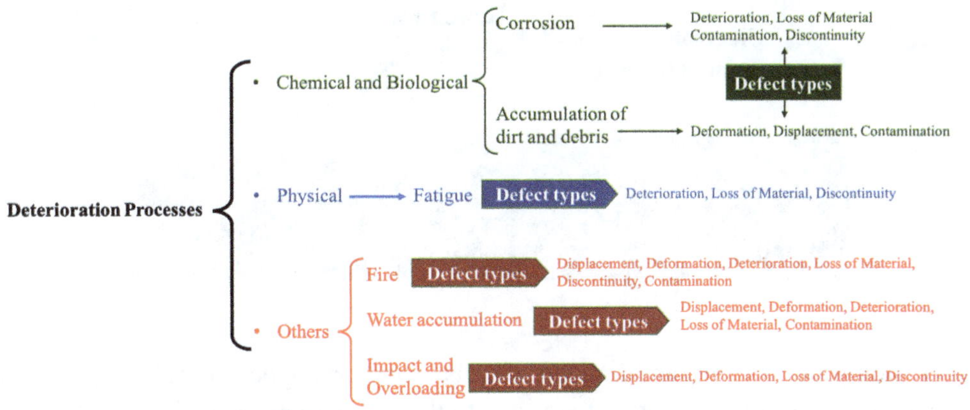

The economic and social impacts due to the impairments caused by these mechanisms on a steel structure, substantiate the implementation of both protective and preventive measures to mitigate their consequences. Corrosion concept can be explained in different ways and can be found in several text books (Bardal, 2004; Chandler & Bayliss, 1985; Davis, 2000; Ghali, Sastri, & Elboujdaini, 2007; Kaesche, 2003; Roberge, 1999). In this chapter, corrosion will be interpreted according to Bardal (2004) and defined as "an attack on a metallic material by reaction with the surrounding environment". Corrosion of metallic materials may be classified in three groups, namely:

1. Wet corrosion (where the corrosive environment is water with dissolved species);
2. Corrosion in other fluids;
3. Dry corrosion (the corrosive environment is a dry gas) and the most representative example is high-temperature corrosion.

Most of corrosion reactions are an electrochemical process and relies on the formation of a corrosion cell consisting on a cathode, an anode and an electrolyte (such as water with dissolved species) that is in contact with the metal. The electrolyte will provide a pathway for the ionic current and for electron flow through the metal. This process is an electrical circuit without any accumulation of charges. The oxidation reaction that occurs at the anode results in the generation of electrons that are consumed at the cathode by reduction reaction. Current flows from the cathode to the anode by the movement of charged species through the electrolyte while the electron go from the anode to the cathode through the electronic path (Figure 3).

The electrochemical corrosion cell controls the corrosion rate. Therefore, the corrosion reaction may be blocked by eliminating one of the above requirements. For example, a dry surface (without any moisture) will not be corroded since there is no ionic current path. This may be an effective way to control the corrosion.

When a metallic material is immersed in fluid with a relative movement its surface is, in many cases, exposed to mechanical wear effects leading to increased corrosion. This mechanism is known as erosion corrosion or flow-induced localized corrosion that takes place in turbulent flowing liquids. A simple way to explain the erosion-corrosion process is that the deposits of corrosion products, or salts precipitated

Figure 3. Schematic representation of an electrochemical corrosion cell
Adapted from Bardal, 2004.

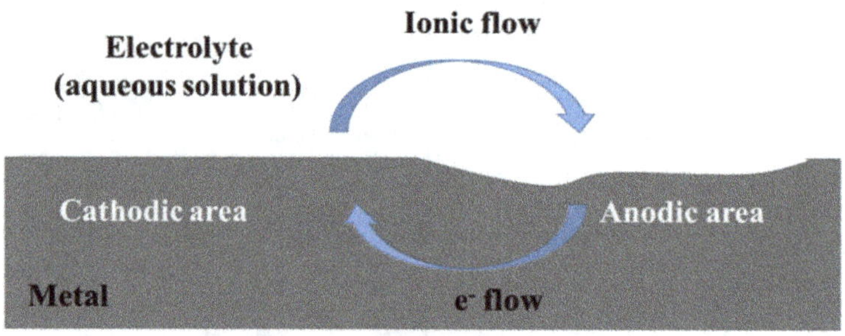

due to the corrosion process itself, are dragged off, dissolved or stopped from being formed. Above a critical flow rate, the domain of erosion initiates due to the action of each erosive event causing rupture or removal of the protective film (Fuchs-Godec, 2010). Therefore, the metallic surface becomes clean and more active. In extreme cases, erosion corrosion process may be complemented by pure mechanical erosion. In this case, solid particles in the fluid may scratch and slash out particles from the material itself causing plastic deformation, which may turn the substrate even more active and prompt to corrosion attack. The pattern results of erosion corrosion are pits determined by the flow direction and the local flow conditions. The components most prone to erosion corrosion are generally pumps, turbine parts, heat exchanger tubes, bends, nozzles, propellers valves and other equipment exposed to liquid sputter or jets (Bardal, 2004).

Corrosion and Environment

The durability of metallic materials and its alloys is determined by the environmental characteristics, the materials properties and their functional conditions. Therefore, the description and classification of the environment, concerning its corrosivity, is essential for the choice and design of metallic structure as well as the choice of the protective measure or the repair method to apply.

Atmospheric exposure is commonly characterized as urban, rural, industrial or marine, based on the existence of major corrosive agents as described in BS EN ISO 9223:2012, BS EN ISO 9224:2012, BS EN ISO 925:2012 and BS EN ISO 9226:2012 (Table 1). Besides the macroclimatic zone, the local and the micro environments (e.g. environment in the interface of the structural element and its surroundings) are the most decisive factors regarding the vulnerability to corrosion and performance of a metal or alloy. Several variables should be considered such as location category (typical exposure conditions of a component, e.g. under shelter, open air, etc.), the meteorological elements, airborne pollutants, operating conditions and technical parameters such as profile, design, manufacture, etc. The most significant factors that influence the corrosion of metals and alloys are the time of wetness (TOW) and pollution level. The TOW is a concept commonly used in atmospheric corrosion of metallic materials and refers to the time when the metal is sufficiently wet for corrosion reaction to occur. The increase of temperature leads to the increase of TOW and corrosion rate and depends on both the climatic conditions and on the properties of the metallic surface (Corvo et al., 2008). BS EN ISO standard 9223 defines TOW as "*The period during which a metallic surface is covered by adsorptive and/or liquid films of electrolyte that are capable of causing atmospheric corrosion*". The wetting of surfaces is caused by many factors, such as dew, rainfall, melting snow and a high humidity level.

Table 1. Type of atmosphere and respective description

Type of Atmosphere	Atmospheric Environments
Rural	In rural areas and small towns without significant contamination by aggressive agents.
Urban	In densely populated areas without industrial concentration and with moderate concentration of pollutants.
Industrial	In areas with heavy industrial activity and high contamination by corrosive pollutants.
Marine	Over the sea, or overland contaminated with airborne salts.
Marine + Industrial (mixed atmospheric environment)	Near the coast and close to industrial concentration areas with moderate to heavy contamination by corrosive agents.

According to BS EN ISO 9223:2012, the TOW of corroding surfaces (τ) can be estimated as the length of time when the relative humidity is higher than 80% at a temperature higher than 0ºC. Information on calculated TOW is helpful for informative atmosphere corrosivity estimation. However, in this standard, it should be kept in mind that TOW is assessed based only on climatic factors (temperature and humidity) regardless of the pollutant level, nature and properties of the metallic surface. Furthermore, the TOW, besides being caused by several factors is also influenced by the type of metal, shape of the component, corrosion products, nature of pollutants and other features. The categories of TOW, according to BS EN ISO 9223:2012, are given in Table 2.

The pollution level may be assessed by performing quantitative measurements of sulphur dioxide (SO_2) and chloride in the air according to the specification of BS EN ISO 9223:2012. SO_2 is a pollutant in the atmosphere that results predominantly from the burning of coal, oil, and gasoline and is one of the most important air pollutants that contributes to the corrosion of metals. It is usually measured in terms of its concentration in air in units of $\mu g/m^3$. However, as it is the SO_2 that deposits on the metal surface, which affects the corrosion, it is also often measured in terms of deposition rate on the surface. The pollution levels can also be measured in terms of the concentration of the dissolved SO_4^{2-} in rain water. Generally, high SO_2 concentration in the air results in a large amount of deposition on an exposed surface and a high concentration of dissolved SO_4^{2-} in rain water. Airborne salinity refers to the content of gaseous and suspended salt in the atmosphere. However, by the same reason previously stated (it is the salt that is deposited on the metal surface that affects the corrosion) it is very often measured in terms of deposition rate. The categories of pollution for outdoor atmospheres are classified according to BS EN ISO 9223:2012 and are described in Table 3.

Table 2. Categories of time of wetness (TOW)

Category	TOW	
	Hours per Year	%
τ_1	$\tau \leq 10$	$\tau \leq 0.1$
τ_2	$10 < \tau \leq 250$	$0.1 < \tau \leq 3$
τ_3	$250 < \tau \leq 2500$	$3 < \tau \leq 30$
τ_4	$2500 < \tau \leq 5500$	$30 < \tau \leq 60$
τ	$5500 <$	$60 < \tau$

Adapted from BS EN ISO 9223:2012.

Table 3. Categories of pollution by sulphur-containing substances represented by SO_2 and by airborne salinity represented NaCl expressed as annual averages for outdoor atmospheres

Classification of SO_2 Pollution in Annual Average				Classification of Pollution by Airborne Salinity Characterized by NaCl in Annual Average		
Category	**Concentration of SO_2 in Air µg m^{-3}**	**Deposition Rate of SO_2 mg m^{-2} day $^{-1}$**	**Typical Environment**	**Category**	**Deposition Rate of Cl^- mg m^{-2} day $^{-1}$**	**Typical Environment**
P_0	$P \leq 12$	$P \leq 10$	Rural	S_0	$S \leq 3$	Non-Coastal
P_1	$12 < P \leq 40$	$10 < P \leq 35$	Urban or industrial	S_1	$3 < S \leq 60$	Coastal Environment
P_2	$40 < P \leq 90$	$35 < P \leq 80$		S_2	$60 < S \leq 300$	
P_3	$90 < P \leq 250$	$80 < P \leq 200$	Severely polluted	S_3	$300 < S \leq 1500$	Within 200 m from Sea

Adapted from BS EN ISO 9223:2012.

The P values (Table 3) are the annual averages determined from continuous measurements for at least one year where P_0 is the background concentration and P_3 is the extreme concentration. Generally, high SO_2 deposition rate tends to result in high rate of corrosion. Nevertheless, it should be kept in mind that two different geographic locations, with similar SO_2 deposition rates, may lead to significantly different corrosion rates. This is mainly due to other factors such as temperature, moisture, etc. that impact significantly on the corrosion rate. The atmospheric corrosion rate appreciably increases in polluted air and for metal surfaces covered by solid particles that retain the moisture.

In the case of deposition rate of Cl^-, the S values (Table 3) correspond to the annual averages determined from continuous measurements for at least one year where S_0 is the background concentration while S_3 represents the salt levels at locations very near the sea. Airborne salinity is predominantly related to the marine climate. It is strongly dependent on many variables such as wind direction and velocity, distance from the shore, local landscape, etc. High salt deposition rate tends to result in high rates of corrosion and for a given geographic location, the corrosion rate decreases with the distance from the sea together with the decrease of salt deposition rate.

The corrosivity of individual metals can be estimated by environmental data, e.g. TOW and pollution categories or based on corrosion rate measurements of standard specimens as specified in BS EN ISO 9226:2012. These measurements allow classifying the corrosivity for different metallic substrates which are described in BS EN ISO 9223:2012 and BS EN ISO 12944-2:1998. Table 4 shows the atmospheric corrosivity categories and respective corrosion rates of carbon steel for the first year of exposure as well as examples of typical environments.

Category C_1 involves generally climate-controlled indoor atmospheres in which the corrosion rate is very low while category C_5 includes atmospheric environments with high corrosivity such as by the sea shore, heavily polluted with SO_2 or extremely humid conditions. The exposure to water from different sources covers a wide range of compositions and properties. The corrosivity of water is affected by several parameters such as O_2 concentration, pH (acidity), hardness, conductivity, temperature, flow rate, the presence of chemical species (e.g. salts) and biological agents. Corrosion in fresh water or briny water (e.g. water with higher salinity than fresh water but lower than seawater) is frequently determined by the diffusion of O_2 (and its concentration), temperature, flow velocity and deposits of calcium and magnesium salts. Organic material and other salts may also affect, negatively or positively the corro-

Table 4. Categories of pollution by sulphur-containing substances represented by SO_2 and by airborne salinity represented by NaCl expressed as annual averages for outdoor atmospheres

Category	Corrosivity	Corrosion Rate of Carbon Steel		Examples of Typical Environments in a Temperate Climate
		g m^{-2}	μm	
C_1	Very low	≤ 10	≤ 1.3	–
C_2	Low	> 10 – 200	> 1.3 – 25	Rural areas (atmospheres with low level of corrosion)
C_3	Medium	> 200 – 400	> 25 – 80	Urban and industrial atmospheres with moderate SO_2 pollution. Coastal areas with low salinity
C_4	High	> 400 – 650	> 50 – 80	Industrial and coastal areas with moderate salinity
C_5	Very high	> 650 – 1500	> 80 – 200	Industrial areas with high humidity and aggressive atmosphere

Adapted from BS EN ISO 9223:2012 and BS EN ISO 12944-2:1998.

sion evolution in water in several ways. For example, acidic salts may favour the H_2 evolution while alkaline salts may act as corrosion inhibitors; some living organisms may produce O_2 while others may consume it and change the environment; under anaerobic conditions sulphate reducing bacteria (SRB) may depolarize hydrogen reduction reaction promoting localized corrosion.

Corrosion in seawater is influenced by the physical, chemical and biological properties of the water and by different sources of pollution as described by (Craig, 2006). Pollution may influence the corrosion kinetics due to toxic effects, depletion of dissolved O_2 and the action of chemical species like sulphides. In offshore environments the structures have long-term exposure to humidity with high salinity, intensive influence of UV light, wave and currents action, gradients (dissolved species, temperature, etc.) and external actions (discharges of industrial wastes). These factors contribute to the behaviour of the metallic elements in different exposure zones with typical damages in a vertical pattern of corrosion rates (Figure 4). The corrosion rate is quite low in the subsoil zone and in adjoining submerged areas. Nevertheless, it increases at the upper regions of the immersed area, until near mean low tide and from mean high tide to the splash zone. In atmospheric zone the corrosion rate of the steel decreases. BS 6349-1-1:2013 provides guidance values for corrosion rates considering the vertical zone classification. Figure 4 shows the vertical zone classification (Davis, 2000), typical average corrosion rates and upper limit for steel in marine environment (BS 6349-1-1:2013).

The differences found are due to the variability effects of bulk and local environmental conditions, based on access of both oxygen and humidity, and influenced by the level of the sea (Figure 4):

- **Atmospheric Zone:** The steel suffers action from marine aerosol, however the structure is not directly attacked by water splashes. The winds carry the salts in the form of solid particles or as droplets of saline solution. The quantity of salt present decreases as a function of height distance from the sea, although wind speed and direction will influence;
- **Splash Zone:** Zone immediately above the maximum level of intertidal variation. The substrate is directly affected by water splash. The height of the splash zone is a function of the wave height, as well as of the speed and wind direction and is subjected to cycles of wetting and drying. This phenomenon becomes more significant as the water evaporates and the salts remain in the surface of the substrate;

Figure 4. Schematic representation of the zones of corrosion for a steel element in seawater, the relative loss of material thickness and the respective corrosion rates
Adapted from Davis, 2000.

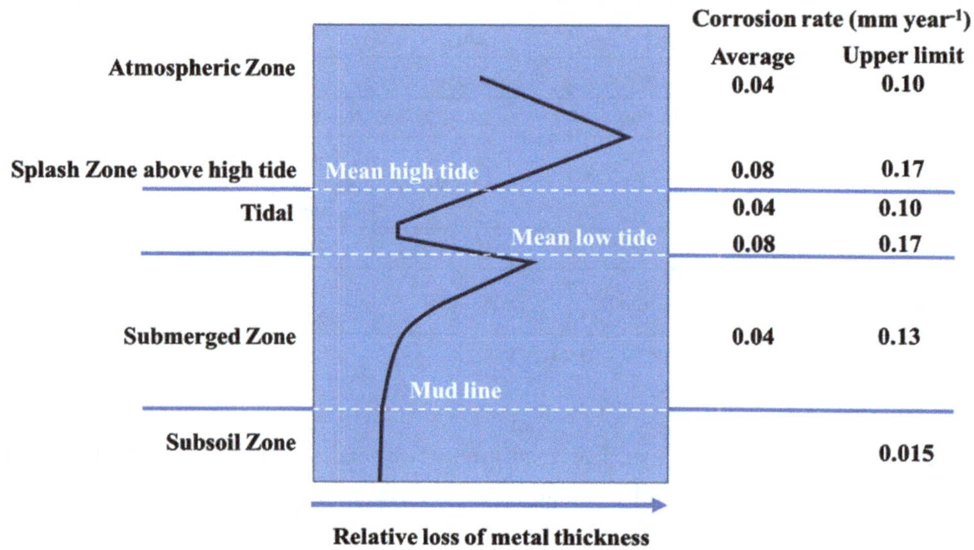

- **Tidal Zone:** The zone between the minimum and the maximum level of tides, governed by their variation. This region is also subjected to wetting and drying cycles. Degradation occurs due to chemical attack, abrasive action of waves and other substances in suspension, and the attack of microorganisms.
- **Submerged Zone:** The accumulation of microorganisms, algae and animals on structures submerged in seawater is certain, regardless of the location and the season. Even if fouling occurs at a slower pace in deep water it can be a significant problem for long-term subsea structures.

Corrosion Prevention

The design of a structure or building may be perfectly satisfactory from the standpoints of structural integrity and aesthetics. However, may include features that accelerate corrosion or influence the cost of controlling it to a significant extent. This is also true for the services and equipment involved with structures as it is for the main structural elements. Welds are a good example of the different requirements of structural stability and corrosion protection. Welds are usually inspected, often to a high level, to ensure structural safety. Though, generally such welds are not in the best condition for the application of a protective coating. For instance, the surface roughness may not be important regarding to structural integrity, however, may have a deleterious effect on the performance of applied coatings. Therefore, welds may require grinding to a smooth contour and possible blast cleaning before coating.

Earlier corrosion problems are generally an outcome of incorrect design, construction mistakes or from poor maintenance of the protective coatings. Corrosion may also be a secondary effect of other deterioration processes (Figure 2). Areas that are poorly drained or ventilated are more vulnerable to accumulation and contamination of dirt, garbage or water and therefore more susceptible to corrosion attack. Corrosion can be prevented by a suitable design, the proper selection of materials and, clearly, by an adequate use of corrosion protection systems. For example, Figure 5 shows typical water traps. This situation may arise from welding sections, in such a way that an "artificial passage" is produced. Therefore, this type of situation should be kept in mind during the design in order to be avoided.

The corrosion arising from connecting two alloys, one of which is more noble (or cathodic) than the other, leads to bimetallic corrosion (or galvanic corrosion) of the less noble (or anodic) alloy. The relative nobility of metals can be found in several galvanic series which are shown and discussed in several fundamental books (Bramfitt, 2007; Chandler & Bayliss, 1985; Ghali et al., 2007; Maaß & Peißker, 2011; Roberge, 1999; Seshan, 2002). The extent of the corrosion depends upon a number of factors:

1. The presence of an electrolyte and its conductivity. Bimetallic corrosion does not arise under dry indoor conditions. However, it is likely to be most severe under immersed conditions, particularly in high conductivity seawater.
2. The relative positions of the alloys in galvanic series. Generally, the effects are likely to be greater when alloys are well separated (higher difference of potential). Corrosion attack will be lower when the alloys are close together (lower difference of potential). For instance, stainless steels (SS) are usually far nobler than carbon steels (CS). Therefore, it will increase the corrosion of the CS to a considerable extension.
3. The relative size of the two alloys has a significant influence on the rate of corrosion. For example, a large area of bare CS in contact with a small area of SS will not be severely attacked. However, when the area of the SS is much higher than the CS, considerable corrosion of the latter may occur. This situation may occur when SS are used with painted steel structures. If there are any small pores or pinholes in the coating, the small area of CS will provide a very large cathode/anode ratio with the more noble bolts and severe pitting of the carbon steel may occur. This is, of course, much severe in harsh environments.
4. The formation of corrosion products. In coastal marine atmospheres, the corrosion products may be hygroscopic, leading to prolonged periods of wetness and increased corrosion attack.

There are a number of methods, which will not be discussed in this chapter, to combat bimetallic corrosion. However, whenever possible it should be avoided by not mixing different alloys particularly in harsh environments. Furthermore, proper access for inspection and eventual maintenance of the protective systems is essential. If, for any reason this is impracticable or extremely expensive (such as offshore environments), then the designers must take this into account when considering the effects of corrosion on the integrity of the structure.

Figure 5. Typical water traps
Adapted from Davis, 2000.

Another usual corrosion problem due to inaccurate design is crevice. Crevices are a problem due to the impediments of cleaning or painting within them. In the sense used here, a crevice is a very small gap between two alloy surfaces which, allow the access of air and moisture. However, do not provide enough space for cleaning and repainting during service life of the structure. Therefore, crevices should be avoided or at least sealed. Sometimes welding in preference to bolted joints may solve the problems. In other cases, the solution may be increasing the gap so that crevice conditions no longer operate. There is clearly an advantage in painting the contacting faces before they are bolted together in a structure. It is preferable to bolt the sections together while the coating is still wet to make sure that any gap will be filled with the coating system.

A good design of a structure is a matter of understanding the principles of corrosion and protection and appreciating situations where changes, sometimes comparatively minor ones, will improve the overall performance of coatings and lead to a reduction in corrosion. If designers, at all levels, develop a concern or an awareness of corrosion, many potential problems can be avoided at an early stage. This approach, generally, will be cheaper and more effective than changes at advanced stages. The maxim "*corrosion control begins at the drawing board*" is true. Before considering the various areas of design, it is worth making a general point. The corrosion problems that may arise must be considered to ensure low operational and maintenance (O&M) costs and to accomplish the expected service life of the structures for which it was dimensioned for. It is unlikely that any conventional organic coating system or a metallic coating can be relied upon for periods over 25 years without maintenance. Therefore, three possible approaches are possible (Bardal, 2004; J. R. Davis, 2000; Ghali et al., 2007; Kaesche, 2003; Roberge, 1999):

1. To build in what is often known as "a corrosion allowance";
2. Use a more resistant alloy;
3. To employ a thick coating.

Each of the mentioned alternatives carries inherent problems. A corrosion allowance is simply the addition of sufficient steel to the section to allow for the anticipated section loss by corrosion. However, when using this approach, the possibilities of enhanced local corrosion or changes in the initial environment must be taken into account. Often, the addition of a coating to the steel even where a corrosion allowance has been added is worth considering, for two reasons:

1. It will add a safety margin;
2. In situations where inspection is possible but access for maintenance is not, it indicates areas of local breakdown.

The use of an alloy that is more corrosion resistant than steel is sometimes a sound approach (such as stainless steel), although in most situations the additional costs prove to be unacceptable. The possibility of monitoring the corrosion of the steel in inaccessible areas is always worth considering since this allows time to decide on a suitable course of action if problems arise. Nevertheless, the design must comply with safety requirements during the service life of the structure. The project detailing is an important measure of corrosion prevention if includes good practices such as:

1. Avoid geometrical configurations that may promote corrosion attack, such as areas that tend to store moisture, water and contaminants (narrow gaps, cavities, crevices) by allowing air circulation and the use of proper drainage (Figure 5).
2. Avoid stress conditions that might lead to environmentally assisted cracking.
3. Avoid bimetallic connections that might lead to galvanic corrosion.
4. Implement an efficient protection system that allows an easy access for inspection and maintenance.

The choice of a protection system, essentially, relies on economic, environmental and safety considerations. The selection of a material is frequently the result of compromises between its function in the structure, its properties and respective needs due to external factors. Concerning corrosion, the features to be taken into account should include the risk of occurrence of different forms of corrosion (and the influencing conditions), the available protection systems that can be used, the environmental conditions and the functional conditions. EN ISO 12944-3:1998 further details the basic design criteria for steel structures in order to minimize the risk of corrosion initiation. The main types of coating defects will be briefly described (Table 5) while other types can be found in the literature (Chandler & Bayliss, 1985).

The occurrence of corrosion is mainly due to loss of the protective effect that should be provided by the coating system. Different causes can promote corrosion such as defects produced during the application of the coating system, reduction of the coating layers' thickness in certain elements e.g. edges, or coating erosion due to environmental conditions. BS EN ISO 4628:2016, provide information concerning the identification, classification and quantification of several coating defects. The assessment of a protection coating system and the definition of the required level of surface preparation are important factors in the choice of the type of intervention that should take place. Coating defects require a choice between total replace of the damage coating by a new one or carry out localized repair of the coating system involving only a renewal of the existing coating system.

Coating systems may integrate several layers of different types of coatings, however, the compatibility between the coats (layers) must be ensured. The coating process involves the application of non-metallic coatings, metallic coatings, or the combination of these two types of coatings on the steel surface. Metallic coatings are generally composed by non-ferrous metals, commonly zinc, aluminum and its alloys (Price & Figueira, 2017).

SOL-GEL METHOD: FUNDAMENTALS

In the past few decades, several text books and reviews devoted to sol-gel methods have been published. Sol-gel method is a low temperature synthetic route, provides excellent control of precursor stoichiometry and ease of composition modification. It also allows the introduction of functional groups, encapsulation of biomolecules and coating substrates of large surface areas (and different forms) involving simple and inexpensive equipment. Generically, sol-gel method includes the formation of a colloidal suspension – *sol* – followed by polymerization leading to the formation of three-dimensional structure dispersed in a solvent – *gel* (Figure 6). The *gel* can be modified with a large number of dopants leading to products with single properties. When drying takes place, under atmospheric conditions of pressure and temperature, typically results in a contraction of the gelled network (C. J. Brinker & Scherer, 1990; Pierre, 1998). The resulting dry *gel* is generally referred as *xerogel* and can be reduced in volume between five to ten times when compared with the initial wet *gel*. Due to its low porosity and large surface area, the *xerogel*

Table 5. Types of coatings defects and respective description

Coatings Defects	Description	Cause	Typical Examples
Cracking	May take several forms and occasionally its detection require some form of view magnifier	The coating is exposed to internal stresses. Depending on the present plasticization, pigmentation, ageing, conditions of application and curing may reduce the resistance to external stresses	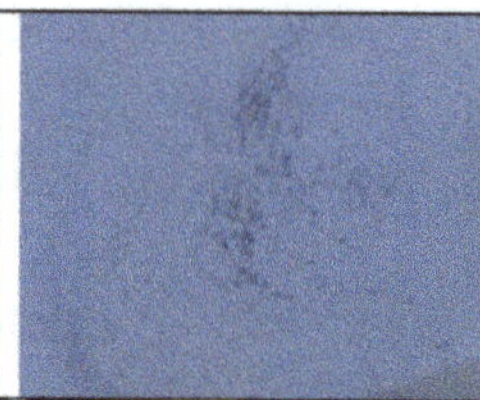
Blistering	Presented as blisters on the surface of the coating	May arise from within the coating itself due to solvent or water entrapment or from corrosion of the substrate.	
Flaking	Detachment of coating from steel substrate in the form of flakes due to loss of adhesion.	Presence of surface contaminants e.g. rust, dirt, oil grease, etc. Lack of roughness of the substrate to achieve mechanical bonding.	
Chalking	Powdery coating on the surface coating	Disintegration of the binder due to atmospheric exposure (moisture and UV)	
Rusting	Formation of rust spots	Blast-cleaning or paint spray	
Filiform corrosion	Particular form of concentration-cell corrosion that occurs on metallic surfaces coated with organic coating.	Appearance of thin filaments in the coating from one or more sources, generally defects or scratches in the coating, in random directions. The filaments are tunnels with corrosion products underneath the swelled and cracked coating and appear as striations with the form of tentacles.	

is suitable for substrate catalysis, filters, among others applications. If the gel is dry in wet supercritical conditions, the shrinkage will be minimal, and the resulting product is known as *aerogel* (Figure 6). If the smallest dimension of the gel is more than a few millimeters, the material is classified as a *monolith* and when the gelation occurs by fast evaporation of the solvent, films and fibers will be produced (Figure

6). However, due to the loss of volatile byproducts formed in the hydrolysis–condensation reactions, it is difficult to control sample shrinkage during three-dimensional network formation (C. J. Brinker & Scherer, 1990; Segal, 1984).

This synthetic process involves the ability to manipulate nano-structural properties by controlling the hydrolysis and condensation reactions, and aggregation processes. The sol-gel process starts with hydrolysis reaction of the metal alkoxide, in which the alkoxide groups (-OR) are replaced by hydroxyl groups (-OH):

$$M(OR) + H_2O \rightarrow M(OH)(OR)_{z-1} + R(OH) \qquad (1)$$

The mechanism of this reaction involves the nucleophilic attack of the hydroxyl group negatively charged ($HO^{\delta-}$) to the metallic atom positively charged ($M^{\delta+}$). Hydroxyl groups are produced followed by the transference of one proton (H^+) and break of the bond M-OR and formation of an alcohol ROH (Equation 1). This process repeats successively until all the alkoxide groups are replaced. Subsequently, the condensation reactions of the resulting hydroxyl groups take place in which the hydrolyzed species $M(OR)_x(OH)_y$ (x = 0, 1, …, z-1; y = z-x) give birth to the formation of a three-dimensional network. The condensation process may occur via *oxolation* or *olation* reactions (Brinker & Scherer, 1990). In the *oxolation* two hydrolyzed species react forming "oxo" bonds M-O-M (Equations 2 and 3) and in olation (Equation 4) one of those species is protonated generating M-OH-M bonds (Brinker & Scherer, 1990).

Oxalation:

$$(OR)_x(OH)_{y-1}\,M\text{-}OH + HO\text{-}M(OR)_x(OH)_{y-1} \rightarrow (OR)_x(OH)_{y-1}\,M\text{-}O\text{-}M(OR)_x(OH)_{y-1} + H_2O \qquad (2)$$

$$(OR)_x(OH)_{y-1}\,M\text{-}OH + RO\text{-}M(OR)_{x-1}(OH)_y \rightarrow (OR)_x(OH)_{y-1}\,M\text{-}O\text{-}M(OR)_{x-1}(OH)_y + ROH \qquad (3)$$

Olation:

$$(OR)_x(OH)_{y-1}\,M\text{-}OH + H_2O\text{-}M(OR)_x(OH)_{y-1} \rightarrow (OR)_x(OH)_{y-1}\,M\text{-}OH\text{-}M(OR)_x(OH)_{y-1} + H_2O \qquad (4)$$

Sol-gel method is influenced by several factors, namely: temperature, pH, catalyst (acid or alkaline), type of solvent, molar ratio between water and alkoxide, nature of the organic group (R), among others (Brinker & Scherer, 1990; P. Innocenzi, Abdirashid, & Guglielmi, 1994; Pierre, 1998). Silicon alkoxide are widely used in the sol-gel process since have kinetics relatively slow in the reactions of the gel formation when compared to titanium, zirconium and aluminium alkoxides. This allows a higher control of the chemical process and end properties, shape of particles, volume and size distribution of the pores and superficial specific area. Additionally, silicon shows a low reactivity relative to chelating groups present in the system and in redox processes which may lead to the formation of undesirable by-products (Brinker & Scherer, 1990; Pierre, 1998). One of the most studied silicon alkoxide is tetraethoxysilane (TEOS). TEOS hydrolysis gives rise to the formation of silanol groups (Si-OH) and the condensations reactions lead to the establishment of siloxane bonds (Si-O-Si) and to the formation of a three-dimensional silica network highly branched (Esteves, Barros-Timmons, & Trindade, 2004). Furthermore, the gelation process of precursors such as TEOS leads to the formation of purely inorganic materials.

Figure 6. Products obtained by sol-gel process
Adapted from Brinker & Scherer, 1990.

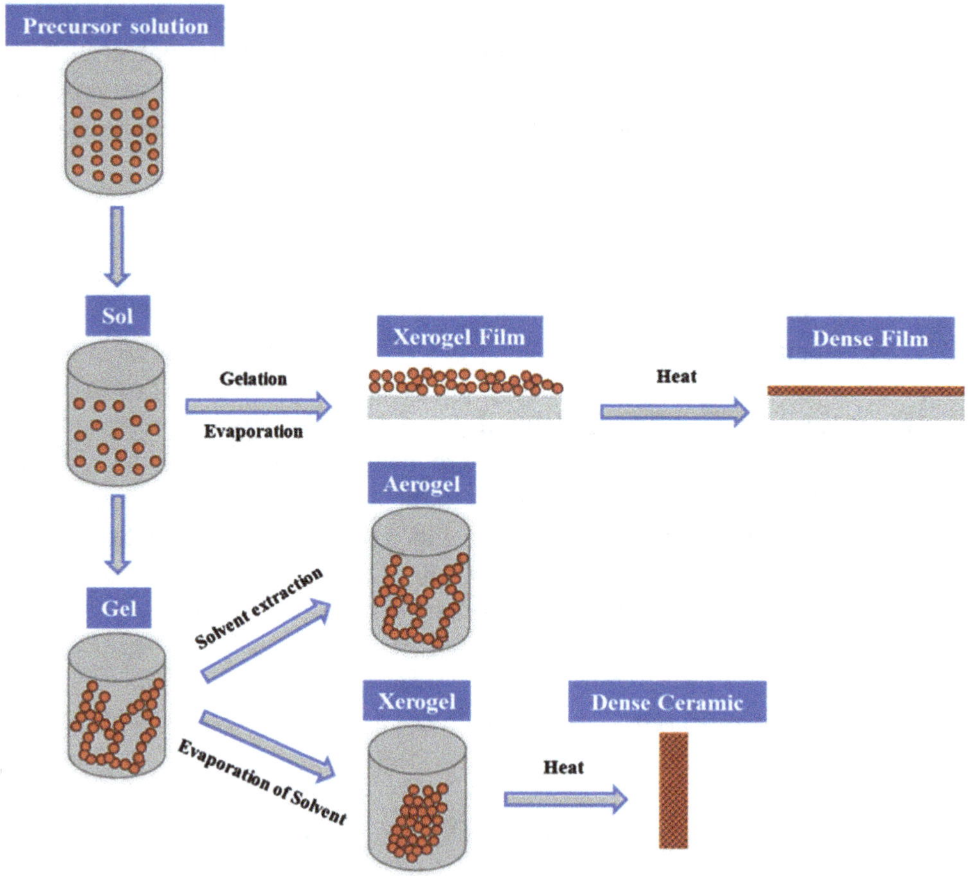

To synthesize O-IH materials it is necessary to introduce molecular precursors (Table 6) allowing for the formation of the organic component. Organic components to produce O-IH materials, using the sol-gel process, may be introduced in two ways (Benvenutti, Moro, Costa, & Gallas, 2009; José & Prado, 2005; Pierre, 1998; Clément Sanchez et al., 2011; Zheludkevich, Salvado, & Ferreira, 2005):

- Addition of organic precursors soluble in the reactional media (where hydrolyse/condensation reactions take place) but do not participate in the gel formation. The O-IH material will show an organic component bonded to the inorganic network by van der Waals forces, ionic or hydrogen bonds and are classified as O-IH class I.
- Addition of organic alkoxydes (R'M(OR)$_x$), which include an organic group *R'* bonded to the element *M* and the binding of which is not hydrolysable. The organic and inorganic components are connected by covalent bonds and this type of O-IH is included in O-IH class II.

The classification of O-IH in two main classes (Figure 7) was first introduced by Sanchez and Ribot in 1994 (Sanchez & Ribot, 1994). Three years later Wojcik and Klein introduced a third class (Wojcik & Klein, 1997) where O-IHs based on the combination of both interactions assigned to O-IHs class I and II are included.

Figure 7. Interactions established between organic and inorganic components for O-IHs class I and class II where ■–■ represents the inorganic network formed from inorganic precursors and ●–● the organic components from organic precursors
Adapted from Benvenutti et al., 2009.

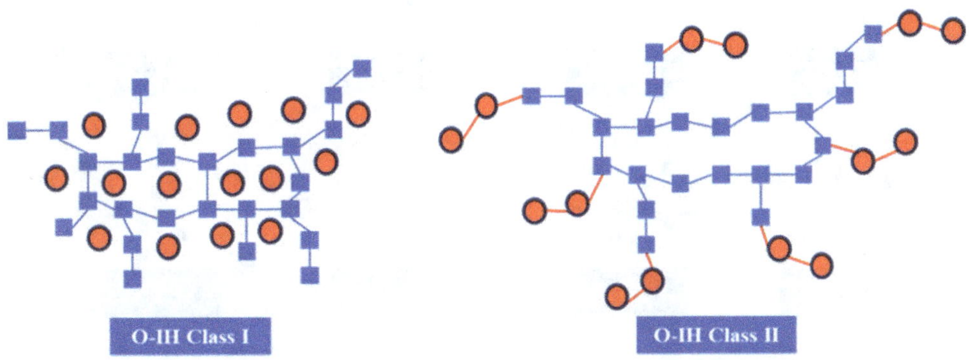

Table 6 and Figure 8 show the most common silicon alkoxides (including organic alkoxydes) and respective functional group used as precursors in the synthesis of O-IH materials. The precursors that contain polymerisable functional groups such as vinyl, epoxide or methacrylate also allow forming an organic network and contributing to additional densification of the final product.

In the nineties (Wen & Wilkes, 1996), the O-IH concept emerged and an increased number of papers in this area were published within a short period of time. Later, Schottner (Schottner, 2001) and Walcarius (Walcarius, 2001) reviewed the perspectives of the O-IH sol–gel polymers and reported the applications of O-IH materials in various fields of electrochemistry. Mitzi (Mitzi, 2001) selected examples of O-IH films and discussed the techniques used on the deposition of O-IH sol–gel materials. Several authors

Table 6. Chemical name and abbreviation of the alkoxysilanes most used as precursors for the synthesis of O-IH materials

Chemical Name	Abbreviation
Tetraethoxysilane	TEOS
Tetramethylorthosilicate	TMOS
Methyltriethoxysilane	MTES
Methyltrimethoxysilane	MTMS
Vinyltrimethoxysilane	VTMS
Phenyltrimethoxysilane	PTMS
Diethylphosphonatoethyltriethoxysilane	PHS
3 – Aminopropyltrimethoxysilane	APTMS
N – (2 – aminoethyl) 3 – aminopropyltrimethoxysilane	AEAPS
3 – Glycidoxypropyltrimethoxysilane	GPTMS
3 – Methacryloxypropyltrimethoxysilane	MAPTS
3 – Mercaptopropyltrimethoxysilane	MPTMS
Bis – [3 – (triethoxysilyl) – propyl]tetrasulfide	BTSTS

Figure 8. Chemical structure of the alkoxysilanes most used as precursors for synthesis of O-IH materials

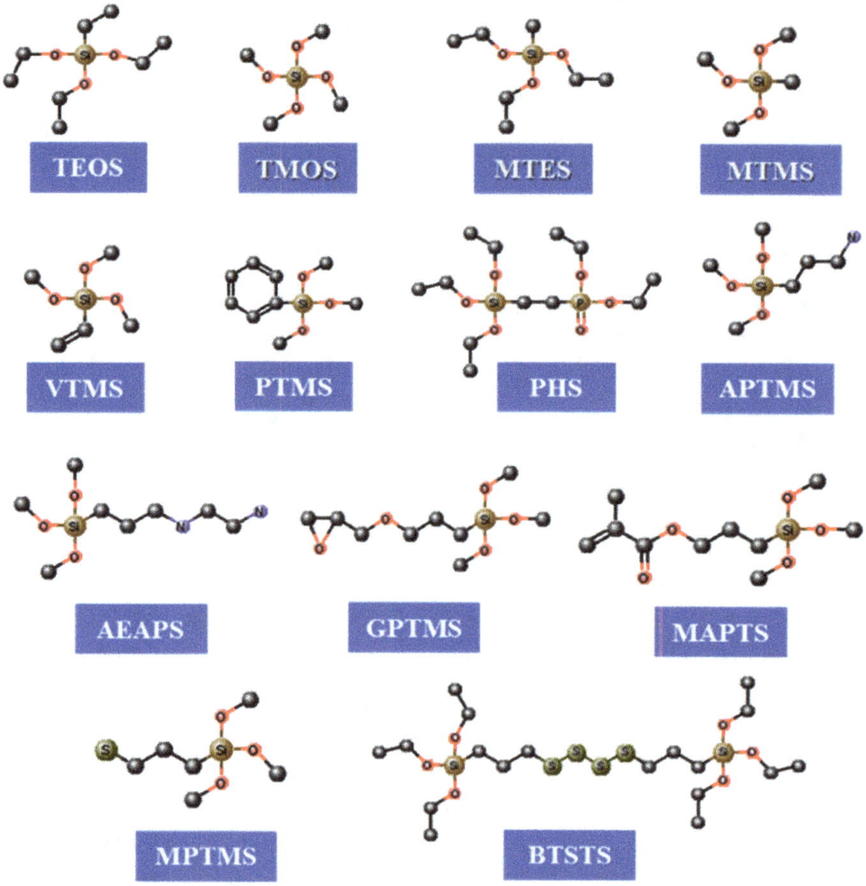

focused review papers on aspects related with gel synthesis processes, such as Hay and Raval (Hay & Raval, 2001) who reported the use of the non-hydrolytic sol–gel route to synthesize O-IHs. Arkhireeva et al. (Arkhireeva et al., 2004) reported the synthesis of O-IHs using both the hydrolytic and non-hydrolytic sol–gel routes. The authors showed that both routes produced O-IHs. However, an improved control of the shape and particle size was obtained using the hydrolytic process. In the same year, Castelvetro et al. (Castelvetro & De Vita, 2004) reviewed and discussed the synthetic approaches to produce nanostructured O-IH materials. The main emphasis was given to the processes and products that can be carried out in an aqueous medium, even though non-aqueous systems have also been discussed. Later in 2005, Sanchez et al. (Clément Sanchez, Julián, Belleville, & Popall, 2005) summarized the chemical pathways to prepare O-IH materials and mentioned outstanding examples of applications of functional O-IHs in distinct areas. The authors highlighted the use of protective coatings based on phenylsilsesquioxanes on structures for naval aircrafts and silsesquioxane–polyester mixtures in coil coating applications. The research efforts on this particular area have justified, almost yearly, a detailed review paper reporting the state-of-art of O-IHs. The most relevant fundamental review papers were written by Innocenzi et al. and Sanchez (Innocenzi, Kidchob, & Yoko, 2005; Innocenzi & Malfatti, 2015; Dash et al., 2008; Lebedev, 2011; Sanchez et al.,2010).

The most effective approach in the production of O-IH materials appears to be the sol–gel technique due to its specific ability to create a spacious network (macromolecular network) containing M–O bonds under mild (room temperature) synthesis conditions either with or without a solvent (José & Prado, 2005; Wang & Bierwagen, 2009). Their properties are highly dependent on the degree of mutual phase dispersion (therefore dependent on the organic/inorganic ratio), the reactivity of crosslinking alkoxide reagents, the processing steps, and the solvents (Mammeri, Bourhis, Rozes, & Sanchez, 2005). The presence of polymeric chains within the O-IH matrix contributes to the material's high electrical resistivity and impermeability, providing similar properties to the usual painting materials. Particular types of O-IH materials have also been termed, either as ORMOSILs (acronym generated from ORganically MOdified SILicates) or OMOCERs (from ORganically MOdified CERamics) depending on the precursors used (Table 6) and the synthesis method employed. The O-IH structure maximizes the preparation of materials that could lead to efficient coatings against corrosion as well as ensuring primary technological needs (Castelvetro & De Vita, 2004).

SOL-GEL COATINGS FOR EROSION CORROSION PROTECTION

Several protection systems are used for erosion-corrosion protection of metallic substrates (e.g. steel, zinc, aluminium, copper, magnesium and respective alloys). The use of organic and inorganic coatings are the most common methods for corrosion protection of industrial components. However, several restrictions have been imposed because of the negative impact on natural environments and human health. These constraints are due to the presence of volatile organic compounds (VOCs), chromium compounds, such as hexavalent chromium (Cr^{6+}) and hazardous air pollutants (HAPs). Therefore, the pursuit for alternative products such as inorganic coatings, paints and other surface treatments, free of VOCs, Cr^{6+} and HAPs has led to the development of a large variety of new coating systems and materials. These novel materials must comply with environmental targets and be effective in mitigate the erosion-corrosion process.

In the last few decades, a large variety of "green" and environmentally friendly processes have been studied. Among them the sol–gel technology has showed to be an efficient alternative for the replacement of the existing technology. Several publications have been devoted to the synthesis of O-IH sol-gel coating systems. The fast evolution of this specific research area and the potential contribution of sol–gel coatings as a corrosion mitigation system for metallic substrates has led to several review and book publications (Arkhireeva et al., 2004; Figueira et al., 2014; Figueira, Fontinha, Silva, & Pereira, 2016; Innocenzi & Malfatti, 2015; Kalidindi & Subasri, 2015; Lutz, Mol, De Graeve, & Terryn, 2016; Mittal, 2014; Montemor, 2016; Sakka & Yoko, 1992; Tiwari & Hihara, 2015).

O-IH sol-gel coatings are complex matrices and the intermolecular interactions between the macromolecular existing structures and metallic surface are particularly pertinent to material properties such as rigidity, low porosity and adhesion to the substrate. The conjugation of these interactions results in materials with enhanced protective properties against corrosion and erosion and with good thermal and electrical insulation features. Sol-gel process is a suitable technique to combine hard and flexible materials. The hydrolysis and condensation reactions of inorganic alkoxides provide an inorganic network after proper thermal treatment. When organosilanes are involved in the process, a highly cross-linked network containing organic and inorganic parts is formed. These materials are of interest since they combine both properties of the organic polymer and the inorganic materials. The hardness and the flexibility of the coating can be tuned by the amount of inorganic compounds (silica, alumina, etc.) and of polymerisable groups (e.g. maethacrylate, epoxy, vinyl, etc.). Furthermore, the chemical compatibility with the organic paints eases the adhesion of the sol-gel coating to the substrate and allows the introduction of corrosion inhibitors, such as cerium (Ce^{3+}), improving coating performances against corrosion (Aramaki, 2002; Cambon, Esteban, et al., 2012; Cambon, Ansart, Bonino, & Turq, 2012; Carneiro et al., 2012; Hammer et al., 2010).

Parkhill et al. (2001) prepared epoxy-silica sol-gel films on 2014-T3 aluminium alloy showing improved barrier properties and enhanced corrosion resistance in salt spray test in comparison to the chromate conversion pre-treatment conventionally applied to aircraft alloys. These results can be considered bottom line for the exponential development of OI-H materials for aircraft alloys. From this point forward, several types of O-IH sol-gel coatings have been studied for the corrosion protection of aluminium alloys (Akid, Gobara, & Wang, 2011; F. Andreatta et al., 2011; Capelossi et al., 2013, p. 201; Dias, Lamaka, Nogueira, Diamantino, & Ferreira, 2012; Fedel et al., 2013; Kozhukharov et al., 2012; Lampke, Darwich, Nickel, & Wielage, 2008; Li, Lu, Wang, & Tran, 2009; Metroke & Apblett, 2004; Rodič, Iskra, & Milošev, 2014; Varma et al., 2010; Voevodin, Kurdziel, & Mantz, 2006). Most of them are based on organo-alkoxysilane precursors containing organo-functional groups, such as epoxy, methacrylic, or vinyl that can be polymerized further increasing the coating's density, and thus improving its barrier effect. Other functional groups were also studied in aluminium alloys and other metallic substrates, for example phenyl (Kumar, Jyothirmayi, Soma Raju, & Subasri, 2012; Sheffer, Groysman, & Mandler, 2003) to increase coatings hydrophobicity or phosphonate (Khramov & Johnson, 2009; D. Wang & Bierwagen, 2009, p. 327), to improve bonding to the substrate. Different organic groups are used to achieve tailored properties of the sol-gel coatings. For example, the amino groups have the ability to react with the epoxy groups and provide compatibility with organic polymer paint systems. Therefore, the inclusion of amine-derived cross-linking agents in the sol-gel synthesis of epoxy-silica based O-IH coatings allows the promotion of the organic network formation at low temperature (Davis, Brough, & Atkinson, 2003), with inherent energy savings. Additionally, the process is essentially water-based which is also environmentally benign. Nevertheless, the amino-cured O-IH sol-gel coatings developed at room temperature show reduced barrier properties in water with time due to low hydrolytic stability (Fontinha, Salta, Zheludkevich, & Ferreira, 2013; Osborne, 2001). The corrosion behaviour of low temperature amine-cured epoxysilane sol-gel coatings has been studied by Vreugdenhil et al. (2001) and by Davis et al. (2003) using diethylenetriamine (DETA). DETA is one of the most common epoxy crosslinkers allowing to obtain a dense protective coating able to provide aluminium corrosion protection. Furthermore, the amino-silanes also contribute to the inorganic network formation. Other amines have been studied as alternative to DETA like di-amines of longer carbon chain (Metroke, Kachurina, & Knobbe, 2002) or branched amines (Pathak & Khanna, 2009; Roussi, Tsetsekou, Tsiourvas, & Karantonis, 2011). The produced coatings showed enhanced corrosion protection properties compared to those derived from formulations containing DETA. Without the addition of amine or other crosslinking agents, epoxy-silane-based sol-gel coatings require thermal curing at temperatures above 100 °C to complete organic network formation. The introduction of organic components in sol-gel films results in thicker crack-free coatings with improved corrosion protection properties. However, it also decreases wear resistance and mechanical properties, making them prone to physical damage. To overcome this drawback several strategies have been studied, such as the addition of inorganic nanoparticles of silica, aluminium oxide, and other oxides to the sols (Conde, Durán, & de Damborenea, 2003; Osborne, 2001; Rosero-Navarro, Pellice, Durán, & Aparicio, 2008; M. Zheludkevich et al., 2005), or the formation of nanostructured films or particles achieved by in situ synthesis during the sol-gel process through controlled hydrolysis/

condensation of sol-gel precursors (Donley et al., 2003; Lamaka et al., 2006; Pirhady Tavandashti et al., 2016; Voevodin et al., 2001). The latter process has the advantage of providing chemical bonds between the O-IH matrix and the nanoparticles.

The hardness and mechanical resistance of the O-IH coatings can be increased by nanoparticle reinforcement (Lavollée et al., 2016). The most common via for nanoparticles introduction within the O-IH sol network are colloidal suspension (Phanasgaonkar & Raja, 2009), synthesis in *situ* (Kiruthika, Subasri, Jyothirmayi, Sarvani, & Hebalkar, 2010) or as nanoscale powder (Kozhukharov, 2009; Kozhukharov et al., 2012). Among the coatings containing oxide nanoparticles (Ciriminna, Sciortino, Alonzo, Schrijver, & Pagliaro, 2011; Liu et al., 2015; Niederberger, 2007; Rodič et al., 2015), some publications dealt with coated metallic substrates with different nanoparticles such as zinc (Zn) (Pepe, Galliano, Ceré, Aparicio, & Durán, 2005), Montmorillonite (Fedel et al., 2013; Montemor, 2016; Zomorodian et al., 2013), alumina (Al_2O_3) (Adraider, Pang, Nabhani, Hodgson, & Zhang, 2011; Balan, Ng, Beng Siang, Singh Raman, & Chan, 2013; Cao, Zhu, Li, Hu, & Liu, 2017; Lakshmi et al., 2017; Wang & Akid, 2008), zirconium oxide (ZrO_2) (Adraider et al., 2013; Rodič & Milošev, 2014, 2015; Romero Pareja, López Ibáñez, Martín, Ramos-Barrado, & Leinen, 2006; Sorochkina, Smotraiev, & Chepurna, 2015; Varma et al., 2010; Yu, Xue, Liu, Li, & Zhang, 2015) and silicon dioxide (SiO_2) (Balan et al., 2013; Gharagozlou, Naderi, & Baradaran, 2016; Frederico Maia et al., 2012; Oldani, Sergi, Pirola, Sacchi, & Bianchi, 2016; Phanasgaonkar & Raja, 2009; Lakshmi et al., 2017; Santana et al., 2015, 2015; Suzana et al., 2016; Turri, Torlaj, Piccinini, & Levi, 2010; Vivar Mora et al., 2017) dispersed in the O-IH matrix.

Other studies to improve the physical properties, such as density, of the organic coatings have been conducted by introducing reinforcing agents such as carbon nanotubes (CNTs). CNTs have been studied as reinforcing agents mainly due to their properties, such as their high density, high tensile strength, high thermal stability and conductivity (Calvert, 1999; Grabowski, Zbyrad, Uhl, Staszewski, & Packo, 2017; Huang et al., 2014; Lekawa-Raus, Gizewski, Patmore, Kurzepa, & Koziol, 2017; Sakka, Antar, Elleuch, & Feller, 2017; Treacy, Ebbesen, & Gibson, 1996). Some studies reported that the incorporation of multiwalled CNTs in epoxy matrices improved the mechanical strength and reduced the friction of the coatings originating composites with improved wear resistance (Baer, Burrows, & El-Azab, 2003; Hammer, dos Santos, Cerrutti, Pulcinelli, & Santilli, 2013; Khun & Frankel, 2016; More & Mhaske, 2016; Wang, Xue, Dong, Wang, & Li, 2008; Zhang, Zarudi, & Xiao, 2006). Recently, Khun and Frankel showed that the introduction of MWCNT significantly decreased the cathodic delamination of the polyurethane composite coatings. The authors concluded that these results were due to the dispersion of denser MWCNTs in the polyurethane matrix and evidenced that the MWCNTs blocked the transport pathways of oxygen and water molecules through the coating during the cathodic delamination progress (Khun & Frankel, 2016).

The main objective of a protective coating is good resistance to corrosion and abrasion/erosion. It is undeniably that mechanical damage may locally result in the reduction of the protective properties compromising the performance of the coating system (designed for a certain period) and therefore, promoting the corrosion of the substrate earlier than expected. The performance of a coating in an abrasion test depends on the properties of the material, on the film thickness and on the adhesion of the coating to the metallic substrate. For instance, the ideal content of corrosion inhibitors incorporated within the O-IH matrix depends on the sol-gel matrix. Several studies have shown that, the selection of an inhibitor and its constituent compound might be critical to the stability of the sol–gel network and therefore the barrier performance of the O-IH coatings may be compromised (Figueira, Silva, Pereira, & Salta, 2013, 2014; Zheludkevich et al., 2005). Other authors studied the impact of the hardness of the coatings on the erosion protection by changing the ratio of organic-rich to inorganic-rich compounds. The mechanical properties of the resulting O-IH coatings were adjusted by nanoparticles (e.g. ZrO_2). These authors demonstrated that the addition of ZrO_2 nanoparticles may have a negative or a positive effect on the erosion performance (Grundwürmer, Nuyken, Meyer, Wehr, & Schupp, 2007). In 2000, (Li, Jordens, & Wilkes, 2000) showed that sol-gel coatings deposited on different metallic substrates (copper, brass and stainless steel) are effective in wear control. However, inferior results were obtained on the plain steel due to poor adhesion to the substrate.

The need to assess the resistance of a coating system to the abrasion is achieved by abrasive wear tests and generally are performed using Taber test according to ASTM D 4060-95. Taber test was initially developed to assess the performance of organic coatings (Cambruzzi, Rossi, & Deflorian, 2005; Rossi, Deflorian, Fontanari, Cambruzzi, & Bonora, 2005; Rossi, Deflorian, & Risatti, 2006) and have not been commonly applied on O-IH sol-gel coatings (Kiruthika et al., 2010). This test has limitations considering that it does not provide an accurate value of the protective capacity. Nevertheless, allows to compare different coating systems (Lavollée et al., 2016) and assess them qualitatively.

Usually, most of the studies performed assess the corrosion and wear properties alone. Thus, the influence of the mechanical degradation on the performance of the coatings for corrosion mitigation is not well established. The damage induced by mechanical action, and considering the thickness decrease and mass loss (according to ASTM D 4060-95), insufficient information can be gathered about the damage of the coatings and its remaining protective properties. A very useful technique that have been widely employed to assess the decrease of the protective barrier properties is Electrochemical Impedance Spectroscopy (EIS) after Taber Abraser Standard test (Lavollée et al., 2016; Rossi et al., 2005). EIS can be extremely helpful for O-IH materials optimization since it is a technique that allows to quantify electric (ionic) resistivity and dielectric properties (Zandi Zand, Verbeken, & Adriaens, 2012).

Nowadays, O-IH sol-gel coatings reinforced with nanoparticles in order to obtain coatings with high abrasion and high corrosion properties have been poorly developed particularly for carbon steel. The main efforts have been devote to aluminium alloys (Andreatta et al., 2011, 2010; Balaskas, Kartsonakis, Tziveleka, & Kordas, 2012; Calabrese, Bonaccorsi, Caprì, & Proverbio, 2014; Dias et al., 2012; Feng, Liu, Thompson, & Skeldon, 2010; Gobara, Kamel, Akid, & Baraka, 2015; Grundwürmer et al., 2007; Iribarren-Mateos et al., 2015; Kalidindi & Subasri, 2015; Maia et al., 2016; Nemeth, 2008; Pirhady Tavandashti et al., 2016; Rahimi, Mozaffarinia, & Hojjati Najafabadi, 2013; Rodič & Milošev, 2015, 2016; Rosero-Navarro et al., 2010; Varma et al., 2010; Varma, Periyat, Oubaha, McDonagh, & Duffy, 2011; Wilson, Hawthorne, Yang, & Troczynski, 2001; Yasakau, Carneiro, Zheludkevich, & Ferreira, 2014; Yu et al., 2015; Zheludkevich et al., 2005) highly promoted by the aeronautic industry. Recently, Lavollée *et al.* reported the synthesis of protection system for mild steel substrate combining both properties, resistant to corrosion and to abrasion. The coating was based on alumino-silicated O-IH containing zirconia nanoparticles. The authors demonstrated that the sol-gel technology offers the possibility to produce innovative systems for corrosion mitigation and wear protection (Lavollée et al., 2016). Rodič et al. (2015) reported the use of ZrTPO and methacrylic acid (MAA), following their combination with sol synthesized from TEOS and 3-methacryloxypropyl trimethoxysilane. The various molar contents of ZrTPO and MAA were studied and the coatings offered high corrosion protection to aluminium substrates under simulated aircraft conditions. Additionally, these nanostructured components can be used to introduce corrosion inhibitors within the sol-gel structure. Some examples of this strategy include the incorporation of cerium salts in nanostructured O-IH sol-gel coating's components synthesized *in situ* such as the examples mentioned above, or the complexation of organic inhibitors with cyclodextrins

(Khramov, Voevodin, Balbyshev, & Donley, 2004). Other approaches to solve the problems linked to long-term storage of inhibitors involve their incorporation in "smart" micro- and nano-containers embedded within the O-IH matrix. Such as functionalization of silica nanoparticles loaded with benzotriazole by a layer-by-layer (LbL) deposition technique coated with poly(ethyleneimine/poly(styrene sulfonate) (PEI/PSS) polyelectrolyte layers that respond to pH change (Shchukin & Grigoriev, 2012; Skorb, Fix, Andreeva, Möhwald, & Shchukin, 2009). Likewise, natural halloysite nanotubes coated with polyelectrolyte shells by LbL were used to host the inhibitor 2-mercaptobenzothiazole (M.L. Zheludkevich, Tedim, & Ferreira, 2012). In both cases a significant increase in corrosion resistance of the doped O-IH coatings was achieved in comparison to the undoped ones. Another type of "smart" reservoirs for corrosion inhibitors are those based on ion-exchange, like the layered double hydroxides (LDHs). LDHs (also known as anionic clays or hydrotalcite-like compounds) consist of stacks of mixed-metal hydroxides, positively-charged, stabilized by anions and solvent molecules located between the positive layers. The release of species from LDHs is based on an anion-exchange mechanism. From the corrosion protection point of view, LDHs have a double role:

1. Release of anionic corrosion inhibitors
2. Entrapment of aggressive anionic species such as chlorides.

Furthermore, the release of inhibitors can also be, indirectly, triggered by pH changes; at very high pH the inhibitors can be exchanged with hydroxyl anions, whereas at very low pH the LDHs dissolve releasing the entrapped inhibitors (Zheludkevich et al., 2012). Álvarez et al. (2010) incorporated different concentrations of hydrotalcite compounds into O-IH ZrTPO-GPTMS sol-gel coatings. The doped films showed increased adherence and corrosion resistance. In two other studies, concerning the incorporation of stimuli-response particles into O-IH epoxy-zirconia-silica sol-gel coatings, calcinated red mud particles were used showing inhibitive action, and modified zeolites were used to host Ce^{3+} compounds, improving the protective performance of the O-IH coatings applied on AA 2014-T3 alloy (Álvarez, Collazo, Covelo, Nóvoa, & Pérez, 2010). Another approach to incorporate and stabilize inhibitor species within O-IH sol-gel coatings is the use of multi-layered coatings. Rosero-Navarro et al. (Rosero-Navarro et al., 2010) developed an active sol-gel coating by sandwiching a Ce^{3+}- doped layer between two O-IH undoped films synthesized from TEOS, MAPTS, ethylene glycol dimethacrylate, and glycidyl methacrylate.

The analysis of the publications in the last two decades shows that it is on aluminium-based alloys where higher search efforts for an efficient coating to replace CCCs has been done. These are followed by steel substrates, zinc and magnesium alloys, with copper alloys in last place (Figueira et al., 2014). Regardless of the type of substrate, the most frequently used precursors were GPTMS, TEOS, and TMOS. Furthermore, the enormous number of studies performed in the field of O-IH sol-gel coatings in the last two decades, shows clearly the potential of these coatings to mitigate corrosion of metallic substrates. Besides, all the advantages linked to sol-gel process these materials can also be deposited on a metallic substrate on an industrial scale. This can be easily achieved by using well established and low-cost methods such as spray, dip and spin coating, flow coating, roll coating, doctor blading, and capillary coating processes or using any combination thereof (Aegerter & Mennig, 2004).

Methodologies of Application of O-IH Sol-Gel Coatings

The most traditional techniques for coatings deposition are physical vapour deposition (PVD) and chemical vapour deposition (CVD). PVD generally uses inorganic compounds or gases. CVD process, spin coating, dip coating and spraying techniques use liquid inorganic compounds, organic compounds and gases (Rout, Reserach & Development Department, Tata Steel Ltd., India, Bera, Udayabhanu, & Narayan, 2016). This section will be focused mainly on the following deposition methods:

- Spin coating;
- Dip coating;
- Spraying.

Spin coating is widely employed for the deposition of O-IH sol-gel coatings (Aoki, Habazaki, & Kunitake, 2009; Cardoso, Minas, & Lanceros-Méndez, 2013; Chiu, Zhang, Tsai, & Wu, 2013; G Carbajal-De La Torre, 2009; Habazaki et al., 2016; Hernández et al., 2016; Lacatusu, Badea, Murariu, Pirvu, & Meghea, 2011; Liu, Wei, Tian, Tong, & Xu, 2014; Norouzi & Afrasiabi Garekani, 2014; Sahu, Parija, & Panigrahi, 2009; Shilova, 2013; Soo et al., 2013; Vorotilov, Petrovsky, & Vasiljev, 1995; Wu, Chang, Yang, & Wang, 2006; Zabihi, Ahmadian-Yazdi, & Eslamian, 2017), allowing obtaining a coating with thickness that may range between nanometres and micrometres. Generically, a typical spin coating process is the deposition of a coating fluid on the surface of a substrate at a high speed spin to spread the fluid. In the end, a drying step is included to evaporate the excess solvent (Rout et al., 2016; Sahu et al., 2009). Commonly, two methods of spin coating (Figure 9) are used; e.g. static dispense method and dynamic dispense method. In the first one, a small drop of fluid is deposited on the centre of the substrate. The drop quantity can range from 1 to 10 mL depending on the fluid viscosity and on the substrate size. In the case of dynamic dispense, the substrate rotates at a low speed (about 500 rpm) and the fluid is dispensed on its surface. After this, the spinning substrate is accelerated to higher spin speed (1500 – 6000 rpm) in order to achieve a certain film thickness. A separate drying step is generally essential to further dry the film without shrinkage (Rout et al., 2016; Sahu et al., 2009; Vorotilov et al., 1995). The main factors that affect spin coating process are the spin speed, spin acceleration and the drying process. The final film thickness is defined by the speed during spin step. For example, a variation of spin speed by $\approx$ 50 rpm can cause a 10% change in film thickness (Rout et al., 2016; Vorotilov et al., 1995). Spin acceleration influences the coated film properties and plays an important role in the coating properties because the sol starts drying during the first part of the spin cycle. Generally, around 50% of the solvent evaporates from sol in the first few seconds of the process (Rout et al., 2016). Therefore, the substrate should be coated uniformly to avoid topographical defects. The drying rate of the sol depends on the nature of the sol and on the volatility of the solvent used. Nevertheless, the drying rate also depends on the surrounding temperature during deposition. It is well known that, the air temperature and humidity are determinant for the coated film properties. Therefore, the airflow and the turbulence above the substrate should minimized or held constant when possible.

Generally, a lengthier drying rate produces coatings with higher thickness and uniform coverage through the substrate. The sol dries gradually from the centre to the edges, where the external areas of the surface dry first and the centre successively. Therefore, the viscosity of the coating sol changes from the centre to the outer, being higher in the centre and lower in the external surface areas. This may lead to coatings with a non-uniform thickness. This can be exceeded by slowing down the rate of drying, which

Figure 9. Sequential steps in a spinning process
Adapted from Rout et al., 2016; Sahu et al., 2009.

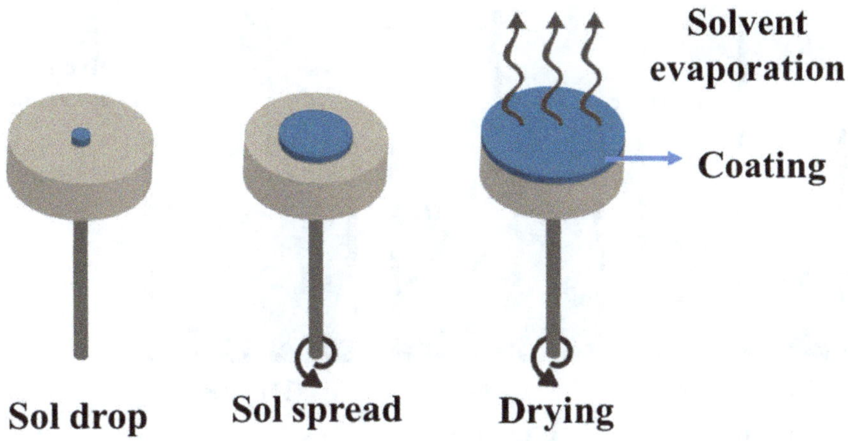

will allow to achieve viscosity, approximately, constant across the substrate surface. Drying rate and final coating thickness is also dependent on the environment humidity. Minor changes in environment humidity may lead to considerable changes in coating thickness. Recently, to prevent this, several spin coater systems are connected with a "closed bowl" in order to provide a controlled environment (Rout et al., 2016). Figure 10 shows how the film thickness changes with the spin time and speed and generally, the film thickness is inversely proportional to the spin speed and time. Nevertheless, it should be kept in mind that these parameters change are highly dependent on the different substrates and coating materials.

Dip coating technique is basically the immersion of a substrate into a fluid (sol) that is withdrawn from the sol in a certain and precise velocity under controlled condition of temperature and pressure (Seshan, 2002). Afterwards, the solvent is evaporated. Figure 11 shows schematically all the steps involved in a dipping process. The coating thickness as well as a uniform coating distribution are strongly dependent on the withdrawal speed and on viscosity of the sol.

Figure 10. Film thickness change with spin time and with spin speed
Adapted from Rout et al., 2016.

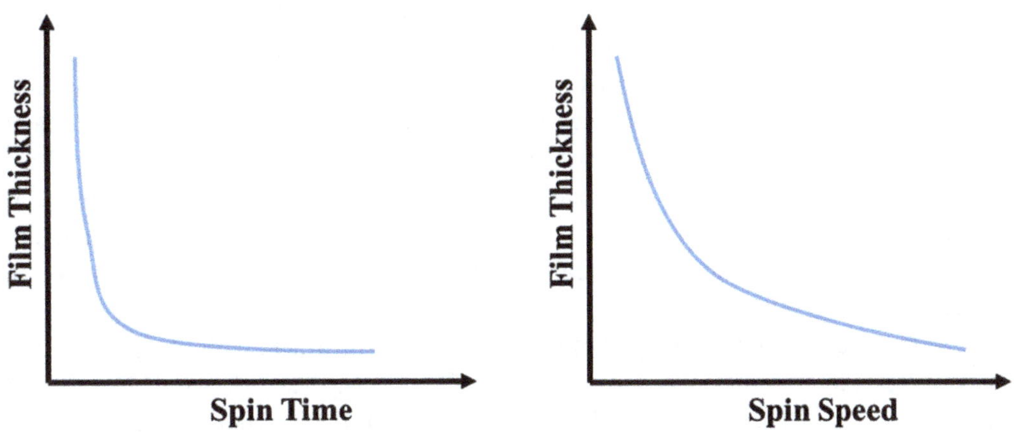

Figure 11. Sequential steps in a dipping process
Adapted from Rout et al., 2016; Sahu et al., 2009.

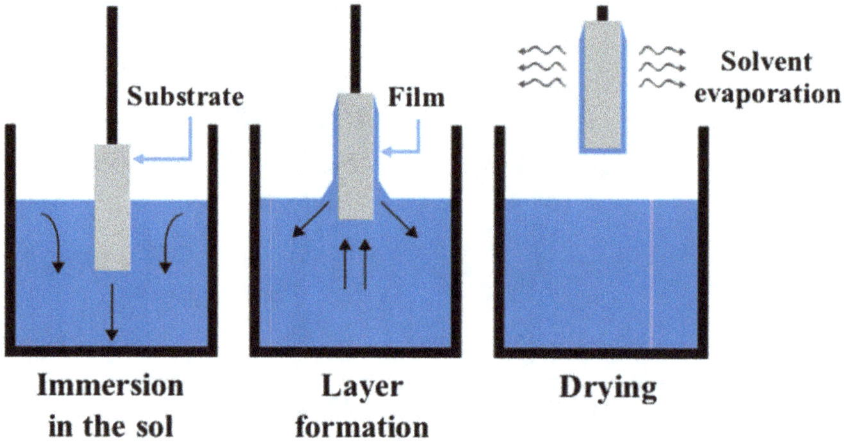

This technique has been widely employed for deposition of O-IH materials in different metallic substrates such as aluminium alloys (Adelkhani, Nasoodi, & Jafari, 2014; Calabrese et al., 2014; Lampke et al., 2008; Sheffer et al., 2003), stainless steel (Hanetho et al., 2014; Hernandez et al., 2013; Messaddeq, Pulcinelli, Santilli, Guastaldi, & Messaddeq, 1999; Nikrooz, Zandrahimi, & Ebrahimifar, 2012; Rahoui, Turq, & Bonino, 2013), carbon steel (Criado, Sobrados, Bastidas, & Sanz, 2015, 2016; Criado, Sobrados, Sanz, & Bastidas, 2014; Hammer et al., 2010, 2013; Hernandez et al., 2013), hot-dip galvanized steel (HDGS) (Fedel et al., 2013; Figueira et al., 2013, 2014, Figueira, Silva, & Pereira, 2015a, 2015b, 2015c, 2016; I. Rute Fontinha et al., 2014; Kunst et al., 2013) and magnesium alloys (Nikrooz & Zandrahimi, 2011; Zomorodian et al., 2013).

Recently, some studies have been devoted in studying the influence of the deposition of several layers during dipping process (Figueira et al., 2015c, 2015a). The authors showed that the use of residence time (time used to let the substrate inside the sol before the withdrawal of the substrate starts), when one layer was deposited, led to improved results compared to samples coated with one layer without any residence time. It was also shown that the time consumed and the cost linked to the deposition of more than one layer (with cures between each deposition) on HDGS was not proportional to the improvement provided concerning the corrosion behaviour of HDGS. Indeed, in certain cases the authors showed that poorer results were obtained with several coating layers than with a single layer. Dip coating method can be considered more advantageous method when compared to the conventional methods such as CVD, evaporation or sputtering, since is a method that requires considerably less equipment and is much cheaper.

CVD is a generic name for a group of processes that involves the deposition of a solid material from a gaseous phase and depending on the nature of the precursor (e.g. gaseous, liquid or solid) a reactor is constructed. CVD major application is to deposit thin film coatings and widely used in microfabrication processes to deposit materials in several forms such as monocrystalline, polycrystalline and amorphous. These materials include for instance silicon, carbon fiber, carbon nanofibers, carbon nanotubes, SiO_2, silicon carbide, silicon oxynitrides, etc.

Spraying is a technique widely employed in industry to coat substrates with irregular shapes and has been extensively used to coat different substrates with O-IH sol-gel coatings (Grbić et al.,2014; Kar, Paul, & Bandyopadhyay, 2016; Liu et al., 2014; Parkhill et al., 2001; Rodriguez et al., 2014; Szymański et al., 2015). Figure 12 compares the main vantages and disadvantages for the dipping, spinning and spraying methods.

Manual deposition can also be performed by brushing and/or rolling however its choice depends on the substrate to be coated. Brushing is particularly useful to coat small substrates or with shapes with difficult access or to retouch surfaces. The main advantage of brushing procedure is related to the low coating loss. Yet, limitations and disadvantages are present such as highly labour costs and obtain a uniform coating thickness (Figueira et al., 2015a, 2015c; Rout et al., 2016). Rolling allows obtaining a film thickness more uniform compared to brushing; achieving smoother surfaces. Other authors reported that electrodeposition (EPD) techniques provided fairly thick crack-free sol–gel-derived coatings when compared either to dip or spin coating techniques (Zheludkevich, Shchukin, Yasakau, Möhwald, & Ferreira, 2007; Zheng & Li, 2010). Castro et al. (Castro, Ferrari, & Moreno, 2003) combined the sol–gel method and the EPD process to prepare thick coatings for metallic substrates obtaining crack-free deposits up to 20 μm in thickness after drying and crack-free glass-like coatings of 12 μm in thickness after sintering at 500 °C for 30 min.

Limitations and Controversies of O-IH Sol-Gel Coatings

Despite the advantages of obtaining O-IH sol-gel materials with different properties, synthesis constraints still remain. The major limitations of O-IH sol–gel processing for coating metals to mitigate corrosion-erosion process are delamination, crackability, adhesion, and thickness limits. Ensuring a uniform distribution on the substrate and thermal treatments (curing/drying) are crucial factors that ensure the quality of the coatings. Cracks affect several properties and compromise barrier properties of the coating in wet corrosive media.

Figure 12. The main vantages and disadvantages for dipping, spinning and spraying methods of deposition
Adapted from Rout et al., 2016; Sahu et al., 2009.

Latella et al. (2005) synthesized and assessed the adhesion behaviour and mechanical properties of O-IH coatings. The authors confirmed that the presence of a thermally grown oxide layer on the substrate, prior to the O-IH sol–gel deposition, had a critical role on the quality of adhesion between the film and the substrate. They also showed that the relation between the structure of the film and the mechanical properties of the coatings were influenced by the nature/characteristics of the organic substituent. Mammeri et al. (2005) investigated and analysed the mechanical properties of O-IH gels reported by several authors and concluded that those properties were dependent on their micro- and nanostructures as well as the nature and extent of the organic–inorganic interfaces within the gel matrix. Other authors showed that when the substrate was subjected to localized plastic deformation due to impact with objects, or extensive plastic deformation due to substrate bending, cracks, and delamination developed easily in the thin films (Guglielmi, 1997; Walcarius, 2001).

Most of the studies reported in the literature were on planar samples and the coatings were deposited by dip or spin coating processes. Objects with complex shapes are far more challenging to coat using these methods, particularly if a uniform thickness is a strict requirement. For these reasons, additional coating methods besides spraying, dipping, spinning, and EPD are expected to be extensively tested and developed in the near future.

The deposition method selected may be a limiting variable since in certain circumstances it could negatively influence the performance of the O-IH coating applied on the substrate. If the deposition method chosen is not the suitable one, even with an O-IH material displaying excellent barrier properties, the coating performance is compromised. The most common methods employed (spinning and dipping) do not ensure fully uniform coatings, particularly when dip-coating systems are used for the deposition of low viscosity O-IH gel materials (Figueira et al., 2015c, 2015a). However, these obstacles can be overcome by the use of intermediate deposition and curing steps. Though, the increase in the number of steps involved in the coating methods turn these procedures extremely complex and time consuming. As such, a compromise between the improvement of the final material properties and the costs involved should be pondered (Figueira et al., 2015a, 2015c).

Another limitation exposed by the consumer industry, which is transversal to all coating materials, is all the trouble linked to the assessment of the erosion-corrosion behaviour of the coatings after being exposed to the environment. In certain cases, such as underground pipelines and storage tanks, visual observation is impossible without extraordinary efforts and costs. In such situations, research should be carried out by pre-evaluating the performance and lifetime of the protective coatings, on lab or at a pilot scale, to plan the adequate maintenance/replacement interventions or by the development of appropriate sensors and monitoring methods based on self-sustainable and remote-control devices.

Until now, it is undeniable that the traditional chromate-based chemical conversion coatings (CCCs) confer the best corrosion protection known at controlled costs. This turns this area even more challenging leading to continuous research for contemporary O-IH materials that fulfil the same properties and performances exhibited by CCCs coatings.

It is noteworthy that no report has yet been found, either in industrial or academic research, of O-IH sol–gel coatings with the same production costs and of equal or superior corrosion performance to traditional conversion layers used processes (chromating and phosphating). Furthermore, as far as the author is aware, no systematic theoretical and experimental approach on how to assess the mid or long-term viability, efficiency, and stability of these OI-H sol–gel coatings have been implemented. And, considering the research evolution on this matter, the author believes that is far from being established.

Therefore, the present situation demands further research and innovation efforts to ensure the success of the application of O-IH sol–gel coatings as metal erosion and corrosion protection.

FUTURE RESEARCH DIRECTIONS AND CHALLENGES OF O-IH SOL-GEL COATINGS

Nowadays, metal corrosion protection is strongly reliant on organic and inorganic coatings technology, as it is a cost-effective mean of providing practical protection against erosion-corrosion for easily corrodible metallic structures and objects. The quality and effectiveness of corrosion control by coatings is, generally, assumed by many users to be low cost and easy to achieve. For these reasons, the consumers of corrosion control coatings often choose these by cost and appearance and not by cost effectiveness, which can be measured by their performance and the lifetime of their behaviour. However, with high labour costs and the difficulties found in recoating large, buried, complex or difficult to reach objects, more coatings consumers are starting to focus on the total costs of corrosion prevention and control. A coating system that increases the lifetime of the product but is more expensive, due to the initial application method, will pay for itself in reduced maintenance costs and reduced need for expensive recoating. Following this line of thought, the cost analysis of a corrosion protection system provided by coatings must lead to further research in order to measure and predict the lifetime of the protective properties of these materials. It is undeniable that future applications and research should be focused primarily on the investigation of more environmentally friendly precursors and industrial colloid particles to replace the traditional precursors in the formation of new O-IH sol–gel protective coatings with improved chemical and physical properties.

Among the most critical constraints identified are the coating's mechanic strength and good adhesion to metallic substrates. In spite of the large number of papers published on this subject, no work was found reporting the use of O-IH sol-gel coatings to stop and prevent corrosion on already damaged substrates as well as studies testing such materials for significant periods of time (years) neither in conditions of natural exposure. The assessment of whether O-IH coatings have the ability to minimize/stop damage caused by corrosion and their durability over time is of extreme importance since O-IHs are known for having a limited shelf life due to sluggish condensation reaction kinetics and the products developed for the protection of metallic substrates against corrosion must have a long lifetime. Another evidence found was that, studies on the barrier properties of the coatings over long time periods (years) were not found. Therefore, a relevant concern related to the ability of these materials to maintain their barrier properties remains. As such, it is vital to undertake long-term tests in the near future as the durability/resistance of the O-IH sol-gel coatings for corrosion prevention/mitigation for long time periods is unknown.

The search for future O-IH gel-based coatings is clearly oriented toward low cost, pollution-free, easily synthesized, and effective corrosion protection as well as O-IH coatings that do not generate hazardous waste during their application and removal. Development of O-IH coatings with self-healing properties, e.g. capacity to recover/repair damages spontaneously and autonomously, that is, without any external intervention (Ghosh, 2009), would be a challenging innovation that will contribute in giving an added high value to the synthesized materials. Inspired by living organisms, materials engineers are focusing on developing materials with self-healing properties. Several authors (Billiet, Van Camp, Hillewaere, Rahier, & Du Prez, 2012; Carneiro et al., 2012; Gao & Suo, 2011; Hughes, 2015; Hughes, Cole, Muster, & Varley, 2010; Khramov et al., 2004; Lutz et al., 2016; Mittal, 2014; Montemor, 2016; Shchukin et al.,

2006; Shchukin & Grigoriev, 2012; Trabelsi, Cecilio, Ferreira, & Montemor, 2005; Yabuki, Yamagami, & Noishiki, 2007; Yoganandan, Pradeep Premkumar, & Balaraju, 2015; Zheludkevich et al., 2010; Zheludkevich et al., 2007), developed and reviewed corrosion protection systems with self-healing abilities. The development of O-IH coatings with those properties by incorporating the release of healing agents, reverting crosslinks or using simultaneous technologies such as conductivity; shape memory effect; nanoparticle migration; and co-deposition might be considered as important achievements. The development of new O-IH coatings with dual behaviour (a protective barrier accumulating the O-IH properties with a self-healing response) is a synergy that will bring benefits and improvements in corrosion prevention, fulfilling the requests needed by the consumers of coatings technology. This concept is based on multilayer coatings where inhibitors and self-healing agents are immobilized in one of the layers being isolated from the external media and metal substrate by intermediate layers (Hughes et al., 2010). The procedure implemented to obtain these new systems should ensures good adhesion and similar thermal expansion coefficients between the layers. This same procedure also relies on an efficient top coating layer barrier protection. Moreover, the immobilization of both self-healing and inhibitor agents within a layer apart from the metallic substrate, by intermediate layers, prevent side reactions and leakage that may affect the substrate.

From the huge variety of studies found very few involved testing their performance with industrial organic paints in order to assess the compatibility of both systems and with the industrial requirements that generally are only partially fulfilled. The introduction of organosilane based O-IH coatings on an industrial scale is still in an early stage of development due to economic reasons and to the limitations imposed for the industrial use of these products related to the long-term stability of the sols. Additionally, some of the higher performing solutions involve complex formulations and conditions of preparation that might be difficult to achieve in large-scale industrial applications. Arkles (2001) published a paper where the commercial applications of the synthesized O-IH gel materials were discussed. According to him, these materials show viability in the market either owing to uncommon properties that allow new end-use applications or significantly improved cost/performance relation when compared with the available materials.

Future demands will involve the development of integrated solutions where O-IH coatings are mixed with conventional components to create coatings that will combine functions of conversion coatings, primer, and topcoat, and certainly, with self-healing ability (Figueira et al., 2014; Figueira, Fontinha, et al., 2016).

CONCLUSION

In the last few decades, the existing knowledge about sol–gel processing and relations between structure and properties achieved an extraordinary development. The research efforts on this area of materials science have grown remarkably, allowing scientists to gain and develop knowledge on how to produce new O-IH sol-gel materials. Since the early 1970s, academic interest in O-IH materials has been accelerating and since the 1990s it has become a fast-growing and very complex subject with an almost exponential increase in the number of the scientific publications. The main driving force in the development of O-IHs to prevent/mitigate corrosion was the search for suitable candidates to replace environmentally unfriendly chromate surface treatments for metallic substrates.

Studies with O-IH coatings were the erosion-corrosion mechanism are involved are scarce. Most of the published papers found studied the corrosion and the erosion/abrasion mechanisms individually. Therefore, studies that take into account the corrosion mechanism and the mechanical properties (erosion/abrasion) in simultaneous should be conducted. These will allow to understand and establish the influence of the mechanical degradation on the performance of coatings for erosion-corrosion mitigation.

Tested O-IH materials present a large diversity of compositions and enhanced physical, mechanical, and morphological properties, among others. O-IH materials show strong potential and are clearly becoming a reality as serious candidates for numerous applications such as smart coatings with "intelligent" self-healing properties and functional protective coatings. Moreover, the O-IH sol–gel process shows high potential for the production of multilayer coating systems which are promising environmentally friendly candidates for replacement of the chromate-based pre-treatments due to a synergistic effect of good barrier properties and effective ''self-healing'' action.

The properties of O-IH sol-gel coatings, such as delamination, crackability, and adhesion, are crucial properties that must be improved to ensure the quality of coatings for corrosion mitigation. Moreover, a systematic theory and procedure on how to evaluate the viability, efficiency and stability of these coatings needs to be implemented. To produce efficient O-IH sol-gel coatings for erosion-corrosion mitigation, these procedures must be considered and applied. O-IH sol-gel materials synthesized and found in the literature, generally, show market viability due to unique properties that allow new end-use applications or show significantly enhanced cost/performance relationship compared to the available materials in the marketplace.

It must be emphasized that there are many epoxy coatings for corrosion commercially available and the continuous research for contemporaneous O-IH polymers has been gaining market niches, leading us to believe that these are the directions for future success. On the other hand, it was also shown that some hybrid materials already exist with significant commercial potential; however, these have not yet reached commercial status. Like all innovative materials, these are no different and are facing many difficulties and challenges for large-scale industrial production. This is due to a number of factors, including capital investment, ease of manufacturing, coating performances, and environmental issues need that must be considered when developing a coating process for industrial application. Therefore, future trends for O-IH sol-gel coatings to be towards low cost, pollution-free, ease of synthesis, and effective in corrosion control. These obstacles and difficulties should be easily overcome by establishing a direct connection between the industry, the users/costumers and the scientists, creating a synergy in which everyone wins. In the end, it can be concluded that the future is bright for the development of more efficient O-IH sol-gel coating systems.

ACKNOWLEDGMENT

The author would like to acknowledge to the National Laboratory for Civil Engineering the physical and technical conditions provided to write part of this chapter.

REFERENCES

Adelkhani, H., Nasoodi, S., & Jafari, A. H. (2014). Corrosion protection properties of silica coatings formed by sol–gel method on Al: The effects of acidity, withdrawal speed, and annealing temperature. *Progress in Organic Coatings*, *77*(1), 142–145. doi:10.1016/j.porgcoat.2013.08.011

Adraider, Y., Pang, Y. X., Nabhani, F., Hodgson, S. N., Sharp, M. C., & Al-Waidh, A. (2013). Fabrication of zirconium oxide coatings on stainless steel by a combined laser/sol–gel technique. *Ceramics International*, *39*(8), 9665–9670. doi:10.1016/j.ceramint.2013.05.089

Adraider, Y., Pang, Y. X., Nabhani, F., Hodgson, S. N., & Zhang, Z. Y. (2011). Laser-induced deposition of sol–gel alumina coating on stainless steel under wet condition. *Surface and Coatings Technology*, *205*(23), 5345–5349. doi:10.1016/j.surfcoat.2011.05.044

Aegerter, M. A., & Mennig, M. (2004). *Sol-gel technologies for glass producers and users*. Boston: Kluwer Academic Publishers. doi:10.1007/978-0-387-88953-5

Akid, R., Gobara, M., & Wang, H. (2011). Corrosion protection performance of novel hybrid polyaniline/sol–gel coatings on an aluminium 2024 alloy in neutral, alkaline and acidic solutions. *Electrochimica Acta*, *56*(5), 2483–2492. doi:10.1016/j.electacta.2010.12.032

Álvarez, D., Collazo, A., Hernández, M., Nóvoa, X. R., & Pérez, C. (2010). Characterization of hybrid sol–gel coatings doped with hydrotalcite-like compounds to improve corrosion resistance of AA2024-T3 alloys. *Progress in Organic Coatings*, *67*(2), 152–160. doi:10.1016/j.porgcoat.2009.10.014

Álvarez, D., Collazo, A., Nóvoa, X. R., & Pérez, C. (2016). Assessment of ZnO nanoparticles as anticorrosive pigment in hybrid sol–gel films. *Progress in Organic Coatings*, *96*, 3–12. doi:10.1016/j.porgcoat.2015.10.017

Álvarez, P., Collazo, A., Covelo, A., Nóvoa, X. R., & Pérez, C. (2010). The electrochemical behaviour of sol–gel hybrid coatings applied on AA2024-T3 alloy: Effect of the metallic surface treatment. *Progress in Organic Coatings*, *69*(2), 175–183. doi:10.1016/j.porgcoat.2010.04.005

Andreatta, F., Paussa, L., Aldighieri, P., Lanzutti, A., Raps, D., & Fedrizzi, L. (2010). Corrosion behaviour of sol–gel treated and painted AA2024 aluminium alloy. *Progress in Organic Coatings*, *69*(2), 133–142. doi:10.1016/j.porgcoat.2010.04.012

Andreatta, F., Paussa, L., Lanzutti, A., Rosero Navarro, N. C., Aparicio, M., Castro, Y., & Fedrizzi, L. (2011). Development and industrial scale-up of ZrO_2 coatings and hybrid organic–inorganic coatings used as pre-treatments before painting aluminium alloys. *Progress in Organic Coatings*, *72*(1–2), 3–14. doi:10.1016/j.porgcoat.2011.01.011

Aoki, Y., Habazaki, H., & Kunitake, T. (2009). Size-scaling of proton conductivity in amorphous aluminosilicate acid thin films. *Journal of the American Chemical Society*, *131*(40), 14399–14406. doi:10.1021/ja904627h PMID:19807185

Aramaki, K. (2002). Self-healing mechanism of an organosiloxane polymer film containing sodium silicate and cerium(III) nitrate for corrosion of scratched zinc surface in 0.5 M NaCl. *Corrosion Science*, *44*(7), 1621–1632. doi:10.1016/S0010-938X(01)00171-8

Arkhireeva, A., Hay, J. N., Lane, J. M., Manzano, M., Masters, H., Oware, W., & Shaw, S. J. (2004). *Synthesis of Organic-Inorganic Hybrid Particles by Sol-Gel Chemistry* (Vol. 31).

Arkles, B. (2001). Commercial Application of Sol-Gel-Derived Hybrid Materials. *MRS Bulletin*, *26*(May), 402–408. doi:10.1557/mrs2001.94

Atkinson, A., & Guppy, R. M. (1991). Mechanical stability of sol-gel films. *Journal of Materials Science*, *26*(14), 3869–3873. doi:10.1007/BF01184984

Baer, D. R., Burrows, P. E., & El-Azab, A. A. (2003). Enhancing coating functionality using nanoscience and nanotechnology. *Progress in Organic Coatings*, *47*(3–4), 342–356. doi:10.1016/S0300-9440(03)00127-9

Balan, P., Ng, A., Beng Siang, C., Singh Raman, R. K., & Chan, E. S. (2013). Effect of Nanoparticle Addition in Hybrid Sol-Gel Silane Coating on Corrosion Resistance of Low Carbon Steel. *Advanced Materials Research*, *686*, 244–249. doi:10.4028/www.scientific.net/AMR.686.244

Balaskas, A. C., Kartsonakis, I. A., Tziveleka, L. A., & Kordas, G. C. (2012). Improvement of anti-corrosive properties of epoxy-coated AA 2024-T3 with TiO2 nanocontainers loaded with 8-hydroxy-quinoline. *Progress in Organic Coatings*, *74*(3), 418–426. doi:10.1016/j.porgcoat.2012.01.005

Bardal, E. (Ed.). (2004). *Corrosion and Protection*. London: Springer London. doi:10.1007/b97510

Benvenutti, E. V., Moro, C. C., Costa, T. M. H., & Gallas, M. R. (2009). Materiais híbridos à base de sílica obtidos pelo método sol-gel. *Quimica Nova*, *32*(7), 1926–1933. doi:10.1590/S0100-40422009000700039

Bhandari, J., Khan, F., Abbassi, R., Garaniya, V., & Ojeda, R. (2015). Modelling of pitting corrosion in marine and offshore steel structures – A technical review. *Journal of Loss Prevention in the Process Industries*, *37*, 39–62. doi:10.1016/j.jlp.2015.06.008

Billiet, S., Van Camp, W., Hillewaere, X. K. D., Rahier, H., & Du Prez, F. E. (2012). Development of optimized autonomous self-healing systems for epoxy materials based on maleimide chemistry. *Polymer*, *53*(12), 2320–2326. doi:10.1016/j.polymer.2012.03.061

Bramfitt, B. L. (2007). Carbon and Alloy Steels. In M. Kutz (Ed.), *Handbook of Materials Selection* (pp. 25–65). John Wiley & Sons, Inc.

Brinker, C., Hurd, A., Schunk, P., Frye, G., & Ashley, C. S. (1992). Review of sol-gel thin film formation. *Journal of Non-Crystalline Solids*, *148*, 424–436. doi:10.1016/S0022-3093(05)80653-2

Brinker, C. J., & Scherer, G. W. (1990). *Sol-gel science: the physics and chemistry of sol-gel processing*. Boston: Academic Press.

BSI British Standards. (2013). BS 6349-1-1:2013 - Maritime works General Code of practice for planning and design for operations.

Calabrese, L., Bonaccorsi, L., Caprì, A., & Proverbio, E. (2014). Adhesion aspects of hydrophobic silane zeolite coatings for corrosion protection of aluminium substrate. *Progress in Organic Coatings*, *77*(9), 1341–1350. doi:10.1016/j.porgcoat.2014.04.025

Calvert, P. (1999). Nanotube composites: A recipe for strength. *Nature*, *399*(6733), 210–211. doi:10.1038/20326

Cambon, J.-B., Ansart, F., Bonino, J.-P., & Turq, V. (2012). Effect of cerium concentration on corrosion resistance and polymerization of hybrid sol–gel coating on martensitic stainless steel. *Progress in Organic Coatings*, *75*(4), 486–493. doi:10.1016/j.porgcoat.2012.06.005

Cambon, J.-B., Esteban, J., Ansart, F., Bonino, J.-P., Turq, V., Santagneli, S. H., & Pulcinelli, S. H. (2012). Effect of cerium on structure modifications of a hybrid sol–gel coating, its mechanical properties and anti-corrosion behavior. *Materials Research Bulletin*, *47*(11), 3170–3176. doi:10.1016/j.materresbull.2012.08.034

Cambruzzi, A., Rossi, S., & Deflorian, F. (2005). Reduction on protective properties of organic coatings produced by abrasive particles. *Wear*, *258*(11), 1696–1705. doi:10.1016/j.wear.2004.11.023

Cao, R., Zhu, L., Li, W., Hu, W., & Liu, H. (2017). The effect of alumina-silica sols on electrodeposited zinc coatings for sintered NdFeB. *Journal of Alloys and Compounds*, *726*, 95–106. doi:10.1016/j.jallcom.2017.06.123

Capelossi, V. R., Poelman, M., Recloux, I., Hernandez, R. P. B., de Melo, H. G., & Olivier, M. G. (2013). Corrosion protection of clad 2024 aluminum alloy anodized in tartaric-sulfuric acid bath and protected with hybrid sol–gel coating. *Electrochimica Acta*, *124*, 69–79. doi:10.1016/j.electacta.2013.09.004

Carbajal-De, G., & La Torre, M. A. E.-M. (2009). Study of Ceramic and Hybrid Coatings Produced by the Sol-Gel Method for Corrosion Protection. *The Open Corrosion Journal*, *2*(1).

Cardoso, V. F., Minas, G., & Lanceros-Méndez, S. (2013). Multilayer spin-coating deposition of poly(vinylidene fluoride) films for controlling thickness and piezoelectric response. *Sensors and Actuators. A, Physical*, *192*, 76–80. doi:10.1016/j.sna.2012.12.019

Carneiro, J., Tedim, J., Fernandes, S. C. M., Freire, C. S. R., Silvestre, A. J. D., Gandini, A., & Zheludkevich, M. L. (2012). Chitosan-based self-healing protective coatings doped with cerium nitrate for corrosion protection of aluminum alloy 2024. *Progress in Organic Coatings*, *75*(1–2), 8–13. doi:10.1016/j.porgcoat.2012.02.012

Castelvetro, V., & De Vita, C. (2004). Nanostructured hybrid materials from aqueous polymer dispersions. *Advances in Colloid and Interface Science*, *108–109*, 167–185. doi:10.1016/j.cis.2003.10.017 PMID:15072940

Castro, Y., Ferrari, B., Moreno, R., & Durán, A. (2003). Silica sol-gel coatings on metals produced by EPD. *Sol-Gel Science and Technology*, *26*(1/3), 735–739. doi:10.1023/A:1020750222322

Chandler, K. A., & Bayliss, D. A. (1985). *Corrosion protection of steel structures*. Elsevier Applied Science Publishers.

Chiu, W.-M., Zhang, Y.-S., Tsai, P.-A., & Wu, J.-H. (2013). Characterization of antireflective coatings on poly(methyl methacrylate) substrate by different process parameters. *Journal of Applied Polymer Science*, *129*(5), 2411–2417. doi:10.1002/app.38975

Ciriminna, R., Fidalgo, A., Pandarus, V., Béland, F., Ilharco, L. M., & Pagliaro, M. (2013). The Sol-Gel Route to Advanced Silica-Based Materials and Recent Applications. *Chemical Reviews*, *113*(8), 6592–6620. doi:10.1021/cr300399c PMID:23782155

Ciriminna, R., Sciortino, M., Alonzo, G., Schrijver, A. D., & Pagliaro, M. (2011). From molecules to systems: Sol-gel microencapsulation in silica-based materials. *Chemical Reviews*, *111*(2), 765–789. doi:10.1021/cr100161x PMID:20726523

Conde, A., Durán, A., & de Damborenea, J. (2003). Polymeric sol–gel coatings as protective layers of aluminium alloys. *Progress in Organic Coatings*, *46*(4), 288–296. doi:10.1016/S0300-9440(03)00014-6

Corvo, F., Pérez, T., Martin, Y., Reyes, J., Dzib, L. R., González-Sánchez, J., & Castañeda, A. (2008). Time of wetness in tropical climate: Considerations on the estimation of TOW according to ISO 9223 standard. *Corrosion Science*, *50*(1), 206–219. doi:10.1016/j.corsci.2007.06.012

Craig, B. D. (2006). Material failure modes, part III: A brief tutorial on corrosion-related material failure modes. *Journal of Failure Analysis and Prevention*, *6*(2), 12–19. doi:10.1361/154770206X99262

Criado, M., Sobrados, I., Bastidas, J. M., & Sanz, J. (2015). Steel corrosion in simulated carbonated concrete pore solution its protection using sol–gel coatings. *Progress in Organic Coatings*, *88*, 228–236. doi:10.1016/j.porgcoat.2015.06.002

Criado, M., Sobrados, I., Bastidas, J. M., & Sanz, J. (2016). Corrosion behaviour of coated steel rebars in carbonated and chloride-contaminated alkali-activated fly ash mortar. *Progress in Organic Coatings*, *99*, 11–22. doi:10.1016/j.porgcoat.2016.04.040

Criado, M., Sobrados, I., Sanz, J., & Bastidas, J. M. (2014). Steel protection using sol–gel coatings in simulated concrete pore solution contaminated with chloride. *Surface and Coatings Technology*, *258*, 485–494. doi:10.1016/j.surfcoat.2014.08.051

Dash, S., Mishra, S., Patel, S., & Mishra, B. K. (2008). Organically modified silica: Synthesis and applications due to its surface interaction with organic molecules. *Advances in Colloid and Interface Science*, *140*(2), 77–94. doi:10.1016/j.cis.2007.12.006 PMID:18321464

Davis, J. R. (2000). *Corrosion understanding the basics*. Materials Park, OH: ASM International.

Davis, S. R., Brough, A. R., & Atkinson, A. (2003). Formation of silica/epoxy hybrid network polymers. *Journal of Non-Crystalline Solids*, *315*(1), 197–205. doi:10.1016/S0022-3093(02)01431-X

Dias, S. A. S., Lamaka, S. V., Nogueira, C. A., Diamantino, T. C., & Ferreira, M. G. S. (2012). Sol–gel coatings modified with zeolite fillers for active corrosion protection of AA2024. *Corrosion Science*, *62*, 153–162. doi:10.1016/j.corsci.2012.05.009

Dimitriev, Y., Ivanova, Y., & Iordanova, R. (2008). History of sol-gel science and technology. *Journal of the University of Chemical Technology and Matallurgy*, *43*(2), 181–192.

Dislich, H. (1986). Sol-gel: Science, processes and products. *Journal of Non-Crystalline Solids*, *80*(1–3), 115–121. doi:10.1016/0022-3093(86)90384-4

Donley, M. S., Mantz, R., Khramov, N., Balbyshev, V. N., Kasten, L. S., & Gaspar, D. J. (2003). The self-assembled nanophase particle (SNAP) process: A nanoscience approach to coatings. *Progress in Organic Coatings*, *47*(3–4), 401–415. doi:10.1016/j.porgcoat.2003.08.017

Esteves, A. C. C., Barros-Timmons, A., & Trindade, T. (2004). Nanocompósitos de matriz polimérica: Estratégias de síntese de materiais híbridos. *Quimica Nova*, *27*(5), 798–806. doi:10.1590/S0100-40422004000500020

Fedel, M., Callone, E., Diré, S., Deflorian, F., Olivier, M.-G., & Poelman, M. (2013). Effect of Na-Montmorillonite sonication on the protective properties of hybrid silica coatings. *Electrochimica Acta*, *124*, 90–99. doi:10.1016/j.electacta.2013.11.006

Feng, Z., Liu, Y., Thompson, G. E., & Skeldon, P. (2010). Sol–gel coatings for corrosion protection of 1050 aluminium alloy. *Electrochimica Acta*, *55*(10), 3518–3527. doi:10.1016/j.electacta.2010.01.074

Figueira, R. B., Fontinha, I. R., Silva, C. J. R., & Pereira, E. V. (2016). Hybrid Sol-Gel Coatings: Smart and Green Materials for Corrosion Mitigation. *Coatings*, *6*(1), 12. doi:10.3390/coatings6010012

Figueira, R. B., Silva, C. J., Pereira, E. V., & Salta, M. M. (2013). Ureasilicate Hybrid Coatings for Corrosion Protection of Galvanized Steel in Cementitious Media. *Journal of the Electrochemical Society*, *160*(10), C467–C479. doi:10.1149/2.033310jes

Figueira, R. B., Silva, C. J., Pereira, E. V., & Salta, M. M. (2014). Alcohol-Aminosilicate Hybrid Coatings for Corrosion Protection of Galvanized Steel in Mortar. *Journal of the Electrochemical Society*, *161*(6), C349–C362. doi:10.1149/2.103406jes

Figueira, R. B., Silva, C. J. R., & Pereira, E. V. (2014). Organic–inorganic hybrid sol–gel coatings for metal corrosion protection: A review of recent progress. *Journal of Coatings Technology and Research*, *12*(1), 1–35. doi:10.1007/s11998-014-9595-6

Figueira, R. B., Silva, C. J. R., & Pereira, E. V. (2015a). Hot-dip galvanized steel dip-coated with ureasilicate hybrid in simulated concrete pore solution: Assessment of coating morphology and corrosion protection efficiency. *Progress in Organic Coatings*, *88*, 245–255. doi:10.1016/j.porgcoat.2015.07.008

Figueira, R. B., Silva, C. J. R., & Pereira, E. V. (2015b). Hybrid sol–gel coatings for corrosion protection of hot-dip galvanized steel in alkaline medium. *Surface and Coatings Technology*, *265*, 191–204. doi:10.1016/j.surfcoat.2015.01.034

Figueira, R. B., Silva, C. J. R., & Pereira, E. V. (2015c). Influence of Experimental Parameters Using the Dip-Coating Method on the Barrier Performance of Hybrid Sol-Gel Coatings in Strong Alkaline Environments. *Coatings*, *5*(2), 124–141. doi:10.3390/coatings5020124

Figueira, R. B., Silva, C. J. R., & Pereira, E. V. (2016). Hybrid sol–gel coatings for corrosion protection of galvanized steel in simulated concrete pore solution. *Journal of Coatings Technology and Research*, *13*(2), 1–19. doi:10.1007/s11998-015-9751-7

Fontinha, I. R., Salta, M. M., Zheludkevich, M. L., Ferreira, M. G., Figueira, R. B., Pereira, E. V., & Silva, C. J. (2014). Influence of pH on the corrosion protection of epoxy-silica-zirconia sol-gel coatings applied on EN AW-6063 aluminium alloy. *ECS Transactions*, *58*(38), 9–16. doi:10.1149/05838.0009ecst

Fontinha, I. R., Salta, M. M., Zheludkevich, M. L., & Ferreira, M. G. S. (2013). EIS Study of Amine Cured Epoxy-silica-zirconia Sol-gel Coatings for Corrosion Protection of the Aluminium Alloy EN AW 6063. *Portugaliae Electrochimica Acta*, *31*(6), 307–319. doi:10.4152/pea.201306307

Fuchs-Godec, R. (2010). The Erosion−Corrosion Inhibition of AISI 431 Martensitic Stainless Steel in 2.0 M H_2SO_4 Solution Using *N* -Alkyl Quaternary Ammonium Salts as Inhibitors. *Industrial & Engineering Chemistry Research*, *49*(14), 6407–6415. doi:10.1021/ie100275a

Gao, J., & Suo, J. (2011). Effects of heating temperature and duration on the microstructure and properties of the self-healing coatings. *Surface and Coatings Technology*, *206*(6), 1342–1350. doi:10.1016/j.surfcoat.2011.08.059

Ghali, E., Sastri, V. S., & Elboujdaini, M. (2007). *Corrosion Prevention and Protection: Practical Solutions*. John Wiley & Sons.

Gharagozlou, M., Naderi, R., & Baradaran, Z. (2016). Effect of synthesized NiFe2O4-silica nanocomposite on the performance of an ecofriendly silane sol–gel coating. *Progress in Organic Coatings*, *90*, 407–413. doi:10.1016/j.porgcoat.2015.08.009

Ghosh, S. K. (2009). *Self-healing Materials: Fundamentals, Design Strategies, and Applications*. Wiley-VCH Verlag GmbH & Co.

Gobara, M., Kamel, H., Akid, R., & Baraka, A. (2015). Corrosion behaviour of AA2024 coated with an acid-soluble collagen/hybrid silica sol–gel matrix. *Progress in Organic Coatings*, *89*, 57–66. doi:10.1016/j.porgcoat.2015.07.026

Grabowski, K., Zbyrad, P., Uhl, T., Staszewski, W. J., & Packo, P. (2017). Multiscale electro-mechanical modeling of carbon nanotube composites. *Computational Materials Science*, *135*, 169–180. doi:10.1016/j.commatsci.2017.04.019

Grbić, B., Radić, N., Stojadinović, S., Vasilić, R., Dohčević-Mitrović, Z., Šaponjić, Z., & Stefanov, P. (2014). TiO_2/WO_3 photocatalytic composite coatings prepared by spray pyrolysis. *Surface and Coatings Technology*, *258*, 763–771. doi:10.1016/j.surfcoat.2014.07.082

Grundwürmer, M., Nuyken, O., Meyer, M., Wehr, J., & Schupp, N. (2007). Sol–gel derived erosion protection coatings against damage caused by liquid impact. *Wear*, *263*(1), 318–329. doi:10.1016/j.wear.2006.12.039

Guglielmi, M. (1997). Sol-gel coatings on metals. *Journal of Sol-Gel Science and Technology*, *8*(1–3), 443–449. doi:10.1007/BF02436880

Habazaki, H., Kimura, T., Aoki, Y., Tsuji, E., Yano, T., Shimizu, K., & Hassel, A. W. (2016). Characterization of Corrosion-resistant, Nanometer-thick, Layer-by-layer Aluminosilicate Coatings Prepared on Stainless Steel. *Electrochimica Acta*, *201*, 311–319. doi:10.1016/j.electacta.2015.11.112

Hammer, P., dos Santos, F. C., Cerrutti, B. M., Pulcinelli, S. H., & Santilli, C. V. (2013). Carbon nanotube-reinforced siloxane-PMMA hybrid coatings with high corrosion resistance. *Progress in Organic Coatings*, *76*(4), 601–608. doi:10.1016/j.porgcoat.2012.11.015

Hammer, P., Schiavetto, M. G., dos Santos, F. C., Benedetti, A. V., Pulcinelli, S. H., & Santilli, C. V. (2010). Improvement of the corrosion resistance of polysiloxane hybrid coatings by cerium doping. *Journal of Non-Crystalline Solids*, *356*(44–49), 2606–2612. doi:10.1016/j.jnoncrysol.2010.05.013

Hanetho, S. M., Kaus, I., Bouzga, A., Simon, C., Grande, T., & Einarsrud, M.-A. (2014). Synthesis and characterization of hybrid aminopropyl silane-based coatings on stainless steel substrates. *Surface and Coatings Technology*, *238*, 1–8. doi:10.1016/j.surfcoat.2013.10.013

Hay, J. N., & Raval, H. M. (2001). Synthesis of Organic–Inorganic Hybrids via the Non-hydrolytic Sol–Gel Process. *Chemistry of Materials*, *13*(10), 3396–3403. doi:10.1021/cm011024n

Hench, L. L., & West, J. K. (1990). The sol-gel process. *Chemical Reviews*, *90*(1), 33–72. doi:10.1021/cr00099a003

Hernandez, M., Barba, A., Genesca, J., Covelo, A., Bucio, E., & Torres, V. (2013). Characterization of Hybrid Sol-Gel Coatings Doped with Hydrotalcite-like Compounds on Steel and Stainless Steel Alloys. *ECS Transactions*, *47*(1), 195–206. doi:10.1149/04701.0195ecst

Hernández, S., Ottone, C., Varetti, S., Fontana, M., Pugliese, D., Saracco, G., & Armandi, M. (2016). Spin-Coated vs. Electrodeposited Mn Oxide Films as Water Oxidation Catalysts. *Materials (Basel)*, *9*(4), 296. doi:10.3390/ma9040296 PMID:28773419

Huang, Q., Yu, D., Xu, B., Hu, W., Ma, Y., Wang, Y., & Tian, Y. (2014). Nanotwinned diamond with unprecedented hardness and stability. *Nature*, *510*(7504), 250–253. doi:10.1038/nature13381 PMID:24919919

Hughes, A. E. (2015). 8 - Self-healing coatings A2 - Li, Guoqiang. In H. Meng (Ed.), *Recent Advances in Smart Self-healing Polymers and Composites* (pp. 211–241). Woodhead Publishing. doi:10.1016/B978-1-78242-280-8.00008-X

Hughes, A. E., Cole, I. S., Muster, T. H., & Varley, R. J. (2010). Designing green, self-healing coatings for metal protection. *NPG Asia Materials*, *2*(4), 143–151. doi:10.1038/asiamat.2010.136

Innocenzi, P., Abdirashid, M. O., & Guglielmi, M. (1994). Structure and Properties of Sol-Gel Coatings from Methytriethoxysilane and Tetraethoxysilane. *Journal of Sol-Gel Science and Technology*, *3*(1), 47–55. doi:10.1007/BF00490148

Innocenzi, P., Kidchob, T., & Yoko, T. (2005). Hybrid Organic-Inorganic Sol-Gel Materials Based on Epoxy-Amine Systems. *Journal of Sol-Gel Science and Technology*, *35*(3), 225–235. doi:10.1007/s10971-005-2290-4

Innocenzi, P., & Malfatti, L. (2015). Processing of Sol–Gel Films from a Top-Down Route. In D. Levy & Marcos Zayat (Eds.), The Sol-Gel Handbook (pp. 165–194). Wiley-VCH Verlag GmbH & Co. KGaA. doi:10.1002/9783527670819.ch05

Iribarren-Mateos, J. I., Buj-Corral, I., Vivancos-Calvet, J., Alemán, C., Iribarren, J. I., & Armelin, E. (2015). Silane and epoxy coatings: A bilayer system to protect AA2024 alloy. *Progress in Organic Coatings, 81*, 47–57. doi:10.1016/j.porgcoat.2014.12.014

José, N. M., & Prado, L. A. S. de A. (2005). Materiais Híbridos Orgânico-Inorgânicos: Preparação e Algumas Aplicações. *Quimica Nova, 28*(2), 281–288. doi:10.1590/S0100-40422005000200020

Kaesche, H. (2003). *Corrosion of Metals: Physicochemical Principles and Current Problems*. Springer. doi:10.1007/978-3-642-96038-3

Kalidindi, R. S. R., & Subasri, R. (2015). 5 - Sol-gel nanocomposite hard coatings. In M. Aliofkhazraei (Ed.), *Anti-Abrasive Nanocoatings* (pp. 105–136). Woodhead Publishing. doi:10.1016/B978-0-85709-211-3.00005-4

Kar, S., Paul, S., & Bandyopadhyay, P. P. (2016). Processing and characterisation of plasma sprayed oxides: Microstructure, phases and residual stress. *Surface and Coatings Technology, 304*, 364–374. doi:10.1016/j.surfcoat.2016.07.043

Khramov, A. N., & Johnson, J. A. (2009). Phosphonate-functionalized ORMOSIL coatings for magnesium alloys. *Progress in Organic Coatings, 65*(3), 381–385. doi:10.1016/j.porgcoat.2009.03.001

Khramov, A. N., Voevodin, N. N., Balbyshev, V. N., & Donley, M. S. (2004). Hybrid organo-ceramic corrosion protection coatings with encapsulated organic corrosion inhibitors. *Thin Solid Films, 447–448*, 549–557. doi:10.1016/j.tsf.2003.07.016

Khun, N. W., & Frankel, G. S. (2016). Cathodic delamination of polyurethane/multiwalled carbon nanotube composite coatings from steel substrates. *Progress in Organic Coatings, 99*, 55–60. doi:10.1016/j.porgcoat.2016.05.002

Kiruthika, P., Subasri, R., Jyothirmayi, A., Sarvani, K., & Hebalkar, N. Y. (2010). Effect of plasma surface treatment on mechanical and corrosion protection properties of UV-curable sol-gel based GPTS-ZrO_2 coatings on mild steel. *Surface and Coatings Technology, 204*(8), 1270–1276. doi:10.1016/j.surfcoat.2009.10.017

Kozhukharov, S. (2009). Relationship between the Conditions of Preparation by the Sol-Gel Route and the Properties of the Obtained Products. *Journal of the University of Chemical Technology and, 44*(2), 143–150.

Kozhukharov, S., Kozhukharov, V., Schem, M., Aslan, M., Wittmar, M., Wittmar, A., & Veith, M. (2012). Protective ability of hybrid nano-composite coatings with cerium sulphate as inhibitor against corrosion of AA2024 aluminium alloy. *Progress in Organic Coatings, 73*(1), 95–103. doi:10.1016/j.porgcoat.2011.09.005

Kumar, N., Jyothirmayi, A., Soma Raju, K. R. C., & Subasri, R. (2012). Effect of functional groups (methyl, phenyl) on organic–inorganic hybrid sol–gel silica coatings on surface modified SS 316. *Ceramics International, 38*(8), 6565–6572. doi:10.1016/j.ceramint.2012.05.040

Kunst, S. R., Cardoso, H. R. P., Oliveira, C. T., Santana, J. A., Sarmento, V. H. V., Muller, I. L., & Malfatti, C. F. (2013). Corrosion resistance of siloxane-poly(methyl methacrylate) hybrid films modified with acetic acid on tin plate substrates: Influence of tetraethoxysilane addition. *Applied Surface Science*, *298*, 1–11. doi:10.1016/j.apsusc.2013.09.182

Lacatusu, I., Badea, N., Murariu, A., Pirvu, C., & Meghea, A. (2011). Vegetal nanoclusters in hybrid silica films prepared by sol-gel spin coating technique. *Journal of Non-Crystalline Solids*, *357*(7), 1716–1723. doi:10.1016/j.jnoncrysol.2011.01.018

Lakshmi, R. V., Aruna, S. T., & Sampath, S. (2017). Ceria nanoparticles vis-à-vis cerium nitrate as corrosion inhibitors for silica-alumina hybrid sol-gel coating. *Applied Surface Science*, *393*, 397–404. doi:10.1016/j.apsusc.2016.10.035

Lamaka, S. V., Zheludkevich, M. L., Yasakau, K. A., Montemor, M. F., Cecílio, P., & Ferreira, M. G. S. (2006). TiOx self-assembled networks prepared by templating approach as nanostructured reservoirs for self-healing anticorrosion pre-treatments. *Electrochemistry Communications*, *8*(3), 421–428. doi:10.1016/j.elecom.2005.12.019

Lampke, T., Darwich, S., Nickel, D., & Wielage, B. (2008). Development and characterization of sol-gel composite coatings on aluminum alloys for corrosion protection. *Materialwissenschaft und Werkstofftechnik*, *39*(12), 914–919. doi:10.1002/mawe.200800396

Latella, B. A., Barbé, C. J., Cassidy, D. J., Li, H., & Ignat, M. (2005). Mechanical Stability and Decohesion of Sol-Gel Hybrid Coatings on Metallic Substrates. *Revista Matéria*, *10*(2), 199–204.

Lavollée, C., Gressier, M., Garcia, J., Sobrino, J.-M., Reby, J., & Menu, M.-J. (2016). New architectured hybrid sol-gel coatings for wear and corrosion protection of low-carbon steel. *Progress in Organic Coatings*, *99*, 337–345. doi:10.1016/j.porgcoat.2016.06.015

Lebedev, E. (2011). Hybrid organo-inorganic polymer systems: Synthesis, Structure and Properties. *Theoretical and Experimental Chemistry*, *46*(6), 391–396. doi:10.1007/s11237-011-9172-3

Lekawa-Raus, A., Gizewski, T., Patmore, J., Kurzepa, L., & Koziol, K. K. (2017). Electrical transport in carbon nanotube fibres. *Scripta Materialia*, *131*, 112–118. doi:10.1016/j.scriptamat.2016.11.027

Li, C., Jordens, K., & Wilkes, G. L. (2000). Abrasion-resistant coatings for plastic and soft metallic substrates by sol–gel reactions of a triethoxysilylated diethylenetriamine and tetramethoxysilane. *Wear*, *242*(1), 152–159. doi:10.1016/S0043-1648(00)00408-7

Li, Y.-S., Lu, W., Wang, Y., & Tran, T. (2009). Studies of (3-mercaptopropyl)trimethoxylsilane and bis(trimethoxysilyl)ethane sol-gel coating on copper and aluminum. *Spectrochimica Acta. Part A: Molecular and Biomolecular Spectroscopy*, *73*(5), 922–928. doi:10.1016/j.saa.2009.04.016 PMID:19464230

Liu, J., Zhang, Y., Yu, M., Li, S., Xue, B., & Yin, X. (2015). Influence of embedded ZnAlCe-NO_3– layered double hydroxides on the anticorrosion properties of sol–gel coatings for aluminum alloy. *Progress in Organic Coatings*, *81*, 93–100. doi:10.1016/j.porgcoat.2014.12.015

Liu, Y., Wei, S., Tian, H., Tong, H., & Xu, B. (2014). Characterization of soft magnetic spinel ferrite coating prepared by plasma spray. *Surface and Coatings Technology*, *258*, 189–199. doi:10.1016/j.surfcoat.2014.09.029

Livage, J. (1997). Sol-gel processes. *Current Opinion in Solid State and Materials Science*, *2*(2), 132–138. doi:10.1016/S1359-0286(97)80057-5

Livage, J. (1999). Sol-gel synthesis of hybrid materials. *Bulletin of Materials Science*, *22*(3), 201–205. doi:10.1007/BF02749920

Lutz, A., Mol, J. M. C., De Graeve, I., & Terryn, H. (2016). 6 - Smart corrosion protection by multi-action self-healing polymeric coatings. In M. F. Montemor (Ed.), *Smart Composite Coatings and Membranes* (pp. 157–181). Woodhead Publishing. doi:10.1016/B978-1-78242-283-9.00006-3

Maaß, P., & Peißker, P. (2011). *Handbook of Hot-dip Galvanization*. John Wiley & Sons. doi:10.1002/9783527636884

Maia, F., Tedim, J., Lisenkov, A. D., Salak, A. N., Zheludkevich, M. L., & Ferreira, M. G. S. (2012). Silica nanocontainers for active corrosion protection. *Nanoscale*, *4*(4), 1287–1298. doi:10.1039/c2nr11536k PMID:22249939

Maia, F., Yasakau, K. A., Carneiro, J., Kallip, S., Tedim, J., Henriques, T., & Ferreira, M. G. S. (2016). Corrosion protection of AA2024 by sol–gel coatings modified with MBT-loaded polyurea microcapsules. *Chemical Engineering Journal*, *283*, 1108–1117. doi:10.1016/j.cej.2015.07.087

Mammeri, F., Bourhis, E. L., Rozes, L., & Sanchez, C. (2005). Mechanical properties of hybrid organic–inorganic materials. *Journal of Materials Chemistry*, *15*(35–36), 3787. doi:10.1039/b507309j

Messaddeq, S. H., Pulcinelli, S. H., Santilli, C. V., Guastaldi, A. C., & Messaddeq, Y. (1999). Microstructure and corrosion resistance of inorganic–organic (ZrO_2–PMMA) hybrid coating on stainless steel1. *Journal of Non-Crystalline Solids*, *247*(1–3), 164–170. doi:10.1016/S0022-3093(99)00058-7

Metroke, T., Kachurina, O., & Knobbe, E. T. (2002). Spectroscopic and corrosion resistance characterization of GLYMO–TEOS Ormosil coatings for aluminum alloy corrosion inhibition. *Progress in Organic Coatings*, *44*(4), 295–305. doi:10.1016/S0300-9440(02)00063-2

Metroke, T. L., & Apblett, A. (2004). Effect of solvent dilution on corrosion protective properties of Ormosil coatings on 2024-T3 aluminum alloy. *Progress in Organic Coatings*, *51*(1), 36–46. doi:10.1016/j.porgcoat.2004.06.002

Mittal, V. (2014). 8 - Self-healing anti-corrosion coatings for applications in structural and petrochemical engineering A2 - Makhlouf, Abdel Salam Hamdy. In *Handbook of Smart Coatings for Materials Protection* (pp. 183–197). Woodhead Publishing. doi:10.1533/9780857096883.2.183

Mitzi, D. B. (2001). Thin-Film Deposition of Organic−Inorganic Hybrid Materials. *Chemistry of Materials*, *13*(10), 3283–3298. doi:10.1021/cm0101677

Montemor, M. F. (2016). 7 - Hybrid nanocontainer-based smart self-healing composite coatings for the protection of metallic assets. In *Smart Composite Coatings and Membranes* (pp. 183–209). Woodhead Publishing. doi:10.1016/B978-1-78242-283-9.00007-5

More, A. P., & Mhaske, S. T. (2016). Anticorrosive coating of polyesteramide resin by functionalized ZnO- Al 2 O 3 -Fly ash composite and functionalized multiwalled carbon nanotubes. *Progress in Organic Coatings*, *99*, 240–250. doi:10.1016/j.porgcoat.2016.06.005

Nemeth, S. (2008). Modelling of the scratch resistance of particle filled sol–gel coatings on aluminium. *Thin Solid Films*, *516*(16), 5355–5359. doi:10.1016/j.tsf.2007.07.044

Niederberger, M. (2007). Nonaqueous sol-gel routes to metal oxide nanoparticles. *Accounts of Chemical Research*, *40*(9), 793–800. doi:10.1021/ar600035e PMID:17461544

Nikrooz, B., & Zandrahimi, M. (2011). Optimization of process variables and corrosion properties of a multi layer silica sol gel coating on AZ91D using the Box–Behnken design. *Journal of Sol-Gel Science and Technology*, *59*(3), 640–649. doi:10.1007/s10971-011-2539-z

Nikrooz, B., Zandrahimi, M., & Ebrahimifar, H. (2012). High temperature oxidation resistance and corrosion properties of dip coated silica coating by sol gel method on stainless steel. *Journal of Sol-Gel Science and Technology*, *63*(3), 286–293. doi:10.1007/s10971-012-2786-7

Norouzi, M., & Afrasiabi Garekani, A. (2014). Corrosion protection by zirconia-based thin films deposited by a sol–gel spin coating method. *Ceramics International*, *40*(2), 2857–2861. doi:10.1016/j.ceramint.2013.10.027

Oldani, V., Sergi, G., Pirola, C., Sacchi, B., & Bianchi, C. L. (2016). Sol-gel hybrid coatings containing silica and a perfluoropolyether derivative with high resistance and anti-fouling properties in liquid media. *Journal of Fluorine Chemistry*, *188*, 43–49. doi:10.1016/j.jfluchem.2016.06.005

Osborne, J. H. (2001). Observations on chromate conversion coatings from a sol–gel perspective. *Progress in Organic Coatings*, *41*(4), 280–286. doi:10.1016/S0300-9440(01)00143-6

Parkhill, R., Knobbe, E., & Donley, M. (2001). Application and evaluation of environmentally compliant spray-coated ormosil films as corrosion resistant treatments for aluminum 2024-T3. *Progress in Organic Coatings*, *41*(4), 261–265. doi:10.1016/S0300-9440(01)00138-2

Pathak, S. S., & Khanna, A. S. (2009). Investigation of anti-corrosion behavior of waterborne organosilane–polyester coatings for AA6011 aluminum alloy. *Progress in Organic Coatings*, *65*(2), 288–294. doi:10.1016/j.porgcoat.2008.12.006

Pepe, A., Galliano, P., Ceré, S., Aparicio, M., & Durán, A. (2005). Hybrid silica sol–gel coatings on Austempered Ductile Iron (ADI). *Materials Letters*, *59*(17), 2219–2222. doi:10.1016/j.matlet.2005.03.001

Phanasgaonkar, A., & Raja, V. S. (2009). Influence of curing temperature, silica nanoparticles- and cerium on surface morphology and corrosion behaviour of hybrid silane coatings on mild steel. *Surface and Coatings Technology*, *203*(16), 2260–2271. doi:10.1016/j.surfcoat.2009.02.020

Pierre, A. C. (1998). Applications of Sol-Gel Processing. In Introduction to Sol-Gel Processing (pp. 347–386). Springer US. doi:10.1007/978-1-4615-5659-6_9

Pirhady Tavandashti, N., Ghorbani, M., Shojaei, A., Gonzalez-Garcia, Y., Terryn, H., & Mol, J. M. C. (2016). pH responsive Ce(III) loaded polyaniline nanofibers for self-healing corrosion protection of AA2024-T3. *Progress in Organic Coatings*, *99*, 197–209. doi:10.1016/j.porgcoat.2016.04.046

Price, S., & Figueira, R. (2017). Corrosion Protection Systems and Fatigue Corrosion in Offshore Wind Structures: Current Status and Future Perspectives. *Coatings*, *7*(2), 25. doi:10.3390/coatings7020025

Rahimi, H., Mozaffarinia, R., & Hojjati Najafabadi, A. (2013). Corrosion and Wear Resistance Characterization of Environmentally Friendly Sol–gel Hybrid Nanocomposite Coating on AA5083. *Journal of Materials Science and Technology*, *29*(7), 603–608. doi:10.1016/j.jmst.2013.03.013

Rahoui, S., Turq, V., & Bonino, J.-P. (2013). Effect of thermal treatment on mechanical and tribological properties of hybrid coatings deposited by sol–gel route on stainless steel. *Surface and Coatings Technology*, *235*, 15–23. doi:10.1016/j.surfcoat.2013.07.008

Roberge, P. (1999). *Handbook of Corrosion Engineering*. McGraw-Hill Professional.

Rodič, P., Iskra, J., & Milošev, I. (2014). Study of a sol–gel process in the preparation of hybrid coatings for corrosion protection using FTIR and 1H NMR methods. *Journal of Non-Crystalline Solids*, *396–397*, 25–35. doi:10.1016/j.jnoncrysol.2014.04.013

Rodič, P., Mertelj, A., Borovšak, M., Benčan, A., Mihailović, D., Malič, B., & Milošev, I. (2015). Composition, structure and morphology of hybrid acrylate-based sol–gel coatings containing Si and Zr composed for protective applications. *Surface and Coatings Technology, 286*, 388-396.

Rodič, P., & Milošev, I. (2014). Corrosion Properties of UV Cured Hybrid Sol-Gel Coatings on AA7075-T6 Determined under Simulated Aircraft Conditions. *Journal of the Electrochemical Society*, *161*(9), C412–C420. doi:10.1149/2.1091409jes

Rodič, P., & Milošev, I. (2015). Electrochemical and Salt Spray Testing of Hybrid Coatings Based on Si and Zr Deposited on Aluminum and Its Alloys. *Journal of the Electrochemical Society*, *162*(10), C592–C600. doi:10.1149/2.0801510jes

Rodič, P., & Milošev, I. (2016). Corrosion Inhibition of Pure Aluminium and Alloys AA2024-T3 and AA7075-T6 by Cerium(III) and Cerium(IV) Salts. *Journal of the Electrochemical Society*, *163*(3), C85–C93. doi:10.1149/2.0431603jes

Rodríguez, M. A., Gil, L., Camero, S., Fréty, N., Santana, Y., & Caro, J. (2014). Effects of the dispersion time on the microstructure and wear resistance of WC/Co-CNTs HVOF sprayed coatings. *Surface and Coatings Technology*, *258*, 38–48. doi:10.1016/j.surfcoat.2014.10.014

Romero Pareja, R., López Ibáñez, R., Martín, F., Ramos-Barrado, J. R., & Leinen, D. (2006). Corrosion behaviour of zirconia barrier coatings on galvanized steel. *Surface and Coatings Technology*, *200*(22–23), 6606–6610. doi:10.1016/j.surfcoat.2005.11.098

Rosero-Navarro, N. C., Paussa, L., Andreatta, F., Castro, Y., Durán, A., Aparicio, M., & Fedrizzi, L. (2010). Optimization of hybrid sol–gel coatings by combination of layers with complementary properties for corrosion protection of AA2024. *Progress in Organic Coatings*, *69*(2), 167–174. doi:10.1016/j.porgcoat.2010.04.013

Rosero-Navarro, N. C., Pellice, S. A., Durán, A., & Aparicio, M. (2008). Effects of Ce-containing sol–gel coatings reinforced with SiO_2 nanoparticles on the protection of AA2024. *Corrosion Science*, *50*(5), 1283–1291. doi:10.1016/j.corsci.2008.01.031

Rossi, S., Deflorian, F., Fontanari, L., Cambruzzi, A., & Bonora, P. L. (2005). Electrochemical measurements to evaluate the damage due to abrasion on organic protective system. *Progress in Organic Coatings*, *52*(4), 288–297. doi:10.1016/j.porgcoat.2004.09.005

Rossi, S., Deflorian, F., & Risatti, M. (2006). Modified Taber apparatus and new test geometry to evaluate the reduction of organic coatings corrosion protective properties induced by abrasive particles. *Surface and Coatings Technology*, *201*(3), 1173–1179. doi:10.1016/j.surfcoat.2006.01.036

Roussi, E., Tsetsekou, A., Tsiourvas, D., & Karantonis, A. (2011). Novel hybrid organo-silicate corrosion resistant coatings based on hyperbranched polymers. *Surface and Coatings Technology*, *205*(10), 3235–3244. doi:10.1016/j.surfcoat.2010.11.037

Rout, T., Bera, S., Udayabhanu, G., & Narayan, R. (2016). Methodologies of Application of Sol-Gel Based Solution onto Substrate: A Review. *Journal of Coating Science and Technology*, *3*(1), 9–22. doi:10.6000/2369-3355.2016.03.01.2

Sahu, N., Parija, B., & Panigrahi, S. (2009). Fundamental understanding and modeling of spin coating process: A review. *Indian Journal of Physics*, *83*(4), 493–502. doi:10.1007/s12648-009-0009-z

Sakka, M. M., Antar, Z., Elleuch, K., & Feller, J. F. (2017). Tribological response of an epoxy matrix filled with graphite and/or carbon nanotubes. *Friction*, *5*(2), 171–182. doi:10.1007/s40544-017-0144-z

Sakka, S., & Yoko, T. (1992). Sol-gel-derived coating films and applications. In R. Reisfeld & C. K. JJørgensen (Eds.), Chemistry, Spectroscopy and Applications of Sol-Gel Glasses (Vol. 77, pp. 89–118). Springer Berlin Heidelberg. doi:10.1007/BFb0036966

Sanchez, C., Belleville, P., Popall, M., & Nicole, L. (2011). Hybrid materials themed issue. *Chemical Society Reviews*, *40*(2), 453–1152.

Sanchez, C., Julián, B., Belleville, P., & Popall, M. (2005). Applications of hybrid organic–inorganic nanocomposites. *Journal of Materials Chemistry*, *15*(35–36), 3559–3592. doi:10.1039/b509097k

Sanchez, C., & Ribot, F. (1994). Design of hybrid organic-inorganic materials synthesized via sol-gel chemistry. *New Journal of Chemistry*, *18*(10), 1007–1047.

Sanchez, C., Rozes, L., Ribot, F., Laberty-Robert, C., Grosso, D., Sassoye, C., & Nicole, L. (2010). "Chimie douce": A land of opportunities for the designed construction of functional inorganic and hybrid organic-inorganic nanomaterials. *Comptes Rendus. Chimie*, *13*(1–2), 3–39. doi:10.1016/j.crci.2009.06.001

Sanchez, C., Soler-Illia, G. J. D. A. A., Ribot, F., Lalot, T., Mayer, C. R., & Cabuil, V. (2001). Designed Hybrid Organic−Inorganic Nanocomposites from Functional Nanobuilding Blocks. *Chemistry of Materials*, *13*(10), 3061–3083. doi:10.1021/cm011061e

Santana, I., Pepe, A., Jimenez-Pique, E., Pellice, S., Milošev, I., & Ceré, S. (2015). Corrosion protection of carbon steel by silica-based hybrid coatings containing cerium salts: Effect of silica nanoparticle content. *Surface and Coatings Technology*, *265*, 106–116. doi:10.1016/j.surfcoat.2015.01.050

Schmidt, H. (1985). New type of non-crystalline solids between inorganic and organic materials. *Journal of Non-Crystalline Solids*, *73*(1–3), 681–691. doi:10.1016/0022-3093(85)90388-6

Schottner, G. (2001). Hybrid Sol–Gel-Derived Polymers: Applications of Multifunctional Materials. *Chemistry of Materials, 13*(10), 3422–3435. doi:10.1021/cm011060m

Segal, D. L. (1984). Sol-gel processing: Routes to oxide ceramics using colloidal dispersions of hydrous oxides and alkoxide intermediates. *Journal of Non-Crystalline Solids, 63*(1–2), 183–191. doi:10.1016/0022-3093(84)90397-1

Seshan, K. (Ed.). (2002). *Handbook of thin-film deposition processes and techniques: principles, methods, equipment, and applications* (2nd ed.). Norwich, N.Y: Noyes Publications.

Shchukin, D. G., & Grigoriev, D. O. (2012). 10 - The use of nanoreservoirs in corrosion protection coatings. In V. S. Saji & R. Cook (Eds.), *Corrosion Protection and Control Using Nanomaterials* (pp. 264–282). Woodhead Publishing. doi:10.1533/9780857095800.2.264

Shchukin, D. G., Zheludkevich, M., Yasakau, K., Lamaka, S., Ferreira, M. G. S., & Möhwald, H. (2006). Layer-by-Layer Assembled Nanocontainers for Self-Healing Corrosion Protection. *Advanced Materials, 18*(13), 1672–1678. doi:10.1002/adma.200502053

Sheffer, M., Groysman, A., & Mandler, D. (2003). Electrodeposition of sol–gel films on Al for corrosion protection. *Corrosion Science, 45*(12), 2893–2904. doi:10.1016/S0010-938X(03)00106-9

Shilova, O. (2013). Synthesis and structure features of composite silicate and hybrid TEOS-derived thin films doped by inorganic and organic additives. *Journal of Sol-Gel Science and Technology, 68*(3), 387–410. doi:10.1007/s10971-013-3026-5

Skorb, E. V., Fix, D., Andreeva, D. V., Möhwald, H., & Shchukin, D. G. (2009). Surface-Modified Mesoporous SiO_2 Containers for Corrosion Protection. *Advanced Functional Materials, 19*(15), 2373–2379. doi:10.1002/adfm.200801804

Soo, M. T., Kawamura, G., Muto, H., Matsuda, A., Lockman, Z., & Cheong, K. Y. (2013). Design of hierarchically meso–macroporous tetragonal ZrO_2 thin films with tunable thickness by spin-coating via sol–gel template route. *Microporous and Mesoporous Materials, 167*, 198–206. doi:10.1016/j.micromeso.2012.09.010

Sorochkina, K., Smotraiev, R., & Chepurna, I. (2015). Zirconium and aluminum oxyhydroxides particles formation during sol–gel process. *Colloids and Surfaces. A, Physicochemical and Engineering Aspects, 484*, 56–61. doi:10.1016/j.colsurfa.2015.07.038

Suzana, A. F., Ferreira, E. A., Benedetti, A. V., Carvalho, H. W. P., Santilli, C. V., & Pulcinelli, S. H. (2016). Corrosion protection of chromium-coated steel by hybrid sol-gel coatings. *Surface and Coatings Technology, 299*, 71–80. doi:10.1016/j.surfcoat.2016.04.075

Szymański, K., Hernas, A., Moskal, G., & Myalska, H. (2015). Thermally sprayed coatings resistant to erosion and corrosion for power plant boilers - A review. *Surface and Coatings Technology, 268*, 153–164. doi:10.1016/j.surfcoat.2014.10.046

Tiwari, A., & Hihara, L. H. (2015). Sol-Gel Route for the Development of Smart Green Conversion Coatings for Corrosion Protection of Metal Alloys. In A. Tiwari, J. Rawlins, & L. H. Hihara (Eds.), *Intelligent Coatings for Corrosion Control* (pp. 363–407). Boston: Butterworth-Heinemann. doi:10.1016/B978-0-12-411467-8.00010-6

Trabelsi, W., Cecilio, P., Ferreira, M. G. S., & Montemor, M. F. (2005). Electrochemical assessment of the self-healing properties of Ce-doped silane solutions for the pre-treatment of galvanised steel substrates. *Progress in Organic Coatings*, *54*(4), 276–284. doi:10.1016/j.porgcoat.2005.07.006

Treacy, M. M. J., Ebbesen, T. W., & Gibson, J. M. (1996). Exceptionally high Young's modulus observed for individual carbon nanotubes. *Nature*, *381*(6584), 678–680. doi:10.1038/381678a0

Turri, S., Torlaj, L., Piccinini, F., & Levi, M. (2010). Abrasion and nanoscratch in nanostructured epoxy coatings. *Journal of Applied Polymer Science*, *118*(3), 1720–1727.

Varma, P. C. R., Colreavy, J., Cassidy, J., Oubaha, M., McDonagh, C., & Duffy, B. (2010). Corrosion protection of AA 2024-T3 aluminium alloys using 3, 4-diaminobenzoic acid chelated zirconium–silane hybrid sol–gels. *Thin Solid Films*, *518*(20), 5753–5761. doi:10.1016/j.tsf.2010.05.088

Varma, P. C. R., Periyat, P., Oubaha, M., McDonagh, C., & Duffy, B. (2011). Application of niobium enriched ormosils as thermally stable coatings for aerospace aluminium alloys. *Surface and Coatings Technology*, *205*(16), 3992–3998. doi:10.1016/j.surfcoat.2011.02.023

Vivar Mora, L., Naik, S., Paul, S., Dawson, R., Neville, A., & Barker, R. (2017). Influence of silica nanoparticles on corrosion resistance of sol-gel based coatings on mild steel. *Surface and Coatings Technology*, *324*, 368–375. doi:10.1016/j.surfcoat.2017.05.063

Voevodin, N. N., Grebasch, N. T., Soto, W. S., Kasten, L. S., Grant, J. T., Arnold, F. E., & Donley, M. S. (2001). An organically modified zirconate film as a corrosion-resistant treatment for aluminum 2024-T3. *Progress in Organic Coatings*, *41*(4), 287–293. doi:10.1016/S0300-9440(01)00156-4

Voevodin, N. N., Kurdziel, J. W., & Mantz, R. (2006). Corrosion protection for aerospace aluminum alloys by Modified Self-assembled NAnophase Particle (MSNAP) sol–gel. *Surface and Coatings Technology*, *201*(3–4), 1080–1084. doi:10.1016/j.surfcoat.2006.01.028

Vorotilov, K., Petrovsky, V., & Vasiljev, V. (1995). Spin coating process of sol-gel silicate films deposition: Effect of spin speed and processing temperature. *Journal of Sol-Gel Science and Technology*, *5*(3), 173–183. doi:10.1007/BF00487014

Vreugdenhil, A. J., Balbyshev, V. N., & Donley, M. S. (2001). Nanostructured silicon sol-gel surface treatments for Al 2024-T3 protection. *Journal of Coatings Technology*, *73*(915), 35–43. doi:10.1007/BF02730029

Walcarius, A. (2001). Electrochemical Applications of Silica-Based Organic−Inorganic Hybrid Materials. *Chemistry of Materials*, *13*(10), 3351–3372. doi:10.1021/cm0110167

Wang, C., Xue, T., Dong, B., Wang, Z., & Li, H.-L. (2008). Polystyrene–acrylonitrile–CNTs nanocomposites preparations and tribological behavior research. *Wear*, *265*(11), 1923–1926. doi:10.1016/j.wear.2008.04.021

Wang, D., & Bierwagen, G. P. (2009). Sol–gel coatings on metals for corrosion protection. *Progress in Organic Coatings*, *64*(4), 327–338. doi:10.1016/j.porgcoat.2008.08.010

Wang, H., & Akid, R. (2008). Encapsulated cerium nitrate inhibitors to provide high-performance anti-corrosion sol–gel coatings on mild steel. *Corrosion Science*, *50*(4), 1142–1148. doi:10.1016/j.corsci.2007.11.019

Wen, J., & Wilkes, G. L. (1996). Organic/inorganic hybrid network materials by the sol-gel approach. *Chemistry of Materials*, *8*(8), 1667–1681. doi:10.1021/cm9601143

Wilkes, Garth L.; Orler, Bruce; Huang, Hao Hsin. (1985). "Ceramers": hybrid materials incorporating polymeric/oligomeric species into inorganic glasses utilizing a sol-gel approach, *26*(2), 300–302.

Wilson, S., Hawthorne, H. M., Yang, Q., & Troczynski, T. (2001). Scale effects in abrasive wear of composite sol–gel alumina coated light alloys. *Wear*, *251*(1), 1042–1050. doi:10.1016/S0043-1648(01)00704-9

Wojcik, A., & Klein, L. C. (1997). Transparent organic/inorganic hybrid gels: A classification scheme. *Applied Organometallic Chemistry*, *11*(November 1995), 129–135.

Wu, K. H., Chang, T. C., Yang, C. C., & Wang, G. P. (2006). Dynamics and corrosion resistance of amine-cured organically modified silicate coatings on aluminum alloys. *Thin Solid Films*, *513*(1–2), 84–89. doi:10.1016/j.tsf.2006.01.022

Yabuki, A., Yamagami, H., & Noishiki, K. (2007). Barrier and self-healing abilities of corrosion protective polymer coatings and metal powders for aluminum alloys. *Materials and Corrosion*, *58*(7), 497–501. doi:10.1002/maco.200604041

Yasakau, K. A., Carneiro, J., Zheludkevich, M. L., & Ferreira, M. G. S. (2014). Influence of sol-gel process parameters on the protection properties of sol–gel coatings applied on AA2024. *Surface and Coatings Technology*, *246*, 6–16. doi:10.1016/j.surfcoat.2014.02.038

Yoganandan, G., Pradeep Premkumar, K., & Balaraju, J. N. (2015). Evaluation of corrosion resistance and self-healing behavior of zirconium–cerium conversion coating developed on AA2024 alloy. *Surface and Coatings Technology*, *270*, 249–258. doi:10.1016/j.surfcoat.2015.02.049

Yu, M., Xue, B., Liu, J., Li, S., & Zhang, Y. (2015). Electrophoretic deposition of hybrid coatings on aluminum alloy by combining 3-aminopropyltrimethoxysilan to silicon–zirconium sol solutions for corrosion protection. *Thin Solid Films*, *590*, 33–39. doi:10.1016/j.tsf.2015.07.055

Zabihi, F., Ahmadian-Yazdi, M.-R., & Eslamian, M. (2017). Photocatalytic Graphene-TiO_2 Thin Films Fabricated by Low-Temperature Ultrasonic Vibration-Assisted Spin and Spray Coating in a Sol-Gel Process. *Catalysts*, *7*(5), 136–152. doi:10.3390/catal7050136

Zandi Zand, R., Verbeken, K., & Adriaens, A. (2012). Corrosion resistance performance of cerium doped silica sol–gel coatings on 304L stainless steel. *Progress in Organic Coatings*, *75*(4), 463–473. doi:10.1016/j.porgcoat.2012.06.008

Zhang, L. C., Zarudi, I., & Xiao, K. Q. (2006). Novel behaviour of friction and wear of epoxy composites reinforced by carbon nanotubes. *Wear*, *261*(7), 806–811. doi:10.1016/j.wear.2006.01.033

Zheludkevich, M., Serra, R., Montemor, M., Yasakau, K., Salvado, I., & Ferreira, M. (2005). Nanostructured sol–gel coatings doped with cerium nitrate as pre-treatments for AA2024-T3Corrosion protection performance. *Electrochimica Acta*, *51*(2), 208–217. doi:10.1016/j.electacta.2005.04.021

Zheludkevich, M. L., Poznyak, S. K., Rodrigues, L. M., Raps, D., Hack, T., Dick, L. F., & Ferreira, M. G. S. (2010). Active protection coatings with layered double hydroxide nanocontainers of corrosion inhibitor. *Corrosion Science*, *52*(2), 602–611. doi:10.1016/j.corsci.2009.10.020

Zheludkevich, M. L., Salvado, I. M., & Ferreira, M. G. S. (2005). Sol–gel coatings for corrosion protection of metals. *Journal of Materials Chemistry*, *15*(48), 5099. doi:10.1039/b419153f

Zheludkevich, M. L., Shchukin, D. G., Yasakau, K. A., Möhwald, H., & Ferreira, M. G. S. (2007). Anticorrosion Coatings with Self-Healing Effect Based on Nanocontainers Impregnated with Corrosion Inhibitor. *Chemistry of Materials*, *19*(3), 402–411. doi:10.1021/cm062066k

Zheludkevich, M. L., Tedim, J., & Ferreira, M. G. S. (2012). "Smart" coatings for active corrosion protection based on multi-functional micro and nanocontainers. *Electrochimica Acta*, *82*, 314–323. doi:10.1016/j.electacta.2012.04.095

Zheng, S., & Li, J. (2010). Inorganic–organic sol gel hybrid coatings for corrosion protection of metals. *Journal of Sol-Gel Science and Technology*, *54*(2), 174–187. doi:10.1007/s10971-010-2173-1

Zomorodian, A., Garcia, M. P., Moura de Silva, T., Fernandes, J. C. S., Fernandes, M. H., & Montemor, M. F. (2013). Corrosion resistance of a composite polymeric coating applied on biodegradable AZ31 magnesium alloy. *Acta Biomaterialia*, *9*(10), 8660–8670. doi:10.1016/j.actbio.2013.02.036 PMID:23454214

KEY TERMS AND DEFINITIONS

Coating: Layer that covers the surface of a metallic substrate, usually referred to as the substrate.

Corrosion: The reaction of a metallic material with its environment and the products of this reaction may be solid, liquid, or gaseous.

Corrosion Rate: The speed at which any metal in a specific environment deteriorates e.g. material loss per area unit and time unit.

Organic-Inorganic Hybrids (O-IH): Multifunctional materials with the presence of both organic and inorganic components within the common matrix support.

Precursor: A substance from which another is formed (sol).

Sol: Stable suspension of colloidal particles within a liquid.

Sol-Gel: Low temperature synthetic route that provides an excellent control of precursor stoichiometry and ease of composition modification.

Time of Wetness (TOW): Time when the metal is sufficiently wet for corrosion reaction to occur.

Chapter 14
Deposition of TiN, TiC, and DLC Coatings by PACVD

Mahboobeh Azadi
Semnan University, Iran

ABSTRACT

In this chapter, the author studied about titanium nitride (TiN), titanium carbide (TiC), diamond like carbon (DLC) single and multilayer coatings that utilize in harsh environments. These hard coatings were usually produced by the plasma assisted chemical vapor deposition (PACVD) method as a modern technique. PACVD is used to deposit thin coatings for different usages such as computer disc drives, automobile and aerospace parts, surgical/medical instruments and the food industry. The author tried to delineate the state of the performance of different coating systems and layer characteristics that suitable either for laboratory -scales or industrial applications. Mechanical features of these coatings contain the hardness, the toughness, the wear resistance and structural properties that were perused. Consequently, this chapter offers a source of information for those who want to familiarize with the knowledge in the area of materials science of functional coatings that was produced by new plasma-based technologies.

INTRODUCTION

In recent years, the growing demand of industries has been increased to improve surface properties of industrial parts (Azadi, 2014). Thus, the usage of various coatings has been spread during last years. Nitrides and carbides of the group IV, V and VI transition metals are important materials which are used for depositing thin coatings in industrial applications (Rie, 1993). While nitride coatings have comparatively good thermal stability and lower interaction tendency with other materials, carbide coatings possess high hardness and better adhesion to metallic substrates (Pfohl, 1992).

Titanium carbide (TiC) and Titanium nitride (TiN) are two materials that belong to the hard material group. TiC and TiN coatings are utilized in various industrial applications such as automobile and aerospace parts, cutting and forming tools, grinding wheels, molding and forging applications due to good mechanical properties including high hardness, low friction coefficient, excellent wear resistance and corrosion resistance in different harsh environments and high resistance to tribo-oxidation and

DOI: 10.4018/978-1-5225-4194-3.ch014

surface fatigue (Shanaghi, 2012, Badisch, 2003). TiC layers are also used as heat shield coatings for atmospheric reentry of spacecraft and biomedical instruments. Graphite substrates which are coated with TiC and TiN layers are promising for using as the container material for pyro- chemical processing of certain rare-earth and nuclear metals, where chemical inertness and a good match of thermal expansion coefficients are required (Maiya, 1998). The deposition of TiC layers on titanium substrates would be useful as the abrasion resistant material for abutment parts of dental implant and sliding parts of artificial joints (Zhu, 2012). Consequently, such coatings are widely used in various industrial applications ranging from machining tools to decorative items (Stoiber, 2003). Similar to TiC layers, Diamond like Carbon (DLC) films were also used for the same applications.

These thin coatings can be made by different methods such as chemical vapor deposition (CVD) and different techniques of physical vapor deposition (PVD) such as ion plating, sputtering and magnetron sputtering pulsed laser deposition (Oguri, 1990, Liepack, 2004). As the temperature of chemical reactions in CVD is still too high and the chlorine content from the metal precursors causes undesired side effects on the layer quality (Rie, 1993), the substitution method is plasma assisted chemical vapor deposition (PACVD). In addition, CVD is not suitable for steel substrates because over-tempering occurs at the temperature about 600 °C (Kim, 1999). Besides, PACVD does not require the high vacuum with respect to PVD techniques. On the other hand, the degradation of mechanical properties of the layer can occur owing to insufficient adhesion for low temperature PVD coatings or to diffuse the binder metal from the substrate into the coating (Rie, 1995). Consequently, PACVD is a suitable technique for depositing thin and hard coatings, due to the possibility of achieving proper properties at low temperatures and high deposition rates (Azadi, 2014, Mofidi, 2014). Indeed, by using PACVD, tools with the complex geometry can be coated and pre-treatment stages like the sputter-cleaning and the plasma nitriding processes can be done in the same deposition chamber (Gammer, 2004). As a result, great advantages that are easily obtained from PACVD processes, are included: the good controllability of constituents of deposited layers, the possibility of not needing to rotate various substrates and the good throwing power (Wang, 199).

The Institute of fur Oberflachentechnik (IOT) developed PACVD process and a pilot system for simultaneous coating of about 240 components (Son, 2007). PACVD apparatus usually contains the reactor of a steel chamber as anode with a cathodic substrate holder and parallel-placed anodic gas admission. The power supply generates pulsed d.c. power or the frequency voltage to supply the proper plasma. An additional heating system which is usually located inside the chamber allows independent adjustment of the substrate temperature (Rie, 1993). Pure gases such as nitrogen, hydrogen and argon are fed into the chamber via mass flow controllers. As different gases are transported through the reactor, distinct coating layers are formed on the tooling substrate. The schematic of PACVD apparatus is depicted in Figure 1. In every PACVD process, gaseous reactants are admitted into a reactor.

TiC COATINGS

Titanium carbide (TiC) is formed thermodynamically as the result of the following chemical reaction at 1030 °C (equation 1):

$$TiCl_4 + CH_4 + H_2 \rightarrow TiC + 4\,HCl + H_2 \quad (1)$$

Figure 1. The schematic of PACVD apparatus

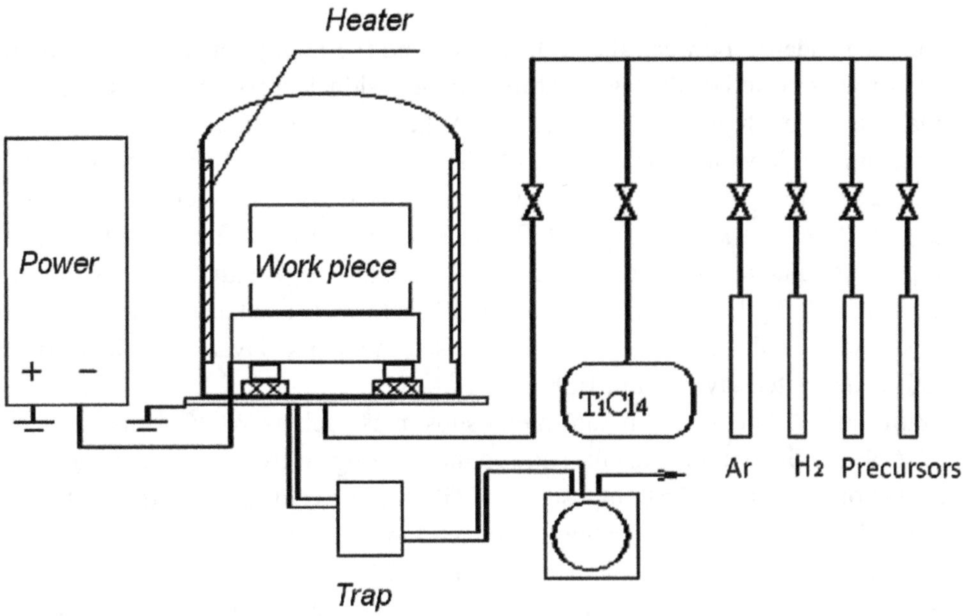

But when the plasma is contributed into the reaction, the required energy and the reaction temperature can be reduced to 450 °C. However, when the TiC coating was deposited in 420 °C, it was not adhesive to the steel substrate. But when the deposition temperature increased to 450 °C, the adhesive coating could be deposited on the hot work tool steel (H13) (Azadi, 2014). In such way, this final product of reaction (1) was the hard and the wear resistant coating that exhibited the chemical and the metallurgical bond to various substrates. The substrate temperature also could be reduced to lower than 300 °C (Rie, 1993), when metal-organic compounds were used as precursors. In this situation, the formation of chlorine-free layers also could be achieved. Sometimes this method was called the metal- organic plasma assisted chemical vapor deposition (MO-PACVD). The deposition temperature in TiC coatings had also an important role in coating characteristic changes such as the structure, the hardness, the wear resistance, the friction coefficient and the surface roughness (Azadi, 2014). In addition, unlike the TiN coating deposition process, it is necessary to control the existence of extra phases, such as the carbon and oxide phases and the chlorine content in the deposited TiC layer for proper industrial applications (Kim, 1999). It is noticeable that TiC phase extended over a wide range of composition in the Ti-C phase diagram (Dahan, 2001). In the absence of proper parameters for depositing the TiC phase, DLC coatings could be created on substrates, meanwhile DLC coatings had their own specific characterizations.

It was noticeable that, the role of hydrogen in such reactions was found to have a good effect on the layer growth process on different substrates. When the hydrogen pressure was low, the TiC layer growth terminated well before the total process duration. As a result, no TiC layer growth was observed without hydrogen addition to the gas mixture (Jonas, 2009).

Besides the deposition temperature, effects of other experimental parameters such as the gas pressure, the duty cycle, the reactor pressure and the precursor pressure on the layer properties has been examined in literatures.

Appearance, Microstructure, and Chemical Composition

There is a strong dependence between the performance and the microstructure of TiC protective hard coatings. The key role is played by the layer texture and the residual stress which, in turn, are determined by the particular grain growth process (Leoni, 1999). When the deposition temperature decreased from 490 °C to 450 °C and the flow rate ratio of CH_4 to $TiCl_4$ was also increased from 1.5 to 6, the TiC coating changed from dark gray to silver color. Such event demonstrated that the stoichiometry of TiC_x coatings would be changed. When the x value in the TiC_x layer was about 0.5 the coating had the silver color and by increasing this value to 0.9 the coating color became darker (Azadi, 2013). Whenever the minimum amount of $TiCl_4$ (3.5 g/h) and CH_4 gas flow of 18 sccm were used as two deposition parameters, the carbon phase and the chlorine content were decreased in the TiC layer and enhanced the chemical composition of the deposited layer (Kim, 1999).

The preparation of hard coatings on hard metal substrates such as WC/Co substrates was done by Taischner et al. (1991). The obtained results showed that coatings were smooth and had the small grain size with the value of 10-30 nm. The composition of TiC coatings could be controlled by the gas phase ratio of reactants. The deposited TiC coating contained amounts of free carbon dependent on the gas phase composition, for example 15 atomic percent at nC/nTi = 25.

Atomic force microscopy is a method that can measure the grain size of such coatings in the nanometric scale. Figure 2 shows the three-dimensional image that was obtained from these techniques. The image demonstrates that the grain shape of TiC layers is spherical and the grain height value in such coatings is about 82 nm (Azadi, 2014).

Figure 2. Three-dimensional AFM image of the surface topography in the TiC coating
Springer, 2014.

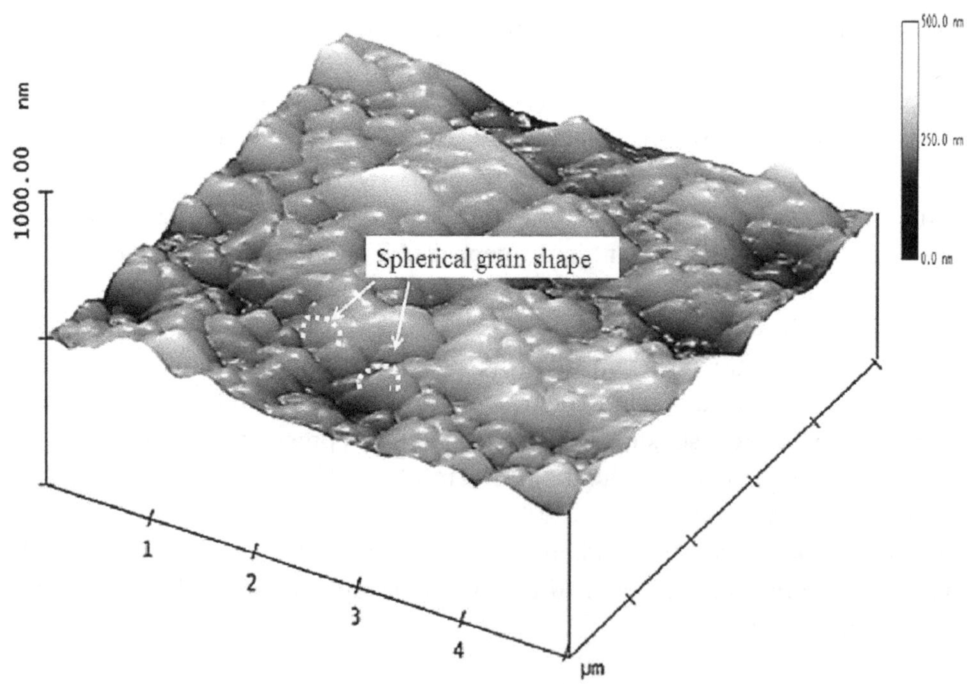

Hardness and Wear Properties

TiC coating is an extremely hard refractory ceramic material, similar to tungsten carbide. The excess carbon content has effects on the hardness and wear properties of TiC coatings. The hardness value was between 2000 to 3000 Hv regardless of the carbon content. In contrast to the hardness value, the friction coefficient of TiC coatings remained constant irrespective of the carbon content. The friction coefficient of TiC coatings was about 0.2. The binding states of excess carbon atoms in such coatings were also investigated by laser Raman spectroscopy technique (Oguri, 1990). The hard layer of low friction consisting of TiC and excess amorphous carbon could be achieved at low process pressures (less than 50 Pa). In this situation, the friction coefficient value lowered to 0.15 (Bartsch, 1997). It was also possible that the nanocrystalline TiC particles were embedded in the matrix of hard amorphous carbon during PACVD process. In this manner, no dependence of hardness on the layer thickness was observed when the thickness of the layer was higher than 300–400 nm (Liepack, 2004). The Vickers microhardness measurement could pass through the maximum value and depended on the gas phase composition (Taischner, 1999). When a single layer of TiC layer was deposited on the stellite 6B substrate at 650°C and 650 V, it would have the smooth surface and the high hardness value (about 3000 HK0.01). In this situation, the layer was bonded to the substrate by the formation of chromium carbides at the interface (Park, 1998)

The hardness of TiC coating was also depended on the deposition temperature. However, it was independent from flux ratio of gases. Besides, the wear mechanism was changed abruptly when the deposition temperature of TiC coatings was changed from 450 °C to 470 °C. The minimum friction coefficient, the surface roughness, the higher hardness, the fracture toughness and the wear resistance of TiC coatings were observed when the deposition temperature was about 450 °C (Azadi, 2014, 2016). Hu et al. (2007) reported that the composition of TiC/C:H had crucial effects on the hardness and other mechanical properties. The hardness and elastic modulus values increased with increasing the titanium concentration up to 39.7 atomic percent. The maximum hardness and elastic modulus values were about 24 GPa and 230 GPa, respectively. However, hardness changes had no obvious effect on tribological properties of the layer. When Ti concentration increased to 39.7 atomic percent, the friction coefficient and the wear rate were about 0.28 and 0.005 $\mu m^3/Nm$. The wear mechanism was depended on the relative amount of TiC and C:H in the layer. In addition, the fretting fatigue and wear properties of titanium alloys could be increased by the duplex plasma treatment combined with shot peening before depositing TiC coatings (Kaestner, 2001).

In Figure 3, the wear surface topography image for the TiC coating can be observed. Due to the speed of the wear (as 0.1 m/s), the mechanical mechanism was a dominant mode of the wear for most hard coatings. In the mechanical wear mechanism, the fractured small ceramic particles from coatings on the surface of specimens could act like the secondary moving indenters (Azadi, 2014). Thus, the small pits could be seen in the wear surfaces. The wear area was separated through the zigzag line from the coating area.

Adhesion Properties to Different Substrates

A clean substrate surface free from oxides and other contaminants plays an important role in creating proper adhesion of layer. Thus, before placing different substrates into the reactor, cleaning procedures such as pickling, grit blasting, degreasing, polishing, etc., are necessary to be done. Besides the clean-

Figure 3. The wear surface topography of the TiC coating
Springer, 2014.

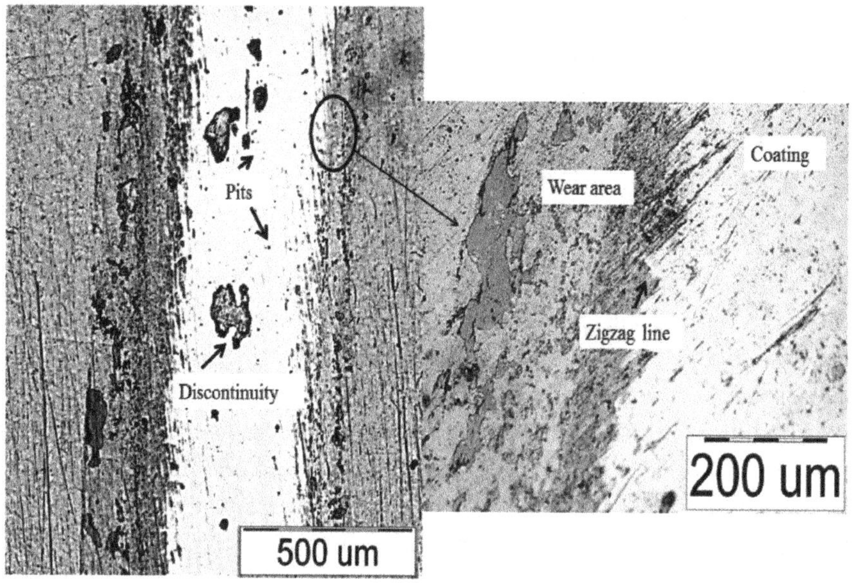

ing process, the chemical composition of the deposited layer has also effects on the adhesive strength of the layer to the substrate. It is noticeable that the layer adhesive force on high-speed steel substrate increased with x in the TiC_x layer and reached to the maximum value when x was about 0.5. Such behavior showed that the adhesive strength would be affected by the stoichiometry of TiC_x layer (Hyakutake, 1998). In addition, the plasma-nitro carburizing pre-treatment led in the best adhesion of the TiC coating to titanium alloys as the substrate. In this way, the combination of plasma treatment and TiC coating enhanced the tribological behavior and extended the applicability of titanium alloys in many industrial sectors (Kaestner, 2001). The plasma-nitriding pre-treatment also enhanced the adhesive strength of TiC coating to hot-working die steel H13 by increasing the hardness of steel substrates (Azadi, 2014). The adhesive energy would be usually increase with increasing the deposition temperature for metallic substrates (Sun, 2017). The adhesive force between the deposited film and the substrate was usually measured by the scratch test.

Thermal Stability and Other Properties

Another remarkable property of TiC coating which is deposited by PACVD is that it had low stress (Liepack, 2004). Indeed, nano-composite layers of TiC/C that composed by small crystallites of hard phases embedded in the amorphous lubricant matrix have been studied as protective coatings. These coatings have to work often in harsh environments, exposed to high temperatures (above 800–900 °C), and oxidizing or corrosive atmospheres. Besides, the grain growth of TiC was not detected up to 1000 °C in the high vacuum, 700 °C in the low vacuum and 350 °C in the air (Liepack, 2009).

Pei et al. (2006) studied about the tribological behavior of the nano-composite coatings by means of ball-on-disk tribo-tests at the elevated temperature in the range of 150-400 °C. Their results showed that the thermal stability of TiC/C: H coating was high up to 350 °C and the oxidation event was not observed. In addition, the origin of failure was attributed to the interfacial delamination due to the discontinuity in mechanical properties between coatings and substrates at the elevated temperature. Other research (Zabransky, 2015) investigated the thermal stability of TiC/C:H coatings by using the thermal desorption spectroscopy in the temperature range from 500 to 1000 °C. It was found that the hardness and the elastic modulus of both Ti-rich and C-rich coatings remained unchanged up to 700 °C.

TiN COATINGS

TiN coatings are deposited on different steel substrates by PACVD using gas mixtures of $TiCl_4$, NH_3, H_2 and Ar. Prior to deposition, the substrate is usually pre-treated by NH_3 or N_2 plasma as reported in the literature (Kim, 1999, Azadi, 2014). The first TiN coating that was deposited by PACVD was reported by Rie et al. (1991). They investigated this coating by the optical emission spectroscopy. Several atomic lines and molecular bands were identified in the emitted spectrum of the discharge. The nitrogen in the discharge was found mostly in the molecular state (N_2^* and N_2^{+*}). Ti was identified in the ionized state that could control the layer growth.

Consequently, TiN is formed thermodynamically as the result of the following chemical reaction at 1000 °C (equation 2):

$$2TiCl_4 + N_2 + 5H_2 \rightarrow 2TiN + 8\ HCl + H_2 \quad (2)$$

When the plasma provided the necessary energy for the chemical reaction, the deposition temperature could be reduced to below 400 °C (Rie, 1991). However, Stockt et al. (1991) reported that whenever the temperature was too high, the decomposition of the metal-organic compound occurred in the gaseous phase to the large extent. Thus, the concentration of the metal-organic compound decreased and the heterogeneous reaction at the substrate surface became slower. When TiN coating deposited on the substrate, the gold-like coating was usually appeared (Stoiber, 2003).

Structure and Chemical Composition

Various chemical compositions of TiN coatings lead to different layer structures. With increasing the chlorine content from 2.6 to 6.3 atomic percent, the change in the preferred orientation from (100) to the random direction would be observed and the grain size and the surface roughness were decreased (Stoiber, 2003)). Columnar grains were also formed with the orientation of (200) at 500 °C, however decreasing the deposition temperature to 400 °C led to form T fibrous grains with the orientation of (200) and the amorphous component (Hilton, 1987). Besides, the gas flow rate can change the preferred orientation in TiN coatings. When the N_2 to $TiCl_4$ flow rate was about 14.3, the preferred orientation and the microstructure changed markedly from (111) to (200), and from columnar to non-oriented. But when this ratio decreased to 8-9, the TiN composition remained unchanged. When N_2 to $TiCl_4$ flow rate was about 3-6, it resulted in the (200) preferred orientation and the fine grained dense structure (Oguri, 1991). Transmission electron microscopy (TEM) results in the research of Taschner et al. (1993) revealed that the small grain size of TiN layer in the initial range followed by the typical columnar growth. Kawata et al. (2001) also demonstrated that all fracture cross-sections of TiN layers that prepared at three deposition temperatures (483, 550 and 600 °C) showed columnar structures. Indeed, all TiN layers with different chlorine concentrations had NaCl structure and the preferred orientation of (200). In all, the grain orientation of TiN coating was depended on various factors, such as the deposition temperature, the flow rate of N_2 to $TiCl_4$ and the chlorine content.

The growth rate of TiN layers seemed to be independent of reactor parameters. It increased nearly proportionally with the current density (Eckel, 1999). However, Son et al. (2007) showed that the deposition rate was dependent on the process parameters such as the flow rate, the deposition temperature and the chamber pressure. The temperature effect was small, whereas the deposition rate sharply increased at the higher pressure in both the pilot plant and the industrial-size scales. The growth of TiN layer with the pulsed D.C. PACVD technique was determined by two parameters: the ion current and the variation of the pause time. The ion current with Ti transfer during the pulse did not lead to any increase of the layer thickness and caused to sputter of material that only weakly incorporated in the growing layer (Hardt, 1999). The variation of the pause time could optimize the growth rate and the process time. Furthermore, the deposition rate was strongly dependent on the flow rate of $TiCl_4$, with a distinct maximum (Mogensen, 1998).

The TiN layer showed a dense and fine-grained structure for every reactor volume. The change of the structure corresponded to increase of the current density (Eckel, 1999). Hilton et al. (1987) also reported that SEM investigations indicated island-like features which fused at increased deposition time. The density of these features was of the order of the TiN grain size.

Mongenson et al. (1998) reported that the plasma intensity, as measured by the light emitted by the plasma, decayed approximately exponentially with the total pressure which was the critical parameter for the quality of TiN layers. The stoichiometric TiN_x layer with low oxygen and chlorine impurities could be prepared at high deposition rates characterized by the sufficient flow rate of $TiCl_4$, (P_{H2}/P_{N2}) >9, the plasma power density between 0.4 and 0.8 Wcm^{-2} and the substrate temperatures of 500-700 °C (Endler, 1995).

In most coatings, the grain size of TiN layer increased with the layer thickness, while 1 to 2 μm-thick layers of TiN was porous. For the 2366 nm-thick layer, sub-grains appeared at grain boundaries and the structure got denser. Sub-grains generated at grain boundaries significantly affected the mechanical properties and the texture (Zhang, 2013). It was also found that the TiN layer and the Ti(C, N) layer had the dense columnar structure while the TiC layer showed the microstructure of nano-crystallite mixed with the amorphous phase. The crystallite size gradually decreased from the TiN to the Ti(C, N) and to the TiC layer. Three-dimensional AFM image of the surface topography in the TiN coating is depicted in Figure 4. The TiN layer had a pyramidal grain shape, since the TiC layer had a spherical grain shape. In addition, sub- grains were seen in some areas of the layer surface (Azadi, 2016).

Wear and Hardness

The wear resistance of TiN layer also increased owing to the increase of the hardness and the adhesion with respect to steel substrates (Kim, 1999). The microhardness values in the range of 25-40 GPa and the high Young's modulus (values between 450 and 590 GPa) were obtained (Taschner, 2001). When the chlorine content was below than 4.0 atomic percent the hardness and the abrasive wear resistance of the coating was not decreased significantly, whereas at higher chlorine content, the decrease in the hardness and the increase in the abrasive wear were observed (Badisch, 2001). Above 200 °C, increased abrasion of the coating occurred, which was primarily the result of the decreasing coating hardness (Badisch, 2003). It was found that the chlorine content below 3.2 atomic percent did not deteriorate the

Figure 4. Three-dimensional AFM image of the surface topography in the TiN coating
Springer, 2014.

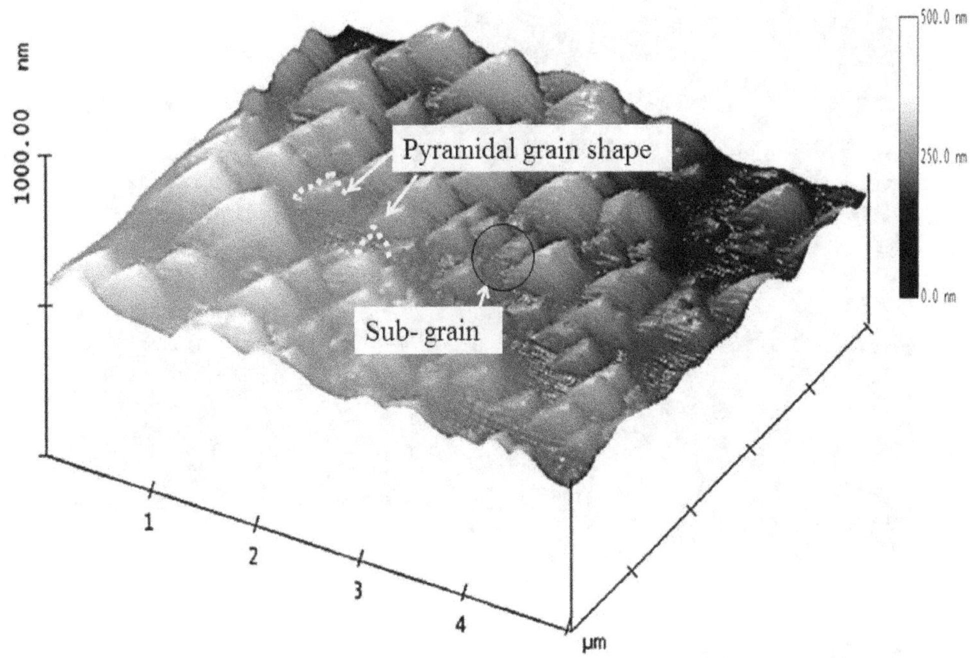

mechanical properties like the hardness and the elastic modulus markedly. For higher chlorine content the decrease in both hardness and the elastic modulus was observed (Stoiber, 2003). The frictional behavior also varied due to incorporate the amount of chlorine into the TiN layer. The wear tract for TiN coating is depicted in Figure 5. Layers of coating seem to be twisted and compressed together. In the wear tract there were some cracks (Azadi, 2014).

Studies (Rie, 1991, Zhu, 1999) on TiN layers indicated that the residual chlorine in the coating deteriorated properties of the coating and led to a reduction in the hardness, the wear resistance, the corrosion resistance and the bonding strength. Whenever TiN layers were deposited at 483, 550 and 600 °C, hardness values of layers were reached to 2070, 2360 and 2140 Hv (0.05 N), respectively (Kawata, 2002). In addition, in such situation, the friction coefficient and the wear factor of TiN layers lowered with increasing the chlorine content. The lower hardness values from metallo-organic compounds as procures were also could be found compared with those with $TiCl_4$ (Rie, 1991). Besides, at the lower bias voltage a uniform composition with the highest hardness could be achieved (Laime, 1990). Park et al. (1998) made the single layer of TiN which had the high hardness value (about 2400 HK0.01) and was formed on the substrate at 650°C and 550 V. No brittle compound was formed at the interface. It has been found that the friction coefficient of TiN coatings with (111) preferred grain orientation was lower than those of TiN coatings with (200) preferred grain orientation. Indeed, the low friction coefficient of TiN coating with (111) preferred grain orientation was due to the production of Ti oxide layer with the good lubricity on the surface of the steel ball, which prevented the Fe adhesion from the steel ball to the surface of the TiN coating (Azushima, 2008).

Figure 5. The wear surface topography of the TiN coating
Elsevier, 2014.

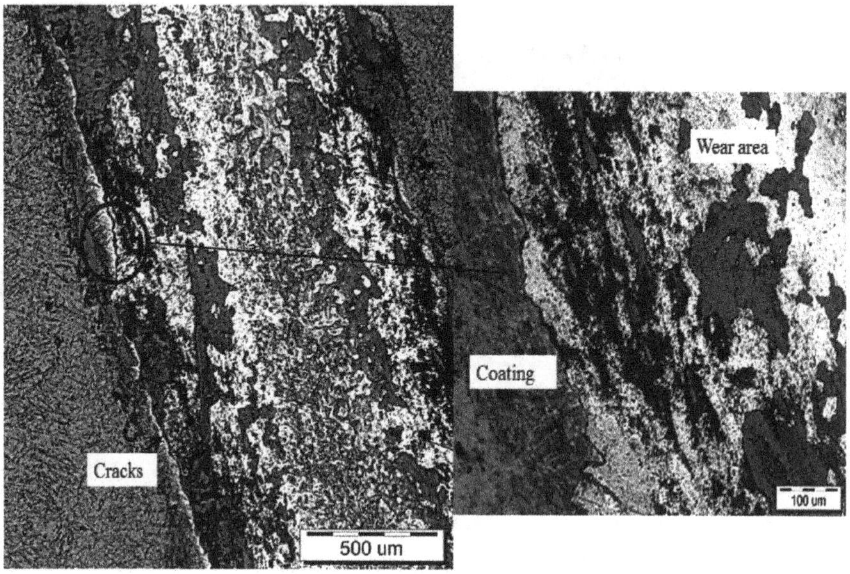

High Oxidation Resistance

When the aluminum element is added to the TiN coating, the coating oxidation resistance would be increased. The high oxidation resistance of such coating was related to the formation of the stable compact aluminum oxide layer on the surface of the substrate when it is exposed to high temperatures like in the cutting tool application (Taschner, 2001). In addition, the incorporation of a small amount of Al or Si to the TiN layer, had the large effect on the improvement in both microhardness and oxidation resistance of TiN layer (Park, 2002). While TiN layer started to oxidize from 500 °C, Ti–Al–N and Ti–Si–N layers showed much improved oxidation resistance up to 700 °C. Comparing Ti–Al–N and Ti–Si–N, the Ti–Si–N layer showed better merit in the mechanical property at the moderate temperature, however Ti–Al–N layer had larger advantages in the resistant property against the high temperature oxidation. However, in recent years, it was found that the TiN layer remained stable up to 800 °C (Yu, 2012). The TiN coating was thermally stable without any layer disintegration or oxidation or phase transformation, up to 24 h at 500 °C. Besides, the nano-multilayers (CrN–TiN) showed better thermal stability as compared to the monolayer, both at room temperature and at 500 °C (Srinivasan, 2007).

Adhesion Properties and Other Properties

Similar to TiC coatings, the cleaning process or other plasma treatments before the layer deposition can enhance the adhesive strength of TiN coatings to substrates. The adhesion of TiN coatings to hot-working die steel H13 as the substrate could be improved by applying the optimized plasma nitriding process conducted at the 25% flow ratio of $N_2/(H_2+ N_2)$ for one hour. The formation of the new compound layer during the nitriding process at the 50% flow ratio of $N_2/(H_2+ N_2)$ deteriorated the adhesion of TiN coatings to the steel substrate. This event was related to premature brittle fracture between the coating and the steel substrate (Ma, 2001). The adhesion strength of the coating also changed markedly with the reactive gases compositions. When the flow rate of N_2 to $TiCl_4$ was about 3-6, it resulted in a better coating adhesion strength to the steel substrate (Oguri, 1991). When the pre-treatment gas mixture contained 40–80% nitrogen, it led to appear the best adhesion properties of PACVD-TiN coatings. The argon addition also had negative effects on the adhesion strength (Gammer, 2004). The adhesion strength of the layer was the best one when the oxygen and chlorine contents were low and therefore it depended on the initial gas phase composition and plasma parameters. The carbon-rich coating also had the lower adhesion than the nitrogen-rich coating (Endler, 1995).

The role played by residual stresses and the texture was highly related both to coating compositions and to the technique and parameters employed in the production of the layer. For example, the compressive in-plane residual stress was usually considered as a positive feature as it tended to improve mechanical properties and the hardness of the TiN layer, but an intense compression could have a negative effect since it would limit the adhesion of protective layers to the underlying metal substrate. In addition, in the fiber textured layer, for instance, the hardness, the mechanical strength and the stiffness could be significantly varied by changing the growth direction (Leoni, 1999). Thomsen et al. (1998) showed that the residual intrinsic stress was related to the preferred orientation of coatings. Total macroscopic stress values were found to range from -1.5 to +1.5 GPa. The low intrinsic strain energies favored the preferred orientation of the TiN layer corresponding to the plane with the lowest surface energy which was (200). Besides, the coating heat treatment improved the adherence of some of TiN coatings, but decreased their wear resistance (Dahan, 2001).

MULTILAYER COATINGS

Multilayer coatings have been found to contain better mechanical properties – i.e. higher hardness, higher fracture toughness and greater strength – than single layer coatings (Azadi, 2014). As an example, although TiC coatings have useful properties, the application as a single layer coating has been limited due to the brittleness of the TiC phase with the strong covalent bond. This limitation can be usually overcomed by increasing the number of layers in coatings (Azadi, 2014). Thus, the improvement of mechanical properties, such as the hardness and the adhesion force of the TiN/TiC multilayer coating could be achieved for practical industrial applications (Kim, 1999). Eroglu et al. (1991) reported that TiN/Ti(C, N)/TiC multilayer coatings, which had been used in the machine tool industry, exhibited performance superior to that of monolithic coatings. Indeed, obtained results demonstrated that graded coatings were receiving considerable interests because of the possibility of tailoring material properties, such as the thermal expansion coefficient or the hardness, by the spatial control of microstructure and composition. Before depositing the multilayer coating, the plasma nitriding process could be done with a careful control of the interlayer thickness and composition. In such situation, the coating performance in tribological applications would be improved (Bull, 1991).

Chemical Composition and Structure

The X-ray diffraction (XRD) pattern showed that TiN/TiC multilayer coating had the crystalline phase and the preferred orientation of (200) thermodynamically that is shown in Figure 6 (Wang, 1993, Azadi 2014). As shown in Figure 6, the preferred orientation of (200) was not changed either for single or multilayer systems. This result was also seen in other research (Rie, 1995) which reported that all deposited layer systems of TiN/TiC multilayer had a preferred orientation in the (200) direction. Besides, this research showed that chlorine contents of 0.9, 1.5 and 3.7 atomic percent were obtained for the TiC, Ti(C,N) and TiN layers, respectively in multilayer systems.

When the multilayer system was deposited on the substrate, sharp boundaries were not seen between various layers as shown in Figure 7. The SEM image of TiN/TiC multilayer coating, deposited with 6 layers is depicted in Figure 7. This coating was a homogeneous and pinhole-free layer.

Azadi et al. (2014) showed that by increasing the number of layers from 2 to 10 in multilayer coatings, it became possible to increase the grain height and the surface roughness of coatings. In addition, it was found that the particle size of TiN–TiC layers also depended on the N_2 flow rate, varying from 100 nm without N_2 flow to 500 nm at higher N_2 flow rates (Takahashi, 2004).

Life Time and Hardness

In recent years, layers with a gradually changing composition, incorporating metal and ceramic components have been reported for wear protective coatings. The sharp interface between layers raises the layer failure, e.g. by interface crack growth, but the potential of the multilayer coating in the functionally graded materials (FGM) concept is also well investigated (Dahan, 2001). The literature showed that TiN coating caused to increase the life time by a factor of 130, whereas the multilayer of TiN/TiC could raise the life time by a factor of 160 (Pfohl, 1999). Azadi et al. (2014) reported that the high hardness with the

Figure 6. XRD pattern results for various specimens including TiN (a), TiC (b), and multilayer coatings with 2 (c) and 10 (d) layers
Elsevier, 2014.

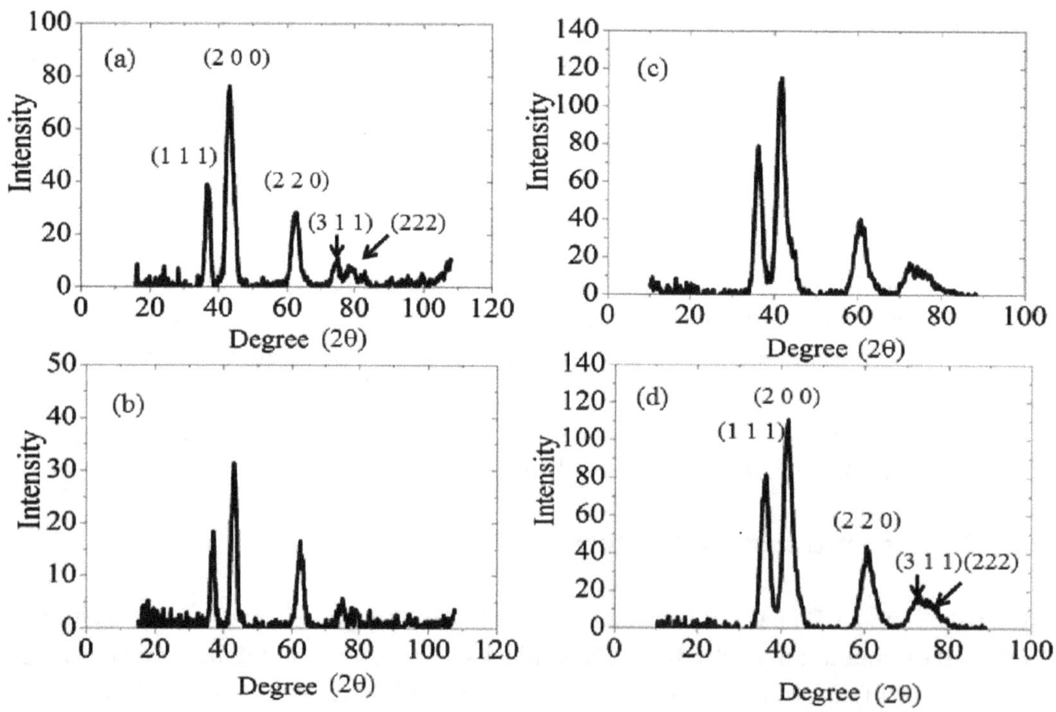

Figure 7. The FESEM image of TiN/TiC multilayer coating, deposited with 6 layers
Elsevier, 2014.

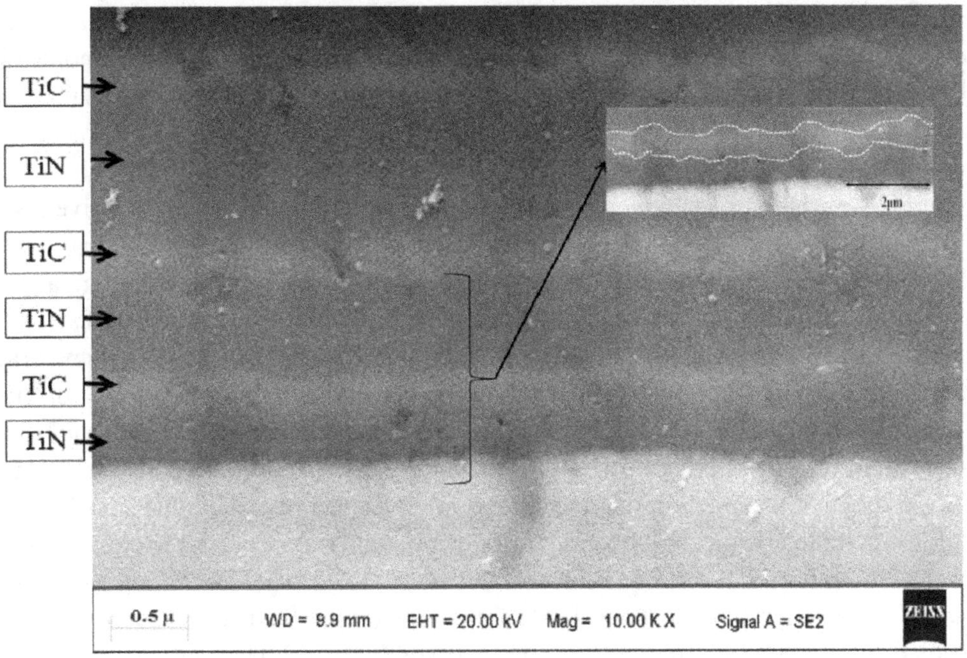

value of about 29 GPa and the high toughness were observed for the TiN/TiC multilayer coating, which had 10 layers. The increase of the hardness of multilayer coatings could be attributed to several interfaces that blocked the dislocation movement across the interface between TiC and TiN layers of multilayer coating. It is noticeable that the hardness of double-layer TiN/TiC coating in comparison to single layer of TiC or TiN coatings was increased about 3 to 31% since barriers for the moving dislocation increased (Azadi, 2014). The highest TiN microhardness values above 3000HV [0.05] were observed for layers which were prepared at 500 °C. The hardness increased up to the maximum value whenever the carbon or silicon was incorporated, to 4200 HV for the titanium carbonitride systems and 3400 HV for (Ti, Si)N_x systems which were deposited at 500 °C (Endler, 1995). Besides, the hardness of the multilayer coating was dependent on the chemical composition of top layer (Park, 1998). The hardness value up to 4100 HK0.01 could be also obtained for layer systems with TiC at the top (Rie, 1995). Consequently, the service lifetime of tools and components were dependent on properties of the near surface layers in the tribological, the corrosive or the mechanical loaded environment. TiC, TiN, Ti(C, N) and multilayer ceramic systems were often considered as the protected coatings of tools due to their high hardness, low friction, high corrosion resistance, good thermal conductivity and high adhesion strength with different substrates. However, the TiN layer could not satisfy the application in specific conditions, especially under extreme conditions because of the relatively low hardness and the high friction coefficient. In order to decrease the friction coefficient and enhance anti-abrasion properties, the TiC layer was considered as the candidate because of incorporation of carbon into the film as a lubrication agent but the high internal stress in the layer often led to spall of layer from the substrate (Zhang, 2013). Thus, Ti(C, N) and multilayer layers are more beneficial to increase the life time of most abrasive tools.

Wear Resistance and Fracture Toughness

When the number of layers in multilayer coating increased, the wear resistance increased and the friction coefficient decreased due to high hardness, high toughness and the surface homogeneity (Azadi, 2014). Mofidi et al. (2014) reported that the fracture toughness of the TiN coating with the thickness of 2 μm increased when the interlayer of Ti was present. For specimens with the lower fracture toughness, the secondary cracking was seen in indentation loads of more than 1250 N. The toughness of TiN/TiC multilayer coating was three times higher than that of the TiN coating. The compressive residual stress was also increased as the number of layers was increased to 10 (Azadi, 2014). The friction coefficient and the volume wear rate of the multilayer composite coatings had the better tribological properties in the vacuum than in the ambient air, especially when compared with those of uncoated steels. Under experimental conditions, the coating wear could be divided into three stages: (a) Voids and crack nucleus formed in the early stage, (b) The coating was pressed up to initiate cracks as a result of plastic deformation of the substrate under the coating. Cracks then propagated under the stress or the heat stress and other influences and (c) The coating broke up in the severe wear condition, and was worn out slowly sheet by sheet in the mild wear condition, depending on the experimental conditions. After the coating had broken down completely, the direct contact between metals was occurred (Renji, 1991).

Zhang et al. (2013) showed that the friction coefficient of TiCN/TiC/TiN coating against the WC ball had the perfect different variation tendencies. At room temperature, the friction coefficient followed the general friction rule from running-in stage to a relatively stable period. The friction coefficient increased from an initial level of about 0.28 to the stable value of about 0.53. However, at 550 °C, the friction coefficient increased sharply from the initial level of about 0.60 to the highest level of approximately 0.7. Then it decreased slowly in the remaining sliding period. In addition, not considerable steady stable stage found in the friction processes at elevated temperature of 550 °C.

Besides, the deposition temperature and flow rates of Ar and CH_4 were two major effective coating parameters for TiC/TiN coatings for wear resistance properties (Wang, 1993). The wear mechanism could be changed from ploughing for Ti(C, N) single layer coatings into cutting for TiC single layer coatings (Azadi, 2016).

DIAMOND LIKE CARBON FILMS (DLC)

The DLC (diamond-like carbon) layers have been studied extensively because of broad applications in various technological areas, especially as passivation layers for fabrication of microelectronic devices in gate dielectrics (Jonas, 2016), when layers were combined with N atoms. Mechanical properties of such coatings are very similar to TiC coating. Whenever it is desirable to create DLC coatings, it is important to know kinetics and the mechanism of the process. This knowledge enables optimization of technological parameters of the depositing process, i.e. obtaining a coating with a proper thickness and the desired structure. However, the appropriately selected doping material, incorporated in specified amounts, allows the DLC layer properties to be modified and provides the new and the desirable layer in particular applications (Batory, 2015).

When the DLC layer combined with other elements, coating properties would be improved more. Batory et al. (2015) investigated mechanical parameters of $C:H:SiO_x$ layers plus DLC layers. The obtained results proved that properties were dependent on various parameters such as the self-bias voltage and the flow ratio of CH_4, the chemical composition and the radio frequency of plasma. Bewilogua et al. (2011) also prepared multilayer coatings consisting of C:H and C:H:Si:O plus C:H and C:H:Si single layers to improve mechanical properties such as the friction coefficient and wear rates. In addition, SiCNH coatings on Ti–6Al–4V (Kyziol, 2014) alloy and Al-Zn alloy (Janus, 2015) surfaces were deposited in order to use in implantology applications. In this situation, having the highest hydrophobicity and the highest hardness are important roles in this particular application.

SUMMARY

The required energy and the reaction temperature for depositing such hard coatings could be reduced from 1000 °C to 450 °C, whenever the plasma was contributed into the reaction. The presence of the plasma had an important role to change the reaction temperature thermodynamically. Results of various researches showed that creating of thin and hard coatings such as TiN and TiC by PACVD established homogeneous and pinhole-free coatings on most substrates with complex shapes. TiC coatings had small grain size with the value of 10 nm and deposited by gas precursors such as the methane or metal- organic materials. Best mechanical properties such as high hardness, wear resistance, fracture toughness and low surface roughness were belonged to the TiC coating that was formed at 450 °C. The challenge in TiN coating was to decrease the chlorine content and other impurities in layers since these two parameters could cause a decreasing effect on mechanical properties like the hardness and the wear resistance. The coatings adhesion with pre-treatment of the substrate such as the plasma nitriding could improve the adhesive strength compared with that without pre-treatment for all coatings. Even a small amount of alloying elements like aluminum or silicon addition to TiN coating had large effects on the improvement in the hardness, the oxidation resistance and the thermal stability of TiN layers. TiC/TiN Multilayer coatings also provided the fine-grained layer structure, with respect to single layers, which mostly revealed the coarse-grained columnar structure. XRD pattern showed the (200) plane as a stable plane for TiN and TiC layers by means of PACVD for both single and multilayer coatings. In multilayer systems, coatings stress decreased comparing with single coatings, which resulted in increasing in the layer adhesive strength. DLC layers with various applications in different industries had the same properties of TiC layers.

CONCLUSION

PACVD offers several advantages in the thin layer deposition and related properties. With the use of metal organic precursors and plasma enhancement deposition temperature can be lowered considerably even to 300 °C. TiN, TiC, DLC and TiN/ TiC multilayer coatings which are deposited by PACVD have various beneficial properties such as high resistance against wear, corrosion, high hardness and fracture toughness and high thermal stability. In this situation these hard coatings are industrially applied for optical, protective, biomedical and other applications. Selective deposition parameters open fascinating prospects for the future, since changing the process factors can enhanced properties of coatings.

REFERENCES

Azadi, M., & Rouhaghdam, A. S. (2014). Nanomechanical properties of TiN/TiC multilayer coatings. *Strength of Materials*, *46*(1), 121–132. doi:10.1007/s11223-014-9523-0

Azadi, M., Rouhaghdam, A. S., & Ahangarani, S. (2013). Properties of TiC Coating by Pulsed DC PACVD. *Journal of Coatings*, *2013*, 1–6. doi:10.1155/2013/712812

Azadi, M., Rouhaghdam, A. S., & Ahangarani, S. (2014). Effect of temperature and gas flux on the mechanical behavior of TiC coating by pulsed DC PECVD. *International Journal of Engineering Transactions B: Application, 27*, 1243-1250.

Azadi, M., Rouhaghdam, A. S., & Ahangarani, S. (2016). A review on titanium nitride and titanium carbide single and multilayer coatings deposited by plasma assisted chemical vapor deposition, *International Journal of Engineering Transactions B: Applications, 29,* 677-687.

Azadi, M., Rouhaghdam, A. S., & Ahangarani, S. (2016). Mechanical Behavior of TiN/TiC-n Multilayer Coatings to Ti(C,N) Multicomponent Coatings by PACVD. *Strength of Materials*, *48*(2), 279–289. doi:10.1007/s11223-016-9763-2

Azadi, M., Rouhaghdam, A. S., Ahangarani, S., & Mofidi, H. H. (2014). Mechanical behavior of TiN/TiC multilayer coatings fabricated by plasma assisted chemical vapor deposition on AISI H13 hot work tool steel. *Surface and Coatings Technology*, *245*, 156–166. doi:10.1016/j.surfcoat.2014.02.055

Azadi, M., Rouhaghdam, A. S., Ahangarani, S., Mofidi, H. H., & Valiei, M. (2014). Effect of number of layers on the toughness of TiN/TiC multilayer coatings. *International Journal of Microstructure and Materials Properties*, *9*(6), 500–516. doi:10.1504/IJMMP.2014.067314

Azadi, M., Rouhaghdam, A. S., Ahangarani, S., Mofidi, H. H., & Valiei, M. (2014). Mechanical behavior and properties of TiN/TiC coating using PACVD. *Advanced Materials Research*, *829*, 476–481. doi:10.4028/www.scientific.net/AMR.829.476

Azushima, A., Tanno, Y., Iwata, H., & Aoki, K. (2008). Coefficients of friction of TiN coatings with preferred grain orientations under dry condition. *Wear*, *265*(7-8), 1017–1022. doi:10.1016/j.wear.2008.02.019

Badisch, E., Fontalvo, G.A., Stoiber, M., Mitterer, C., (2003). Tribological behavior of PACVD TiN coatings in the temperature range up to 500 °C, *Surface and Coatings Technology, 163 –164* 585–590.

Badisch, E., Stoiber, M., Fontalvo, G.A., Mitterer, C., (2003). Low-friction PACVD TiN coatings: influence of Cl-content and testing conditions on the tribological properties, *Surface and Coatings Technology, 174 –175,* 450–454.

Bartsch, K., Leonhardt, A., Langer, U., & Kiinanz, K. (1997). New PACVD-hard material layers for wear protection of high-speed steel. *Surface and Coatings Technology*, *94-95*, 168–173. doi:10.1016/S0257-8972(97)00345-9

Batory, D., Jedrzejczak, A., Szymanski, W., Niedzielski, P., Fijalkowski, M., Louda, P., & Musil, J. et al. (2015). Mechanical characterization of a-C:H:SiOx coatings synthesized using radio-frequency plasma-assisted chemical vapor deposition method. *Thin Solid Films*, *590*, 299–305. doi:10.1016/j.tsf.2015.08.017

Bewilogua, K., Bialuch, I., Ruske, H., & Weigel, K. (2011). Preparation of a-C:H/a-C:H:Si:O and a-C:H/a-C:H:Si multilayer coatings by PACVD. *Surface and Coatings Technology*, *206*(4), 623–629. doi:10.1016/j.surfcoat.2011.02.020

Bull, S. J., Chalker, P. R., Ayres, C. F., & Rickerby, D. S. (1991). The influence of titanium interlayers on the adhesion of titanium nitride coatings obtained by plasma-assisted chemical vapor deposition. *Materials Science and Engineering A*, *139*, 71–78. doi:10.1016/0921-5093(91)90599-I

Dahan, I., Admon, U., Frage, N., Sariel, J., Dariel, M. P., & Moore, J. J. (2001). The development of a functionally graded TiC-Ti multilayer hard coating. *Surface and Coatings Technology*, *137*(2-3), 111–115. doi:10.1016/S0257-8972(00)01122-1

Eckel, M., Hardt, P., & Schmidt, M. (1999). Investigation of TiN deposition in different sized PACVD reactors by means of optical emission spectroscopy. *Surface and Coatings Technology*, *116–119*, 1037–1041. doi:10.1016/S0257-8972(99)00091-2

Endler, I., Wolf, E., Leonhardt, A., & Richter, V. (1995). Preparation, characterization and wear behavior of PACVD cermets. *Surface and Coatings Technology*, *72*(1-2), 37–42. doi:10.1016/0257-8972(94)02338-7

Eroglu, S., & Gallois, B. (1991). Design and chemical vapor deposition of graded TiN/TiC coatings. *Surface and Coatings Technology*, *49*(1-3), 275–278. doi:10.1016/0257-8972(91)90068-8

Gammer, K., Stoiber, M., Wagner, J., Hutter, H., Kullmer, R., & Mitterer, C. (2004). Investigations on the effects of plasma-assisted pre-treatment for PACVD TiN coatings on tool steel. *Thin Solid Films*, *461*, 277–281. doi:10.1016/j.tsf.2004.02.013

Hardt, P., Eckel, M., Schmidt, M., & Wulff, H. (1999). Influence of process parameters on the growth of TiN in different sized plasma-assisted chemical vapor deposition reactors. *Surface and Coatings Technology*, *120–121*, 238–243. doi:10.1016/S0257-8972(99)00496-X

Hilton, M. R., Vandentop, G. J., Salmeron, M., & Somorjai, G. A. (1987). TiN Coatings on M2 Steel produced by PACVD. *Thin Solid Films*, *154*, 377–386. doi:10.1016/0040-6090(87)90380-4

Hu, Y., Li, L., Cai, X., Chen, Q., & Chu, P. K. (2007). Mechanical and tribological properties of TiC/amorphous hydrogenated carbon composite coatings fabricated by D.C magnetron sputtering with and without sample bias. *Diamond and Related Materials*, *16*(1), 181–186. doi:10.1016/j.diamond.2006.05.010

Hyakutake, H., Imada, Y., Honda, F., & Nakajima, K. (1998). Mechanical properties of TiN-TiC mixed crystal films. *Journal of Physics Condensed Matter*, *10*(1), 111–122. doi:10.1088/0953-8984/10/1/013

Janus, M., Kyziol, K., Kluska, S., Konefal-Goral, J., Malek, A., & Jonas, S. (2015). Plasma assisted chemical vapor deposition – Technological design of functional coatings. *Archives of Metallurgy and Materials*, *60*(2), 909–914. doi:10.1515/amm-2015-0228

Jonas, S., Janus, M., Jaglarz, J., & Kyziol, K. (2016). Formation of SixNy(H) and C:N:H layers by Plasma-Assisted Chemical Vapor Deposition method. *Thin Solid Films*, *600*, 162–168. doi:10.1016/j.tsf.2016.01.016

Jonas, S., Kyzioł, K., Lis, J., & Tkacz-Śmiech, K. (2009). Stability of a-C:N:H layers deposited by RF Plasma Enhanced CVD. *Diffusion and Defect Data, Solid State Data. Part B, Solid State Phenomena*, *147-149*, 738–743. doi:10.4028/www.scientific.net/SSP.147-149.738

Kaestner, P., Olfe, J., Hec, J. W., & Rie, K. T. (2001). Improvement in the load-bearing capacity and adhesion of TiC coatings on TiAl6V4 by duplex treatment. *Surface and Coatings Technology*, *142-144*, 928–933. doi:10.1016/S0257-8972(01)01214-2

Kawata, K., Sugimura, H., & Takai, O. (2002). Effects of chlorine on tribological properties of TiN films prepared by pulsed d.c. plasma-enhanced chemical vapor deposition. *Thin Solid Films*, *407*(1-2), 38–44. doi:10.1016/S0040-6090(02)00009-3

Kim, B. J., Kim, Y. C., Lee, D. K., & Lee, J. J. (1999). The effect of NH_3 plasma pre-treatment on the properties of TiN coatings produced by plasma-enhanced chemical vapor deposition (PECVD). *Surface and Coatings Technology*, *111*(1), 56–61. doi:10.1016/S0257-8972(98)00707-5

Kim, D. J., Cho, Y. R., Lee, M. J., Hong, J. M., Kim, Y. K., & Lee, K. H. (1999). Properties of TiN–TiC multilayer coatings using plasma-assisted chemical vapor deposition. *Surface and Coatings Technology*, *116–119*, 906–910. doi:10.1016/S0257-8972(99)00149-8

Kyziol, K., Kaczmarek, L., Brzezinka, G., & Kyziol, A. (2014). Structure, characterization and cytotoxicity study on plasma surface modified Ti–6Al–4V and c-TiAl alloys. *Chemical Engineering Journal*, *240*, 516–526. doi:10.1016/j.cej.2013.10.091

Laimer, J., Stori, H., & Rodhammer, P. (1990). Titanium nitride deposited by PACVD. *Thin Solid Films*, *191*, 77–89. doi:10.1016/0040-6090(90)90276-J

Leoni, M., Scardia, P., Rossi, S., Fedrizzi, L., & Massiani, Y. (1999). (Ti,Cr)N and Ti/TiN PVD coatings on 304 stainless steel substrates: Texture and residual stress. *Thin Solid Films*, *345*(2), 263–269. doi:10.1016/S0040-6090(98)01741-6

Liepack, H., Bartsch, K., Arnold, B., Bauer, H. D., Liu, X., Knupfer, M., & Leonhardt, A. (2004). Characteristics of excess carbon in PACVD TiC-amorphous carbon layers. *Diamond and Related Materials*, *13*(1), 106–110. doi:10.1016/j.diamond.2003.09.007

Liepack, H., Bartsch, K., Bruckner, W., & Leonhardt, A. (2004). Mechanical behavior of PACVD TiC–amorphous carbon composite layers. *Surface and Coatings Technology*, *183*(1), 69–73. doi:10.1016/j.surfcoat.2003.08.083

Ma, S., Li, Y., & Xu, K. (2001). The composite of nitrided steel of H13 and TiN coatings by plasma duplex treatment and the effect of pre-nitriding. *Surface and Coatings Technology*, *137*(2-3), 116–121. doi:10.1016/S0257-8972(00)01073-2

Maiya, P. S., Erck, R. A., & Moon, B. M. (1998). Failure and corrosion resistance of TiN and TiC coatings deposited on graphite by chemical vapor deposition. *Surface and Coatings Technology*, *102*(3), 218–222. doi:10.1016/S0257-8972(97)00690-7

Martinez-Martinez, D., Lopez-Cartes, C., Gago, R. L., Fernandez, A., & Sanchez-Lopez, J. C. (2009). Thermal Stability and Oxidation Resistance of Nanocomposite TiC/a-C Protective Coatings. *Plasma Processes and Polymers*, *6*(S1), 462–467. doi:10.1002/ppap.200931002

Mofidi, H. H., rouhaghdam, A. S., Ahangarani, S., Bozorg, M., Azadi, M., Valiei, M. (2014). Deposition of Titanium Layer on Steel Substrate Using PECVD Method: A Parametric Study. *Materials Sciences and Applications*, *5*, 1–9.

Mofidi, H. H., Rouhaghdam, A. S., Ahangarani, S., Bozorg, M., & Azadi, M. (2014). Fracture toughness of TiN coating as a function of interlayer thickness. *Advanced Materials Research*, *829*, 466–470. doi:10.4028/www.scientific.net/AMR.829.466

Mogensen, K. S., Eskildsen, S. S., Mathiasen, C., & Bottiger, J. (1998). Optical emission spectroscopy on pulsed-DC plasmas used for TiN depositions. *Surface and Coatings Technology*, *102*(1-2), 41–49. doi:10.1016/S0257-8972(97)00529-X

Oguri, K., & Arai, T. (1990). Effect of excess carbon on the hardness of Si-C and TiC coatings formed by plasma-assisted chemical vapor deposition. *Thin Solid Films*, *186*(2), 29–31. doi:10.1016/0040-6090(90)90152-4

Oguri, K., & Arai, T. (1992). Friction coefficient of Si-C, Ti-C and Ge-C coatings with excess carbon formed by plasma-assisted chemical vapor deposition. *Thin Solid Films*, *208*(2), 158–160. doi:10.1016/0040-6090(92)90635-O

Oguri, K., Fujita, H., & Arai, T. (1991). Effect of N_2 to$TiCl_4$ flow rate ratio on the properties of TiN coatings formed by D.C. discharge PACVD. *Thin Solid Films*, *195*, 77–88. doi:10.1016/0040-6090(91)90260-5

Park, I. W., & Kim, K. H. (2002). Coating materials of TiN, Ti–Al–N, and Ti–Si–N by plasma-enhanced chemical vapor deposition for mechanical applications. *Journal of Materials Processing Technology*, *130–131*, 254–259. doi:10.1016/S0924-0136(02)00807-5

Park, J. R., Song, Y. K., Rie, K. T., & Gebauer, A. (1998). Hard coating by plasma-assisted CVD on plasma nitrided stellite. *Surface and Coatings Technology*, *98*(1-3), 1329–1335. doi:10.1016/S0257-8972(97)00264-8

Pei, Y. T., Galvan, D., & De Hosson, J. T. M. (2006). Tribological behavior and thermal stability of TiC/a-C:H nanocomposite coatings. *Journal of Vacuum Science & Technology. A, Vacuum, Surfaces, and Films*, *24*(4), 1448–1454. doi:10.1116/1.2194927

Pfohl, C., Gebauer-Teichmann, A., & Rie, K. T. (1999). Application of wear-resistant PACVD coatings in aluminium diecasting: Economical and ecological aspects. *Surface and Coatings Technology*, *112*(1-3), 347–350. doi:10.1016/S0257-8972(98)00796-8

Renji, Z., Ziwei, L., Zhouping, C., & Qi, S. (1991). Studies on multilayer wear of CVD TiC-TiN multilayer composite coating. *Wear*, *147*(2), 227–251. doi:10.1016/0043-1648(91)90182-T

Rie, K. T., & Gebauer, A. (1991). Plasma-assisted chemical vapor deposition of hard coatings with metallo-organic compounds. *Materials Science and Engineering A*, *139*, 61–66. doi:10.1016/0921-5093(91)90597-G

Rie, K. T., Gebauer, A., Wohle, J., Tonshoff, H. K., & Blawit, C. (1995). Synthesis of TiN/TiCN/TiC layer systems on steel and cermet substrates by PACVD. *Surface and Coatings Technology*, *74*, 375–381. doi:10.1016/0257-8972(95)08243-3

Rie, K. T., & Whole, J. (1991). Spectroscopic investigation of N_2-H_2-Ar-$TiC1_4$-assisted chemical vapor deposition discharge for plasma of TiN. *Materials Science and Engineering A*, *139*, 37–40. doi:10.1016/0921-5093(91)90592-B

Rie, K. T., Wohle, J., & Gebauer, A. (1993). Synthesis of thin coatings by plasma-assisted chemical vapor deposition using metallo-organic compounds as precursors. *Surface and Coatings Technology*, *59*(1-3), 202–206. doi:10.1016/0257-8972(93)90083-Z

Shanaghi, A., Rouhaghdam, A. R. S., Ahangarani, S., Chu, P. K., & Farahani, T. S. (2012). Effects of duty cycle on microstructure and corrosion behavior of tic coatings prepared by dc pulsed plasma CVD. *Applied Surface Science*, *258*(7), 3051–3057. doi:10.1016/j.apsusc.2011.11.036

Son, S. J., Yi, K. W., Mahrholz, J., & Rie, K. T. (2007). Numerical studies on up scaling of metal organic PACVD processes used for tribological coating in automotive industry. *Surface and Coatings Technology*, *201*(16-17), 7318–7326. doi:10.1016/j.surfcoat.2007.01.049

Srinivasan, D., Kulkarni, T. G., & Anand, K. (2007). Thermal stability and high-temperature wear of Ti–TiN and TiN–CrN nanomultilayer coatings under self-mated conditions. *Tribology International*, *40*(2), 266–277. doi:10.1016/j.triboint.2005.09.035

Stockt, H. R., Berndt, H., & Mayr, P. (1991). Plasma-assisted chemical vapor deposition with titanium amides as precursors. *Surface and Coatings Technology*, *46*(1), 15–23. doi:10.1016/0257-8972(91)90146-N

Stoiber, M., Badisch, E., Lugmair, C., Mitterer, C., (2003). Low-friction TiN coatings deposited by PACVD, *Surface and Coatings Technology, 163 –164,* 451-456.

Sun, T., Wu, X., Wang, R., Li, W., & Liu, Q. (2017). First-principles study on the adhesive properties of Al/TiC interfaces: Revisited. *Computational Materials Science*, *126*, 108–120. doi:10.1016/j.commatsci.2016.09.024

Taischner, C., Leonhardt, A., Schonherr, M., Wolf, E., & Henke, J. (1991). Structure and properties of TiCx layers prepared by plasma-assisted chemical vapor deposition methods. *Materials Science and Engineering A*, *139*, 67–70. doi:10.1016/0921-5093(91)90598-H

Takahashi, M., & Shimada, S. (2004). Preparation of composite and compositionally graded TiC–TiN films by liquid injection plasma-enhanced CVD. *Solid State Ionics*, *172*(1-4), 249–252. doi:10.1016/j.ssi.2004.03.015

Taschner, C., Bartsch, K., & Leonhardt, A. (1993). Plasma-assisted deposition of hard material layers from organometallic precursors. *Surface and Coatings Technology*, *59*(1-3), 207–211. doi:10.1016/0257-8972(93)90084-2

Taschner, C., Ljungberg, B., Hoffmann, V., Vogt, C., & Leonhardt, A. (2001). Deposition of TiN, TiC and Ti Al N coatings by pulsed d.c. plasma enhanced chemical vapor deposition methods. *Surface and Coatings Technology*, *142-144*, 823–828. doi:10.1016/S0257-8972(01)01117-3

Thomsen, N. B., Horsewell, A., Mogensen, K. S., Eskildsen, S. S., Mathiasen, C., & Buttiger, J. (1998). Residual stress determination in PECVD TiN coatings by X-ray diffraction: A parametric study. *Thin Solid Films*, *333*(1-2), 50–59. doi:10.1016/S0040-6090(98)00804-9

Wang, H. L., He, J. L., & Hon, M. H. (1993). Sliding wear resistance of TiCN coatings on tool steel made by plasma-enhanced chemical vapor deposition. *Wear*, *169*(2), 195–200. doi:10.1016/0043-1648(93)90298-Z

Yu, H., Tan, T., Wu, W., Tian, C., An, Y., & Sun, F. (2012). Thermal stability of titanium nitride coatings prepared by the mixing technology with laser and plasma. *Current Applied Physics*, *12*(1), 152–154. doi:10.1016/j.cap.2011.05.025

Zabransky, L., Bursikova, V., Daniel, J., Soucek, P., Vasina, P., Dugacek, J., & Perina, V. et al. (2015). Comparative analysis of thermal stability of two different nc-TiC/a-C:H coatings. *Surface and Coatings Technology*, *267*, 32–39. doi:10.1016/j.surfcoat.2015.01.058

Zhang, L., Yang, H., Pang, X., Gao, K., & Volinsky, A. A. (2013). Microstructure, residual stress, and fracture of sputtered TiN films. *Surface and Coatings Technology*, *224*, 120–125. doi:10.1016/j.surfcoat.2013.03.009

Zhu, X., Huang, H., Xu, K., & He, J. (1999). Influence of the interface preparation on properties of PECVD TiN. *Surface and Coatings Technology*, *116–119*, 1029–1032. doi:10.1016/S0257-8972(99)00280-7

Zhu, Y., Wang, W., Jia, X., Akasaka, T., Liao, S., & Watari, F. (2012). Deposition of TiC film on titanium for abrasion resistant implant material by ion-enhanced triode plasma CVD. *Applied Surface Science*, *262*, 156–158. doi:10.1016/j.apsusc.2012.03.152

ADDITIONAL READING

Kaczmarek, L., Sawicki, J., Steglinski, M., Radziszewska, H., Szymanski, W., Kolodziejczyk, L., & Kottfer, D. et al. (2013). The influence of chemical groups on the mechanical properties of SiCNH coatings deposited on 7075 aluminum alloy. *Thin Solid Films*, *534*, 15–21. doi:10.1016/j.tsf.2013.01.055

Kyziol, K., Kluska, S., Janus, M., Sroda, M., Jastrzębski, W., & Kaczmarek, L. (2014). Chemical composition and selected mechanical properties of Al Zn alloy modified in plasma conditions by RF CVD. *Applied Surface Science*, *311*, 33–39. doi:10.1016/j.apsusc.2014.04.193

Kyziol, K., Koper, K., Kaczmarek, L., & Grzesik, Z. (2017). Plasmochemical modification of aluminum-zinc alloys using NH_3 Ar atmosphere with anti-wear coatings deposition. *Materials Chemistry and Physics*, *189*, 198–206. doi:10.1016/j.matchemphys.2016.12.046

Kyziol, K., Koper, K., Sroda, M., Klich, M., & Kaczmarek, L. (2015). Influence of gas mixture during N^+ ion modification under plasma conditions on surface structure and mechanical properties of Al-Zn alloys. *Surface and Coatings Technology*, *278*, 30–37. doi:10.1016/j.surfcoat.2015.07.051

Kyziol, K., Medala, M., & Kaczmarek, L. (2014). Influence of plasmochemical modification of Al-Cu-Mg alloys on surface structure and functional properties. *Vacuum*, *105*, 52–58. doi:10.1016/j.vacuum.2014.02.006

Roach, P., Eglin, D., Rohde, K., & Perry, C. (2007). Modern biomaterials: A review - Bulk properties implications of surface modifications. *Journal of Materials Science*, *18*, 1263–1277. PMID:17443395

KEY TERMS AND DEFINITIONS

AFM: Atomic Force Microscopy (AFM) is a high-resolution type of scanning probe microscopy, which shows the resolution on the order of nanometer.

CVD: Chemical Vapor Deposition (CVD) is a chemical process which is used to produce high quality films, coatings and solid materials.

Fracture Toughness: Fracture Toughness refers to a material property which describes the ability of a material to resist fracture (the nucleation and the growth of cracks). This property has an important role in many design applications for any material.

PACVD: Plasma Assisted Chemical Vapor Deposition (PACVD) is a vacuum based process which is supported by the plasma energy to precipitate several layers from the gas phase. Precursors are contained elements of the layer material.

PVD: Physical Vapor Deposition (PVD) refers to a variety of vacuum deposition methods which are used to deposit thin layer and coatings. Physical reactions are involved in this deposition process.

Thermal Stability: Thermal stability describes the stability of a material against the heat (the high temperature) and its resistance to destruction and decomposition.

Wear Resistance: Wear Resistance refers to a material property which shows the resistance to damage from wear. Wear also describes the loss of material from a surface by means of some mechanical actions.

XRD: X-ray Diffraction (XRD) is a technique which is used for determining the chemical composition of a material. In this technique the atomic or the molecular structure of a crystal is detected since crystalline atoms can cause a beam of incident X-rays to diffract into many special directions.

Chapter 15
Thin Coating Deposition by Magnetron Sputtering

Peter Ifeolu Odetola
Tshwane University of Technology, South Africa

Patricia Popoola
Tshwane University of Technology, South Africa

Philip Oladijo
Botswana International University of Science and Technology, Botswana

ABSTRACT

Advances in thin-film deposition expose new frontiers to structures and phases that are inaccessible by conventional chemical means and have led to innovative modification of existing materials' properties. Thin-film deposition by magnetron sputtering is highly dependent on ion bombardments; coupled with sublimation of solid target unto the substrate through momentum transfer. It is summarily base on phase change of target material under high-energy influence; corresponding controlled condensation of sputtered atoms on substrate material during which process parameters and growth conditions dictate the pace of the atomic scale processes for thin-film formation. Magnetron sputtering is a state-of-the-art thin film deposition technique versatile for several unique applications, especially in the semiconductor industry. Magnetron sputtering is very novel in its use to achieve low-pressure condition that maximizes and conserve stream of electrons for effective knocking of inert atoms into ions. This ensures the high-energy acquired is not dissipated in gas-phase collisions.

INTRODUCTION

This book chapter discusses the basics of different types of sputtering in general and magnetron sputtering in particular as the primary drive for thin film deposition $\left(100nm \leq ThinFilm \leq 10\mu m\right)$. Sputtering is the core of physical vapor deposition (PVD) that shows the cycles of vaporization, transportation, condensation, and nucleation processes used to growing thin films on surfaces. These transformation and phase changes are done in an enabling environment of vacuum under the influence

DOI: 10.4018/978-1-5225-4194-3.ch015

of different process parameters (Thornton, 1974; Alfonso et al., 2012; Gudmundsson et al., 2012; Ponon et al., 2015) such as 1) Power source, 2) Target composition, 3) Target distance to substrate, 4) Pressure, 5) Deposition time, 6) Substrate bias, etc. They determine the course of the entire deposition process from the beginning to the end.

PVD is majorly used to deposit layers of materials by atomic additions in the range of few nanometers to several micrometers (Multilayer). Using PVD technology, material issues, and limitations are easily handle for effective thin film formation as it has a wide spectrum of deposition processes with their specific uniqueness in methods used to generate and deposit material. The common PVD processes (Jameel, 2015; Mattox, 2003, 1998) are 1) Thermal evaporation, 2) Sputtering, 3) Ion plating, 4) Enhanced sputtering, 5) Cathodic arc deposition, 6) Electron beam powder vapor deposition etc.

Despite these diversities, different PVD technologies still utilize the same fundamental steps of vapor-transport-condensation-nucleation approach to grow thin coating. For instance, thermal evaporation uses atomized vapor of coating metal formed by evaporation in a vacuum outfit to coat all the surfaces directly exposed and accessible in the line of sight between the substrate and the target (Mattox, 1998). Sputtering, on the other hand, is also performed in a vacuum environment but uses inert gas and discharge voltage to generate plasma containing a balanced mixture of ionizing inert ions and stream of electrons to vaporize the target material and get them deposited as a thin coat on the substrate. Ion platting is essentially a composite of thermal evaporation and sputtering.

The properties of PVD coatings make them fit for a wide range of applications where corrosion resistance, wear toughness, friction control, and parts lubrication are novel for functionality and efficiency. Because of atomic depositions to form thin films, PVD coatings have high adhesion profile as well as reliable wear and corrosion resistance properties. Application of PVD coatings (Chauhan & Rawal, 2014; Navinsek, 1997; Stefan, 1997) already has a blueprint of satisfaction in aerospace, automotive, surgery/medical, dies, and molds industries. Aside from these, another area of application that has caused a dynamic drive in technology is PVD coatings for semiconductor industry where absolute purity is required.

DEMYSTIFICATION OF PHYSICAL SPUTTERING

From the perspective of definition, sputtering is a physical vapor deposition process of dislodging atoms from a solid target by energetic particle (typically inert Ar^{+}) bombardment. The atoms sputtered from the target are transfer to a strategically placed substrate and become condense there, thereby contributing to forming a thin-film coating.

Sputtering is a unique deposition technique versatile for the formation of thin coating (Chuang et al., 2008) for novel applications and architecture of components and structures. Sputtering can be performed top-down (as shown in Figure one above) giving room for the possibility of advanced processes such as epitaxial growth in the depth of substrate material. The basic building block of sputtering is atomic species of target materials, which build up into a well-formed bonded network of thin films with good adhesion that is not a match with what is obtainable from other conventional coating systems. Nevertheless, the shortcoming of the process is low rates of deposition, overheating of substrate, and insufficient ionization efficiencies. These shortcomings have been overcome through the development of magnetron sputtering system.

Figure 1. Sputtering features and associated parameters

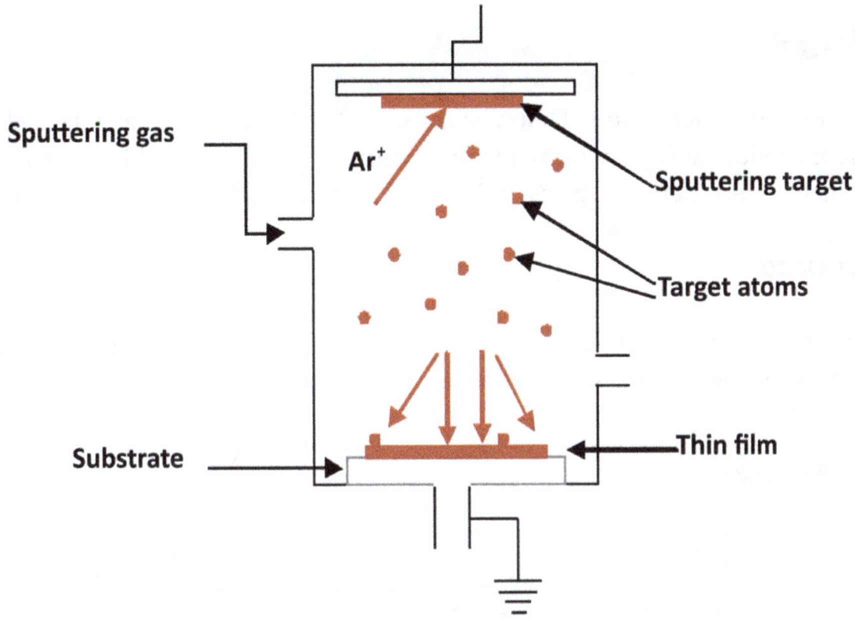

PLASMA TECHNOLOGY

Irving Langmuir, the Nobel laureate who pioneered the scientific study of ionized gases, discovered plasma in 1928. Sputtering is a plasma-based technology. Plasma (Thornton & Greene, 1994; Vasina, 2005; Yagisawa & Makabe, 2006) is the fourth state of matter made up of ionized gas of energy-rich atoms, which collide with one another and knock-off their electrons during the encounter. Plasma also co-exists equal mixture of positively charged inert ions (typically Ar^{+}) and negatively charged electrons to sputter atoms from the surface of the target. Therefore, it is electrically neutral. Electrons in the composition of plasma have the lightest mass compare to positive ions and neutrals, therefore they move much faster. The plasma has two primary functions, first a carrier of inert ions to the target for sputtering and a supplier of ions to aid the growth of films.

Through the years, plasma technology has been the workhorse of continuous development of magnetron sputtering processes to ensure a high degree of metal ionization, improve target usage, increase deposition rates, prevent target poisoning in reactive sputtering, and minimize electrical instabilities such as arcs.

Ionization of sputtered material has a great influence on film characteristics (Lofgran, 2013) such as crystallography, surface morphology, and mechanical properties such as hardness, fatigue, and stress. One of the ways of ensuring a high level of ionization of the sputtered stream of atoms is the application of a negative bias to the substrate.

SPUTTERING PROCESS

Sputtering Target

In sputtering process (Sigmund, 1969; Thompson, 1981; Wehnerand & Anderson, 1970), the target is the coating source; and it is always in a solid form of either low or high melting points. A target material can be pure metal, alloy, ceramic, glass, polymer, semiconductor, composite, or compound based.

Sputtering Sources

The material properties of the target determine the power source (Andersa, 2017; Muhl & Pérez, 2015) that will enable the sputtering process. The two basic power sources are:

Figure 2. Pure metal targets

Figure 3. Schematic diagram of DC and RF sputtering process

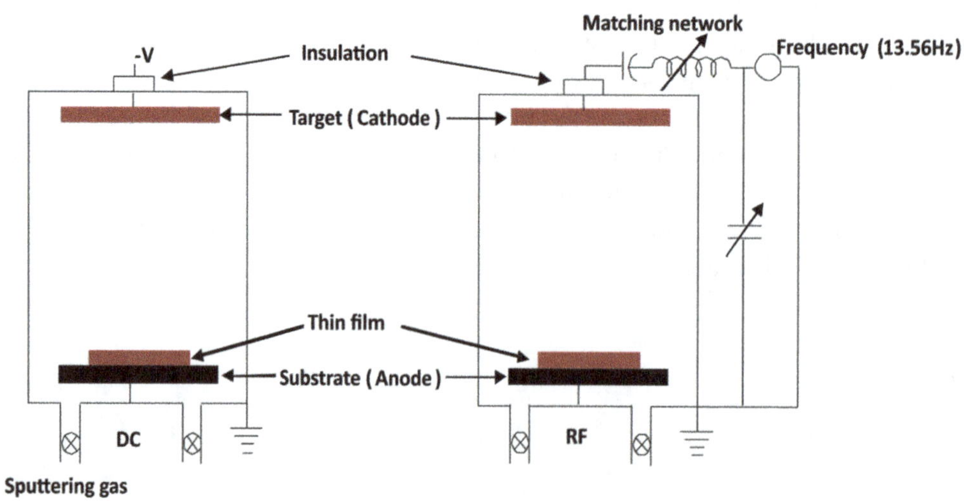

- Direct Current (DC) or Diode source.
- Radio frequency source.

Direct current sputtering is the simplest sputtering process specifically used for metal targets. To generate DC discharge, the cathode has to be electrically conducting material. The insulator cannot be used for DC sputtering due to surface charge that will hinder ion bombardment. In DC sputtering, plasma generated has the capacity to cover a uniform large area of the target with actuation of vaporization.

In DC sputtering, there is a quick drift of electrons from the cathode, which makes them not to be efficiently used to sustain further discharge. Therefore, a magnetron is introduced to confine electrons within the region of the target surface to enhance plasma density and sputtering rate. This often results in improved thin coating deposition.

Radio frequency sputtering is specifically used for sputtering conductors (e.g. metals), nonconductors (e.g. ceramics), and semiconductors (e.g. dielectric materials). As a result, RF is not material limited, and it has a wide scope of target usage. RF sputtering can be adapted to obtain DC-like diode discharge by capacitive coupling it to an electrode as shown in figure 3 above. The RF source comprises two half-cycles. At the first half-cycle, the potential is such that argon ions are accelerated to the target surface with sufficient energy for sputtering atoms from it. In the alternate half-cycle, electrons eventually get to the sputtering sites and prevent any charge accumulation. These alternating cycles make RF suitable for sputtering insulating material, unlike the DC source.

A major disadvantage of RF sputtering of electrical insulating target materials is their inherent properties of poor thermal conductivity, high coefficient of thermal expansion, and brittleness. The plasma generates a huge amount of heat energy that can lead to large thermal gradients and eventual fracture of the target.

Sputtering Gas

Inert gas like argon is specifically used as sputtering gas due to its low reactivity, electrically neutrality, and non-affinity for a chemical reaction. Therefore, the entire process is purely physical. Argon is primarily introduced into the vacuum chamber as background gas to facilitate the ignition of the plasma. As voltage difference is created between the argon gas and the cathode, the argon gas (Ar) become ionized (Ar^+), formed energized plasma, and exhibit a high impact kinetic energy. The ionized argon is propelled under a vacuum condition to sublime and dislodge atoms from the target materials. After impact, these atoms are still carried in plasma stream (Momentum transfer) onto the substrate where they condense and coalesce into films. Sputtering rate depends on the amount of Ar^+ available at the cathode.

Thin-film coatings deposited via sputtering is always a close composition to that of the target materials except in a reactive sputtering where reactive gases like nitrogen or oxygen are introduced to form nitride or oxide compounds of the sputtered materials.

Sputtering Yield

The most prominent parameter for characterizing sputtering process is the yield. In terms of film deposition, sputtering yield (Paruch et al., 2014; Wu et al., 2009; Sigmund, 1969) can be defined as the ratio of the number of target atoms dislodged to the number of inert ions hitting the target.

$$YIELD = \frac{N_{atoms}}{N_{ions}} \tag{1}$$

where N_{atoms} is the number of sputtered atoms, and N_{ions} is the number of incident ion.

Sputtering yield varies; depending on the energy and angle of incidence of bombarding ion particles, relative masses of the bombarding ions and target atoms, and the chemical bond strength of the target surface atoms.

Yield is also a very strong function of the ion energy (Matsunami, 1984; Yamamura & Tawara, 1996; Ziegler et al., 2003). Therefore, if the ion energy is too low within the range of 10-100eV, there will be no sputtering. Thus, yield equals zero. Likewise, in the presence of enormous ion energy above 10keV, the particle penetrates deeper into the target and lose much of their energy (Implantation). Therefore, fewer surface atoms are removed, lowering the sputtering yield. When the ion mass equals the target mass, sputtering yield becomes optimized because this condition maximizes the energy transfer in collisions. There is a need for the ion energy to be optimized between the lower and upper limit of ion energy for effective sputtering to be realized.

Dependence of Sputter Yield on Ion Energy

Sputtering yield is directly proportional to ion energy

$$\left(110eV, 300eV, 600eV, 1000eV, 2000eV\right).$$

The response of materials like

$$\left(Al, Au, Cu, Ni, Pt, Si, Ta, W\right)$$

sputtered with argon ions at various energies is also a function of the atomic weight. Therefore, materials with low atomic weight tend to have higher sputtering yield than those with higher atomic weight. In addition, the sputtering yield of a material depends on the crystal packing which can be single crystal, polycrystalline or amorphous.

The study of sputtering from a single crystal serves as a basis for the theoretical understanding of mechanisms of sputtering. Ions bombardment of single crystals tends to eject atoms preferentially along closed packed directions in the crystal (Alfonso et al., 2012; Sigmund & Jeraj, 2008). In a polycrystalline material (Sigmund, 1969), some crystallographic planes are sputtered faster than the others are. The angular distribution of atoms sputtered from a polycrystalline material can be approximated by a

Figure 4. Ion bombardment of target and characteristic sputtered species

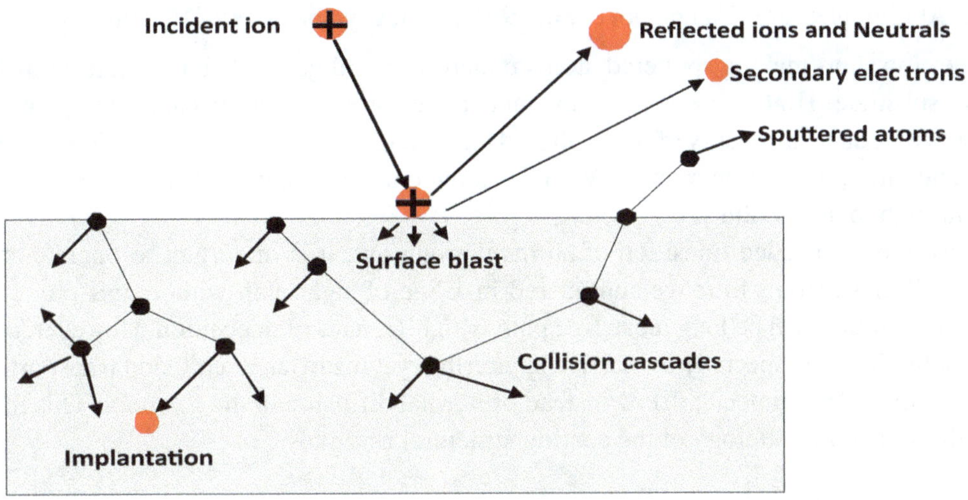

cosine function when the incidence energy is sufficiently above the threshold. The sputtering behavior from polycrystalline targets is similar to the behavior of an amorphous target. In addition, the sputter characteristics of atoms from the alloy's surface are in deposition pattern of their bulk composition ratio.

THEORY OF PARTICLE BOMBARDMENT

Sputtering is energy drive process characterized by collision details of expulsion, propulsion, and deposition of atomized particles. Energy is both needed for target sputtering and to do the momentum transfer of the sputtered particles to the substrate. Inert ions are a carrier of potent energies of several hundred electron volts, which impinge the target and produce collision cascades. If the outermost atom that has been bombarded gains sufficient energy, it will influence other atoms around the surface blast and more collision cascades (Glazov, 2000; Ono et al., 2009; Nakles, 2004) will develop. There can be multiple reflective counter collisions because of action and reaction from the zones of impacts. If the energy gained by a surface atom that is struck from below is enough, it can be physically dislodged from the surface.

During collision cascades, energetic primary ions lose their energy in series of elastic and inelastic collisions with target atoms. The recoiling target atoms, therefore, give rise to secondary ions in the collision encounters. The penetration depth of ions, magnitude of the collision cascade, and a number of surface atoms sputtered are dependent on the mass and energy of ions, and on target's composition and structure. An ion will only dislodge atoms from a target if the energy transferred to the surface atom within the collision cascade exceeds the surface binding energy.

Mean Free Path (MFP)

The deposition rate of films depends on the anode-cathode spacing, which is an equivalence of distance between the target and the substrate. This gap is known as the mean free path. The concept of mean free path (Alfonso et al., 2012; Andersa, 2017; Bhattacharyya et al., 2015; Lofgran, 2013) is used to describe the average distance that atoms can travel without colliding with another gas atom.

The effectiveness of sputtering is dependent on the mean free path, which is inversely proportional to pressure. At high pressure (Van de pol et al., 1991), typically above $1Pa\left(10^{-2}mbar\right)$, the mean free path becomes too short making sputtered atoms undergo several gas collisions resulting in delayed arrival on the substrate. Higher pressure is not innovative for sputtering because gas-phase collisions taking place will reduce the energy of the sputtered atoms and scatter them from a line-of-sight path into the target. Therefore, deposition rates slow down as the energy of sputtered atoms are not thoroughly sustained-through to deposition sites.

To achieve the unimpeded transition of atoms through gas, it is important to operate at low pressures. This will make atoms to move unhindered in a line-of-sight path without gas-phase collisions. The resulting condition will be long mean free path with high rates of deposition. However, this helps to minimize contamination especially in reactive sputtering; yet insufficient collision transport associated with it may result in free molecular flow instead of normal diffusion in the gas pool. This may hamper good formability and morphology of the coating structural assembly.

Ion-Target Interaction

Principally, ion-target (Nakles, 2004; Sigmund & Jeraj, 2008) interaction is characterized by inelastic (electronic energy loss) and elastic (nuclear energy loss) collisions respectively. In the inelastic collision, the ion energy is lost to the electrons in the target material. This results in target ionization and emission of secondary electrons. An elastic collision is the one that actually causes sputtering (Gnaser, 2007). Here, the kinetic energy of the ions is transfer to the stationary atoms. In the presence of sufficient energy, the target gets damaged which eventually caused their atoms to move out of their original sites to another location within or outside the bulk solid giving the possibility of sputtering. Ion reflection and backscattering, electron and ion emission, target damage and heating, atomic sputtering and ion emission are the characteristics features of ion-target interaction.

Modeling the gas phase transport of the sputtering is a rather complex phenomenon owing to several factors (Alfonso et al., 2011; Gudmundsson et al., 2012; Ponon et al., 2015) that the sputtered target atoms depend on; such as the primary ions, target materials and structure, and characteristic behavior of ejected atoms. Factors related to the primary ions include ion mass, ion energy, ion dose rate, and ion angle of incidence. Factors related to target materials and structure are surface binding energy, the atomic weight of target, masses, and fractions of sputtered atoms, crystal orientation, conductivity, and surface curvature. Factors related to the sputtered atoms include their energy distribution, angular distribution, chemical affinity with the target, and clustering.

SPUTTERING MODELS AND MECHANISMS FOR SPUTTERING YIELD

Over the years, a vast number of data has been developed and published on sputtering yield based on analytical models (Kolasinski et al., 2007; Sigmund, 1969, 1981; Saraiva, 2012), semi-empirical formulas (Andersa, 2017; Nakles, 2004; Yamamura & Tawara, 1996), and computer simulation programs (Biersack & Eckstein, 1984; Stoller et al., 2013; Ziegler et al., 2010). Some well-recognized contributors whose findings have made a landmark in sputtering yield are Sigmund, Bohdansky, Yamamura, Wilhelm, Monte-Carlo, Anderson and Bay (1981).

Sigmund reported the most successful model of the sputtering theory, which serves as the foundation for all other findings. Andersen and Bay compiled a complete set of experimental sputtering yields by energetic ions ranging from 0.1 to 100000eV. Yamamura developed semi-empirical equations that led to easier and faster approach to calculating the sputter yield. The other alternative is computer simulation programs that describe the trajectory of the incident particles, the evolution of the cascade processes in the target and the ejection mechanism. In addition, Monte Carlo codes (Stoller et al., 2013; Ziegler et al., 2010) like TRIM (The Transport of Ions in Matter), SRIM, and TRIDYN are adapted to study sputtering yields as well as energy and angular distributions of sputtered atoms. Computer simulations were run at various combinations of ion energy and ion incidence angle. TRIM did not prove to be an adequate model for low-energy sputtering.

Sputtering yield can also be determined experimentally by using both Rutherford backscattering spectrometry (RBS) method, and mass loss methods. In the RBS method, the total sputtering yield was calculated from the integral of differential sputtering yields. Using the mass-loss method, sputtering yield was found by mass-loss technique by using a microbalance to compare the mass of the sample before and after sputtering.

Sigmund's Approach to Sputtering Yield

In 1969, Sigmund pioneered a model successfully used to describe sputter phenomenon is called "Linear Collision Cascade (LCC) model" (Sigmund, 1969). According to this model, energy and momentum are transferred to the target through a series of binary collision cascade. Initially between primary ions and target atoms, and then between recoiled atoms and stationary atoms. Thus, each impinging ion during its collision path can dislodge target atoms that in turn will begin their own collision path, with a probability of displacing several atoms. A 'collision cascade' is generated into the target bulk. Sputtering finally happens after atoms near the surface got enough energy to overcome the sublimation energy of the solid. The primary aim of LCC model is to give quantitative predictions of sputtering yield as a function of ion energy, the angle of incidence, mass and an atomic number of both ions and target atoms.

Sigmund formulated a theory that focuses on the microscopic processes associated with the bulk of the bombarded target material under the assumption of random slowing down throughout the collision cascade in an infinite medium (Yamamura & Tawara, 1996; Yamamura et al., 1983). In this way, Sigmund was able to calculate the sputtering yield of a polycrystalline target (Amorphous material). Sigmund also derives a general expression for the yield as a function of cross-sections of ion and target as well as atomic binding energies both in the bulk of the target and on the surface of a material. One of the important outputs of Sigmund is the deposited energy distribution, which is a function of ion energy and angle of incidence normal to the surface.

A theoretical expression of the sputter yield of amorphous and a polycrystalline target by Sigmund is represented as:

$$Y = \frac{3\alpha}{4\pi^2} \frac{4M_1M_2}{\left(M_1M_2\right)^2} \frac{E_{ion}}{U_{sub}} \tag{2}$$

where Y is the sputtering yield, M_1 is atomic mass of the impinging particle (ion), M_2 is atomic mass of the target atoms, U_{sub} is heat of sublimation of the target material, E_{ion} is the kinetic energy of the incident ion, α is a dimensionless correction factor that depends on mass ratio of target to ion and whose value varies between 0.2 and 1.4.

Thus, from Eq. (2) it is seen that the sputter yield is directly proportional to the ion energy and is inversely proportional to the surface bonding energy of the target material expressed by the sublimation energy. This model has been examined experimentally and found to be in reasonable agreement for low energy ions on various mono-atomic targets.

Sigmund's theory is re-known for success in various aspects of ion sputtering, notwithstanding it has the following notable limitations:

- The theory was designed for polycrystalline materials, and may not provide a reliable description of the crystalline targets.
- The assumption of random slowing down throughout the collision cascade of an infinite medium works satisfactory only at intermediate and high energies and not at low energies.
- The surface binding energy brings in a number of uncertainties like ion erosion that changes the local surface chemistry of near surfaces. Film properties are largely influenced by the energies of depositing atoms.

The target material is known to be available in elemental (Pure metal), compound or composite form. Generally, the sputtering yields of compound materials are lower than that of pure metals. The chemical bonding of the compound materials is stronger than that of pure metals due to their high sublimation energy (Alfonso et al., 2011; Sigmund & Jeraj, 2008). Therefore, the sputtering yield will initially be low until the surface layer is removed.

Semi-Empirical Models of Sputtering

An easier and faster approach to calculate the sputter yield is with semi-empirical formulae (Andersa, 2017; Nakles, 2004; Yamamura & Tawara, 1996) pioneered by Yamamura et al. It is base on Sigmund theory (Sigmund, 1969); uses the Ziegler-Biersack-Littmark (ZBL) screened repulsive potential and include the Lindhard electronic stopping power (Ziegler et al., 1985). The semi-empirical model also includes correction factors for sputtering ion at low energy by including the sputtering threshold energy (Eckstein et al., 2004).

The equations generated from the studies of Yamamura et al. are valid for monatomic solids bombarded at a normal incident angle, and the sputtering yield is giving as:

$$Y\left(E\right) = \frac{0.042\alpha\left(M_2/M_1\right)}{U_S}\frac{Q\left(Z_2\right)S_n\left(E\right)}{1+\Gamma\mathrm{K}\varepsilon^{0.3}}\left[1-\sqrt{E_{th}/E}\right]^s \quad (3)$$

where E is the kinetic energy of the incident ion (projectile), E_{th} is the threshold energy for sputtering, U_S is the surface binding energy, M_1 and M_2 are the atomic masses of the ion and target atom in

A.M.U., respectively, and the numerical factor in units of $\mathring{A}^{-2}$, K is the linhard electronic stopping coefficient, ε is the Linhard-Scharff-Schiott reduced energy, $S_n(E)$ is the collision energy at the surface (nuclear energy loss function), α is a dimensionless coefficient depending on the mass ratio and the ion energy. Q and α are (fitting) parameters as defined in Yamamura and Tawara (1996).

Q Values are the average deviations of best fit of α. The best-fit values α are described as a function of the mass ratio M_2/M_1 in the following way:

$$\text{If } M_1 \leq M_2\text{, then } \alpha = 0.249\left(M_2/M_1\right)^{0.56} + 0.0035\left(M_2/M_1\right)^{1.5} \tag{4}$$

$$\text{If } M_1 \geq M_2\text{, then } \alpha = 0.0875\left(M_2/M_1\right)^{-0.15} + 0.165\left(M_2/M_1\right) \tag{5}$$

The Γ factor now has the form

$$\Gamma = \frac{W\left(Z_2\right)}{1+\left(M_1/7\right)^3} \tag{6}$$

In the previous empirical formula, the factor Γ corresponded to $0.35U_S$, and the power S was fixed at 2.8 for all target materials. In contrast, the S of the new empirical formula is slightly dependent on the target material.

The sputtering threshold $\left(E_{th}\right)$ shown in the semi-empirical formulae dictates the minimum kinetic energy of the bombarding particle for sputtering to occur. A bombarding particle must have a kinetic energy greater than the sputtering threshold for sputtering to occur.

Moreover, the Thompson model (Thompson, 2010), developed from collision cascade theory and is able to provide an empirical formula of the sputtered atom energy distribution function. Sputtered atoms typically have a few electron volts (eV) of kinetic energy due to the ballistic processes in the collision cascade. The energy distribution function depends on the target material because it is essentially determined by the surface binding energy.

$$f_{Thompson}\left(E\right)\alpha\frac{1-\left(E_{SB}+E\Big/\gamma E_i\right)^{1/2}}{E^2\left(1+E_{SB}\Big/E\right)^3} \tag{7}$$

where E_{SB} is the surface binding (cohesive) energy of the target atoms, E is the energy of the ejected atoms, E_i is the kinetic energy of the impinging ions, γ is the kinetic energy mass transfer factor in an elastic collision which is further represented as:

$$\gamma = \frac{m_g m_s}{\left(m_g + m_s\right)^2} \tag{8}$$

where m_s and m_g are the target sputtered atom mass and the gas phase atom mass respectively.

The other alternative is simulations that describe the trajectory of the incident particles, the evolution of the cascade processes in the target and the ejection mechanism. As an example, binary collision Monte Carlo codes like SRIM, and TRIDYN, are available to simulate the sputtering process and through this calculate the sputter yield.

Monte Carlo Simulation Model

A Simulation program is an important and necessary tool to gain a better understanding of the sputtering process in terms of handling the multi-parameters associated with ion-target interactions as well as their control, and optimization. Simulations save both time and money from massive trials and error of experiments.

The overall transport of sputtered atoms is done in the gaseous phase. Monte Carlo Simulation Code computes range profiles of implanted ions, target composition and sputtering rate (Kroese et al, 2011; Brault et al., 2016). Composition changes can be due to the spatial distribution of target atoms deposited in collision cascades, and due to the implanted ions. A comprehensive experimental data for ion-target stopping powers are available at www.srim.org for detail references.

Sputtering yield is already known as one of the most important parameters for analyzing ion-target interaction. Therefore, for description of the sputtering process and estimation of the sputtering yield, a number of advanced simulation programs (Ziegler et al., 2003; Ziegler et al., 2008; & Brault et al., 2016) have been developed in recent years such as: Stopping and Range of Ions in Matter (SRIM), Transport of Ions in Matter (TRIM), and TRIDYN (Moeller et al., 2002). TRIDYN is a TRIM simulation code capable of calculating the dynamic composition changes of the target caused by implanted ions during sputter process.

SRIM is a package in Monte Carlo Simulation software, which is used to give the particle its initial direction and energy. Therefore, the software program probes and monitor each particle trajectory through the chamber and calculates its new position, direction, and energy in every collision encounter until it hits a surface that will finally bring it to rest. All surfaces in the simulation chamber contain several mesh elements that record the number of condensing particles on them calculating their deposition fluxes as well. These are plotted on a graph using a color bar legend for visualization purposes.

MAGNETRON SPUTTERING

Sputtering process using an electron trapping principle is called magnetron sputtering. The advent of magnetron sputtering has opened up a wide prospect for thin film coating applications uncommon with other deposition processes (Andersa, 2017; Brauer et al., 2010; Kelly & Arnell, 2000; Rossnagel, 1990). It is an upgrade to compliment inadequacies of other vacuum coating techniques such as thermal (Wasa et al., 2004) and electron-beam evaporations. Specific distinguishing features of magnetron sputtering are in the following areas:

- **Effective Application of Magnetron:** This device uses a magnetic field to control the motion and path of electrons. The magnetron comprises an assembly of a magnet with a magnetically enhanced glow discharge to effectively trap and utilize energetic electrons for ionization processes. These confined electrons create more ions from the neutral background gas, which are propelled to the target and cause discharge of secondary electrons [Sheridan et al., 1990]. The secondary electrons are used for sustaining the discharge and to compliment for the electrons lost from the trap through neutral collisions.
- **The Novelty of Relative Ease to Form Plasma Under Low-Pressure Application:** This is unique for sustaining high ion energies, and to prevent multiple atom-gas collisions from the target.
- **Different Magnetron Configurations Possible (Planar, Balanced, Unbalanced, or Closed Field Arrangements):** These determine how the electrons will be released, restricted or trapped in the confine of the vacuum system (Rastogi et al., 1987; Waits, 1978) around the cathode.

Advantages of Magnetron Sputtering

- Magnetron sputtering can be powered by both direct current (DC) and radio frequency (RF) sources. Therefore, there is ease of sputtering wide range of materials such as metals, alloys, ceramics, or compounds.
- Easy control of the composition and stoichiometry of deposition layers permit good coating structure and morphology.
- Deposition process at the atomic level with high adhesion thin films' profile on substrate materials is feasible.
- High purity films possible with precise material properties are obtainable.
- There is a possibility of excellent coverage of steps and small features.
- It allows for ease of automation.
- It offers the ability to coat heat-sensitive substrates.

Major Drawback of the Magnetron Sputtering

It is noteworthy that though RF can sputter insulators and alloys, yet have challenges in depositing under thermal conditions because of the high melting point of ceramics and varied vapor pressure of alloy constituents. Therefore, there is need to operate the power source at elevated temperature, which may affect coating morphology.

In addition, plasma confined near the cathode for sputtering is not available for effective momentum transfer of the sputtered atoms. This difficulty can be overcome with an unbalanced magnetron configuration, where the magnetic field permits the escape of some electrons from the cathode region. The major shortcoming of the unbalance magnetron is that the current created by the escaping electrons as well as the plasma generated is non-uniform.

Configuration of Magnetrons

Several types of magnetron cathodes have been developed which include planar, cylindrical, and circular magnetrons such as sputter-gun. These sources have their own strengths and limitations. For instance, planar magnetron setting permits the use of flat sheet or disk of a material, but under-utilized target due to a very small area. This is a great disadvantage especially in case of expensive target materials. In the case of cylindrical and circular magnetrons, specially shaped targets that may not be easily available are required. In addition to the configuration restrictions, magnetic target materials have limited use with these magnetron sources. The magnetic field lines are confined within the target material unless very thin targets are used where magnetic saturation could be achieved.

THIN FILMS FORMATION

Growth Features and Atomic Processes

Generally, deposition sites are characterized by some specific features (Over, 1998; Ploch et al., 2012; Zhang & Lagally, 1997). First is "terrace", known as substrate surface where films are grown from an isolated atom called "ad-atom" into a group of atoms known as "island" as the deposition process proceeds.

Ad-atom mobility (diffusion) on a terrace (flat surface) is by far the most important kinetic process in film growth. It is also known to be a key parameter controlling the surface morphology (Ala-Nissila et al., 2002; Flototto et al., 2012; Luth, 2001). An arriving ad-atom either diffuses randomly on terrace trying to find proper sites to come to rest until it meets another ad-atom to form a stable nucleus (nucleation) or it meets an existing island and sticks to it (growth). At a certain energy level, the diffused atoms cannot move past the surface barrier of a given terrace. Therefore, another terrace layer is activated at a lower energy level for deposition of another film phase. This process continued until the last layer form.

Figure 5. Different growth modes during film formation

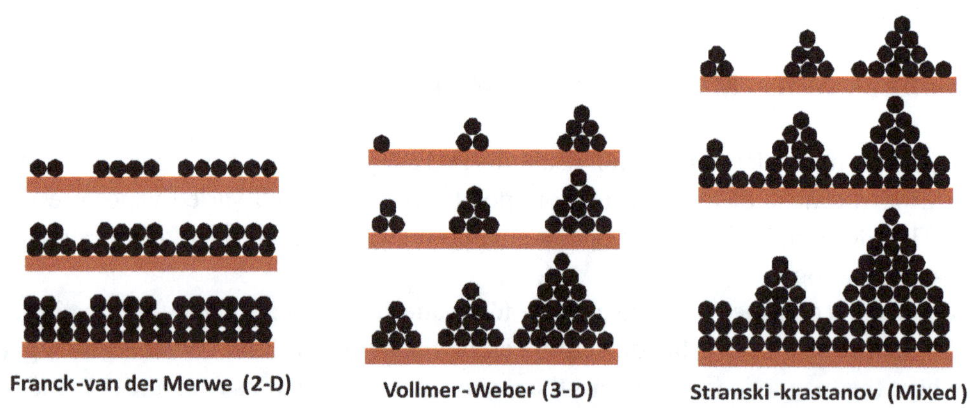

Growth Modes

The deposition energy of an ad-atom is used to form and break a bond with neighboring atoms (Lagally & Zhang, 2002). Nucleation on the substrate depends on the free energy of the ad-atom (Alfonso et al., 2012; Merikoski et al., 1997) and can be group into three growth modes

- **Franck-Van der Merwe Growth Mode:** This is a 2-D or layer-by-layer growth mode whereby a new layer starts to grow only after the preceding one is finished. The ad-atoms fill up into suitable sites as a result of strong adhesion affinity for the substrate atoms until stairs of layers are formed. This mode usually generates a smooth surface because of the long diffusion length of the ad-atoms.
- **Vollmer-Weber Growth Mode:** This is 3-D or an island growth mode whereby the deposited atoms cannot diffuse past the island boundaries. Here, the ad-atoms have a strong cohesive affinity for one another and the 3-D islands grow, interlock, and coalesce in all direction forming network of films. This mode usually produces many isolated islands because of the short diffusion length of the ad-atoms. Therefore, resulting into high dislocation density and rough surface morphology.
- **Stranski-Krastanov Growth Mode:** This is a mode that starts with layer-by-layer growth followed by island growth after one or more monolayer. It is a combination of layer and island growth mode.

Models for Thin Films Growth

The onset of thin film deposition is with condensation drive mechanism of atomic species on the substrate, which is characterized with atomic processes such as impingement, absorption, adsorption, desorption, and aggregation (Flototto et al., 2012; Luth, 2001; Tringides, 1997; Venables, 2000).

Therefore, in-depth understanding of the correlation between these atomic processes and deposition process determine the fundamental basis for the structural evolution of thin coatings during all stages of the growth process. In general, the intrinsic properties of the thin film formed that will dictate the physical properties. Deposition conditions are key factors to crystal growth. The energy profile of the condensed atoms dictates growth mechanism of a thin film. The growth is in direction of lower energy level.

Many models have been established as a guide to an in-depth growth analysis during film growth. Terrace Step Kink (TSK) model is one of the well-established models unique for giving systematic detail of crystal growth at the atomic level. TSK model was pioneered in the early 1950s by Burton, Cabrera and Frank (1951); and this serves as background for many modern technologies to grow thin films. Initially, the study of thin film growth at atomic level was not possible until after the discovery of terrace step kink model. In this model, thermodynamics concepts are used to analyze the initiation, formation, and growth of crystals in a thin coating deposition.

In growing crystal or film, there is a transition barrier to transport atoms between terraces known as Ehrlich-Schwoebel (ES) barrier (Ehrlich & Hudda, 1966; Liu et al., 2002; Schwoebel & Shipsey, 1966). Because of this, the chances for nucleation and growth of a new film layer as more atoms arrive the terrace become possible.

Liu et al. (2002) as rules that govern atoms' transition from one layer to another as films or crystals growth were studied, further stretched the basic understanding of TSK model. In 1991, Villain showed that an ES barrier (which is a two-dimensional phenomenon) produces an effective climbing gradient, leading to unsteady growth and the formation of roughness in the film - an undesirable feature. Therefore, the two-dimensional ES barrier dictates the three-dimensional morphological evolution of growing films. Nevertheless, the concept of the ES barrier also extends to other dimensions.

A step is a link that separates one terrace from another, while a kink is a 'step on a step' representing one lattice unit of lateral displacement of an original straight step (Liu et al., 1991; Pierre-louis et al., 1999; Schwoebel, 1966; Zhong et al., 2001). Villain described that an atom moving along a step edge would encounter a barrier preventing it from crossing over the kink as a result of reduced coordination number (Lagally & Zhang, 2002). The greatest restriction across the kinks occurs at the corner of an island (A two-dimensional crystal) where two-step edges meet. This corner-crossing barrier can induce surface deformities because of growing instability in the morphology of a 3-D film network. If a number of identical terraces were stacked on top of each other, a dividing line separating two crystal facets would be mapped out. According to Liu et al. (2002), there is a resultant increase in height during stacking from transitions over a step edge between adjacent terraces; and this will make the two-dimensional barrier becomes three-dimensional.

Structure Zone Model (SZM)

Another important model for analysis of the thin film is the "Structure Zone Model" which summarizes the relationship between the substrate temperature, the kinetic energy of the sputtered atoms, and their deposition rate. These determine the overall microstructure of a thin coating formed.

Structure formation is the last stage in the sputtering process and is greatly influenced by atomic processes as well as growth conditions. Atomic processes involved are absorption, desorption, nucleation, and growth. The growth conditions are the partial gas pressure, the deposition rate, impurity concentrations, substrate's temperature and bias that actively govern the growth pattern of the film (Flototto et al., 2012).

The microstructure of thin films is highly dependent on the mobility of ad-atoms during growth. The growing atoms are energized from substrate's chemical reactions, thermal effect, and ionic bombardment. The structure zone model (Messier et al., 1984; Thompson, 2000; Petrov et al., 2003; Yam et al., 2011) explains the overall effects produced by these reaction mechanisms. In regard to ad-atoms' mobility, the SZM can determine the morphology and microstructure inherent in any kind of material. The working parameters for analyzing the microstructure of the films using SZM are the final working pressure, target's thermal properties, substrate's temperature and its bias voltage.

According to the research carried out by Movchan and Demchishin (1969), it was established that the microstructure from films of some metals and metal oxides $\left(Ti, Ni, Al_2O_3, ZrO_2\right)$ is related to the normalized temperature with three well define zones.

Zone 1: The controlling parameters here are low temperature $\left(T_S/T_M < 0.3\right)$ and high pressure. Small-elongated grains that form a columnar structure with porous morphology, loose fibrous grain, and voids characterize it. The columnar structure results from the low surface mobility of the atoms adsorbed at substrate's surface.

Zone 2: This region depicts a growing temperature of the substrate with respect to melting temperature of the target in the range $\left(0.3 \le T_S/T_M \le 0.45\right)$. Ad-atoms will experience higher mobility due to more thermal energy and will produce a fully dense columnar grain structure with a higher degree of binding.

Zone 3: This region is characterized by higher temperature $\left(T_S/T_M > 0.45\right)$ as well as enhanced capacity for surface diffusion of ad-atoms that has a great influence on the morphology of the film. There is a transition from columnar grain to equiaxed grain because of grain growth and recrystallization.

Zone T: This is transition region between first and second zones characterized by low-temperature $\left(T_S/T_M \le 0.3\right)$ and low pressure. The zone is formed from grains defined by limits of the low porosity. The surfaces of this zone are more tightly packed (denser) but not fully dense. Therefore, the film is like a fibrous structure.

According to Thornton (1974), the structure zone classification was elaborated considering the final working pressure. This parameter can change both the kinetic energy of the ions that impinge on the surface of a substrate and the mean free path of the sputtered particles. Therefore, it controls surface bombardment of the substrate by dictating the mobility of ad-atoms in that surface.

THIN FILM APPLICATIONS

Thin-film deposition (Bunshah, 1994; Mattox, 1998; Wasa et al., 2004) is novel for producing many functional coatings with exceptional properties and has been integrated into architectures for various operations considered under the subsections.

Electronic Components

This area of application offers the widest and most demanding applications for thin film usage. Innovative fabrication of electronic components especially solid-state devices and microelectronic integrated circuits technologies are effectively validated through thin-film deposition.

- **Areas of Applications:** Semiconductors (solid-state device e.g. flash memory and computer chips), micro-electronic integrated circuits, dielectrics, insulators, conductors, barriers.

Album for thin film materials under electronics (Singh and Shimakawa, 2003; Bindu et al., 2005).

- **Semiconductors:** $Ge, Si, Te, SiC, ZnO, ZnSe, PbS, PbO_2, GaAs, GaP, GaN$
- **Insulators:** $Si_3N_4, Al_2O_3, SiO_2, Ta_2O_5$
- **Resistors:** $Cr, Ta, \mathrm{Re}, TaN, TiN, NiCr, SiCr, SnO_2, In_2O_3$
- **Dielectrics:** $AlN, PbO, MgO, Nb_2O_5, ZrO_2, BaTiO_3, LiNbO_3, PbTiO_3, ZnS, Al_2O_3,$
    - $BeO, SiO, SiO_2, TiO_2, Ta_2O_5, HfO_2, Si_3N_4$
- **Magnetics:** $Fe, Co, Ni - Fe, GdCo$

- **Superconductors:** $Nb, NbN, Nb_3Sn, Nb_3Ge, La - Sr - Cu - O, Y - Ba - Cu - O$
- **Electrodes/Interconnections:** $Au, Al, Cu, Cr, Ti, Pt, Mo, W, Al/Si, Pt/Si, Mo/Si$

Electronic Displays

Thin-film deposition is used to fabricate coating layers with a diversity of display output (Bai et al, 2014) or response to an interface, medium, or systems depending on the characteristics of the films such as; conductivity, luminosity, fluorescence, transparency, insulation, and dielectric. Electronic displays are primarily utilized for interfacing electronic devices with human operators.

- **Areas of Applications:** Liquid-crystal displays (LCDs), Light emitting diodes (LEDs), Electro-luminescent displays (ELDs), Projectors, an organic light emitting diode (OLED), Electrochromic displays, Plasma, and fluorescent displays.

Optical Coatings

Optical coatings (Friz, 1992; Cowell, 1996; Martin et al., 1997; Macleod, 1999; Lee et al., 2011) consist at least one thin layer of material deposited on an optical component to control the reflection and transmission properties of light within an optical system through the mechanism of interference. This improves the efficiency of the system since a limited amount of light is lost. The viability of optical coatings in applications is largely dependent on the following principal factors: wavelength of light being used, the number of layers, individual film's thickness, and on the difference in their refractive index at the layer interfaces (optical properties of metal film).

- **Areas of Applications:** Optical polarizer, beam splitter, antireflection (Hellmich & Deimel, 1992; Macleod & Hobbs, 2004), interference filters on solar panels, plate glass infrared solar reflectors, laser optics, fabrication of filter optics, infrared reflecting coatings, photochromic coatings. The common thin coating materials in optics are: $SiO_2, TiO_2, SnO_2, In_2O_3$

For instance, an antireflective coating (AR) is a type of optical coating widely employed to reduce light reflection that hampers performance in lenses found in optical instruments like cameras, telescope, and digital video disk (DVD) players. On the other hand, it is noteworthy that reflective optical coatings are mirror-like. They are usually made from high lustrous materials like aluminum, gold or silver, and are used on photocopying machines, scanners, barcode, and high-powered lasers.

Magnetic Films for Data Storage

Through thin film coatings, magnetic materials (Kryder, 1992) have found wide commercial applications in data capture devices and control systems as a result of improved wear resistant overcoats for magnetic media; higher coercive multi-layer thin film media with isolated grain structure to achieve low noise; high magnetization, high permeability multilayer thin film materials for advanced recording heads; giant magneto-resistive materials for advanced read heads.

The present generation of magnetic optic disk drives utilizes thin films of ternary amorphous rare earth-transition metal alloy ($R - Fe - Co$ where R is a rare earth element). The substrates can be metal, glass or polymeric materials. Other traditional examples of magnetic films for data storage are

$$NiP, Ni - Fe, Al - Fe - Si, Cu/(Ni - Co), Co - Zr - Nb, Fe - Co - Si - B.$$

Data Storage Devices

Thin films deposition is finding increasing commercial use for optical data storage devices (a high-capacity storage medium e.g. compact disks, hard disk, floppy disk), electronic data storage devices (required electrical power to store and retrieve data e.g. computer chips), magnetic data storage devices (disk drive and magnetic-core memory), solid-state storage devices (flash drive and removable disk), and computer memory applications. Processes for the deposition of organic polymer materials as storage media and as protective overcoats are required for this technology.

A data storage device is a device for storing (recording) data (information). A storage device may hold information, process information or serves the dual functions. It serves as recording medium when it only holds information. On the other hand, it serves as data storage equipment when it can process, access, store or retrieve information from portable (removable) medium or from a permanent component.

Antistatic Coatings

Thin films of an electrically conductive coating are deposited to reduce or prevent a buildup of static-electrical charge and its effects (Shang et al., 2012). The coatings cover a conductive range, which is suitable for dust control, surface shock absorber, and electrostatic painting.

Hard Surface Coatings

These coatings are adhesion providers, wear resistors and friction reducers (Fu et al, 2003; Warcholinski, 2009). They are widely applied on cutting tools (e.g. drill bits), forming tools (e.g. dies), fasteners (e.g. bolts and nuts), injection molding tools, bearings, and machine parts to improve the wear resistance of metal surfaces and service life. Hard surface coatings are generally obtained from metal nitrides, carbides, silicides, and borides ranging from single phase (Cr, Wc, Ti, Ta, W), binary phase

$$(TiN, Zr_2N, TiC, SiC, Zr_2B, TiNi, CrN),$$

and ternary $(TiCN, TiAlN)$ to quaternary $(TiAlCrN, TiNiCrN)$.

Corrosion Resistant Films

This category of thin film coating is used for surface modification, enhancement and barrier purposes (Bouzid et al., 2015; Pillis et al., 2016). They inhibit the formation of corrosion products by acting as a barrier between materials and their environment. For example, in fuel cell system, the metal interconnects of the bipolar plates has an inherent corrosion problem that leads to the development of resistive surface oxides, performance deficiency, and insufficient power output. With the use of novel corrosion resistant coatings, the challenges that had lingered the commercialization of fuel-cell systems can be curbed.

CONCLUSION

In conclusion, material science and engineering have experienced tremendous growth, innovation, and novelty in the field of thin coating deposition over the years through powder vapor technology; of which magnetron sputtering is the key deposition process, well tested and proved for their coating integrity.

The fundamental steps of vapor-transport-condensation-nucleation approach for development of thin coatings of high adhesion profile, increase oxidation resistance, reliable hardness, and good wear resistance, as well as versatility of process parameters' control of the atomic processes and growth conditions, have opened a wide spectrum of getting coatings of several existing materials deposited for novel application.

With magnetron sputtering, so many limitations from the traditional coating systems have been overcome; and the technology now makes it possible to coat materials with reduced operating pressures, low voltage applications, and higher ionization of sputtering ions that improve the overall efficiency of a sputtering process. In addition to this are the different magnetron configurations possible that have enabled ease of fabrication of thin-film coatings.

Established models, mechanisms and calculations that provide a pathway of a better understanding of analysis of the entire process of thin film formation govern all the sputtering of target atoms, the momentum transfers of the sputtered atoms, and the resulting evolution of thin films. Many of these models and mechanism are still work in progress especially in the area of sputtering simulation of the process parameters, growth conditions and the atomic scale processes involved.

Therefore, this book chapter has offered clarity and understanding of necessary details of researches and contributions for further development in the state-of-the-art technology of thin film formation through powder vapor deposition.

REFERENCES

Ala-Nissila, T. F., Ferrando, R., & Ying, S. C. (2002). Collective and single particle. *Advances in Physics*, *51*(3), 949–1078. doi:10.1080/00018730110107902

Alfonso, E. O. (2012). *Thin Film Growth Through Sputtering Technique and Its Applications*. InTech. doi:10.5772/35844

Andersa, A. (2017). Tutorial: Reactive high power impulse magnetron sputtering (R-HiPIMS). American Institute of Physics.

Andersen, H. H., & Bay, H.L. (1981). Sputtering yield measurements. In R. Behrisch (Ed.), Sputtering by Particle Bombardment I (Vol. 47, p. 145). doi:10.1007/3540105212_9

Arnell, R. D., & Kelly, P. J. (1999). Recent advances in magnetron sputtering. *Surface and Coatings Technology*, *112*(1-3), 170–176. doi:10.1016/S0257-8972(98)00749-X

Bai, P. H. (2014). Review of Paper-Like Display Technologies. *Progress In Electromagnetics Research, 147*, 95-116.

Bhattacharyya, A. K. (2015). Sputter Based Epitaxial Growth and Modelling of Cu/Si Thin Films. *Int. J. Thin. Fil. Sci. Tec., 4*(3), 173–177.

Biersack, J. E., & Eckstein, W. (1984). Sputtering Studies with the Monte Carlo Program TRIM.SP. *J. Appl. Phys. A, 34*(2), 73–94. doi:10.1007/BF00614759

Bindu, K. C., Campos, J., Nair, M. T. S., Sánchez, A., & Nair, P. K. (2005). Semiconducting AgSbSe2 thin film and its application in a photovoltaic structure. *Semiconductor Science and Technology, 20*(6), 496–504. doi:10.1088/0268-1242/20/6/004

Bouzid, K. B. (2015). Wear and Corrosion Resistance of CrN-based coatings Deposited by R.F Magnetron Sputtering. *Tribology in Industry, 37*, 60–65.

Brauer, G. S., Szyszka, B., Vergöhl, M., & Bandorf, R. (2010). Magnetron sputtering – Milestones of 30 years. *Vacuum, 84*(12), 1354–1359. doi:10.1016/j.vacuum.2009.12.014

Brault, P. C., Chuon, S., & Bauchire, J.-M. (2016). Molecular Dynamics Simulations of Platinum Plasma Sputtering: A Comparative Case Study. *Frontiers in Physics, 4*, 1–6. doi:10.3389/fphy.2016.00020

Bunshah, R. (1994). *Handbook of Deposition Technologies for Films and Coatings* (2nd ed.). New Jersey: Noyes.

Burton, W. C., Cabrera, N., & Frank, F. C. (1951). The Growth of Crystals and the Equilibrium Structure of their Surfaces. *Trans. R. Soc. London A., 243*(866), 299–358. doi:10.1098/rsta.1951.0006

Chauhan, K. R. (2014). A Review Paper on Tribological and Mechanical Properties of Ternary Nitride Based Coatings. *Procedia Technology*, 430-437. doi:10.1016/j.protcy.2014.08.055

Chuang, C. C. (2008). Effects of Internal Stresses on the Mechanical Properties of Deposition Thin Films. *journal of materials processing technology, 201*, 770–774.

Cowell, R. (1996). Durable optical coatings for energetic UV laser sources. *SPIE, 2703*, 369–374.

Eckstein, W. R., Roth, J., Nagel, W., & Dohmen, R. (2004). Sputtering Mechanisms near the Threshold Energy. *Journal of Nuclear Materials, 328*(1), 55–61. doi:10.1016/j.jnucmat.2004.03.010

Ehrlich, G. H., & Hudda, F. G. (1966). Atomic view of surface self-diffusion: Tungsten on tungsten. *The Journal of Chemical Physics, 44*(3), 1039–1049. doi:10.1063/1.1726787

Flototto, D. W., Wang, Z. M., Jeurgens, L. P. H., Bischoff, E., & Mittemeijer, E. J. (2012). Effect of adatom surface diffusivity on microstructure and intrinsic stress evolutions during Ag film growth. *Journal of Applied Physics, 112*(4), 043503. doi:10.1063/1.4746739

Friz, M. K. (1992). New Materials for Production of Optical Coatings. *Proc. SVC 35th Ann. Techn. Conf.*, (pp. 143-149).

Fu, Y. D. (2003129-136). Deposition of TiN layer on TiNi thin films to improve surface properties. *Surf. Coat. Technol., 167*.

Glazov, L. (2000). *Restricted medium: analytical theory of collision cascades and sputtering*. Izv. Akad. Nauk. Ser. Fiz.

Gnaser, H. (2007). *Energy and angular distributions of sputtered species* (Vol. 110). Berlin: Springer. doi:10.1007/978-3-540-44502-9_5

Gudmundsson, J. B., Brenning, N., Lundin, D., & Helmersson, U. (2012). High Power Impulse Magnetron Sputtering Discharge. *Journal of Vacuum Science & Technology. A, Vacuum, Surfaces, and Films*, *30*(3), 30801. doi:10.1116/1.3691832

Hellmich, W. D. (1992). Optical AR-Coating for Optical Waveguide Devices", Journal of Light wave Technology. *Journal of Lightwave Technology*, *10*(4), 469–476. doi:10.1109/50.134200

Jameel, D. (2015). Thin Film Deposition Processes. *International Journal of Modern Physics A*, *1*(4), 193–199.

Kelly, P. A., & Arnell, R. D. (2000). Magnetron Sputtering: A Review of Recent Developments and Applications. *Vacuum*, *56*(3), 159–172. doi:10.1016/S0042-207X(99)00189-X

Kolasinski, R. P., Polk, J. E., Goebel, D., & Johnson, L. K. (2007). Sputtering Yield Measurements at Glancing Incidence using a Quartz Crystal Microbalance. *Journal of Vacuum Science & Technology. A, Vacuum, Surfaces, and Films*, *25*(2), 236–245. doi:10.1116/1.2435375

Kroese, D. T. (2011). *Handbook of Monte Carlo Methods*. New York: John Wiley & Sons. doi:10.1002/9781118014967

Kryder, M. (1992). Magnetic Thin Films for Data Storage. *Thin Solid Films*, *216*(1), 174–180. doi:10.1016/0040-6090(92)90890-N

Lagally, M. Z., & Zhang, Z. (2002). Thin-film cliffhanger. *Nature*, *417*(6892), 907–910. doi:10.1038/417907a PMID:12087388

Lee, H. K., Lee, H., Park, J.-H., Kim, H.-K., Kong, B. H., & Cho, H. K. (2011). Optical Properties of Ultrathin Copper Thin Films Sandwiched between Nb-Doped TiO2 Films Studied with Spectroscopic Ellipsometry. *Journal of Applied Physics*, *50*(5R), 055805. doi:10.7567/JJAP.50.055805

Liu, J. W., Wu, C., & Tsong, T. T. (1991). Measurement of the binding energy of kink-site atoms of metals and alloys. *Physical Review B: Condensed Matter and Materials Physics*, *43*(14), 11595–11604. doi:10.1103/PhysRevB.43.11595 PMID:9996928

Liu, S. H., Huang, H., & Woo, C. H. (2002). Schwoebel-Ehrlich barrier: From two to three dimension. *Applied Physics Letters*, *80*(18), 3295–3297. doi:10.1063/1.1475774

Lofgran, S. (2013). *Thin film deposition and vacuum technology*. Brigham Young University–Idaho, Department of Physics.

Luth, H. (2001). Solid Surfaces, Interfaces and Thin Film (Forth ed.). New York: Springer-Verlag.

Macleod, A. (1999). The Early Days of Optical Coatings. *Journal of Optics. A, Pure and Applied Optics*, *1*(S), 779–783. doi:10.1088/1464-4258/1/S/305

Macleod, B. H. (2004). Low-Cost Anti-Reflection Technology for Automobile Displays. In *Proceedings of the SID Vehicle Display Conference.*

Martin, P. S., Stewart, D. C., Bennett, W. D., Affinito, J. D., & Gross, M. E. (1997). Multifunctional multilayer optical coatings. *Journal of Vacuum Science & Technology. A, Vacuum, Surfaces, and Films, 15*(3), 1098–1102. doi:10.1116/1.580436

Matsunami, N., Yamamura, Y., Itikawa, Y., Itoh, N., Kazumata, Y., Miyagawa, S., & Tawara, H. et al. (1984). Energy dependence of the yields of ion-induced sputtering of monoatomic solids. *Dat. & Nuc. Dat. Tab, 31*(1), 1–80. doi:10.1016/0092-640X(84)90016-0

Mattox, D. (1998). *Handbook of Physical Vapor Deposition (PVD) Processing.* Westwood, NJ: Noyes Publications.

Mattox, D. (2003). *Foundations of Vacuum Coating Technology.* William Andrew Publishing/Noyes.

Merikoski, J. V.-N., Vattulainen, I., Heinonen, J., & Ala-Nissila, T. (1997). Effect of kinks and concerted diffusion mechanisms on mass transport and growth on stepped metal surfaces. *Surface Science, 387*(1-3), 167–182. doi:10.1016/S0039-6028(97)00349-X

Messier, R. G., Giri, A. P., & Roy, R. A. (1984). Revised structure zone model for thin film physical structure. *Journal of Vacuum Science & Technology. A, Vacuum, Surfaces, and Films, 2*(2), 500–503. doi:10.1116/1.572604

Moeller, W. P., & Rossendorf, F. (2002). *Nano Fabrication using Focused Ion and Electron Beams: Principles and Applications.* Oxford University Press.

Movchan, B. D. (1969). Structure and properties of thick condensates of nickel, titanium, tungsten, aluminum, and zirconium dioxide in vacuum. *Физика металлов и металловедение, 28*, 653–660.

Muhl, S. P., & Pérez, A. (2015). The use of hollow cathodes in deposition processes: A critical review. *Thin Solid Films, 579*, 174–198. doi:10.1016/j.tsf.2015.02.066

Nakles, M. (2004). *Experimental and Modeling Studies of Low-Energy Ion Sputtering for Ion Thrusters.* Blacksburg, Virginia: Virginia Polytechnic Institute and State University.

Navinsek, B. P., Panjan, P., & Milošev, I. (1997). Industrial Applications of CrN (PVD) Coatings, Deposited at High and Low Temperatures. *Surface and Coatings Technology, 97*(1-3), 182–191. doi:10.1016/S0257-8972(97)00393-9

Ono, T., Kenmotsu, T., & Muramoto, T. (2009). *Simulation of the sputtering process.* Berlin: Springer-Verlag.

Over, H. (1998). Crystallographic study of interaction between ad-species on metal surfaces. *Progress in Surface Science, 58*(4), 249–376. doi:10.1016/S0079-6816(98)00029-X

Paruch, R. B., Garrison, B. J., Mlynek, M., & Postawa, Z. (2014). On Universality in Sputtering Yields due to Cluster Bombardment. *The Journal of Physical Chemistry Letters, 5*(18), 3227–3230. doi:10.1021/jz501545t PMID:26276337

Petrov, I. B., Barna, P. B., Hultman, L., & Greene, J. E. (2003). Micro-structural evolution during film growth. *Journal of Vacuum Science & Technology. A, Vacuum, Surfaces, and Films*, *21*(5), S117–S128. doi:10.1116/1.1601610

Pierre-louis, O. D., D'Orsogna, M. R., & Einstein, T. L. (1999). Edge diffusion during growth: The kink Ehrlich-Schwoebel effect and resulting instabilities. *Physical Review Letters*, *82*(18), 3661–3664. doi:10.1103/PhysRevLett.82.3661

Pillis, M. G., Geribola, G. A., Scheidt, G., de Araújo, E. G., de Oliveira, M. C. L., & Antunes, R. A. (2016). Corrosion of thin, magnetron sputtered Nb2O5 films. *Corrosion Science*, *102*, 317–325. doi:10.1016/j.corsci.2015.10.023

Ploch, S. W., Wernicke, T., Dinh, D. V., Pristovsek, M., & Kneissl, M. (2012). Surface diffusion and layer morphology of (11–22) GaN grown by metal-organic vapor phase epitaxy. *Journal of Applied Physics*, *111*(3), 033526. doi:10.1063/1.3682513

Ponon, N. A., Appleby, D. J. R., Arac, E., King, P. J., Ganti, S., Kwa, K. S. K., & O'Neill, A. (2015). Effect of Deposition Conditions and Post Deposition Anneal on Reactively Sputtered Titanium Nitride Thin Films. *Thin Solid Films*, *578*, 31–37. doi:10.1016/j.tsf.2015.02.009

Rastogi, R. V., Vankar, V. D., & Chopra, K. L. (1987). Simple Planar Magnetron Sputtering Source. *The Review of Scientific Instruments*, *58*(8), 1505–1506. doi:10.1063/1.1139388

Rossnagel, S. (1990). *Magnetron Plasma Deposition Processes* (S. C. Rossnagel, Ed.). Park Ridge, New Jersey: Noyes Publications.

Saraiva, M. (2012). *Sputter Deposition of MgO Thin Films : The Effect of Cation Substitution*. Universiteit Gent.

Schwoebel, R. (1969). Step motion on crystal surfaces II. *Journal of Applied Physics*, *40*(2), 614–618. doi:10.1063/1.1657442

Schwoebel, R. S., & Shipsey, E. J. (1966). Step Motion on Crystal Surfaces. *Applied Physics (Berlin)*, *37*(10), 3682–3686. doi:10.1063/1.1707904

Shang, Q. H. (2012). Preparation and characterization of antistatic coatings with modified BaTiO3 powders as conductive fillers. *Journal of Adhesion Science and Technology*, 2642–2652. doi:10.1080/01694243.2013.798926

Sheridan, T. G., Goeckner, M. J., & Goree, J. (1990). Pressure Dependence of Ionization Efficiency in Sputtering Magnetrons. *Applied Physics Letters*, *57*(20), 2080–2082. doi:10.1063/1.103947

Sigmund, P. (1969). Theory of Sputtering I. Sputtering Yield of Amorphous and Polycrystalline Targets. *Physical Review*, *184*(2), 383–416. doi:10.1103/PhysRev.184.383

Sigmund, P. (1981). Sputtering by ion bombardment: theoretical concepts. In R. Behrish (Ed.), *Sputtering by particle bombardment I*. Springer - Verlag. doi:10.1007/3540105212_7

Sigmund, P. J. (2008). Ion Beam Science: Solved and Unsolved Problems (Part I, II). *Medical Physics*, *35*(1), 405.

Singh, J. S. (2003). *Advances in Amorphous Semiconductors*. London, New York: Taylor & Francis. doi:10.1201/9781420023848

Stefan, J. (1997). *Standard Reference Material for Hard Coating, CrN PVD Coating (SRM-HC/2)*. Ljubljana, Slovenia: Institute Report.

Stoller, R. T., Toloczko, M. B., Was, G. S., Certain, A. G., Dwaraknath, S., & Garner, F. A. (2013). On the use of SRIM for computing radiation damage exposure. *Nuclear Instruments & Methods in Physics Research. Section B, Beam Interactions with Materials and Atoms*, *310*, 75–80. doi:10.1016/j.nimb.2013.05.008

Thompson, C. (2000). Structure evolution during processing of polycrystalline films. *Annual Review of Materials Science*, *30*(1), 159–190. doi:10.1146/annurev.matsci.30.1.159

Thompson, M. (1981). Physical mechanisms of sputtering. *Physics Reports*, *69*(4), 335–371. doi:10.1016/0370-1573(81)90106-X

Thompson, M. (2010). The energy spectrum of ejected atoms during the high energy sputtering of gold. *J. Philos. Mag.*, 377-414. doi:10.1080/14786436808227358

Thornton, J. (1974). Influence of apparatus geometry and deposition conditions on structure and topography of thick sputtered coatings. *Journal of Vacuum Science and Technology*, *11*(4), 666–670. doi:10.1116/1.1312732

Thornton, J. G. (1994). Plasmas in Deposition Processes. In R. Bunshah (Ed.), *Deposition Technologies for Films and Coatings*. Park Ridge, New Jersey: Noyes Publications.

Tringides, M. (1997). *Surface Diffusion - Atomistic and Collective Processes*. New York: Plenum Press.

Van de pol, F. B. (1991). R.F Planar Magnetron Sputtering ZnO films I: Structural Properties. *Thin Solid Films, 204*, 349-364.

Vasina, P. (2005). *Plasma Diagnostics Focused on New Magnetron Sputtering Devices for Thin Film Deposition*. Brno, Czech Republic: Masaryk University in Brno, Department of Physical Electronics.

Venables, J. (2000). Surface and Thin Film Processes (1st ed.). Cambridge University Press.

Waits, R. (1978). Planar Magnetron Sputtering. In J. K. Vossenand (Ed.), *Thin Film Processes*. New York: Academic Press. doi:10.1016/B978-0-12-728250-3.50009-X

Warcholinski, B. G. (2009). Tribological properties of TiAlCrN Thin films. *Rev. Mater. Sci.*, *22*, 81–88.

Wasa, K. K. (2004). *Thin Film Materials Technology*. United States: William Andrew Inc. and Springer.

Wehnerand, G. A. (1970). *The Nature of Physical Sputtering" in Handbook of Thin Film Technology* (L. G. Maisseland, Ed.). NY: McGraw-Hill.

Wu, S.-M. K., van de Kruijs, R., Zoethout, E., & Bijkerk, F. (2009). Sputtering yields of Ru, Mo, and Si under low energy Ar+ bombardment. *Journal of Applied Physics*, *106*(5), 1–7. doi:10.1063/1.3149777

Yagisawa, T. M., & Makabe, T. (2006). Modeling of Dc Magnetron Plasma for Sputtering: Transport of Sputtered Copper Atoms. *Journal of Vacuum Science & Technology. A, Vacuum, Surfaces, and Films, 24*(4), 908–913. doi:10.1116/1.2198866

Yam, K. L. (n.d.). Gallium Nitride: An Overview of Structural Defects. *Optoelectronics-Materials and Techniques.*

Yamamura, Y. I. (1983). *Angular dependence of sputtering yields of monatomic solids.* Nagoya University Institute of Plasma Physics Report IPPJ-AM-26.

Yamamura, Y. T., & Tawara, H. (1996). Energy dependence of ion-induced sputtering yields from monoatomic solids at normal incidence. *Atomic Data and Nuclear Data Tables, 62*(2), 149–253. doi:10.1006/adnd.1996.0005

Yasunori, T. (1996). Modification of Thin Film Properties by Sputtered Particles. *R&D Review of Toyota CRDL, 28*(3), 1993.9.

Zhang, Z. L. (1997). Atomistic Processes in the Early Stages of Thin-Film Growth. *Science, 276*(5311), 377–383. doi:10.1126/science.276.5311.377 PMID:9103189

Zhong, J. Z., Zhang, T., Zhang, Z., & Lagally, M. G. (2001). Island-Corner Barrier Effect in two-dimentional Pattern Formation at Surfaces. *Physical Review B: Condensed Matter and Materials Physics, 63*(11), 113403. doi:10.1103/PhysRevB.63.113403

ZieglerJ. (2003). Retrieved from http://www.srim.org

Ziegler, J., Ziegler, M., & Biersack, J. (2010). SRIM – The stopping and range of ions in matter. *Nuclear Instruments & Methods in Physics Research. Section B, Beam Interactions with Materials and Atoms, 268*(11-12), 1818–1823. doi:10.1016/j.nimb.2010.02.091

Ziegler, J. B. (1985). *The Stopping and Range of Ions in Solids.* NY: Pergamon Press.

Ziegler, J. B. (2003). *The Stopping and Range of Ions in Solids.* New York: Pergamon Press.

Ziegler, J. B. (2008). *The Stopping and Range of Ions In Matter.* Retrieved from www.lulu.com/content/1524197

Chapter 16
Oxidation-Protective Coatings for Carbon-Carbon Composites

Omid Ashrafiyan
University of Tehran, Iran

Mohsen Saremi
University of Tehran, Iran

Amirhossein Pakseresht
University of Tehran, Iran & Materials and Energy Research Center, Iran

Ehsan Ghasali
Materials and Energy Research Center, Iran

ABSTRACT

Carbon-carbon composites have attractive properties for high-temperature applications but unfortunately, oxidized at a temperature over 500 °C in the air. Various methods used for protection of C-C composites oxidation such as surface treatment, adding inhibitors in the matrix, and outer coating. High-temperature outer coatings are the most effective method to the protection of C-C composites oxidation. Protective coatings should have low oxygen permeability, compatibility with C-C composites, and adherent to the substrate. SiC is a suitable coating but limited in temperature over 1700 °C and below 1100 °C due to evaporation of silica layer and high viscosity of silica layer respectively and so, multilayer coating system used to solve these problems. SiC coating usually is as a bond layer in coating systems.

INTRODUCTION AND BACKGROUND OF C-C COMPOSITES

Carbon fibers are as reinforcement agent in metals, polymers, and ceramics composites from past decades. Properties of the carbon fibers are a few micrometers in diameter, lightweight, high tensile strength, high stiffness, and high chemical resistance (Chand, 2000). Carbon matrixes reinforced with carbon fiber are called carbon-carbon composites and have a density of 1.6 to 2 $g.cm^{-3}$ which is much lower than many of metals and ceramics density. The carbon fibers can be weaved from one to 3-dimensional as the reinforcement agent in the carbonaceous agent. (Savage, 2012).

DOI: 10.4018/978-1-5225-4194-3.ch016

The carbon can take continuous forms from amorphous and glassy carbon to crystalline graphite. Therefore, although C-C composite made from only one element, both constituents of the composite have different properties that will bring together to produce the carbon-carbon composite. Furthermore, the structure of graphite is consisted hexagonal planes which bonds together by weak van der Waals force, the anisotropy of graphite structure can create different structures based on the orientation and cavities (Buckley & Edie, 1993).

Production of C-C Composite

The carbon fibers are made from PAN, rayon, and pitch and woven in preforms for producing C-C composites. The carbon fibers should place in a carbon matrix. Carbon powders cannot be sintering. For the process of the matrix can use chemical vapor deposition (CVD) and pyrolysis of carbon materials such as pitch. In CVD method, carbon layer deposit on carbon fibers by the organic gas such as methane at high temperature. High densification is difficult and slow in CVD method (Chand, 2000; Deborah, 2010; Shirvanimoghaddam et al., 2017).

The producing of C-C composites by using of pitch or thermosetting resin such as phenolic includes following steps:

- Carbon fibers pre-impregnated with pitch
- Oxidation of Pre-impregnated
- Carbonization of the sample, which production composite with porous structure
- Densification

The final stage is the densification of carbon-carbon composites. This step can be carried out by using the chemical vapor diffusion method, in which carbonaceous gases are pressurized into voids inside the composite or crack formed, then heated up to 1400 °C. Gases tend to fill the cracks of the outer surface of the composite, in resulting the inner cavity can be close and densification incomplete. Another method for densification of carbon-carbon composites is using the liquid phase of pitch and phenolic resin (Savage, 2012).

The condensation process must be repeated to obtain a condensed matter. The pyrolysis process can be carried out both at ambient pressure and high pressure (HIP) in the process of the matrix. The number of saturation cycles much less at high pressure and can improve the composite properties. Therefore, HIP used more than other methods (Raunija & Sharma, 2015)

Properties of C-C Composite

C-C composites have low-density, chemically inert, and high hardness and toughness in temperature around 2000 °C. (Ferreira, Coppini, & Neto, 2001). At high temperatures, C-C composites show 10 to 20 percent increase in mechanical properties (Manocha, 2003). Carbon materials have a thermal shock resistance so that these can be heated from under zero degree of centigrade up to 1500 °C at less than a few seconds. Also, the C-C composite does not severely fracture (Raunija & Babu, 2013).

Oxidation Behavior of C-C Composite

In non-oxidizing environments, the mechanical properties of C-C composites remain up to 2200 °C (H.-J. Li, Fu, Huang, Zeng, & Li, 2005). However, carbon-carbon composites used in the oxidizing environment in many of their applications. Carbon reacts with oxygen in temperatures above 500 °C in air and composite degrades (Westwood, Webster, Day, Hayes, & Taylor, 1996). At low temperature (650-850 °C), the oxidation rate increase with temperature but the oxidation rate decreases as time increases (D. W. McKee, 1991). Also, the rate of oxidation increases with water vapors (Snell, Nelson, & Molian, 2001).

The C-C composite oxidation steps are (Luthra, 1988):

- Diffusion oxygen gas in boundary layers
- Diffusion oxygen gas through the composite crack paths
- The Reaction between carbon and oxygen

At low temperature, the oxygen diffusion rate is faster than its reaction with carbon thus degradation occurs in the bulk of C-C composite and can occur at each part of the composite according to the reactivity of fibers and matrix. At high temperature, the rate of oxygen diffusion is higher than the reaction rate between carbon and oxygen thus degradation occurs in the surface layers of C-C composite (Bacos, 1993). The more oxidation happens in the fiber/matrix interface, leading to the appearance of high energy pores and cracks.

The gas flow has little influence on the oxidation rate at low temperature but at high temperature, oxidation rate increase by increasing the gas flow. Also, external tensile stress increase oxidation rate due to cracks progress but compressive stresses have not effect on C-C composite oxidation (Bacos, 1993).

Applications of C-C Composite

C-C composites have used in a variety of applications in recent decades. These include military aircraft, spacecraft, refractories, furnace elements, biological materials, advanced brakes, and gas turbine blades. In applications which high strength and high ablation resistance need and inability to use cooling systems, C-C composite is a good candidate. Also, this type of composites is used to strengthen the internal parts of broken bones because carbon is compatible with the human body and structurally consistent with the human bone (Kogo, Hatta, Toyoda, & Sugibayashi, 2002). Currently, C-C composite hot pressing molds are in a series of commercial products (Bunker, 2002).

Brake pads should be able to withstand thermal shock, high ablation resistance, high melting temperature and do not lose their performance in high temperatures. Therefore, C-C composites are suitable for this application. For example, C-C composites used in braking linings of Boeing 767 (Buckley & Edie, 1993).

HISTORY AND METHODS OF C-C COMPOSITES OXIDATION PROTECTION

The main work in the field of improving oxidation resistance of graphite and carbon-carbon composites started in 1934 which involved the applying of conversion coating on graphite. In the 1960s, the graphite used in the missile engine was protected by ceramic coatings. The use of iridium coating on a laboratory scale was carried out at this time. Also, during studies in the United States and the United Kingdom, silicon carbide coating was established on C-C composites. These structures were under thermal cycles and could be considered acceptable (Sheehan, 1989). Nowadays, research on improving oxidation resistance of C-C composites focused on using multilayer coating and composite coatings for high-temperature applications and increasing life and durability.

Different methods used to protect the carbon-carbon composite oxidation such as adding inhibitors into the matrix, surface treatment, and outer coating.

Adding Inhibitors to Protection of C-C Composite

There are protective compounds called inhibitors which add to carbon matrix for protecting C-C composite from oxidation. These additives include compounds such as SiC, ZrB_2-Si, ZrH-Si, AlB_{12}, SiB_4, ZrB_2, CrB_2, B, and B_4C that can create glass phases where needed. These glass phases can seal the cracks occurred in the coating during services (Savage, 2012). Adding inhibitors in C-C composite usually improve mechanical properties, but when composites containing glass phase such as boron, zirconium di-boride or heat treatment done over carbonization temperature, mechanical properties are reduced (Bacos, 1993).

Surface Treatment for Protection of C-C Composite Oxidation

Another method for protecting C-C composite oxidation is surface treatment with aqueous solutions (Labruquere, Pailler, Naslain, & Desbat, 1997; Lu & Chung, 2002). Most common materials for surface treatment include halogens such as CCl_4, CCl_2F_2 (X. Hou, Li, Wang, Zhu, & Shen, 2000), $POCl_3$ (Dhami, Bahl, & Awasthy, 1995), boron derivatives such as boric acid (Ehrburger, Baranne, & Lahaye, 1986), and phosphorus derivatives such as oxyfluoride phosphate compounds (Vast, 2009). This method most commonly used for relatively low-temperature applications of C-C composites such as braking systems. The friction coefficient of C-C composite remains constant for a range of 750-1400 K but the braking system made from C-C composite oxidized during repeat braking in contact with air and its lifetime drops significantly (Ehrburger et al., 1986).

In the case of metal phosphate, oxidation decreased by various mechanisms such as blocking active sites, forming a glass coating as a diffusion barrier or deactivating catalytic impurities on the surface by the formation of stable phosphates (Ehrburger et al., 1986). In the case of boron derivatives, the soft layer forms on the carbon surface at a high temperature over 720 K (D. McKee, Spiro, & Lamby, 1984).

Outer Coating for Protection of C-C Composite Oxidation

Oxidation is a surface degradation, so the coating is the logical choice to prevent oxygen diffusion to surface and reaction between carbon and oxygen.

If coating present on C-C composite surface, the oxidation including following steps (Bacos, 1993):

- Oxygen diffusion to surface and boundary layer
- Oxygen diffusion through coatings (If the cracks are in the coating)
- Diffusion through the coating depending on defect structure in coating (If coating is free from cracks)

- Gas diffusion through composite cracks
- The Reaction between carbon and oxygen

Oxides and noble metals have many advantages but have a high thermal expansion. Oxides can reduce by carbon, so they should not establish directly on carbon substrates. Moreover, oxide phases such as ZrO_2 have higher oxygen permeability than C-C composites. High melting refractory metals also can use for oxidation protection but have some issues. Iridium was used in 2000-2200 °C for short time application in the 1960s but have many problems, such as erosion, volatile oxide, relatively high thermal expansion coefficient, poor adhesion to carbon and high cost (Savage, 2012). Rhenium (Re) is only refractory metal which has not carbide, so it can apply on C-C composite surface and act as a bond layer in coating system with iridium outer coating (Yue, Hong-yu, Zeng-rong, & Zhi-xiang, 2015). Also, Different ceramics are candidates for protection against oxidation but due to differences in the thermal expansion coefficient, are used in multilayer coating system. Converted carbides and borate glass have thermal stability and suitable viscosity, so commonly are used as an intermediate layer in coating system (Strife & Sheehan, 1988). SiC is widely applied as a bond layer on C-C composites (Fritze et al., 1998).

REQUIREMENTS OF PROTECTIVE COATING SYSTEMS

The protective coating should have following requirements:

- The coating should prevent oxygen diffusion through coating and substrate and therefore should be; free of defects, and low oxygen permeable (Strife & Sheehan, 1988). These properties should be present in oxidation products too (Corral & Loehman, 2008).
- The coating should be mechanically compatible with the substrate to restrict spallation during extreme thermal heating.
- The coating should be adherent to provide ablation resistance.
- The coating should have reproducibility, and its processing should be accessible and easy (Sheehan, 1989; Strife & Sheehan, 1988).

Protection of C-C composite against oxidation can divide into two categories of limited-life and extended-life protection, each of them having issues and features in their coating system. In the limited-life application, the working temperature is above 1650 °C, the number of thermal cycles is limited and time of operation between several minutes to one hour. In limited-life protection at high temperature and with the oxidizing gas flow, the surface is subject to erosion, evaporation of protective layer and particle impingement. Therefore, coating system especially the outer coating should be protected against erosion and ablation in the limited-life protection. The difference in thermal expansion coefficient between coating and substrate causes coating spallation. If coating system cracked or erode, oxygen can diffuse in coating and substrate (Sheehan, 1989). In extended-life protection, working temperature is less than 1650 °C, operating time is hundreds of hours with thermal cycling. The important subject in this type of protection is thermal expansion coefficient compatibility between coating and substrate. The coating system used in this type of protection should be able to seal occurring cracks (Sheehan, 1989). Secondary phases such as nanowires, nanoparticles, and nanotubes can use for toughening of coatings and prevent cracks which occur due to the difference in thermal expansion coefficient. For example, the Si-Mo-Cr coating has suitable oxidation resistance due to the formation of SiO_2 and Cr_2O_3 oxide layers. The difference in thermal expansion coefficient of Si, Mo, and Cr cause that Si-Mo-Cr cracked (Feng et al., 2015). The performance of Si-Mo-Cr coating improved in high thermal cycles by adding SiC nanowire (J.-P. Zhang, Fu, Qu, Zhou, & Liu, 2016). The SiC nanowire has a small size, high strength, and toughness (Paul, Majumdar, & Suri, 2013). Pre-oxidation treatment can be used to increase the adhesion between the coating and C-C composite (Q. Fu et al., 2014). This method creates a porous structure that provides coating infiltration. The first step in pre-oxidation operations is obtaining the initial temperature of C-C composite oxidation with Thermal gravimeter analysis (TGA) and then oxidize the specimens for short time at this temperature (Q. Fu et al., 2014). In pre-oxidation treatment, the control of oxidized sites is difficult, blasting treatments under the oxyacetylene flame can use as an alternative method (J.-P. Zhang et al., 2016). Figure 1 shows schematic of the pre-oxidation and addition of nanowire on the C-C composite surface.

SOME HIGH-PERFORMANCE COATING SYSTEMS

For the protection of C-C composite oxidation, many of the multilayers coatings developed which SiC is as a bond coat. SiC coating completely and some high-performance coating systems such as SiC/mullite, SiC/$MoSi_2$, SiC/silicate yttrium, and SiC/ZrB_2 described as follow.

Silicon Carbide Coating

Silicon carbide has a three-dimensional structure while silicon atoms surrounded by four atoms of carbon. SiC has more than 250 polymorphs, which many of them amorphous and a large number have crystalline structures that known as polytypes. One of the polytypes is the α-SiC which has the hexagonal crystalline structure as the wurtzite, and another polytype is the ß-SiC crystal with the structure similar to zinc blend. Thermodynamically, the α-SiC type is more stable than ß-SiC. The pure silicon carbide is yellow but usually, has black color due to the formation of thin oxide layer and presence of impurities such as iron (Roy, Chandra, Das, & Maitra, 2014).

Figure 1. Schematic of: (a) pre-oxidation treatment, (b) nanowire adding on the surface of the C-C composites

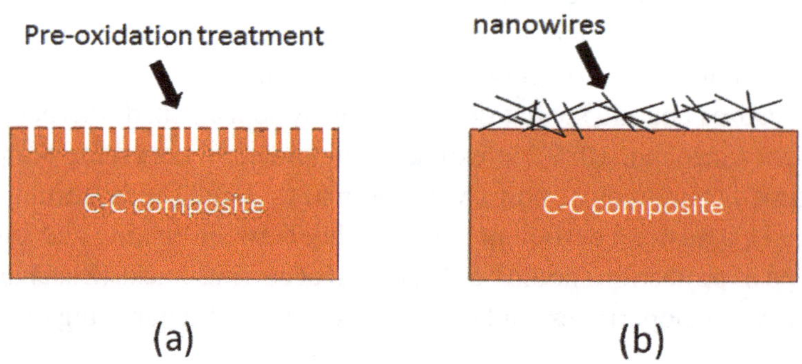

When the oxygen pressure is less than 1 bar, oxidation occurs accordance to below equation (Roy et al., 2014).

However, when the oxygen pressure closes to 1 bar, oxidation occurs according to below equation. By

$$SiC(s) + O_2(g) \rightarrow SiO_2(g) + CO(g). \tag{1}$$

forming SiO_2 on the surface of the SiC, it's weight increase (Roy et al., 2014). During SiC oxidation, a silica layer forms at a temperature above 600 °C with parabolic weight gain and thickness increase with time (Opila, 1994).

$$2SiC(s) + 3O_2(g) \rightarrow 2SiO_2(s) + 2CO(g). \tag{2}$$

Silicon carbide coating widely is used as a protective coating against carbon substrates oxidation. This protection is leading to good physical and chemical compatibility with carbon substrates (Q.-G. Fu, Li, Shi, Li, & Sun, 2005). Different methods are used to apply silicon carbide coating such as pack cementation, chemical vapor deposition (CVD), plasma sintering, and slurry-sintering (H.-J. Li, Jiao, Li, & Wang, 2008). Among these methods, pack cementation is more used leading to gradient interface with the distribution of Si from the surface to the center that giving thermal shock resistance (Jian-Feng, Xie-Rong, He-Jun, Xin-Bo, & Ye-wei, 2004). In this method, the whole of C-C composite surfaces can coat with one process (Xin et al., 2009). The silicon carbide coating applied by two-step pack cementation process; the weight loss of C-C composite with SiC coating after 310 hours at 1500 °C was only 9.9×10^{-5} g/cm²h. Also, the Si element gradually decreases at 18 micrometers below the surface which indicates a gradient interface (Q.-G. Fu et al., 2005).

In the preparation of SiC coating on C-C composite with pack cementation method, the C-C composite samples placed in powder mixture consisting of Si, SiC, Al_2O_3 or Si, C, Al_2O_3 then sealed and heat treatment under the argon or nitrogen atmosphere at temperature 1500 °C to 2200 °C.

To applying the outer ceramic layer in coating systems with SiC as a bond layer, SiC dense or porous layer may need (Shimada & Sato, 2002). In SiC/Mullite coating system (Shimada & Sato, 2002) need a porous SiC layer while a dense layer requires for SiC/Si-W coating (Cheng, Xu, Zhang, & Yin, 1999). Jia-Feng et al. (Jian-Feng et al., 2004) investigated the effect of temperature on the microstructure and the properties of the SiC in pack cementation technique. When the temperature is in the range of 1773-1973 K the coating has a porous ß-SiC structure while at T=2073 K a dense Si/ß-SiC coating is formed. Unfortunately, at 2173 K, the micro cracks in the ß-SiC/α-SiC layer have been formed, so the temperature must be optimized. Pore-forming such as Fe_2O_3 can add in the powder system of pack cementation technique for establishing of the porous structure. In the SiC/ZrB_2 for better adhesion of ZrB_2 to SiC bond layer, the porous coating is obtained by adding Fe_2O_3, due to gas removal according to the equation below (Bai, Yang, Shi, Xu, & Yang, 2012; Feng, Li, Hu, Lin, & Li, 2016).

$$Si(l) + C(s) \rightarrow SiC(s) \tag{3}$$

$$Fe_2O_3(l) + SiC(s) \rightarrow Si(l) + Fe(g) + CO(g) \tag{4}$$

When the SiC coating applied by pack cementation, the type and structure in particular the open porosity and density are effective factors on oxidation. Juan Zhao et al. (Zhao, Liu, Guo, Shi, & Zhai, 2006) showed that the oxidation resistance of C-C composite with SiC coating increased for samples with 15.6% open porosity and 1.8 g/cm^3 density, 12.5% open porosity, and 1.89 g/cm^3 density and 18.8% open porosity and 1.69 g/cm^3 density respectively. In fact, by increasing density along the open porosity of coating, oxidation resistance increased.

Different thermal expansion coefficient between C-C composite (1×10^{-6} K^{-1}) and silicon carbide (5×10^{-6} K^{-1}) creates micro cracks in silicon carbide coating. Also, the silicon carbide coating is heterogeneous due to the difference in reactivity of fiber and matrix in C-C composites (Cairo, Graca, Silva, & Bressiani, 2001). Graphite with SiC coating has homogeneous structure and so, the oxidation resistance of the graphite with SiC coating better than C-C composite with SiC coating (Xin et al., 2009). Figure 2 shows the surface SEM of the SiC coating on the (a) C/C composite and (b) graphite that applied by pack cementation technique at 1650 °C. The SiC coating on the composite is inhomogeneous, while the SiC coating on the graphite is homogeneous and without any cracks due to the uniform structure. If SiO2 glass phase has an appropriate viscosity, the microcrack that forms by thermal shock or due to heterogeneous structure of C-C composite can be sealed (Sheehan, 1989). The graded region with Si distribution in the C-C composite is usually very narrow, and some of the carbon fibers remain without reaction due to the difficult reaction between liquid silicon and carbon fiber. Pure carbon layer can apply on the C-C composite before the gradient SiC coating to solve this problem, (Y.-L. Zhang et al., 2006). SiC coating should be dense in gradient region (Sheehan, 1989). The pack cementation has the ability to coating all layer of surfaces in one process but needs to remove residual material from the C-C composite samples (Xin et al., 2009).

Vacuum plasma can use for applying silicon carbide coating. In this process, the Si deposited on C-C composites, then SiC coating obtained by heat treatment under vacuum (Hu et al., 2015). Also, carbon samples can be immersed in Si liquid to formation SiC at 1410 K under a vacuum or argon atmosphere (Chunhe & Jie, 1995). Another method for Applying SiC coating is the slurry. The slurry method is simple, low cost and suitable but needs to vacuum and high temperature (Xin et al., 2009).

Silicon carbide coatings also have disadvantages which some of them described. Al and Si Oxides react with water or sodium/potassium vapors at high temperatures. Therefore, these oxides shouldn't use in environments with water or Na/K vapors. In temperatures below 1100 °C, the SiO_2 glass phase formed on silicon carbide has a high viscosity so, cracks can't seal (Gee & Little, 1991). SiO_2 evaporates, decomposes and becomes thin at temperatures over 1700 °C (Savage, 2012). Therefore, SiC coating not capable of long-term protection at temperatures up to 1700 °C (Zhu, Qiu, & Ma, 1999). Moreover, SiC has high oxidation rate in high temperature (Sheehan, 1989). All materials based on silicon such as Si_3N_4, $MoSi_2$ have this limitation. The produced gases are entered into the SiO_2 glass phase and create holes. Holes are a place for oxygen diffusion into the C-C composite and so silicon carbide layer degraded (Savage, 2012). Due to expressed problems, SiC single-layer can't provide excellent oxidation and effective for a long time (Sun et al., 2013). Therefore, different multilayer coatings such as SiC/mullite, SiC/silicate yttrium, and SiC/ZrB_2 extended for C-C composites protection and silicon carbide widely is used as an intermediate layer in coating systems (Xin et al., 2009).

The oxidation resistance of C-C composite with SiC coating has improved by adding a boron-rich layer in the 1100-1500 °C. Boron-rich layer can seal the cracks by B_2O_3 glass phase formation (Sheehan, 1993). Boron-rich layer can seal the cracks by formation a B_2O_3 glassy phase (Sheehan, 1993). Borate glasses are suitable for the outer layer in the low-temperature application due to the tendency of B_2O_3

Figure 2. Surface SEM image of; (a SiC-coated C/C composite and (b) SiC-coated graphite

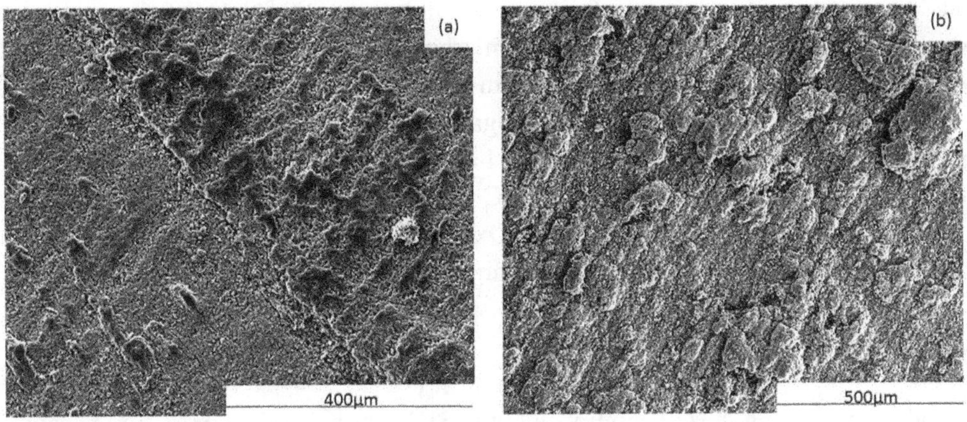

to wet SiC and Si_3N_4. However, this system limited due to the moisture sensitivity of borate glass and incompatibility between SiC coating and borate glass (Savage, 2012). Moisture sensitivity can reduce by adding alkali (Mazurin, Streltsina, & Shvaiko-Shvaikovskaya, 1985). B_2O_3 can use without SiC layer in temperatures below 1100 °C. B_2O_3 has a suitable viscosity and compatibility with carbon composites which allowing to the formation of thin glass layer at 600-1100 °C. If boric oxide contact to carbon, B_2O_3 may reduce to boron carbide due to CO forming. This coating limited at a temperature above 1500 °C because of the CO pressure increase as increase temperature (Sheehan, 1989).

In application above 1800 °C such as space vehicles, the outer coating should have ablation resistant (Levine et al., 2002). Ultra-high-temperature ceramics (UHTCs) are including carbide and borate of transient metal and are a candidate for outer coating (Opeka, Talmy, & Zaykoski, 2004; Tandon, Dumm, Corral, Loehman, & Kotula, 2006). For example, SiC/ZrB_2 coating system can be a good selection for high-temperature operations (Corral & Loehman, 2008).

SiC/ Mullite Coating Systems

The mullite ($3Al_2O_3.2SiO_2$) coating is a suitable candidate for outer coating on SiC bond layer due to high melting temperature, stability, low oxygen permeability, and proper anti-erosion (Bao & Nicholson, 2008; Opeka et al., 2004). Additionally, there is a match in the thermal expansion coefficient and physical and chemical compatibility between silicon carbide and mullite (Damjanović et al., 2007). Also, SiO_2 soft phase formation at the silicon carbide/mullite interface that establishes oxidation protection (Jian-Feng, Xie-Rong, He-Jun, Xin-Bo, & Min, 2003). Generally, the most important method to produce the thick film and high-temperature coating is thermal spray (A. Pakseresht, Javadi, Bahrami, Khodabakhshi, & Simchi, 2016; A. Pakseresht, Javadi, Ghasali, Shahbazkhan, & Shakhesi, 2016; A. Pakseresht, Rahimipour, Alizadeh, Hadavi, & Shahbazkhan, 2016; A. Pakseresht, Rahimipour, Vaezi, & Salehi, 2016; A. H. Pakseresht et al., 2015; A. H. Pakseresht et al., 2014). The mullite coating applied by plasma spraying on a silicon carbide coating which coated on the graphite by pack cementation. Oxidation test shows that the weight loss of graphite sample without coating is 100% after 2 hours, for graphite sample with a silicon carbide coating is 20% after 9 hours, and for a graphite sample with a silicon carbide/mullite coating is 10% after 9 hours at 1450 °C. (Aliakbarpour, Zakeri, Rahimipour, & Irankhah, 2014). For applying mullite outer coating on C-C composite with SiC coating, the precursor solution including mullite powder and additives sprayed and then sintering for 2.5 hours at 1500 °C. One-stage spray and sintering produce a dense and crack-free coating with 178 micrometers thickness. The sample weight loss after 100 hours at 1500 °C is $6.85\times10^{-3}g/cm^2$ (ZhongLiu et al., 2015). Huang Jian-Fenga et al. (Jian-Feng, Xie-Rong, He-Jun, Xin-Bo, & Min, 2003) investigated SiC/mullite coating system by two-step pack cementation which weight loss was 1.86% after 45 hours. Dense mullite coating was also applied by an electrophoretic method on C-C composite with SiC coating. After 322-hour oxidation test at 1500 °C, the weight loss of the sample is $15.76\times10^{-2}g/cm^2$. Oxidation increase with time because of formation of micro-crack and holes in mullite coating (K.-T. Wang, Cao, Huang, & Fei, 2013).

SiC/ $MoSi_2$ Coating Systems

The $MoSi_2$ used for protecting the C-C composite with SiC coating against oxidation because of its high melting point, oxidation and corrosion resistance (Vasudevan & Petrovic, 1992).The SiO_2 glassy phase formation during oxidation of $MoSi_2$ (Sheehan, 1989). $MoSi_2$ coating applied by an electrophoretic process, weight loss after 360 hours of exposure at 1500 °C is 2.49 mg/cm^2 (Huang et al., 2011). Also, The $MoSi_2$ coating applied by the slurry method and could protect C-C composites for 300 hours at 1873 K. The $MoSi_2$ coating applied by pack cementation process have large crystalline particles and cracks and protect C-C composite substrates only for 109 hours at 1773 K (Y.-L. Zhang, Li, Yao, Li, & Qiang, 2011).

ZrB_2 outer coating applied to increase the performance of $MoSi_2$-SiC coating but due to the difference in the thermal expansion coefficient cracked (Niu et al., 2015). So, the SiC nanowires are deposited by CVD before applying the SiC-$MoSi_2$-ZrB_2 coating. The weight loss of this coating system is 1.1 mg/cm^2 after 124 hours at 1773 K (Q.-G. Fu et al., 2016).

$MoSi_2$-WSi_2 Composite coatings have better oxidation resistance than single $MoSi_2$ or WSi_2 outer coating because of formation of intermetallic of $MoSi_2$ and WSi_2 that has high melting point and excellent oxidation resistance at high temperatures (D. Hou et al., 2007).

SiC/ Yttrium Silicate Coating Systems

Yttrium silicate has the best bonding with the SiC bond layer due to the matching in thermal expansion coefficient. Also, yttrium silicate has low evaporation rate and low oxygen penetration (Zeng, Xiong, Guo, & Zhang, 2013). However, the oxidation resistance of yttrium silicate coating depends on its preparation technology (Huang et al., 2007). SiO_2 and Y_2O_3 mixture in slurry dipping method can form $Y_2Si_2O_7$ and $Y_2Si_2O_5$ coatings, but the coating cracked by oxidation of the underlying SiC after 50 hours at 1873 K (Webster et al., 1998). The yttrium silicate gradient multilayer developed by plasma-spray and in the $SiO_2.Y_2O_3$, $1.5SiO_2.Y_2O_3$ and $2SiO_2.Y_2O_3$ form on silicon carbide. This coating system was able to protect C-C composites oxidation for 73 hours at 1773 K (Huang, Li, Zeng, & Li, 2006). Also, yttrium silicate applied by a simple slurry method with a powder mixture of SiO_2 and Y_2O_3 and has oxidation resistance at 1873 K for 202 hours with 2.87×10^{-3} g/cm^2 weight loss (Huang et al., 2007).

SiC/ ZrB_2 Coating Systems

In SiC/ZrB_2 coating systems, the ZrB_2 selected due to the high melting point and silicon carbide because of low-temperature protection. These are synergistic effect and can establish high-temperature protection

(Corral & Loehman, 2008). Zirconium and hafnium-based refractory materials can use at high temperature. These coatings can protect of C-C composites only for a short time at high temperature due to the high oxygen permeability in oxides phases (Savage, 2012). ZrO_2 produced from ZrB_2 oxidation can be reacted with SiO_2 and produce $ZrSiO_4$. $ZrSiO_4$ has excellent oxidation resistance at temperatures over 1973 K for a long time (Kaiser, Lobert, & Telle, 2008). Also, $ZrSiO_4$ decrease consuming of SiO_2 and modify its oxidation properties (P. Wang et al., 2016).

The SiC-Si-ZrB_2 layer on C-C composite with silicon carbide coating applied by pack cementation at 2200 °C. The powder system consisting of Si, ZrB_2, and B_2O_3. The weight loss is 1.53% after 386 hours at 1500 °C. This protection is due to the SiO_2-$ZrSiO_4$ glass layer (Yao, Li, Zhang, Wu, & Qiang, 2012). Gradient coating which preparation by pack cementation process usually hasn't uniform thickness. Therefore, SiC/ZrB_2-SiC gradient coating created by vacuum plasma spray (VPS) process. In this method, the Si and ZrB_2 powder were mixed and after vacuum spraying, coating diffuse in the substrate by heat treatment and ZrB_2-SiC composite coating on silicon carbide gradient coating formed. The specimen with this coating system could protect under H_2-O_2 flame with 2500 °C temperature for at least 60 seconds (Hu et al., 2015). The supersonic atmosphere plasma spray (SAPS) also can be used to create the ZrB_2 protective layer. In the SAPS, ZrB_2 particles with jet impact force are placed on the C-C composite and form a high strength dense layer (Wu et al., 2010). After ZrB_2 coating by SAPS process, the samples were pre-oxidized for 15 minutes at 1373 K. This operation increases resistance to ablation. Samples with ZrB_2 coatings that pre-oxidized have an ablation resistance to 90 seconds under heat flux with 2400 kW/m^2 (Y. Zhang, Hu, Yang, Ren, & Li, 2015). Besides, many of the cavities and oxides formed in SAPS process and the SiC also may decompose (Tului et al., 2008). Creating ZrB_2 coatings with CVD is a challenge, and this method is most commonly used to create carbide coatings (Hu et al., 2015). Ultra-high-temperature ceramics (UHTCs) coatings have high thermal expansion coefficients compared to C/C composite and SiC, which results in poor thermal shock resistance. Nanowires can use to solve this problem. SiC nanoparticles applied by an electrophoretic process on C-C composite with silicon carbide coating and then the SiC-ZrB_2-ZrC coating applied with a pack cementation. The Results Show, the oxidation resistance improved (L. Li et al., 2015). Also, SiC nanowires added to the SiC-ZrB_2 coating. Also, SiC nanowires added to the SiC-ZrB_2 coating. The rate of ablation in the SiC-ZrB_2 coating with SiC nanowires under the oxyacetylene flame is 2.3×10^{-3} g/s and in SiC-ZrB_2 coating without SiC nanowires is equal to 1.4×10^{-3} g/s. SiC nanowires help to increase the bonding strength and toughness thus ablation resistance increased (Y. Zhang et al., 2015). Ferrocene commonly used as a catalyst for the formation of fibers and particles. Ferrocene commonly used as a catalyst for the formation of fibers and particles. Ferrocene used at preparing the ZrB_2-SiC coating by pack cementation process on silicon carbide coated graphite by adding in mixing powders. The results show that by adding 3% ferrocene, the specimen will 96/2% weight after 15 thermal shocks between room temperature and 1500 °C. Also, the 99.8% of weight remains after 19 hours (P. Wang et al., 2016).

CONCLUSION

The fibers and matrix in the carbon-carbon composite are carbon-based but with different structures and properties and various methods such as CVD and HIP are used to make carbon-carbon composites. The carbon-carbon composite has a high melting temperature, high thermal shock resistance, and reliable failure. Carbon-carbon composite suited for aerospace applications, ultrasonic vehicles with high operating temperatures for these unique properties. C-C composites degraded in air and temperatures up to 500 °C due to oxidation but in most applications, C-C composites are in this condition. Therefore, protection of oxidation is necessary. Different methods can use for C-C composite protection such as adding inhibitors in the matrix, surface treatment, and coating. The coating is a most effective method to protect the composite against oxidation. Inhibitors and surface treatment used along coatings. The protective coatings for a carbon substrate should have low oxygen diffusion, thermal expansion coefficient close to the substrate, and thermal stability. The silicon carbide has oxidation resistance and compatibility with carbon substrate, so suitable for C-C composite coating but the results show that this coating can't provide effective protection in long work and temperature above 1700 °C and below 1100 °C. Therefore, multilayer coating systems with SiC bond coat usually used to protect C-C composite. SiC bond coat can use for borate glass, ultra-high temperature ceramics such as ZrB_2, mullite, $MoSi_2$, yttrium silicate.

REFERENCES

Aliakbarpour, S., Zakeri, M., Rahimipour, M., & Irankhah, R. (2014). Effect of SiC-mullite coatings on oxidation resistance of graphite. *Advances in Applied Ceramics*, *113*(6), 358–361. doi:10.1179/1743676114Y.0000000173

Bacos, M. (1993). Carbon-carbon composites: oxidation behavior and coatings protection. *Le Journal de Physique IV, 3*(C7), 1895-1903.

Bai, J., Yang, X., Shi, Y., Xu, S., & Yang, J. (2012). Fabrication of directional SiC porous ceramics using Fe 2 O 3 as pore-forming agent. *Materials Letters*, *78*, 192–194. doi:10.1016/j.matlet.2012.03.046

Bao, Y., & Nicholson, P. S. (2008). AlPO 4-coated mullite/alumina fiber reinforced reaction-bonded mullite composites. *Journal of the European Ceramic Society*, *28*(16), 3041–3048. doi:10.1016/j.jeurceramsoc.2008.05.032

Buckley, J. D., & Edie, D. D. (1993). *Carbon-carbon materials and composites* (Vol. 1254). William Andrew.

Bunker, S. P. (2002). Evaluation of Dynamic Properties of a Carbon-Carbon Composite at Elevated Temperatures.

Cairo, C., Graca, M., Silva, C., & Bressiani, J. (2001). Functionally gradient ceramic coating for carbon–carbon antioxidation protection. *Journal of the European Ceramic Society*, *21*(3), 325–329. doi:10.1016/S0955-2219(00)00191-6

Chand, S. (2000). Review carbon fibers for composites. *Journal of Materials Science*, *35*(6), 1303–1313. doi:10.1023/A:1004780301489

Cheng, L., Xu, Y., Zhang, L., & Yin, X. (1999). Oxidation behavior of carbon–carbon composites with a three-layer coating from room temperature to 1700° C. *Carbon*, *37*(6), 977–981. doi:10.1016/S0008-6223(98)00293-0

Corral, E. L., & Loehman, R. E. (2008). Ultra-High-Temperature Ceramic Coatings for Oxidation Protection of Carbon–Carbon Composites. *Journal of the American Ceramic Society*, *91*(5), 1495–1502. doi:10.1111/j.1551-2916.2008.02331.x

Deborah, D. C. (2010). *Composite materials: Science and applications*. Engineering Materials and Processes.

Dhami, T., Bahl, O., & Awasthy, B. (1995). Oxidation-resistant carbon-carbon composites up to 1700 C. *Carbon*, *33*(4), 479–490. doi:10.1016/0008-6223(94)00173-W

Ehrburger, P., Baranne, P., & Lahaye, J. (1986). Inhibition of the oxidation of carboncarbon composite by boron oxide. *Carbon*, *24*(4), 495–499. doi:10.1016/0008-6223(86)90274-5

Feng, T., Li, H., Fu, Q., Shi, X., Hu, M., & Liu, L. (2015). Erosion resistance of Mo–Si–Cr coating-modified C/C composites in a wind tunnel at 1873K. *Journal of Alloys and Compounds*, *622*, 1049–1054. doi:10.1016/j.jallcom.2014.10.189

Feng, T., Li, H.-J., Hu, M.-H., Lin, H.-J., & Li, L. (2016). Oxidation and ablation resistance of Fe 2 O 3 modified ZrB 2–SiC–Si coating for carbon/carbon composites. *Ceramics International*, *42*(1), 270–278. doi:10.1016/j.ceramint.2015.08.105

Ferreira, J., Coppini, N., & Neto, F. L. (2001). Characteristics of carbon–carbon composite turning. *Journal of Materials Processing Technology*, *109*(1), 65–71. doi:10.1016/S0924-0136(00)00776-7

Fritze, H., Jojic, J., Witke, T., Rüscher, C., Weber, S., Scherrer, S., & Borchardt, G. et al. (1998). Mullite based oxidation protection for SiC–C/C composites in air at temperatures up to 1900K. *Journal of the European Ceramic Society*, *18*(16), 2351–2364. doi:10.1016/S0955-2219(98)00242-8

Fu, Q., Shan, Y., Wen, S., Li, H., Chu, Y., & Liu, X. (2014). Oxidation pre-treatment to improve the mechanical property and oxidation resistance of Si–Mo–Cr coated C/C composites. *Corrosion Science*, *83*, 103–110. doi:10.1016/j.corsci.2014.02.008

Fu, Q.-G., Jing, J.-Y., Tan, B.-Y., Yuan, R.-M., Zhuang, L., & Li, L. (2016). Nanowire-toughened transition layer to improve the oxidation resistance of SiC–MoSi 2–ZrB 2 coating for C/C composites. *Corrosion Science*, *111*, 259–266. doi:10.1016/j.corsci.2016.05.013

Fu, Q.-G., Li, H.-J., Shi, X.-H., Li, K.-Z., & Sun, G.-D. (2005). Silicon carbide coating to protect carbon/carbon composites against oxidation. *Scripta Materialia*, *52*(9), 923–927. doi:10.1016/j.scriptamat.2004.12.029

Gee, S., & Little, J. (1991). Oxidation behaviour and protection of carbon/carbon composites. *Journal of Materials Science*, *26*(4), 1093–1100. doi:10.1007/BF00576792

Hou, D., Li, K., Li, H., Fu, Q., Wei, J., & Zhang, Y. (2007). W-Mo-Si/SiC oxidation protective coating for carbon/carbon composites. *Journal Of Materials Science And Technology-Shenyang*, *23*(4), 559.

Hou, X., Li, H., Wang, C., Zhu, Z., & Shen, J. (2000). Internal friction behavior of carbon–carbon composites. *Carbon*, *38*(15), 2095–2101. doi:10.1016/S0008-6223(00)00069-5

Huang, J.-F., Li, H.-J., Zeng, X.-R., Deng, F., Xiong, X., & Li, K.-Z. (2007). Oxidation resistant yttrium silicates coating for carbon/carbon composites prepared by a novel in-situ formation method. *Ceramics International, 33*(5), 887–890. doi:10.1016/j.ceramint.2006.01.007

Huang, J.-F., Li, H.-J., Zeng, X.-R., & Li, K.-Z. (2006). Yttrium silicate oxidation protective coating for SiC coated carbon/carbon composites. *Ceramics International, 32*(4), 417–421. doi:10.1016/j.ceramint.2005.03.018

Huang, J.-F., Wang, B., Li, H.-J., Liu, M., Cao, L.-Y., & Yao, C.-Y. (2011). A MoSi 2/SiC oxidation protective coating for carbon/carbon composites. *Corrosion Science, 53*(2), 834–839. doi:10.1016/j.corsci.2010.11.024

Jian-Feng, H., Xie-Rong, Z., He-Jun, L., Xin-Bo, X., & Min, H. (2003). Mullite-Al 2 O 3-SiC oxidation protective coating for carbon/carbon composites. *Carbon, 41*(14), 2825–2829. doi:10.1016/S0008-6223(03)00397-X

Jian-Feng, H., Xie-Rong, Z., He-Jun, L., Xin-Bo, X., & Ye-wei, F. (2004). Influence of the preparation temperature on the phase, microstructure and anti-oxidation property of a SiC coating for C/C composites. *Carbon, 42*(8), 1517–1521. doi:10.1016/j.carbon.2004.01.066

Kaiser, A., Lobert, M., & Telle, R. (2008). Thermal stability of zircon (ZrSiO 4). *Journal of the European Ceramic Society, 28*(11), 2199–2211. doi:10.1016/j.jeurceramsoc.2007.12.040

Kogo, Y., Hatta, H., Toyoda, M., & Sugibayashi, T. (2002). Application of three-dimensionally reinforced carbon–carbon composites to dovetail joint structures. *Composites Science and Technology, 62*(16), 2143–2152. doi:10.1016/S0266-3538(02)00157-4

Labruquere, S., Pailler, R., Naslain, R., & Desbat, B. (1997). Enhancement of the oxidation resistance of carbon fibres in C/C composites via surface treatments. *Key Engineering Materials, 132*, 1938–1941. doi:10.4028/www.scientific.net/KEM.132-136.1938

Levine, S. R., Opila, E. J., Halbig, M. C., Kiser, J. D., Singh, M., & Salem, J. A. (2002). Evaluation of ultra-high temperature ceramics foraeropropulsion use. *Journal of the European Ceramic Society, 22*(14), 2757–2767. doi:10.1016/S0955-2219(02)00140-1

Li, H.-J., Fu, Q.-G., Huang, J.-F., Zeng, X.-R., & Li, K.-Z. (2005). Research on the Oxidation-Protective Coatings for Carbon/Carbon Composites. *Carbon letters, 6*(2), 71-78.

Li, H.-J., Jiao, G.-S., Li, K.-Z., & Wang, C. (2008). Multilayer oxidation resistant coating for SiC coated carbon/carbon composites at high temperature. *Materials Science and Engineering A, 475*(1), 279–284. doi:10.1016/j.msea.2007.04.053

Li, L., Li, H., Li, Y., Yin, X., Shen, Q., & Fu, Q. (2015). A SiC-ZrB 2-ZrC coating toughened by electrophoretically-deposited SiC nanowires to protect C/C composites against thermal shock and oxidation. *Applied Surface Science, 349*, 465–471. doi:10.1016/j.apsusc.2015.05.028

Lu, W., & Chung, D. (2002). Oxidation protection of carbon materials by acid phosphate impregnation. *Carbon, 40*(8), 1249–1254. doi:10.1016/S0008-6223(01)00297-4

Luthra, K. L. (1988). Oxidation of carbon/carbon composites—a theoretical analysis. *Carbon*, *26*(2), 217–224. doi:10.1016/0008-6223(88)90040-1

Manocha, L. M. (2003). High performance carbon-carbon composites. *Sadhana*, *28*(1), 349–358. doi:10.1007/BF02717143

Mazurin, O., Streltsina, M., & Shvaiko-Shvaikovskaya, T. (1985). Handbook of Glass Data. Amsterdam: Elsevier.

McKee, D., Spiro, C., & Lamby, E. (1984). The effects of boron additives on the oxidation behavior of carbons. *Carbon*, *22*(6), 507–511. doi:10.1016/0008-6223(84)90083-6

McKee, D. W. (1991). Oxidation protection of carbon materials. *Chemistry and Physics of Carbon*, *23*, 173–232.

Niu, Y., Wang, H., Liu, Z., Hu, C., Wang, X., Zheng, X., & Ding, C. (2015). Microstructure evolution of ZrB 2–MoSi 2 composite coatings at middle and high temperatures. *Surface and Coatings Technology*, *273*, 30–38. doi:10.1016/j.surfcoat.2015.03.029

Opeka, M. M., Talmy, I. G., & Zaykoski, J. (2004). Oxidation-based materials selection for 2000 C+ hypersonic aerosurfaces: Theoretical considerations and historical experience. *Journal of Materials Science*, *39*(19), 5887–5904. doi:10.1023/B:JMSC.0000041686.21788.77

Opila, E. J. (1994). Oxidation Kinetics of Chemically Vapor-Deposited Silicon Carbide in Wet Oxygen. *Journal of the American Ceramic Society*, *77*(3), 730–736. doi:10.1111/j.1151-2916.1994.tb05357.x

Pakseresht, A., Javadi, A., Bahrami, M., Khodabakhshi, F., & Simchi, A. (2016). Spark plasma sintering of a multilayer thermal barrier coating on Inconel 738 superalloy: Microstructural development and hot corrosion behavior. *Ceramics International*, *42*(2), 2770–2779. doi:10.1016/j.ceramint.2015.11.008

Pakseresht, A., Javadi, A., Ghasali, E., Shahbazkhan, A., & Shakhesi, S. (2016). Evaluation of hot corrosion behavior of plasma sprayed thermal barrier coatings with graded intermediate layer and double ceramic top layer. *Surface and Coatings Technology*, *288*, 36–45. doi:10.1016/j.surfcoat.2016.01.012

Pakseresht, A., Rahimipour, M., Alizadeh, M., Hadavi, S., & Shahbazkhan, A. (2016). Concept of advanced thermal barrier functional coatings in high temperature engineering components. In Research perspectives on functional micro-and nanoscale coatings (pp. 396–419). Hershey, PA: IGI Global. doi:10.4018/978-1-5225-0066-7.ch015

Pakseresht, A., Rahimipour, M., Vaezi, M., & Salehi, M. (2016). Thermal plasma spheroidization and spray deposition of barium titanate powder and characterization of the plasma sprayable powder. *Materials Chemistry and Physics*, *173*, 395–403. doi:10.1016/j.matchemphys.2016.02.028

Pakseresht, A. H., Ghasali, E., Nejati, M., Shirvanimoghaddam, K., Javadi, A. H., & Teimouri, R. (2015). Development empirical-intelligent relationship between plasma spray parameters and coating performance of Yttria-Stabilized Zirconia. *International Journal of Advanced Manufacturing Technology*, *76*(5-8), 1031–1045. doi:10.1007/s00170-014-6212-x

Pakseresht, A. H., Javadi, A. H., Nejati, M., Shirvanimoghaddam, K., Ghasali, E., & Teimouri, R. (2014). Statistical analysis and multiobjective optimization of process parameters in plasma spraying of partially stabilized zirconia. *International Journal of Advanced Manufacturing Technology*, *75*(5-8), 739–753. doi:10.1007/s00170-014-6169-9

Paul, B., Majumdar, S., & Suri, A. (2013). Microstructure and mechanical properties of hot pressed Mo–Cr–Si–Ti in-situ composite, and oxidation behavior with silicide coatings. *International Journal of Refractory Metals & Hard Materials*, *38*, 26–34. doi:10.1016/j.ijrmhm.2012.12.005

Raunija, T. S. K., & Babu, S. (2013). Randomly oriented carbon/carbon composite. *Paper presented at the AIP Conference Proceedings*. doi:10.1063/1.4810050

Roy, J., Chandra, S., Das, S., & Maitra, S. (2014). Oxidation behaviour of silicon carbide-a review. *Reviews on Advanced Materials Science*, *38*(1), 29–39.

Savage, E. (2012). *Carbon-carbon composites*. Springer Science & Business Media.

Sheehan, J. E. (1989). Oxidation protection for carbon fiber composites. *Carbon*, *27*(5), 709–715. doi:10.1016/0008-6223(89)90204-2

Sheehan, J. E. (1993). *High-Temperature Coatings on Carbon Fibers and Carbon-Carbon Composites. by JD Buckley and Dan D. Edie, Carbon-Carbon Materials and Composites* (pp. 224–253). Park Ridge, NJ: Noyes Publications.

Shimada, S., & Sato, T. (2002). Preparation and high temperature oxidation of SiC compositionally graded graphite coated with HfO 2. *Carbon*, *40*(13), 2469–2475. doi:10.1016/S0008-6223(02)00159-8

Shirvanimoghaddam, K., Hamim, S. U., Akbari, M. K., Fakhrhoseini, S. M., Khayyam, H., Pakseresht, A. H., & Jia, S. et al. (2017). Carbon fiber reinforced metal matrix composites: Fabrication processes and properties. *Composites. Part A, Applied Science and Manufacturing*, *92*, 70–96. doi:10.1016/j.compositesa.2016.10.032

Snell, L., Nelson, A., & Molian, P. (2001). A novel laser technique for oxidation-resistant coating of carbon–carbon composite. *Carbon*, *39*(7), 991–999. doi:10.1016/S0008-6223(00)00200-1

Strife, J. R., & Sheehan, J. E. (1988). Ceramic coatings for carbon-carbon composites. *American Ceramic Society Bulletin*, *67*(2), 369–374.

Tandon, R., Dumm, H. P., Corral, E. L., Loehman, R. E., & Kotula, P. G. (2006). *Ultra high temperature ceramics for hypersonic vehicle applications.*

Vast, P. (2009). Oxidation protection for carbon materials by oxifluoride phosphate compounds. *Paper presented at the 19th Annual Conference on Composites, Advanced Ceramics, Materials, and Structures-B: Ceramic Engineering and Science Proceedings*. doi:10.2172/887260

Vasudevan, A., & Petrovic, J. (1992). A comparative overview of molybdenum disilicide composites. *Materials Science and Engineering A*, *155*(1-2), 1–17. doi:10.1016/0921-5093(92)90308-N

Wang, K.-T., Cao, L.-Y., Huang, J.-F., & Fei, J. (2013). A mullite/SiC oxidation protective coating for carbon/carbon composites. *Journal of the European Ceramic Society*, *33*(1), 191–198. doi:10.1016/j.jeurceramsoc.2012.08.009

Wang, P., Zhou, C., Zhang, X., Zhao, G., Xu, B., Cheng, Y., & Han, W. et al. (2016). Oxidation protective ZrB 2–SiC coatings with ferrocene addition on SiC coated graphite. *Ceramics International*, *42*(2), 2654–2661. doi:10.1016/j.ceramint.2015.10.072

Webster, J., Westwood, M., Hayes, F., Day, R., Taylor, R., Duran, A., & Vogel, W. et al. (1998). Oxidation protection coatings for C/SiC based on yttrium silicate. *Journal of the European Ceramic Society*, *18*(16), 2345–2350. doi:10.1016/S0955-2219(98)00241-6

Westwood, M., Webster, J., Day, R., Hayes, F., & Taylor, R. (1996). Oxidation protection for carbon fibre composites. *Journal of Materials Science*, *31*(6), 1389–1397. doi:10.1007/BF00357844

Xin, Y., Huang, Q.-Z., Zou, Y.-H., Chang, X., Su, Z.-A., Zhang, M.-Y., & Xie, Z.-Y. (2009). Anti-oxidation behavior of chemical vapor reaction SiC coatings on different carbon materials at high temperatures. *Transactions of Nonferrous Metals Society of China*, *19*(5), 1044–1050. doi:10.1016/S1003-6326(08)60404-5

Yue, G., Hong-yu, X., Zeng-rong, J., & Zhi-xiang, X. (2015). Mechanical Properties and Thermal Shock Resistance of Rhenium Coating in Iridium/Rhenium/Carbon-Carbon Composites. *Procedia Engineering*, *99*, 1407–1414. doi:10.1016/j.proeng.2014.12.676

Zeng, Y., Xiong, X., Guo, S., & Zhang, W. (2013). SiC/SiC–YAG–YSZ oxidation protective coatings for carbon/carbon composites. *Corrosion Science*, *70*, 68–73. doi:10.1016/j.corsci.2013.01.013

Zhang, J.-P., Fu, Q.-G., Qu, J.-L., Zhou, H.-S., & Liu, N.-K. (2016). Surface modification of carbon/carbon composites and in-situ grown SiC nanowires to enhance the thermal cycling performance of Si-Mo-Cr coating under parallel oxyacetylene torch. *Corrosion Science*, *111*, 667–674. doi:10.1016/j.corsci.2016.06.007

Zhang, Y., Yang, B., Zhang, P., Zhang, J., Ren, J., & Hu, Z. (2015). SiC nanowire toughened ZrB 2–SiC ablative coating for SiC coated C/C composites. *Ceramics International*, *41*(10), 14579–14584. doi:10.1016/j.ceramint.2015.07.175

Zhang, Y.-L., Li, H.-J., Yao, X.-Y., Li, K.-Z., & Qiang, X.-F. (2011). Oxidation protection of C/SiC coated carbon/carbon composites with Si–Mo coating at high temperature. *Corrosion Science*, *53*(6), 2075–2079. doi:10.1016/j.corsci.2011.02.024

Zhao, J., Liu, L., Guo, Q., Shi, J., & Zhai, G. (2006). Oxidation protective behavior of SiC/Si–MoSi 2 coating for different graphite matrix. *Materials Letters*, *60*(16), 1964–1967. doi:10.1016/j.matlet.2005.12.072

Zhong Liu, W., Peng, X., Zhuan, L., Wen, H., Heng, L., Xiao Yu, Y., . . . Wen Bo, C. (2015). Microstructure and oxidation behavior of sol–gel mullite coating on SiC-coated carbon/carbon composites. *Journal of the European Ceramic Society, 35*(14), 3789-3796.

KEY TERMS AND DEFINITIONS

Bond Coat: The coating between the substrate and top coat is a bond coat, which protection of oxidation and decrease thermal expansion coefficient between the substrate and another coating.

C/C Composite: The carbon fibers in carbon matrix is carbon-carbon composite.

Multilayer Coating: A system consisting of a layer on each other which each layer has a task.

Oxidation: In oxidation phenomena, the temperature is sufficiently high to the reaction between materials and oxygen and causes failure.

Protective Coating: The layer applied on the surface of materials and protect from failures such as oxidation, corrosion, and wear.

SiC Coating: The coating inserts on C-C composite to prevent oxygen diffusion by SiO_2 glassy forming.

Ultra-High-Temperature cEramics: Ultra-high-temperature ceramics (UHTCs) are borides, carbides, nitrides, and oxides of transition metals and have temperature stability at very high temperature.

Chapter 17
Hardness of Duplex Treated Stainless Steel:
Influence of the Architecture and Composition of Films

Umm-i-Kalsoom
University of Engineering and Technology Lahore – Kala Shah Kaku, Pakistan

Nisar Ali
University of Engineering and Technology Lahore – Faisalabad, Pakistan

Narjis Begum
COMSATS Institute of Information Technology, Pakistan

Ali Hussnain
University of Education, Pakistan

Riaz Ahmad
Government College University, Pakistan

ABSTRACT

This chapter describes how the aim of duplex surface engineering includes chronological application of two surface modification technologies for the production of a surface, with collective properties. Duplex treatment of nitriding and carbonitriding of austenitic stainless steel is of high technical importance owing to its capability to increase hardness, corrosion and wear resistance of treated surface. Duplex treatment has been utilized to enhance the surface mechanical properties of austenitic stainless steel (AISI 304). The microstructure of nitrided surface indicates the development of nitride phases, Fe_4N, Fe_2N, CrN, Cr_2N and γ_N whereas, duplex treated films shows the formation of FeC, Fe_3C, Fe_7C_3, Cr_3C_2, Cr_7C_3, along with nitride phases like Fe_3N. Both nitrided and duplex treated samples show the formation of cauliflower like grains. Surface micro hardness of treated substrates has been dependent on the variation of crystallite size and increased by 1.26 times the hardness of nitrided sample and 4.60 times the hardness of the untreated substrates.

DOI: 10.4018/978-1-5225-4194-3.ch017

INTRODUCTION

Surface hardening of austenitic Stainless Steel (S.S) has been a topic of current research work (Tschiptschin, 2017, Hossary, 2009). S.S material has enhanced oxidation temperature, elevated corrosion and adhesion to substrate but with low wear resistance and surface hardness. These weak properties make them worthless in various applications where good wear resistance is required. These weaknesses can be overcome by improving the surface properties of materials through structuring or coatings (nitriding, carburizing and carbonitriding). Surface treatment can enhance the tribological, mechanical and optical properties of different materials as well as their life time.

SURFACE HARDENING TECHNIQUES

Nitriding and carburizing are two of the best processes, used to increase the surface hardness of some metallic materials without affecting their bulk properties. Nitriding is a surface hardening technique used to enhance the surface corrosion resistance and to improve the surface hardness, wear resistance and fatigue life time of the treated materials.

The plasma nitriding also termed as ion nitriding, is accepted as a cathodic process escorted by bombardment of positive ions. In case of pulsed DC glow discharge the cathode will serve as target and will be bombarded with positive ions, causing the sputtering of the target material, e.g. Fe in S.S. The sputtered Fe atoms will react with the nitrogen atoms and will form the nitrides of Fe i.e., FeN. These nitrides can get deposited on the target surface but as being unstable elements so get further reactions with iron atoms until they form a stable Fe_4N product.

Atomic Nitrogen (N) can diffuse in the target material and forms nitride layer (Figure 1).

Figure 1 shows different steps for the formation of nitride film on the substrate surface. The cathode that is serving as a target is sputtered due to presence of bombarding nitrogen ions e.g. Fe. These sputtered atoms will react in plasma and form FeN. These nitrides will further react with the target (e.g. Fe) atoms and will form Fe_2N, Fe_3N and finally to the stable product (Fe_4N). The atomic nitrogen that is released in all this kind of reactions will diffuse through the target and will cause nitriding of the substrate surface. During plasma nitriding, the material surface gets saturated with the nitrogen that's why plasma nitriding is much faster than other conventional nitriding techniques. This process can be controlled by selecting the suitable parameters like nitriding time, distance between the electrodes, gas pressures and concentrations etc.

Carburizing

Carburizing is achieved by introducing an extra amount of carbon into the surface treated layer. Plasma carburizing is similar to plasma nitriding process. Plasma assisted carburizing is more advantageous to conventional carburizing techniques as we can get more uniform film in less process time (Bogaert, 2002).

Carbonitriding

Carbonitriding is one more method for surface hardening of material. It is basically a combination of plasma carburizing and nitriding process. It can be carried out in presence of mixture of two gases i.e.

Figure 1. Schematic representation of formation of nitride layer on substrate surface

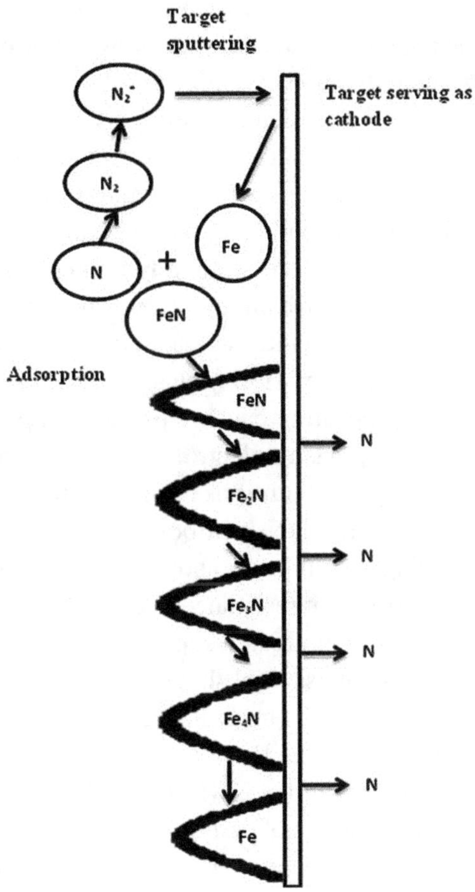

N_2 and Methane, or in the presence of N_2+Ar and graphite target (a carbon source in the present case) (Bogaert, 2002).

Various research groups have worked on improvement in surface properties of various metals through duplex treatment using nitriding, carburizing, or carbonitriding (Kao, 2017, Yang, 2017, Jacquet, 2017) using various techniques like radiofrequency discharges (El-Hossary, 2009), glow discharged plasmas (Selcuk, 2003, Liang, 2001), pulsed DC magnetron sputter system (Zhang, 2017) etc. For the improvement of above mentioned properties. Enhancement in tribomechanical properties of S.S has been observed by Braam et al. (Braam, 1999) after nitriding. Nowadays plasma carbonitriding (Silva, 2017, Catteau, 2016) is most commonly used as an industrial way to enhance the surface tribological and mechanical properties of S.S substrates. During carbonitriding process mixture of carbon, nitrogen and possibly hydrogen at the surface of treated substrates is achieved that adequately diffuse in the instinctive material (Silva, 2017, Catteau, 2016). The carbonitrided metal's case structure generally contains diffusion as well as compound zones. In original metal, diffusion zone is supposed to be the solid solution of carbon and nitrogen and at the surface of material the formed compound zone contains new phases of carbonitrided layer with much higher concentration of nitrogen and carbon. The diffusion zone is generally thick whereas, the compound zone is normally thin, only few tens of microns or may be less than micrometer (Dimitrov, 1996, El-Hossary, 2002).

Ability of treatment of parts of complex shape, reduced process time and removal of negative effects on the smoothness of the surface of treated parts are the key advantages of plasma based carbonitriding process. Furthermore, by this process, the produced layers show better plasticity and more resistance to fatigue and wear (Silva, 2017, Wierzchon, 1993).

Duplex treatment procedure has been considered recently as a fresh industrialized surface processing technique to enhance the properties of the surface of steel and other alloys (Das, 2017, Hariprasad, 2017, Deng, 2017), because of their advantageous effects on adhesion between the coating and substrate and the substrate material's capacity to bear load. This surface modification technique joins two series methods; a hard thin film is coated on the surface of already treated substrate (nitrided or carbided). It has been observed that the substrate should have enough flow strength and hardness to upkeep the thin film with no plastic deformation when applied intense load. Duplex treatment consisting of the process of calorizing following plasma nitriding (Lee, 1997), plasma nitriding resulting from physical vapor deposited coating (Podgornik, 2001), preheating causing physical vapor deposited coating (Pellizzari, 2001) and diamond like carbon followed by plasma electrolytic nitrocarburising (Tsotsos, 2002) are stated here as examples. By controlling the varying conditions of specific techniques we can control the duplex treatment process. Individual coatings of CrN, TiN or carbon deposition were attained on various nitrided targets using several duplex processing techniques exposed the frivolous tribological characteristics, better load-bearing ability, enhanced corrosion resistance in comparison to simply nitrided samples and to the direct CrN, TiN or carbon coating on untreated substrates (Wilson, 1993, Mouri, 2001).

It is observed that carbonitriding of stainless steel shows improved corrosion resistance compared to those prepared by pure carburizing or nitriding. This enhancement in corrosion resistance is attributable to the occurrence of hard FCC phase γ_N (Abd El-Rahman, 2004), which has good corrosion resistance compared to γ- austenite (Zhang, 1985). Blawert et al. (Blawert, 2000) observed that carbide and nitride phases along with the hard solid solution phases of carbon expanded austenite γ_c and nitrogen-expanded austenite γ_N are responsible for substantial enhancement of surface microhardness.

To enhance the mechanical characteristics of AISI 304 stainless steels, duplex treatment of nitriding and carbonitriding has been employed using pulsed dc glow discharge plasmas. A carbonitride film was developed on the nitrided stainless steel substrates by using graphite target as source of carbon and different gas composition of nitrogen and argon w.r.t gas pressure ratio. This chapter will give deep insight to the influence of pulsed dc glow discharge plasma nitrided and duplex treated film's architecture and chemical composition on the hardness of AISI stainless steel 304. Furthermore, this chapter will elucidate and formalize the underlying reasons for the relationships between the chemical composition, crystallinity, film roughness, thickness and hardness.

EXPERIMENTAL SETUP

The Duplex surface treatment process was performed by using dc glow discharge plasma with graphite as target material in nitrogen and argon environment. Austenitic Stainless steel 304 rod having 10 mm diameter, was cut into pieces with dimensions 5×10 mm^2 and these samples were grinded, polished and ultrasonically cleaned for 30 min in water and then washed in acetone. Schematic of experimental setup is revealed in Figure 2. Electrode system comprises of two parallel plate stainless steel electrodes. Both electrodes were circular in shape having thickness of 2 cm and diameter of 9.5 cm.

Figure 2. Schematic diagram of experimental setup

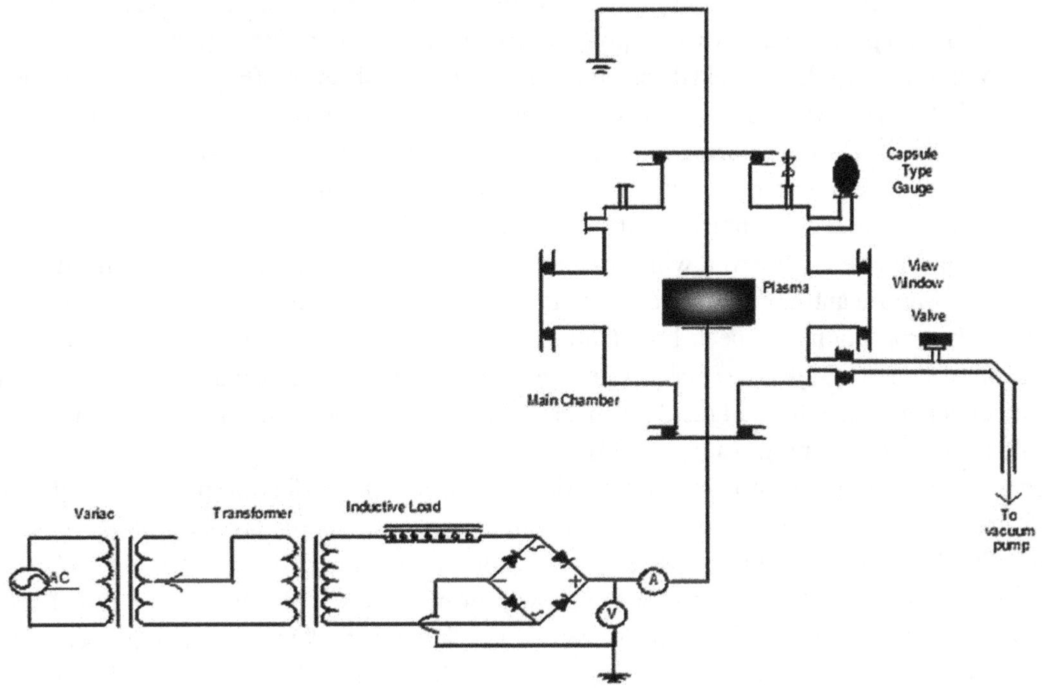

Inter spacing of about 3 cm was kept constant, amongst electrodes throughout experiment. In plasma nitriding, the pre-treatment process, 70% nitrogen and 30% argon were introduced into the chamber at 2 mbar base pressure. The substrates were treated for a processing time of 4 hours at 150 W fixed input power. Nitrided substrates were then carbonitrided by utilizing graphite target as source of carbon and different nitrogen (N_2) to argon (Ar) partial pressure ratio varies from 5% N_2/95% Ar to 30% N_2/70% Ar. The process of carbonitriding was carried out at same plasma power of 150 watt and plasma processing time of 4 hour.

Various characterization methods were employed to examine the nitrided and duplex treated substrates. The structural properties of the deposited films were investigated for the crystallographic structure by using Panalytical X'Pert PRO MPD X-ray diffractometer (XRD) with CuKα (λ=15.406 nm) radiation source, at grazing incidence (6°). The surface morphology of duplex treated substrate is investigated using JEOL (JSM-6480-LV) scanning electron microscope (SEM). Atomic force microscope (AFM) picoplus 2500 is used to characterize RMS roughness and film thickness of the treated samples. Vickers Hardness tester is used to evaluate, the surface hardness of untreated and treated substrates.

RESULTS AND DISCUSSION

Microstructure Analysis

X-ray diffraction analysis is used to measure the variation in cryastallinity and residual stresses by means of X'Pert PRO' (MPD) Xray diffractometer. Operated for voltage and current values of 40 KV & 40 mA, with CuK_α radiation source in air ambient, at room temperature. The pristine sample reveals existence of FCC austenitic stainless steel (γ) and BCC ferritic iron (α-Fe) reflection planes. XRD pattern shows the diffraction peaks related to γ-Fe phases correspond to (111), (200), (220), (311), (222) plane reflections. While α-phase having (110) plane reflection is also observed (Figure 3) (C. Blawart, 1996 , A. M. Abd El-Rahman, 2004). Formation of α-phase is attributable to transformation, induced due to thermal stresses from 'γ' to 'α', during manual polishing (Umm-i-Kalsoom, 2013).

The nitriding process was carried out at 2 mbar pressure for 70% Ar to 30% N_2 ratio, 150 watt power and 4 hours treatment time. After nitriding of stainless steel, reduction in peak intensity of all γ-Fe phases and vanishing of α phase is observed which is attributable to the formation of thicker nitride layer (El-Hossary, 2009). Significantly broader peaks on nitride layer are observed. From Bragg's law crystallite size is inversely proportional to peak broadening so above-mentioned results gives the signatures of the formation of nano-crystallites. During nitriding process ions from plasma causes the formation of surface defect generation enhanced residual stress or gradient of atomic nitrogen (N) in the surface are responsible for peak broadening (Liang, 2003).

Nitrogen absorption in the surface at interstitial cites moves the reflection planes to higher angular positions that causes the peaks to appear at higher angular position that is property of γ_N-Phase (Umm-i-Kalsoom, 2013). This shows that γ austenite phase converts to N expanded γ_N having FCC structure. However, part of N_2 content is utilized for the growth of new phases of nitrides too. Enhancement in surface temperature during nitriding process causes the formation of various new phases of ferrous/chromium nitride, i.e. ferrous nitride Fe_4N (110), (111), (200), Fe_2N (130) and chromium nitride CrN (111), (200), (220), Cr_2N (112).

Figure 3. XRD diffractogram of pristine and nitrided samples of AISI 304 stainless steel when pulsed dc glow discharge for constant power (150 W) and treatment time of (4 hours) with gas composition of 70% N_2 and 30% Ar at a base pressure of 2 mbar.

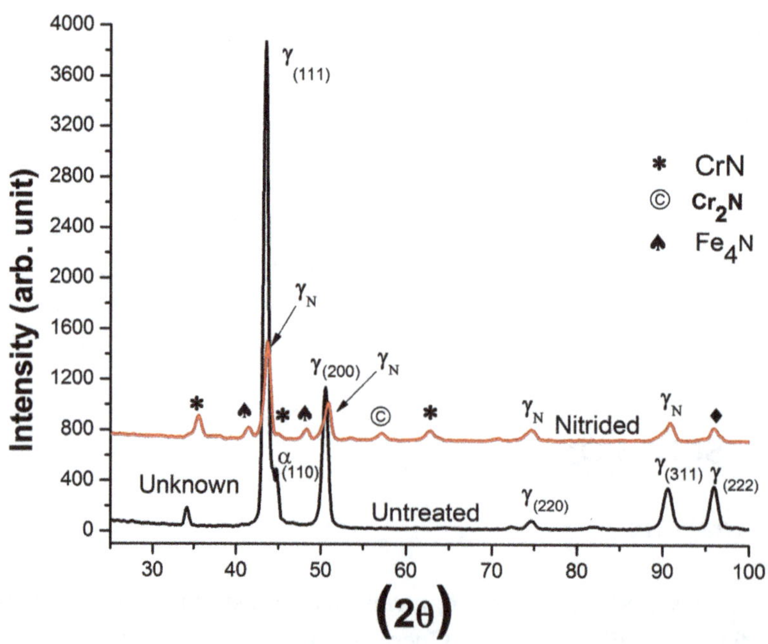

Figure 4 represents the XRD diffractograms and variations in crystallite density and residual stresses of duplex treated stainless steel samples. The nitrided samples were then carbonitrided in the presence of graphite target (carbon source) and varying for various N_2 to Ar ratios (5% N_2/95% Ar to 30% N_2/70% Ar), for a processing time of 4 hours at 150 W fixed input power. Sputtered Fe & Cr ions recombine with carbon ions, to give various iron/chromium carbide plane reflections, such as, Fe_3C (400), FeC (015), Fe_3C (212), Fe_7C_3, Cr_3C_2 (111), Cr_7C_3 (151) along with iron nitride Fe_3N (004), (202), and (200) phases. After duplex treatment γ-phase converts to nitrogen expanded (γ_N) or carbon expanded γ_C phases (Hossary, 2009). Crystallite size is evaluated by using Debye Sherer's formula (Olaya, 2005 , Sanjines, 2006) for Cr_3C_2 (111) plane.

Crystallite Size (C.S)

The C.S of any reflection plane after coating can be calculated using Scherer's formula (Cullity, 1978).

$$C.S\left(D\right) = {\left(0.9\right) * \lambda}\Big/{\left(FWHM\right) * \cos\left(\theta\right)} \tag{1}$$

where λ is incident wavelength of X-rays ($\lambda = 15.4$ nm), θ (rad) is Bragg's angle and FWHM is the full width at half maxima.

XRD data can provide the information regarding the crystallinity, defect density and residual stresses on the deposited surface

Dislocation Density (D.D)

Lattice defects and imperfection in the crystallographic structures is called dislocation. Dislocations are produced when periodicity, of atomic lattice array gets interrupted by some surface treatments along some certain direction in the crystal. Dislocations can be achieved due to aggregation of point defects and interface misfit dislocation networks. The D.D is number of lines of dislocations intersecting unit area in a crystal. It can be defined as length of dislocation lines, in a unit area and can be calculated using formula given in equation 2 (Venkatachalam, 2004).

$$\text{Dislocation Density } (D.D) = {1}\Big/{D^2} \tag{2}$$

Residual Strain

The in-plane residual strain "ε" causes broadening of crystal lattice and therefore a displacement ($d-d_0$) of corresponding XRD reflection planes. For evaluation of strain and the dislocation line density following formulae are used (Ashraf, 2011, Khan, 2008)

$$Strain\left(\varepsilon\right) = \frac{d - do}{do} \tag{3}$$

From above mentioned formulae it is clear that induced surface strain (ε) depends on the observed (d) and standard (d_0) inter planer spacing. The value of Young's modulus (E) for Cr_3C_2 reflection plane is 373.13 GPa (https://www.memsnet.org/material/chromiumcarbidecr3c2bulk/), residual stress induced on the treated substrates can be evaluated from the following relation (Ashraf, 2011, Khan, 2008).

$$Stress\left(\sigma\right) = \varepsilon E \tag{4}$$

Figure 4 (a) shows the XRD diffractograms of duplex treated samples, showing the variation in 2Θ values (peak position) intensity and d-spacing of diffraction planes. The duplex treatment of steel surface changes its microstructure and Bragg's diffraction conditions. These variations are attributable for the changes in d-spacing and intensity of diffraction lines (Mahmood, 2010).

Both the peak intensity (Figure 4a) and crystallinity (Figure 4 b) of Cr_3C_2 (111) plane reflection are found to increase with the increase in concentration of N_2 in Ar up to 20%. The increased intensity and crystallinity of Cr_3C_2(111) reflection plane is due to increased diffusion of carbon into the surface about grain boundaries (Mahmood, 2010) and enhanced diffraction of X-rays from the substrate surface (Khan, 2008, Ali, 2017). The increase in crystallite size can also be due to high energy strain or higher crystallinity of carbonitride layer (Shan, 2009, Ali, 2017) or it may be due to increased adatom mobility caused by higher ion energies, enhancing atomic migrations to grain boundaries and increasing crystallite size (Jelani, 2014). The diffusion of carbon in steel is also exhibited in EDX analysis of table 1. The peak intensity and crystallinity of Cr_3C_2 (111) reflection plane decreases on further increase in concentration of nitrogen in argon up to 30%. Lattice strain and small crystallite size are responsible for this decrease in peak intensity with increase in concentration of nitrogen (Mahmood, 2010, Ali, 2017). Interstitial dissemination of carbon ions/atoms produces microstrain defects which are responsible for reduction in the crystallite size and peak intensity along with an increase in FWHM (Ali, 2013, Meunier, 1998) or it may be due to continuous re-nucleation of growing film due to ion irradiation defects (Sanjines, 2006).

Figure 4 (b) also shows the effect of nitrogen to argon concentration ratio on residual stresses produced within the film. Enough amount of energy delivered to the substrate by striking ions during deposition process that causes the surface temperature to rise locally and results in lattice distortion by changing the d-spacing owing to the variances in coefficients of thermal expansion and inter atomic distances (Kusaka, 2000). The development of residual stresses in the deposited thin film due to energetic nitrogen/carbon ions/atoms is depicted by the displacement of diffraction peaks compared to the stress-free data. Production of residual stresses can also be explained on the basis of strains produced in the lattice due to nitrogen/carbon ions implantation on to the interstitial lattice position and production of defects (ion vacancies).

Inset for Figure 4 (a) shows the lower and higher angle shift of Cr_3C_2 plane reflection. Peak shift to lower angular position confirms the production of tensile residual stresses (Umm-i-Kalsoom, 2013). Whereas, higher angle peak shift from the corresponding stress-free data confirms formation of compressive residual stresses (Ali, 2017).

For 5% nitrogen in argon, less compressive stresses are observed for Cr_3C_2 (111) reflection plane. These stresses are produced due to small crystallite size (Figure 4 a) owing to dissemination of carbon atoms (Ali, 2017, Ali, 2013) into the surface. The reduced crystallite size reduces the slip motion by providing more boundary area and results in appearance of high residual stresses and in turn, strengthens the material. As nitrogen content is increased to 20% relaxation in compressive stresses and their

Figure 4. (a) XRD diffractogram of pristine and duplex treated samples of AISI 304 stainless steel, carbonitriding was carried out by using different nitrogen (N_2) to argon (Ar) partial pressure ratios varying from 5% N_2/95% Ar to 30% N_2/70% Ar for plasma power of 150 watt and plasma processing time of 4 hours, (b) change in crystallite size and surface residual stresses with enhancement in concentration of nitrogen in argon.

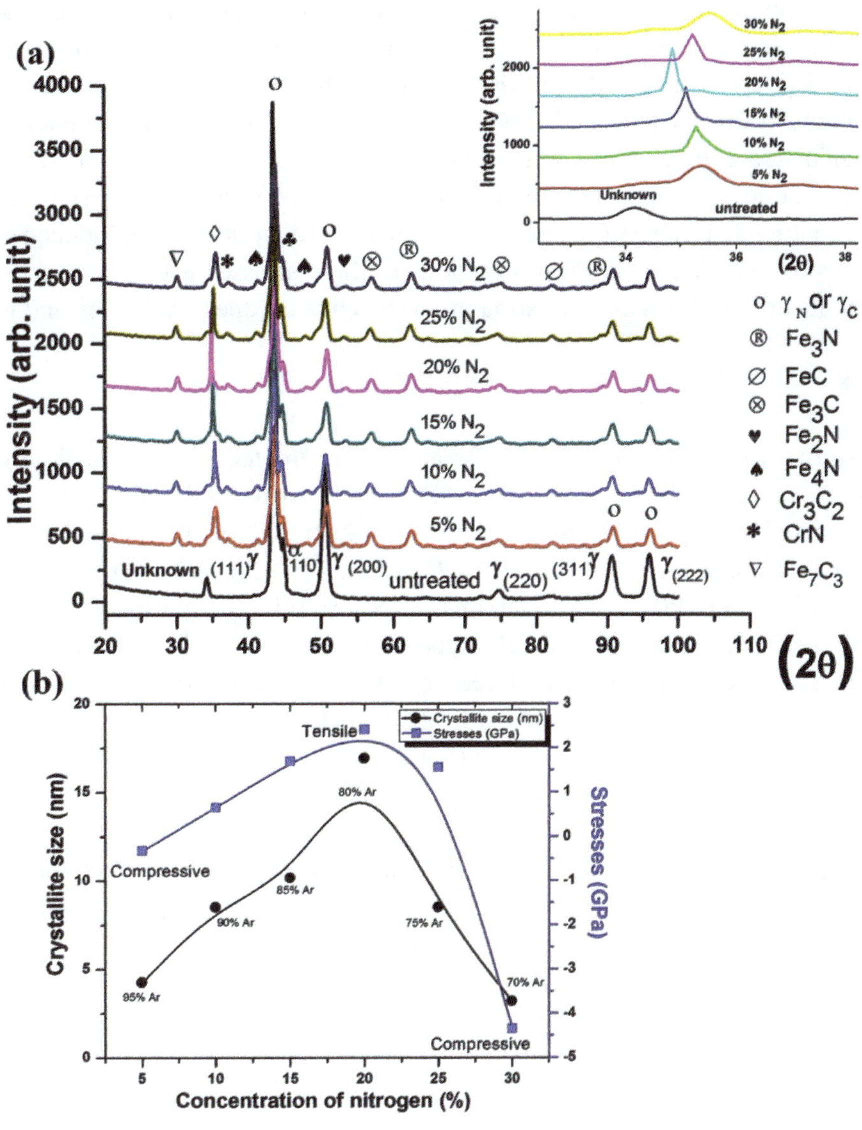

conversion to tensile stresses is achieved. Sputtering of the deposited Cr_3C_2 film by high energy ions may damage the film causing the compressive stresses to relax and transforms in to tensile residual stresses (W. L. Li, 2006, M. Ode´n, 1999). On the other hand Cr_3C_2 phase begins to grow, which results in greater tensile residual stresses. Inset of Figure 4 (a) shows peak shift to lower angular position which confirms the presence of tensile residual stresses. On further increase in nitrogen concentration up to 30% in argon, tensile stresses transform to compressive stresses. This relaxation and transformation of tensile residual stresses to compressive stresses is attributable to reduction in crystallite size. Interstitial carbon diffusion into the surface results in decrease in crystallite size which in turn results in relaxation of tensile residual stresses and their transformation to compressive residual stresses (Ali, 2013). Inset of Figure 4 (a) shows shift of peak to higher angular position which confirms the relaxation of tensile residual stresses and their transformation to compressive residual stresses (Onge, 1998).

Residual stresses in ceramic coatings have values of up to several GPa and are found to depend strongly on input power, pressure, process time and energy of incident ions etc. During film growth, the ion bombardment delivers energy to the atoms reaching at the surface of the growing film, permitting them to find strenuously affirmative site by moving about the surface. Increase in energy of the surface atoms results in an increase in density of defect which will be responsible for increase in residual stresses (Constable, 2004). The development of residual stresses depends on variance in the coefficients of thermal expansion of the coating material and substrate or it may depend on the growth induced stress generated during deposition process (Petrov, 1997). Formations of these stresses within the coating is important due to their pronounced effects on many coating properties like hardness, adhesion and wear resistance.

EDS Analysis

Chemical analysis of the untreated, simple nitrided and duplex treated stainless steel samples treated for varying concentration ratio of nitrogen in argon (5% N_2/95% Ar to 30% N_2/70% Ar), for a processing time of 4 hours at fixed input plasma power of 150 W is illustrated in table 1. Untreated sample shows following contents: Fe ~68.60 wt. %, balanced to 100% by Ni (~9.25 wt.%), Cr (~19.0 wt.%), Si (~1.0 wt.%) Mn (~2.0 wt.%), C (~0.08 wt.%), P (~0.04 wt.%), S (~0.03 wt.%). Nitrided sample shows increase in content of nitrogen up to 30.14 wt.%. Duplex treated samples show diffusion of nitrogen/carbon in to the substrate surface (stainless steel). Sputtered Fe & Cr ions react with nitrogen/carbon ions to give various iron and chromium nitride phases after nitriding, such as, iron nitride Fe_4N (110), (111), (200), Fe_2N (130) and chromium nitride CrN (111), (200), (220), Cr_2N (112).

Table 1. Chemical identification of the untreated, simple nitrided and duplex treated stainless steel samples treated for varying concentration ratio of nitrogen in argon (5% N_2/95% Ar to 30% N_2/70% Ar), for a processing time of 4 hours at fixed input plasma power of 150 W.

Elements	Untreated Wt. %	Simple Nitriding	5% N_2 Wt. %	10% N_2 Wt. %	15% N_2 Wt. %	20% N_2 Wt. %	25% N_2 Wt. %	30% N_2 Wt. %
Ni	9.25	8.21	9.12	8.59	8.24	8.65	8.94	8.62
Cr	19.00	9.35	10.36	10.25	9.56	9.47	7.95	8.56
Fe	68.60	49.58	45.20	46.05	46.72	52.6	46.26	43.36
Si	1.00	0.93	0.82	0.98	1.0	0.95	0.81	0.96
Mn	2.00	1.69	1.52	1.35	1.52	1.67	1.69	1.54
C	0.08	0.06	18.63	17.59	18.24	18.63	19.97	21.71
P	0.04	0.03	0.02	0.03	0.02	0.01	0.03	0.03
S	0.03	0.01	0.01	0.02	0.01	0.01	0.01	0.01
N	----	30.14	14.32	15.14	14.69	13.21	14.34	15.21

After carbonitriding of pre-nitrided samples of stainless steel various carbides of ferrous and chromium are formed along with nitrides, such as, Fe_3C (400), FeC (015), Fe_3C (212), Fe_7C_3, Cr_3C_2 (111), Cr_7C_3 (151) along with ferrous nitride Fe_3N (004), (202), and (200) phases.

Surface Morphological Analysis

The surface morphology of untreated, simple nitrided and duplex treated stainless steel substrates is investigated by employing scanning electron microscope (SEM- JEOL JSM-6480 LV). Figure 5 shows the SEM micrographs of (a) untreated, (b) simple nitrided and (c-h) duplex treated samples treated for various nitrogen concentrations in argon with graphite target as carbon source, i.e. (c) 5% N_2, (d) 10% N_2, (e) 15% N_2, (f) 20% N_2, (g) 25% N_2, (h) 30% N_2 balanced to 100% with argon. Figure 5 (a) demonstrates the surface morphology of the untreated substrate showing the scratches over surface induced owing

Figure 5. SEM images showing variation in surface morphology of (a) untreated, (b) nitrided and (c-h) duplex treated samples of AISI 304 stainless steel treated for different nitrogen (N_2) to argon (Ar) partial pressure ratios varying from(c) 5% N_2 (d) 10% N_2, (e) 15% N_2, (f) 20% N_2, (g) 25% N_2, (h) 30% N_2balanced to 100% by byAr for plasma power of 150 watt and plasma processing time of 4 hours

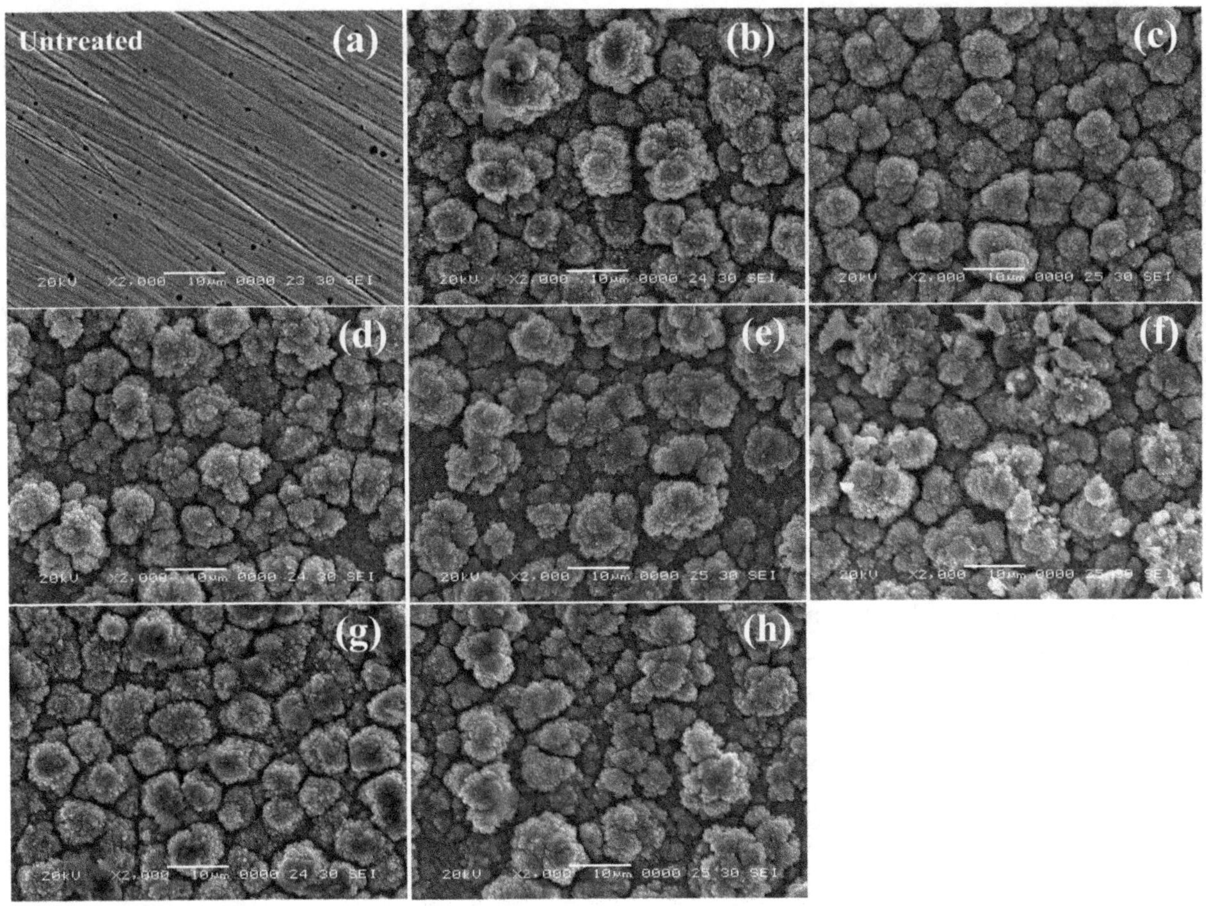

to the friction induced between the substrate surface and carbide paper during manual polishing. After treatment the morphology of the substrate surface changes to granular morphology.

Figure 5 (b) represents the morphology of the surface of nitrided sample treated for 70% nitrogen and 30% argon at base pressure of 2 mbar, input plasma power of 150 W and 4 hours processing time. Homogeneous nucleation and growth of simple nitrided layer is seen which is attributable to energized nitrogen ions bombardment. Nitrided sample shows cauliflower like granular morphology. Grains are randomly packed having wider spaces between them. Figures 5 (c-h) shows the morphology of carbo-nitrided samples treated for various concentrations of nitrogen (5% to 30% N_2) in argon for a processing time of 4 hours at fixed plasma power of 150 watt. Similar to nitrided samples, cauliflower like granular morphology is observed but the size of grains is smaller than nitrided samples.

By increasing nitrogen concentration up to 20% in argon (20% N_2/80% Ar) increase in grain size is observed. Redeposition of the sputtered material is also observed for 20% nitrogen in argon. During the process of deposition, the adatoms of the substrate surface gain K.E from ions which causes an increase in the adatom mobility and results in the growth of grains (Jelani, 2014) owing to presence of tensile residual stresses (presence of enhanced tensile residual stresses is observed from XRD results, Figure 4 b). Atoms at the grain boundary are loosely packed, and have higher values of energy as compare to the atoms in the lattice. The net amount of grain boundary reduces, as the grains grow in size and their numbers decrease, and thereby the free energy of the system reduces. An atom performs tedious thermal motion at the grain boundary. Most of the time, this atom oscillates with small magnitude but sometimes it gains sufficient energy to jumps from the boundary of one atom and get attached to the lattice of the other grain. The probability of jumping of electrons in forward and backward direction will be the same if the two states have the identical energy but the probability will be biased if the two states have different energies. This bias results in macroscopic migration of the grain boundary, in the direction that reduces the energy of the system and causes grain growth. Increase in concentration of nitrogen up to 30% shows reduction in size of grains due to formation of vacancies and vacancy clusters by the bombardment of energetic ions. This enhances the carbon/nitrogen diffusion and results in the reduction of grain size along with enhancement in compressive residual stresses which results in the formation of more compact film over the surface (presence of compressive residual stresses is confirmed by XRD analysis, Figure 4 b) (Garamoon, 2006).

During the process of simple nitriding, in the absence of graphite target, nitrogen atoms are responsible for the nucleation and growth of nitride crystals. However, during carbonitriding (with graphite as target) nucleation and development of carbonitride crystals will be dominated by carbon atoms (confirmed from EDS and XRD results). SEM micrograph of nitrided sample (Figure 3b) having larger and smaller cauliflower like grains. Grains are randomly packed having wider spaces between them. Similar morphology is observed for carbonitrided sample but with reduced cluster size and more compact film. According to Kolbel (1965) and Maliska (2001), high energy neutral particles, energetic nitrogen/argon ions and initially sputtered cathode surface atoms react with nitrogen atoms in the sheath space of cathode and condense on the substrate surface which is negatively charged. Therefore development of preliminary compound layer is considered as the vapor deposition process which is not dominated by the diffusion mechanism, as observed by Inal and Matin (Metin, 1987). Morphology of the surface of plasma treated substrates was influenced strongly by the mechanism of the formation of compound layer. The morphology of the surface is associated directly with the parameters of the growth of film, which in turn depends on the rise in temperature of the substrate surface due to energy deliverance by the energetic ion bombardment (Umm-i-Kalsoom, 2013).

AFM Analysis

AFM is employed to investigate the effect of variation in concentration of N_2 in Ar on the roughness and thickness of thin film of duplex treated samples. Nitrided sample shows rms roughness value of 5.24 nm and film thickness of 410 nm.

Figure 6 (a) represents the variations in rms roughness and film thickness with variation of concentration of N_2 in Ar. Decrease in both rms roughness and film thickness with increase in concentration of N_2 up to 20% is observed. Further increase in concentration of N_2 in Ar up to 30% results in increase in both roughness and thickness of duplex treated film. The reason for the initial decrease in roughness and thickness is the formation of nitrided compound layer (CrN, Cr_2N, Fe_4N) due to the reaction of ambient gas with stainless steel target, which decreases the sputtering yield of metal due to the lower sputtering rate and high emission of secondary electrons of CrN, Cr_2N, Fe_4N phases than metal. Increase in thickness and roughness of the film with increase in concentration of N_2 in Ar up to 30% attributable to more diffusion and deposition during bombardment of highly energetic nitrogen ions on the steel surface (Meunier, 1998).

Hardness Testing

"Hardness is defined as the material's resistance to deformation, indentation or penetration through drilling, scratching or wear that can be measured by using hardness tester."

Vickers hardness tester was employed to investigate the hardness of untreated, nitrided and duplex treated samples. Untreated sample shows hardness value of 250 HV0.5. Hardness of nitrided sample observed by Vicker's hardness tester is 906 HV0.5, which is 3.6 times the hardness of untreated substrate (250 HV0.5). This increase in hardness is attributable to the formation of FCC nitrogen expanded austenite γ_N and hard phase of Fe_4N along with other nitride phases induced during nitriding. In the present work we observed that the hardness of all duplex treated samples is higher than the nitrided samples. This increase in hardness is attributable to the formation of carbon expanded austenite phase (γ_C) and carbide phases (FeC, Fe_3C, Cr_7C_3, Cr_3C_2, Fe_7C_3) during the process of carbonitriding for different N_2/Ar concentration ratios because diffusion of carbon is higher than diffusion of nitrogen in austenitic stainless steels (Rahman, 2004, czerwiec, 2009).

Figure 6 (b) shows decrease in film thickness and hardness with increase in concentration of nitrogen up to 20% in argon. Further increase in nitrogen concentration in argon up to 30% shows an increase in film thickness and hardness. Hardness has its maximum value of 1150 HV0.5 for 30% N_2. Which is 1.26 times the hardness of nitrided and 4.60 times the hardness of the untreated substrates of stainless steel 304. The higher value of microhardness attained for duplex treated layer for different N_2/Ar concentration ratios of 30% is attributable to the combined effect of hard phases of Fe_4N, Fe_2N, Cr_2N, CrN and γ_N along with the formed carbide phases of FeC, Fe_7C_3,Fe_3C, Cr_7C_3, Cr_3C_2 and γc. These nitride and carbide phases produce additional internal stresses to the duplex treated layer.

Figure 6 (c) shows variation in hardness and crystallite size, whereas, Figure 6 (d) shows variation in hardness and residual stresses of Cr_3C_2 phase with increase in concentration of N_2 in Ar. Hardness varies inversely with crystallite size and residual stresses. For 5% N_2 less compressive stresses are attributable to hardness value of 1080 HV which is 4.32 times the hardness of untreated substrate. Increase in concentration of N_2 in Ar up to 20% results in transformation of compressive to increasing tensile residual stresses. Decrease in value of hardness of the treated samples up to 980 HV0.5 can be elucidated

Figure 6. The variation in (a) RMS roughness and film thickness, (b) hardness and film thickness, (c) hardness and crystallite size, (d) hardness and residual stresses for enhancement in concentration of nitrogen in argon.

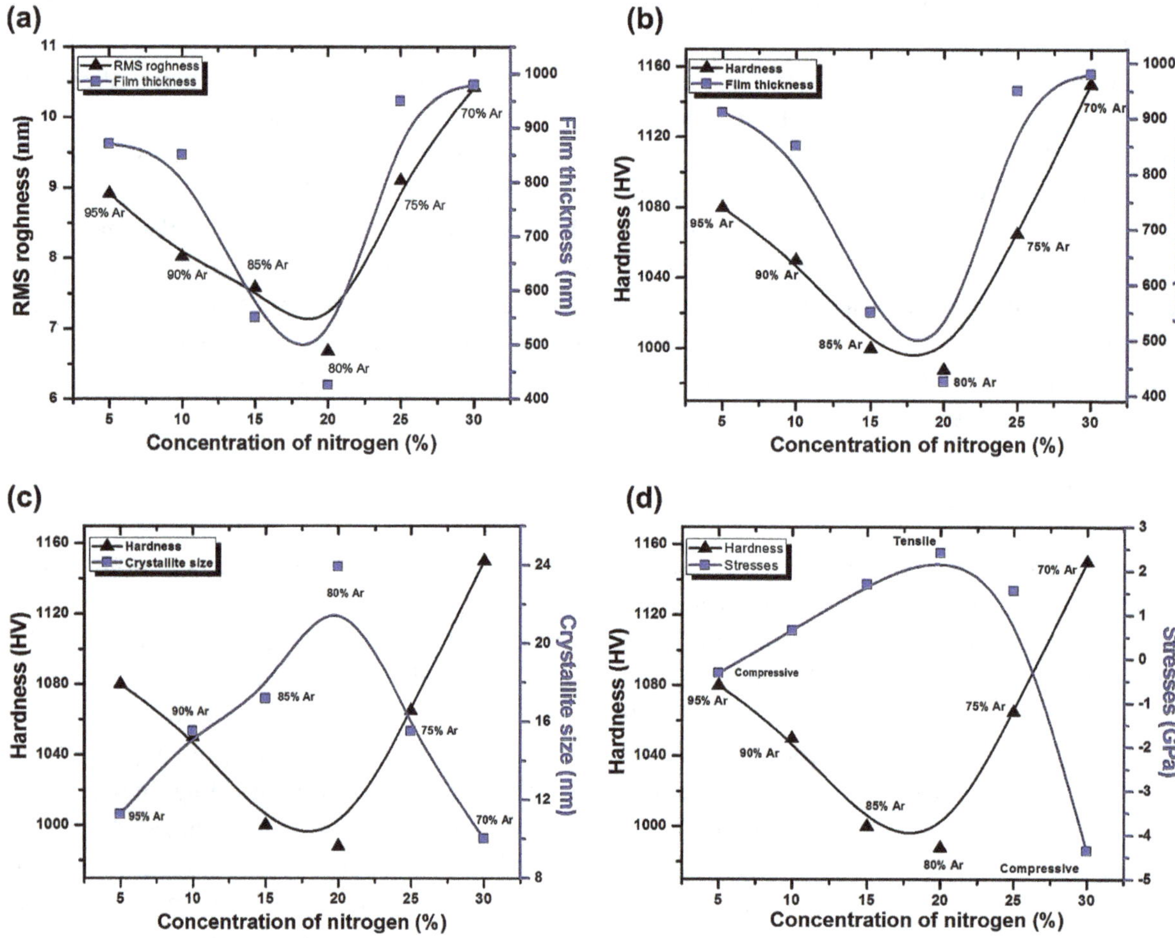

on the basis of increased tensile stresses due to increase in crystallite size. Oxygen diffusion across the boundaries of the grains results in larger crystallites (Figure 6 c) that are more disposed to coarsening and hence lower hardness (Kusaka, 2000 , Umm-i-Kalsoom, 2013, Corengia, 2005). Re-sputtering of the deposited duplex film transforms compressive residual stresses to tensile one, in return reducing the surface hardness (Figure 6 d) (Umm-i-Kalsoom, 2013, Mattox, 1989). Further increase in concentration of N_2 in Ar up to 30% results in an increase in hardness up to 1150 HV0.5. Interstitial diffusion of carbon/nitrogen into the lattice is attributable to the reduction in crystallite size and the transformation of tensile stresses to compressive residual stresses (Figs. 6c-6d) (Kusaka, 2000 , Umm-i-Kalsoom, 2013). Reduced crystallite size hinders the dislocation motion which enhances the strength and hardness of the substrate surface. Development of compressive stresses results in formation of crystallographic textures in which growth of most densely packed planes occur parallel to the surface. The growth of highly compressive stresses can decrease the adhesion between substrate and coating up to a point where spallation occurs during, or subsequent to, deposition. On the other hand, it also causes a substantial increase in hardness (Pan, 1981). Various factors are responsible for the variation in hardness including the phase composition, lattice defects, crystal structure, density of nitride and carbides, size and distribution of grains etc. (Warcholinski, 2009). Variation in values of hardness with crystallite size can be confirmed by employing the Hall-Petch relationship (Lim, 2002):

$$H = H_o + K_H(d)^{-1/2}$$

where K_H and H_o are experimental constants, H is the hardness and d is the average grain size.

CONCLUSION

Stainless steel substrates were duplex treated (nitriding plus carbonitriding) for different gas ratios of N_2 and Ar, in presence of graphite target, a source of carbon at constant time, pressure and power. Here formation of nitrided phases is observed during nitriding and during carbonitriding process nitrides and carbides are perceived at the same time.

EDS analysis of both nitrided and duplex treated substrates shows diffusion of nitrogen during nitriding and diffusion of N and C during duplex treatment of stainless steel. SEM analysis reveals that the growth of duplex treated layers strongly depends on formation mechanisms in presence of plasma. The induced compressive stresses on the substrate surface enhance the compactness of the deposited film (verified from XRD analysis. AFM analysis reveals that both the rms roughness and film thickness show similar behavior with increase in concentration of nitrogen in argon. The surface micro-hardness of nitrided substrate is about 906 HV0.5. The presence of hard nitride phases (Fe_4N & γ_N) enhances the micro hardness of duplex treated substrates as compared to untreated stainless-steel surface. Duplex treated surfaces showed more hardness than nitrided one, due to presence of both nitride and carbide phases (Fe_4N, Fe_2N, CrN, Cr_2N, $\gamma_{N,}$ FeC, Fe_3C, Fe_7C_3, Cr_7C_3, Cr_3C_2 and γc) during the process of carbonitriding. Hardness has its maximum value of 1150 HV0.5 for 30% N_2, with maximum film thickness, minimum crystallite size and maximum compressive residual stresses due to more diffusion of carbon/nitrogen (confirmed by EDS analysis for 30% nitrogen).

REFERENCES

Abd El-Rahman, A. M., El-Hossary, F. M., Fitz, T., Negm, N. Z., Prokert, F., Pham, M. T., & Möller, W. et al. (2004). Effect of N2 to C2H2 ratio on r.f. plasma surface treatment of austenitic stainless steel. *Surface and Coatings Technology*, *183*(2-3), 268–274. doi:10.1016/j.surfcoat.2003.09.057

Ali, N., Bashir, S., Umm-i-Kalsoom, , Akram, M., & Mahmood, K. (2013). Effect of dry and wet ambient environment on the pulsed laser ablation of titanium. *Applied Surface Science*, *270*, 49–57. doi:10.1016/j.apsusc.2012.12.049

Ali, N., Bashir, S., Umm-i-Kalsoom, , Rafique, M. S., Begum, N., & Husinsky, W. (2017). Nanostructuring of zirconium by femtosecond laser irradiation in the ambient environment of air and ethanol. *Optik (Stuttgart)*, *134*, 149–160. doi:10.1016/j.ijleo.2017.01.021

Ali, N., Bashir, S., Umm-i-Kalsoom, , Shahid Rafique, M., Begum, N., Husinsky, W., & Natahala, C. S. R. et al. (2017). Femtosecond laser induced nanostructuring of zirconium in liquid confined environment. *Chinese Physics B*, *26*(1), 015204. doi:10.1088/1674-1056/26/1/015204

Ashraf, M., Akhtar, S. M. J., Khan, A. F., Ali, Z., & Qayyum, A. (2011). Effect of annealing on structural and optoelectronic properties of nanostructured ZnSe films. *Journal of Alloys and Compounds*, *509*(5), 2414–2419. doi:10.1016/j.jallcom.2010.11.032

Blawart, C., Weisheit, A., Mordike, B. L., & Knoop, F. M. (1996). Plasma immersion ion implantation of stainless steel: Austenitic stainless steel in comparison to austenitic-ferritic stainless steel. *Surface and Coatings Technology*, *85*(1-2), 15–27. doi:10.1016/0257-8972(96)02880-0

Blawert, C., Mordike, B. L., Collins, G. A., Short, K. T., Jirásková, Y., Schneeweiss, O., & Perina, V. (2000). Characterisation of duplex layer structures produced by simultaneous implantation of nitrogen and carbon into austenitic stainless steel X5CrNi189. *Surface and Coatings Technology*, *128-129*, 219–225. doi:10.1016/S0257-8972(00)00651-4

Bogaerts, A., Neyts, E., Gijbels, R., & Mullen, J. V. D. (2002). Gas discharge plasmas and their applications. *Spectrochimica Acta. Part B, Atomic Spectroscopy*, *57*(4), 609–658. doi:10.1016/S0584-8547(01)00406-2

Braam, J. J., & Der Zwaag, S. V. (1999). A microstructural model for predicting hardness profiles of Fe-Cr alloys after nitriding. *Philosophical Magazine A*, *79*(5), 1193–1206. doi:10.1080/01418619908210355

Catteau, S. D., Van Landeghem, H. P., Teixeira, J., Dulcy, J., Dehmas, M., Denis, S., & Courteaux, M. et al. (2016). Carbon and nitrogen effects on microstructure and kinetics associated with bainitic transformation in a low-alloyed steel. *Journal of Alloys and Compounds*, *658*, 832–838. doi:10.1016/j.jallcom.2015.11.007

Constable, C. P., Lewis, B. D., Yarwood, J., & Munz, W. D. (2004). Raman microscopic studies of residual and applied stress in PVD hard ceramic coatings and correlation with X-ray diffraction (XRD) measurements. *Surface and Coatings Technology*, *184*(2-3), 291–297. doi:10.1016/j.surfcoat.2003.10.014

Corengia, P., Ybarra, G., Moina, C., Cabo, A., & Broitman, E. (2005). Microstructural and topographical studies of DC-pulsed plasma nitrided AISI 4140 low-alloy steel. *Surface and Coatings Technology*, *200*(7), 2391–2397. doi:10.1016/j.surfcoat.2005.01.060

Cullity, B. D. (1978). *Elements of X-ray Diffraction London*. Addison-Wesley publishing Company, Inc.

Czerwiec, T., Huan, H., Marcos, G., Thiriet, T., Weber, S., & Michel, H. (2009). Fundamental and Innovations in Plasma Assisted Diffusion of Nitrogen and Carbon in Austenitic Stainless Steels and Related Alloys. *Plasma Processes and Polymers*, *6*(6-7), 401–409. doi:10.1002/ppap.200930003

Das, K., Alphonsa, J., Ghosh, M., Ghanshyam, J., Rane, R., & Mukherjee, M. (2017). Influence of pretreatment on surface behavior of duplex plasma treated AISI H13 tool steel. *Surfaces and Interfaces*, *8*, 206–213. doi:10.1016/j.surfin.2017.06.009

Deng, Y., Tan, C., Wang, Y., Chen, L., Cai, P., Kuang, T., & Zhou, K. et al. (2017). Effects of tailored nitriding layers on comprehensive properties of duplex plasma-treated AlTiN coatings. *Ceramics International*, *43*(12), 8721–8729. doi:10.1016/j.ceramint.2017.03.209

Deng, Y., Tan, C., Wang, Y., Chen, L., Cai, P., Kuang, T., & Zhou, K. et al. (2017). Effects of tailored nitriding layers on comprehensive properties of duplex plasma-treated AlTiN coatings. *Ceramics International*, *43*(12), 8721–8729. doi:10.1016/j.ceramint.2017.03.209

El-Hossary, F. M., Negm, N. Z., Abd El-Rahman, A. M., & Hammad, M. (2009). Duplex treatment of 304 AISI stainless steel using rf plasma nitriding and carbonitriding. *Materials Science and Engineering C*, *29*(4), 1167–1173. doi:10.1016/j.msec.2008.09.049

El-Hossary, F. M., Negm, N. Z., Khalil, S. M., & Abd Elrahman, A. M. (2002). Formation and properties of a carbonitrided layer in 304 stainless steel using different radio frequency plasma powers. *Thin Solid Films*, *405*(1-2), 179–185. doi:10.1016/S0040-6090(01)01729-1

Garamoon, A., Rashed, U. M., Abouelela, A., Eissa, M. A., Saudi, A. H., El-Zeer, D. M., & El-Hossary, F. (2006). Hydrogen Effect on Nitriding Process of 304L Austenitic Stainless Steel. *IEEE Transactions on Plasma Science*, *34*(4), 1066–1073. doi:10.1109/TPS.2006.878388

Hariprasad, S., Gowtham, S., Arun, S., Ashok, M., & Rameshbabu, N. (2017). Fabrication of duplex coatings on biodegradable AZ31 magnesium alloy by integrating cerium conversion (CC) and plasma electrolytic oxidation (PEO) processes. *Journal of Alloys and Compounds*, *722*, 698–715. doi:10.1016/j.jallcom.2017.06.119

Jacquet P., Siad A., Nouveau C., Besnard A., & Quesada J. (2015) A study of the tribological behavior of duplex treatment. Annales de Chimie - Science des Matériaux, 39 201-208. doi: 39/3-4 – 20110.3166/acsm.39.201-208

Jelani, M., Bashir, S., Akram, M., Yousaf, D., Afzal, N., & Ahmad, S. (2014). Mechanical behaviour of Excimer laser irradiated polycrystalline Zirconium. *Physica Scripta*, *89*(2), 025703–025710. doi:10.1088/0031-8949/89/02/025703

Kao, W. H., Su, Y. L., Horng, J. H., & Hsieh, Y. T. (2017). Improved tribological properties, electrochemical resistance and biocompatibility of AISI 316L stainless steel through duplex plasma nitriding and TiN coating treatment. *Journal of Biomaterials Applications*, *32*(1), 12–27. doi:10.1177/0885328217712109 PMID:28541124

Khan, I. A., Ahmad, M. H. R., Qayyum, A., Murtaza, G., Zakaullah, A. M., & Rawat, R. J. (2008). Nitridation of zirconium using energetic ions from plasma focus device. *Thin Solid Films*, *516*(23), 8255–8263. doi:10.1016/j.tsf.2008.03.012

Khan, I. A., Hassan, M., Ahmad, R., Qayyum, A., Murtaza, G., Zakaullah, M., & Rawat, R. S. (2008). Nitridation of zirconium using energetic ions from plasma focus device. *Thin Solid Films*, *516*(23), 8255–8263. doi:10.1016/j.tsf.2008.03.012

Kolbel, J., Nordhein-Westfalen, F. D. I., & Verlag, W. D. Koln, & Germany, 1555. (1965).

Kusaka, K., Taniguchi, D., Hanabusa, T., & Tominaga, K. (2000). Effect of input power on crystal orientation and residual stress in AlN film deposited by dc sputtering. *Vacuum*, *59*(2-3), 806–813. doi:10.1016/S0042-207X(00)00351-1

Kusaka, K., Taniguchi, D., Hanabusa, T., & Tominaga, K. (2000). Effect of input power on crystal orientation and residual stress in AlN film deposited by dc sputtering. *Vacuum*, *59*(2-3), 806–813. doi:10.1016/S0042-207X(00)00351-1

Lee, S. Y., Chung, J. W., Kim, C. W., Han, J. G., Kim, S. S., & Lee, J. H. (1997). Effects of relative thickness of the duplex-treated layer on surface properties of AlSl H13 steel. *Surface and Coatings Technology*, *94/95*, 272–278. doi:10.1016/S0257-8972(97)00351-4

Li, W. L., Fei, W. D., & Hanabusa, T. (2006). Effect of deposition condition on residual stress of iron nitride thin films prepared by magnetron sputtering and ion implantation. *Applied Surface Science*, *252*(8), 2847–2852. doi:10.1016/j.apsusc.2005.04.043

Liang, W., Juncai, S., & Xiaolei, X. (2001). Low pressure plasma arc source ion nitriding compared with glow-discharge plasma nitriding of stainless steel. *Surface and Coatings Technology*, *145*(1-3), 31–37. doi:10.1016/S0257-8972(01)01283-X

Lim, Y. Y., & Chaudhri, M. M. (2002). The influence of grain size on the indentation hardness of high-purity copper and aluminium. *Philosophical Magazine A*, *82*(10), 2071–2080. doi:10.1080/01418610208235717

Mahmood, K., Farid, N., Ghauri, I. M., Afzal, N., Idrees, Y., & Mubarik, F. E. (2010). Effects of laser irradiation on the mechanical response of polycrystalline titanium. *Physica Scripta*, *82*(4), 045606–045613. doi:10.1088/0031-8949/82/04/045606

Maliska, A. M., de Oliveira, A. M., Klein, A. N., & Muzart, J. L. R. (2001). Surface porosity sealing effect of plasma nitrocarburizing on sintered unalloyed iron. *Surface and Coatings Technology*, *141*(2-3), 128–134. doi:10.1016/S0257-8972(01)01016-7

Mattox, D. M. (1989). Particle bombardment effects on thin film deposition. A review. *Journal of Vacuum Science & Technology. A, Vacuum, Surfaces, and Films*, *7*(3), 1105–1114. doi:10.1116/1.576238

MEMS & Nanotechnology Exchange. (n.d.) Material: Chromium Carbide (Cr3C2), bulk. Retrieved from https://www.memsnet.org/material/chromiumcarbidecr3c2bulk/

Metin, E., & Inal, O. T. (1987). Formation and growth of iron nitrides during ion-nitriding. *Journal of Materials Science*, *22*(8), 2783–2788. doi:10.1007/BF01086471

Meunier, C., Vives, S., & Bertrand, G. (1998). X-ray diffractometry analysis of r.f.-magnetron-sputtered chromium/chromium nitride coatings. *Surface and Coatings Technology*, *107*(2-3), 149–158. doi:10.1016/S0257-8972(98)00586-6

Meunier, C., Vives, S., & Bertrand, G. (1998). X-ray diffractometry analysis of r.f.-magnetron-sputtered chromium/chromium nitride coatings. *Surface and Coatings Technology*, *107*(2-3), 149–158. doi:10.1016/S0257-8972(98)00586-6

Mouri, L., Mabille, I., Fiaud, C., Amouroux, J., Catillon, G., & Gras, R. (2001). Study of corrosion resistance and surface properties of carbon steel after a duplex plasma treatment. *Thin Solid Films*, *389*(1-2), 153–160. doi:10.1016/S0040-6090(01)00875-6

Ode'n, M., Ericsson, C., Hakansson, G., & Ljungcrantz, H. (1999). Microstructure and mechanical behavior of arc-evaporated Cr–N coatings. *Surface and Coatings Technology*, *114*(1), 39–51. doi:10.1016/S0257-8972(99)00019-5

Olaya, J. J., Rodil, S. E., Muhl, S., & Sanchez, E. (2005). Comparative study of chromium nitride coatings deposited by unbalanced and balanced magnetron sputtering. *Thin Solid Films*, *474*(1-2), 119–126. doi:10.1016/j.tsf.2004.08.067

Pellizzari, M., Molinari, A., & Straffelini, G. (2001). Thermal fatigue resistance of plasma duplex-treated tool steel. *Surface and Coatings Technology*, *142-144*, 1109–1115. doi:10.1016/S0257-8972(01)01223-3

Petrov, I., Losbichler, P., Bergstrom, D., Greene, J. E., Münz, W. D., Hurkmans, T., & Trinh, T. (1997). Ion-assisted growth of Ti1−xAlxN/Ti1−yNbyN multilayers by combined cathodic-arc/magnetron-sputter deposition. *Thin Solid Films*, *302*(1-2), 179–192. doi:10.1016/S0040-6090(96)09524-7

Podgorni k, B., JVižintin, J., Wänstrand, O., Larsson, M., Hogmark, S., Ronkainen, H., & Holmberg, K., (2001). Tribological properties of plasma nitrided and hard coated AISI 4140 steel. *Wear*, *249*, 254-259. doi:10.1016/S0043-1648(01)00564-6

Sanjines, R., Sandu, C. S., Lamni, R., & Levy, F. (2006). Thermal decomposition of Zr1 − xAlxN thin films deposited by magnetron sputtering. *Surface and Coatings Technology*, *200*(22-23), 6308–6312. doi:10.1016/j.surfcoat.2005.11.113

Selcuk, B., Ipek, R., & Karamış, M. B. (2003). A study on friction and wear behaviour of carburized, carbonitrided and borided AISI 1020 and 5115 steels. *Journal of Materials Processing Technology*, *141*(2), 189–196. doi:10.1016/S0924-0136(02)01038-5

Shan, M. S., Khan, N., & Ahmad, R. (2009). Characterization and effect of argon nitrogen plasma on nitridation of aluminum alloy. *CODEN JNSMAC*, *49*, 1–17.

Silva, W. (2017). Carbonitriding of low alloy steels: Mechanical and metallurgical responses. *Materials Science and Engineering A*, *693*, 225–232. doi:10.1016/j.msea.2017.03.077

St-Onge, L., Sabsabi, M., & Cielo, P. (1998). Analysis of solids using laser-induced plasma spectroscopy in double-pulse mode. *Spectrochimica Acta. Part B, Atomic Spectroscopy*, *53*(53), 407–415. doi:10.1016/S0584-8547(98)00080-9

Tschiptschin, A. P., Varela, L. B., Pinedo, C. E., Li, X. Y., & Dong, H. (2017). Development and microstructure characterization of single and duplex nitriding of UNS S31803 duplex stainless steel. *Surface and Coatings Technology*, *327*, 83–92. doi:10.1016/j.surfcoat.2017.08.018

Tsotsos, C., Yerokhin, A. L., Wilson, A. D., Leyland, A., & Matthews, A. (2002). Tribological evaluation of AISI 304 stainless steel duplex treated by plasma electrolytic nitrocarburising and diamond-like carbon coating. *Wear*, *253*(9-10), 986–993. doi:10.1016/S0043-1648(02)00225-9

Umm-i-Kalsoom, , Ahmad, R., Ali, N., Khan, I. A., Saleem, S., Ikhlaq, U., & Khan, N. (2013). Effect of Power and Nitrogen Content on the Deposition of CrN Films by Using Pulsed DC Magnetron Sputtering plasma. *Plasma Science & Technology*, *15*(7), 666–672. doi:10.1088/1009-0630/15/7/12

Umm-i-Kalsoom, R. A., Ali, N., Khan, I. A., Saleem, S., Ikhlaq, U., & Khan, N. (2013). Effect of Power and Nitrogen Content on the Deposition of CrN Films by Using Pulsed DC Magnetron Sputtering Plasma. *Plasma Science & Technology, 15*(7), 667–672. doi:10.1088/1009-0630/15/7/12

Venkatachalam, S., Rajendra Kumar, R. T., Mangalaraj, D., Narayandass, S. K., Kim, K., & Junsin, Y. (2004). Optoelectronic properties of Zn0.52Se 0.48/Si Schottky diodes. *Solid-State Electronics, 48*(12), 2219–2223. doi:10.1016/j.sse.2004.05.082

Warcholinski, B., & Gilewicz, A., (2009). Tribological properties of CrNx coatings. *J. Achiev. Mater. Manuf. Eng.*, 37, 498-504.

Wierzchon, T., & Sobiecki, J. R. (1993). The PACVD method for surface layer deposition using a glow discharge in H_2+N_2+Ti(OC_3H_7)$_4$. *Vacuum, 44*(10), 975–978. doi:10.1016/0042-207X(93)90281-E

Wilson, A., Matthews, A., Housden, J., Turner, R., & Garside, B. (1993). A comparison of the wear and fatigue properties of plasma-assisted physical vapour deposition TiN, CrN and duplex coatings on Ti-6Al-4V. *Surface and Coatings Technology, 62*(1-3), 600–607. doi:10.1016/0257-8972(93)90306-9

Yang, Y., Yan, M. F., Zhang, Y. X., Li, D. Y., Zhang, C. S., Zhu, Y. D., & Wang, Y. X. (2017). Catalytic growth of diamond-like carbon on Fe3C-containing carburized layer through a single-step plasma-assisted carburizing process. *Carbon, 122*, 1–8. doi:10.1016/j.carbon.2017.06.033

Zhang, F., Yan, M., He, J., & Yin, F. (2017). Microstructure evolution and wear resistance of nitride/aluminide coatings on the surface of Ti-coated 2024 Al alloy during plasma nitriding. *Ceramics International, 43*(14), 10832–10839. doi:10.1016/j.ceramint.2017.05.109

Zhang, Z. L., & Bell, T. (1985). Structure and Corrosion Resistance of Plasma Nitrided Stainless Steel. *Surface Engineering, 1*, 131-136. doi:10.1179/sur.1985.1.2.131

Compilation of References

Abd El-Rahman, A. M., El-Hossary, F. M., Fitz, T., Negm, N. Z., Prokert, F., Pham, M. T., & Möller, W. et al. (2004). Effect of N2 to C2H2 ratio on r.f. plasma surface treatment of austenitic stainless steel. *Surface and Coatings Technology, 183*(2-3), 268–274. doi:10.1016/j.surfcoat.2003.09.057

Adachi, N., Endo, M., Menjo, N., & Okuda, T. (2003). Annealing effects of iron nitride thin film grown by plasma-assisted evaporation technique. *Journal of the Ceramic Society of Japan, 111*(6), 369–371. doi:10.2109/jcersj.111.369

Adelkhani, H., Nasoodi, S., & Jafari, A. H. (2014). Corrosion protection properties of silica coatings formed by sol–gel method on Al: The effects of acidity, withdrawal speed, and annealing temperature. *Progress in Organic Coatings, 77*(1), 142–145. doi:10.1016/j.porgcoat.2013.08.011

Adraider, Y., Pang, Y. X., Nabhani, F., Hodgson, S. N., Sharp, M. C., & Al-Waidh, A. (2013). Fabrication of zirconium oxide coatings on stainless steel by a combined laser/sol–gel technique. *Ceramics International, 39*(8), 9665–9670. doi:10.1016/j.ceramint.2013.05.089

Adraider, Y., Pang, Y. X., Nabhani, F., Hodgson, S. N., & Zhang, Z. Y. (2011). Laser-induced deposition of sol–gel alumina coating on stainless steel under wet condition. *Surface and Coatings Technology, 205*(23), 5345–5349. doi:10.1016/j.surfcoat.2011.05.044

Aegerter, M. A., & Mennig, M. (2004). *Sol-gel technologies for glass producers and users*. Boston: Kluwer Academic Publishers. doi:10.1007/978-0-387-88953-5

Ahmadi-Pidani, R., Shoja-Razavi, R., Mozafarinia, R., & Jamali, H. (2012). Evaluation of hot corrosion behavior of plasma sprayed ceria and yttria stabilized zirconia thermal barrier coatings in the presence of Na_2SO_4+V_2O_5 molten salt. *Ceramics International, 38*(8), 6613–6620. doi:10.1016/j.ceramint.2012.05.047

Ahn, H. S., & Kwon, O. K. (1999). Tribological behaviour of plasma-sprayed chromium oxide coating. *Wear, 225*, 814–824. doi:10.1016/S0043-1648(98)00390-1

Ahn, H. S., Lyo, I., & Lim, D. S. (2000). Influence of molybdenum composition in chromium oxide-based coatings on their tribological behavior. *Surface and Coatings Technology, 133*, 351–361. doi:10.1016/S0257-8972(00)00892-6

Ahn, J., Hwang, B., Song, E. P., Lee, S., & Kim, N. J. (2006). Correlation of microstructure and wear resistance of Al_2O_3-TiO_2 coatings plasma sprayed with nanopowders. *Metallurgical and Materials Transactions. A, Physical Metallurgy and Materials Science, 37*(6), 1851–1861. doi:10.1007/s11661-006-0128-5

Ahrens, M., Lampenscherf, S., Vassen, R., & Stover, D. (2012). Deposition and characteristics of submicrometer-structured thermal barrier coatings by suspension plasma spraying. *Journal of Thermal Spray Technology, 21*(3-4), 416–424. doi:10.1007/s11666-012-9762-1

Ahrens, M., Vassen, R., Stoever, D., & Lampenscherf, S. (2004). Sintering and creep processes in plasma-sprayed thermal barrier coatings. *Journal of Thermal Spray Technology*, *13*(3), 432–442. doi:10.1361/10599630420434

Aissa, B., Haddad, E. I., & Jamroz, W. (2014). *Self-Healing Materials: Innovative Materials for Terrestrial & Space Applications*. Smithers Rapra Technology.

Ai, T., Masada, J., & Ito, E. (2014). Development of the High Efficiency and Flexible Gas Turbine M701F5 by Applying "J" Class Gas Turbine Technologies. *Mitsubishi Heavy Industries Technical Review*, *51*(1), 1.

Akid, R., Gobara, M., & Wang, H. (2011). Corrosion protection performance of novel hybrid polyaniline/sol–gel coatings on an aluminium 2024 alloy in neutral, alkaline and acidic solutions. *Electrochimica Acta*, *56*(5), 2483–2492. doi:10.1016/j.electacta.2010.12.032

Ala-Nissila, T. F., Ferrando, R., & Ying, S. C. (2002). Collective and single particle. *Advances in Physics*, *51*(3), 949–1078. doi:10.1080/00018730110107902

Alexander, M. R., Beamson, G., Bailey, P., Noakes, T. C. Q., Skeldon, P., & Thompson, G. E. (2003). The distribution of hydroxyl ions at the surface of anodic alumina. *Surface and Interface Analysis*, *35*(8), 649–657. doi:10.1002/sia.1585

Alexander, M. R., Thompson, G. E., & Beamson, G. (2000). Characterization of the oxide/hydroxide surface of aluminum using X-ray photoelectron spectroscopy: A procedure for curve fitting the O 1s core level. *Surface and Interface Analysis*, *29*(7), 468–477. doi:10.1002/1096-9918(200007)29:7<468::AID-SIA890>3.0.CO;2-V

Alfano, M., Di Girolamo, G., Pagnotta, L., Sun, D., Zekonyte, J., & Wood, R. J. K. (2010). The influence of high-temperature sintering on microstructure and mechanical properties of freestanding APS CeO_2-Y_2O_3-ZrO_2 coatings. *Journal of Materials Science*, *45*(10), 2662–2669. doi:10.1007/s10853-010-4245-6

Alfonso, E. O. (2012). *Thin Film Growth Through Sputtering Technique and Its Applications*. InTech. doi:10.5772/35844

Aliakbarpour, S., Zakeri, M., Rahimipour, M., & Irankhah, R. (2014). Effect of SiC-mullite coatings on oxidation resistance of graphite. *Advances in Applied Ceramics*, *113*(6), 358–361. doi:10.1179/1743676114Y.0000000173

Ali, N., Bashir, S., Umm-i-Kalsoom, , Akram, M., & Mahmood, K. (2013). Effect of dry and wet ambient environment on the pulsed laser ablation of titanium. *Applied Surface Science*, *270*, 49–57. doi:10.1016/j.apsusc.2012.12.049

Ali, N., Bashir, S., Umm-i-Kalsoom, , Rafique, M. S., Begum, N., & Husinsky, W. (2017). Nanostructuring of zirconium by femtosecond laser irradiation in the ambient environment of air and ethanol. *Optik (Stuttgart)*, *134*, 149–160. doi:10.1016/j.ijleo.2017.01.021

Ali, N., Bashir, S., Umm-i-Kalsoom, , Shahid Rafique, M., Begum, N., Husinsky, W., & Natahala, C. S. R. et al. (2017). Femtosecond laser induced nanostructuring of zirconium in liquid confined environment. *Chinese Physics B*, *26*(1), 015204. doi:10.1088/1674-1056/26/1/015204

Allen, A. J., Long, G. G., Boukari, H., Ilavsky, J., Kulkarni, A., Sampath, S., & Goland, A. N. et al. (2001). Microstructure characterization studies to relate the properties of thermal-spray coatings to feedstock and spray conditions. *Surface and Coatings Technology*, *146-147*, 544–552. doi:10.1016/S0257-8972(01)01433-5

Al-Sabouni, O. (1999). *Reactive plasma spraying*. (Unpublished doctrol dissertation). Cranfield University, London, UK.

Álvarez, D., Collazo, A., Hernández, M., Nóvoa, X. R., & Pérez, C. (2010). Characterization of hybrid sol–gel coatings doped with hydrotalcite-like compounds to improve corrosion resistance of AA2024-T3 alloys. *Progress in Organic Coatings*, *67*(2), 152–160. doi:10.1016/j.porgcoat.2009.10.014

Álvarez, D., Collazo, A., Nóvoa, X. R., & Pérez, C. (2016). Assessment of ZnO nanoparticles as anticorrosive pigment in hybrid sol–gel films. *Progress in Organic Coatings*, *96*, 3–12. doi:10.1016/j.porgcoat.2015.10.017

Álvarez, P., Collazo, A., Covelo, A., Nóvoa, X. R., & Pérez, C. (2010). The electrochemical behaviour of sol–gel hybrid coatings applied on AA2024-T3 alloy: Effect of the metallic surface treatment. *Progress in Organic Coatings*, *69*(2), 175–183. doi:10.1016/j.porgcoat.2010.04.005

Amendola, V., & Meneghetti, M. (2011). *Self-healing at the Nanoscale: Mechanisms and Key Concepts of Natural and Artificial Systems*. CRC Press. doi:10.1201/b11484

Andersa, A. (2017). Tutorial: Reactive high power impulse magnetron sputtering (R-HiPIMS). American Institute of Physics.

Andersen, H. H., & Bay, H.L. (1981). Sputtering yield measurements. In R. Behrisch (Ed.), Sputtering by Particle Bombardment I (Vol. 47, p. 145). doi:10.1007/3540105212_9

Andersson, J.-O., & Ågren, J. (1992). Models for numerical treatment of multicomponent diffusion in simple phases. *Journal of Applied Physics*, *72*(4), 1350–1355. doi:10.1063/1.351745

Andreatta, F., Paussa, L., Aldighieri, P., Lanzutti, A., Raps, D., & Fedrizzi, L. (2010). Corrosion behaviour of sol–gel treated and painted AA2024 aluminium alloy. *Progress in Organic Coatings*, *69*(2), 133–142. doi:10.1016/j.porgcoat.2010.04.012

Andreatta, F., Paussa, L., Lanzutti, A., Rosero Navarro, N. C., Aparicio, M., Castro, Y., & Fedrizzi, L. (2011). Development and industrial scale-up of ZrO_2 coatings and hybrid organic–inorganic coatings used as pre-treatments before painting aluminium alloys. *Progress in Organic Coatings*, *72*(1–2), 3–14. doi:10.1016/j.porgcoat.2011.01.011

Anggono, J. (2005). Mullite ceramics: its properties structure and synthesis. *Jurnal Teknik Mesin, 7*(1).

Aoki, Y., Habazaki, H., & Kunitake, T. (2009). Size-scaling of proton conductivity in amorphous aluminosilicate acid thin films. *Journal of the American Chemical Society*, *131*(40), 14399–14406. doi:10.1021/ja904627h PMID:19807185

Aouadi, S. M., Luster, B., Kohli, B., Muratore, C., & Voevodin, A. A. (2009). Progress in the development of adaptive nitride-based coatings for high temperature tribological applications. *Surface and Coatings Technology*, *204*(6-7), 962–968. doi:10.1016/j.surfcoat.2009.04.010

Aouadi, S. M., Paudel, Y., Luster, B., Stadler, S., Kohli, P., Muratore, C., & Voevodin, A. A. et al. (2008). Adaptive $Mo_2N/MoS_2/Ag$ tribological nanocomposite coatings for aerospace applications. *Tribology Letters*, *29*(2), 95–103. doi:10.1007/s11249-007-9286-x

Aouadi, S. M., Paudel, Y., Simonson, W. J., Ge, Q., Kohli, P., Muratore, C., & Voevodin, A. A. (2009). Tribological investigation of adaptive $Mo_2N/MoS_2/Ag$ coatings with high sulfur content. *Surface and Coatings Technology*, *203*(10-11), 1304–1309. doi:10.1016/j.surfcoat.2008.10.040

Aouadi, S. M., Singh, D. P., Stone, D. S., Polychronopoulou, K., Nahif, F., Rebholz, C., & Voevodin, A. A. et al. (2010). Adaptive VN/Ag nanocomposite coatings with lubricious behavior from 25 to 1000°C. *Acta Materialia*, *58*(16), 5326–5331. doi:10.1016/j.actamat.2010.06.006

Aramaki, K. (2002). Self-healing mechanism of an organosiloxane polymer film containing sodium silicate and cerium(III) nitrate for corrosion of scratched zinc surface in 0.5 M NaCl. *Corrosion Science*, *44*(7), 1621–1632. doi:10.1016/S0010-938X(01)00171-8

Archard, J. F. (1953). Contact and rubbing of flat surfaces. *Journal of Applied Physics*, *24*(8), 981–988. doi:10.1063/1.1721448

Ardanuy, M., Velasco, J. I., Maspoch, M. L., Haurie, L., & Fernández, A. I. (2009). Influence of EMAA compatibilizer on the structure and properties of HDPE/hydrotalcite nanocomposites prepared by melt mixing. *Journal of Applied Polymer Science, 113*(2), 950–958. doi:10.1002/app.30133

Arenas, O. L., Porcayo-Calderon, J., Salinas-Bravo, V. M., Martinez-Villafane, A., & Gonzalez-Rodriguez, J. G. (2005). Effect of boron on the hot corrosion resistance of sprayed Fe40Al intermetallics. *High-Temperature Materials and Processes, 24*(2), 93–100. doi:10.1515/HTMP.2005.24.2.93

Arkhireeva, A., Hay, J. N., Lane, J. M., Manzano, M., Masters, H., Oware, W., & Shaw, S. J. (2004). *Synthesis of Organic-Inorganic Hybrid Particles by Sol-Gel Chemistry* (Vol. 31).

Arkles, B. (2001). Commercial Application of Sol-Gel-Derived Hybrid Materials. *MRS Bulletin, 26*(May), 402–408. doi:10.1557/mrs2001.94

Arnell, R. D., & Kelly, P. J. (1999). Recent advances in magnetron sputtering. *Surface and Coatings Technology, 112*(1-3), 170–176. doi:10.1016/S0257-8972(98)00749-X

Ashraf, M., Akhtar, S. M. J., Khan, A. F., Ali, Z., & Qayyum, A. (2011). Effect of annealing on structural and optoelectronic properties of nanostructured ZnSe films. *Journal of Alloys and Compounds, 509*(5), 2414–2419. doi:10.1016/j.jallcom.2010.11.032

ASTM. (2001). *Standard Test Method for Adhesion or Cohesion Strength of Thermal Spray Coatings* (pp. 846–852). ASTM.

ASTM. (2010). *Standard Test Method for Melt Flow Rates of Thermoplastics by Extrusion Plastometer* (pp. 282–296). West Conshohocken, PA: ASTM.

Atkinson, A., & Guppy, R. M. (1991). Mechanical stability of sol-gel films. *Journal of Materials Science, 26*(14), 3869–3873. doi:10.1007/BF01184984

Aygun, A. (2008). *Novel thermal barrier coatings (TBC) that are resistant to high temperature attack by CaO-Mgo-Al_2O_3-SiO_2 (CMAS) glassy deposits.* [Dissertation Doctor of Philosophy]. The Ohio State University, Columbus, OH.

Azadi, M., Rouhaghdam, A. S., & Ahangarani, S. (2014). Effect of temperature and gas flux on the mechanical behavior of TiC coating by pulsed DC PECVD. *International Journal of Engineering Transactions B: Application, 27*, 1243-1250.

Azadi, M., Rouhaghdam, A. S., & Ahangarani, S. (2016). A review on titanium nitride and titanium carbide single and multilayer coatings deposited by plasma assisted chemical vapor deposition, *International Journal of Engineering Transactions B: Applications, 29,* 677-687.

Azadi, M., & Rouhaghdam, A. S. (2014). Nanomechanical properties of TiN/TiC multilayer coatings. *Strength of Materials, 46*(1), 121–132. doi:10.1007/s11223-014-9523-0

Azadi, M., Rouhaghdam, A. S., & Ahangarani, S. (2013). Properties of TiC Coating by Pulsed DC PACVD. *Journal of Coatings, 2013*, 1–6. doi:10.1155/2013/712812

Azadi, M., Rouhaghdam, A. S., & Ahangarani, S. (2016). Mechanical Behavior of TiN/TiC-n Multilayer Coatings to Ti(C,N) Multicomponent Coatings by PACVD. *Strength of Materials, 48*(2), 279–289. doi:10.1007/s11223-016-9763-2

Azadi, M., Rouhaghdam, A. S., Ahangarani, S., & Mofidi, H. H. (2014). Mechanical behavior of TiN/TiC multilayer coatings fabricated by plasma assisted chemical vapor deposition on AISI H13 hot work tool steel. *Surface and Coatings Technology, 245*, 156–166. doi:10.1016/j.surfcoat.2014.02.055

Azadi, M., Rouhaghdam, A. S., Ahangarani, S., Mofidi, H. H., & Valiei, M. (2014). Effect of number of layers on the toughness of TiN/TiC multilayer coatings. *International Journal of Microstructure and Materials Properties, 9*(6), 500–516. doi:10.1504/IJMMP.2014.067314

Azadi, M., Rouhaghdam, A. S., Ahangarani, S., Mofidi, H. H., & Valiei, M. (2014). Mechanical behavior and properties of TiN/TiC coating using PACVD. *Advanced Materials Research, 829*, 476–481. doi:10.4028/www.scientific.net/AMR.829.476

AZo Materials. (July 5, 2012). AISI 1018 Mild/Low Carbon Steel. Retrieved in May 2017 from http://www.azom.com/article.aspx?ArticleID=6115

Azushima, A., Tanno, Y., Iwata, H., & Aoki, K. (2008). Coefficients of friction of TiN coatings with preferred grain orientations under dry condition. *Wear, 265*(7-8), 1017–1022. doi:10.1016/j.wear.2008.02.019

Bacci, T., Bertamini, L., Ferrari, F., Galliano, F. P., & Galvanetto, E. (2000). Reactive plasma spraying of titanium in nitrogen containing plasma gas. *Materials Science and Engineering A, 283*(1-2), 189–195. doi:10.1016/S0921-5093(00)00704-8

Bacos, M. (1993). Carbon-carbon composites: oxidation behavior and coatings protection. *Le Journal de Physique IV, 3*(C7), 1895-1903.

Badisch, E., Fontalvo, G.A., Stoiber, M., Mitterer, C., (2003). Tribological behavior of PACVD TiN coatings in the temperature range up to 500 °C, *Surface and Coatings Technology, 163 –164* 585–590.

Badisch, E., Stoiber, M., Fontalvo, G.A., Mitterer, C., (2003). Low-friction PACVD TiN coatings: influence of Cl-content and testing conditions on the tribological properties, *Surface and Coatings Technology, 174 –175,* 450–454.

Baer, D. R., Burrows, P. E., & El-Azab, A. A. (2003). Enhancing coating functionality using nanoscience and nanotechnology. *Progress in Organic Coatings, 47*(3–4), 342–356. doi:10.1016/S0300-9440(03)00127-9

Bai, P. H. (2014). Review of Paper-Like Display Technologies. *Progress In Electromagnetics Research, 147*, 95-116.

Bai, J., Yang, X., Shi, Y., Xu, S., & Yang, J. (2012). Fabrication of directional SiC porous ceramics using Fe 2 O 3 as pore-forming agent. *Materials Letters, 78*, 192–194. doi:10.1016/j.matlet.2012.03.046

Bakan, E., Mack, D. E., Mauer, G., Mucke, R., & Vassen, R. (2015). Porosity–Property Relationships of Plasma-Sprayed $Gd_2Zr_2O_7$/YSZ Thermal Barrier Coatings. *Journal of the American Ceramic Society, 98*(8), 2647–2654. doi:10.1111/jace.13611

Bakan, E., Mack, D. E., Mauer, G., & Vassen, R. (2014). Gadolinium zirconate/YSZ thermal barrier coatings: Plasma spraying, microstructure and thermal cycling behavior. *Journal of the American Ceramic Society, 97*(12), 4045–4051. doi:10.1111/jace.13204

Bakan, E., & Vaßen, R. (2017). Ceramic Top Coats of Plasma-Sprayed Thermal Barrier Coatings: Materials, Processes, and Properties. *Journal of Thermal Spray Technology, 26*(6), 992–1000. doi:10.1007/s11666-017-0597-7

Baklouti, S., Chartier, T., & Baumard, J. F. (1998). Binder distribution in spray-dried alumina agglomerates. *Journal of the European Ceramic Society, 18*(14), 2117–2121. doi:10.1016/S0955-2219(98)00107-1

Balan, P., Ng, A., Beng Siang, C., Singh Raman, R. K., & Chan, E. S. (2013). Effect of Nanoparticle Addition in Hybrid Sol-Gel Silane Coating on Corrosion Resistance of Low Carbon Steel. *Advanced Materials Research, 686*, 244–249. doi:10.4028/www.scientific.net/AMR.686.244

Balaskas, A. C., Kartsonakis, I. A., Tziveleka, L. A., & Kordas, G. C. (2012). Improvement of anti-corrosive properties of epoxy-coated AA 2024-T3 with TiO2 nanocontainers loaded with 8-hydroxyquinoline. *Progress in Organic Coatings*, *74*(3), 418–426. doi:10.1016/j.porgcoat.2012.01.005

Balint, D. S., & Hutchinson, J. W. (2005). An analytical model of rumpling in thermal barrier coatings. *Journal of the Mechanics and Physics of Solids*, *53*(5), 949–973. doi:10.1016/j.jmps.2004.11.002

Bao, Y., Gawne, D. T., Vesely, D., & Bevis, M. J. (1994a). Formation and microstructure of plasma sprayed polyamide coatings. *Surface Engineering*, *10*(4), 307–313. doi:10.1179/sur.1994.10.4.307

Bao, Y., Gawne, D. T., Vesely, D., & Bevis, M. J. (1994b). Production of polymer matrix composite coatings by thermal spraying. *Transactions of the Institute of Metal Finishing*, *72*(Pt 3), 110–113. doi:10.1080/00202967.1994.11871033

Bao, Y., Gawne, D. T., & Zhang, T. (1995). Effect of feedstock particle size on the heat transfer rates and properties of thermally sprayed polymer coatings. *Transactions of the Institute of Metal Finishing*, *73*(Pt 4), 119–124. doi:10.1080/00202967.1995.11871072

Bao, Y., & Nicholson, P. S. (2008). AlPO 4-coated mullite/alumina fiber reinforced reaction-bonded mullite composites. *Journal of the European Ceramic Society*, *28*(16), 3041–3048. doi:10.1016/j.jeurceramsoc.2008.05.032

Barbooti, M. M., Al-Madfai, S. H., & Nassouri, H. J. (1988). Thermochemical studies on hot ash corrosion of stainless steel 304 and inhibition by magnesium sulphate. *Thermochimica Acta*, *126*, 43–49. doi:10.1016/0040-6031(88)87248-4

Bardal, E. (Ed.). (2004). *Corrosion and Protection*. London: Springer London. doi:10.1007/b97510

Bardal, E., Eggen, T. G., Rogne, T., & Solem, T. (1995). The erosion and corrosion properties of thermal spray and other coatings. In *Proceedings of the Int. Therm. Spray. Conf.*, Kobe, Japan.

Bartsch, K., Leonhardt, A., Langer, U., & Kiinanz, K. (1997). New PACVD-hard material layers for wear protection of high-speed steel. *Surface and Coatings Technology*, *94-95*, 168–173. doi:10.1016/S0257-8972(97)00345-9

Bast, U., & Schumann, E. (2002). Development of novel oxide materials for TBCs. *Ceramic Engineering and Science Proceedings*, *23*, 525–532. doi:10.1002/9780470294758.ch58

Batory, D., Jedrzejczak, A., Szymanski, W., Niedzielski, P., Fijalkowski, M., Louda, P., & Musil, J. et al. (2015). Mechanical characterization of a-C:H:SiOx coatings synthesized using radio-frequency plasma-assisted chemical vapor deposition method. *Thin Solid Films*, *590*, 299–305. doi:10.1016/j.tsf.2015.08.017

Beltran, A. M., & Shores, D. (1972). Hot corrosion. *Superalloys. Compilation of Monographs*, *1972*, 317–339.

Benvenutti, E. V., Moro, C. C., Costa, T. M. H., & Gallas, M. R. (2009). Materiais híbridos à base de sílica obtidos pelo método sol-gel. *Quimica Nova*, *32*(7), 1926–1933. doi:10.1590/S0100-40422009000700039

Berger-Keller, N., Bertrand, G., Filiatre, C., Meunier, C., & Coddet, C. (2003). Microstructure of plasma-sprayed titania coatings deposited from spray dried powder. *Surface and Coatings Technology*, *168*(2-3), 281–290. doi:10.1016/S0257-8972(03)00223-8

Berger, L. M., Herrmann, M., Nebelung, M., Thiele, S., Heimann, R. B., Schnickt, T., & Vuoristo, P. et al. (1998). Investigations on thermal spraying of silicon nitride-based powders. In *Proceeding of 15th International Thermal Spray Conference,* Nice, France (pp. 1149-1154).

Berger, L. M., Stahr, C. C., Saaro, S., Thiele, S., Woydt, M., & Kelling, N. (2009). Dry sliding up to 7.5 m/s and 800°C of thermally sprayed coatings of the TiO_2–Cr_2O_3 system and (Ti, Mo) (C, N)–Ni (Co). *Wear*, *267*(5-8), 954–964. doi:10.1016/j.wear.2008.12.105

Berget, J., Rogne, T., & Bardal, E. (2007). Erosion–corrosion properties of different WC–Co–Cr coatings deposited by the HVOF process—influence of metallic matrix composition and spray powder size distribution. *Surface and Coatings Technology, 201*(18), 7619–7625. doi:10.1016/j.surfcoat.2007.02.032

Bernard, B., Quet, A., Bianchi, L., Joulia, A., Malié, A., Schick, V., & Rémy, B. (2017). Thermal insulation properties of YSZ coatings: Suspension plasma spraying (SPS) versus electron beam physical vapor deposition (EB-PVD) and atmospheric plasma spraying (APS). *Surface and Coatings Technology, 318*, 122–128. doi:10.1016/j.surfcoat.2016.06.010

Bernard-Granger, G., Monchalin, N., & Guizard, C. (2008). Comparisons of grain size-density trajectory during spark plasma sintering and hot-pressing of zirconia. *Materials Letters, 62*(30), 4555–4558. doi:10.1016/j.matlet.2008.08.042

Berndt, C. C., Brogan, J. A., Montavon, G., Claudon, A., & Coddet, C. (1998). Mechanical properties of metal- and ceramic-polymer composites formed via thermal spray consolidation - Extended abstract. *Journal of Thermal Spray Technology, 7*(3), 337–339. doi:10.1361/105996398770350819

Berndt, C. C., & Lavernia, E. J. (1998). Thermal spray processing of nanoscale materials - A conference report with extended abstracts. *Journal of Thermal Spray Technology, 7*(3), 411–440. doi:10.1007/s11666-998-0058-4

Bewilogua, K., Bialuch, I., Ruske, H., & Weigel, K. (2011). Preparation of a-C:H/a-C:H:Si:O and a-C:H/a-C:H:Si multilayer coatings by PACVD. *Surface and Coatings Technology, 206*(4), 623–629. doi:10.1016/j.surfcoat.2011.02.020

Bezzi, F., Fabbri, P., Brentari, A., Mingazzini, C., Burresi, E., Beaulardi, L., & Sangiorgi, S. (2010). Slurry coating of environmental barrier coating (EBC) on silicon carbide based material. In Advances in Science and Technology (Vol. 66, pp. 100-105). Trans Tech Publications.

Bhandari, J., Khan, F., Abbassi, R., Garaniya, V., & Ojeda, R. (2015). Modelling of pitting corrosion in marine and offshore steel structures – A technical review. *Journal of Loss Prevention in the Process Industries, 37*, 39–62. doi:10.1016/j.jlp.2015.06.008

Bhatia, T., Ozturk, A., Xie, L., Jordan, E. H., Cetegen, B., Gell, M., & Padture, N. P. et al. (2002). Mechanisms of ceramic coating deposition in solution precursor plasma spray. *Journal of Materials Research, 17*(09), 2363–2372. doi:10.1557/JMR.2002.0346

Bhattacharyya, A. K. (2015). Sputter Based Epitaxial Growth and Modelling of Cu/Si Thin Films. *Int. J. Thin. Fil. Sci. Tec., 4*(3), 173–177.

Bianchi, L., Grimaud, A., Blein, F., Lucchese, P., & Fauchais, P. (1995). Comparison of plasma-sprayed alumina coatings by RF and DC plasma spraying. *Journal of Thermal Spray Technology, 4*(1), 59–66. doi:10.1007/BF02648529

Biersack, J. E., & Eckstein, W. (1984). Sputtering Studies with the Monte Carlo Program TRIM.SP. *J. Appl. Phys. A, 34*(2), 73–94. doi:10.1007/BF00614759

Billiet, S., Van Camp, W., Hillewaere, X. K. D., Rahier, H., & Du Prez, F. E. (2012). Development of optimized autonomous self-healing systems for epoxy materials based on maleimide chemistry. *Polymer, 53*(12), 2320–2326. doi:10.1016/j.polymer.2012.03.061

Bindu, K. C., Campos, J., Nair, M. T. S., Sánchez, A., & Nair, P. K. (2005). Semiconducting AgSbSe2 thin film and its application in a photovoltaic structure. *Semiconductor Science and Technology, 20*(6), 496–504. doi:10.1088/0268-1242/20/6/004

Birks, N., Meier, G. H., & Pettit, F. S. (2006). *Introduction to the high temperature oxidation of metals* Retrieved from http://swin.eblib.com.au/patron/FullRecord.aspx?p=258481

Birks, N., Meier, G. H., & Pettit, F. S. (2006). *High temperature oxidation of metals*. Cambridge, UK: Cambridge University Press. doi:10.1017/CBO9781139163903

Bjordal, M., Bardal, E., Rogne, T., & Eggen, T. G. (1995a). Erosion and corrosion properties of WC coatings and duplex stainless steel in sand-containing synthetic sea water. *Wear, 186–187*, 508–514. doi:10.1016/0043-1648(95)07148-2

Bjordal, M., Bardal, E., Rogne, T., & Eggen, T. G. (1995b). Combined erosion and corrosion of thermal sprayed WC and CRC coatings. *Surface and Coatings Technology*, *70*(2-3), 215–220. doi:10.1016/0257-8972(94)02266-S

Blawart, C., Weisheit, A., Mordike, B. L., & Knoop, F. M. (1996). Plasma immersion ion implantation of stainless steel: Austenitic stainless steel in comparison to austenitic-ferritic stainless steel. *Surface and Coatings Technology*, *85*(1-2), 15–27. doi:10.1016/0257-8972(96)02880-0

Blawert, C., Mordike, B. L., Collins, G. A., Short, K. T., Jirásková, Y., Schneeweiss, O., & Perina, V. (2000). Characterisation of duplex layer structures produced by simultaneous implantation of nitrogen and carbon into austenitic stainless steel X5CrNi189. *Surface and Coatings Technology*, *128-129*, 219–225. doi:10.1016/S0257-8972(00)00651-4

Boatemaa, L., Kwakernaak, C., van der Zwaag, S., & Sloof, W. G. (2016). Selection of healing agents for autonomous healing of alumina at high temperatures. *Journal of the European Ceramic Society*, *36*(16), 4141–4145. doi:10.1016/j.jeurceramsoc.2016.05.038

Bobzin, K., Bagcivan, N., Brogelmann, T., & Yildirim, B. (2013). Influence of temperature on phase stability and thermal conductivity of single- and double-ceramic-layer EB-PVD TBC top coats consisting of 7YSZ, $Gd_2Zr_2O_7$ and $La_2Zr_2O_7$. *Surface and Coatings Technology*, *237*, 56–64. doi:10.1016/j.surfcoat.2013.08.013

Bogaerts, A., Neyts, E., Gijbels, R., & Mullen, J. V. D. (2002). Gas discharge plasmas and their applications. *Spectrochimica Acta. Part B, Atomic Spectroscopy*, *57*(4), 609–658. doi:10.1016/S0584-8547(01)00406-2

Böhnlein-Mauß, J., Sigmund, W., Wegner, G., Meyer, W. H., Heßel, F., Seitz, K., & Roosen, A. (1992). The function of polymers in the tape casting of alumina. *Advanced Materials*, *4*(2), 73–81. doi:10.1002/adma.19920040203

Bojarski, Z., & Łągiewka, E. (1995). *Rentgenowska analiza strukturalna*. Katowice: Wydawnictwo Uniwersytetu Śląskiego.

Bolcavage, A., Feuerstein, A., Foster, J., & Moore, P. (2004). Thermal shock testing of thermal barrier coating/bondcoat systems. *Journal of Materials Engineering and Performance*, *13*(4), 389–397. doi:10.1361/10599490419883

Bolelli, G., Cannillo, V., Lusvarghi, L., & Manfredini, T. (2006). Wear behavior of thermally sprayed ceramic oxide coatings. *Wear*, *261*(11-12), 1298–1315. doi:10.1016/j.wear.2006.03.023

Boley, B. A., & Weiner, J. H. (2012). *Theory of thermal stresses. New York*. US: Courier Corporation.

Borom, M. P., Johnson, C. A., & Peluso, L. A. (1996). Role of environmental deposits and operating surface temperature in spallation of air plasma sprayed thermal barrier coatings. *Surface and Coatings Technology*, *86-87*, 116–126. doi:10.1016/S0257-8972(96)02994-5

Borovinskaya, I. P., Ignat'eva, T. I., Semenova, V. N., & Chemagina, E. A. (2015). Aluminum oxynitride by SHS in chemical furnace. *International Journal of Self-Propagating High-Temperature Synthesis, 24*(3), 142-147.

Bose, S. (2011). *High temperature coatings*. Butterworth-Heinemann.

Boulos, M. I. (1992). RF induction plasma spraying: State-of-the-art review. *Journal of Thermal Spray Technology*, *1*(1), 33–40. doi:10.1007/BF02657015

Bounazef, M., Guessasma, S., Montavon, G., & Coddet, C. (2004). Effect of APS process parameters on wear behaviour of alumina-titania coatings. *Materials Letters*, *58*(20), 2451–2455. doi:10.1016/j.matlet.2004.02.026

Bourdin, E., Boulos, M., & Fauchais, P. (1983). Transient conduction to a single sphere under plasma conditions. *International Journal of Heat and Mass Transfer*, *26*, 567–582. doi:10.1016/0017-9310(83)90008-X

Bouzid, K. B. (2015). Wear and Corrosion Resistance of CrN-based coatings Deposited by R.F Magnetron Sputtering. *Tribology in Industry*, *37*, 60–65.

Boyce, M. P. (2002). *Gas Turbine Engineering Handbook* (2nd ed.). Portland, US: Gulf Professional Publishing.

Braam, J. J., & Der Zwaag, S. V. (1999). A microstructural model for predicting hardness profiles of Fe-Cr alloys after nitriding. *Philosophical Magazine A*, *79*(5), 1193–1206. doi:10.1080/01418619908210355

Bramfitt, B. L. (2007). Carbon and Alloy Steels. In M. Kutz (Ed.), *Handbook of Materials Selection* (pp. 25–65). John Wiley & Sons, Inc.

Brauer, G. S., Szyszka, B., Vergöhl, M., & Bandorf, R. (2010). Magnetron sputtering – Milestones of 30 years. *Vacuum*, *84*(12), 1354–1359. doi:10.1016/j.vacuum.2009.12.014

Braue, W., & Mechnich, P. (2011). Recession of an EB-PVDYSZ Coated Turbine Blade by $CaSO_4$ and Fe, Ti-Rich CMAS-Type Deposits. *Journal of the American Ceramic Society*, *94*(12), 4483–4489. doi:10.1111/j.1551-2916.2011.04747.x

Brault, P. C., Chuon, S., & Bauchire, J.-M. (2016). Molecular Dynamics Simulations of Platinum Plasma Sputtering: A Comparative Case Study. *Frontiers in Physics*, *4*, 1–6. doi:10.3389/fphy.2016.00020

Brinker, C. J., & Scherer, G. W. (1990). *Sol-gel science: the physics and chemistry of sol-gel processing*. Boston: Academic Press.

Brinker, C., Hurd, A., Schunk, P., Frye, G., & Ashley, C. S. (1992). Review of sol-gel thin film formation. *Journal of Non-Crystalline Solids*, *148*, 424–436. doi:10.1016/S0022-3093(05)80653-2

Brogan, J. A., & Berndt, C. C. (1996). The mechanical properties of combustion-sprayed polymers and blends. *Paper presented at the Thermal Spray: Practical Solutions for Engineering Problems*, Cincinnati, Ohio.

Brogan, J. A., Lamp, R., & Berndt, C. C. (1995). Thermal spraying of polymers. *Paper presented at the Fourth World Congress on Coating Systems for Bridges and Steel Structures: Bridging the Environment*, St. Louis, MO.

Brogan, J. A. (1996). *Processing and property relationships of thermally sprayed polymer systems*. Stony Brook, NY: State University of New York.

Brogan, J. A., & Berndt, C. C. (1997). The coalescence of combustion-sprayed ethylene-methacrylic acid copolymer. *Journal of Materials Science*, *32*(8), 2099–2106. doi:10.1023/A:1018526906452

Browning, J. A. (1992). Hypervelocity impact fusion-a technical note. *Journal of Thermal Spray Technology*, *1*(4), 289–292. doi:10.1007/BF02647155

BSI British Standards. (2013). BS 6349-1-1:2013 - Maritime works General Code of practice for planning and design for operations.

BSI. (2004). *Thermal Spraying - Determination of the Deposition Efficiency for Thermal Spraying* (pp. 1–10). European Committee for Standardization.

Buckley, J. D., & Edie, D. D. (1993). *Carbon-carbon materials and composites* (Vol. 1254). William Andrew.

Bull, S. J. (1998). Modeling of residual stress in oxide scales. *Oxidation of Metals*, *49*(1-2), 1–17. doi:10.1023/A:1018822222663

Bull, S. J., Chalker, P. R., Ayres, C. F., & Rickerby, D. S. (1991). The influence of titanium interlayers on the adhesion of titanium nitride coatings obtained by plasma-assisted chemical vapor deposition. *Materials Science and Engineering A, 139*, 71–78. doi:10.1016/0921-5093(91)90599-I

Bunker, S. P. (2002). Evaluation of Dynamic Properties of a Carbon-Carbon Composite at Elevated Temperatures.

Bunshah, R. (1994). *Handbook of Deposition Technologies for Films and Coatings* (2nd ed.). New Jersey: Noyes.

Bunshah, R. F. (2001). *Handbook of hard coatings, deposition technologies, properties and applications*. New York, USA: Noyes Publication.

Burton, W. C., Cabrera, N., & Frank, F. C. (1951). The Growth of Crystals and the Equilibrium Structure of their Surfaces. *Trans. R. Soc. London A., 243*(866), 299–358. doi:10.1098/rsta.1951.0006

Bussmann, M., Chandra, S., & Mostaghimi, J. (2000). Modeling the splash of a droplet impacting a solid surface. *Physics of Fluids, 12*(12), 3121–3132. doi:10.1063/1.1321258

Cairo, C., Graca, M., Silva, C., & Bressiani, J. (2001). Functionally gradient ceramic coating for carbon–carbon antioxidation protection. *Journal of the European Ceramic Society, 21*(3), 325–329. doi:10.1016/S0955-2219(00)00191-6

Calabrese, L., Bonaccorsi, L., Caprì, A., & Proverbio, E. (2014). Adhesion aspects of hydrophobic silane zeolite coatings for corrosion protection of aluminium substrate. *Progress in Organic Coatings, 77*(9), 1341–1350. doi:10.1016/j.porgcoat.2014.04.025

Calvert, P. (1999). Nanotube composites: A recipe for strength. *Nature, 399*(6733), 210–211. doi:10.1038/20326

Cambon, J.-B., Ansart, F., Bonino, J.-P., & Turq, V. (2012). Effect of cerium concentration on corrosion resistance and polymerization of hybrid sol–gel coating on martensitic stainless steel. *Progress in Organic Coatings, 75*(4), 486–493. doi:10.1016/j.porgcoat.2012.06.005

Cambon, J.-B., Esteban, J., Ansart, F., Bonino, J.-P., Turq, V., Santagneli, S. H., & Pulcinelli, S. H. (2012). Effect of cerium on structure modifications of a hybrid sol–gel coating, its mechanical properties and anti-corrosion behavior. *Materials Research Bulletin, 47*(11), 3170–3176. doi:10.1016/j.materresbull.2012.08.034

Cambruzzi, A., Rossi, S., & Deflorian, F. (2005). Reduction on protective properties of organic coatings produced by abrasive particles. *Wear, 258*(11), 1696–1705. doi:10.1016/j.wear.2004.11.023

Campbell, C. E., Boettinger, W. J., & Kattner, U. R. (2002). Development of a diffusion mobility database for Ni-base superalloys. *Acta Materialia, 50*(4), 775–792. doi:10.1016/S1359-6454(01)00383-4

Cao, L. H., Khor, K. A., Fu, L., & Boey, F. (1999). Plasma spray processing of Al_2O_3/AlN composite powders. *Journal of Materials Processing Technology, 89-90*, 392–398. doi:10.1016/S0924-0136(99)00063-1

Cao, Q., Vessen, R., Schwarz, S., Jungen, W., Tietz, F., & Stoever, D. (2000). Spray-drying of ceramics for plasma-spray coating. *Journal of the European Ceramic Society, 20*(14-15), 2433–2439. doi:10.1016/S0955-2219(00)00112-6

Cao, R., Zhu, L., Li, W., Hu, W., & Liu, H. (2017). The effect of alumina-silica sols on electrodeposited zinc coatings for sintered NdFeB. *Journal of Alloys and Compounds, 726*, 95–106. doi:10.1016/j.jallcom.2017.06.123

Cao, X. Q., Vassen, R., & Stover, D. (2004). Ceramic materials for thermal barrier coatings. *Journal of the European Ceramic Society, 24*(1), 1–10. doi:10.1016/S0955-2219(03)00129-8

Cao, X. Q., Vassen, R., Tietz, F., & Stoever, D. (2006). New double-ceramic-layer thermal barrier coatings based on zirconia-rare earth composite oxides. *Journal of the European Ceramic Society, 26*(3), 247–251. doi:10.1016/j.jeurceramsoc.2004.11.007

Capelossi, V. R., Poelman, M., Recloux, I., Hernandez, R. P. B., de Melo, H. G., & Olivier, M. G. (2013). Corrosion protection of clad 2024 aluminum alloy anodized in tartaric-sulfuric acid bath and protected with hybrid sol–gel coating. *Electrochimica Acta*, *124*, 69–79. doi:10.1016/j.electacta.2013.09.004

Carabat, A. L., Zwaag, S., & Sloof, W. G. (2015). Creating a Protective Shell for Reactive MoSi2 Particles in High-Temperature Ceramics. *Journal of the American Ceramic Society*, *98*(8), 2609–2616. doi:10.1111/jace.13625

Carbajal-De, G., & La Torre, M. A. E.-M. (2009). Study of Ceramic and Hybrid Coatings Produced by the Sol-Gel Method for Corrosion Protection. *The Open Corrosion Journal*, *2*(1).

Cardoso, V. F., Minas, G., & Lanceros-Méndez, S. (2013). Multilayer spin-coating deposition of poly(vinylidene fluoride) films for controlling thickness and piezoelectric response. *Sensors and Actuators. A, Physical*, *192*, 76–80. doi:10.1016/j.sna.2012.12.019

Carim, A. H., Rohrer, G. S., Dando, N. R., Tzeng, S. Y., Rohrer, C. L., & Perotta, A. J. (1997). Conversion of Diaspore to corundum: A new α-alumina transformation sequence. *Communications of the American Ceramic Society*, *80*(10), 2677–2680. doi:10.1111/j.1151-2916.1997.tb03171.x

Carlos, P. (2011). *Bergmann, Juliane Vicenzi, Protection against Erosive Wear Using Thermal Sprayed Cermet-A Review*. Springer-Verlag Berlin Heidelberg; doi:10.1007/978-3-642-21987-0

Carneiro, J., Tedim, J., Fernandes, S. C. M., Freire, C. S. R., Silvestre, A. J. D., Gandini, A., & Zheludkevich, M. L. (2012). Chitosan-based self-healing protective coatings doped with cerium nitrate for corrosion protection of aluminum alloy 2024. *Progress in Organic Coatings*, *75*(1–2), 8–13. doi:10.1016/j.porgcoat.2012.02.012

Carpio, P., Salvador, M. D., Borrell, A., & Sánchez, E. (2017). Thermal behaviour of multilayer and functionally-graded YSZ/Gd 2 Zr 2 O 7 coatings. *Ceramics International*, *43*(5), 4048–4054. doi:10.1016/j.ceramint.2016.11.178

Castelvetro, V., & De Vita, C. (2004). Nanostructured hybrid materials from aqueous polymer dispersions. *Advances in Colloid and Interface Science*, *108–109*, 167–185. doi:10.1016/j.cis.2003.10.017 PMID:15072940

Castro, Y., Ferrari, B., Moreno, R., & Durán, A. (2003). Silica sol-gel coatings on metals produced by EPD. *Sol-Gel Science and Technology*, *26*(1/3), 735–739. doi:10.1023/A:1020750222322

Catteau, S. D., Van Landeghem, H. P., Teixeira, J., Dulcy, J., Dehmas, M., Denis, S., & Courteaux, M. et al. (2016). Carbon and nitrogen effects on microstructure and kinetics associated with bainitic transformation in a low-alloyed steel. *Journal of Alloys and Compounds*, *658*, 832–838. doi:10.1016/j.jallcom.2015.11.007

Cedelle, J., Vardelle, M., & Fauchais, P. (2006). Influence of stainless steel substrate preheating on surface topography and on millimeter- and micrometer-sized splat formation. *Surface and Coatings Technology*, *201*(3-4), 1373–1382. doi:10.1016/j.surfcoat.2006.02.003

Cellard, A., Garnier, V., Fantozzi, G., Baret, G., & Fort, P. (2009). Wear resistance of chromium oxide nanostructured coatings. *Ceramics International*, *35*(2), 913–916. doi:10.1016/j.ceramint.2008.02.022

Cellard, A., Zenati, R., Garnier, V., Fantozzi, G., & Baret, G. (2007). Optimization of chromium oxide nanopowders dispersion for spray-drying. *Journal of the European Ceramic Society*, *27*(2-3), 1021–1027. doi:10.1016/j.jeurceramsoc.2006.04.132

Cernuschi, F., Ahmaniemi, S., Vuoristo, P., & Mantyla, T. J. (2004). Modelling of thermal conductivity of porous materials: application to thick thermal barrier coatings. *Journal of European Ceramic Society, 24*, 2657-2667.

Cernuschi, F., Bianchi, P., Leoni, M., & Scardi, P. (1999). Thermal diffusivity microstructure relationship in Y-PSZ thermal barrier coatings. *Journal of Thermal Spray Technology*, *8*, 102-109.

Cetinel, H., Celik, E., & Kusoglu, M. I. (2008). Tribological behavior of Cr_2O_3 coatings as bearing materials. *Journal of Materials Processing Technology, 196*(1-3), 259–265. doi:10.1016/j.jmatprotec.2007.05.048

Chandler, K. A., & Bayliss, D. A. (1985). *Corrosion protection of steel structures*. Elsevier Applied Science Publishers.

Chandran, R. G., Chandrashekar, B. K., Ganguly, C., & Patil, K. C. (1996). Sintering and microstructural investigations on combustion processed mullite. *Journal of the European Ceramic Society, 16*(8), 843–849. doi:10.1016/0955-2219(96)00001-5

Chandra, S., & Fauchais, P. (2009). Formation of solid splats during thermal spray deposition. *Journal of Thermal Spray Technology, 18*(2), 148–180. doi:10.1007/s11666-009-9294-5

Chand, S. (2000). Review carbon fibers for composites. *Journal of Materials Science, 35*(6), 1303–1313. doi:10.1023/A:1004780301489

Charewicz, W. (1990). *Pierwiastki ziem rzadkich: surowce, technologie, zastosowanie*. Warszawa: Wydawnictwo Naukowo-Techniczne.

Chatha, S. S., Sidhu, H. S., & Sidhu, B. S. (2012). Characterisation and corrosion-erosion behaviour of carbide based thermal spray coatings. *Journal of Minerals & Materials Characterization & Engineering, 11*(6), 569–586. doi:10.4236/jmmce.2012.116041

Chauhan, K. R. (2014). A Review Paper on Tribological and Mechanical Properties of Ternary Nitride Based Coatings. *Procedia Technology*, 430-437. doi:10.1016/j.protcy.2014.08.055

Chen, D., Jordan, E. H., & Gell, M. (2010). The solution precursor plasma spray coatings: Influence of solvent type. *Plasma Chemistry and Plasma Processing, 30*(1), 111–119. doi:10.1007/s11090-009-9200-4

Chen, G. F., & Lou, H. Y. (1999a). Oxidation behavior of sputtered Ni-3Cr-20Al nanocrystalline coating. *Materials Science and Engineering A, 271*(1-2), 360–365. doi:10.1016/S0921-5093(99)00258-0

Chen, G. F., & Lou, H. Y. (1999b). The effect of nanocrystallization on the oxidation resistance of Ni-5Cr-5Al alloy. *Scripta Materialia, 41*(8), 883–887. doi:10.1016/S1359-6462(99)00224-9

Chen, G. F., & Lou, H. Y. (2000a). Effect of nanocrystallization on the oxidation behavior of a Ni–8Cr–3.5Al alloy. *Oxidation of Metals, 54*(1-2), 155–162. doi:10.1023/A:1004610931447

Chen, G. F., & Lou, H. Y. (2000b). The effect nanocrystallization on the oxide formation on Ni-11Cr-8Al alloy. *Scripta Materialia, 43*(2), 119–122. doi:10.1016/S1359-6462(00)00379-1

Chen, G. F., & Rühle, M. (2005). Transient alumina transformation on a sputtered K38G nanocrystalline coating. *Surface and Coatings Technology, 191*(2-3), 263–266. doi:10.1016/j.surfcoat.2004.02.025

Cheng, L., Xu, Y., Zhang, L., & Yin, X. (1999). Oxidation behavior of carbon–carbon composites with a three-layer coating from room temperature to 1700° C. *Carbon, 37*(6), 977–981. doi:10.1016/S0008-6223(98)00293-0

Chen, H., Ding, C., Zhang, P., La, P., & Lee, S. W. (2003). Wear of plasma-sprayed nanostructured zirconia coatings against stainless steel under distilled-water conditions. *Surface and Coatings Technology, 173*(2-3), 144–149. doi:10.1016/S0257-8972(03)00311-6

Chen, H., Gao, Y., Tao, S., Liu, Y., & Luo, H. (2009). Thermophysical properties of lanthanum zirconate coating prepared by plasma spraying and the influence of post-annealing. *Journal of Alloys and Compounds, 486*(1-2), 391–399. doi:10.1016/j.jallcom.2009.06.162

Chen, H., Zhou, X., & Ding, C. (2002). Investigation of the thermomechanical properties of a plasma sprayed nanostructured zirconia coating. *Journal of the European Ceramic Society*, *23*(9), 1449–1455. doi:10.1016/S0955-2219(02)00345-X

Chen, J. H., & Little, J. A. (1997). Degradation of the platinum aluminide coating on CMSX4 at 1100 °C. *Surface and Coatings Technology*, *92*(1-2), 69–77. doi:10.1016/S0257-8972(96)03117-9

Chen, L. B. (2006). Yttria-stabilized zirconia thermal barrier coatings - A review. *Surface Review and Letters*, *13*(05), 535–544. doi:10.1142/S0218625X06008670

Chen, M. H., Li, W. B., Shen, M. L., Zhu, S. L., & Wang, F. H. (2013). Glass–ceramic coatings on titanium alloys for high temperature oxidation protection: Oxidation kinetics and microstructure. *Corrosion Science*, *74*, 178–186. doi:10.1016/j.corsci.2013.04.041

Chen, M. H., Li, W. B., Shen, M. L., Zhu, S. L., & Wang, F. H. (2014). Glass coatings on stainless steels for high-temperature oxidation protection: Mechanisms. *Corrosion Science*, *82*, 316–327. doi:10.1016/j.corsci.2014.01.033

Chen, M. H., Shen, M. L., Wang, X., Zhu, S. L., & Wang, F. H. (2012). Oxidation and thermal shock behavior of a glass-alumina composite coating on K38G superalloy at 1000 °C. *Journal of Materials Science and Technology*, *28*(5), 433–438. doi:10.1016/S1005-0302(12)60079-6

Chen, M. H., Shen, M. L., Wang, X., Zhu, S. L., & Wang, F. H. (2013). Interfacial reaction between SiO_2–Al_2O_3–ZnO–CaO based glass coatings and K38G superalloy substrates. *Surface and Coatings Technology*, *216*, 145–151. doi:10.1016/j.surfcoat.2012.11.045

Chen, M. H., Shen, M. L., Zhu, S. L., & Wang, F. H. (2013). Comparative study of interfacial reaction between superalloy substrate and glass coating with and without alumina particles incorporation. *Applied Surface Science*, *271*, 228–233. doi:10.1016/j.apsusc.2013.01.165

Chen, M. H., Shen, M. L., Zhu, S. L., Wang, F. H., & Niu, Y. (2012). Preparation and thermal shock behavior at 1000 °C of a glass-alumina-NiCrAlY tri-composite coating on K38G superalloy. *Surface and Coatings Technology*, *206*(8-9), 2566–2571. doi:10.1016/j.surfcoat.2011.11.015

Chen, M. H., Shen, M. L., Zhu, S. L., Wang, F. H., & Wang, X. L. (2013). Effect of sand blasting and glass matrix composite coating on oxidation resistance of a nickel-based superalloy at 1000 °C. *Corrosion Science*, *73*, 331–341. doi:10.1016/j.corsci.2013.04.022

Chen, M. H., Zhu, S. L., Shen, M. L., Wang, F. H., & Niu, Y. (2011). Effect of NiCrAlY platelets inclusion on the mechanical and thermal shock properties of glass matrix composites. *Materials Science and Engineering A*, *528*(3), 1360–1366. doi:10.1016/j.msea.2010.10.015

Chen, M. H., Zhu, S. L., Shen, M. L., Wang, F. H., & Niu, Y. (2013). Thermophysical properties of alumina particle reinforced glass matrix composites. *International Journal of Applied Ceramic Technology*, *10*(2), 224–233. doi:10.1111/j.1744-7402.2012.02806.x

Chen, M. H., Zhu, S. L., & Wang, F. H. (2010). Crystallization Behavior of SiO_2–Al_2O_3–ZnO–CaO Glass System at 1123–1273 K. *Journal of the American Ceramic Society*, *93*(10), 3230–3235. doi:10.1111/j.1551-2916.2010.03859.x

Chen, M. H., Zhu, S. L., & Wang, F. H. (2013). Strengthening mechanisms and fracture surface characteristics of silicate glass matrix composites with inclusion of alumina particles of different particle sizes. *Physica B, Condensed Matter*, *413*, 15–20. doi:10.1016/j.physb.2013.01.006

Chen, M. H., Zhu, S. L., & Wang, F. H. (2015). High temperature oxidation of NiCrAlY, nanocrystalline and enamel-metal nano-composite coatings under thermal shock. *Corrosion Science*, *100*, 556–565. doi:10.1016/j.corsci.2015.08.033

Chen, X., Wang, R., Yao, N., Evans, A. G., Hutchinson, J. W., & Bruce, R. W. (2003). Foreign object damage in a thermal barrier system: Mechanisms and simulations. *Materials Science and Engineering A*, *352*(1), 221–231. doi:10.1016/S0921-5093(02)00905-X

Chin, R. J., & Nobe, K. (1972). Electrodissolution kinetics of iron in chloride solutions III. Acidic solutions. *Journal of the Electrochemical Society*, *119*(11), 1457–1461. doi:10.1149/1.2404023

Chiu, W.-M., Zhang, Y.-S., Tsai, P.-A., & Wu, J.-H. (2013). Characterization of antireflective coatings on poly(methyl methacrylate) substrate by different process parameters. *Journal of Applied Polymer Science*, *129*(5), 2411–2417. doi:10.1002/app.38975

Choi, S. R., Zhu, D., & Miller, R. A. (2005). Effect of Sintering on Mechanical Properties of Plasma-Sprayed Zirconia-Based Thermal Barrier Coatings. *Journal of the American Ceramic Society*, *88*(10), 2859–2867. doi:10.1111/j.1551-2916.2005.00504.x

Cho, J., Xiong, Y., Kim, J., Lee, C., & Hwang, S. H. (2008). Tribological behavior of NiCr-base blended and nanostructured composite APS coatings by rig test. *Wear*, *265*(11-12), 1565–1571. doi:10.1016/j.wear.2008.03.008

Chraska, T., & King, A. H. (2001). Transmission electron microscopy study of rapid solidification of plasma sprayed zirconia part I. First splat solidification. *Thin Solid Films*, *397*(1-2), 30–39. doi:10.1016/S0040-6090(01)01360-8

Christoulis, D. K., Pantelis, D. I., Borit, F., Guipont, V., & Jeandin, M. (2004). Effect of substrate roughness and temperature on splat formation in plasma sprayed aluminium bronze. *Paper presented at the 18th International Conference on Surface Modification Technologies*, Dijon, France. Retrieved from http://www.scopus.com/inward/record.url?eid=2-s2.0-33646056723&partnerID=40

Chuang, C. C. (2008). Effects of Internal Stresses on the Mechanical Properties of Deposition Thin Films. *journal of materials processing technology, 201*, 770–774.

Chwa, S. O., Klein, D., Toma, F. L., Bertrand, G., Liao, H., Coddet, C., & Ohmori, A. (2005). Microstructure and mechanical properties of plasma sprayed nanostructured TiO_2–Al composite coatings. *Surface and Coatings Technology*, *194*(2-3), 215–224. doi:10.1016/j.surfcoat.2004.07.080

Cipitria, A., Golosnoy, I. O., & Clyne, T. W. (2009). A sintering model for plasma-sprayed zirconia thermal barrier coatings. Part II: Coatings bonded to a rigid substrate. *Acta Materialia*, *57*(4), 993–1003. doi:10.1016/j.actamat.2008.10.058

Ciriminna, R., Fidalgo, A., Pandarus, V., Béland, F., Ilharco, L. M., & Pagliaro, M. (2013). The Sol-Gel Route to Advanced Silica-Based Materials and Recent Applications. *Chemical Reviews*, *113*(8), 6592–6620. doi:10.1021/cr300399c PMID:23782155

Ciriminna, R., Sciortino, M., Alonzo, G., Schrijver, A. D., & Pagliaro, M. (2011). From molecules to systems: Sol-gel microencapsulation in silica-based materials. *Chemical Reviews*, *111*(2), 765–789. doi:10.1021/cr100161x PMID:20726523

Ciubotariu, C.-R., Frunzăverde, D., Mărginean, G., Şerban, V.-A., & Bîrdeanu, A.-V. (2016). Optimization of the laser remelting process for HVOF-sprayed Stellite 6 wear resistant coatings. *Optics & Laser Technology*, *77*, 98–103. doi:10.1016/j.optlastec.2015.09.005

Clarke, D. R. (2003). Materials selection guidelines for low thermal conductivity thermal barrier coatings. *Surface and Coatings Technology*, *163–164*, 67–74. doi:10.1016/S0257-8972(02)00593-5

Clarke, D. R. (2003). The lateral growth strain accompanying the formation of a thermally grown oxide. *Acta Materialia*, *51*(5), 1393–1407. doi:10.1016/S1359-6454(02)00532-3

Clarke, D. R., & Levi, C. G. (2003). Materials design for the next generation thermal barrier coatings. *Annual Review of Materials Research*, *33*(1), 383–417. doi:10.1146/annurev.matsci.33.011403.113718

Clarke, D. R., & Phillpot, S. R. (2005). Thermal barrier coating materials. *Materials Today*, *8*(6), 22–29. doi:10.1016/S1369-7021(05)70934-2

Clark, H. (1992). The influence of the flow field in slurry erosion. *Wear*, *152*(2), 223–240. doi:10.1016/0043-1648(92)90122-O

Clyne, T. W. (1984). Numerical treatment of rapid solidification. *Metallurgical Transactions. B, Process Metallurgy*, *15*(2), 369–381. doi:10.1007/BF02667341

Conde, A., Durán, A., & de Damborenea, J. (2003). Polymeric sol–gel coatings as protective layers of aluminium alloys. *Progress in Organic Coatings*, *46*(4), 288–296. doi:10.1016/S0300-9440(03)00014-6

Constable, C. P., Lewis, B. D., Yarwood, J., & Munz, W. D. (2004). Raman microscopic studies of residual and applied stress in PVD hard ceramic coatings and correlation with X-ray diffraction (XRD) measurements. *Surface and Coatings Technology*, *184*(2-3), 291–297. doi:10.1016/j.surfcoat.2003.10.014

Corengia, P., Ybarra, G., Moina, C., Cabo, A., & Broitman, E. (2005). Microstructural and topographical studies of DC-pulsed plasma nitrided AISI 4140 low-alloy steel. *Surface and Coatings Technology*, *200*(7), 2391–2397. doi:10.1016/j.surfcoat.2005.01.060

Corral, E. L., & Loehman, R. E. (2008). Ultra-High-Temperature Ceramic Coatings for Oxidation Protection of Carbon–Carbon Composites. *Journal of the American Ceramic Society*, *91*(5), 1495–1502. doi:10.1111/j.1551-2916.2008.02331.x

Corvo, F., Pérez, T., Martin, Y., Reyes, J., Dzib, L. R., González-Sánchez, J., & Castañeda, A. (2008). Time of wetness in tropical climate: Considerations on the estimation of TOW according to ISO 9223 standard. *Corrosion Science*, *50*(1), 206–219. doi:10.1016/j.corsci.2007.06.012

Council, N. R. (1996). *Coatings for high-temperature structural materials: trends and opportunities*. National Academies Press.

Courtin, S., Gardin, C., Bezine, G., & Hamouda, H. B. H. (2005). Advantages of the J-integral approach for calculating stress intensity factors when using the commercial finite element software ABAQUS. *Engineering Fracture Mechanics*, *72*(14), 2174–2185. doi:10.1016/j.engfracmech.2005.02.003

Cowell, R. (1996). Durable optical coatings for energetic UV laser sources. *SPIE*, *2703*, 369–374.

Craig, B. D. (2006). Material failure modes, part III: A brief tutorial on corrosion-related material failure modes. *Journal of Failure Analysis and Prevention*, *6*(2), 12–19. doi:10.1361/154770206X99262

Criado, M., Sobrados, I., Bastidas, J. M., & Sanz, J. (2015). Steel corrosion in simulated carbonated concrete pore solution its protection using sol–gel coatings. *Progress in Organic Coatings*, *88*, 228–236. doi:10.1016/j.porgcoat.2015.06.002

Criado, M., Sobrados, I., Bastidas, J. M., & Sanz, J. (2016). Corrosion behaviour of coated steel rebars in carbonated and chloride-contaminated alkali-activated fly ash mortar. *Progress in Organic Coatings*, *99*, 11–22. doi:10.1016/j.porgcoat.2016.04.040

Criado, M., Sobrados, I., Sanz, J., & Bastidas, J. M. (2014). Steel protection using sol–gel coatings in simulated concrete pore solution contaminated with chloride. *Surface and Coatings Technology*, *258*, 485–494. doi:10.1016/j.surfcoat.2014.08.051

Cullity, B. D. (1978). *Elements of X-ray Diffraction London*. Addison-Wesley publishing Company, Inc.

Curry, N. (2014). Design of Thermal Barrier Coatings. *PhD Thesis Production Technology 2014 No.3*, Trollhattan Sweden: University West.

Curry, N., Markocsan, N., Li, X. A., Tricoire, A., & Dorfman, M. (2011). Next generation thermal barrier coatings for the gas turbine industry. *Journal of Thermal Spray Technology*, *20*(1-2), 108–115. doi:10.1007/s11666-010-9593-x

Czerwiec, T., Huan, H., Marcos, G., Thiriet, T., Weber, S., & Michel, H. (2009). Fundamental and Innovations in Plasma Assisted Diffusion of Nitrogen and Carbon in Austenitic Stainless Steels and Related Alloys. *Plasma Processes and Polymers*, *6*(6-7), 401–409. doi:10.1002/ppap.200930003

Dahan, I., Admon, U., Frage, N., Sariel, J., Dariel, M. P., & Moore, J. J. (2001). The development of a functionally graded TiC-Ti multilayer hard coating. *Surface and Coatings Technology*, *137*(2-3), 111–115. doi:10.1016/S0257-8972(00)01122-1

Dahl, K. V., Hald, J., & Horsewell, A. (2006). Interdiffusion between Ni-based superalloy and MCrAlY coating. *Defect and Diffusion Forum*, 258-260, 73-78.

Dahl, P., Kaus, I., Zhao, Z., Johnsson, M., Nygren, M., Wiik, K., & Einarsrud, M. A. et al. (2007). Densification and properties of zirconia prepared by three different sintering techniques. *Ceramics International*, *33*(8), 1603–1610. doi:10.1016/j.ceramint.2006.07.005

Dai, H., Li, J., Cao, X., Meng, J. (2009). Thermal barrier coating material. *US20097597971*.

Darolia, R., Rigney, J.D., & Grylls, R., (2001). Nickel aluminide alloys with Cr and Zr for adherent coating or interlayer on superalloy turbine parts.

Darolia, R. (2013). Thermal barrier coatings technology: Critical review, progress update, remaining challenges and prospects. *International Materials Reviews*, *58*(6), 315–348. doi:10.1179/1743280413Y.0000000019

Dash, S., Mishra, S., Patel, S., & Mishra, B. K. (2008). Organically modified silica: Synthesis and applications due to its surface interaction with organic molecules. *Advances in Colloid and Interface Science*, *140*(2), 77–94. doi:10.1016/j.cis.2007.12.006 PMID:18321464

Das, K., Alphonsa, J., Ghosh, M., Ghanshyam, J., Rane, R., & Mukherjee, M. (2017). Influence of pretreatment on surface behavior of duplex plasma treated AISI H13 tool steel. *Surfaces and Interfaces*, *8*, 206–213. doi:10.1016/j.surfin.2017.06.009

Davis, J. R. (2000). *Corrosion understanding the basics*. Materials Park, OH: ASM International.

Davis, J. R. (Ed.). (2004). *Handbook of thermal spray technology*. Materials Park, OH, USA: ASM International.

Davis, S. R., Brough, A. R., & Atkinson, A. (2003). Formation of silica/epoxy hybrid network polymers. *Journal of Non-Crystalline Solids*, *315*(1), 197–205. doi:10.1016/S0022-3093(02)01431-X

De Freitas Cunha Lins, V., Cury, F. M., & Moreira, R. (2008). Infrared and ultraviolet-visible spectroscopy study of the degradation of polyester and polyester/ethylene methyl acrylate copolymer blend coatings on steel. *Journal of Applied Polymer Science*, *109*(4), 2103–2112. doi:10.1002/app.28231

De Sansal, C., Devincre, B., & Kubin, L. (2010). Grain size strengthening in microcrystalline copper: A three-dimensional dislocation dynamics simulation. *Key Engineering Materials*, *423*, 25–32. doi:10.4028/www.scientific.net/KEM.423.25

Deborah, D. C. (2010). *Composite materials: Science and applications*. Engineering Materials and Processes.

Delogu, F., & Cocco, G. (2006). Microstructural refinement of ceramic powders under mechanical processing conditions. *Journal of Alloys and Compounds*, *420*(1-2), 246–250. doi:10.1016/j.jallcom.2005.10.037

Deneirjean, A., Lefort, P., & Fauchais, P. (2003). Nitridation process and mechanism of Ti-$_6$Al-$_4$V particles by dc plasma spraying. *Physical Chemistry Chemical Physics*, *5*(22), 5133–5138. doi:10.1039/B311186P

Deng, Y., Tan, C., Wang, Y., Chen, L., Cai, P., Kuang, T., & Zhou, K. et al. (2017). Effects of tailored nitriding layers on comprehensive properties of duplex plasma-treated AlTiN coatings. *Ceramics International*, *43*(12), 8721–8729. doi:10.1016/j.ceramint.2017.03.209

Derelioglu, Z., Carabat, A., Song, G., van der Zwaag, S., & Sloof, W. (2015). On the use of B-alloyed MoSi 2 particles as crack healing agents in yttria stabilized zirconia thermal barrier coatings. *Journal of the European Ceramic Society*, *35*(16), 4507–4511. doi:10.1016/j.jeurceramsoc.2015.08.035

Dhami, T., Bahl, O., & Awasthy, B. (1995). Oxidation-resistant carbon-carbon composites up to 1700 C. *Carbon*, *33*(4), 479–490. doi:10.1016/0008-6223(94)00173-W

Dhineshkumar, S. R., Duraiselvam, M., Natarajan, S., Panwar, S. S., Jena, T., & Khan, M. A. (2016). Enhancement of strain tolerance of functionally graded $LaTi_2Al_9O_{19}$ thermal barrier coating through ultra-short pulse based laser texturing. *Surface and Coatings Technology*, *304*, 263–271. doi:10.1016/j.surfcoat.2016.07.018

Di Girolamo, G., Blasi, C., Brentari, A., & Schioppa, M. (2015). Microstructural, mechanical and thermal characteristics of zirconia-based thermal barrier coatings deposited by plasma spraying. *Ceramics International*, *41*(9), 11776–11785. doi:10.1016/j.ceramint.2015.05.145

Di Girolamo, G., Blasi, C., Pagnotta, L., & Schioppa, M. (2010). Phase evolution and thermophysical properties of plasma sprayed thick zirconia coatings after annealing. *Ceramics International*, *36*(8), 2273–2280. doi:10.1016/j.ceramint.2010.07.035

Di Girolamo, G., Blasi, C., Schioppa, M., & Tapfer, L. (2010). Structure and thermal properties of heat treated plasma sprayed ceria-yttria co-stabilized zirconia coatings. *Ceramics International*, *36*(3), 961–968. doi:10.1016/j.ceramint.2009.10.020

Di Girolamo, G., Marra, F., Blasi, C., Pilloni, L., & Valente, T. (2011). Microstructure, mechanical properties and thermal shock resistance of plasma sprayed nanostructured zirconia coatings. *Ceramics International*, *37*(7), 2711–2717. doi:10.1016/j.ceramint.2011.04.024

Di Girolamo, G., Marra, F., Schioppa, M., Blasi, C., Pulci, G., & Valente, T. (2015). Evolution of microstructural and mechanical properties of lanthanum zirconate thermal barrier coatings at high temperature. *Surface and Coatings Technology*, *268*, 298–302. doi:10.1016/j.surfcoat.2014.07.067

Di Girolamo, G., & Serra, E. (2015). Thermally sprayed nanostructured coatings for anti-wear and TBC applications: state-of-the-art and future perspectives. In M. Aliofkhazarei (Ed.), *Anti-abrasive nanocoatings* (pp. 513–545). Elsevier Ltd. doi:10.1016/B978-0-85709-211-3.00020-0

Dias, S. A. S., Lamaka, S. V., Nogueira, C. A., Diamantino, T. C., & Ferreira, M. G. S. (2012). Sol–gel coatings modified with zeolite fillers for active corrosion protection of AA2024. *Corrosion Science*, *62*, 153–162. doi:10.1016/j.corsci.2012.05.009

DiCarlo, J. A., & Yun, H.-M. (2005). *Non-oxide (silicon carbide) fibers Handbook of ceramic composites* (pp. 33–52). Springer. doi:10.1007/0-387-23986-3_2

Dimitriev, Y., Ivanova, Y., & Iordanova, R. (2008). History of sol-gel science and technology. *Journal of the University of Chemical Technology and Matallurgy*, *43*(2), 181–192.

Dislich, H. (1986). Sol-gel: Science, processes and products. *Journal of Non-Crystalline Solids*, *80*(1–3), 115–121. doi:10.1016/0022-3093(86)90384-4

Doleker, K. M., & Karaoglanli, A. C. (2016). Comparison of oxidation behavior of YSZ and $Gd_2Zr_2O_7$ thermal barrier coatings (TBCs). *Surface and Coatings Technology*, *318*, 198–207. doi:10.1016/j.surfcoat.2016.12.078

Donald, I. W. (1993). Preparation, properties and chemistry of glass and glass-ceramic-to- metal seals and coatings. *Journal of Materials Science*, *28*(11), 2841–2886. doi:10.1007/BF00354689

Dong, H., Yang, G. J., Li, C. X., Luo, X. T., & Li, C. J. (2014). Effect of TGO thickness on thermal cyclic lifetime and failure mode of plasma-sprayed TBCs. *Journal of the American Ceramic Society*, *97*(4), 1226–1232. doi:10.1111/jace.12868

Donley, M. S., Mantz, R., Khramov, N., Balbyshev, V. N., Kasten, L. S., & Gaspar, D. J. (2003). The self-assembled nanophase particle (SNAP) process: A nanoscience approach to coatings. *Progress in Organic Coatings*, *47*(3–4), 401–415. doi:10.1016/j.porgcoat.2003.08.017

Donnet, C., & Erdemir, A. (2004). Historical development and new trends in tribological and solid lubricant coatings. *Surface and Coatings Technology*, *180*, 76–84. doi:10.1016/j.surfcoat.2003.10.022

Drexler, J. M. (2011). *Thermal barrier coatings resistant to glassy deposits*. [Dissertation Doctor of Philosophy]. The Ohio State University, Columbus, OH.

Drexler, J. M., Gledhill, A. D., Shinoda, K., Vasiliev, A. L., Reddy, K. M., Sampath, S., & Padture, N. P. (2011). Jet Engine Coatings for Resisting Volcanic Ash Damage. *Advanced Materials*, *23*(21), 2419–2424. doi:10.1002/adma.201004783 PMID:21484889

Drexler, J. M., Ortiz, A. L., & Padture, N. P. (2012). Composition effects of thermal barrier coating ceramics on their interaction with molten Ca–Mg–Al-silicate (CMAS) glass. *Acta Materialia*, *60*(15), 5437–5447. doi:10.1016/j.actamat.2012.06.053

Drexler, J. M., Shinoda, K., Ortiz, A. L., & Padture, N. (2010). Air-plasma-sprayed thermal barrier coatings that are resistant to high-temperature attack by glass deposits. *Acta Materialia*, *58*(20), 6835–6844. doi:10.1016/j.actamat.2010.09.013

Duvall, D., & Ruckle, D. (1982). Ceramic thermal barrier coatings for turbine engine components. *ASME Paper*, *82*, 332.

Duwez, P., Willen, R. H., & Klement, W. (1960). Continuous series of metastable solid solutions in silver-copper alloys. *Journal of Applied Physics*, *31*(6), 1136–1137. doi:10.1063/1.1735777

Eckardt, T., Mallener, W., & Stover, D. (1996). Development of plasma sprayed silicon/silicon nitride coatings by in situ nitriding. In Proceeding of Thermal Spray '96 (pp. 309-312), German.

Eckardt, T., Mallener, W., & Stover, D. (1994). Reactive plasma spraying of silicon in controlled nitrogen atmosphere. In *Proceeding 7th National Thermal Spray Conference,* Boston, MA (pp. 515-519).

Eckardt, T., Mallener, W., & Stover, D. (1995). Influence of a reactive nitrogen atmosphere on plasma sprayed silicon coatings. *VDI-Berichte*, *1166*, 341–348.

Eckel, M., Hardt, P., & Schmidt, M. (1999). Investigation of TiN deposition in different sized PACVD reactors by means of optical emission spectroscopy. *Surface and Coatings Technology*, *116–119*, 1037–1041. doi:10.1016/S0257-8972(99)00091-2

Eckstein, W. R., Roth, J., Nagel, W., & Dohmen, R. (2004). Sputtering Mechanisms near the Threshold Energy. *Journal of Nuclear Materials*, *328*(1), 55–61. doi:10.1016/j.jnucmat.2004.03.010

Eftis, J., & Liebowitz, H. (1976). On surface energy and the continuum thermodynamics of brittle fracture. *Engineering Fracture Mechanics*, *8*(3), 459–485. doi:10.1016/0013-7944(76)90002-3

Ehrburger, P., Baranne, P., & Lahaye, J. (1986). Inhibition of the oxidation of carboncarbon composite by boron oxide. *Carbon*, *24*(4), 495–499. doi:10.1016/0008-6223(86)90274-5

Ehrlich, G. H., & Hudda, F. G. (1966). Atomic view of surface self-diffusion: Tungsten on tungsten. *The Journal of Chemical Physics*, *44*(3), 1039–1049. doi:10.1063/1.1726787

El Rayes, M. M., Abdo, H. S., & Khalil, K. A. (2013). Erosion - Corrosion of Cermet Coating. *International Journal of Electrochemical Science*, *8(1)*, 1117–1137.

El-Hossary, F. M., Negm, N. Z., Abd El-Rahman, A. M., & Hammad, M. (2009). Duplex treatment of 304 AISI stainless steel using rf plasma nitriding and carbonitriding. *Materials Science and Engineering C*, *29*(4), 1167–1173. doi:10.1016/j.msec.2008.09.049

El-Hossary, F. M., Negm, N. Z., Khalil, S. M., & Abd Elrahman, A. M. (2002). Formation and properties of a carbonitrided layer in 304 stainless steel using different radio frequency plasma powers. *Thin Solid Films*, *405*(1-2), 179–185. doi:10.1016/S0040-6090(01)01729-1

Eliaz, N., Shemesh, G., & Latanision, R. M. (2002). Hot corrosion in gas turbine components. *Engineering Failure Analysis*, *9*(1), 31–43. doi:10.1016/S1350-6307(00)00035-2

Endler, I., Wolf, E., Leonhardt, A., & Richter, V. (1995). Preparation, characterization and wear behavior of PACVD cermets. *Surface and Coatings Technology*, *72*(1-2), 37–42. doi:10.1016/0257-8972(94)02338-7

Erickson, L. C., Hawthorne, H. M., & Troczynski, T. (2001). Correlations between microstructural parameters, micromechanical properties and wear resistance of plasma sprayed ceramic coatings. *Wear*, *250*(1-12), 569–575. doi:10.1016/S0043-1648(01)00608-1

Eroglu, S., & Gallois, B. (1991). Design and chemical vapor deposition of graded TiN/TiC coatings. *Surface and Coatings Technology*, *49*(1-3), 275–278. doi:10.1016/0257-8972(91)90068-8

Escure, C., Vardelle, M., & Fauchais, P. (2003). Experimental and theoretical study of the impact of alumina droplets on cold and hot substrates. *Plasma Chemistry and Plasma Processing*, *23*(2), 185–221. doi:10.1023/A:1022976914185

Espallargas, N., Berget, J., Guilemany, J. M., Benedetti, A. V., & Suegama, P. H. (2008a). Cr3C2–NiCr and WC–Ni thermal spray coatings as alternatives to hard chromium for erosion–corrosion resistance. *Surface and Coatings Technology*, *202*(8), 1405–1417. doi:10.1016/j.surfcoat.2007.06.048

Esteves, A. C. C., Barros-Timmons, A., & Trindade, T. (2004). Nanocompósitos de matriz polimérica: Estratégias de síntese de materiais híbridos. *Quimica Nova*, *27*(5), 798–806. doi:10.1590/S0100-40422004000500020

Ettouil, F. B., Mazhorova, O., Pateyron, B., Ageorges, H., El Ganaoui, M., & Fauchais, P. (2008). Predicting dynamic and thermal histories of agglomerated particles injected within a d. C. plasma jet. *Surface and Coatings Technology*, *202*(18), 4491–4495. doi:10.1016/j.surfcoat.2008.04.032

Evans, H. E., Mitchell, G. P., Lobb, R. C., & Owen, D. R. J. (1993). A numerical-analysis of oxide spallation. *Proceedings of the Royal Society of London Series A*, 440(1908), 1-22.

Evans, A. G., Clarke, D. R., & Levi, C. G. (2008). The influence of oxides on the performance of advanced gas turbines. *Journal of the European Ceramic Society*, *28*(7), 1405–1419. doi:10.1016/j.jeurceramsoc.2007.12.023

Evans, A. G., Crumley, G. B., & Demaray, R. E. (1983). On the mechanical behavior of brittle coatings and layers. *Oxidation of Metals*, *20*(5-6), 193–216. doi:10.1007/BF00656841

Evans, A. G., Fleck, N. A., Faulhalber, S., Vermaak, N., Maloney, M., & Darolia, R. (2006). Scaling laws governing the erosion and impact resistance of thermal barrier coatings. *Wear*, *260*(7-8), 886–894. doi:10.1016/j.wear.2005.05.005

Evans, A. G., He, M. Y., & Hutchinson, J. W. (1997). Effect of interface undulations on the thermal fatigue of thin films and scales on metal substrates. *Acta Materialia*, *45*(9), 3543–3554. doi:10.1016/S1359-6454(97)00062-1

Evans, A. G., Mumm, D. R., Hutchinson, J. W., Meier, G. H., & Petit, F. S. (2001). Mechanisms controlling the durability of thermal barrier coatings. *Progress in Materials Science*, *46*(5), 505–553. doi:10.1016/S0079-6425(00)00020-7

Fan, W., & Bai, Y. (2016). Review of suspension and solution precursor plasma sprayed thermal barrier coatings. *Ceramics International*, *42*(13), 14299–14312. doi:10.1016/j.ceramint.2016.06.063

Fan, X. L., Xu, R., Zhang, W., & Wang, T. J. (2012). Effect of periodic surface cracks on the interfacial fracture of thermal barrier coating system. *Applied Surface Science*, *258*(24), 9816–9823. doi:10.1016/j.apsusc.2012.06.036

Farle, A., Boatemaa, L., Shen, L., Gövert, S., Kok, J., Bosch, M., & Sloof, W. (2016). Demonstrating the self-healing behaviour of some selected ceramics under combustion chamber conditions. *Smart Materials and Structures*, *25*(8), 084019. doi:10.1088/0964-1726/25/8/084019

Fauchais P.L., Heberlein, J.V.R., & Boulos, M. (2014). *Thermal Spray Fundamentals: From Powder to Part*. New York: Springer Science+Business Media. doi:10.1007/978-0-387-68991-3_2

Fauchais, P., & Vardelle, A. (2012). Thermal Sprayed Coatings Used Against Corrosion and Corrosive Wear. In H. Jazi (Ed.), Advanced Plasma Spray Applications. doi:10.5772/34448

Fauchais, P. (2004). Understanding plasma spraying. *Journal of Physics. D, Applied Physics*, *37*(9), R86–R108. doi:10.1088/0022-3727/37/9/R02

Fauchais, P. (2008). Suspension and solution plasma or HVOF spraying. *Journal of Thermal Spray Technology*, *17*(1), 1–3. doi:10.1007/s11666-008-9160-x

Fauchais, P. L., Heberlein, J. V. R., & Boulos, M. I. (2014). *Thermal Spray Fundamentals From Powder to Part*. Springer. doi:10.1007/978-0-387-68991-3

Fauchais, P., & Coudert, J. F. (1996). Measurement of temperatures in thermal plasmas. In R. Benocci, G. Bonizzoni, & E. Sindoni (Eds.), *Thermal Plasmas for Hazardous Waste Treatment* (pp. 39–66). PA: World Scientific. doi:10.1142/9789814447171_0002

Fauchais, P., Etchart-Salas, R., Rat, V., Coudert, J.-F., Caron, N., & Wittmann-Ténèze, K. (2008). Parameters controlling liquid plasma spraying: Solutions, sols, or suspensions. *Journal of Thermal Spray Technology*, *17*(1), 31–59. doi:10.1007/s11666-007-9152-2

Fauchais, P., Fukumoto, M., Vardelle, A., & Vardelle, M. (2004). Knowledge concerning splat formation: An invited review. *Journal of Thermal Spray Technology*, *13*(3), 337–360. doi:10.1361/10599630419670

Fauchais, P., Montavan, G., & Bertrand, G. (2009). Influence of powders on thermal spray coating structures: Recent developments in nano or finely structured coatings and some safety issues. In B. R. Marple, M. M. Hyland, Y.-C. Lau, C.-J. Li, R. S. Lima, & G. Montavon (Eds.), *Thermal Spray 2009: Expanding Thermal Spray Performance to New Markets and Applications* (pp. 799–817). Materials Park, OH, USA: ASM International.

Fauchais, P., Montavan, G., & Bertrand, G. (2010). From powders to thermally sprayed coatings. *Journal of Thermal Spray Technology*, *19*(1-2), 56–80. doi:10.1007/s11666-009-9435-x

Fauchais, P., Rat, V., Coudert, J. F., Etchart-Salas, R., & Montavon, G. (2008). Operating parameters for suspension and solution plasma-spray coatings. *Surface and Coatings Technology*, *202*(18), 4309–4317. doi:10.1016/j.surfcoat.2008.04.003

Fauchais, P., Vardelle, A., & Denoirjean, A. (1997). Reactive thermal plasmas: Ultrafine particles synthesis and coating deposition. *Surface and Coatings Technology*, *97*(1-3), 66–78. doi:10.1016/S0257-8972(97)00294-6

Fauchais, P., Vardelle, A., & Dussoubs, B. (2001). Quo vadis thermal spraying? *Journal of Thermal Spray Technology*, *10*(1), 44–66. doi:10.1361/105996301770349510

Fauchais, P., Vardelle, M., Vardelle, A., Bianchi, L., & Leger, A. C. (1996). Parameters controlling the generation and properties of plasma sprayed zirconia coatings. *Plasma Chemistry and Plasma Processing*, *16*(1), S99–S125.

Fedel, M., Callone, E., Diré, S., Deflorian, F., Olivier, M.-G., & Poelman, M. (2013). Effect of Na-Montmorillonite sonication on the protective properties of hybrid silica coatings. *Electrochimica Acta*, *124*, 90–99. doi:10.1016/j.electacta.2013.11.006

Feng, T., Li, H., Fu, Q., Shi, X., Hu, M., & Liu, L. (2015). Erosion resistance of Mo–Si–Cr coating-modified C/C composites in a wind tunnel at 1873K. *Journal of Alloys and Compounds*, *622*, 1049–1054. doi:10.1016/j.jallcom.2014.10.189

Feng, T., Li, H.-J., Hu, M.-H., Lin, H.-J., & Li, L. (2016). Oxidation and ablation resistance of Fe 2 O 3 modified ZrB 2–SiC–Si coating for carbon/carbon composites. *Ceramics International*, *42*(1), 270–278. doi:10.1016/j.ceramint.2015.08.105

Feng, Z., Liu, Y., Thompson, G. E., & Skeldon, P. (2010). Sol–gel coatings for corrosion protection of 1050 aluminium alloy. *Electrochimica Acta*, *55*(10), 3518–3527. doi:10.1016/j.electacta.2010.01.074

Ferreira, J., Coppini, N., & Neto, F. L. (2001). Characteristics of carbon–carbon composite turning. *Journal of Materials Processing Technology*, *109*(1), 65–71. doi:10.1016/S0924-0136(00)00776-7

Feynman, R. P. (2011). There's plenty of room at the bottom. *Resonance*, *16*(9), 890–905. doi:10.1007/s12045-011-0109-x

Figueira, R. B., Fontinha, I. R., Silva, C. J. R., & Pereira, E. V. (2016). Hybrid Sol-Gel Coatings: Smart and Green Materials for Corrosion Mitigation. *Coatings*, *6*(1), 12. doi:10.3390/coatings6010012

Figueira, R. B., Silva, C. J. R., & Pereira, E. V. (2014). Organic–inorganic hybrid sol–gel coatings for metal corrosion protection: A review of recent progress. *Journal of Coatings Technology and Research*, *12*(1), 1–35. doi:10.1007/s11998-014-9595-6

Figueira, R. B., Silva, C. J. R., & Pereira, E. V. (2015a). Hot-dip galvanized steel dip-coated with ureasilicate hybrid in simulated concrete pore solution: Assessment of coating morphology and corrosion protection efficiency. *Progress in Organic Coatings*, *88*, 245–255. doi:10.1016/j.porgcoat.2015.07.008

Figueira, R. B., Silva, C. J. R., & Pereira, E. V. (2015b). Hybrid sol–gel coatings for corrosion protection of hot-dip galvanized steel in alkaline medium. *Surface and Coatings Technology*, *265*, 191–204. doi:10.1016/j.surfcoat.2015.01.034

Figueira, R. B., Silva, C. J. R., & Pereira, E. V. (2015c). Influence of Experimental Parameters Using the Dip-Coating Method on the Barrier Performance of Hybrid Sol-Gel Coatings in Strong Alkaline Environments. *Coatings*, *5*(2), 124–141. doi:10.3390/coatings5020124

Figueira, R. B., Silva, C. J. R., & Pereira, E. V. (2016). Hybrid sol–gel coatings for corrosion protection of galvanized steel in simulated concrete pore solution. *Journal of Coatings Technology and Research*, *13*(2), 1–19. doi:10.1007/s11998-015-9751-7

Figueira, R. B., Silva, C. J., Pereira, E. V., & Salta, M. M. (2013). Ureasilicate Hybrid Coatings for Corrosion Protection of Galvanized Steel in Cementitious Media. *Journal of the Electrochemical Society*, *160*(10), C467–C479. doi:10.1149/2.033310jes

Figueira, R. B., Silva, C. J., Pereira, E. V., & Salta, M. M. (2014). Alcohol-Aminosilicate Hybrid Coatings for Corrosion Protection of Galvanized Steel in Mortar. *Journal of the Electrochemical Society*, *161*(6), C349–C362. doi:10.1149/2.103406jes

Finnie, I. (1960). Erosion of surfaces by solid particles. *Wear*, *3*(2), 87–103. doi:10.1016/0043-1648(60)90055-7

Finnie, I. (1995). Some reflections on the past and future of erosion. *Wear*, *186-187*(Part 1), 1–10. doi:10.1016/0043-1648(95)07188-1

Fischer, M. (1978). Modellvorstellungen über den Reaktionsverlauf und Diskussion der Funktionen passivierender Zusätze bei der chemischen Passivierung von Eisen im pH-Bereich schwach sauer, neutral, basisch. *Materials and Corrosion*, *29*(3), 188–198. doi:10.1002/maco.19780290305

Flame Spray Technologies. (2007). Thermal spray process. Retrieved from http://www.fst.nl/en/page00031.asp

Flototto, D. W., Wang, Z. M., Jeurgens, L. P. H., Bischoff, E., & Mittemeijer, E. J. (2012). Effect of ad-atom surface diffusivity on microstructure and intrinsic stress evolutions during Ag film growth. *Journal of Applied Physics*, *112*(4), 043503. doi:10.1063/1.4746739

Fontana, M. G. (2005). *Corrosion engineering*. Tata McGraw-Hill Education.

Fontinha, I. R., Salta, M. M., Zheludkevich, M. L., & Ferreira, M. G. S. (2013). EIS Study of Amine Cured Epoxy-silica-zirconia Sol-gel Coatings for Corrosion Protection of the Aluminium Alloy EN AW 6063. *Portugaliae Electrochimica Acta*, *31*(6), 307–319. doi:10.4152/pea.201306307

Fontinha, I. R., Salta, M. M., Zheludkevich, M. L., Ferreira, M. G., Figueira, R. B., Pereira, E. V., & Silva, C. J. (2014). Influence of pH on the corrosion protection of epoxy-silica-zirconia sol-gel coatings applied on EN AW-6063 aluminium alloy. *ECS Transactions*, *58*(38), 9–16. doi:10.1149/05838.0009ecst

Forestier, R., Massoni, E., & Chastel, Y. (2002). Estimation of constitutive parameters using an inverse method coupled to a 3D finite element software. *Journal of Materials Processing Technology*, *125*, 594–601. doi:10.1016/S0924-0136(02)00406-5

Frank, J. (2001). *Hermanek, Thermal Spray Terminology and Company Origins, First Printing*. Materials Park, OH: ASM International.

Frank, W., & Graf, L. (1985). Mechanism of Intergranular Corrosion and Intergranular Stress-Corrosion Cracking in Mild Steels. *Res Mech.*, *13*(4), 251–264.

Fritze, H., Jojic, J., Witke, T., Rüscher, C., Weber, S., Scherrer, S., & Borchardt, G. et al. (1998). Mullite based oxidation protection for SiC–C/C composites in air at temperatures up to 1900K. *Journal of the European Ceramic Society*, *18*(16), 2351–2364. doi:10.1016/S0955-2219(98)00242-8

Friz, M. K. (1992). New Materials for Production of Optical Coatings. *Proc. SVC 35th Ann. Techn. Conf.*, (pp. 143-149).

Frommhertz, M., Scholz, A., Oechsner, M., Bakan, E., & Vassen, R. (2016). Gadolinium zirconate/YSZ thermal barrier coatings: Mixed-mode interfacial fracture toughness and sintering behavior. *Surface and Coatings Technology*, *286*, 119–128. doi:10.1016/j.surfcoat.2015.12.012

Fu, Y. D. (2003129-136). Deposition of TiN layer on TiNi thin films to improve surface properties. *Surf. Coat. Technol.*, *167*.

Fuchs-Godec, R. (2010). The Erosion−Corrosion Inhibition of AISI 431 Martensitic Stainless Steel in 2.0 M H_2SO_4 Solution Using *N* -Alkyl Quaternary Ammonium Salts as Inhibitors. *Industrial & Engineering Chemistry Research, 49*(14), 6407–6415. doi:10.1021/ie100275a

Fujii, T., Yoshida, K., Suzuki, K., & Ito, S. (2001). Direct nitriding of large grains of aluminum with 2 mm size. *Solid State Ionics, 141-142*, 593–598. doi:10.1016/S0167-2738(01)00773-1

Fukumoto, M., Nagai, H., & Yasui, T. (2006a). Influence of surface character change of substrate due to heating on flattening behavior of thermal sprayed particles. *Paper presented at the 2006 International Thermal Spray Conference: Building on 100 Years of Success*, Seattle, WA. Retrieved from http://www.scopus.com/inward/record.url?eid=2-s2.0-34548752700&partnerID=40&md5=38ae31c0418098efc96220bff9c2022f

Fukumoto, M., Nishiyama, H., & Nishioka, A. (2002). Effect of surface morphology of substrate on flattening behavior of freely fallen metal droplet. *Paper presented at the International Thermal Spray Conference*, Essen, Germany.

Fukumoto, M., Yamada, M., Yasui, T., & Takahashi, K. (2004). Fabrication of aluminum nitride coating by reactive RF plasma spray process. *In Proceeding of the International Thermal Spray Conference* (CD). Osaka, Japan.

Fukumoto, M., Yamaguchi, T., Yamada, M., & Yasui, T. (2007b). Splash splat to disk splat transition behavior in plasma-sprayed metallic materials. *Paper presented at the International Thermal Spray Conference on Thermal Spray 2007: Global Coating Solutions*, Beijing, China. Retrieved from http://www.scopus.com/inward/record.url?eid=2-s2.0-59349099839&partnerID=40&md5=8c8c7aaaafea2cc271b8687671fea83a

Fukumoto, M., & Huang, Y. (1999). Flattening mechanism in thermal sprayed nickel particle impinging on flat substrate surface. *Journal of Thermal Spray Technology, 8*(3), 427–432. doi:10.1361/105996399770350386

Fukumoto, M., Itoh, S., & Itoh, S. (1997). Fabrication of functionally graded TiN coating by reactive plasma spraying. *Surface Engineering, 13*(4), 315–319. doi:10.1179/sur.1997.13.4.315

Fukumoto, M., Nagai, H., & Yasui, T. (2006b). Influence of surface character change of substrate due to heating on flattening behavior of thermal sprayed particles. *Journal of Thermal Spray Technology, 15*(4), 759–764. doi:10.1361/105996306X146776

Fukumoto, M., Nishioka, E., & Matsubara, T. (2002). Effect of interface wetting on flattening of freely fallen metal droplet onto flat substrate surface. *Journal of Thermal Spray Technology, 11*(1), 69–74. doi:10.1361/105996302770348998

Fukumoto, M., Yamaguchi, T., Yamada, M., & Yasui, T. (2007a). Splash splat to disk splat transition behavior in plasma-sprayed metallic materials. *Journal of Thermal Spray Technology, 16*(5-6), 905–912. doi:10.1007/s11666-007-9083-y

Fundamentals of Gas Turbine Engines. (n.d.) Retrieved in May 2017 from http://www.skybrary.aero/bookshelf/books/1621.pdf

Fu, Q.-G., Jing, J.-Y., Tan, B.-Y., Yuan, R.-M., Zhuang, L., & Li, L. (2016). Nanowire-toughened transition layer to improve the oxidation resistance of SiC–MoSi 2–ZrB 2 coating for C/C composites. *Corrosion Science, 111*, 259–266. doi:10.1016/j.corsci.2016.05.013

Fu, Q.-G., Li, H.-J., Shi, X.-H., Li, K.-Z., & Sun, G.-D. (2005). Silicon carbide coating to protect carbon/carbon composites against oxidation. *Scripta Materialia, 52*(9), 923–927. doi:10.1016/j.scriptamat.2004.12.029

Fu, Q., Shan, Y., Wen, S., Li, H., Chu, Y., & Liu, X. (2014). Oxidation pre-treatment to improve the mechanical property and oxidation resistance of Si–Mo–Cr coated C/C composites. *Corrosion Science, 83*, 103–110. doi:10.1016/j.corsci.2014.02.008

Gadow, R., Killinger, A., Friedrich, C., & Voss, A. (1998). Combined metallurgical and ceramic coating in the development of tubular ozone generators. In *Proceedings of 15th International Thermal Spray Conference*, Nice, France (pp. 1081-1090).

Gadow, R., & Lischka, M. (2002). Lanthanum hexaaluminate – novel thermal barrier coatings for gas turbine applications – materials and process development. *Surface and Coatings Technology*, *151*, 392–399. doi:10.1016/S0257-8972(01)01642-5

Gammer, K., Stoiber, M., Wagner, J., Hutter, H., Kullmer, R., & Mitterer, C. (2004). Investigations on the effects of plasma-assisted pre-treatment for PACVD TiN coatings on tool steel. *Thin Solid Films*, *461*, 277–281. doi:10.1016/j.tsf.2004.02.013

Gao, J., & Suo, J. (2011). Effects of heating temperature and duration on the microstructure and properties of the self-healing coatings. *Surface and Coatings Technology*, *206*(6), 1342–1350. doi:10.1016/j.surfcoat.2011.08.059

Garamoon, A., Rashed, U. M., Abouelela, A., Eissa, M. A., Saudi, A. H., El-Zeer, D. M., & El-Hossary, F. (2006). Hydrogen Effect on Nitriding Process of 304L Austenitic Stainless Steel. *IEEE Transactions on Plasma Science*, *34*(4), 1066–1073. doi:10.1109/TPS.2006.878388

Gawne, D. T., Bao, Y., & Zhang, T. (2001). Plasma-spray deposition of silicon nitride-based coatings. In *Proceeding of International Thermal Spray Conference*, Singapore.

Gee, S., & Little, J. (1991). Oxidation behaviour and protection of carbon/carbon composites. *Journal of Materials Science*, *26*(4), 1093–1100. doi:10.1007/BF00576792

Gell, M., Jordan, E. H., Sohn, Y. H., Goberman, D., Shaw, L., & Xiao, T. D. (2001). Development and implementation of plasma sprayed nanostructured ceramic coatings. *Surface and Coatings Technology*, *146*, 48–54. doi:10.1016/S0257-8972(01)01470-0

Gell, M., Jordan, E. H., Teicholz, M., Cetegen, B. M., Padture, N. P., Xie, L., & Roth, J. et al. (2008). Thermal barrier coatings made by the solution precursor plasma spray process. *Journal of Thermal Spray Technology*, *17*(1), 124–135. doi:10.1007/s11666-007-9141-5

Geng, S. J., Wang, F. H., & Zhu, S. L. (2002). High-temperature oxidation behavior of sputtered IN 738 nanocrystalline coating. *Oxidation of Metals*, *57*(3-4), 231–243. doi:10.1023/A:1014870101143

Ghali, E., Sastri, V. S., & Elboujdaini, M. (2007). *Corrosion Prevention and Protection: Practical Solutions*. John Wiley & Sons.

Gharagozlou, M., Naderi, R., & Baradaran, Z. (2016). Effect of synthesized NiFe2O4-silica nanocomposite on the performance of an ecofriendly silane sol–gel coating. *Progress in Organic Coatings*, *90*, 407–413. doi:10.1016/j.porgcoat.2015.08.009

Ghasemi, R., Shoja-Razavi, R., Mozafarinia, R., & Jamali, H. (2010). Laser glazing of plasma-sprayed nanostructured yttria stabilized zirconia thermal barrier coatings. *Ceramics International*, *39*(8), 9483–9490. doi:10.1016/j.ceramint.2013.05.066

Ghosh, S. K. (2009). *Self-healing Materials: Fundamentals, Design Strategies, and Applications*. Wiley-VCH Verlag GmbH & Co.

Ghosh, S. (2015). Thermal Barrier Ceramic Coatings - A Review. In A. Mohamed (Ed.), *Advanced ceramic processing* (pp. 11–138). Rijeka, Croatia: InTech; doi:10.5772/61346

Glass, T. W., & Depay, J. A. (1991). Thermal spray coatings: properties, processes and applications. *Paper presented at the Fourth National Thermal Spray Conference*, Pittsburgh, PA.

Glazov, L. (2000). *Restricted medium: analytical theory of collision cascades and sputtering*. Izv. Akad. Nauk. Ser. Fiz.

Gledhill, A. D., Reddy, K. M., Drexler, J. M., Shinoda, K., Sampath, S., & Padture, N. P. (2011). Mitigation of damage from molten fly ash to air-plasma-sprayed thermal barrier coatings. *Materials Science and Engineering A*, *528*(24), 7214–7221. doi:10.1016/j.msea.2011.06.041

Gleeson, B. (2006). Thermal barrier coatings for aeroengine applications. *Journal of Propulsion and Power*, *22*(2), 375–383. doi:10.2514/1.20734

Gnaser, H. (2007). *Energy and angular distributions of sputtered species* (Vol. 110). Berlin: Springer. doi:10.1007/978-3-540-44502-9_5

Gobara, M., Kamel, H., Akid, R., & Baraka, A. (2015). Corrosion behaviour of AA2024 coated with an acid-soluble collagen/hybrid silica sol–gel matrix. *Progress in Organic Coatings*, *89*, 57–66. doi:10.1016/j.porgcoat.2015.07.026

Goberman, D., Sohn, Y. H., Shaw, L., Jordan, E., & Gell, M. (2002). Microstructure development of Al_2O_3-13wt.% TiO_2 plasma sprayed coatings derived from nanocrystalline powders. *Journal of Acta Materialia*, *50*(5), 1141–1152. doi:10.1016/S1359-6454(01)00414-1

Gonzales-Rodrigues, J. G., Haro, S., Martinez-Villafañe, A., Salinas-Bravo, V. M., & Porcayo-Calderon, J. (2006). Corrosion performance of heat resistant alloy in Na2SO4-V2O5 molten salts. *Materials Science and Engineering A*, 435, 258-265.

Gonzalez, J. E. G., & Mirza-Rosca, J. C. (1999). Study of the corrosion behavior of titanium and some of its alloys for biomedical and dental implant applications. *Journal of Electroanalytical Chemistry*, *471*(2), 109–115. doi:10.1016/S0022-0728(99)00260-0

Gorlach, I. A. (2008). Development of the Thermal Spraying Process for Anticorrosion Surface Protection. In *Proceedings of the International MultiConference of Engineers and Computer Scientists IMECS '08*, Hong Kong, March 19-21.

Goward, G. W. (1998). Progress in coatings for gas turbine airfoils. *Surface and Coatings Technology*, *108*, 73–79. doi:10.1016/S0257-8972(98)00667-7

Grabowski, K., Zbyrad, P., Uhl, T., Staszewski, W. J., & Packo, P. (2017). Multiscale electro-mechanical modeling of carbon nanotube composites. *Computational Materials Science*, *135*, 169–180. doi:10.1016/j.commatsci.2017.04.019

Grachev, S. Y., Borsa, D. M., & Boerma, D. O. (2002). On the growth of magnetic Fe_4N films. *Surface Science*, *516*(1-2), 159–168. doi:10.1016/S0039-6028(02)02029-0

Graziani, G., Bianchi, M., Sassoni, E., Russo, A., & Marcacci, M. (2017). Ion-substituted calcium phosphate coatings deposited by plasma-assisted techniques: A review. *Materials Science and Engineering C*, *74*, 219–229. doi:10.1016/j.msec.2016.12.018 PMID:28254288

Grbić, B., Radić, N., Stojadinović, S., Vasilić, R., Dohčević-Mitrović, Z., Šaponjić, Z., & Stefanov, P. (2014). TiO_2/WO_3 photocatalytic composite coatings prepared by spray pyrolysis. *Surface and Coatings Technology*, *258*, 763–771. doi:10.1016/j.surfcoat.2014.07.082

Greenwood, R., & Kendall, K. (1999). Selection of suitable dispersants for aqueous suspensions of zirconia and titania powders using acoustophoresis. *Journal of the European Ceramic Society*, *19*(4), 479–488. doi:10.1016/S0955-2219(98)00208-8

Grenier, S., Shanker, K., Tsantrizos, P., & Ajersch, F. (1996). Deposition of TiN films using a thermal plasma reactive forming technology. *Surface and Coatings Technology*, *82*(3), 311–316. doi:10.1016/0257-8972(95)02744-0

Gromov, A., Ilyin, A., Ditts, A., & Vereshchagin, V. (2005). Combustion of Al–Al_2O_3 mixtures in air. *Journal of the European Ceramic Society*, *25*(9), 1575–1579. doi:10.1016/j.jeurceramsoc.2004.05.017

Grosvenor, A. P., Kobe, B. A., & McIntyre, N. S. (2004a). Examination of the oxidation of iron by oxygen using X-ray photoelectron spectroscopy and QUASES™. *Surface Science*, *565*(2-3), 151–162. doi:10.1016/j.susc.2004.06.210

Grosvenor, A. P., Kobe, B. A., & McIntyre, N. S. (2004b). Studies of the oxidation of iron by water vapour using X-ray photoelectron spectroscopy and QUASES™. *Surface Science*, *572*(2-3), 217–227. doi:10.1016/j.susc.2004.08.035

Grundwürmer, M., Nuyken, O., Meyer, M., Wehr, J., & Schupp, N. (2007). Sol–gel derived erosion protection coatings against damage caused by liquid impact. *Wear*, *263*(1), 318–329. doi:10.1016/j.wear.2006.12.039

Gualtieri, M. L., Prudenziati, M., & Gualtieri, A. F. (2006). Quantitative determination of the amorphous phase in plasma sprayed alumina coatings using the Rietveld method. *Surface and Coatings Technology*, *201*(6), 2984–2989. doi:10.1016/j.surfcoat.2006.06.009

Gudmundsson, J. B., Brenning, N., Lundin, D., & Helmersson, U. (2012). High Power Impulse Magnetron Sputtering Discharge. *Journal of Vacuum Science & Technology. A, Vacuum, Surfaces, and Films*, *30*(3), 30801. doi:10.1116/1.3691832

Guessasma, S., Bounazef, M., & Coddet, C. (2004). Note on POD test parameters to study wear behaviour of alumina–titania coatings. *Materials Characterization*, *52*(3), 237–241. doi:10.1016/j.matchar.2004.05.002

Guessasma, S., Bounazef, M., Nardin, P., & Sahraoui, T. (2006). Wear behavior of alumina–titania coatings: Analysis of process and parameters. *Ceramics International*, *32*(1), 13–19. doi:10.1016/j.ceramint.2004.11.014

Guglielmi, M. (1997). Sol-gel coatings on metals. *Journal of Sol-Gel Science and Technology*, *8*(1–3), 443–449. doi:10.1007/BF02436880

Guilemany, J. M., Nutting, J., & Dougan, M. J. (1997). A Transmission Electron Microscopy Study of the Microstructures Present in Alumina Coatings Produced by Plasma Spraying. *Journal of Thermal Spray Technology*, *6*(4), 425–429. doi:10.1007/s11666-997-0025-5

Günthner, M., Schütz, A., Glatzel, U., Wang, K., Bordia, R. K., Greißl, O., & Motz, G. et al. (2011). High performance environmental barrier coatings, Part I: Passive filler loaded SiCN system for steel. *Journal of the European Ceramic Society*, *31*(15), 3003–3010. doi:10.1016/j.jeurceramsoc.2011.05.027

Guo, X., Lu, Z., Jung, Y. G., Li, L., Knapp, J., & Zhang, J. (2016). Thermal properties, thermal shock, and thermal cycling behavior of lanthanum zirconate-based thermal barrier coatings. *Metallurgical and Materials Transactions E*, 3(2), 64-70.

Guo, L., Li, M., He, S., Zhang, C., Wang, Q., & Ye, F. (2017). Preparation and hot corrosion behavior of plasma sprayed nanostructured $Gd_2Zr_2O_7$-$LaPO_4$ thermal barrier coatings. *Journal of Alloys and Compounds*, *698*, 13–19. doi:10.1016/j.jallcom.2016.12.241

Guo, L., Zhang, Y., & Ye, F. X. (2015). Phase Structure Evolution and Thermo-Physical Properties of Nonstoichiometry $Nd_{2-x}Zr_{2+x}O_{7+x/2}$ Pyrochlore Ceramics. *Journal of the American Ceramic Society*, *98*(3), 1013–1018. doi:10.1111/jace.13374

Gupta, M. (2014). Design of Thermal Barrier Coatings. A modelling approach. *PhD Thesis Production Technology 2014 No.3*, Trollhattan Sweden: University West.

Gurrappa, I. (2001). Identification of hot corrosion resistant MCrAlY based bond coatings for gas turbine engine applications. *Surface and Coatings Technology*, *139*(2), 272–283. doi:10.1016/S0257-8972(00)01156-7

Habazaki, H., Kimura, T., Aoki, Y., Tsuji, E., Yano, T., Shimizu, K., & Hassel, A. W. (2016). Characterization of Corrosion-resistant, Nanometer-thick, Layer-by-layer Aluminosilicate Coatings Prepared on Stainless Steel. *Electrochimica Acta, 201*, 311–319. doi:10.1016/j.electacta.2015.11.112

Habibi, M. H. (May 2014). *Hot corrosion behaviour of new candidates for thermal barrier coatings application in turbine simulated environments* [Dissertation]. Louisiana State University.

Habibi, M. H., Wang, L., & Guo, S. (2012). An investigation on hot corrosion resistance of plasma sprayed composite YSZ- $Gd_2Zr_2O_7$ and $Gd_2Zr_2O_7$ thermal barrier coatings in simulated turbine environment at 1050 °C. *Journal of the European Ceramic Society, 32*, 1635–1642. doi:10.1016/j.jeurceramsoc.2012.01.006

Hajizadeh-Oghaz, M., Shoja-Razavi, R., Ghasemi, A., & Valefi, Z. (2016). Na_2SO_4 and V_2O_5 molten salts corrosion resistance of plasma-sprayed nanostructured ceria and yttria co-stabilized zirconia thermal barrier coatings. *Ceramics International, 42*(4), 5433–5446. doi:10.1016/j.ceramint.2015.12.085

Hamdy, A. S., Sa'eh, A. G., Shoeib, M. A., & Barakat, Y. (2007). Evaluation of corrosion and erosion–corrosion resistances of mild steel in sulfide-containing NaCl aerated solutions. *Electrochimica Acta, 52*(24), 7068–7074. doi:10.1016/j.electacta.2007.05.034

Hammer, P., dos Santos, F. C., Cerrutti, B. M., Pulcinelli, S. H., & Santilli, C. V. (2013). Carbon nanotube-reinforced siloxane-PMMA hybrid coatings with high corrosion resistance. *Progress in Organic Coatings, 76*(4), 601–608. doi:10.1016/j.porgcoat.2012.11.015

Hammer, P., Schiavetto, M. G., dos Santos, F. C., Benedetti, A. V., Pulcinelli, S. H., & Santilli, C. V. (2010). Improvement of the corrosion resistance of polysiloxane hybrid coatings by cerium doping. *Journal of Non-Crystalline Solids, 356*(44–49), 2606–2612. doi:10.1016/j.jnoncrysol.2010.05.013

Hanetho, S. M., Kaus, I., Bouzga, A., Simon, C., Grande, T., & Einarsrud, M.-A. (2014). Synthesis and characterization of hybrid aminopropyl silane-based coatings on stainless steel substrates. *Surface and Coatings Technology, 238*, 1–8. doi:10.1016/j.surfcoat.2013.10.013

Harada, H. (2003, November). High temperature materials for gas turbines: the present and future. *Paper presented at the International Gas Turbine Congress*, Tokyo, Japan.

Hardt, P., Eckel, M., Schmidt, M., & Wulff, H. (1999). Influence of process parameters on the growth of TiN in different sized plasma-assisted chemical vapor deposition reactors. *Surface and Coatings Technology, 120–121*, 238–243. doi:10.1016/S0257-8972(99)00496-X

Hardwicke, C., & Lau, Y. C. (2013). Advances in Thermal Spray Coatings for Gas Turbines and Energy Generation: A Review. *Journal of Thermal Spray Technology, 22*(5), 564–576. doi:10.1007/s11666-013-9904-0

Hariprasad, S., Gowtham, S., Arun, S., Ashok, M., & Rameshbabu, N. (2017). Fabrication of duplex coatings on biodegradable AZ31 magnesium alloy by integrating cerium conversion (CC) and plasma electrolytic oxidation (PEO) processes. *Journal of Alloys and Compounds, 722*, 698–715. doi:10.1016/j.jallcom.2017.06.119

Harvey, D. (1996). The tough truth - Wear-resistant coatings using high velocity oxyfuel. *Industrial Lubrication and Tribology, 48*(2), 11–16. doi:10.1108/00368799610781230

Hass, D. D., Slifka, A. J., & Wadley, H. N. G. (2001). Low thermal conductivity vapor deposited zirconia microstructures. *Acta Materialia, 49*(6), 973–983. doi:10.1016/S1359-6454(00)00403-1

Hasselman, D. P. H. (1985). Thermal stress resistance of engineering ceramics. *Materials Science and Engineering, 71*, 251–264. doi:10.1016/0025-5416(85)90235-6

Hayashi, Y., & Morinaga, S. (1995). Evaluation of iron nitride film resistors deposited by reactive sputtering method [in Japanese]. *Journal of the Vacuum Society of Japan, 38*(3), 350–353. doi:10.3131/jvsj.38.350

Hay, J. N., & Raval, H. M. (2001). Synthesis of Organic–Inorganic Hybrids via the Non-hydrolytic Sol–Gel Process. *Chemistry of Materials, 13*(10), 3396–3403. doi:10.1021/cm011024n

Haynes, J. A., Ferber, M. K., & Porter, W. D. (1999). Thermal cycling behavior of plasma-sprayed thermal barrier coatings with various MCrAlX bond coats. *Journal of Thermal Spray Technology, 9*(1), 38–48. doi:10.1361/105996300770350041

Haynes, J. A., Pint, B. A., Porter, W. D., & Wright, I. G. (2004). Comparison of thermal expansion and oxidation behavior of various high temperature coating materials and superalloys. *Materials at High Temperatures, 21*(2), 87–94. doi:10.1179/mht.2004.012

Heimann, R. B., & Lehmann, H. D. (2008). Recently Patented Work on Thermally Sprayed Coatings for Protection Against Wear and Corrosion of Engineered Structures. *Recent Patents on Materials Science, 1*(1), 41–55. doi:10.2174/1874464810801010041

Heimann, R. B., Thiele, S., Berger, L. M., Herrmann, M., Nebelung, M., Wielage, B., & Vuoristo, P. et al. (1998). Thermally sprayed silicon nitride-based coatings on steel for application in severe operation environments: Preliminary results. *Microstructural Science, 26*, 389–394.

Heiman, R. B. (1996). *Plasma Spray Coating: Principle and Application* (1st ed.). Weinheim, Germany: VCH. doi:10.1002/9783527614851

Heitz, E. (1991). Chemo-Mechanical Effects of Flow on Corrosion. *Corrosion, 47*(2), 135–145. doi:10.5006/1.3585229

He, J., & Schoenung, J. M. (2002). Nanostructured coatings. *Materials Science and Engineering A, 336*(1-2), 274–319. doi:10.1016/S0921-5093(01)01986-4

Hellmich, W. D. (1992). Optical AR-Coating for Optical Waveguide Devices", Journal of Light wave Technology. *Journal of Lightwave Technology, 10*(4), 469–476. doi:10.1109/50.134200

Hench, L. L., & West, J. K. (1990). The sol-gel process. *Chemical Reviews, 90*(1), 33–72. doi:10.1021/cr00099a003

Herman, H. (1998). Plasma sprayed coatings. *Scientific American,* 259, 112-117.

Herman, H., Sampath, S., & McCune, R. (2000). Thermal spray: Current status and future trends. *MRS Bulletin, 25*(7), 17–25. doi:10.1557/mrs2000.119

Hernandez, M., Barba, A., Genesca, J., Covelo, A., Bucio, E., & Torres, V. (2013). Characterization of Hybrid Sol-Gel Coatings Doped with Hydrotalcite-like Compounds on Steel and Stainless Steel Alloys. *ECS Transactions, 47*(1), 195–206. doi:10.1149/04701.0195ecst

Hernández, S., Ottone, C., Varetti, S., Fontana, M., Pugliese, D., Saracco, G., & Armandi, M. (2016). Spin-Coated vs. Electrodeposited Mn Oxide Films as Water Oxidation Catalysts. *Materials (Basel), 9*(4), 296. doi:10.3390/ma9040296 PMID:28773419

Hetmanczyk, M., Swadzba, L., & Mendala, B. (2007). Advanced materials and protective coatings in aero-engines applications. *Journal of Achievements of Materials and Manufacturing Engineering, 24*(1), 372–381.

High, M. S., Painter, P. C., & Coleman, M. M. (1992). Polymer-polymer mutual diffusion using transmission FTIR spectroscopy. *Macromolecules, 25*(2), 797–801. doi:10.1021/ma00028a045

Hille, T. S., Suiker, A. S. J., & Turteltaub, S. (2009). Microcrack nucleation in thermal barrier coating systems. *Engineering Fracture Mechanics, 76*(6), 813–825. doi:10.1016/j.engfracmech.2008.12.010

Hilton, M. R., Vandentop, G. J., Salmeron, M., & Somorjai, G. A. (1987). TiN Coatings on M2 Steel produced by PACVD. *Thin Solid Films*, *154*, 377–386. doi:10.1016/0040-6090(87)90380-4

Hiraoka, K., Shiwaku, T., & Ohji, T. (1996). Temperature distributions of gas tungsten arc plasma by spectroscopic methods. *Quarterly Journal of the Japan welding* [in Japanese]. *Society*, *14*(4), 641–648.

Hirose, J., Ito, S., Akashi, K., & Koura, N. (1994). Preparation of Fe_4N sintered compact by HIP. [in Japanese]. *Journal of the Japan Society of Powder and Powder Metallurgy*, *41*(8), 989–993. doi:10.2497/jjspm.41.989

Hongsong, Z., Kun, P., Qiang, X., Fuchi, W., & Ling, L. (2009). Thermal conductivity of $(Sm_{1-x}La_x)Zr_2O_7$ (x=0, 0.25, 0.5, 0.75 and 1) oxides for advanced thermal barrier coatings. *Journal of Rare Earths*, *27*(2), 222–226. doi:10.1016/S1002-0721(08)60223-9

Hoshiyama, Y., Miyake, H., Murakami, K., & Nakajima, H. (2002). Titanium aluminide-based composite depositions with dispersed nitride particles produced by reactive thermal spraying. *Journal of Japan Institute of Metals*, *66*(7), 784–791. doi:10.2320/jinstmet1952.66.7_784

Hou, D., Li, K., Li, H., Fu, Q., Wei, J., & Zhang, Y. (2007). W-Mo-Si/SiC oxidation protective coating for carbon/carbon composites. *Journal Of Materials Science And Technology-Shenyang*, *23*(4), 559.

Hou, P. Y., & Smialek, J. L. (1995). Effect of sulfur removal by H_2-annealing on the oxidation behavior of a Ni-25Cr alloy. *Scripta Metallurgica et Materialia*, *33*(9), 1409–1416. doi:10.1016/0956-716X(95)00385-9

Hou, X., Li, H., Wang, C., Zhu, Z., & Shen, J. (2000). Internal friction behavior of carbon–carbon composites. *Carbon*, *38*(15), 2095–2101. doi:10.1016/S0008-6223(00)00069-5

Huang, H., Eguchi, K., & Yoshida, T. (2003). Novel structured yttria-stabilized zirconia coatings fabricated by hybrid thermal plasma spraying. *Science and Technology of Advanced Materials*, *4*(6), 617–622. doi:10.1016/j.stam.2004.01.004

Huang, H., Liu, C., Ni, I. Y., & Zhou, C. G. (2011). Evaluation of microstructural evolution of thermal barrier coatings exposed to Na_2SO_4 using impedance spectroscopy. *Corrosion Science*, *53*(4), 1369–1374. doi:10.1016/j.corsci.2010.12.023

Huang, J.-F., Li, H.-J., Zeng, X.-R., Deng, F., Xiong, X., & Li, K.-Z. (2007). Oxidation resistant yttrium silicates coating for carbon/carbon composites prepared by a novel in-situ formation method. *Ceramics International*, *33*(5), 887–890. doi:10.1016/j.ceramint.2006.01.007

Huang, J.-F., Li, H.-J., Zeng, X.-R., & Li, K.-Z. (2006). Yttrium silicate oxidation protective coating for SiC coated carbon/carbon composites. *Ceramics International*, *32*(4), 417–421. doi:10.1016/j.ceramint.2005.03.018

Huang, J.-F., Wang, B., Li, H.-J., Liu, M., Cao, L.-Y., & Yao, C.-Y. (2011). A MoSi 2/SiC oxidation protective coating for carbon/carbon composites. *Corrosion Science*, *53*(2), 834–839. doi:10.1016/j.corsci.2010.11.024

Huang, Q., Yu, D., Xu, B., Hu, W., Ma, Y., Wang, Y., & Tian, Y. (2014). Nanotwinned diamond with unprecedented hardness and stability. *Nature*, *510*(7504), 250–253. doi:10.1038/nature13381 PMID:24919919

Hudis, M. (1973). Study of ion-nitriding. *Journal of Applied Physics*, *44*(4), 1489–1496. doi:10.1063/1.1662398

Hughes, A. E. (2015). 8 - Self-healing coatings A2 - Li, Guoqiang. In H. Meng (Ed.), *Recent Advances in Smart Self-healing Polymers and Composites* (pp. 211–241). Woodhead Publishing. doi:10.1016/B978-1-78242-280-8.00008-X

Hughes, A. E., Cole, I. S., Muster, T. H., & Varley, R. J. (2010). Designing green, self-healing coatings for metal protection. *NPG Asia Materials*, *2*(4), 143–151. doi:10.1038/asiamat.2010.136

Hu, J., Zhao, H., Tao, S., Zhou, X., & Ding, C. (2010). Thermal conductivity of plasma sprayed $Sm_2Zr_2O_7$ coating. *Journal of the European Ceramic Society*, *30*(3), 799–804. doi:10.1016/j.jeurceramsoc.2009.09.010

Humphrey, J. A. C. (1990). Fundamentals of fluid motion in erosion by solid particle impact. *Int. J. of Heat and Flow*, *11*(3), 170–195. doi:10.1016/0142-727X(90)90036-B

Hurst, C. (March 2010). China's Rare Earth Elements Industry: What Can the West Learn? *Institute for the Analysis of Global Security (IAGS)*. Retrieved from http://fmso.leavenworth.army.mil/documents/rareearth.pdf

Hutchings, I. M. (1992). *Friction and Wear of Engineering Materials*. In Tribology.

Hutchinson, J. W., & Evans, A. G. (2002). On the delamination of thermal barrier coatings in a thermal gradient. *Surface and Coatings Technology*, *149*(2-3), 179–184. doi:10.1016/S0257-8972(01)01451-7

Hutchinson, R. G., Fleck, N. A., & Cocks, A. C. F. (2006). A sintering model for thermal barrier coatings. *Acta Materialia*, *54*(5), 1297–1306. doi:10.1016/j.actamat.2005.10.056

Hu, Y., Li, L., Cai, X., Chen, Q., & Chu, P. K. (2007). Mechanical and tribological properties of TiC/amorphous hydrogenated carbon composite coatings fabricated by D.C magnetron sputtering with and without sample bias. *Diamond and Related Materials*, *16*(1), 181–186. doi:10.1016/j.diamond.2006.05.010

Hyakutake, H., Imada, Y., Honda, F., & Nakajima, K. (1998). Mechanical properties of TiN-TiC mixed crystal films. *Journal of Physics Condensed Matter*, *10*(1), 111–122. doi:10.1088/0953-8984/10/1/013

Ibrahim, A., Abdel Hamid, Z., & Abdel Aal, A. (2010). Investigation of nanostructured and conventional alumina–titania coatings prepared by air plasma spray process. *Materials Science and Engineering A*, *527*(3), 663–668. doi:10.1016/j.msea.2009.08.054

Ichinose, N., & Kuwabara, H. (1998). *Nitride Ceramics*. Tokyo, Japan: Nikkan-Kogyo-Shinbunsha.

Ilhan, Z., Hornes, A., Sata, N., Freitag, T. O., Ansar, A., Schiller, G., Krebs, A. (2015). Self-Healing Plasma Sprayed Ceramic Coatings.

Ingham, H. S., & Fabel, A. J. (1975). Comparison of plasma flame spray gases. *Welding Journal*, 101–105.

Innocenzi, P., & Malfatti, L. (2015). Processing of Sol–Gel Films from a Top-Down Route. In D. Levy & Marcos Zayat (Eds.), The Sol-Gel Handbook (pp. 165–194). Wiley-VCH Verlag GmbH & Co. KGaA. doi:10.1002/9783527670819.ch05

Innocenzi, P., Abdirashid, M. O., & Guglielmi, M. (1994). Structure and Properties of Sol-Gel Coatings from Methytriethoxysilane and Tetraethoxysilane. *Journal of Sol-Gel Science and Technology*, *3*(1), 47–55. doi:10.1007/BF00490148

Innocenzi, P., Kidchob, T., & Yoko, T. (2005). Hybrid Organic-Inorganic Sol-Gel Materials Based on Epoxy-Amine Systems. *Journal of Sol-Gel Science and Technology*, *35*(3), 225–235. doi:10.1007/s10971-005-2290-4

International, A. S. M. (1987). Corrosion. [Metals park, Ohio.]. *ASM Metals Handbook.*, *13*, 38.

Iribarren-Mateos, J. I., Buj-Corral, I., Vivancos-Calvet, J., Alemán, C., Iribarren, J. I., & Armelin, E. (2015). Silane and epoxy coatings: A bilayer system to protect AA2024 alloy. *Progress in Organic Coatings*, *81*, 47–57. doi:10.1016/j.porgcoat.2014.12.014

Irissou, E., Legoux, J. G., Ryabinin, A. N., Jodoin, B., & Moreau, C. (2008). Review on cold spray process and technology: Part I - Intellectual property. *Journal of Thermal Spray Technology*, *17*(4), 495–516. doi:10.1007/s11666-008-9203-3

Ishikawa, K., & Tobe, M. S. S. (1993). Application of thermal sprayed coatings to prevent corrosion of construction in Japan. In *Proceeding of 5th National Thermal Spray Conference,* Anaheim, CA (pp. 679-684). .

Itoh, Y., & Saitoh, M. (2005). Mechanical properties of overaluminized MCrAlY coatings at room temperature. *Journal of Engineering for Gas Turbines and Power*, *127*(4), 23–30. doi:10.1115/1.1688766

Ito, S., Shinmura, S., Hosokoshi, T., & Akashi, K. (1996). HIP sintering and saturation magnetization of Fe_4N [in Japanese]. *Journal of the Japan Society of Powder and Powder Metallurgy*, *43*(12), 1415–1419. doi:10.2497/jjspm.43.1415

Jack, K. H. (1951). The iron-nitrogen system: The preparation and the crystal structures of nitrogen-austenite (γ) and nitrogen-martensite (α′). *Proceedings of the Royal Society of London. Series A, Mathematical and Physical Sciences*, *208*(1093), 200–215. doi:10.1098/rspa.1951.0154

Jacquet P., Siad A., Nouveau C., Besnard A., & Quesada J. (2015) A study of the tribological behavior of duplex treatment. Annales de Chimie - Science des Matériaux, 39 201-208. doi: 39/3-4 – 20110.3166/acsm.39.201-208

Jadhav, A.D., Padture, N.P., Jordan, E.H., Gell, M., Miranzo, P., & Fuller, Jr., E.R. (2006). Low thermal conductivity plasma-sprayed thermal barrier coatings with engineered microstructures. *Acta Materiallica*, *54*, 3343-3349.

Jadhav, A. D., & Padture, N. P. (2008). Mechanical properties of solution-precursor plasma sprayed thermal barrier coatings. *Surface and Coatings Technology*, *202*(20), 4976–4979. doi:10.1016/j.surfcoat.2008.04.091

Jamali, H., Mozafarinia, R., Shoja-Razavi, R., & Ahmadi-Pidani, R. (2014). Comparison of hot corrosion behaviors of plasma-sprayed nanostructured and conventional YSZ thermal barrier coatings exposure to molten vanadium pentoxide and sodium-sulfate. *Journal of the European Ceramic Society*, *34*(2), 185–192. doi:10.1016/j.jeurceramsoc.2013.08.006

Jameel, D. (2015). Thin Film Deposition Processes. *International Journal of Modern Physics A*, *1*(4), 193–199.

Janus, M., Kyziol, K., Kluska, S., Konefal-Goral, J., Malek, A., & Jonas, S. (2015). Plasma assisted chemical vapor deposition – Technological design of functional coatings. *Archives of Metallurgy and Materials*, *60*(2), 909–914. doi:10.1515/amm-2015-0228

Jarligo, M. O., Mack, D. E., Mauer, G., Vassen, R., & Stöver, D. (2009). Atmospheric Plasma Spraying of High Melting Temperature Complex Perovskites for TBC Application. *Journal of Thermal Spray Technology*, *19*(1-2), 303–310. doi:10.1007/s11666-009-9377-3

Jelani, M., Bashir, S., Akram, M., Yousaf, D., Afzal, N., & Ahmad, S. (2014). Mechanical behaviour of Excimer laser irradiated polycrystalline Zirconium. *Physica Scripta*, *89*(2), 025703–025710. doi:10.1088/0031-8949/89/02/025703

Jian, C. Y., Hashida, T., Takahashi, H., & Saito, M. (1995). Thermal shock and fatigue resistance evaluation of functionally graded coating for gas turbine blades by laser heating method. *Composites Engineering*, *5*(7), 879–889. doi:10.1016/0961-9526(95)00041-K

Jian-Feng, H., Xie-Rong, Z., He-Jun, L., Xin-Bo, X., & Min, H. (2003). Mullite-Al 2 O 3-SiC oxidation protective coating for carbon/carbon composites. *Carbon*, *41*(14), 2825–2829. doi:10.1016/S0008-6223(03)00397-X

Jian-Feng, H., Xie-Rong, Z., He-Jun, L., Xin-Bo, X., & Ye-wei, F. (2004). Influence of the preparation temperature on the phase, microstructure and anti-oxidation property of a SiC coating for C/C composites. *Carbon*, *42*(8), 1517–1521. doi:10.1016/j.carbon.2004.01.066

Jiang, S. M., Li, H. Q., Ma, J., Xu, C. Z., Gong, J., & Sun, C. (2010). High temperature corrosion behavior of a gradient NiCoCrAlYSi coating II: Oxidation and hot corrosion. *Corrosion Science*, *52*(7), 2316–2322. doi:10.1016/j.corsci.2010.03.032

Jiang, S. M., Xu, C. Z., Li, H. Q., Ma, J., Gong, J., & Sun, C. (2010). High temperature corrosion behaviour of a gradient NiCoCrAlYSi coating I: Microstructure evolution. *Corrosion Science*, *52*(5), 1746–1752. doi:10.1016/j.corsci.2010.01.025

Jiang, X., Wan, Y., Herman, H., & Sampath, S. (2001). Role of condensates and adsorbates on substrate surface on fragmentation of impinging molten droplets during thermal spray. *Thin Solid Films*, *385*(1-2), 132–141. doi:10.1016/S0040-6090(01)00769-6

Jin, L., Yu, Q., Ni, L., & Zhou, C. (2012). Microstructure and thermal properties of nanostructured 8 wt.% CeO_2 doped YSZ coatings prepared by atmospheric plasma spraying. *Journal of Thermal Spray Technology, 21*(5), 928–934. doi:10.1007/s11666-012-9744-3

Jin, Y., & Yang, Y. (1996). Tribological behavior of various plasma sprayed ceramic coatings. *Surface and Coatings Technology, 88*(1-3), 248–254. doi:10.1016/S0257-8972(96)02918-0

Jin, Z. H., & Feng, Y. Z. (2008). Thermal fracture resistance of a functionally graded coating with periodic edge cracks. *Surface and Coatings Technology, 202*(17), 4189–4197. doi:10.1016/j.surfcoat.2008.03.009

Jonas, S., Janus, M., Jaglarz, J., & Kyziol, K. (2016). Formation of SixNy(H) and C:N:H layers by Plasma-Assisted Chemical Vapor Deposition method. *Thin Solid Films, 600*, 162–168. doi:10.1016/j.tsf.2016.01.016

Jonas, S., Kyzioł, K., Lis, J., & Tkacz-Śmiech, K. (2009). Stability of a-C:N:H layers deposited by RF Plasma Enhanced CVD. *Diffusion and Defect Data, Solid State Data. Part B, Solid State Phenomena, 147-149*, 738–743. doi:10.4028/www.scientific.net/SSP.147-149.738

Jones, R. L. (1997). Vanadate-sulfate melt thermochemistry relating to hot corrosion of thermal barrier coatings. Naval Research Laboratory.

Jones, D. A. (1992). *Principles and prevention of corrosion.* Macmillan.

Jones, R. L. (1989). Scandia-stabilized zirconia for resistance to molten vanadate-sulfate corrosion. *Surface and Coatings Technology, 39*, 89–96. doi:10.1016/0257-8972(89)90044-3

Jordan, E. H., Gell, M., Sohn, Y. H., Goberman, D., Shaw, L., Jiang, S., & Strutt, P. et al. (2001). Fabrication and evaluation of plasma sprayed nanostructured alumina-titania coatings with superior properties. *Materials Science and Engineering A, 301*(1), 80–89. doi:10.1016/S0921-5093(00)01382-4

Jordan, E. H., Jiang, C., & Gell, M. (2015). The solution precursor plasma spray (SPPS) process: A review with energy considerations. *Journal of Thermal Spray Technology, 24*(7), 1153–1165. doi:10.1007/s11666-015-0272-9

José, N. M., & Prado, L. A. S. de A. (2005). Materiais Híbridos Orgânico-Inorgânicos: Preparação e Algumas Aplicações. *Quimica Nova, 28*(2), 281–288. doi:10.1590/S0100-40422005000200020

Juang, R. C., Lee, C. J., & Chen, C. C. (2003). Combustion synthesis of hexagonal aluminum nitride powders under low nitrogen pressure. *Materials Science and Engineering A, 357*(1-2), 219–227. doi:10.1016/S0921-5093(03)00163-1

Kaesche, H. (2003). *Corrosion of Metals: Physicochemical Principles and Current Problems.* Springer. doi:10.1007/978-3-642-96038-3

Kaestner, P., Olfe, J., Hec, J. W., & Rie, K. T. (2001). Improvement in the load-bearing capacity and adhesion of TiC coatings on TiAl6V4 by duplex treatment. *Surface and Coatings Technology, 142-144*, 928–933. doi:10.1016/S0257-8972(01)01214-2

Kaiser, A., Lobert, M., & Telle, R. (2008). Thermal stability of zircon (ZrSiO 4). *Journal of the European Ceramic Society, 28*(11), 2199–2211. doi:10.1016/j.jeurceramsoc.2007.12.040

Kalidindi, R. S. R., & Subasri, R. (2015). 5 - Sol-gel nanocomposite hard coatings. In M. Aliofkhazraei (Ed.), *Anti-Abrasive Nanocoatings* (pp. 105–136). Woodhead Publishing. doi:10.1016/B978-0-85709-211-3.00005-4

Kamal, S., Jayaganthan, R., & Prakash, S. (2009). Mechanical and microstructural characterisations of NiCrAoAlYTa coatings on superalloys deposited by detonation gun technique. *Surface and Coatings Technology, 25*, 303–312.

Kao, W. H., Su, Y. L., Horng, J. H., & Hsieh, Y. T. (2017). Improved tribological properties, electrochemical resistance and biocompatibility of AISI 316L stainless steel through duplex plasma nitriding and TiN coating treatment. *Journal of Biomaterials Applications, 32*(1), 12–27. doi:10.1177/0885328217712109 PMID:28541124

Karch, J., Birringer, R., & Gleiter, H. (1987). Ceramics ductile at low-temperature. *Nature, 330*(6148), 556–558. doi:10.1038/330556a0

Kar, S., Paul, S., & Bandyopadhyay, P. P. (2016). Processing and characterisation of plasma sprayed oxides: Microstructure, phases and residual stress. *Surface and Coatings Technology, 304*, 364–374. doi:10.1016/j.surfcoat.2016.07.043

Karthikeyan, J., Berndt, C. C., Tikkanen, J., Reddy, S., & Herman, H. (1997). Plasma spray synthesis of nanomaterial powders and deposits. *Materials Science and Engineering a-Structural Materials Properties Microstructure and Processing, 238*(2), 275-286.

Karthikeyan, J., Berndt, C. C., Tikkanen, J., Wang, J. Y., King, A. H., & Herman, H. (1997b). Preparation of nanophase materials by thermal spray processing of liquid precursors. *Nanostructured Materials, 9*(1-8), 137-140.

Karthikeyan, J., Kay, C. M., Lindeman, J., Lima, P. S., & Berndt, C. C. (2000). *Cold spray processing of titanium powder.* Paper presented at the International Thermal Spray Conference, Montreal, QUE, Canada. Retrieved from http://www.scopus.com/inward/record.url?eid=2-s2.0-1442353812&partnerID=40

Karthikeyan, J., Berndt, C. C., Tikkanen, J., Wang, J. Y., King, A. H., & Herman, H. (1997a). Nanomaterial powders and deposits prepared by flame spray processing of liquid precursors. *Nanostructured Materials, 8*(1), 61–74. doi:10.1016/S0965-9773(97)00066-4

Karunaratne, M. S. A., Ogden, S. L., Kenny, S. D., & Thomson, R. C. (2009). A multicomponent diffusion model for prediction of microstructural evolution in coated Ni based superalloy systems. *Materials Science and Technology, 25*(2), 287–299. doi:10.1179/174328408X355415

Kassabji, F., Tourenne, F., Derradji, A., & Fauchais, P. (1983). Aluminium and aluminium nitride deposition by low pressure nitrogen arc plasma spraying. In *Proceeding of 10th International Thermal Spray Conference*, Düsseldorf, Germany (Vol. 80, pp. 82-84).

Kassner, H., Stuke, A., Rodig, M., Vassen, R., & Stover, D. (2009). Influence of porosity on thermal conductivity and sintering in suspension plasma sprayed thermal barrier coatings. *Ceramic Engineering and Science Proceedings, 29*(4), 147–158. doi:10.1002/9780470456323.ch13

Kawasaki, A., Watanabe, R., Yuki, M., Nakanishi, Y., & Onabe, H. (1996). Effect of microstructure on thermal shock cracking of functionally graded thermal barrier coatings studied by burner heating test. *Materials Transactions*, 37(4), 788-795.

Kawasaki, A., & Watanabe, R. (2002). Thermal fracture behavior of metal/ceramic functionally graded materials. *Engineering Fracture Mechanics, 69*(14-16), 1713–1728. doi:10.1016/S0013-7944(02)00054-1

Kawata, K., Sugimura, H., & Takai, O. (2002). Effects of chlorine on tribological properties of TiN films prepared by pulsed d.c. plasma-enhanced chemical vapor deposition. *Thin Solid Films, 407*(1-2), 38–44. doi:10.1016/S0040-6090(02)00009-3

Kelly, P. A., & Arnell, R. D. (2000). Magnetron Sputtering: A Review of Recent Developments and Applications. *Vacuum, 56*(3), 159–172. doi:10.1016/S0042-207X(99)00189-X

Kembaiyan, K.T., & Keshavan, K. (1995). *Combating severe fluid erosion and corrosion of drill bits using thermal spray coatings.*

Keyvani, A., Saremi, M., & Sohi, M. H. (2011). Oxidation resistance of YSZ-alumina composites compared to normal YSZ TBC coatings at 1100°C. *Journal of Alloys and Compounds*, *509*(33), 8370–8377. doi:10.1016/j.jallcom.2011.05.029

Khan, A. U., Briscoe, B. J., & Luckham, P. F. (2000). Interaction of binders with dispersant stabilised alumina suspensions. *Colloids and Surfaces. A, Physicochemical and Engineering Aspects*, *161*(2), 243–257. doi:10.1016/S0927-7757(99)00374-X

Khan, I. A., Ahmad, M. H. R., Qayyum, A., Murtaza, G., Zakaullah, A. M., & Rawat, R. J. (2008). Nitridation of zirconium using energetic ions from plasma focus device. *Thin Solid Films*, *516*(23), 8255–8263. doi:10.1016/j.tsf.2008.03.012

Khor, K. A., Boey, F. Y. C., Zhao, X. L., & Cao, L. H. (2001). Aluminum nitride by plasma spraying of an Al_2O_3-C-Sm_2O_3 system. *Materials Science and Engineering A*, *300*(1-2), 203–210. doi:10.1016/S0921-5093(00)01658-0

Khor, K. A., & Li, Y. (2001). Crystallization bahaviors of in the plasma-spheroidized alumina/zirconia mixtures. *Materials Letters*, *48*(2), 57–63. doi:10.1016/S0167-577X(00)00280-9

Khramov, A. N., & Johnson, J. A. (2009). Phosphonate-functionalized ORMOSIL coatings for magnesium alloys. *Progress in Organic Coatings*, *65*(3), 381–385. doi:10.1016/j.porgcoat.2009.03.001

Khramov, A. N., Voevodin, N. N., Balbyshev, V. N., & Donley, M. S. (2004). Hybrid organo-ceramic corrosion protection coatings with encapsulated organic corrosion inhibitors. *Thin Solid Films*, *447–448*, 549–557. doi:10.1016/j.tsf.2003.07.016

Khun, N. W., & Frankel, G. S. (2016). Cathodic delamination of polyurethane/multiwalled carbon nanotube composite coatings from steel substrates. *Progress in Organic Coatings*, *99*, 55–60. doi:10.1016/j.porgcoat.2016.05.002

Kieback, B., Neubrand, A., & Riedel, H. (2003). Processing techniques for functionally graded materials. *Materials Science and Engineering A*, *362*(1), 81–106. doi:10.1016/S0921-5093(03)00578-1

Kim, B. J., Kim, Y. C., Lee, D. K., & Lee, J. J. (1999). The effect of NH_3 plasma pre-treatment on the properties of TiN coatings produced by plasma-enhanced chemical vapor deposition (PECVD). *Surface and Coatings Technology*, *111*(1), 56–61. doi:10.1016/S0257-8972(98)00707-5

Kim, D. J., Cho, Y. R., Lee, M. J., Hong, J. M., Kim, Y. K., & Lee, K. H. (1999). Properties of TiN–TiC multilayer coatings using plasma-assisted chemical vapor deposition. *Surface and Coatings Technology*, *116–119*, 906–910. doi:10.1016/S0257-8972(99)00149-8

Kim, G., Choi, H., Han, C., Uhm, S., & Lee, C. (2005). Characterization of atmospheric plasma spray NiCr–Cr_2O_3–Ag–CaF_2/BaF_2 coatings. *Surface and Coatings Technology*, *195*(1), 107–115. doi:10.1016/j.surfcoat.2004.11.041

Kimura, I., Ichiya, K., Ishii, M., Hotta, N., & Kitamura, T. (1989). Synthesis of fine AlN powder by a floating nitridation technique using an N_2/NH_3 Gas Mixture. *Journal of Materials Science Letters*, *8*(3), 303–304. doi:10.1007/BF00725505

Kippax, P. (2005). Measuring particle size using modern diffraction techniques. *Paint & Coatings*, *8*.

Kiruthika, P., Subasri, R., Jyothirmayi, A., Sarvani, K., & Hebalkar, N. Y. (2010). Effect of plasma surface treatment on mechanical and corrosion protection properties of UV-curable sol-gel based GPTS-ZrO_2 coatings on mild steel. *Surface and Coatings Technology*, *204*(8), 1270–1276. doi:10.1016/j.surfcoat.2009.10.017

Kittle, C. (1996). *Introduction to Solid State Physics*. New York: Wiley.

Klemens, P. G. (1991). Thermal conductivity of inhomogeneous media. *High Temperatures-High Pressures, 23*, 241-248.

Klug, H. P., & Alexander, L. E. (1974). *X-ray Diffraction Procedures for Polycrystalline and Amorphous Materials* (2nd ed.). New York: Wiley.

Klupa, A. (2012). Cenniejsze niż złoto? Metale ziem rzadkich w światowej strategii gospodarczej. *Przegląd strategiczny, 1*, 239-250.

Kogo, Y., Hatta, H., Toyoda, M., & Sugibayashi, T. (2002). Application of three-dimensionally reinforced carbon–carbon composites to dovetail joint structures. *Composites Science and Technology, 62*(16), 2143–2152. doi:10.1016/S0266-3538(02)00157-4

Kohzaki, M., Tsujita, Y., Takizawa, A., & Kinoshita, T. (1987). Crystallization and formation of cluster of ethylene ionomer during physical aging. *Journal of Applied Polymer Science, 33*(7), 2393–2402. doi:10.1002/app.1987.070330710

Kokini, K., DeJonge, J., Rangaraj, S., & Beardsley, B. (2002). Thermal shock of functionally graded thermal barrier coatings with similar thermal resistance. *Surface and Coatings Technology, 154*(2), 223–231. doi:10.1016/S0257-8972(02)00031-2

Kokini, K., Takeuchi, Y. R., & Choules, B. D. (1996). Surface thermal cracking of thermal barrier coatings owing to stress relaxation: Zirconia vs. mullite. *Surface and Coatings Technology, 82*(1-2), 77–82. doi:10.1016/0257-8972(95)02647-9

Kolasinski, R. P., Polk, J. E., Goebel, D., & Johnson, L. K. (2007). Sputtering Yield Measurements at Glancing Incidence using a Quartz Crystal Microbalance. *Journal of Vacuum Science & Technology. A, Vacuum, Surfaces, and Films, 25*(2), 236–245. doi:10.1116/1.2435375

Kolbel, J., Nordhein-Westfalen, F. D. I., & Verlag, W. D. Koln, & Germany, 1555. (1965).

Kolman, B., Forman, J., Dubsky, J., & Chraska, P. (1994). Homogeneity studies of powders and plasma sprayed deposits. *Mikrochimica Acta, 114*(1), 335–342. doi:10.1007/BF01244560

Komaki, M., Mimura, T., Kusumoto, Y., Kurahasi, R., Kouzaki, M., & Yamasaki, T. (2010). Formation of Fe-Based Amorphous Coating Films by Thermal Spraying Technique. Materials Transactions, 51(9), 1581-1585.

Kong, J., Ye, Z., & Lv, F. (2009). Non-equilibrium solidification character of $Zr_{56.2}Ti_{13.8}Nb_{5.0}Cu_{6.9}Ni_{5.6}Be_{12.5}$ bulk metallic glass composites containing ductile dendrite phase. *Journal of Alloys and Compounds, 478*(1-2), 202–205. doi:10.1016/j.jallcom.2008.11.130

Konyashin, I. (2001). A technique for fabrication of coated TiCN-based cermets with functionally graded structure. *International Journal of Refractory Metals & Hard Materials, 19*(4), 523–526. doi:10.1016/S0263-4368(01)00058-0

Kozhukharov, S. (2009). Relationship between the Conditions of Preparation by the Sol-Gel Route and the Properties of the Obtained Products. *Journal of the University of Chemical Technology and, 44*(2), 143–150.

Kozhukharov, S., Kozhukharov, V., Schem, M., Aslan, M., Wittmar, M., Wittmar, A., & Veith, M. (2012). Protective ability of hybrid nano-composite coatings with cerium sulphate as inhibitor against corrosion of AA2024 aluminium alloy. *Progress in Organic Coatings, 73*(1), 95–103. doi:10.1016/j.porgcoat.2011.09.005

Kramer, M. J., Mecco, H., Dennis, K. W., Vargonova, E., McCallum, R. W., & Napolitano, R. E. (2007). Rapid solidification and metallic glass formation – Experimental and theoretical limits. *Journal of Non-Crystalline Solids, 353*(32-40), 3633–3639. doi:10.1016/j.jnoncrysol.2007.05.172

Kramer, S., Yang, J., & Levi, C. G. (2008). Infiltration-Inhibiting Reaction of Gadolinium Zirconate Thermal Barrier Coatings with CMAS Melts. *Journal of the American Ceramic Society, 91*(2), 576–583. doi:10.1111/j.1551-2916.2007.02175.x

Kramer, S., Yang, J., Levi, C. G., & Jonhson, C. A. (2006). Thermochemical interaction of thermal barrier coatings with molten CaO-MgO-Al_2O_3-SiO_2 (CMAS) deposits. *Journal of the American Ceramic Society, 89*(10), 3167–3175. doi:10.1111/j.1551-2916.2006.01209.x

Krishna, L. R., Sen, D., Rao, Y. S., Rao, G. V. N., & Sundararajan, G. (2002). Thermal spray coating of aluminum nitride utilizing the detonation spray technique. *Journal of Materials Research*, *17*(10), 2514–2523. doi:10.1557/JMR.2002.0366

Krishnan, R., Dash, S., Kesavamoorthy, R., Babu Rao, C., Tyagi, A. K., & Raj, B. (2006). Laser surface modification and characterization of air plasma sprayed alumina coatings. *Surface and Coatings Technology*, *200*(8), 2791–2799. doi:10.1016/j.surfcoat.2005.05.002

Kroese, D. T. (2011). *Handbook of Monte Carlo Methods*. New York: John Wiley & Sons. doi:10.1002/9781118014967

Kryder, M. (1992). Magnetic Thin Films for Data Storage. *Thin Solid Films*, *216*(1), 174–180. doi:10.1016/0040-6090(92)90890-N

Kuang, J. C., Zhang, C., Zhou, X., Liu, Q., & Ye, C. (2005). Formation and characterization of cubic AlN crystalline in a carbothermal reduction reaction. *Materials Letters*, *59*(16), 2006–2010. doi:10.1016/j.matlet.2005.02.039

Kulkarni, A., Wang, Z., Nakamura, T., Sampath, S., Goland, A., Herman, H., … Steinbrecht, R.W. (2003). Comprehensive microstructural characterization and predictive property modeling of plasma-sprayed zirconia coatings. *Acta Materiallia*, *5*(1), 2457-2475.

Kumar, N., Jyothirmayi, A., Soma Raju, K. R. C., & Subasri, R. (2012). Effect of functional groups (methyl, phenyl) on organic–inorganic hybrid sol–gel silica coatings on surface modified SS 316. *Ceramics International*, *38*(8), 6565–6572. doi:10.1016/j.ceramint.2012.05.040

Kumar, V., & Kandasubramanian, B. (2016). Processing and design methodologies for advanced and novel thermal barrier coatings for engineering applications. *Particuology*, *27*, 1–28. doi:10.1016/j.partic.2016.01.007

Kunioshi, C. T., Correa, O. V., & Ramanathan, L. V. (2005). Erosion-Oxidation Behavior of Thermal Sprayed Ni20Cr Alloy and WC and Cr_3C_2 Cermet Coatings. *Materials Research*, *8*(2), 125–129. doi:10.1590/S1516-14392005000200005

Kunst, S. R., Cardoso, H. R. P., Oliveira, C. T., Santana, J. A., Sarmento, V. H. V., Muller, I. L., & Malfatti, C. F. (2013). Corrosion resistance of siloxane-poly(methyl methacrylate) hybrid films modified with acetic acid on tin plate substrates: Influence of tetraethoxysilane addition. *Applied Surface Science*, *298*, 1–11. doi:10.1016/j.apsusc.2013.09.182

Kuroda, S., Kawakita, J., Watanabe, M., & Katanoda, H. (2008). Warm spraying-a novel coating process based on high-velocity impact of solid particles. *Science and Technology of Advanced Materials*, *9*(3). doi:10.1088/1468-6996/9/3/033002

Kusaka, K., Taniguchi, D., Hanabusa, T., & Tominaga, K. (2000). Effect of input power on crystal orientation and residual stress in AlN film deposited by dc sputtering. *Vacuum*, *59*(2-3), 806–813. doi:10.1016/S0042-207X(00)00351-1

Kyziol, K., Kaczmarek, L., Brzezinka, G., & Kyziol, A. (2014). Structure, characterization and cytotoxicity study on plasma surface modified Ti–6Al–4V and c-TiAl alloys. *Chemical Engineering Journal*, *240*, 516–526. doi:10.1016/j.cej.2013.10.091

Labruquere, S., Pailler, R., Naslain, R., & Desbat, B. (1997). Enhancement of the oxidation resistance of carbon fibres in C/C composites via surface treatments. *Key Engineering Materials*, *132*, 1938–1941. doi:10.4028/www.scientific.net/KEM.132-136.1938

Lacatusu, I., Badea, N., Murariu, A., Pirvu, C., & Meghea, A. (2011). Vegetal nanoclusters in hybrid silica films prepared by sol-gel spin coating technique. *Journal of Non-Crystalline Solids*, *357*(7), 1716–1723. doi:10.1016/j.jnoncrysol.2011.01.018

Lagally, M. Z., & Zhang, Z. (2002). Thin-film cliffhanger. *Nature*, *417*(6892), 907–910. doi:10.1038/417907a PMID:12087388

Laimer, J., Stori, H., & Rodhammer, P. (1990). Titanium nitride deposited by PACVD. *Thin Solid Films*, *191*, 77–89. doi:10.1016/0040-6090(90)90276-J

Lajtai, E. Z. (1971). A theoretical and experimental evaluation of the Griffith theory of brittle fracture. *Tectonophysics*, *11*(2), 129–156. doi:10.1016/0040-1951(71)90060-6

Lakshmi, R. V., Aruna, S. T., & Sampath, S. (2017). Ceria nanoparticles vis-à-vis cerium nitrate as corrosion inhibitors for silica-alumina hybrid sol-gel coating. *Applied Surface Science*, *393*, 397–404. doi:10.1016/j.apsusc.2016.10.035

Lamaka, S. V., Zheludkevich, M. L., Yasakau, K. A., Montemor, M. F., Cecílio, P., & Ferreira, M. G. S. (2006). TiOx self-assembled networks prepared by templating approach as nanostructured reservoirs for self-healing anticorrosion pre-treatments. *Electrochemistry Communications*, *8*(3), 421–428. doi:10.1016/j.elecom.2005.12.019

Lampke, T., Darwich, S., Nickel, D., & Wielage, B. (2008). Development and characterization of sol-gel composite coatings on aluminum alloys for corrosion protection. *Materialwissenschaft und Werkstofftechnik*, *39*(12), 914–919. doi:10.1002/mawe.200800396

Lamuta, C., Di Girolamo, G., & Pagnotta, L. (2015). Microstructural, mechanical and tribological properties of nano-structured YSZ coatings produced with different APS process parameters. *Ceramics International*, *41*(7), 8904–8914. doi:10.1016/j.ceramint.2015.03.148

Lance, M. J., Unocic, K. A., Haynes, J. A., & Pint, B. A. (2013). Effect of water vapor on thermally-grown alumina scales on Pt-modified and simple aluminide bond coatings. *Surface and Coatings Technology*, *237*, 2–7. doi:10.1016/j.surfcoat.2013.06.041

Langer, S. A., Carter, W. C., & Fuller, E. R. (1997). Object Oriented Finite Element Analysis for Materials Science. *Center for Theoretical and computational Materials Science and the Information Lab.* NIST. Retrieved April 25, 2017 from www.ctcms.nist.gov/oof/

Latella, B. A., Barbé, C. J., Cassidy, D. J., Li, H., & Ignat, M. (2005). Mechanical Stability and Decohesion of Sol-Gel Hybrid Coatings on Metallic Substrates. *Revista Matéria*, *10*(2), 199–204.

Latka, L., Cattini, A., Chicot, D., Pawlowski, L., Kozerski, S., Petit, F., & Denoirjean, A. (2013). Mechanical properties of yttria-and ceria-stabilized zirconia coatings obtained by suspension plasma spraying. *Journal of Thermal Spray Technology*, *22*(2-3), 125–130. doi:10.1007/s11666-012-9874-7

Lavollée, C., Gressier, M., Garcia, J., Sobrino, J.-M., Reby, J., & Menu, M.-J. (2016). New architectured hybrid sol-gel coatings for wear and corrosion protection of low-carbon steel. *Progress in Organic Coatings*, *99*, 337–345. doi:10.1016/j.porgcoat.2016.06.015

Lebedev, E. (2011). Hybrid organo-inorganic polymer systems: Synthesis, Structure and Properties. *Theoretical and Experimental Chemistry*, *46*(6), 391–396. doi:10.1007/s11237-011-9172-3

Lee, D. H., Kang, N. K., Lee, K. S., Moon, H. S., Kim, H. T., & Kim, C. (2017). Evaluation of Thermal Durability of Thermal Barrier Coating and Change in Mechanical Behavior. *Journal of the Korean Ceramic Society*, *54*(4), 314–322. doi:10.4191/kcers.2017.54.4.05

Lee, H. K., Lee, H., Park, J.-H., Kim, H.-K., Kong, B. H., & Cho, H. K. (2011). Optical Properties of Ultrathin Copper Thin Films Sandwiched between Nb-Doped TiO2 Films Studied with Spectroscopic Ellipsometry. *Journal of Applied Physics*, *50*(5R), 055805. doi:10.7567/JJAP.50.055805

Lee, H. Y., & Baik, K. H. (2009). Comparison of corrosion resistance between Al 2 O 3 and YSZ coatings against high temperature LiCl-Li 2 O molten salt. *Metals and Materials International*, *15*(5), 783–787. doi:10.1007/s12540-009-0783-8

Lee, S. Y., Chung, J. W., Kim, C. W., Han, J. G., Kim, S. S., & Lee, J. H. (1997). Effects of relative thickness of the duplex-treated layer on surface properties of AlSl H13 steel. *Surface and Coatings Technology*, *94/95*, 272–278. doi:10.1016/S0257-8972(97)00351-4

Lee, W. E., & Rainforth, M. (1994). *Ceramic microstructures: property control by processing*. Springer Science & Business Media.

Leger, A. C., Vardelle, M., Vardelle, A., Dussoubs, B., & Fauchais, P. (1995). Splat formation: ceramic carticles on ceramic substrates. *Paper presented at the Advances in Thermal Spray Science & Technology*, Houston, TX.

Lehmann, H., Pitzer, D., Pracht, G., Vassen, R., & Stover, D. (2003). Thermal Conductivity and Thermal Expansion Coefficients of the Lanthanum Rare-Earth-Element Zirconate System. *Journal of the American Ceramic Society*, *86*(8), 1338–1344. doi:10.1111/j.1151-2916.2003.tb03473.x

Lekawa-Raus, A., Gizewski, T., Patmore, J., Kurzepa, L., & Koziol, K. K. (2017). Electrical transport in carbon nanotube fibres. *Scripta Materialia*, *131*, 112–118. doi:10.1016/j.scriptamat.2016.11.027

Leoni, M., Scardia, P., Rossi, S., Fedrizzi, L., & Massiani, Y. (1999). (Ti,Cr)N and Ti/TiN PVD coatings on 304 stainless steel substrates: Texture and residual stress. *Thin Solid Films*, *345*(2), 263–269. doi:10.1016/S0040-6090(98)01741-6

Levi, C. G. (2004). Emerging materials and processes for thermal barrier systems. *Current Opinion in Solid State and Materials Science*, *8*(1), 77–91. doi:10.1016/j.cossms.2004.03.009

Levi, C. G., Hutchinson, J. W., Vidal-Sétif, M.-H., & Johnson, C. A. (2012). Environmental degradation of thermal-barrier coatings by molten deposits. *MRS Bulletin*, *37*(10), 932–941. doi:10.1557/mrs.2012.230

Levine, S. R., Opila, E. J., Halbig, M. C., Kiser, J. D., Singh, M., & Salem, J. A. (2002). Evaluation of ultra-high temperature ceramics foraeropropulsion use. *Journal of the European Ceramic Society*, *22*(14), 2757–2767. doi:10.1016/S0955-2219(02)00140-1

Levy, A. V. (1995). *Solid Particle Erosion and Erosion-corrosion of Materials*. USA: ASM International.

Li, H.-J., Fu, Q.-G., Huang, J.-F., Zeng, X.-R., & Li, K.-Z. (2005). Research on the Oxidation-Protective Coatings for Carbon/Carbon Composites. *Carbon letters*, *6*(2), 71-78.

Li, M., & Christofides, P. D. (2009). Modeling and control of high-velocity oxygen-fuel (HVOF) thermal spray: A tutorial review. *Journal of Thermal Spray Technology*.

Li, R., Ashgriz, N., & Chandra, S. (2010). Maximum spread of droplet on solid surface: Low reynolds and weber numbers. *Journal of Fluids Engineering-Transactions of the Asme*, *132*(6).

Li, Y., Li, C. J., Zhang, Q., Yang, G. J., Li, C. X. (2010). Influence of TGO composition on the thermal shock lifetime of thermal barrier coatings with cold-sprayed MCrAlY bond coat. *Journal of thermal spray technology*, *19*(1), 168-177.

Liang, B., Liao, H., Ding, C., & Coddet, C. (2005). Nanostructured zirconia-30 vol.% alumina composite coatings deposited by atmospheric plasma spraying. *Thin Solid Films*, *484*(1-2), 225–231. doi:10.1016/j.tsf.2005.02.040

Liang, W., Juncai, S., & Xiaolei, X. (2001). Low pressure plasma arc source ion nitriding compared with glow-discharge plasma nitriding of stainless steel. *Surface and Coatings Technology*, *145*(1-3), 31–37. doi:10.1016/S0257-8972(01)01283-X

Liao, H., Beche, E., Berger, F., & Coddet, C. (1998). On the microstructures of thermally sprayed "PEEK" polymer. *Paper presented at the Thermal Spray: Meeting the Challenges of the 21st Century*, Nice, France. Retrieved from http://www.scopus.com/inward/record.url?eid=2-s2.0-1442354654&partnerID=40&md5=b271e3c393176e0a41ae34b584c15782

Li, C. J., & Li, J. L. (2004). Evaporated-gas-induced splashing model for splat formation during plasma spraying. *Surface and Coatings Technology*, *184*(1), 13–23. doi:10.1016/j.surfcoat.2003.10.048

Li, C. J., Li, W. Y., & Liao, H. (2006). Examination of the critical velocity for deposition of particles in cold spraying. *Journal of Thermal Spray Technology*, *15*(2), 212–222. doi:10.1361/105996306X108093

Li, C., Jordens, K., & Wilkes, G. L. (2000). Abrasion-resistant coatings for plastic and soft metallic substrates by sol–gel reactions of a triethoxysilylated diethylenetriamine and tetramethoxysilane. *Wear*, *242*(1), 152–159. doi:10.1016/S0043-1648(00)00408-7

Liepack, H., Bartsch, K., Arnold, B., Bauer, H. D., Liu, X., Knupfer, M., & Leonhardt, A. (2004). Characteristics of excess carbon in PACVD TiC-amorphous carbon layers. *Diamond and Related Materials*, *13*(1), 106–110. doi:10.1016/j.diamond.2003.09.007

Liepack, H., Bartsch, K., Bruckner, W., & Leonhardt, A. (2004). Mechanical behavior of PACVD TiC–amorphous carbon composite layers. *Surface and Coatings Technology*, *183*(1), 69–73. doi:10.1016/j.surfcoat.2003.08.083

Li, G., & Meng, H. (2015). *Recent Advances in Smart Self-healing Polymers and Composites*. Elsevier Science.

Li, H.-J., Jiao, G.-S., Li, K.-Z., & Wang, C. (2008). Multilayer oxidation resistant coating for SiC coated carbon/carbon composites at high temperature. *Materials Science and Engineering A*, *475*(1), 279–284. doi:10.1016/j.msea.2007.04.053

Li, L., Li, H., Li, Y., Yin, X., Shen, Q., & Fu, Q. (2015). A SiC-ZrB 2-ZrC coating toughened by electrophoretically-deposited SiC nanowires to protect C/C composites against thermal shock and oxidation. *Applied Surface Science*, *349*, 465–471. doi:10.1016/j.apsusc.2015.05.028

Lima, R. S., & Marple, B. R. (2007). Thermal spray coatings engineered from nanostructured ceramic agglomerated powders for structural, thermal barrier and biomedical applications: A review. *Journal of Thermal Spray Technology*, *16*(1), 40–63. doi:10.1007/s11666-006-9010-7

Lima, R. S., & Marple, B. R. (2008). Nanostructured YSZ thermal barrier coatings engineered to counteract sintering effects. *Materials Science and Engineering A*, *485*(1-2), 182–193. doi:10.1016/j.msea.2007.07.082

Lima, R., Kucuk, A., & Berndt, C. (2002). Bimodal distribution of mechanical properties on plasma sprayed nanostructured partially stabilized zirconia. *Materials Science and Engineering A*, *327*(2), 224–232. doi:10.1016/S0921-5093(01)01530-1

Lim, Y. Y., & Chaudhri, M. M. (2002). The influence of grain size on the indentation hardness of high-purity copper and aluminium. *Philosophical Magazine A*, *82*(10), 2071–2080. doi:10.1080/01418610208235717

Lin, C., & Chung, S. (2001). Combustion synthesis of aluminum nitride powder using additives. *Journal of Materials Research*, *16*(8), 2200–2208. doi:10.1557/JMR.2001.0302

Lin, X., Zeng, Y., Lee, S. W., & Ding, C. (2004). Characterization of alumina-3 wt.% titania coating prepared by plasma spraying of nanostructured powders. *Journal of the European Ceramic Society*, *24*(4), 627–634. doi:10.1016/S0955-2219(03)00254-1

Lin, X., Zeng, Y., Zhou, X., & Ding, C. (2003). Microstructure of alumina-3wt.% titania coatings by plasma spraying with nanostructured powders. *Materials Science and Engineering A*, *357*(1-2), 228–234. doi:10.1016/S0921-5093(03)00164-3

Liu, Y., Hang, Z., Chen, H., Ceng, S., Gou, G., Wang, X., ... & Wu, X. (2016). Erosion–Corrosion Property of CeO2-Modified HVOF WC-Co Coating. *Journal of Thermal Spray Technology*, *25*(4), 815-822. doi:.10.1007/s11666-016-0391-y

Liu, H. F., Xiong, X., Li, X. B., & Wang, Y. I. (2014). Hot corrosion behavior of Sc_2O_3-Y_2O_3-ZrO_2 thermal barrier coatings in presence of Na_2SO_4+V_2O_5 molten salt. *Corrosion Science*, *85*, 87–93. doi:10.1016/j.corsci.2014.04.001

Liu, H., Li, S., Li, Q., & Li, Y. (2010). Investigation on the phase stability, sintering, and thermal conductivity of Sc_2O_3-Y_2O_3-ZrO_2 for thermal barrier coating application. *Materials & Design, 31*(6), 2972–2977. doi:10.1016/j.matdes.2009.12.019

Liu, H., Tao, J., Xu, J., Chen, Z., & Gao, Q. (2009). Corrosion and tribological behaviors of chromium oxide coatings prepared by the glow-discharge plasma technique. *Surface and Coatings Technology, 204*(1-2), 28–36. doi:10.1016/j.surfcoat.2009.06.020

Liu, J. W., Wu, C., & Tsong, T. T. (1991). Measurement of the binding energy of kink-site atoms of metals and alloys. *Physical Review B: Condensed Matter and Materials Physics, 43*(14), 11595–11604. doi:10.1103/PhysRevB.43.11595 PMID:9996928

Liu, J., Zhang, Y., Yu, M., Li, S., Xue, B., & Yin, X. (2015). Influence of embedded ZnAlCe-NO_3– layered double hydroxides on the anticorrosion properties of sol–gel coatings for aluminum alloy. *Progress in Organic Coatings, 81*, 93–100. doi:10.1016/j.porgcoat.2014.12.015

Liu, S. H., Huang, H., & Woo, C. H. (2002). Schwoebel-Ehrlich barrier: From two to three dimension. *Applied Physics Letters, 80*(18), 3295–3297. doi:10.1063/1.1475774

Liu, T., Chen, X., Yang, G. J., & Li, C. J. (2016). Properties evolution of plasma-sprayed $La_2Zr_2O_7$ coating induced by pore structure evolution during thermal exposure. *Ceramics International, 42*(14), 15485–15492. doi:10.1016/j.ceramint.2016.06.201

Liu, T., Luo, X. T., Chen, X., Yang, C. X., Li, C. X., & Li, J. (2015). Morphology and size evolution of interlamellar two-dimensional pores in plasma-sprayed $La_2Zr_2O_7$ coatings during thermal exposure at 1300 °C. *Journal of Thermal Spray Technology, 24*(5), 739–748. doi:10.1007/s11666-015-0236-0

Liu, X., Zhang, B., & Deng, Z. (2002). Grinding of nanostructured ceramic coatings: Surface observations and material removal mechanisms. *International Journal of Machine Tools & Manufacture, 42*(15), 1665–1676. doi:10.1016/S0890-6955(02)00115-3

Liu, Y., Vidal, V., Le Roux, S., Blas, F., Ansart, F., & Lours, P. (2015). Influence of isothermal and cyclic oxidation on the apparent interfacial toughness in thermal barrier coating systems. *Journal of the European Ceramic Society, 35*(15), 4269–4275. doi:10.1016/j.jeurceramsoc.2015.07.018

Liu, Y., Wei, S., Tian, H., Tong, H., & Xu, B. (2014). Characterization of soft magnetic spinel ferrite coating prepared by plasma spray. *Surface and Coatings Technology, 258*, 189–199. doi:10.1016/j.surfcoat.2014.09.029

Liu, Z. G., Zhang, W. H., Ouyang, J. H., & Zhou, Y. (2015). Novel thermal barrier coatings based on rare-earth zirconates/YSZ double-ceramic-layer system deposited by plasma spraying. *Journal of Alloys and Compounds, 647*, 438–444. doi:10.1016/j.jallcom.2015.05.189

Livage, J. (1997). Sol-gel processes. *Current Opinion in Solid State and Materials Science, 2*(2), 132–138. doi:10.1016/S1359-0286(97)80057-5

Livage, J. (1999). Sol-gel synthesis of hybrid materials. *Bulletin of Materials Science, 22*(3), 201–205. doi:10.1007/BF02749920

Li, W. B., Chen, M. H., Wang, C., Zhu, S. L., & Wang, F. H. (2013). Preparation and oxidation behavior of SiO_2–Al_2O_3–glass composite coating on Ti–47Al–2Cr–2Nb alloy. *Surface and Coatings Technology, 218*, 30–38. doi:10.1016/j.surfcoat.2012.12.022

Li, W. B., Chen, M. H., Wu, M. Y., Zhu, S. L., Wang, C., & Wang, F. H. (2014). Microstructure and oxidation behavior of a SiC–Al_2O_3–glass composite coating on Ti–47Al–2Cr–2Nb alloy. *Corrosion Science*, *87*, 179–186. doi:10.1016/j.corsci.2014.06.015

Li, W. B., Zhu, S. L., Chen, M. H., Wang, C., & Wang, F. H. (2014). Development of an oxidation resistant glass–ceramic composite coating on Ti–47Al–2Cr–2Nb alloy. *Applied Surface Science*, *292*, 583–590. doi:10.1016/j.apsusc.2013.12.012

Li, W. B., Zhu, S. L., Wang, C., Chen, M. H., Shen, M. L., & Wang, F. H. (2013). SiO_2–Al_2O_3–glass composite coating on Ti–6Al–4V alloy: Oxidation and interfacial reaction behavior. *Corrosion Science*, *74*, 367–378. doi:10.1016/j.corsci.2013.05.010

Li, W. L., Fei, W. D., & Hanabusa, T. (2006). Effect of deposition condition on residual stress of iron nitride thin films prepared by magnetron sputtering and ion implantation. *Applied Surface Science*, *252*(8), 2847–2852. doi:10.1016/j.apsusc.2005.04.043

Li, Y. S., Niu, Y., & Wu, W. T. (2003). Accelerated corrosion of pure Fe, Ni, Cr and several Fe-based alloys induced by ZnCl 2–KCl at 450 C in oxidizing environment. *Materials Science and Engineering A*, *345*(1), 64–71. doi:10.1016/S0921-5093(02)00386-6

Li, Y.-S., Lu, W., Wang, Y., & Tran, T. (2009). Studies of (3-mercaptopropyl)trimethoxylsilane and bis(trimethoxysilyl) ethane sol-gel coating on copper and aluminum. *Spectrochimica Acta. Part A: Molecular and Biomolecular Spectroscopy*, *73*(5), 922–928. doi:10.1016/j.saa.2009.04.016 PMID:19464230

Lofgran, S. (2013). *Thin film deposition and vacuum technology*. Brigham Young University–Idaho, Department of Physics.

Loghman-Estarki, M. R., Nejati, M., Edris, H., Shoja Razavi, R., Jamali, H., & Pakseresht, A. H. (2015). Evaluation of hot corrosion behavior of plasma sprayed Scandia and yttria co-stabilized nanostructured thermal barrier coatings in the presence of molten sulfate and vanadate salt. *Journal of the European Ceramic Society*, *35*(2), 693–702. doi:10.1016/j.jeurceramsoc.2014.08.029

Lorbeer, P., & Lorenz, W. J. (1980). The kinetics of iron dissolution and passivation in solutions containing oxygen. *Electrochimica Acta*, *25*(4), 375–381. doi:10.1016/0013-4686(80)87026-5

Lou, H. Y., Wang, F. H., Xia, B. J., & Zhang, L. X. (1992). High temperature oxidation resistance of sputtered micrograin superalloy K38G. *Oxidation of Metals*, *38*(3-4), 299–307. doi:10.1007/BF00666917

Lou, H. Y., Wang, F. H., Zhu, S. L., Xia, B. J., & Zhang, L. X. (1994). Oxide formation of K38G superalloy and its sputtered micrograined coating. *Surface and Coatings Technology*, *63*(2), 105–114. doi:10.1016/S0257-8972(05)80014-3

Lou, H. Y., Zhu, S. L., & Wang, F. H. (1995). Rehealing ability of oxide scales formed on microcrystalline K38G coatings. *Oxidation of Metals*, *43*(3-4), 317–328. doi:10.1007/BF01047033

Loustanau, P. J., & Horton, D. (1994). EMAA thermoplastic powder coatings in shop and field applications. *Materials Performance*, *33*(7), 32–36.

Lugscheider, E., Limbach, R., Liden, A., & Lodin, J. (1990). Plasma spraying of silicon nitride (Si_3N_4). *High Temperature Materials Powder Engineering*, *1*, 877–880.

Lugscheider, E., Remer, P., & Zhao, L. (1996). Reactive plasma spraying of wear resistant coatings of Ti composites. In C. C. Berndt (Ed.), *Thermal spray: practical solutions for engineering problems* (pp. 927–932). PA: ASM International, Materials Park, OH.

Lu, J. T., Zhu, S. L., & Wang, F. H. (2011). High temperature corrosion behavior of an AIP NiCoCrAlY coating modified by aluminizing. *Surface and Coatings Technology*, *205*(21-22), 5053–5058. doi:10.1016/j.surfcoat.2011.05.005

Lu, L., Sui, M. L., & Lu, K. (2000). Superplastic extensibility of nanocrystalline copper at room temperature. *Science*, *287*(5457), 1463–1466. doi:10.1126/science.287.5457.1463 PMID:10688789

Luo, H., Goberman, D., Shaw, L., & Gell, M. (2003). Indentation fracture behaviour of plasma-sprayed nanostructured Al_2O_3-13wt%TiO_2 coatings. *Materials Science and Engineering A*, *346*(1-2), 237–245. doi:10.1016/S0921-5093(02)00523-3

Luth, H. (2001). Solid Surfaces, Interfaces and Thin Film (Forth ed.). New York: Springer-Verlag.

Luthra, K. L. (1988). Oxidation of carbon/carbon composites—a theoretical analysis. *Carbon*, *26*(2), 217–224. doi:10.1016/0008-6223(88)90040-1

Lutz, A., Mol, J. M. C., De Graeve, I., & Terryn, H. (2016). 6 - Smart corrosion protection by multi-action self-healing polymeric coatings. In M. F. Montemor (Ed.), *Smart Composite Coatings and Membranes* (pp. 157–181). Woodhead Publishing. doi:10.1016/B978-1-78242-283-9.00006-3

Lu, W. M., Pan, T. J., Zhang, K., & Niu, Y. (2008). Accelerated corrosion of five commercial steels under a ZnCl 2–KCl deposit in a reducing environment typical of waste gasification at 673–773K. *Corrosion Science*, *50*(7), 1900–1906. doi:10.1016/j.corsci.2008.03.004

Lu, W., & Chung, D. (2002). Oxidation protection of carbon materials by acid phosphate impregnation. *Carbon*, *40*(8), 1249–1254. doi:10.1016/S0008-6223(01)00297-4

Lu, Y., & Liaw, P. K. (2001). The mechanical properties of nanostructured materials. *Journal of Metals*, *53*, 31–35.

Lv, B., Fan, X., Xie, H., & Wang, T. J. (2017). Effect of neck formation on the sintering of air-plasma-sprayed thermal barrier coatings. *Journal of the European Ceramic Society*, *37*(2), 811–821. doi:10.1016/j.jeurceramsoc.2016.08.033

Lyo, I. W., Ahn, H. S., & Lim, D. S. (2003). Microstructure and tribological properties of plasma-sprayed chromium oxide–molybdenum oxide composite coatings. *Surface and Coatings Technology*, *163*, 413–421. doi:10.1016/S0257-8972(02)00613-8

Maaß, P., & Peißker, P. (2011). *Handbook of Hot-dip Galvanization*. John Wiley & Sons. doi:10.1002/9783527636884

Macleod, A. (1999). The Early Days of Optical Coatings. *Journal of Optics. A, Pure and Applied Optics*, *1*(S), 779–783. doi:10.1088/1464-4258/1/S/305

Macleod, B. H. (2004). Low-Cost Anti-Reflection Technology for Automobile Displays. In *Proceedings of the SID Vehicle Display Conference*.

Mahade, S., Curry, N., Bjorklund, S., Markocsan, N., Nylen, P., & Vassen, R. (2016). Functional performance of $Gd_2Zr_2O_7$/YSZ multi-layered thermal barrier coatings deposited by suspension plasma spray. *Surface and Coatings Technology*, *318*, 208–216. doi:10.1016/j.surfcoat.2016.12.062

Mahmood, K., Farid, N., Ghauri, I. M., Afzal, N., Idrees, Y., & Mubarik, F. E. (2010). Effects of laser irradiation on the mechanical response of polycrystalline titanium. *Physica Scripta*, *82*(4), 045606–045613. doi:10.1088/0031-8949/82/04/045606

Maia, F., Tedim, J., Lisenkov, A. D., Salak, A. N., Zheludkevich, M. L., & Ferreira, M. G. S. (2012). Silica nanocontainers for active corrosion protection. *Nanoscale*, *4*(4), 1287–1298. doi:10.1039/c2nr11536k PMID:22249939

Maia, F., Yasakau, K. A., Carneiro, J., Kallip, S., Tedim, J., Henriques, T., & Ferreira, M. G. S. (2016). Corrosion protection of AA2024 by sol–gel coatings modified with MBT-loaded polyurea microcapsules. *Chemical Engineering Journal*, *283*, 1108–1117. doi:10.1016/j.cej.2015.07.087

Maiya, P. S., Erck, R. A., & Moon, B. M. (1998). Failure and corrosion resistance of TiN and TiC coatings deposited on graphite by chemical vapor deposition. *Surface and Coatings Technology, 102*(3), 218–222. doi:10.1016/S0257-8972(97)00690-7

Maliska, A. M., de Oliveira, A. M., Klein, A. N., & Muzart, J. L. R. (2001). Surface porosity sealing effect of plasma nitrocarburizing on sintered unalloyed iron. *Surface and Coatings Technology, 141*(2-3), 128–134. doi:10.1016/S0257-8972(01)01016-7

Malkin, S., & Hwang, T. W. (1996). Grinding mechanisms for ceramics. *Annals of the CIRP, 45*(2), 569–580. doi:10.1016/S0007-8506(07)60511-3

Malony, M. J. (2001). U.S. Patent 6, 284,323.

Malzbender, J., & Steinbrech, R. (2003). Determination of the stress-dependent stiffness of plasma-sprayed thermal barrier coatings using depth-sensitive indentation. *Journal of Materials Research, 18*(08), 1975–1984. doi:10.1557/JMR.2003.0274

Mammeri, F., Bourhis, E. L., Rozes, L., & Sanchez, C. (2005). Mechanical properties of hybrid organic–inorganic materials. *Journal of Materials Chemistry, 15*(35–36), 3787. doi:10.1039/b507309j

Manocha, L. M. (2003). High performance carbon-carbon composites. *Sadhana, 28*(1), 349–358. doi:10.1007/BF02717143

Marchand, O., Girardot, L., Planche, M. P., Bertrand, P., Bailly, Y., & Bertrand, G. (2011). An insight into suspension plasma spray: Injection of the suspension and its interaction with the plasma flow. *Journal of Thermal Spray Technology, 20*(6), 1310–1320. doi:10.1007/s11666-011-9682-5

Markocsan, N., Nylén, P., Wigren, J., & Li, X.-H. (2007). Low thermal conductivity coatings for gas turbine applications. *Journal of thermal spray technology, 16*(4), 498-505.

Marple, B. R., Voyer, J., Thibodeau, M., Nagy, D. R., & Vassen, R. (2006). Hot corrosion of lanthanum zirconate and partially stabilized zirconia thermal barrier coatings. *Transactions of ASME, 128*, 144-152. doi:10.1115/1.1924534

Martinez-Martinez, D., Lopez-Cartes, C., Gago, R. L., Fernandez, A., & Sanchez-Lopez, J. C. (2009). Thermal Stability and Oxidation Resistance of Nanocomposite TiC/a-C Protective Coatings. *Plasma Processes and Polymers, 6*(S1), 462–467. doi:10.1002/ppap.200931002

Martin, P. S., Stewart, D. C., Bennett, W. D., Affinito, J. D., & Gross, M. E. (1997). Multifunctional multilayer optical coatings. *Journal of Vacuum Science & Technology. A, Vacuum, Surfaces, and Films, 15*(3), 1098–1102. doi:10.1116/1.580436

Ma, S., Li, Y., & Xu, K. (2001). The composite of nitrided steel of H13 and TiN coatings by plasma duplex treatment and the effect of pre-nitriding. *Surface and Coatings Technology, 137*(2-3), 116–121. doi:10.1016/S0257-8972(00)01073-2

Masoule, S. T., Valefi, Z., Ehsani, N., & Lavasani, H. Q. (2016). Thermal insulation and thermal shock behavior of conventional and nanostructured plasma-sprayed TBCs. *Journal of Thermal Spray Technology, 25*(8), 1684–1691. doi:10.1007/s11666-016-0476-7

Matejicek, J., Sampath, S., Gilmore, D., & Neiser, R. (2003). In situ measurement of residual stresses and elastic moduli in thermal sprayed coatings: Part 2: processing effects on properties of Mo coatings. *Acta Materialia, 51*(3), 873–885. doi:10.1016/S1359-6454(02)00477-9

Matikainen, V., Koivuluoto, H., Milanti, A., & Vuoristo, P. (2015). Advanced coatings by novel high-kinetic thermal spray processes. Tiede & Tekniikka.

Matsumoto, M., Takayama, H., Yokoe, D., Mukai, K., Matsubaru, H., Kagyia, Y., & Sugita, Y. (2006). Thermal cycle behavior of plasma sprayed La_2O_3, Y_2O_3 stabilized ZrO_2 coatings. *Scripta Materialia*, *54*(12), 2035–2039. doi:10.1016/j.scriptamat.2006.03.015

Matsumoto, M., Yamaguchi, N., & Matsubara, H. (2004). Low thermal conductivity and high temperature stability of ZrO_2–Y_2O_3–La_2O_3 coatings produced by electron beam PVD. *Scripta Materialia*, *50*(6), 867–871. doi:10.1016/j.scriptamat.2003.12.008

Matsunami, N., Yamamura, Y., Itikawa, Y., Itoh, N., Kazumata, Y., Miyagawa, S., & Tawara, H. et al. (1984). Energy dependence of the yields of ion-induced sputtering of monoatomic solids. *Dat. & Nuc. Dat. Tab*, *31*(1), 1–80. doi:10.1016/0092-640X(84)90016-0

Mattox, D. (2003). *Foundations of Vacuum Coating Technology.* William Andrew Publishing/Noyes.

Mattox, D. (1998). *Handbook of Physical Vapor Deposition (PVD) Processing.* Westwood, NJ: Noyes Publications.

Mattox, D. M. (1989). Particle bombardment effects on thin film deposition. A review. *Journal of Vacuum Science & Technology. A, Vacuum, Surfaces, and Films*, *7*(3), 1105–1114. doi:10.1116/1.576238

Ma, W., Dong, H., Guo, H., Gong, P., & Zheng, X. (2010). Thermal cycling behavior of $La_2Ce_2O_7$/8YSZ double-ceramic-layer thermal barrier coatings prepared by atmospheric plasma spraying. *Surface and Coatings Technology*, *204*(21-22), 3336–3370. doi:10.1016/j.surfcoat.2010.03.053

Ma, W., Mack, D. E., Vassen, R., & Stover, D. (2008). Perovskite-type strontium zirconate as a new material for thermal barrier coatings. *Journal of the American Ceramic Society*, *91*(8), 2630–2635. doi:10.1111/j.1551-2916.2008.02472.x

Ma, X., Wu, F., Roth, J., Gell, M., & Jordan, E. H. (2006). Low thermal conductivity thermal barrier coating deposited by the solution plasma spray process. *Surface and Coatings Technology*, *201*(7), 4447–4452. doi:10.1016/j.surfcoat.2006.08.095

Maxwell, J. C. (1904). *A treatise on electricity and magnetism.* Oxford: Clarendon Press.

Mazurin, O., Streltsina, M., & Shvaiko-Shvaikovskaya, T. (1985). Handbook of Glass Data. Amsterdam: Elsevier.

McDonald, A., Moreau, C., & Chandra, S. (2007). Effect of substrate oxidation on spreading of plasma-sprayed nickel on stainless steel. *Surface and Coatings Technology*, *202*(1), 23–33. doi:10.1016/j.surfcoat.2007.04.041

McFadden, S. X., Mishra, R. S., Valiev, R. Z., Zhilyaev, A. P., & Mukherjee, A. K. (1999). Low-temperature superplasticity in nanostructured nickel and metal alloys. *Nature*, *398*(6729), 684–686. doi:10.1038/19486

McKee, D. W. (1991). Oxidation protection of carbon materials. *Chemistry and Physics of Carbon*, *23*, 173–232.

McKee, D., Spiro, C., & Lamby, E. (1984). The effects of boron additives on the oxidation behavior of carbons. *Carbon*, *22*(6), 507–511. doi:10.1016/0008-6223(84)90083-6

McPherson, R. (1973). The structure of Al_2O_3-Cr_2O_3 powders condensed from a plasma. *Journal of Materials Science*, *8*(6), 859–862. doi:10.1007/BF02397915

Mehdizadeh, N. Z., Chandra, S., & Mostaghimi, J. (2002). Effect of substrate temperature and roughness on coating formation. *Paper presented at the International Thermal Spray Conference*, Essen, Germany.

MEMS & Nanotechnology Exchange. (n.d.) Material: Chromium Carbide (Cr3C2), bulk. Retrieved from https://www.memsnet.org/material/chromiumcarbidecr3c2bulk/

Merikoski, J. V.-N., Vattulainen, I., Heinonen, J., & Ala-Nissila, T. (1997). Effect of kinks and concerted diffusion mechanisms on mass transport and growth on stepped metal surfaces. *Surface Science*, *387*(1-3), 167–182. doi:10.1016/S0039-6028(97)00349-X

Mesa, D. H., Toro, A., Sinatora, A., & Tschiptschin, A. P. (2003). The effect of testing temperature on corrosion–erosion resistance of martensitic stainless steels. *Wear*, *255*(1-6), 139–145. doi:10.1016/S0043-1648(03)00096-6

Messaddeq, S. H., Pulcinelli, S. H., Santilli, C. V., Guastaldi, A. C., & Messaddeq, Y. (1999). Microstructure and corrosion resistance of inorganic–organic (ZrO_2–PMMA) hybrid coating on stainless steel1. *Journal of Non-Crystalline Solids*, *247*(1–3), 164–170. doi:10.1016/S0022-3093(99)00058-7

Messier, R. G., Giri, A. P., & Roy, R. A. (1984). Revised structure zone model for thin film physical structure. *Journal of Vacuum Science & Technology. A, Vacuum, Surfaces, and Films*, *2*(2), 500–503. doi:10.1116/1.572604

Metco, O. (2014). Thermal spray equipment guide. Retrieved from https://www.oerlikon.com/metco/en/products-services/coating-equipment/thermal-spray/?rd=1

Metikos-Hukovic, M., Pilic, Z., Babic, R., & Omanovic, D. (2006). Influence of alloying elements on the corrosion stability of CoCrMo implant alloy in Hank's solution. *Acta Biomaterialia*, *2*(6), 693–700. doi:10.1016/j.actbio.2006.06.002 PMID:16884967

Metin, E., & Inal, O. T. (1987). Formation and growth of iron nitrides during ion-nitriding. *Journal of Materials Science*, *22*(8), 2783–2788. doi:10.1007/BF01086471

Metroke, T. L., & Apblett, A. (2004). Effect of solvent dilution on corrosion protective properties of Ormosil coatings on 2024-T3 aluminum alloy. *Progress in Organic Coatings*, *51*(1), 36–46. doi:10.1016/j.porgcoat.2004.06.002

Metroke, T., Kachurina, O., & Knobbe, E. T. (2002). Spectroscopic and corrosion resistance characterization of GLYMO–TEOS Ormosil coatings for aluminum alloy corrosion inhibition. *Progress in Organic Coatings*, *44*(4), 295–305. doi:10.1016/S0300-9440(02)00063-2

Meunier, C., Vives, S., & Bertrand, G. (1998). X-ray diffractometry analysis of r.f.-magnetron-sputtered chromium/chromium nitride coatings. *Surface and Coatings Technology*, *107*(2-3), 149–158. doi:10.1016/S0257-8972(98)00586-6

Meure, S., Wu, D. Y., & Furman, S. A. (2010). FTIR study of bonding between a thermoplastic healing agent and a mendable epoxy resin. *Vibrational Spectroscopy*, *52*(1), 10–15. doi:10.1016/j.vibspec.2009.09.005

Michlik, P., & Berndt, C. (2006). Image-based extended finite element modeling of thermal barrier coatings. *Surface and Coatings Technology*, *201*(6), 2369–2380. doi:10.1016/j.surfcoat.2006.04.003

Miguel, J. M., Guilemany, J. M., & Vizcaino, S. (2003). Tribological study of NiCrBSi coatings obtained by different processes. *Tribology International*, *36*(3), 181–187. doi:10.1016/S0301-679X(02)00144-5

Miller, R. A. (1997). Thermal barrier coatings for aircraft engines: history and directions. *Journal of thermal spray technology, 6*(1), 35-42.

Miller, R. A. (1997). Thermal barrier coatings for aircraft engines: History and directions. *Journal of Thermal Spray Technology*, *6*(1), 35–42. doi:10.1007/BF02646310

Mishra, S. B., Chandra, K., & Prakash, S. (2013). Erosion–corrosion performance of NiCrAlY coating produced by plasma spray process in a coal-fired thermal power plant. *Surface and Coatings Technology*, *216*, 23–34. doi:10.1016/j.surfcoat.2012.09.044

Mittal, V. (2014). 8 - Self-healing anti-corrosion coatings for applications in structural and petrochemical engineering A2 - Makhlouf, Abdel Salam Hamdy. In *Handbook of Smart Coatings for Materials Protection* (pp. 183–197). Woodhead Publishing. doi:10.1533/9780857096883.2.183

Mittemeijer, E. J., & Scardi, P. (2004). *Diffraction Analysis of the Microstructure of Materials*. Berlin: Springer-Verlag. doi:10.1007/978-3-662-06723-9

Mitzi, D. B. (2001). Thin-Film Deposition of Organic–Inorganic Hybrid Materials. *Chemistry of Materials*, *13*(10), 3283–3298. doi:10.1021/cm0101677

Mizuta, H., Oda, K., Shibasaki, Y., Maeda, M., Michida, M., & Ohshima, K. (1992). Preparation of high strength and Translucent Alumina by Hot Isostatic Pressing. *Journal of the American Ceramic Society*, *75*(2), 469–473. doi:10.1111/j.1151-2916.1992.tb08203.x

Moeller, W. P., & Rossendorf, F. (2002). *Nano Fabrication using Focused Ion and Electron Beams: Principles and Applications*. Oxford University Press.

Mofidi, H. H., rouhaghdam, A. S., Ahangarani, S., Bozorg, M., Azadi, M., Valiei, M. (2014). Deposition of Titanium Layer on Steel Substrate Using PECVD Method: A Parametric Study. *Materials Sciences and Applications*, *5*, 1–9.

Mofidi, H. H., Rouhaghdam, A. S., Ahangarani, S., Bozorg, M., & Azadi, M. (2014). Fracture toughness of TiN coating as a function of interlayer thickness. *Advanced Materials Research*, *829*, 466–470. doi:10.4028/www.scientific.net/AMR.829.466

Mogensen, K. S., Eskildsen, S. S., Mathiasen, C., & Bottiger, J. (1998). Optical emission spectroscopy on pulsed-DC plasmas used for TiN depositions. *Surface and Coatings Technology*, *102*(1-2), 41–49. doi:10.1016/S0257-8972(97)00529-X

Mohan, P., Yuan, B., Patetrson, T., Desai, V. H., & Sohn, Y. H. (2007). Degradation of yttria-stabilized zirconia thermal barrier coatings by vanadium pentoxide, phoshorous pentoxide, and sodium sulfate. *Journal of the American Ceramic Society*, *90*(11), 3601–3607. doi:10.1111/j.1551-2916.2007.01941.x

Montanaro, L., Tulliani, J. M., Perrot, C., & Negro, A. (1997). Sintering of industrial mullites. *Journal of the European Ceramic Society*, *17*(14), 1715–1723. doi:10.1016/S0955-2219(97)00043-5

Montemor, M. F. (2016). 7 - Hybrid nanocontainer-based smart self-healing composite coatings for the protection of metallic assets. In *Smart Composite Coatings and Membranes* (pp. 183–209). Woodhead Publishing. doi:10.1016/B978-1-78242-283-9.00007-5

More, A. P., & Mhaske, S. T. (2016). Anticorrosive coating of polyesteramide resin by functionalized ZnO- Al 2 O 3 -Fly ash composite and functionalized multiwalled carbon nanotubes. *Progress in Organic Coatings*, *99*, 240–250. doi:10.1016/j.porgcoat.2016.06.005

Moreau, C., Gougeon, P., & Lamontagne, M. (1995). Influence of substrate preparation on the flattening and cooling of plasma-sprayed particles. *Journal of Thermal Spray Technology*, *4*(1), 25–33. doi:10.1007/BF02648525

Moriyama, M., Tahara, H., & Yoshikawa, T. (2003). Silicon nitride coating using electromagnetic acceleration plasma generators [in Japanese]. *Koon Gakkaishi*, *29*(4), 13.

Moriyama, M., Tahara, H., & Yoshikawa, T. (2004). Silicon nitride coatings by magneto-plasma-dynamic plasma spraying [in Japanese]. *Koon Gakkaishi*, *30*(1), 40–41.

Morks, M. F., Tsunekawa, Y., Okumiya, M., & Shoeib, M. A. (2003). Splat microstructure of plasma sprayed cast iron with different chamber pressures. *Journal of Thermal Spray Technology*, *12*(2), 282–289. doi:10.1361/105996303770348393

Moskal, G., Rozmysłowska, A., Gazda, A., & Homa, M. (2009). Wybrane termofizyczne właściwości proszków cyrkonianowych na bazie pierwiastków ziem rzadkich typu RE2Zr2O7 (RE- Gd, La, Sm, Nd) przeznaczonych do natryskiwania cieplnego powłokowych warstw barierowych. *Prace Instytutu Odlewnictwa*, Tom XLIX. *Zeszyt*, *4*, 15–26.

Mostaghimi, J., & Chandra, S. (2007). Heat transfer in plasma spray coating processes. *Advances in Heat Transfer*, *40*(C), 143–204. doi:10.1016/S0065-2717(07)40002-8

Mouri, L., Mabille, I., Fiaud, C., Amouroux, J., Catillon, G., & Gras, R. (2001). Study of corrosion resistance and surface properties of carbon steel after a duplex plasma treatment. *Thin Solid Films*, *389*(1-2), 153–160. doi:10.1016/S0040-6090(01)00875-6

Movchan, B. D. (1969). Structure and properties of thick condensates of nickel, titanium, tungsten, aluminum, and zirconium dioxide in vacuum. *Физика металлов и металловедение*, *28*, 653–660.

Mudgal, D., Singh, S., & Prakash, S. (2015). Cyclic Hot corrosion behavior of Superni 718, Superni 600, and Superco 605 in sulfate and chloride containing environment at 900°C. *Metallography Microstructure and Analysis*, *4*(1), 13–25. doi:10.1007/s13632-014-0182-0

Muhl, S. P., & Pérez, A. (2015). The use of hollow cathodes in deposition processes: A critical review. *Thin Solid Films*, *579*, 174–198. doi:10.1016/j.tsf.2015.02.066

Mulligan, C. P., Blanchet, T. A., & Gall, D. (2010). CrN–Ag nanocomposite coatings: Tribology at room temperature and during a temperature ramp. *Surface and Coatings Technology*, *204*(9-10), 1388–1394. doi:10.1016/j.surfcoat.2009.09.018

Mulligan, C. P., & Gall, D. (2005). CrN-Ag self-lubricating hard coatings. *Surface and Coatings Technology*, *200*(5-6), 1495–1500. doi:10.1016/j.surfcoat.2005.08.063

Muratore, C., Hu, J. J., & Voevodin, A. A. (2009). Tribological coatings for lubrication over multiple thermal cycles. *Surface and Coatings Technology*, *203*(8), 957–962. doi:10.1016/j.surfcoat.2008.08.073

N'Gandu-Muamba, J. M., & Streiff, R. (1993). The reactive element effect (R.E.E.): A tentative classification. *Journal de Physique. IV*, *03*(C9), 281–290.

Nagano, M., Nagashima, S., Maeda, H., & Kato, A. (1999). Sintering behaviour of Al_2TiO_5 base ceramics and their thermal properties. *Ceramics International*, *25*(8), 681–687. doi:10.1016/S0272-8842(98)00083-2

Nakajima, K., Taki, K., Takata, M., & Okamoto, S. (1991). An effort on preparing sintered bodies of Fe_4N [in Japanese]. *Journal of the Ceramic Society of Japan*, *99*(4), 311–314. doi:10.2109/jcersj.99.311

Nakao, W., Takahashi, K., & Ando, K. (2009). Self-healing of Surface Cracks in Structural Ceramics. In *Self-healing materials: fundamentals, design strategies, and applications*.

Nakles, M. (2004). *Experimental and Modeling Studies of Low-Energy Ion Sputtering for Ion Thrusters*. Blacksburg, Virginia: Virginia Polytechnic Institute and State University.

Nandiyanto, A. B. D., & Okuyama, K. (2011). Progress in developing spray-drying methods for the production of controlled morphology particles: From the nanometer to submicrometer size ranges. *Advanced Powder Technology*, *22*(1), 1–19. doi:10.1016/j.apt.2010.09.011

Narita, T. (2009). A view of compatible heat-resistant alloy and coating systems at high-temperatures. *Paper presented at the International Workshop on Advanced Material for New and Renewable Energy*, Jakarta, Indonesia. doi:10.1063/1.3243272

Naveena, B. E., Keshavamurthy, R., & Sekhar, N. (2017). Slurry erosive wear behaviour of plasma-sprayed flyash–Al_2O_3 coatings. In *Surface Engineering*. Taylor & Francis. doi:10.1080/02670844.2017.1288341

Navinsek, B. P., Panjan, P., & Milošev, I. (1997). Industrial Applications of CrN (PVD) Coatings, Deposited at High and Low Temperatures. *Surface and Coatings Technology*, *97*(1-3), 182–191. doi:10.1016/S0257-8972(97)00393-9

Nemeth, S. (2008). Modelling of the scratch resistance of particle filled sol–gel coatings on aluminium. *Thin Solid Films*, *516*(16), 5355–5359. doi:10.1016/j.tsf.2007.07.044

Neville, A. (1995). An Investigation of the Corrosion Behaviour of a Range of Engineering Materials in Marine Environments [PhD Thesis]. University of Glasgow, UK.

Neville, A., Hodgkiess, T. & Dallas, J. T, (1995). A study of the erosion-corrosion behaviour of engineering steels for marine pumping applications. *Wear,* 186-187, 497-507.

Nicholls, J. (2003). Advances in coating design for high-performance gas turbines. *MRS Bulletin*, *28*(09), 659–670. doi:10.1557/mrs2003.194

Niederberger, M. (2007). Nonaqueous sol-gel routes to metal oxide nanoparticles. *Accounts of Chemical Research*, *40*(9), 793–800. doi:10.1021/ar600035e PMID:17461544

Niizuma, K. (2001). Synthesis of iron nitride by high energy plasma [in Japanese]. *Collaboration Report in Joining and Welding Research Institute Osaka University*, *2000*, 194–195.

Niizuma, K., Utsushikawa, Y., & Kobayashi, A. (2001). Magnetic properties of Fe coating prepared by gas tunnel plasma spraying [in Japanese]. *Journal of IAPS*, *9*, 136–138.

Nikrooz, B., & Zandrahimi, M. (2011). Optimization of process variables and corrosion properties of a multi layer silica sol gel coating on AZ91D using the Box–Behnken design. *Journal of Sol-Gel Science and Technology*, *59*(3), 640–649. doi:10.1007/s10971-011-2539-z

Nikrooz, B., Zandrahimi, M., & Ebrahimifar, H. (2012). High temperature oxidation resistance and corrosion properties of dip coated silica coating by sol gel method on stainless steel. *Journal of Sol-Gel Science and Technology*, *63*(3), 286–293. doi:10.1007/s10971-012-2786-7

Niu, Y., Wang, H., Liu, Z., Hu, C., Wang, X., Zheng, X., & Ding, C. (2015). Microstructure evolution of ZrB 2–MoSi 2 composite coatings at middle and high temperatures. *Surface and Coatings Technology*, *273*, 30–38. doi:10.1016/j.surfcoat.2015.03.029

Norouzi, M., & Afrasiabi Garekani, A. (2014). Corrosion protection by zirconia-based thin films deposited by a sol–gel spin coating method. *Ceramics International*, *40*(2), 2857–2861. doi:10.1016/j.ceramint.2013.10.027

Nosonovsky, M., & Rohatgi, P. K. (2011). *Biomimetics in materials science: self-healing, self-lubricating, and self-cleaning materials* (Vol. 152). Springer Science & Business Media.

Ode'n, M., Ericsson, C., Hakansson, G., & Ljungcrantz, H. (1999). Microstructure and mechanical behavior of arc-evaporated Cr–N coatings. *Surface and Coatings Technology*, *114*(1), 39–51. doi:10.1016/S0257-8972(99)00019-5

Oguma, H., Okada, I., Torigoe, T., & Takahashi, K. (2004). High temperature corrosion resistant alloy, thermal barrier coating material, and gas turbine using high temperature corrosion resistant alloy.

Oguri, K., & Arai, T. (1990). Effect of excess carbon on the hardness of Si-C and TiC coatings formed by plasma-assisted chemical vapor deposition. *Thin Solid Films*, *186*(2), 29–31. doi:10.1016/0040-6090(90)90152-4

Oguri, K., & Arai, T. (1992). Friction coefficient of Si-C, Ti-C and Ge-C coatings with excess carbon formed by plasma-assisted chemical vapor deposition. *Thin Solid Films*, *208*(2), 158–160. doi:10.1016/0040-6090(92)90635-O

Oguri, K., Fujita, H., & Arai, T. (1991). Effect of N_2 toTiCl_4 flow rate ratio on the properties of TiN coatings formed by D.C. discharge PACVD. *Thin Solid Films*, *195*, 77–88. doi:10.1016/0040-6090(91)90260-5

Ohmori, A., Hirano, S., & Kamada, K. (1993). Spraying TiN by a combined laser and low-pressure plasma spray system. *Journal of Thermal Spray Technology*, *2*(2), 137–144. doi:10.1007/BF02652021

Ohmori, A., Wakamatsu, M., & Kamada, K. (1993). Synthesis of Al2O3-AlN coatings by low pressure plasma spraying and nitriding. *Transactions of JWRI*, *22*(2), 227-232.

Ohmori, A., Zhang, Y. C., Hirano, S., & Arata, Y. (1991). Reaction spray by laser-plasma jet synthesis of TiN coatings. *Transactions of JWRI*, *20*(2), 27-33.

Okada, T., Hamatani, H., & Yoshida, T. (1989). Radio-Frequency plasma spraying of ceramics. *Journal of the American Ceramic Society*, *72*(11), 2111–2116. doi:10.1111/j.1151-2916.1989.tb06040.x

Oksa, M., Turunen, E., Suhonen, T., Varis, T., & Hannula, S.-P. (2011). Optimization and Characterization of High Velocity Oxy-fuel Sprayed Coatings: Techniques, Materials, and Applications. *Coatings*, *1*(2), 17–52. doi:10.3390/coatings1010017

Olaya, J. J., Rodil, S. E., Muhl, S., & Sanchez, E. (2005). Comparative study of chromium nitride coatings deposited by unbalanced and balanced magnetron sputtering. *Thin Solid Films*, *474*(1-2), 119–126. doi:10.1016/j.tsf.2004.08.067

Oldani, V., Sergi, G., Pirola, C., Sacchi, B., & Bianchi, C. L. (2016). Sol-gel hybrid coatings containing silica and a perfluoropolyether derivative with high resistance and anti-fouling properties in liquid media. *Journal of Fluorine Chemistry*, *188*, 43–49. doi:10.1016/j.jfluchem.2016.06.005

Olszak, D., & Olszyna, A. (2004). Określanie wielkości krystalitów i odkształceń sieciowych w nanokompozycie NiAl-Al_2O_3 na podstawie poszerzenia rentgenowskich linii dyfrakcyjnych. *Kompozyty*, *4/11*, 284–288.

Ono, H., Teramoto, T., & Shinoda, T. (1993). Development of plasma sprayed lubricant free slide plates with ceramic coat for railway use. *Materials and Manufacturing Processes*, *8*(4-5), 451–463. doi:10.1080/10426919308934850

Ono, T., Kenmotsu, T., & Muramoto, T. (2009). *Simulation of the sputtering process*. Berlin: Springer-Verlag.

Opeka, M. M., Talmy, I. G., & Zaykoski, J. (2004). Oxidation-based materials selection for 2000 C+ hypersonic aerosurfaces: Theoretical considerations and historical experience. *Journal of Materials Science*, *39*(19), 5887–5904. doi:10.1023/B:JMSC.0000041686.21788.77

Opila, E. J. (1994). Oxidation Kinetics of Chemically Vapor-Deposited Silicon Carbide in Wet Oxygen. *Journal of the American Ceramic Society*, *77*(3), 730–736. doi:10.1111/j.1151-2916.1994.tb05357.x

Osborne, J. H. (2001). Observations on chromate conversion coatings from a sol–gel perspective. *Progress in Organic Coatings*, *41*(4), 280–286. doi:10.1016/S0300-9440(01)00143-6

Ouyang, J. H., & Sasaki, S. (2001). Effects of different additives on microstructure and high-temperature tribological properties of plasma-sprayed Cr_2O_3 ceramic coatings. *Wear*, *249*(1-2), 56–67. doi:10.1016/S0043-1648(01)00530-0

Ouyang, T., Fang, X., Zhang, Y., Liu, D., Wang, Y., Feng, S., & Suo, J. (2016). Enhancement of high temperature oxidation resistance and spallation resistance of SiC-self-healing thermal barrier coatings. *Surface and Coatings Technology*, *286*, 365–375. doi:10.1016/j.surfcoat.2015.12.054

Over, H. (1998). Crystallographic study of interaction between ad-species on metal surfaces. *Progress in Surface Science*, *58*(4), 249–376. doi:10.1016/S0079-6816(98)00029-X

Padture, N. P., Gell, M., & Jordan, E. H. (2002). Thermal barrier coatings for gas-turbine engine applications. *Science*, *296*(5566), 280–284. doi:10.1126/science.1068609 PMID:11951028

Padture, N. P., Schlichting, K. W., Bhatia, T., Ozturk, A., Cetegen, B., Jordan, E. H., & Osendi, M. I. et al. (2001). Toward durable thermal barrier coatings with novel microstructure deposited by solution precursors plasma spray. *Acta Materialia*, *49*(12), 2251–2257. doi:10.1016/S1359-6454(01)00130-6

Pakseresht, A., Rahimipour, M., Alizadeh, M., Hadavi, S., & Shahbazkhan, A. (2016). Concept of advanced thermal barrier functional coatings in high temperature engineering components. In Research perspectives on functional micro- and nanoscale coatings (pp. 396–419). Hershey, PA: IGI Global. doi:10.4018/978-1-5225-0066-7.ch015

Pakseresht, A. H., Ghasali, E., Nejati, M., Shirvanimoghaddam, K., Javadi, A. H., & Teimouri, R. (2015). Development empirical-intelligent relationship between plasma spray parameters and coating performance of Yttria-Stabilized Zirconia. *International Journal of Advanced Manufacturing Technology*, *76*(5-8), 1031–1045. doi:10.1007/s00170-014-6212-x

Pakseresht, A. H., Javadi, A. H., Nejati, M., Shirvanimoghaddam, K., Ghasali, E., & Teimouri, R. (2014). Statistical analysis and multiobjective optimization of process parameters in plasma spraying of partially stabilized zirconia. *International Journal of Advanced Manufacturing Technology*, *75*(5-8), 739–753. doi:10.1007/s00170-014-6169-9

Pakseresht, A., Javadi, A., Bahrami, M., Khodabakhshi, F., & Simchi, A. (2016). Spark plasma sintering of a multilayer thermal barrier coating on Inconel 738 superalloy: Microstructural development and hot corrosion behavior. *Ceramics International*, *42*(2), 2770–2779. doi:10.1016/j.ceramint.2015.11.008

Pakseresht, A., Javadi, A., Ghasali, E., Shahbazkhan, A., & Shakhesi, S. (2016). Evaluation of hot corrosion behavior of plasma sprayed thermal barrier coatings with graded intermediate layer and double ceramic top layer. *Surface and Coatings Technology*, *288*, 36–45. doi:10.1016/j.surfcoat.2016.01.012

Pakseresht, A., Rahimipour, M., Vaezi, M., & Salehi, M. (2016). Thermal plasma spheroidization and spray deposition of barium titanate powder and characterization of the plasma sprayable powder. *Materials Chemistry and Physics*, *173*, 395–403. doi:10.1016/j.matchemphys.2016.02.028

Panat, R., Hsia, K. J., & Oldham, J. (2005). Rumpling instability in thermal barrier systems under isothermal conditions in vacuum. *Philosophical Magazine*, *85*(1), 45–64. doi:10.1080/14786430412331309344

Pan, J., Leygraf, C., Jargelius Pettersson, R. F. A., & Linden, J. (1998). Characterization of High-Temperature Oxide Films on Stainless Steels by Electrochemical-Impedance Spectroscopy. *Oxidation of Metals*, 50(5), 431-455.

Pantelis, D. I., Psyllaki, P., & Alexo-poulos, N. (2000). Tribological behaviour of plasma sprayed Al_2O_3 coatings under severe wear conditions. *Wear*, *237*(2), 197–204. doi:10.1016/S0043-1648(99)00324-5

Pan, W., Wan, C. L., Xu, Q., Wang, J. D., & Qu, Z. X. (2007). Thermal diffusivity of samarium–gadolinium zirconate solid solutions. *Thermochimica Acta*, *455*(1-2), 16–20. doi:10.1016/j.tca.2006.12.001

Papyrin, A., Kosarev, V., Klinkov, S., Alkhimov, A., & Fomin, V. (2007). *Cold spray technology*.

Parkhill, R., Knobbe, E., & Donley, M. (2001). Application and evaluation of environmentally compliant spray-coated ormosil films as corrosion resistant treatments for aluminum 2024-T3. *Progress in Organic Coatings*, *41*(4), 261–265. doi:10.1016/S0300-9440(01)00138-2

Park, I. W., & Kim, K. H. (2002). Coating materials of TiN, Ti–Al–N, and Ti–Si–N by plasma-enhanced chemical vapor deposition for mechanical applications. *Journal of Materials Processing Technology*, *130–131*, 254–259. doi:10.1016/S0924-0136(02)00807-5

Park, J. R., Song, Y. K., Rie, K. T., & Gebauer, A. (1998). Hard coating by plasma-assisted CVD on plasma nitrided stellite. *Surface and Coatings Technology*, *98*(1-3), 1329–1335. doi:10.1016/S0257-8972(97)00264-8

Paruch, R. B., Garrison, B. J., Mlynek, M., & Postawa, Z. (2014). On Universality in Sputtering Yields due to Cluster Bombardment. *The Journal of Physical Chemistry Letters*, *5*(18), 3227–3230. doi:10.1021/jz501545t PMID:26276337

Pathak, S. S., & Khanna, A. S. (2009). Investigation of anti-corrosion behavior of waterborne organosilane–polyester coatings for AA6011 aluminum alloy. *Progress in Organic Coatings*, *65*(2), 288–294. doi:10.1016/j.porgcoat.2008.12.006

Paul, B., Majumdar, S., & Suri, A. (2013). Microstructure and mechanical properties of hot pressed Mo–Cr–Si–Ti in-situ composite, and oxidation behavior with silicide coatings. *International Journal of Refractory Metals & Hard Materials*, *38*, 26–34. doi:10.1016/j.ijrmhm.2012.12.005

Paul, S., Cipitria, A., Tsipas, S. A., & Clyne, T. W. (2009). Sintering characteristics of plasma sprayed zirconia coatings containing different stabilisers. *Surface and Coatings Technology*, *203*(8), 1069–1074. doi:10.1016/j.surfcoat.2008.09.037

Pawlowski, L. (2008). The science and engineering of thermal spray coatings (2nd ed.). Chichester, UK: John Wiley & Sons. doi:10.1002/9780470754085

Pawlowski, L., & Dépôts Physiques. (2003). *Techniques, microstructures et propriétés*. Lausanne, Switzerland Presse Polytechnique et Universitaires Romandes.

Pawlowski, L. (2007). *The science and engineering of thermal spray coatings* (2nd ed.). Wiley.

Pawlowski, L. (2008). Finely grained nanometric and submicrometric coatings by thermal spraying: A review. *Surface and Coatings Technology*, *202*(18), 4318–4328. doi:10.1016/j.surfcoat.2008.04.004

Pawlowski, L. (2008). Finely grained nanometric and submicrometric coatings by thermal spraying: A review. *Surface and Coatings Technology*, *203*(3-4), 397–397. doi:10.1016/j.surfcoat.2008.09.004

Pawlowski, L. (2009). Suspension and solution thermal spray coatings. *Surface and Coatings Technology*, *203*(19), 2807–2829. doi:10.1016/j.surfcoat.2009.03.005

Pei, Y. T., Galvan, D., & De Hosson, J. T. M. (2006). Tribological behavior and thermal stability of TiC/a-C:H nanocomposite coatings. *Journal of Vacuum Science & Technology. A, Vacuum, Surfaces, and Films*, *24*(4), 1448–1454. doi:10.1116/1.2194927

Pellizzari, M., Molinari, A., & Straffelini, G. (2001). Thermal fatigue resistance of plasma duplex-treated tool steel. *Surface and Coatings Technology*, *142-144*, 1109–1115. doi:10.1016/S0257-8972(01)01223-3

Peng, H., Wang, L., Guo, L., Miao, W., Guo, H., & Gong, S. (2012). Degradation of EB-PVD thermal barrier coatings caused by CMAS deposits. Progress in Natural Science. *Materials International*, *22*, 461–467.

Pennefather, R. C., & Boone, D. H. (1995). Mechanical degradation of coating systems in high-temperature cyclic oxidation. *Surface and Coatings Technology*, *76*(1-3), 47–52. doi:10.1016/0257-8972(95)02551-0

Pepe, A., Galliano, P., Ceré, S., Aparicio, M., & Durán, A. (2005). Hybrid silica sol–gel coatings on Austempered Ductile Iron (ADI). *Materials Letters*, *59*(17), 2219–2222. doi:10.1016/j.matlet.2005.03.001

Pershin, V., Lufitha, M., Chandra, S., & Mostaghimi, J. (2003). Effect of substrate temperature on adhesion strength of plasma-sprayed nickel coatings. *Journal of Thermal Spray Technology*, *12*(3), 370–376. doi:10.1361/105996303770348249

Petrov, I. B., Barna, P. B., Hultman, L., & Greene, J. E. (2003). Micro-structural evolution during film growth. *Journal of Vacuum Science & Technology. A, Vacuum, Surfaces, and Films*, *21*(5), S117–S128. doi:10.1116/1.1601610

Petrov, I., Losbichler, P., Bergstrom, D., Greene, J. E., Münz, W. D., Hurkmans, T., & Trinh, T. (1997). Ion-assisted growth of Ti1−xAlxN/Ti1−yNbyN multilayers by combined cathodic-arc/magnetron-sputter deposition. *Thin Solid Films*, *302*(1-2), 179–192. doi:10.1016/S0040-6090(96)09524-7

Petrovicova, E., Knight, R., Smith, R. W., & Schadler, L. S. (1997). Structure and properties of HVOF sprayed ceramic/polymer nano-composite coatings. *Paper presented at the Thermal Spray: A United Forum for Scientific and Technological Advances*, Indianapolis, IN.

Petrovicova, E., & Schadler, L. S. (2002). Thermal spraying of polymers. *International Materials Reviews*, *47*(4), 169–190. doi:10.1179/095066002225006566

Petrovicova, E., Smith, R. W., Knight, R., & Schadler, L. S. (1998). Particulate nanoreinforced polymer composites processing and properties. *J. Therm. Spray Techn.*, *7*(3), 426–427.

Pfohl, C., Gebauer-Teichmann, A., & Rie, K. T. (1999). Application of wear-resistant PACVD coatings in aluminium diecasting: Economical and ecological aspects. *Surface and Coatings Technology*, *112*(1-3), 347–350. doi:10.1016/S0257-8972(98)00796-8

Phanasgaonkar, A., & Raja, V. S. (2009). Influence of curing temperature, silica nanoparticles- and cerium on surface morphology and corrosion behaviour of hybrid silane coatings on mild steel. *Surface and Coatings Technology*, *203*(16), 2260–2271. doi:10.1016/j.surfcoat.2009.02.020

Pidani, R. (2014). The influence of laser treatment on hot corrosion behavior of plasma-sprayed nanostructured yttria stabilized zirconia thermal barrier coatings. *Journal of the European Ceramic Society*, *34*(8), 2013–2021. doi:10.1016/j.jeurceramsoc.2014.01.031

Pierre, A. C. (1998). Applications of Sol-Gel Processing. In Introduction to Sol-Gel Processing (pp. 347–386). Springer US. doi:10.1007/978-1-4615-5659-6_9

Pierre-louis, O. D., D'Orsogna, M. R., & Einstein, T. L. (1999). Edge diffusion during growth: The kink Ehrlich-Schwoebel effect and resulting instabilities. *Physical Review Letters*, *82*(18), 3661–3664. doi:10.1103/PhysRevLett.82.3661

Pierson, H. O. (1996). *Handbook of refractory carbides and nitrides –Properties, characteristics, processing and applications*. New Jersey, USA: Noyes Publications.

Pilling, N. B., & Bedworth, R. E. (1923). The oxidation of metals at high temperatures. *Journal of the Institute of Metals*, *29*, 529–582.

Pillis, M. G., Geribola, G. A., Scheidt, G., de Araújo, E. G., de Oliveira, M. C. L., & Antunes, R. A. (2016). Corrosion of thin, magnetron sputtered Nb2O5 films. *Corrosion Science*, *102*, 317–325. doi:10.1016/j.corsci.2015.10.023

Pirhady Tavandashti, N., Ghorbani, M., Shojaei, A., Gonzalez-Garcia, Y., Terryn, H., & Mol, J. M. C. (2016). pH responsive Ce(III) loaded polyaniline nanofibers for self-healing corrosion protection of AA2024-T3. *Progress in Organic Coatings*, *99*, 197–209. doi:10.1016/j.porgcoat.2016.04.046

Ploch, S. W., Wernicke, T., Dinh, D. V., Pristovsek, M., & Kneissl, M. (2012). Surface diffusion and layer morphology of (11–22) GaN grown by metal-organic vapor phase epitaxy. *Journal of Applied Physics*, *111*(3), 033526. doi:10.1063/1.3682513

PN-EN 23923-2:1998: Proszki metaliczne -- Oznaczenie gęstości nasypowej -- Metoda wolumetru Scotta.

PN-EN ISO 4490:2014-11. Proszki metaliczne -- Oznaczanie czasu przesypywania przez kalibrowany lejek (sypkościomierz Halla).

Podgornik, B., JVižintin, J., Wänstrand, O., Larsson, M., Hogmark, S., Ronkainen, H., & Holmberg, K., (2001). Tribological properties of plasma nitrided and hard coated AISI 4140 steel. *Wear, 249*, 254-259. doi:10.1016/S0043-1648(01)00564-6

Poerschke, D. L., & Levi, C. G. (2015). Effects of cation substitution and temperature on the interaction between thermal barrier oxides and molten CMAS. *Journal of the European Ceramic Society, 35*(2), 681–691. doi:10.1016/j.jeurceramsoc.2014.09.006

Ponon, N. A., Appleby, D. J. R., Arac, E., King, P. J., Ganti, S., Kwa, K. S. K., & O'Neill, A. (2015). Effect of Deposition Conditions and Post Deposition Anneal on Reactively Sputtered Titanium Nitride Thin Films. *Thin Solid Films, 578*, 31–37. doi:10.1016/j.tsf.2015.02.009

Porcayo-Calderon, J., Brito-Figueroa, E., & González-Rodríguez, J. G. (1999). Oxidation behaviour of Fe–Si thermal spray coatings. *Materials Letters, 38*(1), 45–53. doi:10.1016/S0167-577X(98)00130-X

Porcayo-Calderón, J., Sotelo-Mazón, O., Casales-Diaz, M., Ascencio-Gutierrez, J. A., Salinas-Bravo, V. M., & Martinez-Gomez, L. (2014). Electrochemical study of Ni20Cr coatings applied by HVOF process in ZnCl2-KCl at high temperatures. *Journal of Analytical Methods in Chemistry*. PMID:25210645

Porcayo-Calderón, J., Sotelo-Mazon, O., Salinas-Bravo, V. M., Arrieta-Gonzalez, C. D., Ramos-Hernandez, J. J., & Cuevas-Arteaga, C. (2012). Electrochemical Performance of Ni20Cr coatings applied by combustion powder spray in ZnCl2-KCl molten salts. *International Journal of Electrochemical Science, 7*(2), 1134–1148.

Poulson, B. (1990). Mass transfer from rough surfaces. *Corrosion Science, 30*(6/7), 743–746. doi:10.1016/0010-938X(90)90037-6

Pourbaix, M. (1974). *Atlas of electrochemical equilibria in aqueous solutions.*

Price, S., & Figueira, R. (2017). Corrosion Protection Systems and Fatigue Corrosion in Offshore Wind Structures: Current Status and Future Perspectives. *Coatings, 7*(2), 25. doi:10.3390/coatings7020025

Qin, E., Yin, S., Ji, H., Huang, Q., Liu, Z., Wu, S. (2017). Hot Corrosion Behavior of Arc-Sprayed Highly Dense NiCr-Based Coatings in Chloride Salt Deposit. *Journal of Thermal Spray Technology.*

Qu, Z., Wan, C., & Pan, W. (2007). Thermal expansion and defect chemistry of MgO doped $Sm_2Zr_2O_7$. *Chemistry of Materials, 19*(20), 4913–4918. doi:10.1021/cm071615z

Rabiei, A. G. E. A., & Evans, A. G. (2000). Failure mechanisms associated with the thermally grown oxide in plasma-sprayed thermal barrier coatings. *Acta Materialia, 48*(15), 3963–3976. doi:10.1016/S1359-6454(00)00171-3

Radwan, M., & Miyamoto, Y. (2007). Growth of quasi-aligned AlN nanofibers by nitriding combustion synthesis. *Journal of the American Ceramic Society, 90*(8), 2347–2351. doi:10.1111/j.1551-2916.2007.01747.x

Rahimi, H., Mozaffarinia, R., & Hojjati Najafabadi, A. (2013). Corrosion and Wear Resistance Characterization of Environmentally Friendly Sol–gel Hybrid Nanocomposite Coating on AA5083. *Journal of Materials Science and Technology, 29*(7), 603–608. doi:10.1016/j.jmst.2013.03.013

Rahman, M., Ibrahim, T. K., & Abdalla, A. N. (2011). Thermodynamic performance analysis of gas-turbine power-plant. *International Journal of Physical Sciences, 6*(14), 3539–3550.

Rahoui, S., Turq, V., & Bonino, J.-P. (2013). Effect of thermal treatment on mechanical and tribological properties of hybrid coatings deposited by sol–gel route on stainless steel. *Surface and Coatings Technology, 235*, 15–23. doi:10.1016/j.surfcoat.2013.07.008

Ramachandran, C. S., Balasubramanian, V., & Ananthapadmanabhan, P. V. (2012). Synthesis, spheroidization and spray deposition of lanthanum zirconate using thermal plasma process. *Surface and Coatings Technology*, *206*(13), 3017–3035. doi:10.1016/j.surfcoat.2011.10.052

Ramachandran, C. S., Balasubramanian, V., Ananthapadmanabhan, P. V., & Viswabaskaran, V. (2012). Influence of the intermixed interfacial layers on the thermal cycling behavior of atmospheric plasma sprayed lanthanum zirconate based coatings. *Ceramics International*, *38*(5), 4081–4096. doi:10.1016/j.ceramint.2012.01.066

Ramesh, C. S., & Keshavamurthy, R. (2011). Slurry erosive wear behavior of Ni-P coated Si3N4 reinforced Al6061 composites. *Materials & Design*, *32*(4), 1833–1843. doi:10.1016/j.matdes.2010.12.024

Rapp, R. A. (2002). Hot corrosion of materials: A fluxing mechanism? *Corrosion Science*, *44*(2), 209–221. doi:10.1016/S0010-938X(01)00057-9

Rastogi, R. V., Vankar, V. D., & Chopra, K. L. (1987). Simple Planar Magnetron Sputtering Source. *The Review of Scientific Instruments*, *58*(8), 1505–1506. doi:10.1063/1.1139388

Rauf, A., Yu, Q. H., Jin, L., & Zhou, C. G. (2012). Microstructure and thermal properties of nanostructured lanthana-doped yttria-stabilized zirconia thermal barrier coatings by air plasma spraying. *Scripta Materialia*, *66*(2), 109–112. doi:10.1016/j.scriptamat.2011.10.017

Raunija, T. S. K., & Babu, S. (2013). Randomly oriented carbon/carbon composite. *Paper presented at the AIP Conference Proceedings*. doi:10.1063/1.4810050

Reed, R. C. (2006). *The superalloys: fundamentals and applications*. Cambridge, UK: Cambridge University Press. doi:10.1017/CBO9780511541285

Remesh, K., Ng, H. W., & Yu, S. C. M. (2003). Influence of process parameters on the deposition footprint in plasma-spray coating. *Journal of Thermal Spray Technology*, *12*(3), 377–392. doi:10.1361/105996303770348258

Renji, Z., Ziwei, L., Zhouping, C., & Qi, S. (1991). Studies on multilayer wear of CVD TiC-TiN multilayer composite coating. *Wear*, *147*(2), 227–251. doi:10.1016/0043-1648(91)90182-T

Ren, P., Zhu, S. L., & Wang, F. H. (2015a). Characterization and oxidation behavior of a sputtered nanocomposite Ni+CrAlYSiHfN coating. *Corrosion*, *71*(4), 523–535. doi:10.5006/1492

Ren, P., Zhu, S. L., & Wang, F. H. (2015b). Microstructural stability of AlN diffusion barrier for nanocomposite Ni + CrAlYSiHfN coating on single crystal superalloy at high temperatures. *Applied Surface Science*, *359*, 420–425. doi:10.1016/j.apsusc.2015.10.096

Ren, P., Zhu, S. L., & Wang, F. H. (2016a). Microstructure and oxidation behavior of a Ni+CrAlYSiHfN/AlN multilayer coating fabricated by reactive magnetron sputtering. *Corrosion Science*, *104*, 197–206. doi:10.1016/j.corsci.2015.12.012

Ren, P., Zhu, S. L., & Wang, F. H. (2016b). TEM study of the evolution of sputtered Ni+CrAlYSiHfN nanocomposite coating with an AlN diffusion barrier at high temperature. *Surface and Coatings Technology*, *286*, 262–267. doi:10.1016/j.surfcoat.2015.12.046

Rhines, F. N., & Wolf, J. S. (1970). The role of oxide microstructure and growth stresses in the high-temperature scaling of nickel. *Metallurgical Transactions*, *1*(6), 1701–1710. doi:10.1007/BF02642020

Rhys-Jones, T. (1989). Coatings for blade and vane applications in gas turbines. *Corrosion Science*, *29*(6), 623–646. doi:10.1016/0010-938X(89)90104-2

Richer, R., Yandouzi, M., Beauvais, L., & Jodoin, B. (2010). Oxidation behavior of CoNiCrAlY bond coats produced by plasma, HVOF and cold gas dynamic spraying. *Surface and Coatings Technology*, *204*(24), 3962–3974. doi:10.1016/j.surfcoat.2010.03.043

Rico, A., Rodriguez, J., Otero, E., Zeng, P., & Rainforth, W. M. (2009). Wear behaviour of nanostructured alumina-titania coatings deposited by atmospheric plasma spray. *Wear*, *267*(5-8), 1191–1197. doi:10.1016/j.wear.2009.01.022

Rie, K. T., & Gebauer, A. (1991). Plasma-assisted chemical vapor deposition of hard coatings with metallo-organic compounds. *Materials Science and Engineering A*, *139*, 61–66. doi:10.1016/0921-5093(91)90597-G

Rie, K. T., Gebauer, A., Wohle, J., Tonshoff, H. K., & Blawit, C. (1995). Synthesis of TiN/TiCN/TiC layer systems on steel and cermet substrates by PACVD. *Surface and Coatings Technology*, *74*, 375–381. doi:10.1016/0257-8972(95)08243-3

Rie, K. T., & Whole, J. (1991). Spectroscopic investigation of N_2-H_2-Ar-$TiC1_4$-assisted chemical vapor deposition discharge for plasma of TiN. *Materials Science and Engineering A*, *139*, 37–40. doi:10.1016/0921-5093(91)90592-B

Rie, K. T., Wohle, J., & Gebauer, A. (1993). Synthesis of thin coatings by plasma-assisted chemical vapor deposition using metallo-organic compounds as precursors. *Surface and Coatings Technology*, *59*(1-3), 202–206. doi:10.1016/0257-8972(93)90083-Z

Roberge, P. (1999). *Handbook of Corrosion Engineering*. McGraw-Hill Professional.

Rodič, P., Mertelj, A., Borovšak, M., Benčan, A., Mihailović, D., Malič, B., & Milošev, I. (2015). Composition, structure and morphology of hybrid acrylate-based sol–gel coatings containing Si and Zr composed for protective applications. *Surface and Coatings Technology*, *286*, 388-396.

Rodič, P., Iskra, J., & Milošev, I. (2014). Study of a sol–gel process in the preparation of hybrid coatings for corrosion protection using FTIR and 1H NMR methods. *Journal of Non-Crystalline Solids*, *396–397*, 25–35. doi:10.1016/j.jnoncrysol.2014.04.013

Rodič, P., & Milošev, I. (2014). Corrosion Properties of UV Cured Hybrid Sol-Gel Coatings on AA7075-T6 Determined under Simulated Aircraft Conditions. *Journal of the Electrochemical Society*, *161*(9), C412–C420. doi:10.1149/2.1091409jes

Rodič, P., & Milošev, I. (2015). Electrochemical and Salt Spray Testing of Hybrid Coatings Based on Si and Zr Deposited on Aluminum and Its Alloys. *Journal of the Electrochemical Society*, *162*(10), C592–C600. doi:10.1149/2.0801510jes

Rodič, P., & Milošev, I. (2016). Corrosion Inhibition of Pure Aluminium and Alloys AA2024-T3 and AA7075-T6 by Cerium(III) and Cerium(IV) Salts. *Journal of the Electrochemical Society*, *163*(3), C85–C93. doi:10.1149/2.0431603jes

Rodriguez-Diaz, R. A., Banos, L., Novelo, O., Flores, C., Colin, J., Arenas-Alatorre, J., & Juarez-Islas, J. A. (2009). Microstructural characterization of $Fe_{40}Al_{15}Cr$ (% at.) intermetallic alloy produced by rapid solidification. *Acta Microscopica*, *18*(2), 169–173.

Rodríguez, M. A., Gil, L., Camero, S., Fréty, N., Santana, Y., & Caro, J. (2014). Effects of the dispersion time on the microstructure and wear resistance of WC/Co-CNTs HVOF sprayed coatings. *Surface and Coatings Technology*, *258*, 38–48. doi:10.1016/j.surfcoat.2014.10.014

Romero Pareja, R., López Ibáñez, R., Martín, F., Ramos-Barrado, J. R., & Leinen, D. (2006). Corrosion behaviour of zirconia barrier coatings on galvanized steel. *Surface and Coatings Technology*, *200*(22–23), 6606–6610. doi:10.1016/j.surfcoat.2005.11.098

Ropp, R. C., & Libowitz, G. G. (1978). The nature of the alumina-rich phase in the System La_2O_3-$A1_2O_3$. *Journal of the American Ceramic Society*, *61*(11-12), 473–475. doi:10.1111/j.1151-2916.1978.tb16119.x

Rosenband, V., & Gany, A. (2004). Activation of combustion synthesis of aluminium nitride powder. *Journal of Materials Processing Technology*, *147*(2), 179–203. doi:10.1016/j.jmatprotec.2003.12.017

Rosenflanz, A., Frey, M., Endres, B., Anderson, T., Richards, E., & Schardt, C. (2004). Bulk glasses and ultra hard nanoceramic based on alumina and rare-earth oxides. *Nature*, *430*(7001), 761–764. doi:10.1038/nature02729 PMID:15306804

Rosero-Navarro, N. C., Paussa, L., Andreatta, F., Castro, Y., Durán, A., Aparicio, M., & Fedrizzi, L. (2010). Optimization of hybrid sol–gel coatings by combination of layers with complementary properties for corrosion protection of AA2024. *Progress in Organic Coatings*, *69*(2), 167–174. doi:10.1016/j.porgcoat.2010.04.013

Rosero-Navarro, N. C., Pellice, S. A., Durán, A., & Aparicio, M. (2008). Effects of Ce-containing sol–gel coatings reinforced with SiO_2 nanoparticles on the protection of AA2024. *Corrosion Science*, *50*(5), 1283–1291. doi:10.1016/j.corsci.2008.01.031

Rossi, S., Deflorian, F., Fontanari, L., Cambruzzi, A., & Bonora, P. L. (2005). Electrochemical measurements to evaluate the damage due to abrasion on organic protective system. *Progress in Organic Coatings*, *52*(4), 288–297. doi:10.1016/j.porgcoat.2004.09.005

Rossi, S., Deflorian, F., & Risatti, M. (2006). Modified Taber apparatus and new test geometry to evaluate the reduction of organic coatings corrosion protective properties induced by abrasive particles. *Surface and Coatings Technology*, *201*(3), 1173–1179. doi:10.1016/j.surfcoat.2006.01.036

Rossnagel, S. (1990). *Magnetron Plasma Deposition Processes* (S. C. Rossnagel, Ed.). Park Ridge, New Jersey: Noyes Publications.

Roukes, M. (2001). Plenty of room, indeed. *Scientific American*, *285*(3), 42. doi:10.1038/scientificamerican0901-48 PMID:11524969

Roussi, E., Tsetsekou, A., Tsiourvas, D., & Karantonis, A. (2011). Novel hybrid organo-silicate corrosion resistant coatings based on hyperbranched polymers. *Surface and Coatings Technology*, *205*(10), 3235–3244. doi:10.1016/j.surfcoat.2010.11.037

Rout, T., Bera, S., Udayabhanu, G., & Narayan, R. (2016). Methodologies of Application of Sol-Gel Based Solution onto Substrate: A Review. *Journal of Coating Science and Technology*, *3*(1), 9–22. doi:10.6000/2369-3355.2016.03.01.2

Roy, J., Chandra, S., Das, S., & Maitra, S. (2014). Oxidation behaviour of silicon carbide-a review. *Reviews on Advanced Materials Science*, *38*(1), 29–39.

Ruder, A., Buchkremer, H. P., Jansen, H., Mallener, W., & Stöver, D. (1992). Wet powder spraying—a process for the production of coatings. *Surface and Coatings Technology*, *53*(1), 71–74. doi:10.1016/0257-8972(92)90105-J

Ruijin, W., Tianjian, X. L. Z., & Xiaoou, H. (2004). The properties of high productive high velocity arc sprayed coatings and its applications. In *Proceedings of International Thermal Spray Conference*. Osaka, Japan.

Saber-Samandari, S., & Berndt, C.C. (2010). *Thermal spray coatings – A technology review,* International Heat Treatment and Surface Engineering, 4(1). doi:.10.1179/174951410X12572442577381

Sabiruddin, K. (2011). Properties of splats and coatings obtained from plasma sprayed alumina based powders [Ph.D Thesis]. Department of Mechanical Engineering, Indian Institute of Technology Kharagpur, India.

Sadowski, T., & Golewski, P. (2016). *Protective Thermal Barrier Coatings Loadings in Thermal Barrier Coatings of Jet Engine Turbine Blades* (pp. 5–11). Springer. doi:10.1007/978-981-10-0919-8_2

Sadri, E. (2012). Plasma spraying and tribological evaluation of Cr_2O_3-Ag nanostructured composite coating. Unpublished master's thesis, Isfahan University of Technology, Isfahan, Iran.

Sadri, E., & Ashrafizadeh, F. (2013). Structural characterization and mechanical properties of plasma sprayed nanostructured Cr_2O_3-Ag composite coatings. *Surface and Coatings Technology, 236*, 91–101. doi:10.1016/j.surfcoat.2013.09.033

Saha, G. C., Khan, T. I., & Zhang, G. A. (2011). Erosion–corrosion resistance of microcrystalline and near-nanocrystalline WC–17Co high velocity oxy-fuel thermal spray coatings. *Corrosion Science, 53*(6), 2106–2114. doi:10.1016/j.corsci.2011.02.028

Sahu, N., Parija, B., & Panigrahi, S. (2009). Fundamental understanding and modeling of spin coating process: A review. *Indian Journal of Physics, 83*(4), 493–502. doi:10.1007/s12648-009-0009-z

Sakka, S., & Yoko, T. (1992). Sol-gel-derived coating films and applications. In R. Reisfeld & C. K. JJørgensen (Eds.), Chemistry, Spectroscopy and Applications of Sol-Gel Glasses (Vol. 77, pp. 89–118). Springer Berlin Heidelberg. doi:10.1007/BFb0036966

Sakka, M. M., Antar, Z., Elleuch, K., & Feller, J. F. (2017). Tribological response of an epoxy matrix filled with graphite and/or carbon nanotubes. *Friction, 5*(2), 171–182. doi:10.1007/s40544-017-0144-z

Salak, A., Rundnayova, E., & Selecka, M. (1999). Internal nitriding of sintered iron in nitrogen-low ammonia atmospheres. *Powder Metallurgy, 42*(1), 45–50. doi:10.1179/pom.1999.42.1.45

Salhi, Z., Guessasma, S., & Fenineche, N. (2009). Yttria-Stabilized zirconia in-flight particle characteristics under vacuum plasma spray conditions. *Vacuum, 83*(11), 1382–1387. doi:10.1016/j.vacuum.2009.04.051

Samal, S. (2017). Thermal plasma technology: The prospective future in material processing. *Journal of Cleaner Production, 142*, 3131–3150. doi:10.1016/j.jclepro.2016.10.154

Sampath, S., Jiang, X. Y., Matejicek, J., Leger, A. C., & Vardelle, A. (1999). Substrate temperature effects on splat formation, microstructure development and properties of plasma sprayed coatings Part I: Case study for partially stabilized zirconia. *Materials Science and Engineering A, 272*(1), 181–188. doi:10.1016/S0921-5093(99)00459-1

Sanchez, C., Belleville, P., Popall, M., & Nicole, L. (2011). Hybrid materials themed issue. *Chemical Society Reviews, 40*(2), 453–1152.

Sanchez, C., Julián, B., Belleville, P., & Popall, M. (2005). Applications of hybrid organic–inorganic nanocomposites. *Journal of Materials Chemistry, 15*(35–36), 3559–3592. doi:10.1039/b509097k

Sanchez, C., & Ribot, F. (1994). Design of hybrid organic-inorganic materials synthesized via sol-gel chemistry. *New Journal of Chemistry, 18*(10), 1007–1047.

Sanchez, C., Rozes, L., Ribot, F., Laberty-Robert, C., Grosso, D., Sassoye, C., & Nicole, L. (2010). "Chimie douce": A land of opportunities for the designed construction of functional inorganic and hybrid organic-inorganic nanomaterials. *Comptes Rendus. Chimie, 13*(1–2), 3–39. doi:10.1016/j.crci.2009.06.001

Sanchez, C., Soler-Illia, G. J. D. A. A., Ribot, F., Lalot, T., Mayer, C. R., & Cabuil, V. (2001). Designed Hybrid Organic–Inorganic Nanocomposites from Functional Nanobuilding Blocks. *Chemistry of Materials, 13*(10), 3061–3083. doi:10.1021/cm011061e

Sanjines, R., Sandu, C. S., Lamni, R., & Levy, F. (2006). Thermal decomposition of Zr1 – xAlxN thin films deposited by magnetron sputtering. *Surface and Coatings Technology, 200*(22-23), 6308–6312. doi:10.1016/j.surfcoat.2005.11.113

Santana, I., Pepe, A., Jimenez-Pique, E., Pellice, S., Milošev, I., & Ceré, S. (2015). Corrosion protection of carbon steel by silica-based hybrid coatings containing cerium salts: Effect of silica nanoparticle content. *Surface and Coatings Technology*, *265*, 106–116. doi:10.1016/j.surfcoat.2015.01.050

Saraiva, M. (2012). *Sputter Deposition of MgO Thin Films : The Effect of Cation Substitution*. Universiteit Gent.

Saremi, M., & Habibi, M. H. (2011). An investigation on hot corrosion resistance of plasma sprayed YSZ-ceria TBC in Na2SO4-V2O5 at 1050°C. In *TMS 2011- 140th annual meeting and exhibition, supplemental proceedings*. (pp. 429-437). San Diego, CA: John Wiley & Sons Inc.

Saremi, M., Afrasiabi, A., & Kobayashi, A. (2007). Bond coat oxidation and hot corrosion behavior of plasma sprayed YSZ coating on Ni superalloy. *Transactions of JWRI (Joining and Welding Research Institute)*, *36*, 41-45.

Saremi, M., Afrasiabi, A., & Kobayashi, A. (2008). Microstructural analysis of YSZ and YSZ/Al_2O_3 plasma sprayed thermal barrier coatings after high temperature oxidation. *Surface and Coatings Technology*, *202*(14), 3233–3238. doi:10.1016/j.surfcoat.2007.11.029

Sarikaya, O. (2005). Effect of some parameters on microstructure and hardness of alumina coatings prepared by the air plasma spraying process. *Surface and Coatings Technology*, *190*(2-3), 388–393. doi:10.1016/j.surfcoat.2004.02.007

Saunders, S. R. J., Monteiro, M., & Rizzo, F. (2008). The oxidation behaviour of metals and alloys at high temperatures in atmospheres containing water vapour: A review. *Progress in Materials Science*, *53*(5), 775–837. doi:10.1016/j.pmatsci.2007.11.001

Savage, E. (2012). *Carbon-carbon composites*. Springer Science & Business Media.

Scharf, T. W., Kotula, P. G., & Prasad, S. V. (2010). Friction and wear mechanisms in MoS_2/Sb_2O_3/Au nanocomposite coatings. *Acta Materialia*, *58*(12), 4100–4109. doi:10.1016/j.actamat.2010.03.040

Schiefler Filho, M. F. O., Buschinelli, A. J. A. B., Gärtner, F., Kirsten, A., Voyer, J., & Kreye, H. (2004). Influence of process parameters on the quality of thermally sprayed X46Cr13 stainless steel coatings. *Journal of the Brazilian Society of Mechanical Sciences and Engineering*, *26*(1), 98–106. doi:10.1590/S1678-58782004000100016

Schmidt, H. (1985). New type of non-crystalline solids between inorganic and organic materials. *Journal of Non-Crystalline Solids*, *73*(1–3), 681–691. doi:10.1016/0022-3093(85)90388-6

Schottner, G. (2001). Hybrid Sol−Gel-Derived Polymers: Applications of Multifunctional Materials. *Chemistry of Materials*, *13*(10), 3422–3435. doi:10.1021/cm011060m

Schulz, U. (2000). Phase transformation in EB-PVD yttria partially stabilized zirconia thermal barrier coatings during annealing. *Journal of the American Ceramic Society*, *83*(4), 904–910. doi:10.1111/j.1151-2916.2000.tb01292.x

Schulz, U., & Braue, W. (2013). Degradation of La2Zr2O7 and other novel EB-PVD thermal barrier coatings by CMAS (CaO–MgO–Al2O3–SiO2) and volcanic ash deposits. *Surface and Coatings Technology*, *235*, 165–173. doi:10.1016/j.surfcoat.2013.07.029

Schütze, M. (1995). Mechanical properties of oxide scales. *Oxidation of Metals*, *44*(1-2), 29–61. doi:10.1007/BF01046722

Schwoebel, R. (1969). Step motion on crystal surfaces II. *Journal of Applied Physics*, *40*(2), 614–618. doi:10.1063/1.1657442

Schwoebel, R. S., & Shipsey, E. J. (1966). Step Motion on Crystal Surfaces. *Applied Physics (Berlin)*, *37*(10), 3682–3686. doi:10.1063/1.1707904

Segal, D. L. (1984). Sol-gel processing: Routes to oxide ceramics using colloidal dispersions of hydrous oxides and alkoxide intermediates. *Journal of Non-Crystalline Solids*, *63*(1–2), 183–191. doi:10.1016/0022-3093(84)90397-1

Selcuk, B., Ipek, R., & Karamış, M. B. (2003). A study on friction and wear behaviour of carburized, carbonitrided and borided AISI 1020 and 5115 steels. *Journal of Materials Processing Technology, 141*(2), 189–196. doi:10.1016/S0924-0136(02)01038-5

Selezneva, S. E., & Boulos, M. (2001). Supersonic induction plasma jet modeling. *Nuclear Instruments & Methods in Physics Research. Section B, Beam Interactions with Materials and Atoms, 180*(1-4), 306–311. doi:10.1016/S0168-583X(01)00436-0

Semenov, S., & Cetegen, B. (2001). Spectroscopic temperature measurements in direct current arc plasma jets used in thermal spray processing of materials. *Journal of Thermal Spray Technology, 10*(2), 326–336. doi:10.1361/105996301770349411

Seraffon, M., Simms, N. J., Sumner, J., & Nicholls, J. R. (2011). The development of new bond coat compositions for thermal barrier coating systems operating under industrial gas turbine conditions. *Surface and Coatings Technology, 206*(7), 1529–1537. doi:10.1016/j.surfcoat.2011.06.023

Seshan, K. (Ed.). (2002). *Handbook of thin-film deposition processes and techniques: principles, methods, equipment, and applications* (2nd ed.). Norwich, N.Y: Noyes Publications.

Seviostanov, I., & Kachanov, M. (2001). Plasma-sprayed ceramic coatings: Anisotropic elastic and conductive properties in relation to the microstructure; Cross-property correlations. *Materials Science Engineering* A, *297*, 235-243.

Shabnam, H. (2012). Studies on the role of heavy metal ions on the corrosion behaviour of iron base alloys.

Shafiro, B., & Kachanov, M. (2000). Anisotropic effective conductivity of materials with nonrandomly oriented inclusions of diverse ellipsoidal shapes. *Journal of Applied Physic,* 87, 8561-8569.

Shahien, M., Yamada, M., Yasui, T., & Fukumoto, M. (2011d). Controlling of Nitriding Process on Reactive Plasma Spraying of Al Particles. In *IOP Conference Series: Mater. Sci. Eng.* doi:10.1088/1757-899X/18/20/202006

Shahien, M., Yamada, M., Yasui, T., & Fukumoto, M. (2010a). Cubic Aluminum Nitride Coating through Atmospheric Reactive Plasma Nitriding. *Journal of Thermal Spray Technology, 19*(3), 635–641. doi:10.1007/s11666-010-9469-0

Shahien, M., Yamada, M., Yasui, T., & Fukumoto, M. (2010b). Fabrication of AlN coatings by reactive atmospheric plasma spray nitriding of Al powders. *Materials Transactions, 51*(5), 957–961. doi:10.2320/matertrans.T-M2010804

Shahien, M., Yamada, M., Yasui, T., & Fukumoto, M. (2010c). Influence of Plasma Gases on Fabrication of AlN Coatings through Atmospheric Plasma Nitriding Process. In S. Ikezawa (Ed.), *Industrial Application of Plasma Process* (Vol. 3, pp. 1–10). India: Transworld Research Network.

Shahien, M., Yamada, M., Yasui, T., & Fukumoto, M. (2011a). Influence of NH_4Cl Powder Addition for Fabrication of Aluminum Nitride Coating in Reactive Atmospheric Plasma Spray Process. *Journal of Thermal Spray Technology, 20*(1-2), 205–212. doi:10.1007/s11666-010-9584-y

Shahien, M., Yamada, M., Yasui, T., & Fukumoto, M. (2011b). Reactive Atmospheric Plasma Spraying of AlN coatings: Influence of Aluminum Feedstock Particle Size. *Journal of Thermal Spray Technology, 20*(3), 580–589. doi:10.1007/s11666-010-9582-0

Shahien, M., Yamada, M., Yasui, T., & Fukumoto, M. (2011c). Fabrication of AlN/Al_2O_3 coatings via atmospheric reactive plasma nitriding of A l_2O_3 powders. *Key Engineering Materials, 484*, 166–171. doi:10.4028/www.scientific.net/KEM.484.166

Shahien, M., Yamada, M., Yasui, T., & Fukumoto, M. (2011e). In situ fabrication of AlN coating by reactive plasma spraying of Al/AlN powder. *Coatings, 1-2*(2), 88–107. doi:10.3390/coatings1020088

Shahien, M., Yamada, M., Yasui, T., & Fukumoto, M. (2012). Reactive Plasma Nitriding of Al_2O_3 Powder in Thermal Spray. *International Journal of Modern Physics, 6*, 546–551.

Shahien, M., Yamada, M., Yasui, T., & Fukumoto, M. (2013a). Synthesis of Cubic Aluminum Nitride Coating from Al_2O_3 Powder in Reactive Plasma Spray process. *Materials Transactions, 54*(2), 207–214. doi:10.2320/matertrans.T-M2012831

Shahien, M., Yamada, M., Yasui, T., & Fukumoto, M. (2013b). N_2 and H_2 Plasma Gases' Effects in Reactive Plasma Spraying of Al_2O_3 Powder. *Surface and Coatings Technology, 216*, 308–317. doi:10.1016/j.surfcoat.2012.11.062

Shahien, M., Yamada, M., Yasui, T., & Fukumoto, M. (2013c). Reactive Plasma Spraying of Fine Al_2O_3/AlN Feedstock Powder. *Journal of Thermal Spray Technology, 22*(8), 1283–1293. doi:10.1007/s11666-013-9992-x

Shanaghi, A., Rouhaghdam, A. R. S., Ahangarani, S., Chu, P. K., & Farahani, T. S. (2012). Effects of duty cycle on microstructure and corrosion behavior of tic coatings prepared by dc pulsed plasma CVD. *Applied Surface Science, 258*(7), 3051–3057. doi:10.1016/j.apsusc.2011.11.036

Shang, Q. H. (2012). Preparation and characterization of antistatic coatings with modified BaTiO3 powders as conductive fillers. *Journal of Adhesion Science and Technology*, 2642–2652. doi:10.1080/01694243.2013.798926

Shan, M. S., Khan, N., & Ahmad, R. (2009). Characterization and effect of argon nitrogen plasma on nitridation of aluminum alloy. *CODEN JNSMAC, 49*, 1–17.

Shaw, L. L., Goberman, D., Ren, R., Gell, M., Jiang, S., Wang, Y., & Strutt, P. R. et al. (2000). The dependency of microstructure and properties of nanostructured coatings on plasma spray conditions. *Surface and Coatings Technology, 130*(1), 1–8. doi:10.1016/S0257-8972(00)00673-3

Shchukin, D. G., & Grigoriev, D. O. (2012). 10 - The use of nanoreservoirs in corrosion protection coatings. In V. S. Saji & R. Cook (Eds.), *Corrosion Protection and Control Using Nanomaterials* (pp. 264–282). Woodhead Publishing. doi:10.1533/9780857095800.2.264

Shchukin, D. G., Zheludkevich, M., Yasakau, K., Lamaka, S., Ferreira, M. G. S., & Möhwald, H. (2006). Layer-by-Layer Assembled Nanocontainers for Self-Healing Corrosion Protection. *Advanced Materials, 18*(13), 1672–1678. doi:10.1002/adma.200502053

Sheehan, J. E. (1989). Oxidation protection for carbon fiber composites. *Carbon, 27*(5), 709–715. doi:10.1016/0008-6223(89)90204-2

Sheehan, J. E. (1993). *High-Temperature Coatings on Carbon Fibers and Carbon-Carbon Composites. by JD Buckley and Dan D. Edie, Carbon-Carbon Materials and Composites* (pp. 224–253). Park Ridge, NJ: Noyes Publications.

Sheffer, M., Groysman, A., & Mandler, D. (2003). Electrodeposition of sol–gel films on Al for corrosion protection. *Corrosion Science, 45*(12), 2893–2904. doi:10.1016/S0010-938X(03)00106-9

Shen, M. L., Zhu, S. L., Chen, M. H., & Wang, F. H. (2011). The oxidation and oxygen permeation resistance of quartz particle-reinforced aluminosilicate glass coating on titanium alloy. *Journal of the American Ceramic Society, 94*(8), 2436–2441. doi:10.1111/j.1551-2916.2011.04587.x

Shen, M. L., Zhu, S. L., & Wang, F. H. (2015). Formation kinetics of multi-layered interfacial zone between γ-TiAl and glass–ceramic coatings via interfacial reactions at 1000 °C. *Corrosion Science, 91*, 341–351. doi:10.1016/j.corsci.2014.11.036

Shen, Y. G., Mai, Y. W., Zhang, Q. C., McKenzie, D. R., McFall, W. D., & McBride, W. E. (2000). Residual stress, microstructure, and structure of tungsten thin films deposited by magnetron sputtering. *Journal of Applied Physics, 87*(1), 177–187. doi:10.1063/1.371841

Sheridan, T. G., Goeckner, M. J., & Goree, J. (1990). Pressure Dependence of Ionization Efficiency in Sputtering Magnetrons. *Applied Physics Letters*, *57*(20), 2080–2082. doi:10.1063/1.103947

Shi, J., Darzens, S., & Karlsson, A. M. (2005). Aspects of the morphological evolution in thermal barrier coatings and the intrinsic thermal mismatch therein. *Materials Science and Engineering A*, *392*(1), 301–312. doi:10.1016/j.msea.2004.09.062

Shilova, O. (2013). Synthesis and structure features of composite silicate and hybrid TEOS-derived thin films doped by inorganic and organic additives. *Journal of Sol-Gel Science and Technology*, *68*(3), 387–410. doi:10.1007/s10971-013-3026-5

Shimada, S., & Sato, T. (2002). Preparation and high temperature oxidation of SiC compositionally graded graphite coated with HfO 2. *Carbon*, *40*(13), 2469–2475. doi:10.1016/S0008-6223(02)00159-8

Shipway, P. H., & Hogg, J. J. (2005). Dependence of microscale mechanisms of WC-Co hard metals. *Wear*, *259*, 44–51. doi:10.1016/j.wear.2004.12.021

Shirvanimoghaddam, K., Hamim, S. U., Akbari, M. K., Fakhrhoseini, S. M., Khayyam, H., Pakseresht, A. H., & Jia, S. et al. (2017). Carbon fiber reinforced metal matrix composites: Fabrication processes and properties. *Composites. Part A, Applied Science and Manufacturing*, *92*, 70–96. doi:10.1016/j.compositesa.2016.10.032

Shreir, L. L. (1963b). [Newnes-Butterworths, London.]. *Corrosion*, *2*.

Shrestha, S., Hodgkiess, T., & Neville, A. (2005). Erosion–corrosion behaviour of high-velocity oxy-fuel Ni–Cr–Mo–Si–B coatings under high-velocity seawater jet impingement. *Wear*, *259*(1-6), 208–218. doi:10.1016/j.wear.2005.01.038

Sidhu, T.S., Agrawal, R.D., Prakash, S. (2006) Hot corrosion of some superalloys and role of high-velocity oxy-fuel spray coatings—a review, *Surface & Coatings Technology, 198*, 441-446.

Sidhu, T. S., Agrawal, R. D., & Prakash, S. (2005). Hot corrosion of some superalloys and role of high-velocity oxy-fuel spray coatings - a review. *Surface and Coatings Technology*, *198*(1-3), 441–446. doi:10.1016/j.surfcoat.2004.10.056

Sidhu, T. S., Prakash, S., & Agrawal, R. D. (2006). Evaluation of hot corrosion resistance of HVOF coatings on a Ni-based superalloy in molten salt environment. *Materials Science and Engineering A*, *430*(1-2), 64–78. doi:10.1016/j.msea.2006.05.099

Sidhu, T. S., Prakash, S., & Agrawa, R. D. (2005). Studies on the properties of high-velocity oxy–fuel thermal spray coatings for higher temperature applications. *Materials Science*, *41*(6), 805–823. doi:10.1007/s11003-006-0047-z

Siegel, R. W. (1996). Creating nanophase materials. *Scientific American*, *275*(6), 74–79. doi:10.1038/scientificamerican1296-74

Sigmund, P. (1969). Theory of Sputtering I. Sputtering Yield of Amorphous and Polycrystalline Targets. *Physical Review*, *184*(2), 383–416. doi:10.1103/PhysRev.184.383

Sigmund, P. (1981). Sputtering by ion bombardment: theoretical concepts. In R. Behrish (Ed.), *Sputtering by particle bombardment I*. Springer - Verlag. doi:10.1007/3540105212_7

Sigmund, P. J. (2008). Ion Beam Science: Solved and Unsolved Problems (Part I, II). *Medical Physics*, *35*(1), 405.

Silva, W. (2017). Carbonitriding of low alloy steels: Mechanical and metallurgical responses. *Materials Science and Engineering A*, *693*, 225–232. doi:10.1016/j.msea.2017.03.077

Silverman, D. C. (1982). Technical Note: Presence of Solid Fe (OH) 2 in EMF-pH Diagram for Iron. *Corrosion*, *38*(8), 453–455. doi:10.5006/1.3578222

Silverson. (2017). Welcome. Retrieved in May 2017 from http://www.silverson.com/images/uploads/documents/CPVA-Solutions.pdf

Singh, J. S. (2003). *Advances in Amorphous Semiconductors*. London, New York: Taylor & Francis. doi:10.1201/9781420023848

Singh, K. (2014). Advanced materials for land based gas turbines. *Transactions of the Indian Institute of Metals*, *67*(5), 601–615. doi:10.1007/s12666-014-0398-3

Singh, V. P., Sil, A., & Jayaganthan, R. (2011). Tribological behavior of plasma sprayed Cr_2O_3-3%TiO_2 coatings. *Wear*, *272*(1), 149–158. doi:10.1016/j.wear.2011.08.004

Skapin, S., Kolar, D., & Suvorov, D. (1993). X-ray diffraction and microstructural investigation of the Al_2O_3-La_2O_3-TiO_2 system. *Journal of the American Ceramic Society*, *76*(9), 2359–2362. doi:10.1111/j.1151-2916.1993.tb07777.x

Skorb, E. V., Fix, D., Andreeva, D. V., Möhwald, H., & Shchukin, D. G. (2009). Surface-Modified Mesoporous SiO_2 Containers for Corrosion Protection. *Advanced Functional Materials*, *19*(15), 2373–2379. doi:10.1002/adfm.200801804

Sliney, H. E. (2001). Solid lubricants. In P.J. Blau (Ed.), ASM Handbook (Vol. 18, pp. 193-216) ASM International.

Sliney, H. E. (1979). Wide temperature spectrum self-lubricating coatings prepared by plasma spraying. *Thin Solid Films*, *64*(2), 211–217. doi:10.1016/0040-6090(79)90511-X

Sloof, W. G. (2007). Self Healing in Coatings at High Temperatures. In S. van der Zwaag (Ed.), *Self Healing Materials: An Alternative Approach to 20 Centuries of Materials Science* (pp. 309–321). Dordrecht: Springer Netherlands. doi:10.1007/978-1-4020-6250-6_14

Smialek, J. L., & Auping, J. V. (2002). COSP for windows—Strategies for rapid analyses of cyclic-oxidation behavior. *Oxidation of Metals*, *57*(5-6), 559–581. doi:10.1023/A:1015308606869

Smith, R. W. (1993). Reactive plasma spray forming for advanced materials synthesis. *Powder Metallurgy International*, *25*(1), 9–16.

Smith, R. W., Lugcheider, E., Jokiel, P., Mueller, U., Merz, J., & Wilbert, M. (1993). Synthesis of composite materials by reactive plasma spray processing. In C. C. Berndt & T. F. Bernecki (Eds.), *Thermal Spray: Research, Design and Applications* (pp. 439–444). Materials Park, OH, USA: ASM International.

Snell, L., Nelson, A., & Molian, P. (2001). A novel laser technique for oxidation-resistant coating of carbon–carbon composite. *Carbon*, *39*(7), 991–999. doi:10.1016/S0008-6223(00)00200-1

Sodeoka, S., Suzuki, M., Ueno, K., Sakuramoto, H., Shibata, T., & Ando, M. (1997). Thermal and mechanical properties of ZrO_2-CeO_2 plasma-sprayed coatings. *Journal of Thermal Spray Technology*, *6*(3), 361–367. doi:10.1007/s11666-997-0071-z

Soleimanipour, Z., Baghshahi, S., & Shoja-razavi, R. (2017). Improving the Thermal Shock Resistance of Thermal Barrier Coatings Through Formation of an In Situ YSZ/Al2O3 Composite via Laser Cladding. *Journal of Materials Engineering and Performance*, *4*(26), 1890–1899. doi:10.1007/s11665-017-2591-0

Song, E. P., Ahn, J., Lee, S., & Kim, N. J. (2006). Microstructure and wear resistance of nanostructured Al_2O_3-8wt.%TiO_2 coatings plasma-sprayed with nanopowders. *Surface and Coatings Technology*, *201*(3-4), 1309–1315. doi:10.1016/j.surfcoat.2006.01.052

Son, S. J., Yi, K. W., Mahrholz, J., & Rie, K. T. (2007). Numerical studies on up scaling of metal organic PACVD processes used for tribological coating in automotive industry. *Surface and Coatings Technology*, *201*(16-17), 7318–7326. doi:10.1016/j.surfcoat.2007.01.049

Soo, M. T., Kawamura, G., Muto, H., Matsuda, A., Lockman, Z., & Cheong, K. Y. (2013). Design of hierarchically meso–macroporous tetragonal ZrO_2 thin films with tunable thickness by spin-coating via sol–gel template route. *Microporous and Mesoporous Materials*, *167*, 198–206. doi:10.1016/j.micromeso.2012.09.010

Sorochkina, K., Smotraiev, R., & Chepurna, I. (2015). Zirconium and aluminum oxyhydroxides particles formation during sol–gel process. *Colloids and Surfaces. A, Physicochemical and Engineering Aspects*, *484*, 56–61. doi:10.1016/j.colsurfa.2015.07.038

Sourmail, T. (2004). *Coatings for high temperature applications*. University of Cambridge, UK. Retrieved from http://Thomas-sourmail.net/coatings/index.html

Souza, V. A. D., & Neville, A. (2007). Aspects of microstructure on the synergy and overall material loss of thermal spray coatings in erosion–corrosion environments. *Wear*, *263*(1-6), 339–346. doi:10.1016/j.wear.2007.01.071

Sreedhar, G., Alam, M. D. M., & Raja, V. S. (2009). Hot corrosion behavior of plasma sprayed YSZ/Al_2O_3 dispersed NiCrAlY coatings on Inconel-718 superalloy. *Surface and Coatings Technology*, *204*(3), 291–299. doi:10.1016/j.surfcoat.2009.07.026

Sreedhar, G., & Raja, V. S. (2010). Hot corrosion of YSZ/Al_2O_3 dispersed NiCrAlY plasma-sprayed coatings in Na_2SO_4-10 wt. % NaCl melt. *Corrosion Science*, *52*(8), 2592–2602. doi:10.1016/j.corsci.2010.04.007

Srinivasan, D., Kulkarni, T. G., & Anand, K. (2007). Thermal stability and high-temperature wear of Ti–TiN and TiN–CrN nanomultilayer coatings under self-mated conditions. *Tribology International*, *40*(2), 266–277. doi:10.1016/j.triboint.2005.09.035

Stack, M., & Bray, L. (1995). M and Bray L, *Interpretation of wastage mechanisms of materials exposed to elevated temperature erosion-corrosion using erosion-corrosion maps and computer graphics*. *Wear*, *186-187*, 273–283. doi:10.1016/0043-1648(95)07174-1

Stecura, S. (1985). Optimization of the NiCrAl-Y/ZrO_2-Y_2O_3 Thermal Barrier System. *NASA Report Technical Memorandum* 86905, Cleveland.

Stefan, J. (1997). *Standard Reference Material for Hard Coating, CrN PVD Coating (SRM-HC/2)*. Ljubljana, Slovenia: Institute Report.

Steffens, H.-D., Wilden, J., & Buchmann, C. (1995). Laser beam nitriding of thermally sprayed titanium coatings. In C.C. Berndt & S. Sampath (Ed.), Thermal Spray Science & Technology (pp. 981-986). Materials Park, OH: ASM International.

Stern, K. H. (1996). *Metallurgical and ceramic protective coatings*. Springer Science & Business Media. doi:10.1007/978-94-009-1501-5

Stockt, H. R., Berndt, H., & Mayr, P. (1991). Plasma-assisted chemical vapor deposition with titanium amides as precursors. *Surface and Coatings Technology*, *46*(1), 15–23. doi:10.1016/0257-8972(91)90146-N

Stoiber, M., Badisch, E., Lugmair, C., Mitterer, C., (2003). Low-friction TiN coatings deposited by PACVD, *Surface and Coatings Technology, 163 –164*, 451-456.

Stoller, R. T., Toloczko, M. B., Was, G. S., Certain, A. G., Dwaraknath, S., & Garner, F. A. (2013). On the use of SRIM for computing radiation damage exposure. *Nuclear Instruments & Methods in Physics Research. Section B, Beam Interactions with Materials and Atoms*, *310*, 75–80. doi:10.1016/j.nimb.2013.05.008

St-Onge, L., Sabsabi, M., & Cielo, P. (1998). Analysis of solids using laser-induced plasma spectroscopy in double-pulse mode. *Spectrochimica Acta. Part B, Atomic Spectroscopy*, *53*(53), 407–415. doi:10.1016/S0584-8547(98)00080-9

Stott, F. H., & Rocchini, G. (1991). *Electrochemical Aspect of high temperatura corrosion reaction in combustión systems, Colombo, UK corrosion* (Vol. 2). Manchester: Europan Federation of Corrosion.

Stöver, D., & Funke, C. (1999). Directions of the development of thermal barrier coatings in energy applications. *Journal of Materials Processing Technology*, *92*, 195–202. doi:10.1016/S0924-0136(99)00244-7

Stover, D., Pratch, G., Lehmann, H., Dietrich, M., Doring, J. E., & Vassen, R. (2004). New materials concepts for the next generation of plasma-sprayed thermal barrier coatings. *Journal of Thermal Spray Technology*, *13*(1), 76–83. doi:10.1007/s11666-004-0052-4

Strangman, T. E. (1986). Thermal barrier coatings for turbine airfoils. *Thin Solid Films*, *127*(1-2), 93–105. doi:10.1016/0040-6090(85)90215-9

Strife, J. R., & Sheehan, J. E. (1988). Ceramic coatings for carbon-carbon composites. *American Ceramic Society Bulletin*, *67*(2), 369–374.

StruttP. R.KearB. H.RolandR. F. (2000).

Subramanian, R. (2001). Thermal barrier coatings having high phase stability.

Subramanian, R. (2002). U.S. Patent 6,387,539.

Suh, N. P. (1973). The delamination theory of wear. *Wear*, *25*, 111–124. doi:10.1016/0043-1648(73)90125-7

Sumner, J., Encinas-Oropesa, A., Symms, N. J., & Oakey, E. J. (2013). High temperature oxidation and corrosion of gas turbine materials in burner rig exposures. *Materials Science and Technology*, *29*(7), 813–821. doi:10.1179/17432 84712Y.0000000097

Sundararajan, G., Joshi, S. V., & Krishna, L. R. (2016). Engineered surfaces for automotive engine and power train components. *Current Opinion in Chemical Engineering*, *11*, 1–6. doi:10.1016/j.coche.2015.10.001

Sundarrajan, G., & Roy, M. (1997). Solid particle erosion of metallic materials at room and elevated temperatures. *Tribology International*, *30*(5), 339–359. doi:10.1016/S0301-679X(96)00064-3

Sundgren, J.-E. (1985). Structure and properties of TiN coatings. *Thin Solid Films*, *128*(1-2), 21–44. doi:10.1016/0040-6090(85)90333-5

Sun, T., Wu, X., Wang, R., Li, W., & Liu, Q. (2017). First-principles study on the adhesive properties of Al/TiC interfaces: Revisited. *Computational Materials Science*, *126*, 108–120. doi:10.1016/j.commatsci.2016.09.024

Sun, Y., & Bell, T. (1997). A numerical model of plasma nitriding of low alloy steels. *Materials Science and Engineering A*, *224*(1-2), 33–47. doi:10.1016/S0921-5093(96)10561-X

Suo, Z., Kubair, D. V., Evans, A. G., Clarke, D. R., & Tolpygo, V. K. (2003). Stresses induced in alloys by selective oxidation. *Acta Materialia*, *51*(4), 959–974. doi:10.1016/S1359-6454(02)00499-8

Suresh, G., Seenivsan, G., Krishnaiah, M. V., & Murti, P. S. (1997). Investigation of the thermal conductivity of selected compounds of gadolinium and lanthanum. *Journal of Nuclear Materials*, *249*(2-3), 259–261. doi:10.1016/S0022-3115(97)00235-3

Suryanarayana, C. (2001). Mechanical Alloying and Milling. *Progress in Materials Science*, *46*(1-2), 1–184. doi:10.1016/S0079-6425(99)00010-9

Suzana, A. F., Ferreira, E. A., Benedetti, A. V., Carvalho, H. W. P., Santilli, C. V., & Pulcinelli, S. H. (2016). Corrosion protection of chromium-coated steel by hybrid sol-gel coatings. *Surface and Coatings Technology, 299*, 71–80. doi:10.1016/j.surfcoat.2016.04.075

Sweet, G. K. (1993). Applying thermoplastic/thermoset powder with a modified plasma system. *Paper presented at the Thermal Spray Coatings: Research, Design and Applications*, Anaheim, CA.

Szymański, K., Hernas, A., Moskal, G., & Myalska, H. (2015). Thermally sprayed coatings resistant to erosion and corrosion for power plant boilers - A review. *Surface and Coatings Technology, 268*, 153–164. doi:10.1016/j.surfcoat.2014.10.046

Tahara, H., Moriyama, M., & Fujiuchi, K. (2004). Ceramic spraying using electromagnetically accelerated plasma. In *Proceeding of the International Thermal Spray Conference* (CD), Osaka, Japan.

Taischner, C., Leonhardt, A., Schonherr, M., Wolf, E., & Henke, J. (1991). Structure and properties of TiCx layers prepared by plasma-assisted chemical vapor deposition methods. *Materials Science and Engineering A, 139*, 67–70. doi:10.1016/0921-5093(91)90598-H

Takada, S., Okumura, Y., & Sasaki, T. (1988). Preparation of iron-nitride by reduction sintering method. *Research reports of the Faculty of Engineering* [in Japanese]. *Gifu University, 38*, 43–48.

Takagi, K., Kudo, D., Kawasaki, A., & Harada, Y. (2011). Microstructural dependency of thermal expansion and sintering shrinkage in plasma-sprayed zirconia coatings. *Surface and Coatings Technology, 205*(19), 4411–4417. doi:10.1016/j.surfcoat.2011.03.053

Takahashi, M., & Shimada, S. (2004). Preparation of composite and compositionally graded TiC–TiN films by liquid injection plasma-enhanced CVD. *Solid State Ionics, 172*(1-4), 249–252. doi:10.1016/j.ssi.2004.03.015

Talib, R. J., Saad, S., Toff, M. R. M., & Hashim, H. (2003). Thermal spray coating technology – A review. *Solid State Science and Technology, 11*(1), 109–117.

Tamaki, M., Tomii, Y., Yamamoto, N., Kuwabara, H., & Nishikawa, I. (1997). Behavior and role of hydrogen during plasma nitriding [in Japanese]. *Journal of the Japan Society of Powder and Powder Metallurgy, 44*(7), 712–715. doi:10.2497/jjspm.44.712

Tamilselvi, S., Raman, V., & Rajendran, N. (2006). Corrosion behaviour of Ti–6Al–7Nb and Ti–6Al–4V ELI alloys in the simulated body fluid solution by electrochemical impedance spectroscopy. *Electrochimica Acta, 52*(3), 839–846. doi:10.1016/j.electacta.2006.06.018

Tamura, M., Takahashi, M., Ishii, J., Suzuki, K., Sato, M., & Shimomura, K. (1999). Multilayered thermal barrier coating for land-based gas turbines. *Journal of Thermal Spray Technology, 8*(1), 68–72. doi:10.1361/105996399770350584

Tanaka, K., & Hayashi, K. (2000). Preparation of Fe_4N sintered compact by heating green and sintered compacts of Fe powder in flowing NH_3+H_2 mixed gas. *Journal of the Japan Society of Powder and Powder Metallurgy, 47*(6), 600–606. doi:10.2497/jjspm.47.600

Tanaka, S., Pin, C. C., & Uematsu, K. (2006). Effect of Organic Binder Segregation on Sintered Strength of Dry-Pressed Alumina. *Journal of the American Ceramic Society, 89*(6), 1903–1907. doi:10.1111/j.1551-2916.2006.01057.x

Tanaka, Y., & Fukumoto, M. (1999). Investigation of dominating factors on flattening behavior of plasma sprayed ceramic particles. *Surface and Coatings Technology, 120-121*, 124–130. doi:10.1016/S0257-8972(99)00348-5

Tandon, R., Dumm, H. P., Corral, E. L., Loehman, R. E., & Kotula, P. G. (2006). *Ultra high temperature ceramics for hypersonic vehicle applications.*

Tan, K. S., Wharton, J. A., & Wood, R. J. K. (2005). Solid particle erosion–corrosion behaviour of a novel HVOF nickel aluminium bronze coating for marine applications—correlation between mass loss and electrochemical measurements. *Wear*, *258*(1-4), 629–640. doi:10.1016/j.wear.2004.02.019

Tao, S., Liang, B., Ding, C., Lao, H., & Coddet, C. (2005). Wear Characteristics of Plasma-Sprayed Nanostructured Yttria Stabilized Zirconia Coatings. *Journal of Thermal Spray Technology*, *14*(4), 518–523. doi:10.1361/105996305X76540

Taschner, C., Bartsch, K., & Leonhardt, A. (1993). Plasma-assisted deposition of hard material layers from organometallic precursors. *Surface and Coatings Technology*, *59*(1-3), 207–211. doi:10.1016/0257-8972(93)90084-2

Taschner, C., Ljungberg, B., Hoffmann, V., Vogt, C., & Leonhardt, A. (2001). Deposition of TiN, TiC and Ti Al N coatings by pulsed d.c. plasma enhanced chemical vapor deposition methods. *Surface and Coatings Technology*, *142-144*, 823–828. doi:10.1016/S0257-8972(01)01117-3

Taylor, M. L., Murphy, J. G., & King, H. W. (1987). Brittle/ductile erosion of plasma sprayed ceramic and metallic coatings. In D. L. Houck (Ed.), *Thermal Spray: Advances in Coatings Technology* (pp. 143–149). Materials Park, OH: ASM International.

Tekmen, C., Tsunekawa, Y., & Okumiya, M. (2008). Effect of plasma spray parameters on in-flight particle characteristics and in-situ alumina formation. *Surface and Coatings Technology*, *203*(3-4), 223–228. doi:10.1016/j.surfcoat.2008.08.057

Tekmen, C., Tsunekawa, Y., & Okumiya, M. (2009). In-situ TiB_2-Al_2O_3 formed composite coatings by atmospheric plasma spraying: Influence of process parameters and in-flight particle characteristics. *Surface and Coatings Technology*, *203*(12), 1649–1655. doi:10.1016/j.surfcoat.2008.12.016

TEKNA. (2009). Plasma state technology. Retrieved from http://www.tekna.com/technology/plasma-state/

Tezuka, N., Low, I. M., Davies, I. J., Alecu, I. D., Stead, R. J., Mehrtens, E. G., & Latella, B. A. (2009). Effect of fluoride and oxide additives on the phase transformations in alumina/clay ceramics. *J. Aust. Ceram. Soc*, *45*(1), 19–27.

Thiele, S., Heimann, R. B., Berger, L. M., Herrmann, M., Nebelung, M., Schnick, T., & Vuoristo, P. et al. (2002). Microstructure and porperties of thermally sprayed silicon nitride-based coatings. *Journal of Thermal Spray Technology*, *11*(2), 218–225. doi:10.1361/105996302770348871

Thiel, S., Heimann, R. B., Herrmann, M., Berger, L. M., Nebelung, M., Zschunke, M., & Wielage, B. (1996). Thermal spraying of silicon nitride – based powders. In C. C. Berndt (Ed.), *Thermal spray: practical solutions for engineering problems* (pp. 325–331). Materials Park, OH: ASM International.

Thompson, M. (2010). The energy spectrum of ejected atoms during the high energy sputtering of gold. *J. Philos. Mag.*, 377-414. doi:10.1080/14786436808227358

Thompson, C. (2000). Structure evolution during processing of polycrystalline films. *Annual Review of Materials Science*, *30*(1), 159–190. doi:10.1146/annurev.matsci.30.1.159

Thompson, J., & Clyne, T. (2001). The effect of heat treatment on the stiffness of zirconia top coats in plasma-sprayed TBCs. *Acta Materialia*, *49*(9), 1565–1575. doi:10.1016/S1359-6454(01)00065-9

Thompson, M. (1981). Physical mechanisms of sputtering. *Physics Reports*, *69*(4), 335–371. doi:10.1016/0370-1573(81)90106-X

Thomsen, N. B., Horsewell, A., Mogensen, K. S., Eskildsen, S. S., Mathiasen, C., & Buttiger, J. (1998). Residual stress determination in PECVD TiN coatings by X-ray diffraction: A parametric study. *Thin Solid Films*, *333*(1-2), 50–59. doi:10.1016/S0040-6090(98)00804-9

Thornton, J. (1974). Influence of apparatus geometry and deposition conditions on structure and topography of thick sputtered coatings. *Journal of Vacuum Science and Technology*, *11*(4), 666–670. doi:10.1116/1.1312732

Thornton, J. G. (1994). Plasmas in Deposition Processes. In R. Bunshah (Ed.), *Deposition Technologies for Films and Coatings*. Park Ridge, New Jersey: Noyes Publications.

Thorpe, M. L., & Richter, H. J. (1992). A pragmatic analysis and comparison of HVOF processes. *Journal of Thermal Spray Technology*, *1*(2), 161–170. doi:10.1007/BF02659017

Tian, W., Wang, Y., & Yang, Y. (2008). Fretting wear behaviour of conventional and nanostructured Al_2O_3-13wt%TiO_2 coatings fabricated by plasma spray. *Wear*, *256*(11-12), 1700–1707. doi:10.1016/j.wear.2008.04.012

Tian, W., Wang, Y., Yang, Y., & Li, C. (2009). Toughening and strengthening mechanism of plasma sprayed nanostructured Al_2O_3–13 wt. %TiO_2 coatings. *Surface and Coatings Technology*, *204*(5), 642–649. doi:10.1016/j.surfcoat.2009.08.045

Tian, W., Wang, Y., Zhang, T., & Yang, Y. (2009). Microstructure, spallation and corrossion of plasma sprayed Al_2O_3-13% TiO_2 coatings. *Corrosion Science*, *51*(12), 2924–2931. doi:10.1016/j.corsci.2009.08.005

Tilly, D. J., Lofvander, J. P. A., & Levi, C. G. (1996). Solidification paths and carbide morphologies in melt processed $MoSi_2$-SiC in situ composites. *Metallurgical and Materials Transactions*, *28A*, 1889–1900.

Tiwari, A., & Hihara, L. H. (2015). Sol-Gel Route for the Development of Smart Green Conversion Coatings for Corrosion Protection of Metal Alloys. In A. Tiwari, J. Rawlins, & L. H. Hihara (Eds.), *Intelligent Coatings for Corrosion Control* (pp. 363–407). Boston: Butterworth-Heinemann. doi:10.1016/B978-0-12-411467-8.00010-6

TMR-Transparency Market Research. (2017). Global thermal barrier coatings market. Retrieved from http://www.transparencymarketresearch.com

Tolpygo, V. K., & Clarke, D. R. (2000). Surface rumpling of a (Ni, Pt)Al bond coat induced by cyclic oxidation. *Acta Materialia*, *48*(13), 3283–3293. doi:10.1016/S1359-6454(00)00156-7

Tolpygo, V. K., & Clarke, D. R. (2004a). On the rumpling mechanism in nickel-aluminide coatings Part I: An experimental assessment. *Acta Materialia*, *52*(17), 5115–5127.

Tolpygo, V. K., & Clarke, D. R. (2004b). Rumpling induced by thermal cycling of an overlay coating: The effect of coating thickness. *Acta Materialia*, *52*(3), 615–621. doi:10.1016/j.actamat.2003.10.001

Toma, D., Brandl, W., & Marginean, G. (2001). Wear and corrosion behaviour of thermally sprayed cermet coatings. *Surface and Coatings Technology*, *138*(2-3), 149–158. doi:10.1016/S0257-8972(00)01141-5

Tomiczek, B., & Dobrzański, L.A. (2013). Określanie struktury i wielkości krystalitów nanokompozytów wzmacnianych haloizytem, *XLI Szkoła Inżynierii Materiałowej Kraków – Krynica*, 24-27 IX 2013, 314-318.

Tortorelli, P. F., Specht, E. D., More, K. L., & Hou, P. Y. (2012). Oxide growth stress measurements and relaxation mechanisms for alumina scales grown on FeCrAlY. *Materials and Corrosion*, *63*(10), 857–861. doi:10.1002/maco.201206760

Toth, L. E. (1971). *Transition metal carbides and nitrides*. New York, London: Academic Press.

Trabelsi, W., Cecilio, P., Ferreira, M. G. S., & Montemor, M. F. (2005). Electrochemical assessment of the self-healing properties of Ce-doped silane solutions for the pre-treatment of galvanised steel substrates. *Progress in Organic Coatings*, *54*(4), 276–284. doi:10.1016/j.porgcoat.2005.07.006

Tran, A. T. T., & Hyland, M. M. (2009). The role of substrate surface chemistry on splat formation during plasma spray deposition by experiments and simulations. *Paper presented at the International Thermal Spray Conference*, Las Vegas, NV. Retrieved from http://www.scopus.com/inward/record.url?eid=2-s2.0-76749126654&partnerID=40&md5=830a0ccf95426d84effca4bd27fbb1f0

Tran, A. T. T. (2010). *Effects of surface chemistry on splat formation during plasma spraying. (Ph.D. Ph.D.)*. Auckland: The University of Auckland.

Tran, A. T. T., & Hyland, M. M. (2010). The role of substrate surface chemistry on splat formation during plasma spray deposition by experiments and simulations. *Journal of Thermal Spray Technology*, *19*(1-2), 11–23. doi:10.1007/s11666-009-9414-2

Tran, A. T. T., Hyland, M. M., Qiu, T., Withy, B., & James, B. J. (2008). Effects of surface chemistry on splat formation during plasma spraying. *Journal of Thermal Spray Technology*, *17*(5-6), 637–645. doi:10.1007/s11666-008-9237-6

Tran, A. T. T., Hyland, M. M., Shinoda, K., & Sampath, S. (2011). Influence of substrate surface conditions on the deposition and spreading of molten droplets. *Thin Solid Films*, *519*(8), 2445–2456. doi:10.1016/j.tsf.2010.11.047

Treacy, M. M. J., Ebbesen, T. W., & Gibson, J. M. (1996). Exceptionally high Young's modulus observed for individual carbon nanotubes. *Nature*, *381*(6584), 678–680. doi:10.1038/381678a0

Trethewey, K. R., & Chamberlain, J. (1995). *Corrosion for Science and Engineering* (2nd ed.). England: Longman Group Ltd.

Trice, R. W., Su, Y. J., Mawdsley, J. R., Faber, K. T., Arellano-Lopez, A. R., Wang, H., & Porter, W. D. (2002). Effect of heat treatment on phase stability, microstructure, and thermal conductivity of plasma-sprayed YSZ. *Journal of Materials Science*, *37*(11), 2359–2365. doi:10.1023/A:1015310509520

Tringides, M. (1997). *Surface Diffusion - Atomistic and Collective Processes*. New York: Plenum Press.

Tripathi, H. S., & Banerjee, G. (1998). Synthesis and mechanical properties of mullite from beach sand sillimanite: Effect of TiO 2. *Journal of the European Ceramic Society*, *18*(14), 2081–2087. doi:10.1016/S0955-2219(98)00149-6

Tripathi, H. S., Das, S. K., & Banerjee, G. (2000). Thermal shock behaviour of high alumina aggregates derived from sillimanite beach sand with and without Fe2O3 doping. *Ceramics International*, *26*(1), 1–6. doi:10.1016/S0272-8842(99)00010-3

Tschiptschin, A. P., Varela, L. B., Pinedo, C. E., Li, X. Y., & Dong, H. (2017). Development and microstructure characterization of single and duplex nitriding of UNS S31803 duplex stainless steel. *Surface and Coatings Technology*, *327*, 83–92. doi:10.1016/j.surfcoat.2017.08.018

TSET. (2008). Plasma spray process. Retrieved from http://www.tsetinc.com/ceramicoat.htm

Tsotsos, C., Yerokhin, A. L., Wilson, A. D., Leyland, A., & Matthews, A. (2002). Tribological evaluation of AISI 304 stainless steel duplex treated by plasma electrolytic nitrocarburising and diamond-like carbon coating. *Wear*, *253*(9-10), 986–993. doi:10.1016/S0043-1648(02)00225-9

Tsui, Y. C., & Clyne, T. W. (1997). An analytical model for predicting residual stresses in progressively deposited coatings Part 1: Planar geometry. *Thin Solid Films*, *306*(1), 23–33. doi:10.1016/S0040-6090(97)00199-5

Tsujita, Y., Shibayama, K., Takizawa, A., Kinoshita, T., & Uematsu, I. (1987). Thermal properties of ethylene ionomers. *Journal of Applied Polymer Science*, *33*(4), 1307–1314. doi:10.1002/app.1987.070330421

Tsunashima, K., Nishioji, H., Hirasawa, E., & Yano, S. (1992). EXAFS studies on Zn(II) complexes with 1,3-bis(aminomethyl) cyclohexane of ethylene-methacrylic acid copolymer. *Polymer*, *33*(9), 1809–1812. doi:10.1016/0032-3861(92)90475-C

Tsunekawa, Y., Hiromura, M., & Okumiya, M. (1998). Nitriding of Metal Droplets in Synthesis of Intermetallic Matrix Composite Coatings by Reactive Plasma Spraying. In C. Coddet (Ed.), *Thermal Spray: Meeting the Challenges of the 21st Century* (pp. 1143–1148). PA: ASM International, Materials Park, OH, USA.

Tsunekawa, Y., Hiromura, M., & Okumiya, M. (2000). Nitride formation in synthesis of titanium aluminide matrix composite coatings by reactive RF plasma spraying. *Journal of Thermal Spray Technology*, *9*(1), 83–89. doi:10.1361/105996300770350096

Tsunekawa, Y., Okumiya, M., Kobayashi, T., Okuda, M., & Fukumoto, M. (1996). Chromium-nitride in situ composites with a compositional gradient formed by reactive DC plasma spraying. *Journal of Thermal Spray Technology*, *5*(2), 139–144. doi:10.1007/BF02646427

Tsunekawa, Y., Okumiya, M., Okuda, M., & Fukumoto, M. (1997). Reaction of metal droplets during flight and on substrate in reactive low pressure DC plasma spraying. In *Proceeding of the Eleventh International Conference on Surface Modification Technologies,* Paris, France (pp. 234-245).

Tuler, F. R., & Butcher, B. M. (1968). A criterion for the time dependence of dynamic fracture. *International Journal of Fracture*, *4*(4), 431–437.

Turri, S., Torlaj, L., Piccinini, F., & Levi, M. (2010). Abrasion and nanoscratch in nanostructured epoxy coatings. *Journal of Applied Polymer Science*, *118*(3), 1720–1727.

Twardowski, T., Reilly, M., & Knight, R. (2001). Properties of HVOF sprayed multi-scale polymer/silica nanocomposite coatings. *Paper presented at the Thermal Spray 2001: New Surfaces for a New Millennium,* Singapore.

Umm-i-Kalsoom, , Ahmad, R., Ali, N., Khan, I. A., Saleem, S., Ikhlaq, U., & Khan, N. (2013). Effect of Power and Nitrogen Content on the Deposition of CrN Films by Using Pulsed DC Magnetron Sputtering plasma. *Plasma Science & Technology*, *15*(7), 666–672. doi:10.1088/1009-0630/15/7/12

University of Toronto. (2006). DC plasma modeling. Retrieved from http://www.mie.utoronto.ca/labs/cact/Research/DCPlasma/DCPlasma.html

Uusitalo, M. A., Vuoristo, P. M. J., & Mäntylä, T. A. (2002). Elevated temperature erosion–corrosion of thermal sprayed coatings in chlorine containing environments. *Wear*, *252*(7-8), 586–594. doi:10.1016/S0043-1648(02)00014-5

Vaben, R., Traeger, F., & Stover, D. (2004). New thermal barrier coatings based on pyrochlore/YSZ double layer systems. *International Journal of Applied Ceramic Technology*, *1*, 351–356.

Valefi, Z., & Saremi, M. (2017). The effects of plasma spray parameters on the microstructure and phase composition of thermal barrier coatings made by spps process. *Iranian Journal of Materials Science and Engineering*, *14*(2), 11–23.

Valero-Vidal, C., & Igual-Muñoz, A. (2011). Electrochemical Aspects in Biomedical Alloy Characterization: Electrochemical Impedance Spectroscopy. In A. N. Laskovski (Ed.), *Biomedical Engineering, Trends in Materials Science.* InTech. doi:10.5772/13039

Van de pol, F. B. (1991). R.F Planar Magnetron Sputtering ZnO films I: Structural Properties. *Thin Solid Films, 204*, 349-364.

Van der Zwaag, S. (2010). Routes and mechanisms towards self healing behaviour in engineering materials. *Bulletin of the Polish Academy of Sciences. Technical Sciences*, *58*(2), 227–236. doi:10.2478/v10175-010-0022-6

Varacalle, D. J. Jr, Guillen, D. P., Deason, D. M., Rhodaberger, W., & Sampson, E. (2006). Effect of grit-blasting on substrate roughness and coating adhesion. *Journal of Thermal Spray Technology, 15*(3), 348–355. doi:10.1361/105996306X124347

Vardelle, A., Moreau, C., & Fauchais, P. (2000). Dynamics of deposit formation in thermal-spray processes. *MRS Bulletin, 25*(7), 32–37. doi:10.1557/mrs2000.121

Vardelle, M., Vardelle, A., & Fauchais, P. (1993). Spray parameter and particle behavior relationships during plasma spraying. *Journal of Thermal Spray Technology, 2*(1), 79–91. doi:10.1007/BF02647426

Vardelle, M., Vardelle, A., Fauchais, P., Li, K.-I., Dussoubs, B., & Themelis, N. J. (2001). Controlling particle injection in plasma spraying. *Journal of Thermal Spray Technology, 10*(2), 267–284. doi:10.1361/105996301770349367

Varma, P. C. R., Colreavy, J., Cassidy, J., Oubaha, M., McDonagh, C., & Duffy, B. (2010). Corrosion protection of AA 2024-T3 aluminium alloys using 3, 4-diaminobenzoic acid chelated zirconium–silane hybrid sol–gels. *Thin Solid Films, 518*(20), 5753–5761. doi:10.1016/j.tsf.2010.05.088

Varma, P. C. R., Periyat, P., Oubaha, M., McDonagh, C., & Duffy, B. (2011). Application of niobium enriched ormosils as thermally stable coatings for aerospace aluminium alloys. *Surface and Coatings Technology, 205*(16), 3992–3998. doi:10.1016/j.surfcoat.2011.02.023

Vasina, P. (2005). *Plasma Diagnostics Focused on New Magnetron Sputtering Devices for Thin Film Deposition.* Brno, Czech Republic: Masaryk University in Brno, Department of Physical Electronics.

Vassen, R., Cao, X., Tietz, F., Basu, D., & Stover, D. (2000). Zirconates as new materials for thermal barrier coatings. *Journal of American Ceramic Society*, 83, 2023-2028.

Vassen, R., Stuke, A., & Stöver, D. (2009). Recent developments in the field of thermal barrier coatings. *Journal of thermal spray technology, 18*(2), 181-186.

Vassen, R., Cao, X., Tietz, F., Basu, D., & Stover, D. (2000). Zirconates as new materials for thermal barrier coatings. *Journal of the American Ceramic Society, 83*(8), 2023–2028. doi:10.1111/j.1151-2916.2000.tb01506.x

Vassen, R., Jarligo, M. O., Steinke, T., Mack, D. E., & Stoever, D. (2010). Overview of advanced thermal barrier coatings. *Surface and Coatings Technology, 205*(4), 938–942. doi:10.1016/j.surfcoat.2010.08.151

Vassen, R., Tietz, F., Kerkhoff, G., & Stoever, D. (November 1998). New materials for advanced thermal barrier coatings. In J. Lecomte-Beckers, F. Schuber, & P. J. Ennis, (Ed.), *Proceedings of the 6 th Liége Conference on Materials for Advanced Power Engineering*, Universite de Liége, Belgium (pp. 1627-1635).

Vassen, R., Traeger, F., & Stoever, D. (2004). New thermal barrier coatings based on pyrochlore/YSZ double-layer systems. *International Journal of Applied Ceramic Technology, 1*(4), 351–361. doi:10.1111/j.1744-7402.2004.tb00186.x

Vast, P. (2009). Oxidation protection for carbon materials by oxifluoride phosphate compounds. *Paper presented at the 19th Annual Conference on Composites, Advanced Ceramics, Materials, and Structures-B: Ceramic Engineering and Science Proceedings.* doi:10.2172/887260

Vasudevan, A., & Petrovic, J. (1992). A comparative overview of molybdenum disilicide composites. *Materials Science and Engineering A, 155*(1-2), 1–17. doi:10.1016/0921-5093(92)90308-N

Veal, B. W., Paulikas, A. P., Gleeson, B., & Hou, P. Y. (2007). Creep in α-Al_2O_3 thermally grown on β-NiAl and NiAlPt alloys. *Surface and Coatings Technology, 202*(4-7), 608–612. doi:10.1016/j.surfcoat.2007.05.093

Venables, J. (2000). Surface and Thin Film Processes (1st ed.). Cambridge University Press.

Venkatachalam, S., Rajendra Kumar, R. T., Mangalaraj, D., Narayandass, S. K., Kim, K., & Junsin, Y. (2004). Optoelectronic properties of Zn0.52Se 0.48/Si Schottky diodes. *Solid-State Electronics*, *48*(12), 2219–2223. doi:10.1016/j.sse.2004.05.082

Verdian, M. M., Salehi, M., & Raeissi, K. (2010). Effect of feedstock particle size on microstructure of APS coatings prepared from mechanically alloyed nickel-titanium powders. *Surface Engineering*, *26*(6), 447–452. doi:10.1179/026708409X12490360425927

Verma, R., Kant, S., & Suri, N. M. (2016a). Adhesion strength optimization of slurry sprayed mullite-based coating using Taguchi method. *Proceedings of the Institution of Mechanical Engineers. Part E, Journal of Process Mechanical Engineering*, *230*(2), 87–96. doi:10.1177/0954408915595948

Verma, R., Suri, N. M., & Kant, S. (2016b). Effect of parameters on adhesion strength for slurry spray coating technique. *Materials and Manufacturing Processes*.

Verma, R., Suri, N. M., & Kant, S. (2016c). Parametric Appraisal of Slurry-Sprayed Mullite Coatings for Coating Thickness. *Journal of Thermal Spray Technology*, *25*(7), 1289–1301. doi:10.1007/s11666-016-0437-1

Viswanath, B., & Vijayarangan, S. (2012). Characterization Studies of Mullite Coatings on Cast Aluminum. *Journal of thermal spray technology, 21*(2), 325-334.

Viswanathan, V., Dwivedi, G., & Sampath, S. (2014). Engineered multilayer thermal barrier coatings for enhanced durability and functional performance. *Journal of the American Ceramic Society*, *97*(9), 2770–2778. doi:10.1111/jace.13033

Vivar Mora, L., Naik, S., Paul, S., Dawson, R., Neville, A., & Barker, R. (2017). Influence of silica nanoparticles on corrosion resistance of sol-gel based coatings on mild steel. *Surface and Coatings Technology*, *324*, 368–375. doi:10.1016/j.surfcoat.2017.05.063

Voevodin, A. A., & Zabinski, J. S. (2006). Laser surface texturing for adaptive solid lubrication. *Wear*, *261*(11-12), 1285–1292. doi:10.1016/j.wear.2006.03.013

Voevodin, N. N., Grebasch, N. T., Soto, W. S., Kasten, L. S., Grant, J. T., Arnold, F. E., & Donley, M. S. (2001). An organically modified zirconate film as a corrosion-resistant treatment for aluminum 2024-T3. *Progress in Organic Coatings*, *41*(4), 287–293. doi:10.1016/S0300-9440(01)00156-4

Voevodin, N. N., Kurdziel, J. W., & Mantz, R. (2006). Corrosion protection for aerospace aluminum alloys by Modified Self-assembled NAnophase Particle (MSNAP) sol–gel. *Surface and Coatings Technology*, *201*(3–4), 1080–1084. doi:10.1016/j.surfcoat.2006.01.028

Vorotilov, K., Petrovsky, V., & Vasiljev, V. (1995). Spin coating process of sol-gel silicate films deposition: Effect of spin speed and processing temperature. *Journal of Sol-Gel Science and Technology*, *5*(3), 173–183. doi:10.1007/BF00487014

Vreugdenhil, A. J., Balbyshev, V. N., & Donley, M. S. (2001). Nanostructured silicon sol-gel surface treatments for Al 2024-T3 protection. *Journal of Coatings Technology*, *73*(915), 35–43. doi:10.1007/BF02730029

Vuoristo, P. (2014). Thermal spray coating processes. In Comprehensive materials processing (1st ed., Vol. 4, pp. 229-276). doi:10.1016/B978-0-08-096532-1.00407-6

Waits, R. (1978). Planar Magnetron Sputtering. In J. K. Vossenand (Ed.), *Thin Film Processes*. New York: Academic Press. doi:10.1016/B978-0-12-728250-3.50009-X

Walcarius, A. (2001). Electrochemical Applications of Silica-Based Organic–Inorganic Hybrid Materials. *Chemistry of Materials*, *13*(10), 3351–3372. doi:10.1021/cm0110167

Wallwork, G. R., & Hed, A. Z. (1971). Some limiting factors in the use of alloys at high temperatures. *Oxidation of Metals, 3*(2), 171–184. doi:10.1007/BF00603485

Walston, W. S. (2004). Coating and surface technologies for turbine airfoils. *Superalloys*, 579-588.

Wang, C. (2006, September). Experimental and Computational phase studies of the ZrO_2-based systems for thermal barrier coatings [Dissertation an der Universität Stuttgart]. Universität Stuttgart, Stuttgart, Germany.

Wang, Z., Kulkarni, A., Deshpande, S., Nakamura, T., & Herman, H. (2003). Effects of pores and interfaces on effective properties of plasma sprayed zirconia coatings. *Acta Materiallia, 51*, 5319-5334.

Wang, B. (1996). Erosion-corrosion of thermal sprayed coatings in FBC boilers. *Wear, 199*(1), 24–32. doi:10.1016/0043-1648(96)06972-4

Wang, C. M., Guo, L., & Ye, F. X. (2016). $LaPO_4$ as a toughening agent for rare earth zirconate ceramics. *Materials & Design, 111*, 389–393. doi:10.1016/j.matdes.2016.09.014

Wang, C., Wang, J., Wang, L., Hao, G., Sun, X., Shan, F., & Zou, Z. (2014). Nanocomposite lanthanum zirconate thermal barrier coating deposited by suspension plasma spray process. *Journal of Thermal Spray Technology, 23*(7), 1030–1036. doi:10.1007/s11666-014-0068-3

Wang, C., Xue, T., Dong, B., Wang, Z., & Li, H.-L. (2008). Polystyrene–acrylonitrile–CNTs nanocomposites preparations and tribological behavior research. *Wear, 265*(11), 1923–1926. doi:10.1016/j.wear.2008.04.021

Wang, D., & Bierwagen, G. P. (2009). Sol–gel coatings on metals for corrosion protection. *Progress in Organic Coatings, 64*(4), 327–338. doi:10.1016/j.porgcoat.2008.08.010

Wang, D., Tian, Z., Shen, L., Liu, Z., & Huang, Y. (2009). Microstructural characteristics and formation mechanism of Al_2O_3-13wt.% TiO_2 coatings plasma-sprayed with nanostructured agglomerated powders, *Journal of Surface and coatings. Technology (Elmsford, N.Y.), 203*, 1298–1303.

Wang, F. H., Lou, H. Y., Zhu, S. L., & Wu, W. T. (1996). The mechanism of scale adhesion on sputtered microcrystalline CoCrAl films. *Oxidation of Metals, 45*(1-2), 39–50. doi:10.1007/BF01046819

Wang, H. L., He, J. L., & Hon, M. H. (1993). Sliding wear resistance of TiCN coatings on tool steel made by plasma-enhanced chemical vapor deposition. *Wear, 169*(2), 195–200. doi:10.1016/0043-1648(93)90298-Z

Wang, H., & Akid, R. (2008). Encapsulated cerium nitrate inhibitors to provide high-performance anti-corrosion sol–gel coatings on mild steel. *Corrosion Science, 50*(4), 1142–1148. doi:10.1016/j.corsci.2007.11.019

Wang, J. L., Chen, M. H., Yang, L. L., Zhu, S. L., & Wang, F. H. (2015). Comparative study of oxidation and interdiffusion behavior of AIP NiCrAlY and sputtered nanocrystalline coatings on a nickel-based single-crystal superalloy. *Corrosion Science, 98*, 530–540. doi:10.1016/j.corsci.2015.05.062

Wang, J. L., Chen, M. H., Zhu, S. L., & Wang, F. H. (2015). Ta effect on oxidation of a nickel-based single-crystal superalloy and its sputtered nanocrystalline coating at 900–1100oC. *Applied Surface Science, 345*, 194–203. doi:10.1016/j.apsusc.2015.03.157

Wang, J., Li, S. H. P., & Stevens, R. (1992). Effects of organic binders on the sintering of isostatically compacted zirconia powders. *Journal of Materials Science, 27*(1), 63–67. doi:10.1007/BF02403644

Wang, K.-T., Cao, L.-Y., Huang, J.-F., & Fei, J. (2013). A mullite/SiC oxidation protective coating for carbon/carbon composites. *Journal of the European Ceramic Society, 33*(1), 191–198. doi:10.1016/j.jeurceramsoc.2012.08.009

Wang, L., Li, D. C., Yang, J. S., Shao, F., Zhong, X. H., Zhao, H. Y., & Wang, Y. (2016). Modeling of thermal properties and failure of thermal barrier coatings with the use of finite element methods: A review. *Journal of the European Ceramic Society, 36*(6), 1313–1331. doi:10.1016/j.jeurceramsoc.2015.12.038

Wang, L., Wang, Y., & Sun, X. G. (2011). Influence of pores on the thermal insulation behavior of thermal barrier coatings prepared by APS. *Materials & Design, 32*, 36–47. doi:10.1016/j.matdes.2010.06.040

Wang, L., Wang, Y., Sun, X. G., He, J. Q., Pan, Z. Y., & Wang, C. H. (2012). Thermal shock behavior of 8YSZ and double-ceramic-layer $La_2Zr_2O_7$/8YSZ thermal barrier coatings fabricated by atmospheric plasma spraying. *Ceramics International, 38*(5), 3595–3606. doi:10.1016/j.ceramint.2011.12.076

Wang, M., & Shaw, L. L. (2007). Effect of the powder manufacturing method on microstructure and wear performance of plasma sprayed Alumina-Titania coatings. *Surface and Coatings Technology, 202*(1), 34–44. doi:10.1016/j.surfcoat.2007.04.057

Wang, P., Zhou, C., Zhang, X., Zhao, G., Xu, B., Cheng, Y., & Han, W. et al. (2016). Oxidation protective ZrB 2–SiC coatings with ferrocene addition on SiC coated graphite. *Ceramics International, 42*(2), 2654–2661. doi:10.1016/j.ceramint.2015.10.072

Wang, X., Chen, M. H., Zhu, S. L., & Wang, F. H. (2013). Interfacial reactions between a SiO_2-Al_2O_3-ZnO-CaO based glass and alpha alumina. *Surface and Coatings Technology, 232*, 6–12. doi:10.1016/j.surfcoat.2013.04.044

Wang, X., Chen, M. H., Zhu, S. L., & Wang, F. H. (2015). Oxidation behavior of glass-based composite thermal barrier coating on K417G superalloy with a NiCoCrAlY bond coat at 1000 °C. *Surface and Coatings Technology, 270*, 314–323. doi:10.1016/j.surfcoat.2015.02.014

Wang, X., Shen, M. L., Chen, M. H., Zhu, S. L., & Wang, F. H. (2015). Microstructure stabilization of a novel glass/YSZ composite coating material by adding alumina particles. *Ceramics International, 41*(8), 9753–9762. doi:10.1016/j.ceramint.2015.04.047

Wang, Y., Guo, H., & Gong, S. (2009). Thermal physical properties and thermal shock resistance of $La_2Ce_2O_7$ thermal barrier coatings with segmented structure. *Ceramics International, 35*, 2639–2644. doi:10.1016/j.ceramint.2009.02.025

Wang, Y., Jiang, S., Wang, M., Wang, S., Xiao, T. D., & Strutt, P. R. (2000). Abrasive wear characteristics of plasma sprayed nanostructured alumina/titania coatings. *Wear, 237*(2), 176–185. doi:10.1016/S0043-1648(99)00323-3

Wang, Y., Zheng, Y. G., Ke, W., Sun, W. H., Hou, W. L., Chang, X. C., & Wang, J. Q. (2011). Slurry erosion–corrosion behaviour of high-velocity oxy-fuel (HVOF) sprayed Fe-based amorphous metallic coatings for marine pump in sand-containing NaCl solutions. *Corrosion Science, 53*(10), 3177–3185. doi:10.1016/j.corsci.2011.05.062

Wang, Y., & Zhou, C. (2016). Microstructure and thermal properties of nanostructured gadolinia doped yttria-stabilized zirconia thermal barrier coatings produced by air plasma spraying. *Ceramics International, 42*(11), 13047–13052. doi:10.1016/j.ceramint.2016.05.084

Wang, Z., Kulkarni, A., Deshpande, S., Nakamura, T., & Herman, H. (2003). Effects of pores and interfaces on effective properties of plasma sprayed zirconia coatings. *Acta Materialia, 51*(18), 5319–5334. doi:10.1016/S1359-6454(03)00390-2

Wank, A. (n.d.). *Basics of thermal spray technology*. GmbH, Luckenbach, Germany: GTV Verschleiss-Schutz.

Warcholinski, B., & Gilewicz, A., (2009). Tribological properties of CrNx coatings. *J. Achiev. Mater. Manuf. Eng.*, 37, 498-504.

Warcholinski, B. G. (2009). Tribological properties of TiAlCrN Thin films. *Rev. Mater. Sci., 22*, 81–88.

Wasa, K. K. (2004). *Thin Film Materials Technology*. United States: William Andrew Inc. and Springer.

Webster, J., Westwood, M., Hayes, F., Day, R., Taylor, R., Duran, A., & Vogel, W. et al. (1998). Oxidation protection coatings for C/SiC based on yttrium silicate. *Journal of the European Ceramic Society*, *18*(16), 2345–2350. doi:10.1016/S0955-2219(98)00241-6

Wehnerand, G. A. (1970). *The Nature of Physical Sputtering" in Handbook of Thin Film Technology* (L. G. Maisseland, Ed.). NY: McGraw-Hill.

Wellman, R. G., & Nicholls, J. R. (2004, May). High Temperature Erosion-Oxidation Mechanisms, Maps and Models. *Wear*, *256*(9-10), 907–917. doi:10.1016/j.wear.2003.04.003

Wellman, R. G., & Nicholls, J. R. (2008). Erosion, corrosion and erosion–corrosion of EB PVD thermal barrier coatings. *Tribology International*, *41*(7), 657–662. doi:10.1016/j.triboint.2007.10.004

Wen, J., & Wilkes, G. L. (1996). Organic/inorganic hybrid network materials by the sol-gel approach. *Chemistry of Materials*, *8*(8), 1667–1681. doi:10.1021/cm9601143

Westwood, M., Webster, J., Day, R., Hayes, F., & Taylor, R. (1996). Oxidation protection for carbon fibre composites. *Journal of Materials Science*, *31*(6), 1389–1397. doi:10.1007/BF00357844

Wierderhorn, S. M., & Hockey, B. J. (1983). Effect of material parameters on the erosion resistance of brittle materials. *Materials Science*, *18*(3), 766–780. doi:10.1007/BF00745575

Wierzchon, T., & Sobiecki, J. R. (1993). The PACVD method for surface layer deposition using a glow discharge in H_2+N_2+$Ti(OC_3H_7)_4$. *Vacuum*, *44*(10), 975–978. doi:10.1016/0042-207X(93)90281-E

Wiese, W. L., Smith, M. W., & Miles, B. M. (1996). Atomic Transition Probabilities, Vol. II: Sodium through Calcium. Washington, DC: U.S. Government.

Wilkes, Garth L.; Orler, Bruce; Huang, Hao Hsin. (1985). "Ceramers": hybrid materials incorporating polymeric/oligomeric species into inorganic glasses utilizing a sol-gel approach, *26*(2), 300–302.

Wilson, A., Matthews, A., Housden, J., Turner, R., & Garside, B. (1993). A comparison of the wear and fatigue properties of plasma-assisted physical vapour deposition TiN, CrN and duplex coatings on Ti-6Al-4V. *Surface and Coatings Technology*, *62*(1-3), 600–607. doi:10.1016/0257-8972(93)90306-9

Wilson, S., Hawthorne, H. M., Yang, Q., & Troczynski, T. (2001). Scale effects in abrasive wear of composite sol–gel alumina coated light alloys. *Wear*, *251*(1), 1042–1050. doi:10.1016/S0043-1648(01)00704-9

Wojcik, A., & Klein, L. C. (1997). Transparent organic/inorganic hybrid gels: A classification scheme. *Applied Organometallic Chemistry*, *11*(November 1995), 129–135.

Wood, R. J. K. & Fry, S.A. (1989). The Synergistic Effect of Cavitation Erosion and Corrosion for Copper and Cupro-Nickel in Seawater. *Journal of Fluids Engineering*, *111*, 271-276.

Wriedt, H. A. (1986). The Al-N (aluminum-nitrogen) system. *Bulletin of Alloy Phase Diagrams*, *7*(4), 329–333. doi:10.1007/BF02873001

Wu, D. Y., Meure, S., & Solomon, D. (2008). Self-healing polymeric materials: A review of recent developments. *Progress in Polymer Science*, *33*(5), 479–522. doi:10.1016/j.progpolymsci.2008.02.001

Wu, J., Guo, H. B., Gao, Y. Z., & Gong, S. (2011). Microstructure and thermophysical properties of yttria stabilized zirconia coatings with CMAS deposits. *Journal of the European Ceramic Society*, *31*(10), 1881–1888. doi:10.1016/j.jeurceramsoc.2011.04.006

Wu, J., Guo, H., Zhou, L., Wang, L., & Gong, S. (2010). Microstructure and thermal properties of plasma sprayed thermal barrier coatings from nanostructured YSZ. *Journal of Thermal Spray Technology*, *19*(6), 1186–1194. doi:10.1007/s11666-010-9535-7

Wu, J., Wei, X. Zh., Padture, N. P., Klemens, P. G., Gell, M., & Garcia, E. (2002). Thermal conductivity of ceramics in the ZrO2-GdO1.5system. *Journal of Materials Research*, *17*(12), 3193–3200. doi:10.1557/JMR.2002.0462

Wu, J., Wei, X., Padture, N. P., Klemens, P. G., Gell, M., Garcia, E., & Osendi, M. I. et al. (2002). Low-thermal conductivity rare-earth zirconates for potential thermal-barrier-coating applications. *Journal of the European Ceramic Society*, *85*(12), 3031–3035. doi:10.1111/j.1151-2916.2002.tb00574.x

Wu, K. H., Chang, T. C., Yang, C. C., & Wang, G. P. (2006). Dynamics and corrosion resistance of amine-cured organically modified silicate coatings on aluminum alloys. *Thin Solid Films*, *513*(1–2), 84–89. doi:10.1016/j.tsf.2006.01.022

Wu, M. Y., Chen, M. H., Zhu, S. L., & Wang, F. H. (2016). Protection mechanism of enamel–alumina composite coatings on a Cr-rich nickel-based superalloy against high-temperature oxidation. *Surface and Coatings Technology*, *285*, 57–67. doi:10.1016/j.surfcoat.2015.11.026

Wu, S.-M. K., van de Kruijs, R., Zoethout, E., & Bijkerk, F. (2009). Sputtering yields of Ru, Mo, and Si under low energy Ar+ bombardment. *Journal of Applied Physics*, *106*(5), 1–7. doi:10.1063/1.3149777

Xie, W. (2013). *Thermal Spray Processing of Ethylene Methacrylic Acid and Polymer-Ceramic Composites. (PhD Doctor of Phyloshophy)*. Australia: Swinburne University of Technology.

Xie, W., Wang, J., & Berndt, C. C. (2012). Spreading Behavior and Morphology of Ethylene Methacrylic Acid (EMAA) Deposits via the Flame Spray Process. *Coatings*, *2*(2), 76–93. doi:10.3390/coatings2020076

Xie, W., Wang, J., & Berndt, C. C. (2013). Ethylene Methacrylic Acid (EMAA) Single Splat Morphology. *Coatings*, *3*(2), 82–97. doi:10.3390/coatings3020082

Xie, W., Wang, J., & Berndt, C. C. (2014). Analysis of EMAA Splats on Glass and Mild Steel Substrates. *Journal of Thermal Spray Technology*, *23*(3), 317–324. doi:10.1007/s11666-013-9989-5

Xie, X., Guo, H., Gonga, P., & Xua, H. (2011). Lanthanum-titanium-aluminum oxide: A novel thermal barrier coating material for applications at 1300 °C. *Journal of the European Ceramic Society*, *31*(9), 1677–1683. doi:10.1016/j.jeurceramsoc.2011.03.036

Xie, X., Guo, H., & Gong, S. (2010). Mechanical properties of $LaTi_2Al_9O_{19}$ and thermal cycling behaviours of plasma-sprayed $LaTi_2Al_9O_{19}$/YSZ thermal barrier coatings. *Journal of Thermal Spray Technology*, *19*(6), 1179–1185. doi:10.1007/s11666-010-9529-5

Xin, Y., Huang, Q.-Z., Zou, Y.-H., Chang, X., Su, Z.-A., Zhang, M.-Y., & Xie, Z.-Y. (2009). Anti-oxidation behavior of chemical vapor reaction SiC coatings on different carbon materials at high temperatures. *Transactions of Nonferrous Metals Society of China*, *19*(5), 1044–1050. doi:10.1016/S1003-6326(08)60404-5

Xiong, Y. M., Guan, C. H., Zhu, S. L., & Wang, F. H. (2006). Effect of enamel coating on oxidation and hot corrosion behaviors of Ti-24Al-14Nb-3V alloy. *Journal of Materials Engineering and Performance*, *15*(5), 564–569. doi:10.1361/105994906X136151

Xiong, Y. M., Zhu, S. L., & Wang, F. H. (2005). The oxidation behavior and mechanical performance of Ti60 alloy with enamel coating. *Surface and Coatings Technology*, *190*(2-3), 195–199. doi:10.1016/j.surfcoat.2004.09.005

Xiong, Y. M., Zhu, S. L., & Wang, F. H. (2008). Synergistic corrosion behavior of coated Ti60 alloys with NaCl deposit in moist air at elevated temperature. *Corrosion Science*, *50*(1), 15–22. doi:10.1016/j.corsci.2007.06.007

Xu, H., & Guo, H. (2011). *Thermal Barrier Coatings*. Cambridge, UK: Woodhead Publishing Limited. doi:10.1533/9780857090829

Xu, Y., & Hutchings, I. M. (2006). Cold spray deposition of thermoplastic powder. *Surface and Coatings Technology, 201*(6), 3044–3050. doi:10.1016/j.surfcoat.2006.06.016

Xu, Y., Hu, X., Xu, F., & Li, K. (2017). Rare earth silicate environmental barrier coatings: Present status and prospective. *Ceramics International, 43*(8), 5847–5855. doi:10.1016/j.ceramint.2017.01.153

Xu, Z. H., He, L. M., Tang, Z. H., Mu, R. D., & Cao, X. Q. (2012). Evolution of high temperature corrosion behavior of $La_2(Zr_{0.7}Ce_{0.3})_2O_7$ with the addition of Y_2O_3 thermal barrier coatings in contacts with vanadate–sulfate salts. *Journal of Alloys and Compounds, 536*, 106–112. doi:10.1016/j.jallcom.2012.04.074

Xu, Z., He, L., Mu, R., He, S., Huang, G., & Cao, X. (2010). Hot corrosion behavior of rare earth zirconates and yttria partially stabilized zirconia thermal barrier coatings. *Surface and Coatings Technology, 204*(21-22), 3652–3661. doi:10.1016/j.surfcoat.2010.04.044

Yabuki, A., Yamagami, H., & Noishiki, K. (2007). Barrier and self-healing abilities of corrosion protective polymer coatings and metal powders for aluminum alloys. *Materials and Corrosion, 58*(7), 497–501. doi:10.1002/maco.200604041

Yagisawa, T. M., & Makabe, T. (2006). Modeling of Dc Magnetron Plasma for Sputtering: Transport of Sputtered Copper Atoms. *Journal of Vacuum Science & Technology. A, Vacuum, Surfaces, and Films, 24*(4), 908–913. doi:10.1116/1.2198866

Yam, K. L. (n.d.). Gallium Nitride: An Overview of Structural Defects. *Optoelectronics-Materials and Techniques*.

Yamada, M. (2006). *Fabrication of nitride ceramic coatings by reactive plasma spraying*. Unpublished doctoral dissertation, Toyohashi University of Technology, Toyohashi, Japan.

Yamada, K., Tomono, Y., Morimoto, J., Sasaki, Y., & Ohmori, A. (2002). Hot corrosion behavior of boiler tube materials in refuse incineration environment. *Vacuum, 65*(3), 533–540. doi:10.1016/S0042-207X(01)00468-7

Yamada, M., Inamoto, T., Fukumoto, M., & Yasui, T. (2004). Fabrication of silicon nitride thick coatings by reactive RF plasma spraying. *Materials Transactions, 45*(12), 3304–3308. doi:10.2320/matertrans.45.3304

Yamada, M., Inamoto, T., Fukumoto, M., & Yasui, T. (2005). Spectroscopic measurement of plasma in the reactive plasma spray process [in Japanese]. *Journal of the Vacuum Society of Japan, 48*(3), 241–243. doi:10.3131/jvsj.48.241

Yamada, M., Kouzaki, Y., Yasui, T., & Fukumoto, M. (2006). Fabrication of iron nitride coatings by reactive RF plasma spraying. *Surface and Coatings Technology, 201*(3-4), 1745–1751. doi:10.1016/j.surfcoat.2006.02.046

Yamada, M., Nakamura, H., Yasui, T., Fukumoto, M., & Takahashi, K. (2006). Influence of substrate materials upon fabrication of aluminum nitride coatings by reactive RF plasma spraying. *Materials Transactions, 47*(7), 1671–1676. doi:10.2320/matertrans.47.1671

Yamada, M., Yasui, T., Fukumoto, M., & Takahashi, K. (2007). Nitridation of aluminum particles and formation process of aluminum nitride coatings by reactive RF plasma spraying. *Thin Solid Films, 515*(9), 4166–4171. doi:10.1016/j.tsf.2006.02.054

Yamamura, Y. I. (1983). *Angular dependence of sputtering yields of monatomic solids.* Nagoya University Institute of Plasma Physics Report IPPJ-AM-26.

Yamamura, Y. T., & Tawara, H. (1996). Energy dependence of ion-induced sputtering yields from monoatomic solids at normal incidence. *Atomic Data and Nuclear Data Tables, 62*(2), 149–253. doi:10.1006/adnd.1996.0005

Yan, F. Y., Gross, K. A., Simon, G. P., & Berndt, C. C. (2004). Mechanical and erosion properties of $CaCO_3$-EMAA thermal sprayed coatings. *Polymer Engineering and Science*, *44*(8), 1448–1459. doi:10.1002/pen.20141

Yang, L. L., Chen, M. H., Cheng, Y. X., Wang, J. L., Liu, L., Zhu, S. L., & Wang, F. H. (2016). Effects of surface finish of single crystal superalloy substrate on cyclic thermal oxidation of its nanocrystalline coating. *Corrosion Science*, *111*, 313–324. doi:10.1016/j.corsci.2016.04.023

Yang, L. L., Chen, M. H., Wang, J. L., Bao, Z. B., Zhu, S. L., & Wang, F. H. (2017). Diffusion of Ta and its influence on oxidation behavior of nanocrystalline coatings with different Ta, Y and Al contents. *Corrosion Science*, *126*, 344–355. doi:10.1016/j.corsci.2017.07.017

Yang, L. L., Chen, M. H., Wang, J. L., Zhu, S. L., & Wang, F. H. (2016). A duplex nanocrystalline coating for high-temperature applications on single-crystal superalloy. *Corrosion Science*, *102*, 72–83. doi:10.1016/j.corsci.2015.09.020

Yang, S. L., & Wang, F. H. (2006). Effect of nanocrystallization on sulfur segregation in Fe–Cr–Al alloy during oxidation at 1000 °C. *Oxidation of Metals*, *65*(3-4), 195–205. doi:10.1007/s11085-006-9014-1

Yang, Y. F., Jiang, C. Y., Yao, H. R., Bao, Z. B., Zhu, S. L., & Wang, F. H. (2016). Preparation and enhanced oxidation performance of a Hf-doped single-phase Pt-modified aluminide coating. *Corrosion Science*, *113*, 17–25. doi:10.1016/j.corsci.2016.09.014

Yang, Y., & Urban, M. W. (2013). Self-healing polymeric materials. *Chemical Society Reviews*, *42*(17), 7446–7467. doi:10.1039/c3cs60109a PMID:23864042

Yang, Y., Yan, M. F., Zhang, Y. X., Li, D. Y., Zhang, C. S., Zhu, Y. D., & Wang, Y. X. (2017). Catalytic growth of diamond-like carbon on Fe3C-containing carburized layer through a single-step plasma-assisted carburizing process. *Carbon*, *122*, 1–8. doi:10.1016/j.carbon.2017.06.033

Yasakau, K. A., Carneiro, J., Zheludkevich, M. L., & Ferreira, M. G. S. (2014). Influence of sol-gel process parameters on the protection properties of sol–gel coatings applied on AA2024. *Surface and Coatings Technology*, *246*, 6–16. doi:10.1016/j.surfcoat.2014.02.038

Yasunori, T. (1996). Modification of Thin Film Properties by Sputtered Particles. *R&D Review of Toyota CRDL*, *28*(3), 1993.9.

Yoganandan, G., Pradeep Premkumar, K., & Balaraju, J. N. (2015). Evaluation of corrosion resistance and self-healing behavior of zirconium–cerium conversion coating developed on AA2024 alloy. *Surface and Coatings Technology*, *270*, 249–258. doi:10.1016/j.surfcoat.2015.02.049

Yoshioka, S., Boatemaa, L., van der Zwaag, S., Nakao, W., & Sloof, W. G. (2016). On the use of TiC as high-temperature healing particles in alumina based composites. *Journal of the European Ceramic Society*, *36*(16), 4155–4162. doi:10.1016/j.jeurceramsoc.2016.06.008

Yue, G., Hong-yu, X., Zeng-rong, J., & Zhi-xiang, X. (2015). Mechanical Properties and Thermal Shock Resistance of Rhenium Coating in Iridium/Rhenium/Carbon-Carbon Composites. *Procedia Engineering*, *99*, 1407–1414. doi:10.1016/j.proeng.2014.12.676

Yugeswaran, S., Selvarajan, V., Vijay, M., Ananthapadmanababhan, P. V., & Sreekumar, K. P. (2010). Influence of critical plasma spraying parameter (CPSP) on plasma sprayed alumina-titania coatings. *Ceramics International*, *36*(1), 141–149. doi:10.1016/j.ceramint.2009.07.012

Yu, H., Tan, T., Wu, W., Tian, C., An, Y., & Sun, F. (2012). Thermal stability of titanium nitride coatings prepared by the mixing technology with laser and plasma. *Current Applied Physics*, *12*(1), 152–154. doi:10.1016/j.cap.2011.05.025

Yu, M., Xue, B., Liu, J., Li, S., & Zhang, Y. (2015). Electrophoretic deposition of hybrid coatings on aluminum alloy by combining 3-aminopropyltrimethoxysilan to silicon–zirconium sol solutions for corrosion protection. *Thin Solid Films*, *590*, 33–39. doi:10.1016/j.tsf.2015.07.055

Yu, Q. H., Rauf, A., & Zhou, C. G. (2010). Microstructure and thermal properties of nanostructured 4 wt.%Al_2O_3-YSZ coatings produced by atmospheric plasma spraying. *Journal of Thermal Spray Technology*, *19*(6), 1294–1300. doi:10.1007/s11666-010-9524-x

Yu, Z., Huang, J., Wang, W., Yu, J., & Wu, L. (2016). Deposition and properties of a multilayered thermal barrier coating. *Surface and Coatings Technology*, *288*, 126–134. doi:10.1016/j.surfcoat.2016.01.001

Zabihi, F., Ahmadian-Yazdi, M.-R., & Eslamian, M. (2017). Photocatalytic Graphene-TiO_2 Thin Films Fabricated by Low-Temperature Ultrasonic Vibration-Assisted Spin and Spray Coating in a Sol-Gel Process. *Catalysts*, *7*(5), 136–152. doi:10.3390/catal7050136

Zabransky, L., Bursikova, V., Daniel, J., Soucek, P., Vasina, P., Dugacek, J., & Perina, V. et al. (2015). Comparative analysis of thermal stability of two different nc-TiC/a-C:H coatings. *Surface and Coatings Technology*, *267*, 32–39. doi:10.1016/j.surfcoat.2015.01.058

Zandi Zand, R., Verbeken, K., & Adriaens, A. (2012). Corrosion resistance performance of cerium doped silica sol–gel coatings on 304L stainless steel. *Progress in Organic Coatings*, *75*(4), 463–473. doi:10.1016/j.porgcoat.2012.06.008

Zavareh, M. A., Sarhan, A. A. D. M., Bushroa, B. A. R., & Basirun, W. J. (2014). Plasma thermal spray of ceramic oxide coating on carbon steel with enhanced wear and corrosion resistance for oil and gas applications. *Ceramics International*, *40*(9), 14267–14277. doi:10.1016/j.ceramint.2014.06.017

Zeng, L. C., Wang, W., & Wu, T. W. (2001). Electrochemical impedance models for molten salt corrosión. *Corrosion Science*, *43*(4), 787–801. doi:10.1016/S0010-938X(00)00108-6

Zeng, Y., Xiong, X., Guo, S., & Zhang, W. (2013). SiC/SiC–YAG–YSZ oxidation protective coatings for carbon/carbon composites. *Corrosion Science*, *70*, 68–73. doi:10.1016/j.corsci.2013.01.013

Zerfass, H. (2006). *ArC Spraying*. Osu Gunter Hessler GmbH & Co.KG. Retrieved from http://www.osu.de/e/arcspraying.html

Zhang, Z. L., & Bell, T. (1985). Structure and Corrosion Resistance of Plasma Nitrided Stainless Steel. *Surface Engineering*, *1*, 131-136. doi:10.1179/sur.1985.1.2.131

Zhang, F., Yan, M., He, J., & Yin, F. (2017). Microstructure evolution and wear resistance of nitride/aluminide coatings on the surface of Ti-coated 2024 Al alloy during plasma nitriding. *Ceramics International*, *43*(14), 10832–10839. doi:10.1016/j.ceramint.2017.05.109

Zhang, H., Guo, L., Ma, Y., Peng, H., Guo, H., & Gong, S. (2014). Thermal cycling behavior of $(Gd_{0.9}Yb_{0.1})_2Zr_2O_7$/8YSZ gradient thermal barrier coatings deposited on Hf-doped NiAl bond coat by EB-PVD. *Surface and Coatings Technology*, *258*, 950–955. doi:10.1016/j.surfcoat.2014.07.051

Zhang, J.-P., Fu, Q.-G., Qu, J.-L., Zhou, H.-S., & Liu, N.-K. (2016). Surface modification of carbon/carbon composites and in-situ grown SiC nanowires to enhance the thermal cycling performance of Si-Mo-Cr coating under parallel oxyacetylene torch. *Corrosion Science*, *111*, 667–674. doi:10.1016/j.corsci.2016.06.007

Zhang, L. C., Zarudi, I., & Xiao, K. Q. (2006). Novel behaviour of friction and wear of epoxy composites reinforced by carbon nanotubes. *Wear*, *261*(7), 806–811. doi:10.1016/j.wear.2006.01.033

Zhang, L., Yang, H., Pang, X., Gao, K., & Volinsky, A. A. (2013). Microstructure, residual stress, and fracture of sputtered TiN films. *Surface and Coatings Technology*, *224*, 120–125. doi:10.1016/j.surfcoat.2013.03.009

Zhang, Y.-L., Li, H.-J., Yao, X.-Y., Li, K.-Z., & Qiang, X.-F. (2011). Oxidation protection of C/SiC coated carbon/carbon composites with Si–Mo coating at high temperature. *Corrosion Science*, *53*(6), 2075–2079. doi:10.1016/j.corsci.2011.02.024

Zhang, Y., Yang, B., Zhang, P., Zhang, J., Ren, J., & Hu, Z. (2015). SiC nanowire toughened ZrB 2–SiC ablative coating for SiC coated C/C composites. *Ceramics International*, *41*(10), 14579–14584. doi:10.1016/j.ceramint.2015.07.175

Zhang, Z. L. (1997). Atomistic Processes in the Early Stages of Thin-Film Growth. *Science*, *276*(5311), 377–383. doi:10.1126/science.276.5311.377 PMID:9103189

Zhao, B., Yadian, B. L., Li, Z. J., Liu, P., & Zhang, Y. F. (2009). Improvement on wettability between carbon nanotubes and Sn. *Surface Engineering*, *25*(1), 31–35. doi:10.1179/026708408X334104

Zhao, H., Begley, M. R., Heuer, A., Shargi-Mostaghimi, R., & Wadley, H. N. G. (2011). Reaction, transformation and delamination of samarium zirconate thermal barrier coatings. *Surface and Coatings Technology*, *205*(19), 4355–4365. doi:10.1016/j.surfcoat.2011.03.028

Zhao, H., Levi, C. G., & Wadley, H. N. G. (2009). Vapor deposited samarium zirconate thermal barrier coatings. *Surface and Coatings Technology*, *203*(20-21), 3157–3167. doi:10.1016/j.surfcoat.2009.03.048

Zhao, H., Levi, C. G., & Wadley, H. N. G. (2014). Molten silicate interactions with thermal barrier coatings. *Surface and Coatings Technology*, *251*, 74–86. doi:10.1016/j.surfcoat.2014.04.007

Zhao, J., Liu, L., Guo, Q., Shi, J., & Zhai, G. (2006). Oxidation protective behavior of SiC/Si–MoSi 2 coating for different graphite matrix. *Materials Letters*, *60*(16), 1964–1967. doi:10.1016/j.matlet.2005.12.072

Zhao, S., Gu, L., Zhao, Y., Huang, W., Zhu, L., Fan, X., & Cao, X. et al. (2013). Thermal cycling behavior and failure mechanism of $La_2(Zr_{0.7}Ce_{0.3})_2O_7/Eu^{3+}$ - doped 8YSZ thermal barrier coating prepared by atmospheric plasma spraying. *Journal of Alloys and Compounds*, *580*, 101–107. doi:10.1016/j.jallcom.2013.05.070

Zhao, W., Wang, Y., Liu, C., Dong, L., Yu, H., & Ai, H. (2010). Erosion–corrosion of thermally sprayed coatings in simulated splash zone. *Surface and Coatings Technology*, *205*(7), 2267–2272. doi:10.1016/j.surfcoat.2010.09.011

Zheludkevich, M. L., Poznyak, S. K., Rodrigues, L. M., Raps, D., Hack, T., Dick, L. F., & Ferreira, M. G. S. (2010). Active protection coatings with layered double hydroxide nanocontainers of corrosion inhibitor. *Corrosion Science*, *52*(2), 602–611. doi:10.1016/j.corsci.2009.10.020

Zheludkevich, M. L., Salvado, I. M., & Ferreira, M. G. S. (2005). Sol–gel coatings for corrosion protection of metals. *Journal of Materials Chemistry*, *15*(48), 5099. doi:10.1039/b419153f

Zheludkevich, M. L., Shchukin, D. G., Yasakau, K. A., Möhwald, H., & Ferreira, M. G. S. (2007). Anticorrosion Coatings with Self-Healing Effect Based on Nanocontainers Impregnated with Corrosion Inhibitor. *Chemistry of Materials*, *19*(3), 402–411. doi:10.1021/cm062066k

Zheludkevich, M. L., Tedim, J., & Ferreira, M. G. S. (2012). "Smart" coatings for active corrosion protection based on multi-functional micro and nanocontainers. *Electrochimica Acta*, *82*, 314–323. doi:10.1016/j.electacta.2012.04.095

Zheludkevich, M., Serra, R., Montemor, M., Yasakau, K., Salvado, I., & Ferreira, M. (2005). Nanostructured sol–gel coatings doped with cerium nitrate as pre-treatments for AA2024-T3Corrosion protection performance. *Electrochimica Acta*, *51*(2), 208–217. doi:10.1016/j.electacta.2005.04.021

Zheng, D. Y., Xiong, Y. M., Zhu, S. L., Wang, F. H., & Wu, W. T. (2008). Solid reactions between enamel and O-phase Ti-Al-Nb intermetallics at 800 °C. *Materials Scienc Forum*, 595-598, 233-238.

Zheng, D. Y., Zhu, S. L., & Wang, F. H. (2006). Oxidation and hot corrosion behavior of a novel enamel-Al_2O_3 composite coating on K38G superalloy. *Surface and Coatings Technology*, *200*(20-21), 5931–5936. doi:10.1016/j.surfcoat.2005.09.004

Zheng, S., & Li, J. (2010). Inorganic–organic sol gel hybrid coatings for corrosion protection of metals. *Journal of Sol-Gel Science and Technology*, *54*(2), 174–187. doi:10.1007/s10971-010-2173-1

Zhong Liu, W., Peng, X., Zhuan, L., Wen, H., Heng, L., Xiao Yu, Y., . . . Wen Bo, C. (2015). Microstructure and oxidation behavior of sol–gel mullite coating on SiC-coated carbon/carbon composites. *Journal of the European Ceramic Society*, *35*(14), 3789-3796.

Zhong, J. Z., Zhang, T., Zhang, Z., & Lagally, M. G. (2001). Island-Corner Barrier Effect in two-dimentional Pattern Formation at Surfaces. *Physical Review B: Condensed Matter and Materials Physics*, *63*(11), 113403. doi:10.1103/PhysRevB.63.113403

Zhong, X., Zhao, H., Zhou, X., Liu, C., Wang, L., Shao, F., & Ding, C. et al. (2014). Thermal shock behavior of toughened gadolinium zirconate/YSZ double-ceramic-layered thermal barrier coating. *Journal of Alloys and Compounds*, *593*, 50–55. doi:10.1016/j.jallcom.2014.01.060

Zhou, C. G., Yu, J. S., Gong, S. K., & Xu, H. B. (2002). Influence of water vapor on the isothermal oxidation behavior of low pressure plasma sprayed NiCrAlY coating at high temperature. *Surface and Coatings Technology*, *161*(1), 86–91. doi:10.1016/S0257-8972(02)00398-5

Zhu, D. M., & Miller, R. A. (2000). Thermal barrier coatings for advanced gas-turbine engines. *Materials Research Society*, *25*(07), 43–47. doi:10.1557/mrs2000.123

Zhu, D., Nesbitt, J. A., Barrett, C. A., McCue, T. R., & Miller, R. A. (2004). Furnace cyclic oxidation behavior of multi-component low conductivity thermal barrier coatings. *Journal of Thermal Spray Technology*, *13*(1), 84–92. doi:10.1007/s11666-004-0053-3

Zhu, L. J., Zhu, S. L., & Wang, F. H. (2012). Preparation and oxidation behaviour of nanocrystalline Ni + CrAlYSiN composite coating with AlN diffusion barrier on Ni-based superalloy K417. *Corrosion Science*, *60*, 265–274. doi:10.1016/j.corsci.2012.03.027

Zhu, L. J., Zhu, S. L., Wang, F. H., & Zhang, J. Q. (2014). Comparison of the cyclic oxidation behavior of a low expansion Ni + CrAlYSiN nanocomposite and a NiCrAlYSi coating. *Corrosion Science*, *80*, 393–401. doi:10.1016/j.corsci.2013.11.052

Zhu, S. L., Wang, F. H., Lou, H. Y., & Wu, W. T. (1995). Reactive sputter deposition of alumina films on superalloys and their high-temperature corrosion resistance. *Surface and Coatings Technology*, *71*(1), 9–15. doi:10.1016/0257-8972(94)02289-3

Zhu, S. L., Wei, Z. W., & Liu, J. H. (2014). Kinetics modeling of parabolic growth competing to linear declining: A theoretical tool to interpret corrosion data in real world. *Advanced Materials Research*, *860-863*, 899–902. doi:10.4028/www.scientific.net/AMR.860-863.899

Zhu, X., Huang, H., Xu, K., & He, J. (1999). Influence of the interface preparation on properties of PECVD TiN. *Surface and Coatings Technology*, *116–119*, 1029–1032. doi:10.1016/S0257-8972(99)00280-7

Zhu, Y., Wang, W., Jia, X., Akasaka, T., Liao, S., & Watari, F. (2012). Deposition of TiC film on titanium for abrasion resistant implant material by ion-enhanced triode plasma CVD. *Applied Surface Science*, *262*, 156–158. doi:10.1016/j.apsusc.2012.03.152

Ziegler, J. B. (2008). *The Stopping and Range of Ions In Matter.* Retrieved from www.lulu.com/content/1524197

ZieglerJ. (2003). Retrieved from http://www.srim.org

Ziegler, J. B. (1985). *The Stopping and Range of Ions in Solids.* NY: Pergamon Press.

Ziegler, J., Ziegler, M., & Biersack, J. (2010). SRIM – The stopping and range of ions in matter. *Nuclear Instruments & Methods in Physics Research. Section B, Beam Interactions with Materials and Atoms*, *268*(11-12), 1818–1823. doi:10.1016/j.nimb.2010.02.091

Zois, D., Lekatou, A., & Vardavoulias, M. (2009). A microstructure and mechanical property investigation on thermally sprayed nanostructured ceramic coatings before and after a sintering treatment. *Surface and Coatings Technology*, *204*(1-2), 15–27. doi:10.1016/j.surfcoat.2009.06.013

Zomorodian, A., Garcia, M. P., Moura de Silva, T., Fernandes, J. C. S., Fernandes, M. H., & Montemor, M. F. (2013). Corrosion resistance of a composite polymeric coating applied on biodegradable AZ31 magnesium alloy. *Acta Biomaterialia*, *9*(10), 8660–8670. doi:10.1016/j.actbio.2013.02.036 PMID:23454214

Zwaag, S. (2008). *Self healing materials: an alternative approach to 20 centuries of materials science*: Springer Science+ Business Media BV.

About the Contributors

Amir Hossein Pakseresht has worked as a researcher at Tehran University (Postdoc position with Iran National Science Foundation: INSF, Since 2016) and Materials and Energy Research Center (MERC) Karaj- Iran, since 2010. Pakseresht completed his Ph.D. in materials science and engineering at MERC in 2015. He received his undergraduate study in materials science and engineering from Teheran University. His research interests lie in the area of plasma spraying, with a focus on splat morphology and new thermal barrier coatings. He, also, has collaborated actively with researchers in several other disciplines of materials science, particularly film, coating and metal matrix composite. Dr. Pakseresht has also published a book that has been published 5 times in 7 years. Totally, he has published more than 40 journal papers, 5 patents and one international book chapter (till 2017). He is an industry expert with more than 10 years of successful experience. He also, teaches physical chemistry of materials, thermodynamics of materials, non-destructive testing, and physical properties of materials.

* * *

Nisar Ali completed his masters, M.Phil and Ph.D from GC University Lahore, Pakistan. He won Higher Education Commission of Pakistan (HEC) Indigenous scholarship for M.Phil and Ph.D studies. During his M.Phil he worked on the research project based upon thin film deposition by using cold plasma technique. During his Ph.D his research project was "nanosecond laser ablation of transition metals in dry and liquid confined environments". During his Ph.D research he won HEC scholarship under the IRSIP project (International Research Support Initiative Program) to visit abroad for six months. Under IRSIP project he visited Technical University, Vienna, Austria, where he worked on the project of "femtosecond laser ablation of transition metals in dry and liquid confined environments." After completion of his Ph.D he joined the University of Engineering and Technology Lahore, Faisalabad Campus, Faisalabad Pakistan as an Assistant Professor in the Department of Basic Sciences and Humanities. At present, he am working on three different projects, the first one is based upon Ion beam modification of pulsed laser deposited thin films, the second one about the laser ablation and laser induced breakdown spectroscopy and third one is synthesis and application of Piezoceramic materials for biomedical applications.

Omid Ashrafiyan was born in Tehran, Iran, in 1993. He completed his bachelor of materials engineering at the Semnan University, Iran, in 2015 and master of materials engineering at the University of Tehran, Iran, in 2017. His research interests lie in the corrosion, protection of the materials, and surface engineering. His master's project is the protection of carbon-carbon composite against oxidation by coating.

Fakhreddin Ashrafizadeh is a professor of materials engineering at Isfahan University of Technology. He was graduated from Department of Metallurgy and Materials, the University of Birmingham, in 1988 and, since that time, he has been dealt with education and research in the fields of materials engineering and surface technology. Prof. Ashrafizadeh has published more than 330 scientific papers in national and international journals and conference proceedings, as well as 45 successful research and industrial projects. He was selected as the National Distinguished Professor in 2008.

Mahboobeh Azadi received her BS and MS in Material Engineering from Shiraz University, Iran, in 2007 and 2009. She then received her PhD in corrosion, coating and surface engineering from Tarbiat Modares University (TMU), Iran, in 2014. She is assistance professor at Semnan university from 2014 until now. Azadi's research interests include multilayer coatings, ceramics, mechanical properties of coating, corrosion and wear.

Naveena B. E. is an Assistant Professor, Department of Automobile Engineering Areas of Interest: Thermal Spray Coatings, Automotive, Surface Engineering.

Partha Bandyopadhyay has been working on thermal spraying for last 20 years. At present he leads a thermal spray research team in IIT Kharagpur, India. He has already graduated six PhD students and around 25 master's students. At present, seven students are working for their PhD with him. He has published around 50 papers in peer reviewed journals and investigated a number of research projects.

Giovanni Di Girolamo is an Italian researcher with long expertise in project management and researches on materials for multifunctional applications. He focused part of his researches on the development, fabrication and testing of materials and coatings for components operating at high-temperature in gas turbines and aircraft engines.

Rita Figueira finished her PhD in Science (Chemistry) in February 2015. She has worked more than 10 years as researcher fellowship at National Laboratory for Civil Engineering – Department of Materials. During her PhD, she developed organic-inorganic hybrid sol-gel materials, doped and undoped with corrosion inhibitors; to mitigate corrosion on galvanized steel when in contact with concrete. Dr. Figueira has published more than 15 papers and 1 book chapter, where she was both the first and corresponding author. Dr. Figueira is also an active member of several scientific societies such as The Royal Society of Chemistry and an active reviewer for 30 journals overall. Currently, she is working at Engiprojects, Lda as consultant for research projects in the area of corrosion.

Ehsan Ghasali was born in Tehran, Iran, in 1987. He received the B.E. degree in Material science engineering from the University of Semnan, Iran, in 2010, and the MSc. and Ph.D. degrees in Material science from the Materials and Energy research center, Karaj, Iran 2013 and 2018, respectively. Dr. Ghasali has contributed greatly to the understanding of mechanical and microstructure properties of composites and also high temperature coatings. Dr. Ghasali has published eighteen journal articles. His publications reflect his research interests on the composite science and high-temperature coatings.

M. Hassanzadeh is currently a Ph.D. candidate in Material Science at Tehran University, Tehran, Iran Since 2015. He completed his MSc in materials science and engineering at MUT in 2013. He received his undergraduate study in materials science and engineering from Shahid Bahonar University of Kerman. His primary research interests are the high-temperature coatings, with a focus on plasma spraying thermal barrier coating. He is currently working on a Ph.D. thesis on self-healing thermal barrier coating (The research is supervised by Professor Mohsen Saremi) and also, teaching corrosion and coating lab.

John Henao was born in Colombia, in 1988. He completed his Ph.D at the Universitat de Barcelona, Spain and his Master degree at the Université de Limoges, France. He did his undergraduate studies at the Universidad de Antioquia, Colombia. Dr. Henao is specialist in Materials Science, having experience in the field of thermal spray, energy storage and characterization of materials.

Sebastian Jucha is a Ph.D. student of the Materials Science on the Faculty of Materials Science and Metallurgy of the Silesian University of Technology. Graduated a major of Materials Science in the field of engineering and masters studies. His research interests include surface engineering especially heat-resistant coatings, in it thermal barrier coatings.

Suman Kant serves as Assistant Professor in Production Engineering Department of PEC University of Technology, Chandigarh, where he supervises graduate and postgraduate students in areas of production engineering. His research has been centered on industrial engineering, operations research, casting technology, metal matrix composites and coating technology. Dr. Kant authored or co-authored more than 30 papers in archival journals and 20 papers in national/ international conferences. Dr. Kant received his Ph.D. in Mechanical Engineering from IIT Roorkee.

R. Keshavamurthy received his Bachelor of Engineering in Mechanical Engineering, post-graduation in Manufacturing Science and Engineering and PhD in Metal Matrix Composites from Visvesvaraya Technological University, Belgaum, India in the year 2005, 2007 and 2011 respectively. Dr. Keshavamurthy started his career as Research Associate under DST Sponsored Research Project in 2006. Later in 2010, he joined as Assistant Professor in the Department of Mechanical Engineering, PES Institute of Technology, Bangalore. In 2012 he has been appointed as Scientist in Rapid Prototyping Division, at Central Manufacturing Technology Institute (CMTI), Bangalore, India. Since August 2013, He is with Dayananda Sagar College of Engineering, Bangalore. His areas of interest are Metal Matrix Composites, Thermal Spray Coatings and Non-Traditional Machining. He has published over 90 research papers in international journals and conferences and has filed one Indian patent in the area of Thermal spray coatings.

Grzegorz Moskal is an employee of the Faculty of Materials Science and Metallurgy of the Silesian University of Technology. Graduated a major of Materials Science in the field of masters and doctoral studies. Since 2004 he is an employee of the Silesian University of Technology. In 2013 he defended his habilitation dissertation, and since 2017 has been the Head of the Department of Ceramics, Composites and Foundry Technologies at the Institute of Materials Science. His research interests include nickel- and cobalt-based high temperature creep resisting materials, steels for energetic industries and heat-resistant coatings, especially thermal barrier coatings. Is the author and co-author of numerous scientific publications in this field, published in internationally recognized journals. Participated in numerous material professional evaluations on the range of assessment the processes of destruction, degradation and assessment of the durability of high temperature creep resisting materials under real operating conditions. Reviewer of numerous scientific journals indexed in WoS database. Author of monograph: „ Microstructure and properties of plasma sprayed thermal barrier coating on the base of gadolinium zirconate", and co-author of the chapter: "Microstructural degradation in boiler steels: materials developments, properties and assessment" in "Power plant life management and performance improvement", Edited by J.E. Oakey, Cranfield University, UK, Woodhead Publishing Series in Energy No. 23, (2011).

Sekhar N., Professor, Department of Automobile Engineering Areas of interest: Thermal spray, Surface engineering of automotive components Industrial experience: 31 years in M/s BOSCH, Bangalore, as divisional manager (Engineering and Manufacturing).

Peter Ifeolu Odetola holds BSc and MSc degrees in metallurgical and Materials Engineering from University of Lagos (UNILAG), Nigeria. He is currently a PhD researcher in Chemical, Metallurgical, and Materials Engineering at Tshwane University of Technology (TUT), Pretoria, South Africa. His research focuses on the development of nanocomposite coatings for corrosion protection, deposition of metals and other materials using several techniques, and development of novel multi-faceted coating on bipolar plate for fuel cell applications. His accomplishments include two published book chapters on electrodeposition of composite materials.

Philip Oladijo has joined the Department of Chemical, Material and Metallurgical Engineering at Botswana International University of Science and Technology (BIUST), Palapye, Botswana. Prior to that, he completed his postdoctoral fellowship under the tutelage of Professor Lesley Cornish in School of Chemical and Metallurgical Engineering at University of the Witwatersrand, South Africa. He received his PhD degree in Material Science and Engineering field from the University of the Witwatersrand, South Africa in 2013. Until now, he has published several peer-reviewed publication and research project. His research interest focused on hard materials, coatings, residual stress analyses using diffraction techniques, tribology, corrosion, self-healing of a material and material characterization.

Patricia Popoola is an expert in the field of Advanced Materials Processing with focus on Surface Technologies. Her research cum team findings have been presented at both local and international conferences. Patricia has over 200 publications (journals and conference papers) apart from book chapters. Currently, she has produced fifteen Masters' and seven Doctoral degrees graduates. She is employed at the Tshwane University of Technology, Pretoria, South Africa.

Pritee Purohit received her PhD in Metallurgy and Materials Science from the Savitribai Phule Pune University in 2017. Presently working as an Assistant Professor in the department of Mechanical engineering in Army Institute of Technology Pune.

Ehsan Sadri is currently a Ph.D candidate in Materials engineering at Isfahan University of Technology. He completed his MSc in materials engineering at Isfahan University of Technology under the supervision of Professor F. Ashrafizadeh in 2012. His research interests include surface engineering, thermal spray coatings, tribology and nanomaterials. He is an industry expert specially in materials selection, casting and corrosion of materials with more than 7 years of experience. He also teaches physical chemistry and thermodynamics of materials, surface engineering and physical metallurgy.

Mohsen Saremi is a member of academic staff in The Faculty of Engineering, Tehran University, Tehran, Iran since 1985. Dr Saremi completed his Ph.D. in Metallurgy & Materials Engineering at Birmingham University in 1984. His research interests lie in the area of Applied Electrochemistry, Corrosion of Metals, Coatings (High Temperature Coatings, TBC and coatings on body implants) and SCC of Metals. Prof. Saremi has published more than 150 papers in international journals and conferences.

Mohammed Shahien is part of Central Metallurgical Research and Development Institute (CMRDI), Egypt and; Toyohashi University of Technology (TUT), Japan. Dr Shahien received his Bachelor degree in Chemistry from El-Monofia University Egypt, then he received his M.Sc and Ph.D degrees from Toyohashi University of Technology, Japan, in 2004, 2009 and 2012 respectively. In 2005 he joined Central Metallurgical Research and Development Institute (CMRDI), Egypt as Assistant Researcher. In 2013 he has been appointed as Researcher in Refractory and Ceramic Materials Division, Advanced Materials Department, CMRDI, Egypt. His research interest includes fabrication of advanced ceramic materials, thermal spray coatings, Surface Engineering and development of ceramic coatings for several applications. Dr Shahien had numerous scientific publications in the field including 3 book-chapters, and till 2017 he received more than 22 awards and honors. He had wide academic experiences in several Japanese organizations including Osaka University, Toyohashi University of Technology (TUT) and National Institute of Advanced Industrial Science and Technology, (AIST). Dr Shahien is an editorial board member of the Journal of Advanced Thermal Science Research and he is an outstanding reviewer in more than 10 international scientific journals.

Oscar Sotelo is a Postdoctoral Researcher in the Division of Materials Science within the Insititute of Physical Sciences at the Universidad Nacional Autónoma de México, Campus Morelos where he has been a faculty member since 2012.

Narendra Mohan Suri serves as Professor and Head of Production Engineering Department of PEC University of Technology, Chandigarh, where he supervises graduate and postgraduate students in areas of production engineering. His research has been centered on welding engineering, casting technology, metal matrix composites, and coating technology. Dr. N.M. Suri authored or co-authored more than 25 papers in archival journals and 25 papers in national/ international conferences. He supervised an in-house project on the development of SAW flux for stainless steel. Dr. Suri received his PhD in Mechanical Engineering from IIT Delhi.

Shashikant Vagge is currently working as Professor in the department of Metallurgy and Materials Science, College of Engineering Pune.

B. Venkateshwarlu is currently working Associate Professor in VIT University, Vellore, India. He obtained his doctorate in Manufacturing in Mechanical Engineering department of IIT Kharagpur, West Bengal, India. He previously worked as Assistant Professor at different institutes in Hyderabad, India, after post-graduation from Osmania University, Hyderabad. His research interests are in surface engineering and Advanced Machining Technologies.

Rajeev Verma is faculty in the Department of Production & Industrial Engineering, PEC University of Technology, Chandigarh, India. He received his Ph.D. and Masters in Engineering degrees from Production & Industrial Engineering Department of PEC University of Technology, Chandigarh. He has more than 5 years of industrial and teaching experience. He has authored and co-authored 7 papers in international journals, 7 papers in international conferences and 3 papers in national conferences. His current area of research includes surface engineering and optimization.

Wei Xie is the CEO of Seaway Advanced Materials Co. Ltd. He has been appointed as a researcher and project manager in Qinghua Redbud Innovation Institute since 2015. He received his PhD in materials science and engineering in 2013 from Swinburne University of Technology in Australia. Immediately after that he was a postdoctoral research associate from Pacific Northwest National Laboratory in USA for one year. He has four years of academic experience in Europe and two years of academic experience in the Chinese Academy of Science. He has twenty-two publications and five patents. He is the outstanding reviewer for Journal of Alloy and Compounds, the reviewer for Physics Letter A and the reviewer for Surface and Coatings Technology. His research interest is permanent and soft magnetic materials research and application, metamaterials, computational materials, additive manufacturing (3D printing), thermal spray coatings, powder metallurgy and semiconductors thin film coating.

Index

E

F

G

H

I

L

M

N

O

P

R

S

T

U

V

W

X

Y

Z

Purchase Print, E-Book, or Print + E-Book

IGI Global books can now be purchased from three unique pricing formats: Print Only, E-Book Only, or Print + E-Book. Shipping fees apply.

www.igi-global.com

Recommended Reference Books

ISBN: 978-1-4666-9975-5
© 2016; 539 pp.
List Price: $225

ISBN: 978-1-4666-9811-6
© 2016; 441 pp.
List Price: $215

ISBN: 978-1-4666-9619-8
© 2016; 1,670 pp.
List Price: $2,250

ISBN: 978-1-4666-8473-7
© 2015; 1,702 pp.
List Price: $2,295

ISBN: 978-1-4666-9911-3
© 2016; 410 pp.
List Price: $310

ISBN: 978-1-4666-8291-7
© 2015; 743 pp.
List Price: $335

Looking for free content, product updates, news, and special offers?
Join IGI Global's mailing list today and start enjoying exclusive perks sent only to IGI Global members.
Add your name to the list at **www.igi-global.com/newsletters.**

Publishing Information Science and Technology Research Since 1988

IGI Global
DISSEMINATOR OF KNOWLEDGE

www.igi-global.com Sign up at www.igi-global.com/newsletters facebook.com/igiglobal twitter.com/igiglobal

www.ingramcontent.com/pod-product-compliance
Lightning Source LLC
LaVergne TN
LVHW060252011026
849161LV00011B/2701